Electrocatalysis of Direct Methanol Fuel Cells

Edited by
Hansan Liu and Jiujun Zhang

Further Reading

Kolb, G.
Fuel Processing
for Fuel Cells
2008
ISBN: 978-3-527-31581-9

Elvers, B. (ed.)
Handbook of Fuels
Energy Sources for Transportation
2008
ISBN: 978-3-527-30740-1

Sundmacher, K., Kienle, A., Pesch, H. J., Berndt, J. F., Huppmann, G. (eds.)
Molten Carbonate Fuel Cells
Modeling, Analysis, Simulation, and Control
2007
ISBN: 978-3-527-31474-4

Logan, B. E.
Microbial Fuel Cells
2008
ISBN: 978-0-470-23948-3

Mitsos, A., Barton, P. I. (eds.)
Microfabricated Power Generation Devices
Design and Technology
2009
ISBN: 978-3-527-32081-3

Garcia-Martinez, J. (ed.)
Nanotechnology for the Energy Challenge
2010
ISBN: 978-3-527-32401-9

Ozawa, K. (ed.)
Lithium Ion Rechargable Batteries
Materials, Technology, and New Applications
2009
ISBN: 978-3-527-31983-1

Electrocatalysis of Direct Methanol Fuel Cells

From Fundamentals to Applications

Edited by
Hansan Liu and Jiujun Zhang

WILEY-VCH Verlag GmbH & Co. KGaA

The Editors

Dr. Hansan Liu
National Research Council Canada
Institute for Fuell Cell Innovation
4250 Westbrook Mall
Vancouver, BC V6T 1W5
Canada

Dr. Jiujun Zhang
National Research Council Canada
Institute for Fuel Cell Innovation
4250 Wesbrook Mall
Vancouver, BC V6T 1W5
Canada

All books published by Wiley-VCH are carefully produced. Nevertheless, authors, editors, and publisher do not warrant the information contained in these books, including this book, to be free of errors. Readers are advised to keep in mind that statements, data, illustrations, procedural details or other items may inadvertently be inaccurate.

Library of Congress Card No.: applied for

British Library Cataloguing-in-Publication Data
A catalogue record for this book is available from the British Library.

Bibliographic information published by the Deutsche Nationalbibliothek
The Deutsche Nationalbibliothek lists this publication in the Deutsche Nationalbibliografie; detailed bibliographic data are available on the Internet at http://dnb.d-nb.de.

© 2009 WILEY-VCH Verlag GmbH & Co. KGaA, Weinheim

All rights reserved (including those of translation into other languages). No part of this book may be reproduced in any form – by photoprinting, microfilm, or any other means – nor transmitted or translated into a machine language without written permission from the publishers. Registered names, trademarks, etc. used in this book, even when not specifically marked as such, are not to be considered unprotected by law.

Composition Thomson Digital, Noida, India
Printing Betz-Druck GmbH, Darmstadt
Binding Litges & Dopf GmbH, Heppenheim
Cover Design Schulz Grafik-Design, Fußgönheim

Printed in the Federal Republic of Germany
Printed on acid-free paper

ISBN: 978-3-527-32377-7

Contents

Preface *XV*
List of Contributors *XIX*

1	**Direct Methanol Fuel Cells: History, Status and Perspectives** *1*	
	Antonino Salvatore Aricò, Vincenzo Baglio, and Vincenzo Antonucci	
1.1	Introduction *1*	
1.2	Concept of Direct Methanol Fuel Cells *2*	
1.2.1	Principles of Operation *2*	
1.2.1.1	DMFC Components *2*	
1.2.1.2	DMFC Operation Mode *3*	
1.2.1.3	Fuel Cell Process *3*	
1.2.2	Performance, Efficiency and Energy Density *4*	
1.2.2.1	Polarization Curves and Performance *4*	
1.2.2.2	Fuel Utilization *5*	
1.2.2.3	Cathode Operating Conditions *7*	
1.2.2.4	Heat Production *8*	
1.2.2.5	Cell Efficiency *9*	
1.2.2.6	Energy Density *9*	
1.3	Historical Aspects of Direct Methanol Fuel Cell Development and State-of-the-Art *10*	
1.3.1	Historical Development of Methanol Oxidation Catalysts *10*	
1.3.2	Status of Knowledge of Methanol Oxidation Process and State-of-the-Art Anode Catalysts *14*	
1.3.2.1	Oxidation Mechanism *14*	
1.3.2.2	Pt-Ru Catalysts *15*	
1.3.2.3	Alternative Anode Formulations *19*	
1.3.2.4	Practical Anode Catalysts *19*	
1.3.2.5	Anode Catalysts for Alkaline DMFC Systems *20*	
1.3.3	Technological Advances in Electrolyte Development for DMFCs *21*	
1.3.4	State-of-the-Art DMFC Electrolytes *24*	
1.3.4.1	General Aspects of DMFC Electrolyte Development *24*

Electrocatalysis of Direct Methanol Fuel Cells. Edited by Hansan Liu and Jiujun Zhang
Copyright © 2009 WILEY-VCH Verlag GmbH & Co. KGaA, Weinheim
ISBN: 978-3-527-32377-7

1.3.4.2	Proton Conducting Membranes	24
1.3.4.3	Membranes for High Temperature Applications	26
1.3.4.4	Alkaline Membranes	29
1.3.4.5	Effects of Crossover on DMFC Performance and Efficiency	32
1.3.5	Historical Development of Oxygen Electroreduction in DMFCs	33
1.3.6	Status of Knowledge of Oxygen Reduction Electrocatalysis and State-of-the-Art Cathode Catalysts	33
1.3.6.1	Oxygen Reduction Process	33
1.3.6.2	Pt-Based Catalysts and Non-noble Metal Electrocatalysts	35
1.3.6.3	Alternative Cathode Catalysts	37
1.3.7	DMFC Power Sources in the 'Pre-1990 Era'	39
1.4	Current Status of DMFC Technology for Different Fields of Application	40
1.4.1	Portable Power Sources	40
1.4.2	Transportation	53
1.4.3	Technology Development	60
1.4.3.1	DMFC Technology	60
1.4.3.2	Catalyst Preparation	61
1.4.3.3	Electrode Manufacturing and Membrane Electrode Assemblies Membrane Electrode Assembly (MEAs)	61
1.4.3.4	Stack Hardware and Design	62
1.4.3.5	DMFC Systems	64
1.5	Perspectives of Direct Methanol Fuel Cells and Techno-Economical Challenges	65
	References	70
2	**Nanostructured Electrocatalyst Synthesis: Fundamental and Methods**	**79**
	Nitin C. Bagkar, Hao Ming Chen, Harshala Parab, and Ru-Shi Liu	
2.1	Introduction	79
2.2	Fundamental Understanding of the Structure–Activity Relationship	80
2.2.1	Particle Size Effect	81
2.2.2	Crystal Facet Effect	82
2.3	Synthetic Methods of Conventional Carbon-Supported Catalysts	85
2.3.1	Impregnation Method	87
2.3.2	Colloidal Method	88
2.3.2.1	Polyol Synthesis	89
2.3.3	Microemulsion Method	93
2.4	Synthetic Methods of Novel Unsupported Pt Nanostructures	95
2.4.1	Template-Based Synthesis	95
2.4.1.1	Soft Template	96
2.4.1.2	Hard Template	96
2.4.2	Galvanic Replacement Reaction	99
2.4.2.1	Platinum Nanospheres	99
2.4.2.2	Platinum Nanotubes	99

2.4.2.3	Nanoporous Platinum	*101*
2.4.2.4	Hollow Platinum Nanochannels	*102*
2.4.2.5	Bimetallic Clusters	*103*
2.4.3	Electrochemical Synthesis	*104*
2.4.3.1	Spherical Platinum Clusters	*104*
2.4.3.2	Nanoporous Platinum	*105*
2.4.3.3	Tetrahexahedral Platinum Nanostructures	*106*
2.5	Conclusions	*108*
	References	*110*
3	**Electrocatalyst Characterization and Activity Validation – Fundamentals and Methods**	*115*
	Loka Subramanyam Sarma, Fadlilatul Taufany, and Bing-Joe Hwang	
3.1	Introduction	*115*
3.2	Direct Methanol Fuel Cells – Role of Electrocatalysts	*117*
3.2.1	Role of Anode Electrocatalysts – Methanol Oxidation Reaction (MOR)	*118*
3.2.2	Role of Cathode Electrocatalysts – Oxygen Reduction Reaction (ORR)	*119*
3.3	Characterization Techniques for Anode and Cathode Catalysts	*121*
3.3.1	Fundamental Aspects	*121*
3.3.2	Evaluation of Catalyst Crystallite Size, Elemental Composition, Morphology, Dispersion	*122*
3.3.2.1	X-ray Diffraction (XRD)	*122*
3.3.2.2	Transmission Electron Microscopy (TEM)	*124*
3.3.2.3	Scanning Electron Microscopy (SEM)	*125*
3.3.2.4	Atomic Force Microscopy (AFM)	*125*
3.3.2.5	Energy Dispersive X-ray Spectroscopy (EDS/EDX)	*125*
3.3.3	Elucidation of Nanostructural Characteristics of Electrocatalysts – Alloying Extent or Atomic Distribution, Electronic Structural Features, State of Catalyst Species, Surface Composition, Segregation Phenomena	*126*
3.3.3.1	X-ray Absorption Spectroscopy	*126*
3.3.3.2	X-ray Photoelectron Spectroscopy (XPS)	*132*
3.3.3.3	Electrochemical-Nuclear Magnetic Resonance Spectroscopy (EC-NMR)	*133*
3.3.3.4	Low-Energy Electron Diffraction (LEED)	*134*
3.3.3.5	Auger Electron Spectroscopy (AES)	*134*
3.3.3.6	Low-Energy Ion Scattering (LEIS)	*135*
3.3.3.7	Temperature Programmed Reduction (TPR)	*135*
3.4	Evaluation of Electrocatalyst Activity, Electrochemical Active Surface Area, Catalyst – Adsorbate Interactions, and Activity Validation Techniques	*136*
3.4.1	Cyclic Voltammetry (CV)	*136*
3.4.1.1	Applications	*137*

3.4.1.2	State of the Pt-Based Catalysts *137*
3.4.1.3	Qualitative Indicator for the Activity of Pt-Based Catalysts Towards ORR *139*
3.4.1.4	Qualitative Indicator for the Activity of Pt-Based Catalysts Towards MOR *139*
3.4.1.5	Reversibility of $H_{ads/des}$ Reaction and Roughness Factor (RF) of Pt-Based Catalysts *140*
3.4.1.6	Electrochemical Active Surface Area (ECASA) and Particle Size of Pt-Based Catalysts *140*
3.4.2	Adsorptive CO-Stripping Voltammetry (CO_{ads}-SV) *141*
3.4.3	Underpotential Deposition (UPD) *143*
3.4.3.1	Surface Composition of Pt-Based Catalysts Through the Cu-UPD method *145*
3.4.4	Rotating Disk Electrode (RDE) *146*
3.4.5	Rotating Ring-Disk Electrode (RRDE) Method *148*
3.4.6	Linear Sweep Voltammetry (LSV) *149*
3.5	Conclusions and Outlook *155*
	References *156*

4 Combinatorial and High Throughput Screening of DMFC Electrocatalysts *165*

Rongzhong Jiang and Deryn Chu

4.1	Introduction *165*
4.2	Common Procedures for the Development of DMFC Catalysts *168*
4.3	General Methods for Combinatorial and High Throughput Screening *169*
4.4	Methods of Combinatorial Synthesis *171*
4.4.1	Chemical Synthesis *171*
4.4.2	Electrochemical Synthesis *173*
4.4.3	Physical Synthesis *174*
4.5	Electrode Arrays for High Throughput Screening *175*
4.5.1	Direct Electrode Array and Automated Screening Method *175*
4.5.2	Material Spot Array on a Single Electrode and Optical Screening Method *177*
4.5.3	Indirect Electrode Array on a Single Conductive Substrate and Electrolyte Probe Screening Methods *179*
4.6	Other Screening Methods for Catalyst Discovery *182*
4.6.1	Infrared (IR) Thermography *183*
4.6.2	Scanning Mass Spectrometry *183*
4.6.3	Scanning Electrochemical Microscope (SECM) *185*
4.7	Combinatorial Methods for DMFC Evaluation and Data Analysis *187*
4.7.1	Micro Fuel Cell Array *187*
4.7.2	A Method for High Throughput Screening of DMFC Single Cells *189*
4.7.3	Data Analysis of Combinatorial Results *190*

4.8	Challenge and Perspective *190*	
	References *193*	
5	**State-of-the-Art Electrocatalysts for Direct Methanol Fuel Cells** *197*	
	Hanwei Lei, Paolina Atanassova, Yipeng Sun, and Berislav Blizanac	
5.1	Introduction *197*	
5.2	Electrocatalysis and Electrocatalysts for DMFC *198*	
5.2.1	Electrocatalysis for Methanol Oxidation *198*	
5.2.2	Electrocatalyst Development *199*	
5.2.3	Spray Conversion Reaction Platform for Electrocatalyst Manufacturing *202*	
5.3	DMFC Electrocatalyst Characterization and Evaluation *203*	
5.3.1	Physical Characterization *204*	
5.3.2	Electrochemical Evaluation *208*	
5.3.2.1	Thin Film Rotating Disc Electrode (TFRDE) Catalyst Characterization *208*	
5.3.2.2	Fuel Cell Evaluation *210*	
5.3.3	Durability Study *212*	
5.4	DMFC Performance Advancement via MEA Design *215*	
5.4.1	Electrocatalyst Layer Design *215*	
5.4.2	Hydrocarbon Membrane for DMFC Performance Improvement *217*	
5.4.3	Other Aspects of DMFC Catalyst Development *220*	
5.5	Prospects for DMFC *222*	
5.6	Conclusions *222*	
	References *224*	
6	**Platinum Alloys as Anode Catalysts for Direct Methanol Fuel Cells** *227*	
	Ermete Antolini	
6.1	Introduction *227*	
6.2	Phase Diagram vs. Activity: New Chances for DMFC Anodes *229*	
6.2.1	PtRu Catalysts: The Effect of Alloying and Ru Oxide Presence *229*	
6.2.2	PtSn Catalysts: Activity of PtSn Alloys and Non-Alloyed Pt-SnO$_x$ *233*	
6.2.3	Pt-Co and Pt-Ni Catalysts: Effect of Alloying and CoO$_x$ and NiO$_x$ Presence *235*	
6.3	Preparation Methods of Pt Alloys *238*	
6.3.1	Unsupported Catalysts *238*	
6.3.2	Supported Catalysts *240*	
6.4	Activity Evaluation of Pt Alloys *242*	
6.4.1	Pt-Based Binary Catalysts *243*	
6.4.1.1	Pt-W *243*	
6.4.1.2	Pt-Mo *244*	
6.4.1.3	Pt-Au *245*	
6.4.2	Ternary Pt-Ru-Based Catalysts *246*

6.5	Stability of Pt-Ru Catalysts in DMFC Environment *248*
6.6	Conclusions *250*
	References *251*
7	**Methanol-Tolerant Cathode Catalysts for DMFC** *257*
	Claude Lamy, Christophe Coutanceau, and Nicolas Alonso-Vante
7.1	Introduction *257*
7.2	Thermodynamics and Kinetics of the Oxygen Reduction Reaction (ORR) *258*
7.2.1	The ORR at a Platinum Electrode in a DMFC *258*
7.2.2	Concepts for Novel Oxygen Reduction Electrocatalysts *261*
7.3	Experimental Details *264*
7.3.1	Determination of the Methanol Crossover of Proton Exchange Membranes *264*
7.3.1.1	Experimental Procedures *264*
7.3.1.2	Results *266*
7.3.2	Electrochemical Measurements (Voltammetry, RDE, RRDE, etc.) *269*
7.3.2.1	Experimental Set-Up *269*
7.3.2.2	Analysis of the Data *269*
7.4	Synthesis and Characterizations of Nanostructured Catalysts for the ORR *272*
7.4.1	Platinum-Based Catalysts and Electrodes *272*
7.4.1.1	Synthesis of Platinum-Based Catalysts by the Carbonyl Complex Route *273*
7.4.1.2	Synthesis of Platinum-Based Catalysts by the Colloidal Route *273*
7.4.1.3	Physicochemical Characterizations of the Catalysts *274*
7.4.1.4	Electrochemical Characterization of the Catalysts *275*
7.4.2	Syntheses and Characterization of Transition Metal Macrocycles *278*
7.4.2.1	Syntheses of Transition Metal Phthalocyanines *279*
7.4.2.2	Syntheses of Transition Metal Porphyrins *280*
7.4.2.3	Synthesis of Transition Metal Tetraazaannulenes *281*
7.4.2.4	Characterization of the Macrocycles *281*
7.4.2.5	Preparation of Macrocycle Electrodes and Characterization of their Activity *282*
7.4.3	Transition Metal Chalcogenide Catalysts and Electrodes *289*
7.4.3.1	Synthesis of Metal Chalcogenides *289*
7.4.3.2	Physicochemical Characterizations *292*
7.4.4	Fuel Cell Tests *294*
7.5	Catalyst Tolerance in the Presence of Methanol *296*
7.5.1	Behavior of PtM/C Catalysts for the ORR in the Presence of Methanol *296*
7.5.2	Behavior of Transition Metal Macrocycles for the ORR in the Presence of Methanol *300*
7.5.3	Transition Metal Chalcogenides *302*
7.5.4	Other Non Pt-Based Catalysts *304*

7.6	Summary and Outlook 306
	References 308

8	**Carbon Nanotube-Supported Catalysts for the Direct Methanol Fuel Cell** 315
	Chen-Hao Wang, Li-Chyong Chen, and Kuei-Hsien Chen
8.1	Introduction 315
8.2	Preparation of Carbon Nanotube-Supported Catalysts 316
8.2.1	Functionalization of Carbon Nanotubes 316
8.2.2	Polymer-Modified CNTs 317
8.2.3	Impregnation Method 318
8.2.4	Colloidal Method 322
8.2.5	Electrodeposition Method 323
8.3	Characteristics of the Carbon Nanotube Electrode 325
8.3.1	Electrochemical Behavior of the CNT Electrode 325
8.3.2	Durability of the CNT Electrode 329
8.4	Electrochemical Behavior of Carbon Nanotube-Supported Catalysts 331
8.4.1	Methanol Oxidation Reaction 331
8.4.2	Oxygen Reduction Reaction 337
8.4.3	Electrochemical Impedance Analysis 340
8.5	Direct Growth of Carbon Nanotubes as Catalyst Supports 341
8.5.1	Direct Growth of CNTs on Carbon Cloth (CNT-CC) 342
8.5.2	Appearance of CNT-CC-Supported Catalysts 344
8.5.3	Electrochemical Behavior of CNT-CC-Supported Catalysts 344
8.6	Conclusion 348
	References 348

9	**Mesoporous Carbon-Supported Catalysts for Direct Methanol Fuel Cells** 355
	Chanho Pak, Ji Man Kim, and Hyuk Chang
9.1	Introduction 355
9.2	Mesoporous Carbon 356
9.2.1	General Aspects of Mesoporous Carbon 356
9.2.2	Synthesis of Mesoporous Carbon 357
9.2.2.1	Synthesis of OMC Materials via Nano-Casting Method 357
9.2.2.2	Synthesis of OMC Materials via Direct Self-Assembly Approach 359
9.2.3	Characteristics of Mesoporous Carbon 360
9.3	Mesoporous Carbon-Supported Catalyst 363
9.3.1	Concepts of Mesoporous Carbon-Supported Catalyst 363
9.3.2	Preparation Methods for Mesoporous Carbon-Supported Catalyst 364
9.3.3	Characterization of Mesoporous Carbon-Supported Catalyst 366
9.4	Fuel Cell Performance of Mesoporous Carbon-Supported Catalyst 367

9.5	Summary and Prospect 373	
	References 375	
10	**Proton Exchange Membranes for Direct Methanol Fuel Cells** 379	
	Dae Sik Kim, Michael D. Guiver, and Yu Seung Kim	
10.1	Introduction 379	
10.2	Synthesis of Polymer Electrolyte Membranes for DMFC 380	
10.2.1	Synthesis and Properties of PEMs Containing Aliphatic Polymers 380	
10.2.2	Synthesis of Sulfonated Poly(aryl ether) Copolymers 385	
10.2.2.1	Post-Sulfonation of Polymers 385	
10.2.2.2	Direct Copolymerization of Sulfonated Monomers 388	
10.2.2.3	Other Synthetic Strategies: Introducing Sulfonic Acid Groups 397	
10.2.2.4	Properties of Sulfonated Poly(arylene ether) Copolymers 402	
10.2.3	Single Cell Performances 403	
10.3	Conclusions 412	
	References 412	
11	**Fabrication and Optimization of DMFC Catalyst Layers and Membrane Electrode Assemblies** 417	
	Liang Ma, Yunjie Huang, Ligang Feng, Wei Xing, and Jiujun Zhang	
11.1	Introduction 417	
11.2	Components for DMFC Catalyst Layer Optimization 419	
11.2.1	Catalysts for the Methanol Oxidation Reaction (MOR) 419	
11.2.2	Catalysts for the Oxygen Reduction Reaction (ORR) 423	
11.2.3	Ionomer in the Catalyst Layer 424	
11.2.4	Components Related to Mass Transport 430	
11.3	Catalyzed DMFC Electrode Structure and Fabrication Process 433	
11.3.1	Fabrication Process of DMFC Electrode 433	
11.3.2	Novel Structures with Extended Reaction Zone 436	
11.4	Other Electrode Fabrication Methods for DMFCs 438	
11.4.1	Electrodeposition 438	
11.4.2	Sputtering 438	
11.4.3	Chemical Reduction Method 439	
11.4.4	Dry Production Techniques 439	
11.4.5	Glue Method 440	
11.4.6	Sedimentation Method 440	
11.4.7	Breathing Process 441	
11.5	Summary 441	
	References 441	
12	**Local Current Distribution in Direct Methanol Fuel Cells** 449	
	Andrei A. Kulikovsky and Klaus Wippermann	
12.1	Introduction 449	
12.2	Model 451	
12.2.1	General Description 451	

12.2.2	Basic Assumptions	452
12.2.3	Through-Plane Relations	453
12.2.4	Equations Along the Channel	454
12.2.5	Large Methanol Stoichiometry, Small Current	456
12.3	The Bifunctional Regime of DMFC Operation	459
12.3.1	Experimental	459
12.3.2	Experimental Results	461
12.3.3	Polarization Curves at Constant Oxygen Stoichiometry	462
12.3.4	Critical Air Flow Rate	464
12.4	Direct Methanol–Hydrogen Fuel Cells (DMHFCs)	466
12.4.1	Experiment	466
12.4.2	DMHFC: The Mechanism of Functioning	470
12.4.2.1	Potentials Across the Cell	470
12.4.2.2	Potentials Along the Channel	471
12.4.2.3	Potentials in the Galvanic Domain	473
12.4.2.4	The Transition Region: Hydrogen Cell	474
12.5	Bifunctional Activation of DMFC	476
12.5.1	Single Channel Cell	476
12.5.2	Activation of Square-Shaped Cells	480
12.6	Conclusions	482
12.7	List of symbols	483
12.7.1	Superscripts	484
12.7.2	Subscripts	484
12.7.3	Greek Symbols	485
	References	485

13 Electrocatalysis in the Direct Methanol Alkaline Fuel Cell 487
Keith Scott and Eileen Yu

13.1	Introduction	487
13.2	History of Alkaline Methanol Fuel Cells	488
13.3	Electrocatalysis of Methanol Oxidation in Alkaline Media	490
13.3.1	Mechanism of Methanol Oxidation in Alkaline Media	491
13.3.2	Precious Metal Catalysts	491
13.3.3	Non-Precious Metal Catalysts	494
13.3.4	Effect of pH and Electrolyte	496
13.4	Oxygen Reduction and Methanol Tolerant Electrocatalysts	498
13.4.1	Oxygen Reduction Mechanism in Alkaline Media	498
13.4.1.1	Cyclic Voltammetry of Pt/C and Pt/Ru/C Catalysts in 1 M NaOH	500
13.4.2	Non-platinum Electrocatalysts	502
13.4.3	Mixed Reduction Potential of Methanol with Oxygen	503
13.5	Direct Methanol Fuel Cells in Alkaline Media	506
13.5.1	Aqueous Electrolyte Media	506
13.5.2	Cationic Exchange Membranes	508
13.5.3	Invariant Electrolyte Media	511
13.6	Direct Alkaline Polymer Electrolyte Membrane Fuel Cells	511

13.6.1	Anion Exchange Membrane for Methanol Fuel Cells	*511*
13.6.2	Direct Methanol Alkaline Membrane Fuel Cell Performance	*513*
13.6.3	Membraneless Fuel Cell	*517*
13.7	Alkaline Fuel Cells with other Direct Liquid Fuels	*518*
13.8	Conclusions	*520*
	References	*521*
14	**Electrocatalysis in Other Direct Liquid Fuel Cells**	*527*
	Sharon L. Blair and Wai Lung (Simon) Law	
14.1	Introduction	*527*
14.1.1	Fuel Characteristics and Theoretical Comparison of Various Fuels	*527*
14.2	Electrocatalysis of Direct Formic Acid Fuel Cells	*528*
14.2.2	Anode Catalysts for DFAFCs	*530*
14.2.2.1	Pt-based Anode Catalysts	*531*
14.2.2.2	Alternative Pt Modifiers	*536*
14.2.2.3	Pd-Based Anode Catalysts	*536*
14.2.3	Cathode Catalysts for DFAFCs	*541*
14.2.4	Direct Formic Acid Fuel Cell Performance	*541*
14.2.5	Summary of Electrocatalysis in DFAFCs	*545*
14.3	Electrocatalysis of Direct Ethanol Fuel Cells	*545*
14.3.1	Anode Catalysts for DEFCs	*546*
14.3.1.1	Binary Catalysts	*547*
14.3.1.2	Ternary Catalysts	*549*
14.3.1.3	Anode Catalysts for Alkaline Electrolytes	*550*
14.3.2	Cathode Catalysts for DEFCs	*551*
14.3.3	Direct Ethanol Fuel Cell Performance	*552*
14.3.4	Summary of Electrocatalysis in DEFCs	*553*
14.4	Electrocatalysis of Direct Hydrazine Fuel Cells	*554*
14.4.1	Anode Catalysts for DHFCs	*555*
14.4.1.1	Noble Metal-based Anode Catalysts	*556*
14.4.1.2	Non-Noble Metal-based Anode Catalysts	*557*
14.4.2	Cathode Catalysts for DHFCs	*558*
14.4.3	Direct Hydrazine Fuel Cell Performance	*558*
14.4.4	Summary of Electrocatalysis in DHFCs	*560*
14.5	Other Direct Liquid Fueled Fuel Cells	*561*
14.5.1	2-Propanol	*561*
14.5.2	Ethylene Glycol	*561*
14.5.3	Other Liquid Organic Fuels for Fuel Cells	*562*
14.6	Summary	*562*
	References	*563*

Index *567*

Preface

In today's world, human energy demands are continually on the rise, and the prospect of an energy shortage, or even crisis, is likely in the near future. Among the various efforts being made to meet global energy needs, the energy carried by hydrogen and liquid biofuels such as methanol and ethanol has become an attractive option in terms of sustainability and low environmental impact. Within this category of energy conversion technologies, the fuel cell is one of the most environmentally friendly and sustainable possibilities. In the past several decades, governments, along with academic communities and industries, have made enormous investments to develop fuel cells, especially proton exchange membrane fuel cells (PEMFCs), into a viable technology for stationary, portable, and transportation applications. As a result, great advances have been achieved in fuel cell technology in recent years. To date, several kinds of fuel cells, including direct methanol fuel cells (DMFCs), have shown great promise for near-term commercialization.

DMFCs, using renewable liquid methanol as fuel, have some advantages over the hydrogen-fuelled PEMFCs, such as easy fuel production, storage, and transportation, simple feed strategy, as well as system simplification. In comparison with other advanced energy devices such as lithium ion batteries, DMFCs show one order of magnitude higher energy density, making them a better choice for meeting the high energy density requirement of traditionally battery-powered devices. In the past two decades, great progress has been made in DMFC technology. For example, a few pre-commercial products based on this technology are currently emerging in the portable device category, including laptop computers and cell phones. However, several challenges still hinder commercialization, such as inadequate performance and high cost. The sluggish electrocatalytic reactions, that is, the anodic methanol oxidation reaction (MOR) and the cathodic oxygen reduction reaction (ORR), are largely responsible for this low performance. Methanol crossover from anode to cathode through the membrane can also cause performance deterioration. In order to overcome these technical difficulties, high-loading and high-cost Pt and Pt alloy catalysts must currently be used in DMFCs to mitigate the negative effects of the sluggish MOR and ORR, as well as the methanol crossover. In the past few years, numerous studies have been devoted to developing DMFC electrocatalysis, in an effort to further understand the reaction mechanisms, develop cost-effective and

Electrocatalysis of Direct Methanol Fuel Cells. Edited by Hansan Liu and Jiujun Zhang
Copyright © 2009 WILEY-VCH Verlag GmbH & Co. KGaA, Weinheim
ISBN: 978-3-527-32377-7

high-performance or methanol-tolerant catalysts, and optimize the fabrication of catalyst layers and membrane electrode assemblies membrane electrode assembly (MEA).

This book is designed to draw a clear picture of the current status of DMFC technology, with a particular focus on the technical progress, challenges, and perspectives in the field of DMFC electrocatalysis. A group of top fuel cell scientists and engineers, who have not only excellent academic records but also strong industrial fuel cell expertise, were invited to contribute chapters. These leading researchers from universities, government laboratories, and fuel cell industry companies in North America, Europe, and Asia share in this volume their knowledge, information, and insights on recent advances in the fundamental theories, experimental methodologies, and research achievements in DMFC electrocatalysis.

In the first chapter, Arico and his colleagues provide a comprehensive review of DMFC history, the current status of fundamental studies, and the technical advances that have been made in catalyst preparation, MEA fabrication, as well as stack/system design. They also give a global summary of DMFC prototypes in portable and transportation applications, and discuss the techno-economic challenges still confronting DMFC technology. In Chapter 2, Liu and his co-workers describe various methods of synthesizing nanostructured electrocatalysts for DMFCs, including both conventional carbon-supported, Pt-based catalysts and novel unsupported catalyst nanostructures. In Chapter 3, Sarma and Hwang give an overview of versatile characterization techniques for elucidating nanoscale characteristics and activity validation methodologies for DMFC electrocatalysts. In Chapter 4, Jiang and Chu introduce the principles and modern technologies of combinatorial methods used in the synthesis and high throughput screening of electrocatalyst libraries for DMFCs. In Chapter 5, Lei and his co-workers introduce DMFC catalyst development in industry, particularly focusing on Cabot's spray conversion reaction (SCR) platform for the discovery and manufacturing of high-performance fuel cell electrocatalysts. In Chapter 6, Antolini surveys the progress in preparation methods, activity validation, and stability studies of Pt binary and ternary alloys for DMFC anodes. In Chapter 7, Lamy and his colleagues discuss the thermodynamics and kinetics of the oxygen reduction reaction on DMFC cathodes, and introduce strategies to mitigate the methanol-crossover effect; they also describe the progress made in developing methanol-tolerant catalysts. In Chapter 8, Chen and his co-authors review recent advances in carbon nanotube (CNT)-supported DMFC catalysts. In Chapter 9, Pak and Chang cover studies of mesoporous carbon-supported DMFC catalysts. In Chapter 10, Kim and Guiver discuss strategies for making proton exchange membranes with high proton conductivity and lower methanol permeability, mainly describing their own work on developing hydrocarbon-based copolymers containing sulfonic acid groups for DMFC membranes. In Chapter 11, Xing and his co-authors review the progress that has been made in the fabrication and optimization of DMFC catalyst layers and membrane electrode assemblies, with an emphasis on the effects of catalyst layer components, structures, and fabrication methods on DMFC performance. In Chapter 12, Kulikovsky and Wippermann present their experimental and modeling efforts to understand local current distribution during

DMFC operation. In Chapter 13, Scott and Yu provide a detailed description of direct methanol alkaline fuel cells in terms of MOR catalysts, ORR catalysts, and membranes in alkaline circumstances. In Chapter 14, Blair and Law introduce electrocatalysis in other direct liquid fuel cells, mainly focusing on direct ethanol fuel cells and on their own work with direct formic acid fuel cells at Tekion Inc.

It is our hope that this volume will prove to be a good resource for electrochemists, chemical engineers, material scientists, students, and the public, providing up-to-date information on DMFC principles, the current status of DMFC electrocatalysis, and the future prospects for DMFC technology. We anticipate that this book will also be used as a reference by undergraduate and graduate post-secondary students, as well as scientists and engineers who work in the areas of energy, electrochemistry science/technology, fuel cells, and electrocatalysis.

We would like to express our appreciation to Wiley-VCH for inviting us to lead this book project, and we thank Dr. Elke Maase and Dr. Heike Noethe for their guidance and support in smoothing the book preparation process. We also extend our thanks to all our colleagues at the National Research Council of Canada's Institute for Fuel Cell Innovation (IFCI), and especially to the members of the catalysis team, for their support and help. We gratefully acknowledge all the chapter authors for their enthusiastic, collaborative, and reliable contributions. Special thanks go to Prof. Arico, Dr. Lei, Dr. Chang, and Dr. Kulikovsky for providing raw materials for the book's cover design. (Kulikovsky's current distribution picture and IFCI's mesoporous catalysts pictures were incorporated into the cover art.) Finally, our special appreciation goes to our families, for their understanding and their ongoing support of our work.

Vancouver, Canada *Hansan Liu* and *Jiujun Zhang*
February 2009

List of Contributors

Nicolas Alonso-Vante
CNRS-Université de Poitiers
Laboratory of Electrocatalysis
LACCO, UMR 6503
40 avenue du Recteur Pineau
86022 Poitiers Cedex
France

Ermete Antolini
Scuola di Scienza dei Materiali
Via 25 aprile 22
16016 Cogoleto
Genova
Italy

Vincenzo Antonucci
CNR-ITAE Institute
Via Salita S. Lucia sopra Contesse 5
98126 Messina
Italy

Antonino Salvatore Aricò
CNR-ITAE Institute
Via Salita S. Lucia sopra Contesse 5
98126 Messina
Italy

Paolina Atanassova
Cabot Corporation
5401 Venice Ave NE
Albuquerque, NM 87113
USA

Nitin C. Bagkar
National Taiwan University
Department of Chemistry
Taipei 106
Taiwan

Vincenzo Baglio
CNR-ITAE Institute
Via Salita S. Lucia sopra Contesse 5
98126 Messina
Italy

Sharon L. Blair
Tekion Inc.
8602 Commerce Court
Burnaby, BC V5A 4N6
Canada

Berislav Blizanac
Cabot Corporation
5401 Venice Ave NE
Albuquerque, NM 87113
USA

Hyuk Chang
Samsung Electronics Co. Ltd.
Energy Group
Emerging Center
Corporate Technology Operations SAIT
San #14-1 Nongseo-dong, Giheung-gu
Yongin-si 446-712
Korea

Electrocatalysis of Direct Methanol Fuel Cells. Edited by Hansan Liu and Jiujun Zhang
Copyright © 2009 WILEY-VCH Verlag GmbH & Co. KGaA, Weinheim
ISBN: 978-3-527-32377-7

List of Contributors

Hao Ming Chen
National Taiwan University
Department of Chemistry
Taipei 106
Taiwan

Kuei-Hsien Chen
Academia Sinica
Institute of Atomic and Molecular Sciences
Taipei 10617
Taiwan

and

National Taiwan University
Center for Condensed Matter Sciences
Taipei 10617
Taiwan

Li-Chyong Chen
National Taiwan University
Center for Condensed Matter Sciences
Taipei 10617
Taiwan

Deryn Chu
U.S. Army Research Laboratory
Sensors and Electron Devices Directorate
2800 Powder Mill Road
Adelphi, MD 20783-1197
USA

Christophe Coutanceau
CNRS-Université de Poitiers
Laboratory of Electrocatalysis
LACCO, UMR 6503
40 avenue du Recteur Pineau
86022 Poitiers Cedex
France

Ligang Feng
Chinese Academy of Sciences
Changchun Institute of Applied Chemistry
5625 Renmin Street
Changchun, Jilin Province 130022
China

Michael D. Guiver
National Research Council
Institute for Chemical Process and Environmental Technology
Ottawa, ON K1A 0R6
Canada

Yunjie Huang
Chinese Academy of Sciences
Changchun Institute of Applied Chemistry
5625 Renmin Street
Changchun, Jilin Province 130022
China

Bing-Joe Hwang
National Taiwan University of Science
Department of Chemical Engineering
43, Keelung Rd., Sec. 4
Taipei
Taiwan

Rongzhong Jiang
U.S. Army Research Laboratory
Sensors and Electron Devices Directorate
2800 Powder Mill Road
Adelphi, MD 20783-1197
USA

Dae Sik Kim
Los Alamos National Laboratory
Materials Physics & Applications
Sensors and Electrochemical Devices Group
Los Alamos, NM 87545
USA

Ji Man Kim
Sungkyunkwan University
BK21 School of Chemical Materials
Science and SKKU Advanced Institute
of Nanotechnology
Department of Chemistry
Functional Materials Lab
Suwon 440-746
Korea

Yu Seung Kim
Los Alamos National Laboratory
Materials Physics & Applications
Sensors and Electrochemical Devices
Group
Los Alamos, NM 87545
USA

Andrei A. Kulikovsky
Research Centre 'Juelich'/
Forschungszentrum 'Jülich'
Institute for Energy Research –
Fuel Cells (IEF-3)
52425 Jülich
Germany

and

Moscow State University
Research Computing Center
119991 Moscow
Russia

Claude Lamy
CNRS-Université de Poitiers
Laboratory of Electrocatalysis
LACCO, UMR 6503
40 avenue du Recteur Pineau
86022 Poitiers Cedex
France

Wai Lung (Simon) Law
Tekion Inc.
8602 Commerce Court
Burnaby, BC V5A 4N6
Canada

Hanwei Lei
Cabot Corporation
5401 Venice Ave NE
Albuquerque, NM 87113
USA

Ru-Shi Liu
National Taiwan University
Department of Chemistry
Taipei 106
Taiwan

Liang Ma
Chinese Academy of Sciences
Changchun Institute of Applied
Chemistry
5625 Renmin Street
Changchun, Jilin Province 130022
China

Chanho Pak
Samsung Electronics Co. Ltd.
Energy Group
Emerging Center
Corporate Technology Operations SAIT
San #14-1 Nongseo-dong, Giheung-gu
Yongin-si 446-712
Korea

Harshala Parab
National Taiwan University
Department of Chemistry
Taipei 106
Taiwan

Loka Subramanyam Sarma
National Taiwan University of Science
and Technology
Department of Chemical Engineering
Nanoelectrochemistry Laboratory
Taipei 106
Taiwan

Keith Scott
University of Newcastle upon Tyne
School of Chemical Engineering and
Advanced Materials
Newcastle upon Tyne NE1 7RU
UK

Yipeng Sun
Cabot Corporation
5401 Venice Ave NE
Albuquerque, NM 87113
USA

Fadlilatul Taufany
National Taiwan University of Science
and Technology
Department of Chemical Engineering
Nanoelectrochemistry Laboratory
Taipei 106
Taiwan

Chen-Hao Wang
Academia Sinica
Institute of Atomic and Molecular
Sciences
Taipei 10617
Taiwan

Klaus Wippermann
Research Centre 'Juelich'/
Forschungszentrum 'Jülich'
Institute for Energy Research – Fuel
Cells (IEF-3)
52425 Jülich
Germany

Wei Xing
Chinese Academy of Sciences
Changchun Institute of Applied
Chemistry
5625 Renmin Street
Changchun, Jilin Province 130022
China

Eileen Yu
University of Newcastle upon Tyne
School of Chemical Engineering and
Advanced Materials
Newcastle upon Tyne NE1 7RU
UK

Jiujun Zhang
National Research Council Canada
Institute for Fuel Cell Innovation
4250 Wesbrook Mall
Vancouver, BC V6T 1W5
Canada

1
Direct Methanol Fuel Cells: History, Status and Perspectives
Antonino Salvatore Aricò, Vincenzo Baglio, and Vincenzo Antonucci

1.1
Introduction

Significant efforts in recent decades have been focused on the direct electrochemical oxidation of alcohol and hydrocarbon fuels. Organic liquid fuels are characterized by high energy density, whereas the electromotive force associated with their electrochemical combustion to CO_2 is comparable to that of hydrogen combustion to water [1–3]. Among the liquid organic fuels, methanol has promising characteristics in terms of reactivity at low temperatures, storage and handling. Accordingly, a methanol-feed proton exchange membrane fuel cell (PEMFC) would help to alleviate some of the issues surrounding fuel storage and processing for fuel cells. Technological improvements in DMFCs are, thus, fuelled by their perspectives on applications in portable, transportation and stationary systems especially with regard to the remote and distributed generation of electrical energy [4, 5]. Methanol is cheap and can be distributed by using the present infrastructure for liquid fuels. It can be obtained from fossil fuels, such as natural gas or coal, as well as from sustainable sources through fermentation of agricultural products and from biomasses. Compared with ethanol, methanol has the significant advantage of high selectivity to CO_2 formation in the electrochemical oxidation process [1–3]. However, despite these practical system benefits, DMFCs are characterized by a significantly lower power density and lower efficiency than a PEMFC operating with hydrogen because of the slow oxidation kinetics of methanol and methanol crossover from the anode to the cathode [1–3].

This chapter deals with an analysis of the history, current status of technology, potential applications and techno-economic challenges of DMFCs. The basic aspects of DMFC operation are presented with particular regard to thermodynamics, performance, efficiency and energy density characteristics. The historical development of DMFC devices and components is analyzed with special regard to the study of catalysts and electrolytes. The section on fundamentals is focused on the electrocatalysis of the methanol oxidation reaction (MOR) and oxygen electroreduction. The

Electrocatalysis of Direct Methanol Fuel Cells. Edited by Hansan Liu and Jiujun Zhang
Copyright © 2009 WILEY-VCH Verlag GmbH & Co. KGaA, Weinheim
ISBN: 978-3-527-32377-7

current knowledge in the basic research areas is presented and particular emphasis is given to required breakthroughs. The technology section deals with the fabrication methodologies for the manufacturing of membrane electrode assemblies membrane electrode assembly (MEA), stack hardware and system design. Recent efforts in developing DMFC stack for both portable and electro-traction applications are reported.

1.2 Concept of Direct Methanol Fuel Cells

1.2.1 Principles of Operation

1.2.1.1 DMFC Components

The core of the present DMFCs is a polymer electrolyte ion exchange membrane. The electrodes (anode and cathode) are in intimate contact with the membrane faces (Figure 1.1). The electrodes usually consist of three layers: catalytic layer, diffusion layer and backing layer. The catalytic layer is composed of a mixture of catalyst and ionomer and it is characterized by a mixed electronic-ionic conductivity. The catalysts are often based on carbon supported or unsupported PtRu and Pt materials at the anode and cathode, respectively. The membrane as well as the ionomer consist, in most cases, of a perfluorosulfonic acid polymer. The diffusion layer is usually a mixture of carbon and polytetrafluoroethylene (Teflon®) with hydrophobic properties necessary to transport oxygen molecules to the catalytic sites at the cathode or to favor the escape of CO_2 from the anode. The overall thickness of a 'membrane and electrode assembly' (MEA) is generally smaller than one millimeter. Several cells are

Figure 1.1 SEM micrograph of a DMFC membrane and electrode assembly equipped with Nafion 112 membrane.

usually connected in series to form a fuel cell stack that is integrated in a system which contains the auxiliaries, allowing stack operation and delivery of the electrical power to the external load.

1.2.1.2 DMFC Operation Mode

In the literature, a distinction is usually made between 'active' and 'passive' operation mode [5]. In the active mode, the auxiliaries such as pumps, blowers, sensors, and so on are used to supply reactants and to control the stack operation in order to optimize working conditions. This allows the achievement of the most appropriate electrical characteristics. In the passive mode, there are no energy consuming auxiliaries (excluding step-up DC/DC converters) and the reactants reach the catalytic sites by natural convection, the effect of the capillary forces or due to the concentration/partial pressure gradient. The system is simpler than in the active mode; no significant amount of power from the stack is dissipated on auxiliaries, but, the operating conditions may not be optimal to achieve the best efficiency and performance.

1.2.1.3 Fuel Cell Process

Protonic electrolyte based DMFCs are directly fed with a methanol/water mixture at the anode. Methanol is directly oxidized to carbon dioxide although the possible formation of compounds such as formaldehyde, formic acid or other organic molecules is not excluded. The formation of such organic molecules decreases the fuel use.

A scheme of the overall reaction process occurring in a DMFC equipped with a proton conducting electrolyte is outlined below:

$$CH_3OH + H_2O \rightarrow CO_2 + 6H^+ + 6e^- \quad \text{(anode)} \tag{1.1}$$

$$3/2\, O_2 + 6H^+ + 6e^- \rightarrow 3H_2O \quad \text{(cathode)} \tag{1.2}$$

$$CH_3OH + 3/2\, O_2 \rightarrow CO_2 + 2H_2O \quad \text{(overall)} \tag{1.3}$$

In the presence of an alkaline electrolyte, this process can be written as follows:

$$CH_3OH + 6OH^- \rightarrow CO_2 + 5H_2O + 6e^- \quad \text{(anode)} \tag{1.4}$$

$$3/2O_2 + 3H_2O + 6e^- \rightarrow 6OH^- \quad \text{(cathode)} \tag{1.5}$$

$$CH_3OH + 3/2O_2 \rightarrow CO_2 + 2H_2O \quad \text{(overall)} \tag{1.3'}$$

The thermodynamic efficiency of the process is given by the ratio between the Gibbs free energy, that is, the maximum value of electrical work ($\Delta G°$) that can be obtained, and the total available energy for the process, that is, the enthalpy ($\Delta H°$). Under standard conditions:

$$\eta_{rev} = \Delta G° / \Delta H°; \text{ reversible energy efficiency} \tag{1.6}$$

with

$$\Delta G° = \Delta H° - (T \times \Delta S°); \tag{1.7}$$

and

$$\Delta G° = -nF \times \Delta E_{rev} \qquad (1.8)$$

ΔE_{rev} is the electromotive force. At 25 °C, 1 atm and with pure oxygen feed, the reversible potential for methanol oxidation is 1.18 V [3]. It does not vary significantly in the operating range 20–130 °C and 1–3 bar abs. pressure.

1.2.2
Performance, Efficiency and Energy Density

1.2.2.1 Polarization Curves and Performance

Usually, the open circuit voltage of a polymer electrolyte DMFC is significantly lower than the thermodynamic or reversible potential for the overall process. This is mainly due to methanol crossover that causes a mixed potential at the cathode and to the irreversible adsorption of intermediate species at electrode potentials close to the reversible potential [6–19]. The coverage of methanolic species is large at high cell potentials, that is, at low anode potentials. This determines a strong anode activation control that reflects on the overall polarization curve (Figure 1.2). In Figure 1.2, the terminal voltage of the cell is deconvoluted into the anode and cathode polarizations:

$$E_{cell} = E_{cathode} - E_{anode}. \qquad (1.9)$$

Anode, cathode and cell potentials can be measured simultaneously by a dynamic hydrogen electrode (DHE). Alternatively, the anode polarization can be measured in the driven mode and the cathode curve is calculated from Equation 1.9. In the driven mode, hydrogen is fed to the cathode that acts as both counter and reference

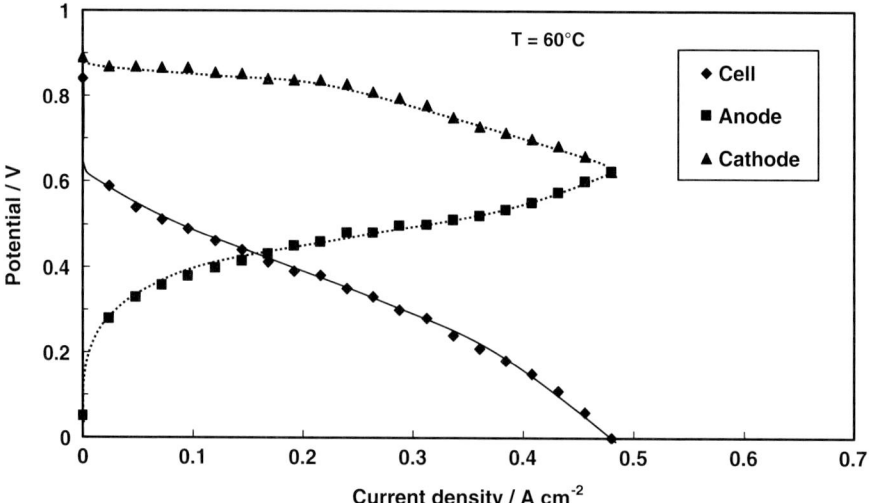

Figure 1.2 Single cell and *in situ* half-cell electrode polarizations for a DMFC operating at 60 °C, ambient pressure, with 1 M methanol at the anode and air feed at the cathode.

electrode. This is also the usual mode to carry out *in situ* cyclic voltammetry experiments for the anode.

Besides the strong activation control at the anode, the effect of the mixed potential on the cathode polarization curve is clearly observed in Figure 1.2. The onset potential for oxygen reduction in the presence of methanol crossover is below 0.9 V versus the reversible hydrogen electrode (RHE). This is much lower than the reversible potential for oxygen reduction in the absence of methanol, which is, 1.23 V vs. RHE. The methanol adsorption on the cathode mainly influences the region of activation control for oxygen reduction. In fact, at high cathode potentials, oxygen reduction is slow and oxidation of methanol permeated through the membrane is enhanced by the elevated potential. The two opposite reactions compete with each other and no spontaneous current is registered above 0.9 V (Figure 1.2).

At high currents (Figure 1.2), both anodic and cathodic polarization curves show the onset of mass transport constraints due to the removal of CO_2 from the anode and the effect of flooding at the cathode. In the protonic electrolyte methanol fuel cell, the flooding of the cathode is not only due to the water formed by the electrochemical process; it occurs, especially, as a consequence of the fact that a water/methanol mixture permeates the hydrophilic membrane.

The polarization curves of a DMFC device can be registered at different temperatures in order to study in detail the activation behavior (Figure 1.3), which clearly shows the presence of a strong effect of the temperature on the activation process. Temperature, pressure and methanol concentration are the most important variables determining performance and efficiency. Performance is often reported in terms of maximum power density under defined operating conditions.

1.2.2.2 Fuel Utilization

In a polarization diagram, beside the terminal voltage and the power density, it is also useful to report variations of the ohmic resistance and the crossover current (equivalent current density) as functions of the electrical current density. Usually, internal resistance does not vary significantly in the current density range of a DMFC whereas the equivalent current density is quite important for the methanol fuel cell because it determines the fuel use and influences the overall performance. It represents the current corresponding to the methanol permeation rate. A direct comparison with the effective (measured) electrical current permits evaluation of the fuel lost due to the crossover (Figure 1.4). The crossover or permeation rate of methanol can be determined *in situ* by the so-called CO_2 sensor method. Alternatively, chromatographic analyses can be used. In the presence of a Pt based catalyst, almost all the methanol that is permeated to the cathode is oxidized to CO_2 at high electrochemical potentials. From the CO_2 flow rate, the MOR and Faraday law, the equivalent current density is calculated according to the following equation:

$$I_{\text{cross over}} = \text{mol}_{\text{MeOH cross-over}} \cdot 6 \cdot F \tag{1.10}$$

Where $\text{mol}_{\text{MeOH crossover}}$ is the rate of methanol permeation to the cathode per unit of time and geometric electrode area (moles min^{-1} cm^{-2}).

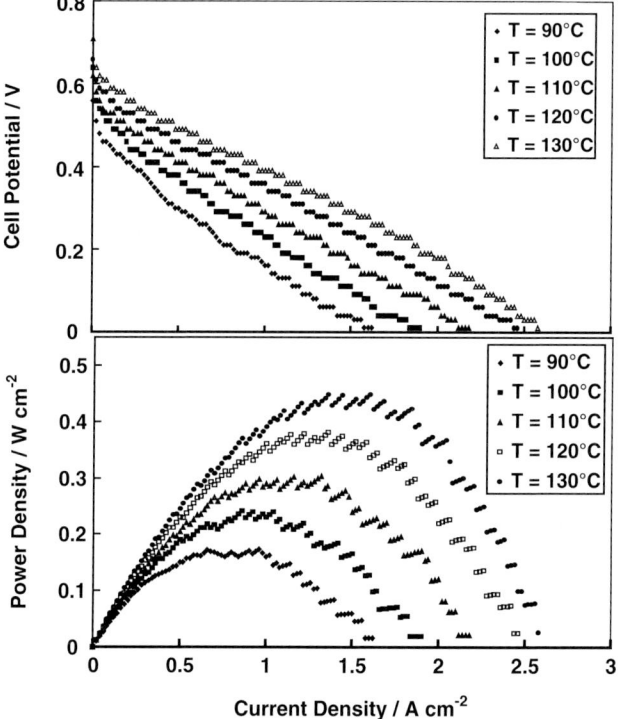

Figure 1.3 Galvanostatic polarization data for the DMFC equipped with CNR-ITAE Pt-Ru (anode) and 90% Pt/C (cathode) catalysts; 2 M CH$_3$OH, oxygen feed, interdigitated flow field [75].

The CO$_2$-sensor based crossover measurement is instantaneous and it is carried out simultaneously with the polarization experiment. It takes into proper account the effect of the electro-osmotic drag. However, possible CO$_2$ permeation from the anode compartment through the membrane may cause some interference at high electrical current densities in the presence of thin membranes. Alternatively, the permeation can be measured in a separate experiment. An inert gas is fed to the cathode compartment and the electrode is polarized anodically while the methanol electrode is polarized cathodically. The measured anodic current is related to the methanol permeation rate through the membrane. This procedure discards two relevant phenomena that usually occur during practical fuel cell operation: electro-osmotic drag and methanol concentration gradient at the anode-electrolyte interface. Such effects are not reproduced in the driven mode. By using the CO$_2$ sensor method, it is observed that the equivalent current density usually decreases as a function of the electrical current density due to the methanol consumption at the anode/electrolyte interface, which reduces the methanol concentration gradient between the anode and the cathode (Figure 1.4).

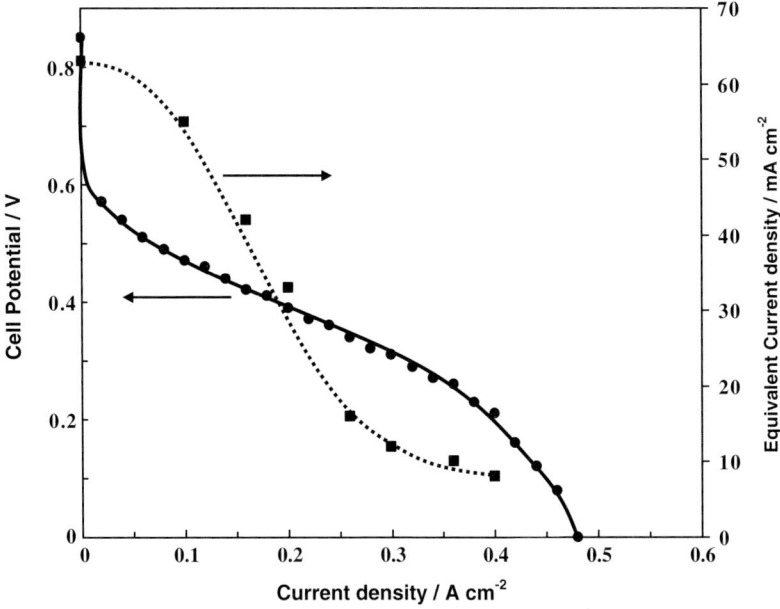

Figure 1.4 Cell potential and equivalent current density (due to methanol crossover) as a function of electrical current density for a DMFC operating at 60 °C.

1.2.2.3 Cathode Operating Conditions

Air is usually fed at stoichiometry of 2 in the active mode. If the system operates at ambient pressure, the power consumption by the blower, in the case of a large air flow, is not as significant as in the case of a compressor (pressurized DMFC). An increase of air flow produces better performance and it is not unusual to see DMFC experiments reported with an air flow corresponding to a stoichiometry of 5 or even higher.

A similar effect is produced by the cathode pressure. The negative effects of methanol poisoning at the cathode can be counteracted by an increase of oxygen partial pressure. The Temkin adsorption isotherm is often used to model the adsorption of oxygen at the cathode in PEMFCs [20]. Accordingly, an increase of oxygen partial pressure significantly influences the coverage of adsorbed oxygen species. This process is in competition with the adsorption of methanol permeated through the membrane, on the cathode surface. It appears that the increase of air stoichiometry especially favors the physical removal from the cathode of the liquid mixture of water/methanol that permeates through the membrane or is formed by the reaction (water) avoiding the electrode flooding. The flooding of the cathode is more significant in a protonic electrolyte DMFC than in a PEMFC due to the supply of plenty of liquid water together with methanol to the anode that permeates to the cathode through the hydrophilic membrane. This effect is less dramatic in a vapor-fed DMFC.

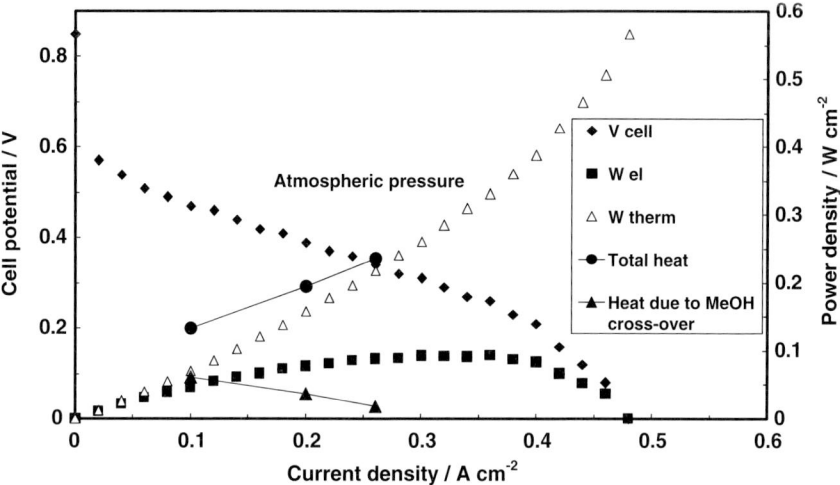

Figure 1.5 Electrical and thermal characteristics of a DMFC operating at 60 °C under atmospheric pressure.

1.2.2.4 Heat Production

For several applications, it is quite important to know how much heat is released during fuel cell operation. Thus, another useful polarization diagram to qualify the behavior of the fuel cell should include the heat released in the process deconvoluted into the heat produced by both the electrochemical process and that produced by the chemical reaction associated with methanol crossover (Figure 1.5). We are not aware of the use of such a plot in the literature. We have reported in Figure 1.5 the results of the heat release calculation that we carried out in the framework of European Community project called Morepower, which dealt with the development of a DMFC system for portable applications [21]. For sake of simplicity, the calculation reported here concerns single cell behavior.

The heat produced per unit of time by the electrochemical process only is derived from the reaction enthalpy and the methanol consumption rate from the following formula:

$$Q = \text{mol}_{\text{MeOH}}(\Delta H_r - nF\Delta E_{\text{cell}}); \tag{1.11}$$

mol_{MeOH} are the number of moles of methanol which are consumed per unit of time. This term is calculated from the electrical current density and the Faraday law; ΔE_{cell} is the cell voltage at the operating current density.

The redox process occurring at the cathode, associated with the methanol crossover, can be assumed as a chemical oxidation to CO_2 by the effect of the oxygen molecules, mediated by the Pt catalyst, since there is no electrical work produced:

$$Q = \text{mol}_{\text{MeOH}}\Delta H_r; \tag{1.12}$$

In Figure 1.5 the effect of heat release due to the electrochemical process is compared with that related to methanol crossover for a device operating at 60 °C with 1 M methanol and air feed at the cathode. The crossover decreases significantly as a function of electrical current density; accordingly, the amount of heat released diminishes. The heat due to the crossover is comparable to the heat due to the electrochemical process at 0.1 A cm^{-2} whereas the effect of crossover is much less significant above 250 mA cm^{-2}. The chemical energy that is dissipated as heat represents a net loss of efficiency. The heat released increases exponentially as a function of current and it reaches a maximum at the short circuit.

1.2.2.5 Cell Efficiency

At a defined current density, the voltage efficiency is thus defined as the ratio between the terminal cell voltage and the reversible potential for the process at the same temperature and pressure.

$$\eta_v = \Delta V / \Delta E_{rev} \qquad (1.13)$$

As a consequence of crossover, the current delivered by the DMFC device is smaller than that calculated on the basis of overall methanol consumption. The ratio between the measured electrical current (I) and that calculated from the Faraday law on the basis of the total methanol consumption (I_{total}) is defined as fuel efficiency:

$$\eta_f = I / I_{total} \qquad (1.14)$$

A determination of the fuel efficiency based only on the methanol crossover may represent a source of error if there are other side effects such as a loss of methanol by evaporation. For a passive DMFC, the overall efficiency can thus be expressed as:

$$\eta = \eta_{rev} \times \eta_v \times \eta_f \qquad (1.15)$$

In the active mode, the amount of energy consumed by the auxiliaries (pumps, blowers etc.) as compared with that delivered by the stack must be taken into consideration when the system efficiency is calculated.

1.2.2.6 Energy Density

Besides the performance and efficiency of the DMFC device, the energy density of the fuel plays a significant role in several practical applications, including transportation and portable power sources. It is also a relevant factor for stationary generation since it determines which infrastructure is appropriate for fuel distribution [22].

The energy density of a fuel is defined with respect to the weight (kWh/kg) or volume (kWh/l) as

$$W_e = (-\Delta G / 3600\, M) \text{ or } W_s = (-\Delta G\, \rho / 3600\, M); \qquad (1.16)$$

where M is the molecular weight (g/moles) and ρ the density (g/l).

Table 1.1 summarizes the energy density for various fuels. The gravimetric energy density of pure methanol is about one order of magnitude larger than that of H_2 stored in a pressurized tank (e.g., at 200 bar) and in a metal hydride system (4–5%). Similar considerations can be made with regard to the volume. The energy density of pure methanol is also much higher than Li-Ion batteries but lower than conventional liquid fuels used in transportation, such as gasoline and diesel (Figure 1.6).

To use all the potential energy density associated with methanol combustion, a tank with pure methanol should be used. When considering the range, that is, the driving

1 Direct Methanol Fuel Cells: History, Status and Perspectives

Table 1.1 Volumetric and gravimetric energy density for various fuels of technical interest for low temperature fuel cells.

Fuels	Volumetric Energy density (kWh l^{-1})	Gravimetric Energy density (kWh kg^{-1})
Diluted Hydrogen (1.5%)	—	0.49
Hydrogen	0.18 (@ 1000 psi, 25 °C)	—
Methanol	4.82 (100 wt.%)	6.1
Ethanol	6.28 (100 wt.%)	8
Formic acid	1.75 (88 wt.%)	—
Dimethyl ether (DME)	5.61 (in liquid of 100 wt.%)	8.4
Ethylene glycol	5.87 (100 wt.%)	5.3

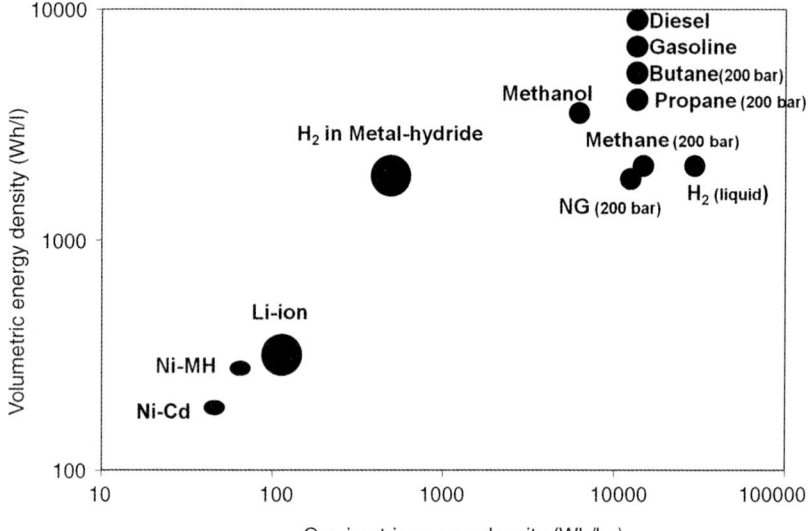

Figure 1.6 Gravimetric and volumetric energy density of various fuels/devices.

distance, of a fuel cell car compared with an internal combustion engine or the operating time of a methanol portable power source compared with a Li-battery, besides the energy density, the overall efficiency of the process should also be considered.

1.3
Historical Aspects of Direct Methanol Fuel Cell Development and State-of-the-Art

1.3.1
Historical Development of Methanol Oxidation Catalysts

The methanol electro-oxidation process was explored for the first time by E. Muller in 1922 [23]. However, the concept of methanol fuel cells started to be investigated in the early 1950s by Kordesch and Marko [24] as well as by Pavela [25]. Accordingly, much research on the anode and cathode electrocatalysts for such an application was initiated at the same time [1, 26]. Alkaline electrolytes were initially used for methanol fuel cells; the search for the active anode and cathode catalysts mainly regarded nickel or platinum for the MOR and silver for the oxygen reduction process [25, 26]. Parallel investigations of the MOR were conducted in acidic electrolytes such as sulfuric acid in the same period [3, 23]. It was observed that the kinetics of methanol electro-oxidation was slower in an acidic environment compared with the alkaline electrolyte. However, better perspectives were envisaged for the acid electrolyte based DMFCs. The main issue of a liquid alkaline electrolyte, such as KOH, was its chemical interaction with the reaction product of methanol oxidation, that is, carbon dioxide, to give rise to the formation of carbonate.

Among the pioneering studies carried out on catalysts for methanol oxidation in acidic media, the work of Cathro [27] that investigated the Pt-Sn system and that of Jansen and Molhuysen [28] represented the first attempts to make a screening of bimetallic catalysts using a systematic approach. Pt-Sn and Pt-Ru were isolated as the most promising anode formulations [28]. In effect, PtSn was initially a better catalytic system than PtRu [28]; this was essentially due to the use of the ad-atoms approach for both the formulations. Successive studies by Watanabe and Motoo [29] in the 1960s showed the large potentialities of the Pt-Ru system especially when Pt and Ru were combined in a solid solution (face centered cubic (fcc) alloy).

Research on DMFCs initially addressed the search of active anode formulations; half-cell studies proved that the methanol oxidation process was slower than the oxygen reduction; thus, the anode reaction attracted more interest as the rate determining step (r.d.s.) of the overall DMFC process [1, 2]. The first decades of activity on anode catalysts mainly addressed the investigation of the mechanism and the search for new or improved catalyst formulations. One of the first attempts to rationalize the methanol oxidation process was by Bagotzky and Vassiliev [30]. Their work was essentially carried out on pure platinum; they proposed some relevant kinetic equations for the methanol electro-oxidation rate as a function of the coverage of methanolic residues and oxygen species adsorbed on the electrodes. These studies served as a basis for the successive formulation of the bifunctional theory [29] for bimetallic catalysts. Also worth mentioning is the work of Shibata and Motoo on the effect of ad-atoms [31] that individuated the influence of steric effects. Of relevant interest also were the successive attempts of McNicol [32], Parsons and Vander-Noot [33] and Aramata [34] to further rationalize the mechanism of methanol oxidation by electrochemical studies. However, an in-depth analysis of the methanol oxidation process, initially at smooth electrode surfaces, was made possible by the use of spectro-electrochemical methods. This work was carried out by several groups including those of Lamy [35], Bockris [36] and Christensen [37]. These studies essentially investigated adsorbed methanolic residues by infrared spectroscopy whereas, for the adsorbed oxygen species, ellipsometry gave appropriate results [38]. Further knowledge of the methanol oxidation process was provided by the use of *in situ* mass spectrometry. This method allowed detection of the anode potentials at which CO_2 was formed. In the late 1980s and beginning of the 1990s, a relevant amount of work addressed the amelioration of catalyst formulations and investigated further the structural, surface and electronic properties of the most promising formulations, essentially Pt-Ru. The work carried out by Goodenough, Hamnett and Shukla in the 1980s and 1990s [39, 40] was of relevant interest in this regard. They focused their attention not only on the catalyst but also on the electrode structure (including diffusion and backing layers). Mc Breen and Mukerjee [41], Ross et al. [42] used advanced physico-chemical tools such as extended X-ray absorption fine structure (EXAFS), Low-energy ion scattering spectroscopy, Auger and X-ray photoelectron spectroscopy (XPS) to characterize model and practical anode and cathode catalysts.

The 1990s opened a new era for the DMFCs; the investigation of catalysts formulations in polymer electrolyte single cells progressively replaced the half-cell studies in liquid electrolytes. The number of anodic formulations investigated reduced consistently. More attention addressed the behavior of the catalyst inserted in a practical MEA in a single cell. It was observed that operation at high anode potentials often caused Ru dissolution and migration through the membrane to the cathode [43]. The extensive use of combinatorial catalyst discovery studies in the 1990s suggested that multifunctional catalysts could be of significant importance [44].

It had been established in the 1970s and 1980s that the activity of a methanol oxidation catalyst depends on several factors, including catalyst formulation, support, electrode structure and operating conditions. Most of the work was concentrated on examining the effect of changing the catalyst formulation as a means of enhancing catalytic activity; alloys of various compositions were used as electrode materials although most of these alloys were based on Pt. It was evident that the reaction rate was improved by electrocatalysts adsorbing water and/or oxygen species at potentials similar to the reversible potential of the CH_3OH oxidation reaction and/or able to minimize poisoning by the methanolic residue [45, 46].

It was recognized that the presence of an alloying metal or ad-atom either: (i) modified the electronic nature of the surface; (ii) modified the physical structure; (iii) blocked the poison formation reactions; (iv) adsorbed oxygen/hydroxyl species which take part in the main oxidation reaction.

The following analysis tries to bring to a 'rationale' the main features of different bimetallic Pt-based electrocatalysts, including ad-atoms [46] (Figures 1.7–1.9). The following aspects are considered: (a) the influence of the interaction energies of CO on transition metals; (b) the influence of electronegativity; (c) possible steric and electronic effects, reflected by the influence of atomic radius and ionic potential values. As for the involvement of different poisons in the oxidation mechanism, spectroelectrochemical methods in the 1980s established that the two most probable adsorbed species were −CHO and CO with evidence to support both possibilities [33]. On the whole, the evidence for CO as the poisoning species appeared to be, however, more conclusive, coming from various 'in situ' spectroscopic techniques. Accordingly, both linear and bridge-bonded CO species have been detected on the electrode surface. Furthermore, the potentials where the poison was oxidized also corresponded very closely to those where adsorbed species from pure CO were oxidized in separate experiments [33]. A plot of several electro-catalytic activity results [28, 46–48] as a function of the adsorption heat of CO on transition metals taken from

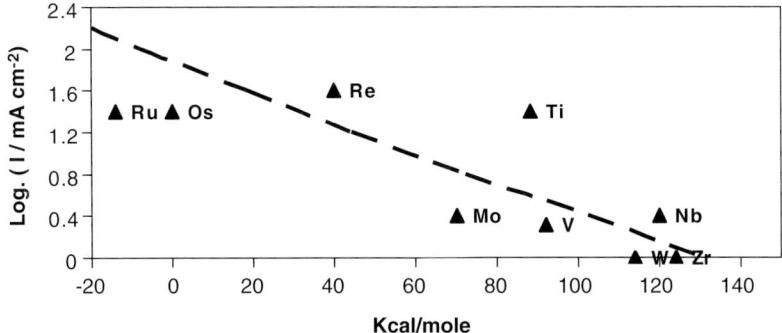

Figure 1.7 Methanol electrocatalytic activity vs. calculated interaction energy of CO adsorption on various Pt alloyed transition metals.

Miyazaki [49] is reported in Figure 1.7. It appears that higher electro-oxidation activity is found in metals having low interaction energies with the CO molecule. The experiments of Shibata et al. [31] on the enhancement of CO oxidation on Pt by the electronegativity of ad-atoms showed that a strong interaction between the ad-atom site and either the hydrated hydrogen ion or the adsorbed CO molecule was required to obtain high enhancement effects. Yet, such evidence was not observed for the CH_3OH oxidation reaction (Figure 1.8); in our opinion, this can be explained by the detrimental effects of non-transition elements upon the methanol dehydrogenation which, as is well known, constitutes the first step in the overall electro-oxidation reaction [2, 50]. Accordingly, the positive influence of the more electronegative elements still holds for transition metals which are known to favor CH_3OH dehydrogenation. According to the ad-atoms theory, two distinct effects were identified from a catalytic view-point in enhancing the electro-oxidation of organic molecules; the first was related to the modification of the electronic environment of the adsorption site, the other was linked to the steric factor which also influenced both the extent and the strength of the adsorption process. As for the latter, a correlation

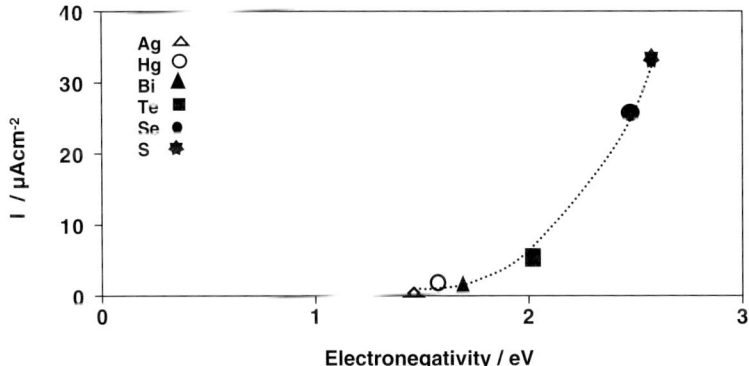

Figure 1.8 Methanol electro-oxidation activity vs. electronegativity of the ad-atoms used for Pt modification.

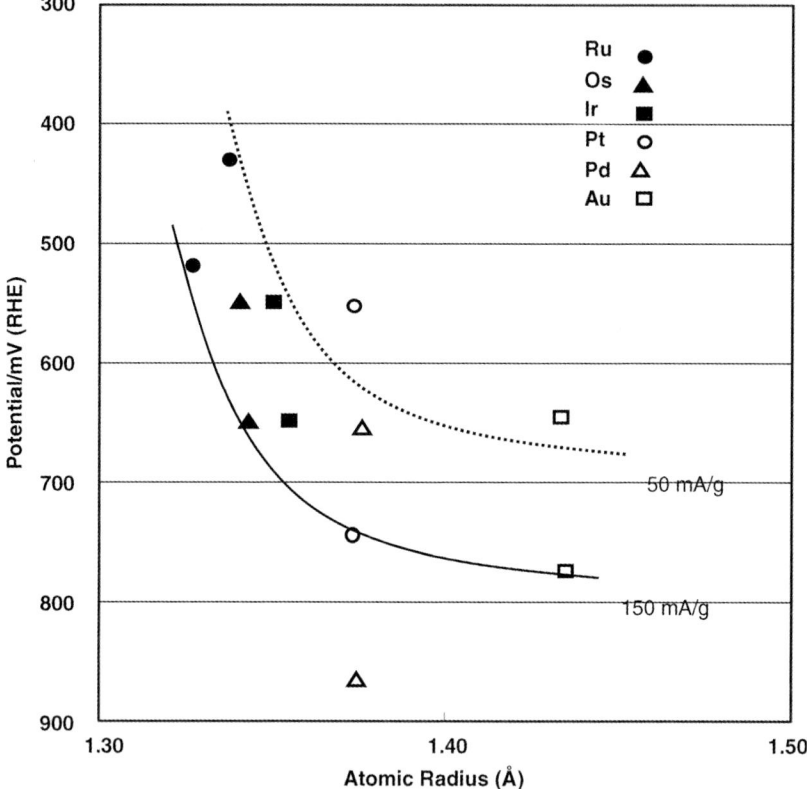

Figure 1.9 Methanol electro-oxidation activity vs. atomic radius of various metals used to form Pt-alloys.

between the overpotential at which a sustained current is obtained for CH_3OH electro-oxidation and the atomic radius of the alloyed metal with Pt was envisaged (Figure 1.9). The combination of labile adsorption intermediates with a large metal area for use was recognized to be favored by the small dimensions of the alloyed element. Regarding the influence of ad-atoms on the modification of the electronic environment of Pt, the literature predicted that the positive catalytic effects of Pt-Ru and Pt-Sn formulations were due to the adsorption, by these elements, of active oxygen on the catalyst surface at low potentials. These species were identified in the late 1980s as adsorbed OH species by using spectro-electrochemical methods.

1.3.2
Status of Knowledge of Methanol Oxidation Process and State-of-the-Art Anode Catalysts

1.3.2.1 Oxidation Mechanism

The detailed mechanism of methanol oxidation has been elucidated in the last three to four decades by using a variety of experimental procedures [1, 2, 8, 33, 51–63]. This mechanism is discussed in detail in several reviews [1, 2]. Studies found that the electrochemical oxidation of methanol on Pt involves several intermediate steps: dehydrogenation, CO-like species chemisorption, OH (or H_2O) species adsorption, chemical interaction between adsorbed CO and OH compounds, and CO_2 evolution [2]. One of these steps is the rate determining step (r.d.s.) depending on the operation temperature and particular catalyst surface (crystallographic orientation, presence of defects, etc.) [37, 64–66]. The state-of-the-art electrocatalysts for the electro-oxidation of methanol in fuel cells are generally based on Pt alloys supported on carbon black [51, 52, 67] or high surface area unsupported catalysts [68]. The electrocatalytic activity of Pt is known to be promoted by the presence of a second metal, such as Ru or Sn, acting either as an ad-atom or a bimetal [2]. According to the bifunctional theory, water discharging occurs on Ru sites with formation of Ru–OH groups on the catalyst surface [38]:

$$Ru + H_2O \rightarrow Ru - OH + H^+ + 1e^-$$

The final step is the reaction of Ru–OH groups with neighboring methanolic residues adsorbed on Pt to produce carbon dioxide [2]:

$$Ru - OH + Pt - CO \rightarrow Ru + Pt + CO_2 + H^+ + 1e^-$$

1.3.2.2 Pt-Ru Catalysts

The Pt-Ru binary alloy electrocatalyst appears as the most promising formulation. Pt sites in Pt-Ru alloys are especially involved in both the methanol dehydrogenation step and strong chemisorption of methanol residues. Although, the subject still remains controversial [29, 54, 55, 63, 69], an optimal Ru content of 50 at.% in carbon supported Pt-Ru catalysts for the MOR at high temperatures (90–130 °C) was found [70]. The optimum Ru surface composition is referable to the relevant synergism accomplished by a Pt-Ru surface with 50% atomic Ru in maximizing the product of θ_{OH} (OH coverage) and k (intrinsic rate constant), assuming the surface reaction between CO_{ads} and OH_{ads} as r.d.s. At low temperatures, adsorption of methanol on Pt requiring an ensemble of three neighboring atoms appears as the r.d.s. Gasteiger et al. [69] have observed that methanol oxidation occurs more readily at room temperature on pure Pt-Ru alloys having low Ru content ($\approx$10%) whereas at intermediate temperatures (60 °C) the reaction is faster on alloys with increased Ru content ($\approx$33%). At both intermediate and high temperatures, the removal of strongly adsorbed carbon monoxide by OH species is usually considered the r.d.s.

The synergistic promotion exerted by Pt-Ru alloys is supported by X-ray absorption analysis [41]. Accordingly, an increase of Pt d-band vacancies is produced by alloying with Ru; possibly, this modifies the adsorption energy of methanolic residues on Pt. Such evidence suggests that the reaction rate is not only dictated by the bifunctional mechanism but it is also influenced by electronic effects occurring on account of the interaction between Pt and Ru [41, 71].

The promoting effect of the RuO_x species for the MOR has been extensively investigated by several authors [72–74]; a very high performance was obtained in a DMFC with unsupported Pt-RuO_x anode electrocatalyst [9]. It was suggested that facile oxygen transfer from Ru to Pt rich regions where adsorption of CO-like residues preferentially occurs could enhance the catalytic oxidation of methanol [74, 75].

The formation of oxidized species of Pt and Ru as well as the electronic properties of the active phase are also influenced by the metal-support interaction. Various catalysts characterized by different concentrations of metal phase on carbon have been investigated [76]. A comparison of the *in situ* stripping behavior of adsorbed methanolic residues for three Pt-Ru/C catalysts at various temperatures is shown in Figure 1.10 [76]. As the temperature increased above 90 °C, the stripping area of the

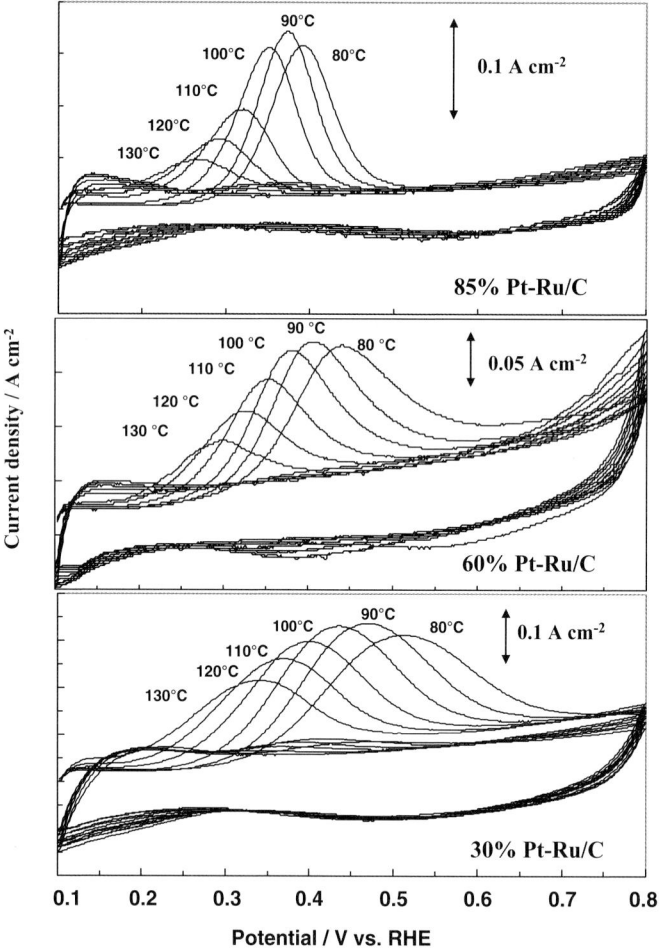

Figure 1.10 *In situ* stripping voltammetry of methanol residues at the various carbon-supported Pt-Ru catalyst/Nafion 117 membrane interfaces, at various temperatures under the DMFC configuration. Anode: 1 M methanol, 1 atm rel. adsorbed for 30 min; cathode: H_2 feed 1 atm rel. [76].

methanolic residues decreased progressively for all catalysts, whereas the peak shifted towards lower potentials on account of the decrease of the activation energy for CO removal. By comparing the behavior of the various catalysts, it was observed that the 30% Pt-Ru/C sample was characterized by the largest stripping area per unit of weight. Yet, the stripping peak potential at each temperature was shifted towards negative values for the 85% PtRu/C catalyst. XPS analysis of Pt 4f spectra (Figure 1.11)

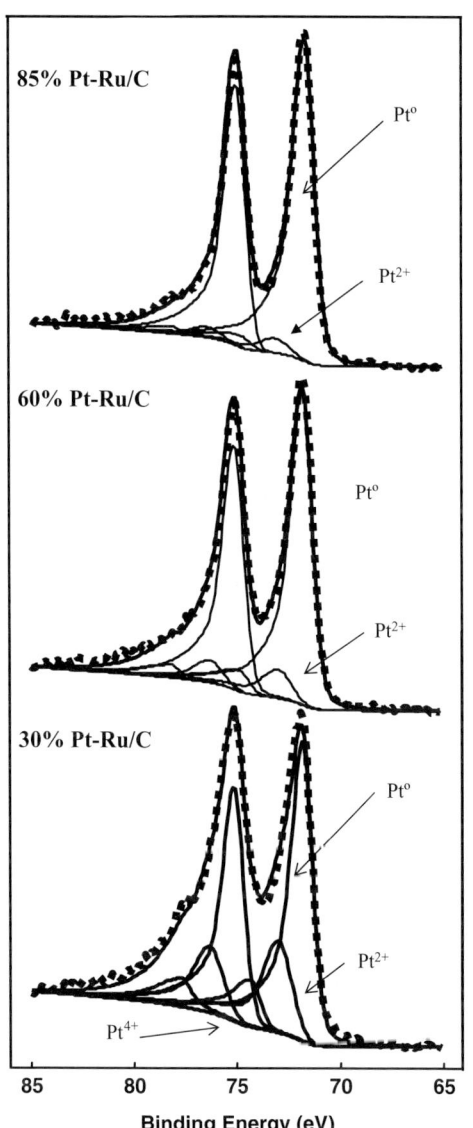

Figure 1.11 X-ray photoelectron spectra of Pt-Ru catalysts (Pt 4f doublet) [76].

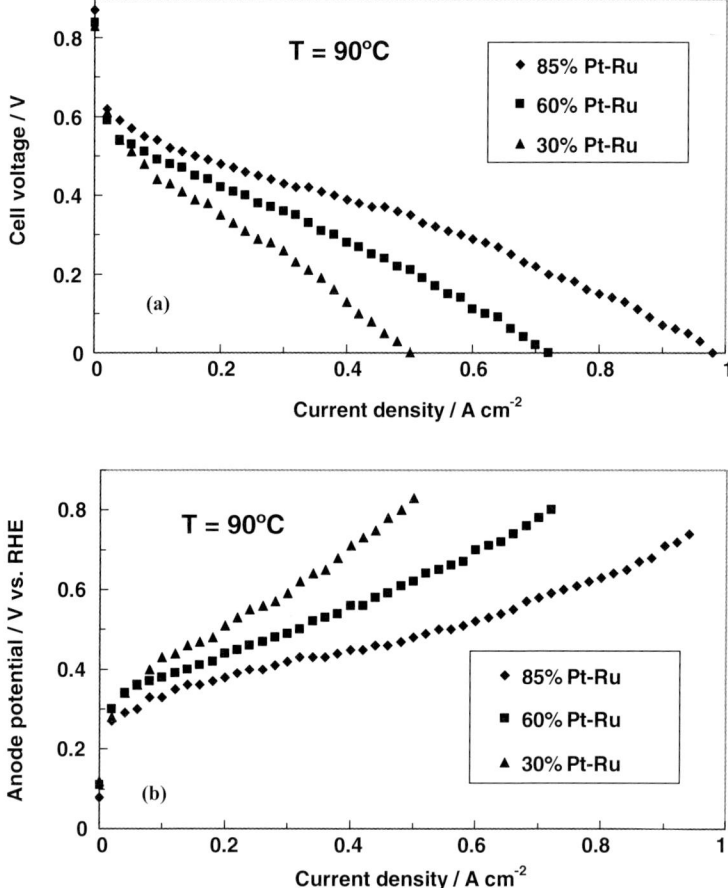

Figure 1.12 DMFC single cell (a) and anodic half-cell polarization behavior (b) at 90 °C for various Pt-Ru/C catalysts. Anode: 1 M methanol, 1 atm rel.; cathode: air feed 2.5 atm (a), H_2 feed 1 atm rel. (b) [76].

showed a shift to higher binding energies with a larger fraction of oxidized species for the catalysts with lower concentrations of metallic phase on carbon, that is, 30 and 60% Pt-Ru/C as an effect of metal support interaction. By comparing the behavior of the three different catalysts either in terms of single cell and half-cell polarization curves at 90 °C (Figure 1.12) [76], better performance was achieved for the catalyst showing both lower stripping peak potentials (as expected) but also lower coverage of methanolic residues. Thus, the higher intrinsic catalytic activity (lower activation barrier) appears to be more relevant than catalyst dispersion.

Significant interest has recently addressed the development of decorated catalysts [77, 78]. Pt-nanoparticles on the surface of a less expensive metal, which participates in the reaction, may represent a useful approach to reduce the incidence

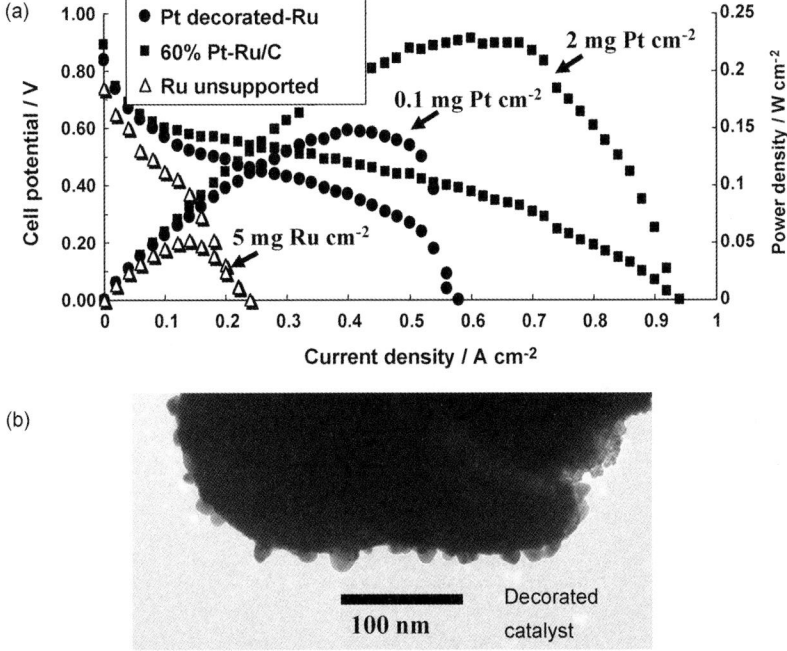

Figure 1.13 (a) DMFC single cell polarizations at 130 °C for commercial Pt-Ru/C, Pt-decorated and bare unsupported Ru catalysts. Anode: 1 M methanol, 2 atm rel.; cathode: air feed 2.5 atm., (b) transmission electron micrograph of Pt-decorated Ru catalyst [80].

of the catalyst cost in DMFC devices. Figure 1.13 shows a comparison of the DMFC polarization curves in the potential range of technical interest for the decorated (0.1 mg Pt cm^{-2}), carbon supported Pt-Ru alloy (2 mg Pt cm^{-2}) and bare unsupported Ru catalysts [79, 80]. The carbon supported Pt-Ru alloy based electrode shows lower potential losses in the activation controlled region than the decorated catalyst. However, the Pt loading in the decorated catalyst based anode is 20 times lower.

1.3.2.3 Alternative Anode Formulations

Only a few electrocatalyst formulations, alternative to Pt, have been proposed for methanol electro-oxidation in an acidic environment. These are mainly based on non-noble transition metal alloys like NiZr [81], transition metal oxides and tungsten-based compounds [82–84]. All these materials showed lower reaction rates than Pt-based electrocatalysts and, thus, such unsatisfactory preliminary results have not stimulated much work in these directions.

1.3.2.4 Practical Anode Catalysts

In the case of methanol electro-oxidation on carbon-supported Pt electrocatalysts, two different trends were observed. For what concerns anode catalyst morphology,

McNicol et al. [8] observed in their Pt electrocatalysts maximum activity at about 80 m^2/g surface area. Another group has shown that the specific activity increases as a function of particle size [85]. Thus, a maximum in mass activity vs. particle size should be observed as in the case of oxygen reduction [86]. On the other hand, Watanabe et al. [87] found that the specific activity for methanol oxidation on a carbon supported Pt electrocatalyst does not change for a particle size above 2 nm (Pt fcc structure); thus, the mass activity increases as the dispersion of the metal phase is increased [87]. These latter findings have been in part confirmed for the Pt-Ru system for a particle size above 3 nm [88]. A poor catalytic performance was observed for catalysts with mean size of about 1–1.5 nm compared with the conventional catalysts; it was also observed that the structure is mainly amorphous in that range of particle size [89].

A high Pt wt.% on the carbon substrate will significantly decrease the anode thickness for the same Pt loading per geometric electrode area (e.g., 1 mg m^{-2}). Thus, it is possible to enhance mass transport through the electrode and, at the same time, reduce the ohmic drop. However, it has been found that an increase in Pt loading (above 40 wt.%) on the carbon support decreases the dispersion of the electrocatalyst, due to some particle agglomeration.

The synthesis of a highly dispersed electrocatalyst phase in conjunction with a high metal loading on carbon support is one of the goals of the recent activity in the field of DMFCs. The mostly used carbon blacks were: Acetylene Black (BET Area: 50 m^2/g), Vulcan XC-72 (BET Area: 250 m^2/g) and Ketjen Black (BET Area: ∼900 m^2/g) [40].

1.3.2.5 Anode Catalysts for Alkaline DMFC Systems

The methanol oxidation rate is accelerated at high pH values. Thus, from a kinetic viewpoint, it is advantageous to carry out methanol oxidation in alkaline electrolytes [90]. Furthermore, because the corrosion constraints are less significant in alkaline media, in principle, a wider number of catalyst formulations can be investigated for methanol oxidation than for proton conducting electrolytes. Despite these promising aspects, studies on the development of anode catalysts for alkaline DMFCs are less numerous than in acidic electrolytes. Due to the enhanced reaction rate at high pHs, alkaline DMFCs can employ non-precious transition metals, for example, Ni [90], which are characterized by low intrinsic activity. Although the reaction rates are usually faster for Pt-Ru than Ni, the increase of reaction kinetics due to the increased pH compensates, in part, for this gap in intrinsic activity [90]. The Ni-based catalysts can operate suitably in combination with a liquid electrolyte containing a concentrated base such as 5 M KOH or NaOH, characterized by a high pH. For practical purposes, anion exchange membranes have recently been preferred to the liquid electrolyte [91]. However, due to carbonation occurring during steady-state operation, the electrolyte in the anode compartment turns progressively into a carbonate/bicarbonate mixture with corresponding lower pH than the KOH solution. Furthermore, the conductivity decreases. Pt-electrocatalysts have mainly been considered for operation in conjunction with anion exchange membranes. These include the conventional Pt/C catalyst, platinized Ti electrodes [92] and Pt-Ru alloys [90]. Platinized mesh anodes have shown higher catalytic activities than conventional Pt/C electrodes. Due to the open area of the mesh, the liquid can easily reach the interface reducing mass transport resistance [92]. The typical activation losses recorded for methanol oxidation at PtRu in acidic systems are less accentuated in alkaline media. The methanol oxidation rate at PtRu electrodes in a carbonate/bicarbonate mixture is about 8 and 2.5 times larger than in sulfuric acid, at 0.35 and 0.45 V RHE, respectively [90]. However, the reaction kinetics in the presence of a Nafion®-type ionomer are faster than in sulfuric acid due to the absence of sulfate anion adsorption on the catalyst surface.

1.3.3
Technological Advances in Electrolyte Development for DMFCs

The electrolytes that were first used in DMFC devices in the 1950s consisted of a concentrated alkaline solution for example, KOH which contained dissolved methanol [3, 25, 26, 93, 94]. Although some attempts were also carried out with anion exchange membranes onto which the electrodes were pressed [95], using an approach similar to the modern PEMFCs, the best results in terms of output power were achieved with 5 M KOH [94].

Alkaline instead of protonic electrolytes were initially selected because methanol oxidation in alkaline media is faster than in the presence of acidic electrolytes [95]. This made possible the use of low cost catalysts, for example, Ni at the anode and silver at the cathode [94]. Unfortunately, there were several practical constraints that convinced most of the DMFC developers to abandon the alkaline electrolyte approach [2, 3, 90, 96] and focus their attention on protonic electrolytes. This approach was undertaken despite the fact that the number of possible catalyst formulations was restricted to those stable in acidic environments [90, 96]. The main problem of alkaline electrolytes consisted of the acid–base reaction between CO_2 and the alkaline solution with carbonate precipitation in the catalyst pores. The lack of adequate alkaline polymer electrolyte membranes with conductivity comparable to Nafion [97, 98] retarded the development of a new generation of anionic membrane direct methanol fuel cells (AMDMFC) [95] Most of the anion-exchange membranes used in the past required KOH recirculation at the anode [90, 95]. Regeneration of this electrolyte due to carbonation and dilution was considered a significant constraint for practical applications. The use of carbonate/bicarbonate media instead of KOH partly reduced the kinetic advantage of alkaline media over protonic electrolytes for methanol electro-oxidation [95]. Proton conducting solid polymer electrolyte membranes for hydrogen-fed fuel cells were initially developed at General Electric for the Gemini Earth-orbiting program in the early 1960s [99]. These devices were based on a polystyrene sulfonic acid membrane that exhibited poor oxidative stability, thus no suitable long-term performance [99]. The poor stability was due to the oxidation of the C—H bonds occurring at high potentials at the cathode, especially in the presence of hydrogen peroxide-type radicals. A large improvement in terms of stability was achieved when Nafion replaced the sulfonated polystyrene-divinylbenzene membrane in the late 1960s [97, 98]. Nafion was originally developed for chloro-alkali electrolyzers; its peculiar characteristics rely on the excellent chemical, electrochemical stability and high proton conductivity that is derived from its unique chemical structure [99].

Despite the large methanol crossover shown by Nafion membranes, these membranes were the most used electrolyte in DMFCs in the 1990s [97]. Nafion 117 is still considered a standard electrolyte to compare the performance, conductivity, and methanol crossover of alternative or newly developed membranes for DMFCs [1, 2]. As alternative to Nafion, Hyflon® was employed with success in DMFCs especially at high temperatures [100]. Hyflon is characterized by the presence of short side chains and an equivalent weight smaller than Nafion [100, 101]. The peculiar characteristic of Hyflon is a glass transition temperature higher than Nafion, which makes this polymer more stable at high temperatures [100]. Prior to the advent of Nafion in DMFCs, which essentially occurred in the late 1980s, various acid electrolytes were used for DMFCs, such as sulfuric acid and phosphoric acid. In these devices, the anode and cathode were separated by a ceramic matrix, for example, a porous silicon carbide separator impregnated with the acidic electrolyte. The electrodes were impregnated with the same acid. Yet, due to extremely high levels of methanol crossover, this concept was abandoned in favor of Nafion. However, in the late 1990s, a similar approach was developed by Peled et al. at Tel Aviv University using a nanoporous membrane based on poly(ethylene oxide) (PEO) filled with silica nanoparticles and impregnated with sulfuric acid or trifluoromethanesulfonic acid (TFMFSA) [102]. The projected cost of this nanoporous membrane filled with sulfuric acid was considered quite low compared with Nafion. However, the best performance was achieved with TFMFSA due to the lower adsorption of anionic species from the electrolyte on the electrode [103].

The strong activation control of the MOR indicated in the high temperature operation, the most useful strategy to improve performance. High temperature operation allowed the achievement of high current densities with consequent fast methanol consumption at the anode/electrolyte interface. This effect reduces the concentration gradient, allowing a decrease in methanol crossover. In this regard, the use of thin membranes like Nafion 112 was sometimes adopted for high temperature operation [9, 104]. The most promising strategies to increase the operating temperatures concerned the use of phosphoric acid-doped polybenzoimidazole membranes operating at about 180–200 °C [105] and composite perfluorosulfonic acid membranes operating up to 145 °C, containing inorganic fillers such as silica, zirconium phosphate, heteropolyacid-doped silica, and titanium oxide [106, 107]. In several attempts, the filler was formed in situ, for example, silica was synthesized inside the membrane by using a sol-gel type procedure using tetraethyl orthosilicate (TEOS) as precursor [108]. Although both approaches were demonstrated to be appropriate for extending the operating temperature range, the main constraint of phosphoric acid-doped polybenzoimidazole was represented by the leaching of acid molecules from the membrane in the presence of hot methanol whereas a composite membrane operated properly at 145 °C in the presence of 3 bar abs. pressure [107]. Subsequently, it was shown that the water retention properties in composite membranes were promoted by the presence of acidic functionalities on the filler surface [107]. Operation at high temperature (145 °C) and reasonable pressure (1.5 bar abs) with acceptable level of performance was made possible by amelioration of membrane properties [107–109]. The composite membrane approach was also extended to membranes alternative to Nafion such as sulfonated poly(ether ether ketone) (S-PEEK) and polysulfone-type [110].

Regarding membrane stability, some aspects related to the operating conditions appear less critical in a DMFC than in a hydrogen-fed fuel cell. For example, in a methanol fuel cell, the cathode never experiences electrochemical potentials above 1 V. Furthermore, the formation of hydrogen peroxide radicals, which can cause significant membrane degradation, occurs mainly in a PEMFC by effect of hydrogen crossover to the cathode. On the other hand, it is observed that the presence of hot concentrated methanol in DMFCs may increase membrane swelling. However, due to the lower electrochemical stability requirements, the range of membranes explored for DMFCs appears larger than that for PEMFCs [2, 101]. Recently, partially fluorinated, non-fluorinated aromatic polymers, radiation grafted ethylene tetra-fluoroethylene (ETFE)-based membranes, acid–base blends and so on [2, 101], have been explored as alternatives to Nafion. Several excellent reviews have already been published on this topic [96, 101]. Most of these alternative electrolytes have the characteristics of lower methanol crossover but also less conductivity than Nafion (as mentioned above these aspects are often interrelated), and, especially, the projected costs appear quite promising compared with classic perfluorosulfonic membranes. Among the various proposed membranes, S-PEEK [111], despite its promising properties in terms of conductivity, fuel permeation and costs, still seems affected by significant swelling; the properties of sulfonated poly(aryl-ether)-type membranes such as polysulfone or polyimide ionomer membranes as well as acid–base blends appear more promising [99, 110, 112].

More recently, several attempts have been carried out to develop a new generation of alkaline anion exchange membranes (AAEMs). The availability of new anionic polymers, with conductivity approaching values that are half of the conductivity of Nafion [91, 92], but characterized by much lower methanol crossover, has given new emphasis to the development of alkaline methanol fuel cells. The new membranes significantly reduced the drawbacks associated with conventional aqueous KOH electrolyte fuel cells that is, carbonate formation and the need to frequently regenerate the electrolyte. OH^- ions, necessary for ion conduction, are formed at the cathode by the water added to humidify the oxidant. These ions migrate to the anode reducing methanol crossover by the electro-osmotic drag. The alkaline environment allows the use of non-noble metal catalysts (usually unstable in the acidic environment); in fact, catalyst corrosion and membrane degradation problems are significantly mitigated due to the high pH [91]. Accordingly, cheap catalysts and hydrocarbon-only membranes have been explored [90, 91, 95, 96]. There are, however, some drawbacks which concern the formation of a pH gradient between anode and cathode [95], the need of cathode humidification (protonic membranes based DMFCs are usually fed by dry air) and the need to increase the operating temperature to enhance conductivity, which may be not useful for portable applications.

1.3.4
State-of-the-Art DMFC Electrolytes

1.3.4.1 General Aspects of DMFC Electrolyte Development

It is widely recognized that the electrolyte is a key component in DMFCs. The electrolyte determines the fuel permeation rate and the choice of the catalysts, and influences the reaction rate. The standard electrolyte membrane for DMFCs is usually a perfluorosulfonic acid membrane such as Nafion, which is also widely used in PEMFCs. Most of the electrolyte alternatives to Nafion, both the proton-conducting and alkaline type, are cheaper than the classic perfluorosulfonic membranes used in PEMFCs; in some cases, they are also characterized by lower methanol crossover; however, life-time characteristics similar to those shown by Nafion-type membranes in fuel cells (60 000 h of operation) have not been achieved yet with the alternative membranes [97, 98]. Concerning conductivity, only recently membrane alternatives to Nafion-type have shown similar levels of performance. One critical aspect is related to the fact that the presence of water is a requirement of low-temperature DMFCs for the occurrence of the electrochemical reactions and to promote ion conductivity. As methanol is highly soluble in water, the transport of water through the membrane is commonly associated with methanol permeation. This effect is more critical with protonic membranes because, besides methanol transport due to the concentration gradient (diffusion), there is an effect due to the electro-osmotic drag. High ionic conductivity is often associated with the presence of high levels of water uptake by the membrane whereas what is required is a low water uptake. These aspects are mainly related to polymer electrolyte membrane DMFCs. No drawbacks in terms of methanol crossover with consequent cathode poisoning and poor anode reaction kinetics are envisaged in intermediate temperature solid oxide fuel cells (IT-SOFCs) which employ dense ceramic anionic electrolytes and operate at 500°–750°C. However, such devices are less suitable for most of the applications of methanol fuel cells including portable and assisted power units (APU). Thus, the research efforts on low temperature methanol fuel cell membranes have addressed improving the conductivity, reducing the crossover of methanol and catalyst degradation. The latter regards, for example, dissolution of Ru and Ru ion migration from the anode to the cathode [43], Co or Fe dissolution from the cathode into the membrane (e.g., in the case of a PtCo or PtFe alloy cathode) and so on. Other important aspects are the increase of chemical and electrochemical stability, reduction of water uptake and swelling, extension of the operating temperature range and finally cost reduction for market application.

1.3.4.2 Proton Conducting Membranes

In the present section, we have restricted our discussion to the characteristics of membranes that are actually considered for commercial DMFC systems. Several excellent reviews have been published on this topic [101, 113–116]. Perfluorosulfonic polymer electrolyte membranes are currently used in H_2/air and methanol/air fuel cells because of their excellent conductivity and electrochemical stability [98]. Unfortunately, they suffer several drawbacks such as methanol crossover and membrane dehydration. The latter severely hinders fuel cell operation above 100 °C, which is a prerequisite for high rate oxidation of small organic molecules involving the formation of strongly adsorbed reaction intermediates such as CO-like species [2]. Since methanol is rapidly transported across perfluorinated membranes and is chemically oxidized to CO_2 and H_2O at the cathode, there is a significant decrease in coulombic efficiency for methanol consumption, as much as 20% under practical operating conditions [117]. Thus, it is very important to modify these membranes by, for example, developing composites or finding alternative proton conductors with the capability of inhibiting methanol transport. It is generally accepted that a solid-state proton conductor is preferable for liquid fuel fed DMFCs because it hinders corrosion and rejects carbon dioxide (produced during the methanol oxidation). However, there are some prerequisites that should be properly considered. The polymer electrolyte should have a high ionic conductivity (5×10^{-2} ohm^{-1} cm^{-1}) under working conditions and low permeability to methanol (less than 10^{-6} moles min^{-1} cm^{-2}). Furthermore, it must be chemically and electrochemically stable under operating conditions. These requirements appear, potentially, to be met by new classes of solid polymer electrolytes that show promising properties. Alternative membranes based on poly(arylene ether sulfone) [118], sulfonated poly(ether ketone) [119] or block co-polymer ion-channel-forming materials as well as acid-doped polyacrylamide and polybenzoimidazole have been suggested [116, 118–120]. Various relationships between membrane nanostructure and transport characteristics, including conductivity, diffusion, permeation and electro-osmotic drag, have been observed [121]. Interestingly, the presence of less connected hydrophilic channels and the wider separation of sulfonic groups in sulfonated poly(ether ketone) reduces water/methanol permeation and electro-osmotic drag with respect to Nafion while maintaining high protonic conductivity [121]. Furthermore, an improvement in thermal and mechanical stability has been shown in nano-separated acid–base polymer blends obtained by combining polymeric N-bases and polymeric sulfonic acids [119].

The perfluorinated polymers such as polysulfones, polyetherketones, and polyimides usually combine reduced crossover and appropriate conductivity levels. However, the conduction mechanism in these systems is not very different from that in Nafion. Thus, methanol crossover cannot be eliminated completely. It appears that the main advantage of these polymers is cost reduction with respect to Nafion. Alternatively, membranes such as phosphoric acid impregnated-polybenzoimidazole (PBI) which do not need water transport to maintain high proton conductivity may represent a valid approach [105, 122]. However, these electrolytes still present methanol crossover effects; moreover, suitable DMFC life-time for such membranes has not been yet demonstrated. Other issues are cold start-up and low temperature operation.

In principle, the water uptake properties of sulfonic acid-based membranes may be modulated by selecting the proper concentration and distribution of sulfonic groups inside the polymer. Such an objective is generally pursued in the preparation of grafted polymer membranes [123]. The application of the radiochemical grafting technique to the production of DMFC membranes has been explored in the framework of the Nemecel European program (see Section 1.4.2). In this procedure the material properties may be properly tailored by varying a few parameters in the synthesis while maintaining the process characteristics and plants for large scale production. The main efforts are addressed to reduce the cost of production through a flexible preparation process and the proper selection of cheap base materials. For this purpose films of ETFE (which has C_2H_2 and C_2F_2 groups with 1/1 ratio and nearly perfect alternance) have been selected as substrate material; the films were radio-chemically grafted with styrene and subsequently sulfonated in order to obtain sulfonic acid anchored groups. The present cost of the base irradiated ETFE material compares favorably with the average industrial cost of the commercial perfluorinated sulfonic membranes. In order to improve the mechanical strength properties of the polymer, increase the thermal resistance and reduce the crossover of gases or liquids (such as methanol) through it, while maintaining suitable conductivity, appropriate crosslinking was made during the grafting step by adding a crosslinking agent [123].

Considerable efforts in the last decade have addressed the development of composite membranes. These include ionomeric membranes modified by dispersing insoluble acids, oxides, zirconium phosphate, and so on, inside their polymeric matrix; other examples are ionomers or inorganic solid acids with high proton conductivity embedded in porous non-proton-conducting polymers [120]. In an attempt to reduce the drawbacks of perfluorosulfonic membranes, nanoceramic fillers have been included in the polymer electrolyte network. Stonehart, Watanabe and co-workers [124] have successfully reduced the humidification constraints in PEMFCs by the inclusion of small amounts of SiO_2 and Pt/TiO_2 (~7 nm) nanoparticles to retain the electrochemically produced water inside the membrane. This approach was used in DMFCs to increase operating temperatures (up to 145 °C) and reduce methanol crossover by increasing the tortuosity factor for methanol permeation [106]. Although it has been hypothesized [116, 120] that the inorganic filler induces structural changes in the polymer matrix, the water retention mechanism and protonic conductivity appear favored in the presence of acidic functional groups on the surface of nanoparticle fillers [107].

1.3.4.3 Membranes for High Temperature Applications

A rational analysis of filler effects on structural, proton transport properties and the electrochemical characteristics of composite perfluorosulfonic membranes for DMFCs was reported [107, 125]. It was observed that a proper tailoring of the surface acid–base properties of the inorganic filler for application in composite Nafion membranes allows appropriate DMFC operation at high temperatures and reduced pressure [107]. An increase in both the strength and amount of acidic surface functional groups in the fillers enhances water retention inside the composite membranes through an electrostatic interaction in the presence of humidification constraints, in the same way as for the adsorption of hydroxyl ions in solution [107, 109, 125].

The DMFC performance of various MEAs based on composite membranes that contain fillers with different acid–base characteristics improves as the pH of the slurry of the inorganic filler decreases (Figure 1.14). As expected, the surface

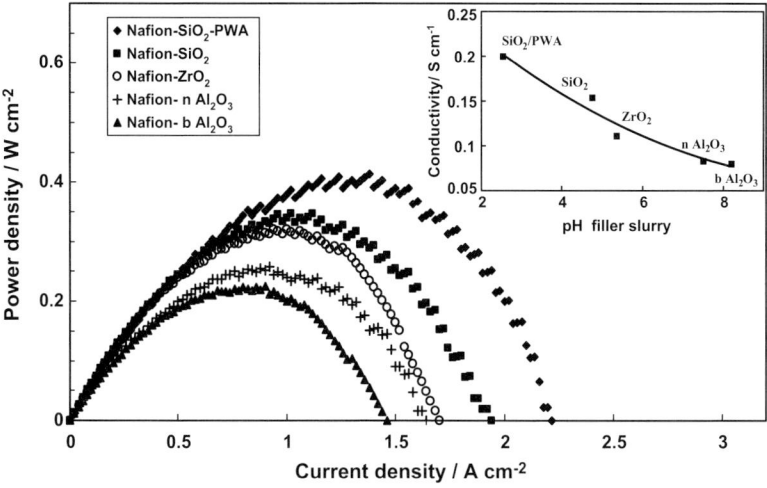

Figure 1.14 DMFC power density curves at 145 °C for MEAs containing different inorganic fillers. Methanol feed 2 M, 2.5 atm; oxygen feed 2.5 atm. Pt loading 2 ± 0.1 mg cm^{-2}. The inset shows the variation of membrane conductivity at 145 °C as a function of the pH of slurry of the filler [109].

properties play a more important role than the crystalline structure of the filler, since the water molecules, acting as promoters towards the proton migration, are effectively coordinated by the surface groups. The conductivity and performance of composite perfluorosulfonic membranes in DMFCs are strongly related to the surface acidity, which, in turn, influences the characteristics of the water physically adsorbed on the inorganic filler surface. It has been observed that the more acidic the filler surface, the larger its capability of undergoing a strong interaction with water through the formation of hydrogen bonds (Figure 1.15). This latter effect produces a decrease in the O—H stretching and bending frequencies in the physically adsorbed water. Furthermore, an increase in the water uptake in the composite membrane and an enhancement of proton conductivity are observed in the presence of acidic fillers [107, 125]. The proton migration inside the membrane appears to be assisted by the water molecules on the surface of the nanofiller particles and could also be promoted by the formation and breaking of hydrogen bonds [109, 125].

Conventional ion-exchange perfluoropolymer membranes such as the well-known Nafion membrane are based on long-side-chain polymers (LSC). In the last few decades, Solvay Solexis has developed a new short-side-chain (SSC) proton conducting perfluoropolymer membrane, that is, Hyflon Ion, characterized by excellent chemical stability and equivalent weight (850 g/eq.) lower than conventional Nafion 117 (1100 g/eq.) [100]. Besides the improved conductivity related to the higher degree of sulfonation, the short-side-chain Hyflon Ion ionomer is characterized, in the protonic form, by a primary transition at around 160 °C whereas the conventional Nafion shows this transition at about 110 °C. This characteristic of the Hyflon Ion membrane ensures proper operation at high temperatures (100–150 °C) provided

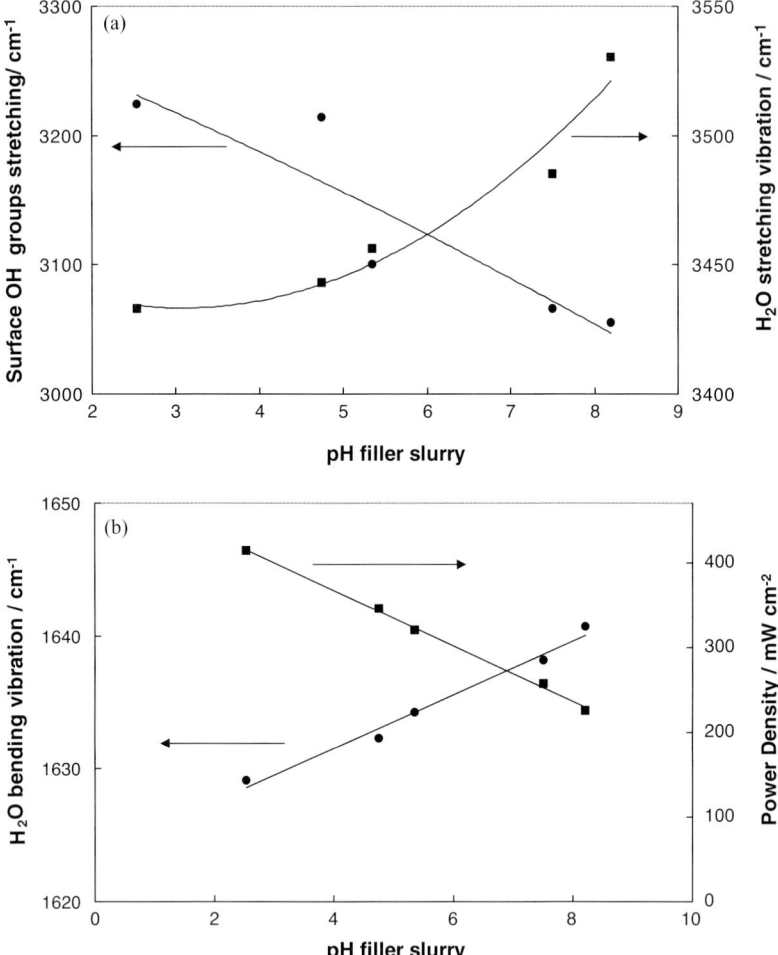

Figure 1.15 Variation of O—H stretching vibration frequencies for surface OH functionalities and physically adsorbed water versus the pH of slurry of the fillers (a); variation of O—H bending vibration frequencies of physically adsorbed water and DMFC maximum power density versus the pH of slurry of the fillers (b) [109].

that a sufficient amount of water is supplied to the membrane or retained inside the polymer under these conditions.

Hyflon Ion membranes have been investigated for applications in DMFCs operating between 90° and 140 °C in the European FP5 Dreamcar project, which is described in the transportation section. DMFC assemblies based on these membranes showed low cell resistance and promising performances compared with conventional membranes. The peak power density reached about 290 mW cm^{-2} at 140 °C and 3 bars abs. with 1 M methanol and air feed (Figure 1.16).

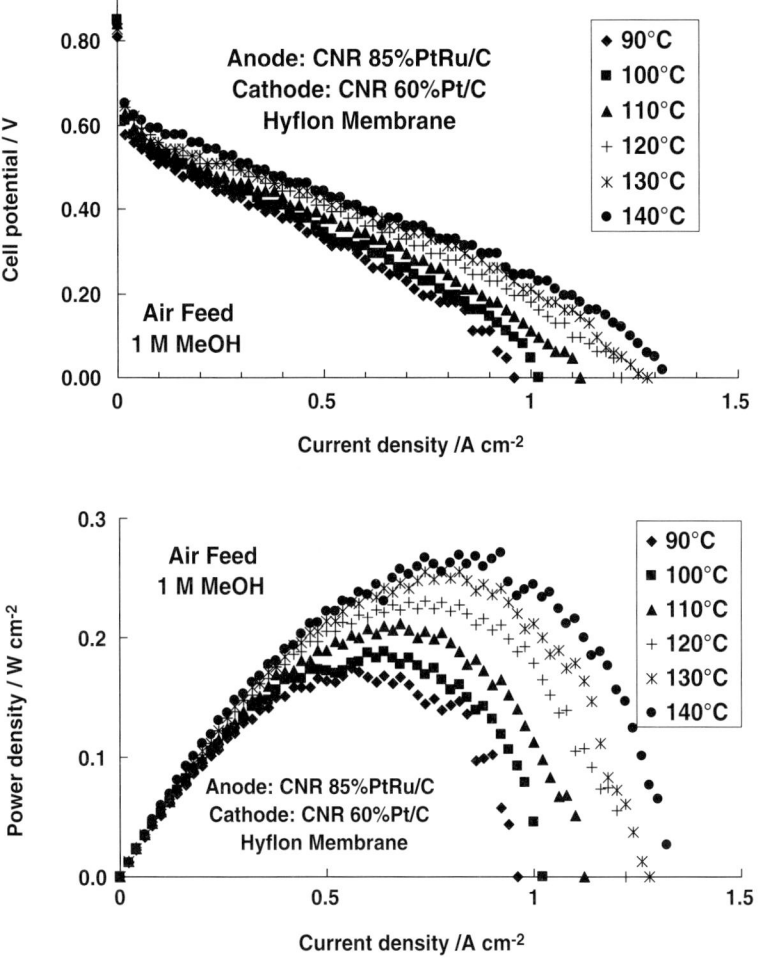

Figure 1.16 DMFC polarization and power density curves for a Hyflon Ion membrane-based MEA [100].

1.3.4.4 Alkaline Membranes

As is well known, the major drawbacks of proton-conducting electrolyte-based DMFCs concern slow reaction kinetics and fuel crossover. A large amount of precious catalysts is necessary and this has a considerable impact on the cost of these devices. Furthermore, catalyst corrosion and membrane degradation at low pH values limit the number of materials that can be selected for long-term stability [101]. On the other hand, the liquid alkaline electrolytes that were initially preferred for DMFCs [94] are affected by practical constraints such as potassium or sodium carbonate formation and precipitation in the catalyst pores, the need to frequently regenerate the electrolyte and liquid electrolyte leakage through the electrode. Some of the drawbacks associated with the behavior of liquid alkaline electrolytes in fuel cells can be solved by using new anion exchange membranes [90, 96]. In anionic polymer electrolytes, OH^- ions, responsible for ionic conduction, are formed at the cathode by the water fed to humidify the oxidant stream according to Equation 1.5. The OH^- ions are transported through the membrane to the anode where they react with methanol to form CO_2 (Equation 1.4). The CO_2 reaction product reacts easily with OH^- ions to form carbonate/bicarbonate (CO_3^{2-}/HCO_3^-) anions [90]. In the membrane region in contact with the anode, the CO_3^{2-}/HCO_3^- ions neutralize the positive charge fixed on the polymeric membrane, for example, quaternary ammonium functionalities, affecting conductivity and causing a variation of the local pH in the anode compartment with respect to the cathode [90]. The decrease of pH at the anode causes a positive shift of the redox potential (in the absolute potential scale) for the oxidation process, thus diminishing the electromotive force. In other words, the pH difference will reduce the voltage thermodynamically. The thermodynamic voltage loss can be 290 mV in the presence of a pH difference of about 4 at 80 °C [90]. This loss decreases by increasing the operating temperature. High temperature operation restricts the number of anionic polymers that can be used. Furthermore, the high temperature approach might not be appropriate for portable applications. Besides these aspects, there are several advantages in using anion exchange membranes. As discussed above, the main advantage concerns with the favorable reaction rates in alkaline media with respect to acidic electrolytes for both oxygen electro-reduction and methanol electro-oxidation reactions [90, 96]. Methanol is oxidized to carbon dioxide in both acid and alkaline electrolytes in the presence of proper catalysts for example, PtRu [90]. The significant reduction of activation overpotential in alkaline media can compensate for the voltage loss due to the thermodynamic effects associated to the pH gradient [90]. However, the enhanced reaction kinetics may allow the use of cheaper materials and, possibly, non-noble metals. In alkaline media, Ni anodes and Ag cathodes represent a suitable compromise in terms of activity and cost [25, 26, 50]. In general, a significant reduction of the catalyst cost may be envisaged.

The OH^- ions migrate from the cathode to the anode; this pathway is the opposite direction to the electro-osmotic drag in proton exchange membrane DMFCs. Thus, the electro-osmotic drag does not contribute to the methanol crossover in alkaline systems. However, it should be pointed out that even at high current densities (in the cell voltage region of technical interest), the contribution of the electro-osmotic drag to methanol crossover, in protonic electrolyte-based DMFCs is quite small with respect to the concentration gradient (diffusion). OH^- migration in the membrane is assisted by water as it occurs for protons in the analogous acidic polymer electrolyte. Thus, methanol crossover cannot be completely eliminated by the anion exchange membranes; however, in principle, it can be reduced. In general, the alkaline electrolyte causes lower corrosion problems. This allows the investigation of a large number of catalyst formulations, especially with regard to methanol tolerance characteristics.

Another important aspect is membrane stability. One of the main degradation mechanisms of proton exchange membranes during fuel cell operation is caused by the hydrogen peroxide-type radicals formation during oxygen electroreduction in an acidic environment. This process is less effective at high pH values [90, 96]. Thus, as occurs for the catalysts, there is a wider range of polymers that can be potentially used as anionic electrolytes in DMFCs. Cheap hydrocarbon-only membranes may be selected [96]; yet, these systems are less appropriate than the fluorinated membranes in terms of high temperature stability. Besides the above mentioned thermodynamic effect due to the pH gradient, another significant problem associated with the use of anion exchange membranes is the low ionic conductivity. This is essentially caused by the lower mobility of anions such as OH^- or the carbonate-type with respect to the protons. The lack of appropriate anionic membranes with conductivity characteristics similar to Nafion has practically hindered the development of alkaline DMFCs for several decades. It should be mentioned that several researchers such as Ogumi et al. [126] and Yu and Scott [92] have reported interesting results with anionic membranes but alkaline solutions were indeed used in those studies to enhance membrane conductivity. Recirculation of the liquid electrolyte through the device not only enhances conductivity but significantly reduces the pH gradient. Thus, electrolyte recirculation eliminates the thermodynamic constraints and enables an extension of the three-phase reaction zone from the electrode/membrane interface to the bulk of the electrode, favoring the presence of a mixed conductivity in the catalytic layer. This aspect is quite important in the absence of a suitable ionomer solution. In this regard, there is considerable interest in the synthesis of perfluorinated anion exchange membranes that can be dissolved in non-volatile solvents, enabling the preparation of mixed conductivity (ionic and electronic) catalytic layers as occurs for conventional PEMFCs and DMFCs. Perfluorinated anion-exchange membranes are also of interest for their perspectives to extend the operating range to higher temperatures [90]. Of course, these are significantly more expensive than hydrocarbon membranes containing C–H bonds in the backbone.

One of the reasons why an increasing number of DMFC developers have recently been showing interest in alkaline membranes is especially due to the development of new polymers that show appropriate conductivities in a range from ambient temperature to 80 °C [91, 92]. Although these conductivity values are still lower than those of conventional perfluorosulfonic membranes, since the alkaline membranes are less affected by methanol permeation, they can be used in a thinner form. This compensates, in part, for the effect of high cell resistance [90, 96]. Among the various types of anion-exchange membranes recently proposed, significant interest has been stimulated by radiation grafted alkaline membranes [91]. Radiation-grafting of styrene into non-fluorinated (LDPE), partially fluorinated (PVDF) and fully fluorinated (FEP) films has been considered widely for PEMFCs and DMFCs [91, 101]. Lower methanol crossover with respect to Nafion and the issue of high temperature operation has been demonstrated in DMFCs with grafted protonic membranes [123]. The radiation grafting process is also appealing for anionic membranes. This process is generally carried out on commercial preformed films; several parameters can be modulated such as the process temperature, the level of grating, radiation dose, and thickness to obtain particular membrane properties [96]. Methanol crossover can be further diminished by using appropriate crosslinking procedures.

Interesting results were obtained by Yu and Scott using Morgane ADP, a commercial anion exchange membrane produced by Solvay (Belgium) [92]. The main drawbacks concerned with conductivity and stability. Another interesting commercial membrane is produced by Tokyzama (Japan) [90, 96]. Recently, high conductivity anion exchange membranes were prepared by the radiation grafting of vinylbenzylchloride (VBC) into FEP with a subsequent amination with triethylamine and an ion exchange with KOH [91]. These membranes were characterized by a degree of grafting of about 20% and conductivities of $2 \cdot 10^{-2}$ S cm^{-1} at 50 °C in aqueous solutions. The conductivities are 20% of the values obtained for state-of-the-art perfluorosulfonic acid membranes. The activation energy for OH$^-$ migration in the membrane was twice compared with that observed for protons in fully hydrated Nafion indicating that OH$^-$ mobility is strongly temperature dependent [92].

1.3.4.5 Effects of Crossover on DMFC Performance and Efficiency

Methanol crossover through the polymer membrane is known to be one of the most challenging problems affecting the performance of DMFCs [88]. The overall efficiency of a methanol fuel cell is determined by both voltage and faradaic efficiency for the consumption of methanol [12]. The faradaic efficiency is influenced mainly by methanol crossover through the membrane. The methanol crossover is usually measured indirectly by determining the amount of CO_2 produced at the cathode by the oxidation of methanol on the Pt surface [88]. This CO_2 can be monitored on line by using an IR-detector. A more accurate method consists of a chromatographic analysis of aliquot samples of the cathode outlet stream [106, 117].

The crossover of methanol is influenced by both membrane characteristics and temperature, as well as by the operating current density [88, 117]. In general, an increase in temperature causes an increase in the diffusion coefficient of methanol and determines a swelling of the polymer membrane. Both effects contribute to an increase in the methanol crossover rate. The crossover includes both methanol permeability due to a concentration gradient and molecular transport caused by electro-osmotic drag in the presence of a proton conducting electrolyte. The latter is directly related to the proton migration through the membrane and it increases with the current density [127]. For DMFCs equipped with an alkaline electrolyte, the electro-osmotic drag is directed towards the anode. Thus, it does not contribute to the crossover. Methanol permeability, caused by the concentration gradient at the anode-electrolyte interface, depends on the operating current density. In a polarization curve, the onset of diffusional limitations occurs when the rate of reactant supply is lower than the rate of its electrochemical consumption. Thus, if the anode is sufficiently active to oxidize methanol electrochemically to CO_2 at a rate comparable to or higher than the rate of the methanol supply, the methanol concentration gradient between anode/electrolyte and cathode/electrolyte interfaces could be reduced significantly [11, 88, 117]. Membranes that are very thick are effective barriers for reducing methanol crossover; conversely, an increase in thickness causes an increase of ohmic overpotentials. In some cases, it may be more productive to use a thinner membrane with reduced ohmic limitations and select appropriate operating conditions which limit the methanol crossover.

1.3.5
Historical Development of Oxygen Electroreduction in DMFCs

Regarding the development of catalysts for the oxygen reduction process in methanol fuel cells during the last few decades, it should be mentioned that silver was considered mainly at the beginning, especially for operation in conjunction with alkaline electrolytes [25, 26]. Silver is used presently in oxygen depolarized cathodes for industrial chloro-alkali cells which use a liquid electrolyte (KOH) [128] whereas Pt and its alloys were employed for oxygen reduction in the presence of acidic electrolytes both in unsupported and carbon supported form. The development of cathode catalysts for proton conducting electrolyte based DMFCs was initially influenced by similar studies carried out on phosphoric acid fuel cells (PAFCs). The catalytic layer was essentially a mixture of Pt/C catalysts and PTFE binder sintered at around 350 °C. These hydrophobic electrodes were useful in reducing the flooding caused by liquid electrolytes. However, the need to use a methanol tolerant cathode catalyst was quickly discerned. Accordingly, the research was also directed towards alternative catalytic systems including non-noble metals.

In the late 1980s, the development of methanol tolerant oxygen reduction catalysts became of practical interest with the development of metal chalcogenides [129], phthalocyanines and phorphyrins [130] -based cathodes with catalytic activities approaching those of Pt in the presence of methanol poisoning.

As for the development of non-noble metal catalysts for the oxygen reduction reaction (ORR), essentially three classes of materials were investigated in the first decades of DMFC development. These included oxides, chalcogenides other than oxides and organometallic compounds. Metal oxides were initially considered being these materials the most obvious candidates as Pt substitutes in electrocatalysis. Oxides are present on the surfaces of all non-noble metals at potentials useful for oxygen reduction. It was assumed that oxygen reduction occurred by the exchange of oxygen atoms between molecular oxygen, surface oxide, and water (referred to as a regenerative mechanism). Several non-noble metal oxides with metallic conductivity are available; a few are stable in acid electrolytes. These include mainly tungsten oxides as a possible candidate. Only recently, Co-oxides with a perovskite structure similar to that used in intermediate temperature solid oxide fuel cells (IT-SOFCs) have been considered [131].

1.3.6
Status of Knowledge of Oxygen Reduction Electrocatalysis and State-of-the-Art Cathode Catalysts

1.3.6.1 Oxygen Reduction Process

Although Pt/C electrocatalysts are, at present, the most widely used materials as cathodes in proton conducting electrolyte-based low temperature fuel cells, due to their intrinsic activity and stability in acidic solutions, there is still great interest in developing more active, selective and less expensive electrocatalysts for oxygen reduction. However, there are a few directions that can be investigated to reduce the costs and to improve the electrocatalytic activity of Pt, especially in the presence of methanol crossover. One is to increase Pt use; this can be achieved by increasing its dispersion on carbon and the interfacial region with the electrolyte. Another successful approach to enhance the electrocatalysis of O_2 reduction is by alloying Pt with transition metals. This enhancement in electrocatalytic activity has been interpreted differently, and several studies were done to analyze in depth the surface properties of the proposed alloy combinations [132–134]. Although a comprehensive understanding of numerous reports has not been reached yet, the observed electrocatalytic effects have been ascribed to several factors (interatomic spacing, preferred orientation, electronic interactions) which play, under fuel cell conditions, a favorable role in enhancing the ORR rate [135–143].

Also, there is an increasing interest in developing methanol tolerant catalyst alternatives to Pt for oxygen reduction; however, it should be taken into account that if the methanol which permeates through the membrane is not completely oxidized at the cathode surface to CO_2, it would contaminate the water at the outlet of the cathode compartment. This could cause several environmental problems in the absence of a proper technical solution, which could be a single chamber DMFC with highly selective anode and cathode catalysts. Alternatively, a proper catalytic burner should be used at the cathode outlet.

As discussed in a previous section, various studies carried out in the past, especially on carbon supported Pt electrocatalysts for oxygen reduction in phosphoric acid fuel cells, showed that the electrocatalytic activity (mass activity, $mA\,g^{-1}$ Pt, and specific activity, $mA\,cm^{-2}$ Pt) depends on the mean particle size. The mass activity for oxygen reduction reaches a maximum at a dimension of about 3 nm, corresponding closely with the particle size at which there is a maximum in the fraction of (1 1 1) and (1 0 0) surface atoms on Pt particles of cubo-octahedral geometry [144]. Platinum atoms at edge and corner sites are considered less active than Pt atoms on the crystal faces. Accordingly, both mass and specific activity should decrease significantly as the relative fraction of atoms at edge and corner sites approach unity [144]. This situation occurs with Pt particles smaller than 1–2 nm in diameter.

In the case of DMFCs, an additional aspect should be considered, which is methanol crossover through the membrane. Methanol oxidation and oxygen reduction in the cathode compartment compete for the same sites, producing a mixed potential which reduces the cell open circuit potential.

In the case of methanol oxidation at the cathode, three neighboring Pt sites in a proper crystallographic arrangement will favor methanol chemisorption. Since at high cathodic potentials the water discharging reaction is largely favored, oxidation of the methanolic residues adsorbed on the surface proceeds very fast producing a parasitic anodic current on this electrode.

When the particle size of the electrocatalyst is very small or one has an amorphous Pt electrocatalyst for the oxygen reduction, methanol chemisorption energy could be lower and hence the cathode less poisonable. At the same time, however, due to the fact that only the inactive edge and corner atoms will be present and dual sites of the proper orientation will not be available, the activity of such an electrocatalyst for oxygen reduction will be lower. The best compromise is to modulate the structure and the particle size between amorphous and crystalline in order to decrease the poisoning by methanol and enhance oxygen reduction. A second possibility is to use a promoting element for oxygen reduction that simultaneously hinders the methanol chemisorption still maintaining the proper structure and particle size.

1.3.6.2 Pt-Based Catalysts and Non-noble Metal Electrocatalysts

The intrinsic electrocatalytic activity of Pt alloys (Pt-Cr, Pt-Ni, Pt-Co, Pt-Cu, Pt-Fe) with a lattice parameter smaller than that of Pt was found to be higher than on the base metal [132–140, 145, 146]. This effect is related to the nearest neighboring distance of Pt-Pt atoms on the surface of the fcc crystals. The r.d.s. involves the rupture of the O—O bond through a dual site mechanism; a decrease of the Pt—Pt distance favors the dual site O_2 adsorption. Leaching of non-noble elements produces a roughening of the surface with a corresponding increase of the Pt surface area.

Many investigations in PEMFCs have shown that enhanced electrocatalytic activity for the ORR of some binary Pt based alloy catalysts, such as Pt—M, (where M = Co, Fe, etc.), in comparison with pure Pt [147–152] can be also interpreted in terms of increased Pt d-band vacancy (electronic factor) and its relative effect on the OH chemisorption from the electrolyte [153].

Methanol chemisorption and ORRs require an appropriate geometrical arrangement of Pt atoms. Both processes are favored on a Pt (1 1 1) surface, which possesses the reasonable nearest Pt—Pt interatomic distance. Thus, the poisoning effect of methanol crossover should be more significant on the Pt (1 1 1) surface. However, it is difficult to quantify the compensation effect due to the increased methanol oxidation rate at the sites where oxygen reduction is favored. Beside these aspects, a recent work has also taken into consideration the role played by the promoting element (Co, Cr) for the removal of strongly bonded oxygenated species on Pt through an intra-alloy electron transfer [142]. Chemisorption of oxygen molecules occurs more easily on oxide-free Pt surfaces [142], but, as reported in cyclic voltammetry studies [64], methanol adsorption and oxidation are favored on a reduced Pt surface rather than on platinum oxide. The addition of Co and Cr to Pt appears to simultaneously favor both ORRs and MORs. For example, the promoting effect of Cr on Pt for methanol electrooxidation has already been reported [98, 141]. Furthermore, the presence of electropositive elements alloyed to Pt favors the chemisorption of OH species on neighboring Pt sites. In the absence of oxygen, a small but noticeable promoting effect for methanol oxidation in a wide range of anodic overpotentials has been observed by Cr, Fe and other elements usually selected as catalytic enhancers for the ORR in PEFCs [6]. At present, it is difficult to establish if the beneficial effect on oxygen reduction is prevailing with respect to the promoting effect on methanol oxidation at the DMFC cathode. For the Pt-Fe system, Watanabe et al. reported that after electrochemical testing of a Pt-Fe alloy the catalyst was covered by a thin Pt skin of less than 1 nm in thickness [150, 151]. Moreover, they suggested that during the adsorption step, a p orbital of O_2 interacts with empty d orbitals of Pt and, consequently, back donation occurs from the partially filled orbital of Pt to the p* (antibonding) molecular orbital of O_2. The increase in d-band vacancies on Pt by alloying produces a strong metal–O_2 interaction. This interaction weakens the O—O bonds,

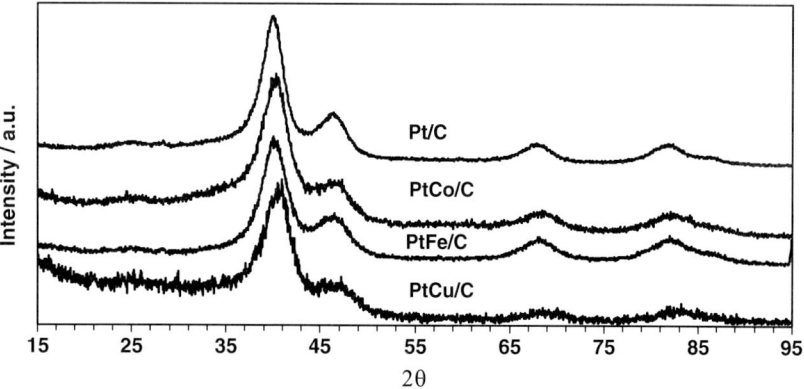

Figure 1.17 XRD patterns of Pt and Pt-based bimetallic catalysts for the oxygen reduction reaction in DMFCs [154].

resulting in bond cleavage and bond formation between O and H^+ of the electrolyte, thus improving the ORR. The leaching of Fe ions into the membrane may cause an increase in resistance and accelerate the degradation reactions of the polymer.

Recently, cathode catalysts synthesized by a low-temperature colloidal-incipient wetness route characterized by a high concentration of metallic phase on carbon black and a particle size smaller than 3 nm have been investigated [154, 155]. The new approach allowed carbon-supported bimetallic nanoparticles with a particle size of about 2–2.5 nm and a suitable degree of alloying, to be obtained. X-ray diffraction (XRD) patterns of these Pt/C and Pt-M/C catalysts are reported in Figure 1.17. These showed the typical fcc crystallographic structure of Pt. A moderate degree of alloying was found in Pt-Fe catalysts whereas the degree of alloying was slightly higher for Pt-Co/C and significantly larger for Pt-Cu/C compared with Pt-Fe, using the same procedure. These findings were derived from a decrease of the lattice parameter.

In terms of polarization behavior, the Pt-Fe/C (2.4 nm) performed better than the Pt/C, Pt-Cu/C and Pt-Co/C catalysts with similar particle sizes (2.1–2.8 nm) at 60 °C (Figure 1.18). This was also confirmed by cathode polarization curves (Figure 1.18). Methanolic residues stripping analysis of these catalysts (Figure 1.19) showed that this enhanced activity was possibly derived from better methanol tolerance and higher intrinsic catalytic activity for oxygen reduction. The presence of a significant current density in the hydrogen desorption region ($E < 0.4$ V RHE) for the Pt-Fe even after methanol adsorption, which was not observed for the other catalysts, indicated suitable methanol tolerance properties. The positive shift of the potential for Pt-Oxide reduction was associated with better intrinsic catalytic activity. The electrochemical active surface area derived from the methanolic residues stripping analysis was larger for catalysts with a smaller particle size, for example, PtCu/C. Yet, it appeared that the small increase of electrochemically active surface area in PtCu did not play the same role of the increase of methanol tolerance and intrinsic catalytic activity in PtFe.

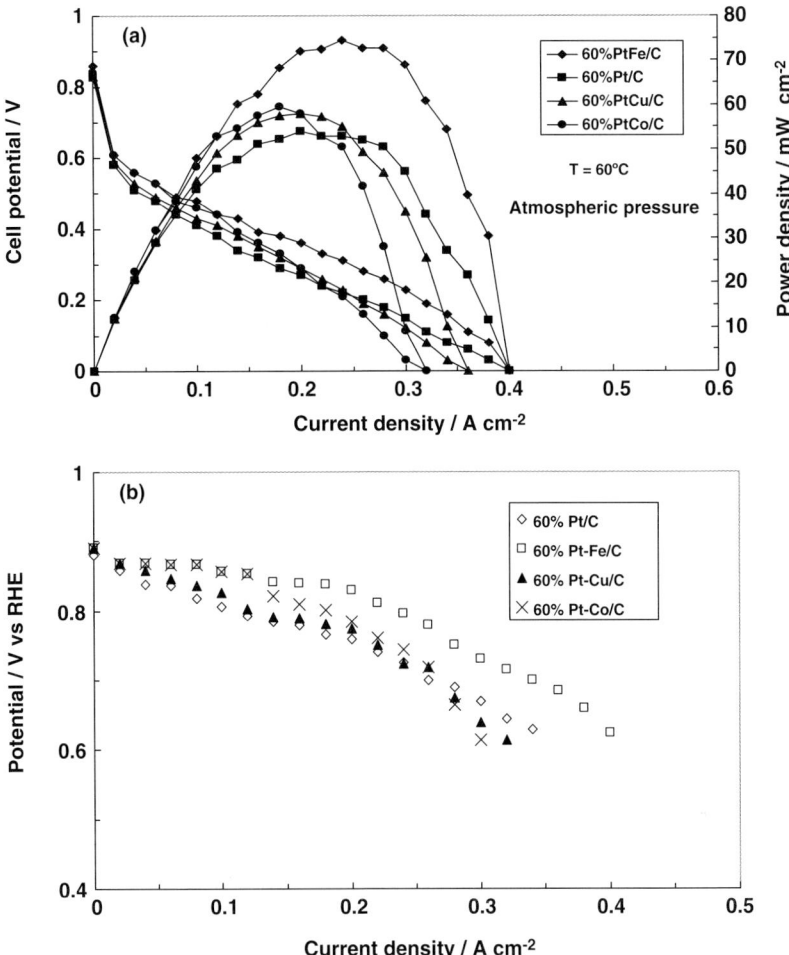

Figure 1.18 (a) Polarization and power density curves for the DMFCs equipped with the various cathode catalysts at 60 °C under atmospheric pressure, and (b) cathodic polarization curves for DMFCs based on the different cathode catalysts recorded at the same operating conditions [154].

1.3.6.3 Alternative Cathode Catalysts

Alternatively to platinum, organic transition metal complexes are known to be good electrocatalysts for the ORR. Transition metals, such as iron or cobalt organic macrocycles from the families of phenylporphyrins, phthalocyanines and azoannulenes have been tested as O_2-reduction electrocatalysts in fuel cells [130, 156–159]. One major problem with these metal organic macrocyclics is their chemical stability under fuel cell operation at high potentials. In many cases, the metal ions dissolve irreversibly in the acid electrolyte. However, if the metal-organic macrocyclic is

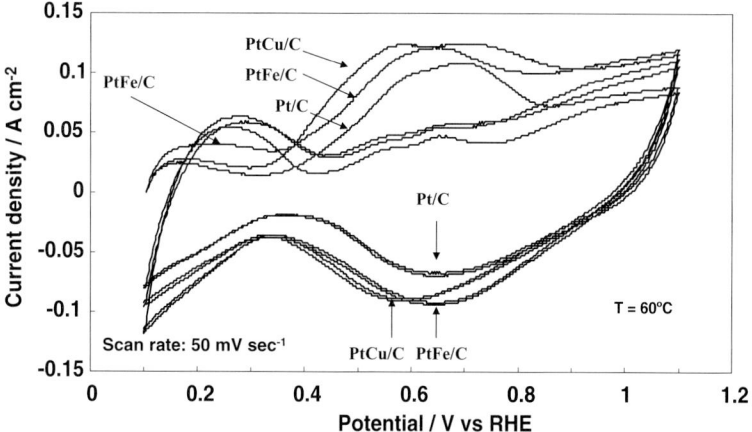

Figure 1.19 Adsorbed methanolic residues stripping voltammetry at a scan rate of 50 mVs^{-1} for the different cathode catalysts at 60 °C [154].

supported on high surface area carbon and treated at high temperatures (from 500 to 800 °C), the residue exhibits promising electrocatalytic activity without any degradation in performance, from which one may infer the good stability of the metal in the electrocatalyst [158].

In some other studies, a few inorganic materials have been proposed as suitable substitutes for platinum in methanol fuel cells due to their selectivity for oxygen reduction, even in the presence of methanol. These materials consist mainly of the Chevrel-phase type ($Mo_4Ru_2Se_8$), transition metal sulfides ($Mo_xRu_yS_z$, $Mo_xRh_yS_z$) or other transition metal chalcogenides (($Ru_{1-x}Mo_x)SeO_z$) [129, 160]. Some of these possess semiconducting properties, thus, in theory, they could introduce an additional ohmic drop in the electrode. However, their activity for oxygen reduction is significantly lower than Pt [2]. Carbon supported Ru electrocatalysts are reported to exhibit high selectivity for oxygen reduction in the presence of methanol but their activities are significantly lower [161]. The tolerance of these materials to methanol is due to the absence of adsorption sites for methanol dehydrogenation. In the case of Ru/carbon, at high potentials, the surface is covered mainly by Ru oxides on which methanol chemisorption is hampered [161]. Regarding the development of cathode catalysts for alkaline DMFCs, it should be pointed out, as for the anode catalyst, that corrosion problems are minimized by the operation at high pH values. Thus, due to the large variety of catalytic formulations that may be screened, it should be easier to discover a methanol tolerant cathode catalyst. The reaction kinetics for oxygen reduction at the cathode are more favorable in alkaline media. This allows the replacement of Pt with less noble or non-precious catalysts with significant advantages in terms of cost reduction. Among the various cathode formulations, Ag and MnO_2 catalysts have shown suitable methanol tolerance and catalytic activity for oxygen reduction [90].

1.3.7
DMFC Power Sources in the 'Pre-1990 Era'

Even though from the 1960s to the 1980s the development of DMFCs was a 'Fuel Cell Researcher's Dream' [97], only in the last two decades have the attractive features of DMFC power sources (portable liquid fuel with an energy density of about half that of gasoline, environmentally friendly technology, a 10-fold increase in power density with a proton exchange membrane electrolyte) clearly indicated their possible application in transportation, portable power and power generation/cogeneration applications.

There were only a few attempts to develop DMFC stacks/systems in the decades preceding the 1990s. The first attempts to develop methanol fuel cells [23, 24] were carried out by Kordesch and Marko in 1951 on the basis of earlier studies by E. Muller. The DMFC devices initially developed were based on alkaline electrolytes, Ni-based or Pt-Pd-based anodes, and silver cathodes [25, 26]. One of the first DMFC stacks of reasonable power was based on alkaline electrolytes and developed in the 1960s by Murray and Grimes at Allis-Chalmers in 1963 [94]. It was operating at 50 °C and consisted of an aqueous alkaline electrolyte (5 M KOH) Pt-Pd anode and Ag cathode catalysts. A porous Ni sheet was used as the backing layer for the electrode. The stack was composed of 40 cells and provided maximum electrical power of 750 W at 9 V with an average cell power density of about 40 mW/cm^2. The approach of using concentrated KOH as electrolyte was similar to that of hydrogen-fed alkaline fuel cells developed in the same period for space applications mainly [97–99]. However, the problem of an acid–base reaction between the electrolyte and the reaction product at the anode, that is, CO_2 with formation of potassium carbonate, was quickly recognized. This caused carbonate precipitation inside catalyst pores with occlusion and consequent increase of mass transport constraints. The increase of resistance over time and the need to regenerate the cell (excess of carbonate removal) induced most of the DMFC developers to address their efforts towards the development of DMFCs based on proton conducting electrolytes. Some attempts addressed the use of carbonate electrolytes working at high temperature or anion exchange membranes [95]. Unfortunately, the performance achieved by using the latter approach was not satisfying in the past [95]. In recent years, however, the approach of anion exchange membranes for DMFCs has been reconsidered. The new anionic membranes show proper conductivity values even in the absence of KOH recirculation [91].

DMFC devices based on acidic electrolytes were initially developed in the mid-1960s by leading laboratories such as Shell, Exxon and Hitachi [2]. In all these cases, 1–2 M sulfuric acid was used as the electrolyte and unsupported platinum black was initially used as electrocatalyst. However, studies conducted by researchers at Shell in 1968 selected Pt-Ru as one of the most effective anode electro-catalysts and developed a 300 W prototype [23]. Esso developed a 100 W stack for communication applications [23]. In terms of stack development, another highlight in terms of performance was the development of a 50 W DMFC stack at Hitachi [2]. Interest in developing DMFCs was stimulated in the early 1990s when the sulfuric acid electrolyte was replaced by a solid-state proton conductor (Nafion). There were two significant effects, (a) an increase in electro-catalytic activity of the electrodes, and (b) improved open circuit potential of the cell due to reduced methanol crossover. In addition, an enhanced oxygen electrode performance was observed because of the replacement of the liquid electrolyte with the perfluorosulfonic acid solid polymer.

Interest in stack development initially focused on transportation applications. Recently, due to the lower efficiency and power densities of DMFCs compared with PEMFCs and the higher projected costs of DMFC power sources (mainly because of the significantly higher noble metal loading), short-term projected applications were directed towards portable applications [4].

Due to the poor anode reaction kinetics and cathode poisoning by methanol crossover, high noble metal loading was initially used in both electrodes (about 10 mg/cm^2 unsupported catalysts); this decreased progressively up to reach 2 mg cm^{-2} and even lower.

After the use of unsupported catalysts, high concentration carbon supported catalysts (e.g., 85% PtRu and 60%Pt) were used in practical stacks [162]. One recent approach involves the use of decorated catalysts with ultra-low Pt loadings [80].

1.4
Current Status of DMFC Technology for Different Fields of Application

1.4.1
Portable Power Sources

The potential market for portable fuel cell systems deals mainly with the energy supply for electronic devices, but it also includes remote and micro-distributed electrical energy generation. Accordingly, DMFC power sources can be used in mobile phones, lap-top computers, as well as energy supply systems for weather stations, medical devices, auxiliary power units (APU) and so on. Direct methanol fuel cells (DMFCs) are promising candidates for these applications because of their high energy density, light weight, compactness, and simplicity as well as their easy and fast recharging [24, 163–165]. Theoretically, methanol has a superior specific energy density (6000 Wh/kg) in comparison with the best rechargeable battery, lithium polymer and lithium ion polymer (theoretical, 600 Wh/kg) systems. This performance advantage translates into more conversation time using cell phones, more time for the use of laptop computers between the replacement of fuel cartridges, and more power available on these devices to support consumer demand. In relation to consumer convenience, another significant advantage of the DMFC over the rechargeable battery is its potential for instantaneous refueling. Unlike rechargeable batteries that require hours to charge a depleted power pack, a DMFC can have its fuel replaced in minutes. These significant advantages make DMFCs an exciting development in the portable electronic devices market.

Several organizations (Table 1.2) are actively engaged in the development of low power DMFCs for cellular phone, laptop computer, portable camera and electronic game applications [104, 163–166]. The primary goal of this research is to develop

1.4 Current Status of DMFC Technology for Different Fields of Application | 41

Table 1.2 DMFC power sources for portable applications.

Developer	Number/area of cells	Power density	Temperature (°C)	Oxidant	Methanol concentration (M)	Anode catalyst and loading	Electrolyte	Cathode catalyst and loading
Motorola Labs	4 cells (planar stack)/13–15 cm^2	12–27 mW cm^{-2}	21	Ambient air[a]	1	PtRu alloy, 6–10 mg cm^{-2}	Nafion 117	6–10 mg cm^{-2}
Energy Related Devices	Planar stack	3–5 mW cm^{-2}	25	Ambient air[a]	1	PtRu alloy	Nafion	Pt
Jet Propulsion Lab	6 cells (flat pack)/ 6–8 cm^2	6–10 mW cm^{-2}	20–25	Ambient air[a]	1	PtRu alloy, 4–6 mg cm^{-2}	Nafion 117	Pt, 4–6 mg cm^{-2}
Los Alamos National Labs	5 cells/45 cm^2	300 W/l	60	Air (3–5 times stoichiometry)	0.5	PtRu alloy, 0.8–16.6 mg cm^{-2}	Nafion	Pt, 0.8–16.6 mg cm^{-2}
Forschungszentrum Julich GmbH	40 cells/100 cm^2	45–55 mW cm^{-2}	50–70	O$_2$ (3 atm)	1	PtRu, 2 mg cm^{-2}	Nafion 115	Pt, 2 mg cm^{-2}
Samsung advanced Institute of Technology	12 cells (monopolar)/ 2 cm^2	23 mW cm^{-2}	25	Ambient air[a]	5 Passive mode	PtRu, 3–8 mg cm^{-2}	Hybrid membrane	Pt, 3–8 mg cm^{-2}
Korea Institute of Energy Research	6 cells (bipolar)/ 52 cm^2	121–207 mW cm^{-2}	25–50	O$_2$ (300 ml min^{-1}), ambient pressure	2.5 Active mode	PtRu/C	Nafion 115 & 117	Pt black
Korea Institute of Science & Technology	6 cells (monopolar)/ 6 cm^2	40 W cm^{-2}	25	Ambient air[a]	4 Passive mode	PtRu	Nafion 115	Pt
More Energy Ltd.	20 cm^2	60–100 mW cm^{-2}	25	Ambient air[a]	30–5%	PtRu	Liquid electrolyte	Pt
Institute for Fuel Cell Innovation, Canada	3 cells (monopolar)	8.6 mW cm^{-2}	25	Ambient air[a]	2 Passive mode	80% PtRu, 4 mg cm^{-2}	Nafion 117	Pt black, 4 mg cm^{-2}

(Continued)

Table 1.2 (Continued)

Developer	Number/area of cells	Power density	Temperature (°C)	Oxidant	Methanol concentration (M)	Anode catalyst and loading	Electrolyte	Cathode catalyst and loading
University of Connecticut, USA	4 cells/18–36 cm^2	30 mW cm^{-2}	25	Ambient air[a]	2–5 Passive mode	PtRu alloy, 7 mg cm^{-2}	Nafion 117	Pt, 6.5 mg cm^{-2}
Honk Kong University	Single cell/4 cm^2	28 mW cm^{-2}	22	Ambient air[a]	4 Passive mode	PtRu, 4 mg cm^{-2}	Nafion 115	40% Pt/C, 2 mg cm^{-2}
The Pennsylvania State University, USA	Single cell/5 cm^2	93 mW cm^{-2}	85	Air (700 ml min^{-1} and 15 psig)	2 Active mode	PtRu, 4 mg cm^{-2}	Nafion 112	40% Pt/C, 1.3 mg cm^{-2}
Harbin Institute of Technology	Single cell	9 mW cm^{-2}	30	Ambient air[a]	2 Passive mode	40% PtRu/C, 2 mg cm^{-2}	Nafion 117	40% Pt/C, 2 mg cm^{-2}
Tel-Aviv University, Israel	Flat fuel cell/6 cm^2	12.5 mW cm^{-2}	25	Ambient air[a]	1–6 in H$_2$SO$_4$ Passive mode	PtRu, 5–7 mg cm^{-2}	NP-PCM	Pt, 4–7 mg cm^{-2}
Tekion Inc., USA	Single cell/5 cm^2	65 mW cm^{-2}	60	Ambient air[a]	2 Active mode	PtRu	Nafion	Pt
University of California, USA	μ-Single cell/1.625 cm^2	16–50 mW cm^{-2}	25–60	Air (88 ml min^{-1})	2 Active mode	PtRu, 4–6 mg cm^{-2}	Nafion 112	40% Pt/C, 1.3 mg cm^{-2}
Waseda University, Japan	μ-Single cell/0.018 cm^2	0.8 mW cm^{-2}	25	O$_2$ (10 μl min^{-1}) sat. in H$_2$SO$_4$	2	PtRu, 2.85 mg cm^{-2}	Nafion 112	Pt, 2.4 mg cm^{-2}
Institute of Microelectronic of Barcelona-CNM, Spain	μ-Single cell	11 mW cm^{-2}	25	Ambient air[a]	4–5 Passive mode	PtRu, 4 mg cm^{-2}	Nafion 117	Pt, 4 mg cm^{-2}
Yonsei University, Korea	Multi-cell structure (monopolar)	33 mW cm^{-2}	80	O$_2$ (30 ml min^{-1})	2	60% PtRu/C, 4 mg cm^{-2}	Nafion 117	60% Pt/C, 4 mg cm^{-2}
CNR-ITAE, Italy	3 cells (monopolar)/4 cm^2	20 mW cm^{-2}	21	Ambient air[a]	5 Passive mode	PtRu, 4 mg cm^{-2}	Nafion 117	Pt, 4 mg cm^{-2}

[a] Ambient air usually refers to the air breathing mode.

proof of concept DMFCs capable of replacing high performance rechargeable batteries in the US$ 6-billion portable electronic devices market.

Motorola Labs—Solid State Research Center, USA, [4] in collaboration with Los Alamos National Laboratory (LANL), USA, is actively engaged in the development of low power DMFCs (greater than 300 mW) for cellular phone applications [167]. Motorola has recently demonstrated a prototype of a miniature DMFC based on a MEA set between ceramic fuel delivery substrates [4]. Motorola used their proprietary low temperature co-fired ceramic (LTCC) technology to create a ceramic structure with embedded microchannels for mixing and delivering methanol/water to the MEA and exhausting the by-product CO_2. The active electrode area for a single cell was approximately 3.5–3.6 cm^2. In the stack assembly, four cells were connected in series in a planar configuration with an MEA area of 13–14 cm^2; the cells exhibited average power densities between 15–22 $mW\,cm^{-2}$. Four cells (each cell operating at 0.3 V) were required for portable power applications because DC/DC converters typically require 1 V to efficiently step up to the operating voltage for electronic devices. Improved assembly and fabrication methods have led to peak power densities greater than 27 $mW\,cm^{-2}$. Motorola is currently improving their ceramic substrate design to include micro-pumps, methanol concentration sensors and supporting circuitry for second generation systems.

Energy Related Devices Inc. (ERD), USA, is working in alliance with Manhattan Scientific Inc., USA) to develop miniature fuel cells for portable electronic applications [163, 168]. A relatively low-cost sputtering method, similar to the one used by the semiconductor industry for the production of microchips, was used for the deposition of electrodes (anode and cathode) on either side of a microporous plastic substrate; the micropores (15 nm to 20 μm) are etched into the substrate using nuclear particle bombardment. Micro-fuel arrays with external connections in series were fabricated precisely and had a thickness of about a millimeter. The principal advantages of the cell include the high use of catalyst, controlled pore geometry, low-cost materials and minimum cell thickness and weight. A MicroFuel Cell® was reported to have achieved a specific energy density of 300 Wh/kg using methanol/water and air as the anodic and cathodic reactants, respectively [4].

The anode design that was developed by MicroFuel Cell has represented a critical new advance in the development of a cost-effective pore-free electrode that is permeable to only hydrogen ions [4]. This increases the efficiency of a methanol fuel cell because it blocks the deleterious effect of methanol crossover across the membrane. The first layer of the anode electrode formed a plug in the pore of the porous membrane; an example is a 20 nm thick palladium metal film on a Nuclepore® filter membrane with 15 nm diameter pores. The second layer (platinum) was deposited to mitigate the hydration induced cracking that occurs in many of these films. The third layer was deposited over the structural metal film and was the most significant layer because it needed to be catalytically active to methanol and capable of accepting hydrogen ions. An alternative method of forming the electrode was to include powder catalyst particles (Pt/Ru on activated carbon) on the surface of the metal films to enhance the catalytic properties of the electrode. Between the anode electrode and the cathode electrode was the electrolyte filled pore, the cell interconnect and the cell break. In the pores of the membrane the electrolyte (Nafion) was immobilized and ERD claims this collimated structure results in improved protonic conductivity. Each of the cells was electrically separated from the adjacent cells by cell breaks, useless space occupying the central thickness of the etched nuclear particle track plastic membrane. The cathode was formed by sputter depositing a conductive gold film onto the porous substrate first, followed by a platinum catalyst film. The electrode was subsequently coated with a Nafion film. Alternatively, platinum powder catalyst particles were added to the surface of the electrode via an ink slurry of 5% Nafion solution. A hydrophobic coating was then deposited onto this Nafion layer in order to prevent liquid product water from condensing on the surface of the air electrodes. ERD developed a novel configuration to use their fuel cell as a simple charger in powering a cellular phone. The fuel cell was configured into a plastic case that was in close proximity to a rechargeable battery. Methanol was delivered to the fuel cell via fuel needle and fuel ports, which allowed methanol to wick or evaporate into the fuel manifold, and be delivered to the fuel electrodes.

The Jet Propulsion Laboratory (JPL), USA, has been actively engaged in the development of 'miniature' DMFCs for cellular phone applications over the last 2 years [165, 169]. According to their analysis, the power requirement of cellular phones during the standby mode is small and steady at 100–150 mW. However, under operating conditions the power requirement fluctuates between 800–1800 mW. In the JPL DMFC, the anode was formed from Pt-Ru alloy particles, either as fine metal powders (unsupported) or dispersed on high surface area carbon. Alternatively, a bimetallic powder made up of submicron Pt and Ru particles was reported to give better results than the Pt-Ru alloy. Another method describes the sputter-deposition of a Pt-Ru catalyst onto the carbon substrate. The preferred electrolyte was Nafion 117; however, other materials may be used to form proton-conducting membranes. Air was delivered to the cathode by natural convection and the cathode is prepared by applying a platinum ink to a carbon substrate. Another component of the cathode was the hydrophobic Teflon polymer used to create a three-phase boundary and to achieve efficient removal of water produced by the electroreduction of oxygen. Sputtering techniques can also be used to apply the platinum catalyst to the carbon support. The noble metal loading in both electrodes was 4–6 mg cm^{-2}. The MEA was prepared by pressing the anode, electrolyte and cathode at 8.62×10^6 Pa and 146 °C. JPL opted for a 'flat-pack' instead of the conventional bipolar plate design, but this resulted in higher ohmic resistance and non-uniform current distribution. In this design the cells were externally connected in series on the same membrane, with through membrane interconnect and air electrodes on the stack exterior. Two 'flat packs' were deployed in a back to back configuration with a common methanol feed to form a 'twin-pack' [4]. Three 'twin-packs' in series were needed to power a cellular phone. In the stack assembly, six cells were connected in series in a planar configuration, which exhibited average power densities between 6–10 mW cm^{-2}. The fuel cell was typically run at ambient air, 20–25 °C with 1 M methanol. Improvements in the configuration and interconnect design have resulted in improved performance characteristics of the six cell 'flat-pack' DMFC. Based on the results of current technology, the JPL researchers predict that a 1 W DMFC power source with the desired specifications for weight and volume and an efficiency of 20% for fuel consumption can be developed for a 10 h operating time, prior to replacement of methanol cartridges.

As stated earlier, Los Alamos National Laboratory (LANL) has been in collaboration with Motorola Labs—Solid State Research Center to produce a ceramic based DMFC, which provides better than 10 mW cm^{-2} power density. LANL researchers have also been engaged in a project to develop a portable DMFC power source capable of replacing the 'BA 5590' primary lithium battery used by the US Army in communication systems [170]. A 30-cell DMFC stack with electrodes having an active area of 45 cm^2 was constructed, an important feature of which was the narrow width (i.e., 2 mm) of each cell. MEAs were made by the decal method, that is, thin film catalysts bonded to the membrane resulting in superior catalyst use and overall cell performance. An anode catalyst loading of Pt between 0.8–16.6 mg cm^{-2} in unsupported PtRu and carbon supported PtRu were used. A highly effective flow field for air made it possible to use a dry air blower to operate the cathode at three to five times stoichiometry. The stack temperature was limited to 60 °C and the air pressure was 0.76 atm, which is the atmospheric pressure at Los Alamos (altitude of 2500 m). To reduce the crossover rate, methanol was fed into the anode chamber at a concentration of 0.5 M. Since water management becomes more difficult at such low methanol concentrations, a proposed solution was to return water from the cathode exhaust to the anode inlet, while using a pure methanol source and a methanol concentration sensor to maintain the low methanol concentration feed to the anode. The peak power attained in the stack near ambient conditions was 80 W at a stack potential of 14 V and approximately 200 W near 90 °C. From this result, it was predicted that this tight packed stack could have a power density of 300 W/l. An energy density of 200 Wh/kg was estimated for a 10 h operation, assuming that the weight of the auxiliaries is twice the weight of the stack.

Forschungszentrum Julich GmbH (FJG), Germany, has developed and successfully tested a 40-cell 50 W DMFC stack [171]. The FJG system consisted of the cell stack, a water/methanol tank, a pump, and ventilators as auxiliaries. The stack was designed in the traditional bipolar plate configuration, which results in lower ohmic resistances but heavier material requirements. To circumvent the weight limitations current collectors were manufactured from stainless steel (MEAs were mounted between the current collectors) and were inserted into plastic frames to reduce the stack's weight. The 6 mm distance between MEAs (cell pitch) revealed a very tight packaging of the stack design. Each frame carried two DMFC single cells that were connected in series by external wiring [4]. MEAs were fabricated in house with an anode loading of 2 mg cm^{-2} PtRu black, catalyst loading of 2 mg cm^{-2} Pt black and cell area of 100 cm^2 for each of the 40 cells. At the anode a novel construction allowed the removal of CO_2 by convection forces at individual cell anodes. The conditions for running the stack were 1 M methanol, 60 °C and 3 bar O_2 which led to peak energy densities of 45–55 mW cm^{-2}. The cathode used air at ambient or elevated pressures; when the stack operated at temperatures above 60 °C the air was fed into the cathode by convection forces. Recent developments include a three-cell short stack desig which has reduced the cell pitch to only 2 mm. The individual cell area of this design is larger, 145 cm^2, than the previous prototype's and although it is not air breathing, it works with low air stoichiometric rates (more efficient cathodic flow distribution structure).

Samsung Advanced Institute of Technology (SAIT), South Korea, has developed a small monopolar DMFC cell pack (2 cm^2, 12 cells, CO_2 removal path, 5–10 M methanol, air breathing and room temperature) of 600 mW for mobile phone applications [172, 173]. Unsupported PtRu and Pt catalysts were coated onto a diffusion electrode of porous carbon substrate of anode and cathode, respectively. In order to allow methanol wicking and air breathing, short and capillary paths were designed as the diffusion layer. Catalyst loading was around 3–8 mg cm^{-2}. Ternary alloys with low binding energy for CO adsorption were investigated with the aid of quantum chemical methods. Inorganic phase dispersed hybrid membranes based on Nafion or Co-PTFS were prepared and applied to the MEA to attain high fuel efficiency and prevent a voltage loss on the cathode. A monopolar structure was investigated; 12 cells of 2 cm^2 were connected in series within a flat cell pack. Fuel storage was attached to the cell pack and power characteristics were measured on the free-standing basis without any fuel and air supply systems. A power density of 50 mW cm^{-2} at 0.3 V was achieved in the normal diffusion electrode design. For application in portable electronic devices, methanol wicking and air breathing electrodes were required. A monopolar design consisting of 12-cell flat pack was assembled and tested. Each cell had an active area of 2 cm^2 and the pack was equipped with a path for CO_2 removal at the anode. The maximum power output was 560 mW at 2.8 V, close to that required by the cellular phone. For this cell pack condition with small active area, the unit cell power density was 23 mW cm^{-2}.

The Korea Institute of Energy Research (KIER, South Korea) has developed a 10 W DMFC stack (bipolar plate, graphite construction) fabricated with six single cells with a 52 cm^2 electrode area [174]. The stack was tested at 25–50 °C using 2.5 M methanol, supplied without a pumping system, and O_2 at ambient pressure, at a flow rate of 300 cc min^{-1}. The maximum power densities obtained in this system were 6.3 W (121 mW cm^{-2}) at 87 mA cm^{-2} at 25 °C and 10.8 W (207 mW cm^{-2}) at 99 mA cm^{-2} at 50 °C. MEAs using Nafion 115 and 117 were formed by hot pressing and the electrodes were produced from carbon supported Pt-Ru metal powders and Pt-black for anode and cathode electrodes, respectively.

More Energy Ltd. (MEL), Israel, a subsidiary of Medis Technologies Ltd. (MDTL, USA), is developing direct liquid methanol (DLM) fuel cells (a hybrid PEM/DMFC system) for portable electronic devices [175]. The key features of the DLM fuel cell are as follows: (i) the anode catalyst extracts hydrogen from methanol directly, (ii) the DLM fuel cell uses a proprietary liquid electrolyte that acts as the membrane in place of a solid polymer electrolyte (Nafion) and (iii) novel polymers and electrocatalysts enable the fabrication of more effective electrodes. The company's fuel cell module delivers approximately 0.9 V and 0.24 W at 60% of its nominal capacity for eight hours. This translates into energy densities of approximately 60 mW cm^{-2} with efforts underway to improve that result to 100 mW cm^{-2}. The high power capacity of the cell is attributed to the proprietary electrode's ability to efficiently oxidize methanol. In addition Medis claims the use of high concentrations of methanol (30%) in its fuel stream with plans for increasing that concentration to 45% methanol.

At the Institute for Fuel Cell Innovation in Vancouver, Canada, a passive (air breathing) planar three-cell DMFC stack was designed, fabricated and tested [176]. In order to maintain design flexibility, polycarbonate was chosen for the plate material whereas 304 stainless steel mesh current collectors were used. In order to test the DMFC in different electrical cell configurations (single cell, multiple cells connected in series or in parallel), a stainless threaded rod was attached to each mesh current collector on the anode and cathode sides to allow for an external electrical connection. Commercial electrodes from E-TEK were used. The catalyst loading was 4 mg cm^{-2} and consisted of an 80% Pt: Ru alloy on optimized carbon. Unsupported Pt black with a 4 mg cm^{-2} loading was used for the cathode. A Nafion 117 membrane was used as the electrolyte. A power density of 8.6 mW cm^{-2} was achieved at ambient temperatures and under passive operation. Stacks with a parallel connection of the single cells showed a significantly lower performance than in a series configuration. High electrical resistance proved to be the dominant factor in the low performance as a result of the stainless steel hardware and poor contact between the electrodes and current collectors.

At the University of Connecticut, USA, the group of Z. Guo and A. Faghri developed a design for planar air breathing DMFC stacks [177]. This design incorporated a window-frame structure that provided a large open area for more efficient mass transfer with modular characteristics, making it possible to fabricate components separately. The current collectors had a niobium expanded metal mesh core with a platinum coating. Two four-cell stacks, one with a total active area of 18 cm^2 and the other with 36 cm^2, were fabricated by inter-connecting four identical cells in series. These stacks were suitable for portable passive power source application. Peak power outputs of 519 and 870 mW were achieved in the stacks with active areas of 18 and 36 cm^2, respectively. A study of the effects of methanol concentration and fuel cell self-heating on fuel cell performance was carried out. Power density reached its highest value in this investigation when 2 and 3 M methanol solutions were used.

At the Honk Kong University of Science and Technology, China, the group of R. Chen and T.S. Zhao [178–181] studied the effect of methanol concentration on the performance of a passive DMFC single cell. They found that cell performance improved substantially with an increase in methanol concentration; a maximum of power density of 20 mW cm^{-2} was achieved with 5.0 M methanol solution. The measurements indicated that better performance with higher methanol concentrations was attributed mainly to the increase in the cell's operating temperature, a result of the exothermic reaction between permeated methanol and oxygen on the cathode. This finding was subsequently confirmed by the fact that cell performance decreased when the cell that was running with higher methanol concentrations, cooled down to room temperature. Moreover, they proposed a new MEA, in which the conventional cathode gas diffusion layer (GDL) is eliminated while using a porous metal structure made of a metal foam for transporting oxygen and collecting current. They showed theoretically that the new MEA [180] and the porous current collector enabled a higher mass transfer rate of oxygen and, thus, better performance. The improved performance of the porous current collector was attributed to the increased operating temperature, a result of the lower effective thermal conductivity of the porous structure and its fast water removal, a result of the capillary action [181].

Another group at the Honk Kong University, H.F. Zhang *et al.* [182], reported on a flexible graphite-based integrated anode plate for DMFCs operating at high methanol feed concentrations under active mode. This anode structure made of flexible graphite materials not only played a dual role for the liquid diffusion layer and flow field plate but also served as a methanol blocker by decreasing methanol flux at the interface of the catalyst and membrane electrolyte. DMFCs incorporating this new anode structure exhibited a much higher open circuit voltage (OCV) (0.51 V) than that (0.42 V) of a conventional DMFC at a 10 M methanol feed. Cell polarization data showed that this new anode structure significantly improved cell performance at high methanol concentrations (e.g., 12 M or above).

M.A. Abdelkareem and N. Nakagawa from Gunma University, Japan, [183] studied the effect of oxygen and methanol supply modes (passive and active supplies of methanol, and air-breathing and flowing supplies of oxygen) on the performance of a DMFC. The experiments were carried out with and without a porous carbon plate (PCP) under ambient conditions using methanol concentrations of 2 M for the MEA without PCP and 16 M for that with PCP. For the conventional MEA, flowing oxygen and methanol were essential to stabilize the cell's performance, avoiding flooding at the cathode and depletion of methanol at the anode. As a result of flowing oxygen, methanol and water fluxes, the conventional MEAs performance increased more than double compared with that obtained from the air-breathing cell. For the MEA with a porous plate, MEA/PCP, the flow of oxygen and methanol had no significant effect on cell performance, because porous carbon plate, PCP, prevented the cathode from flooding by reducing the mass transport through the MEA.

The effect of operating conditions on the energy efficiency of a small passive DMFC was analyzed by D. Chu and R. Jiang from the US Army Research Laboratory, Adelphi, USA [184]. Both faradaic and energy conversion efficiencies decreased significantly with increasing methanol concentration and environmental temperatures. The faradaic conversion efficiency was as high as 94.8%, and the energy conversion efficiency was 23.9% an environmental temperature low enough (10 °C) and under a constant voltage discharge at 0.6 V with 3 M methanol for a DMFC bi-cell using Nafion 117 as the electrolyte. Although higher temperatures and higher methanol concentrations allowed higher discharge power, they resulted in considerable losses of faradaic and energy conversion efficiencies using the Nafion electrolyte membrane.

Various research groups have focused their attention on the critical aspects which need to be addressed for the design a high-performance DMFC. These are CO_2 bubble flow at the anode [185] and water flooding at the cathode [186]. Lu and Wang from Pennsylvania State University, USA, [187] developed a 5 cm^2 transparent cell to visualize these phenomena *in situ*. Two types of MEA based on Nafion 112 were used to investigate the effects of backing pore structure and wettability on cell polarization characteristics and two-phase flow dynamics. One employed carbon paper backing material and the other, carbon cloth. Experiments were performed with various methanol feed concentrations. The transparent fuel cell reached a peak power of 93 mW cm^{-2} at 0.3 V, using a Toray carbon-paper based MEA with 2 M methanol solution preheated to 85 °C. For the hydrophobic carbon paper backing, it was observed that CO_2 bubbles nucleated at certain locations and formed large and discrete bubble slugs in the channels. For the hydrophilic carbon cloth backing, the bubbles were produced more uniformly and were smaller in size. It was thus shown that the anode backing layer of uniform pore size and more hydrophilicity was preferable for gas management in the anode. Flow visualization of water flooding on the cathode side of the DMFC was also carried out. It showed that the liquid droplets appeared more easily on the surface of carbon paper due to its reduced hydrophobicity at elevated temperatures. For the single-side ELAT carbon cloth, liquid droplets tended to form in the corner between the current collecting rib and GDL since ELAT is highly hydrophobic and the rib (stainless steel) surface is hydrophilic. Even if this study was performed at a relatively high temperature (85 °C), a basic understanding of its results is indispensable for portable DMFC design and optimization.

Lai et al. [188] investigated the long-term discharge performance of passive DMFCs at different currents with different cell orientations. Water produced in the cathode was observed from the photographs taken by a digital camera. The results revealed that the passive DMFCs with anodes facing upward showed the best long-term discharge performance at high currents. A few independent water droplets accumulated in the cathode when the anode faced upward. Instead, in the passive DMFC with vertical orientation, a large amount of the water produced flowed down along the surface of current collector. The passive DMFC with vertical orientation performed relatively well at low currents. It was concluded that the cathode produced less water in a certain period of time at lower currents. In addition, the rate of methanol crossover in the passive DMFC with the anode facing upward was relatively high, which lead to a more rapid decrease of methanol concentration in the anode. The passive DMFC with the anode facing downward resulted in the worst performance because it was very difficult to remove CO_2 bubbles produced in the anode.

Water loss and water recycling in direct-methanol fuel cells (DMFCs) are significant issues that affect the complexity, volume and weight of the system and become of greater concern as the size of the DMFC decreases. A research group at Tel-Aviv University, Israel, [189] developed a flat micro DMFC in a plastic housing with a water-management system that controlled the flux of liquid water through the membrane and the loss of water during operation. These cells contained a nanoporous proton-conducting membrane (NP PCM). Methanol consumption and water loss were measured during operation in static air at room temperature for up to 900 h. Water flux through the membrane varied from negative to zero to positive values as a function of the thickness and the properties of the water-management system. The loss of water molecules (to the air) per molecule of methanol consumed in the cell reaction (defined as the w factor) varied from 0.5 to 7. When w was equal to 2 (water flux through the membrane was equal to zero), there was no need to add water to the DMFC and the cell was operating under water-neutral conditions. On the other hand, when W was smaller than 2, it was necessary to remove water from the cell and when it was larger than 2, water was added. The cell showed stable operation up to 900 h and its maximum power was 12.5 mW cm^{-2}.

At the Korea Institute of Science and Technology (KIST), Kim et al. [190] developed passive micro-DMFCs with capacities under 5 W to be used as portable power sources. Research activities were focused on the development of MEAs and the design of monopolar stacks operating under passive and air-breathing conditions. The passive cells showed many unique features, much different from the active ones. Single cells with an active area of 6 cm^2 showed a maximum power density of 40 mW cm^{-2} at 4 M of methanol concentration at room temperature. A six-cell stack with a total active area of 27 cm^2 was constructed in a monopolar configuration and it produced a power output of 1000 mW (37 mW cm^{-2}). Effects of experimental parameters on performance were also examined to investigate the operational characteristics of single cells and monopolar stacks.

Tekion Inc., Champaign, USA, [191] has developed an advanced air breathing DMFC for portable applications. A novel MEA was fabricated to improve the performance of air-breathing DMFCs. A diffusion barrier on the anode side was designed to control methanol transport to the anode catalyst layer, thus suppressing methanol crossover. A catalyst coated membrane with a hydrophobic gas diffusion layer on the cathode side was employed to improve the oxygen mass transport. The advanced DMFC achieved a maximum power density of 65 mW cm^{-2} at 60 °C with 2 M methanol solution. The value was nearly twice that of a commercial MEA. At 40 °C, the power densities operating with 1 and 2 M methanol solutions were over 20 mW cm^{-2} with a cell potential at 0.3 V.

Pennsylvania State University together with the University of California in Los Angeles, USA, [192] developed a silicon-based micro-DMFC for portable applications. Anode and cathode flow-fields with a channel and rib width of 750 µm and a channel depth of 400 µm were fabricated on Si wafers using microelectromechanical system (MEMS) technology. A MEA was specially fabricated to mitigate methanol crossover. This MEA features a modified anode backing structure in which a compact microporous layer is added to create an additional barrier to methanol transport, thereby reducing the rate of methanol crossing over the polymer membrane. The cell with the active area of 1.625 cm^2 was assembled by sandwiching the MEA between two micro-fabricated Si wafers. Extensive cell polarization testing demonstrated a maximum power density of 50 mW cm^{-2} using 2 M methanol feed at 60 °C. When the cell operated at room temperature, the maximum power density was about 16 mW cm^{-2} with both a 2 and 4 M methanol feed. It was further observed that the present µDMFC still performed reasonably with 8 M methanol solution at room temperature.

The Waseda University, Japan, proposed a new concept for µDMFC (0.018 cm^2 active area) based on MEMS technology [193]. The µDMFC was prepared using a series of fabrication steps from a micro-machined silicon wafer including photolithography, deep reactive ion etching, and electron beam deposition. The novelty of this structure is that anodic and cathodic micro-channels arranged in plane were fabricated, dissimilar to the conventional bipolar structure. The first objective of the experimental trials was to verify the feasibility of this novel structure on the basis of

MEMS technology. The methanol anode and oxidant cathode were prepared by electroplating either Pt-Ru or Pt and Pt, respectively, onto the Ti/Au electrodes. A Nafion 112 membrane was used as the electrolyte. The performance of the μDMFC was assessed at room temperature using 2 M CH_3OH/0.5 M H_2SO_4/H_2O as the fuel and O_2-sat./0.5 M H_2SO_4/H_2O as the oxidant. The fuel supply was by means of a microsyringe pump connected to the μDMFC unit. The OCV for the Pt cell was 300 mV while it was 400 mV for the Pt-Ru cell. The maximum power density was 0.44 mW cm^{-2} at 3 mA cm^{-2} at the Pt electrode while, maximum power density reached 0.78 mW cm^{-2} at 3.6 mA cm^{-2} for the cell with the Pt-Ru anode. The reason for this low performance could be due to a non-optimal composition of Pt-Ru anode catalyst.

The Institute of Microelectronics of Barcelona-CNM (CSIC), Spain, presented a passive and silicon-based micro DMFC [194]. The device was based on a hybrid approach composed of a commercial MEA consisting of a Nafion 117 membrane with a 4.0 mg cm^{-2} Pt-Ru catalyst loading on the anode side and 4.0 mg cm^{-2} Pt on the cathode (E-TEK ELAT) sandwiched between two microfabricated silicon current collectors. The silicon plates were provided with an array of vertical squared channels, 300 micrometers in depth, that covered an area of 5.0 × 5.0 mm. In order to provide the current collectors with an appropriate electrical conductivity, a 150 nm Ti/Ni sputtered layer was deposited covering the front side of the wafer. This conductive layer was used as a seed layer for the 4 mm thick Ni layer that was electrodeposited afterwards. This layer enhanced the electrical conductivity of the current collector; it was then covered by a thin Au layer to prevent oxidation. The cell was equipped with a 100 ml methanol reservoir. The cell was tested at room temperature and different methanol concentrations. It was found that methanol concentration had little impact on the fuel cell's maximum power density, which reached a value of around 11 mW cm^{-2} and was comparable to values reported in the literature for larger passive and stainless-steel fuel cells.

A research group at Yonsei University, Korea, developed a DMFC on printed circuit board (PCB) substrates by means of a photolithography process [195]. The effects of the channel pattern, channel width and methanol flow rate on the performance of the fabricated DMFC were evaluated over a range of flow-channel widths from 200 to 400 μm and flow rates of methanol from 2 to 80 ml min^{-1}. A μDMFC with a cross-stripe channel, zig-zag and serpentine-type patterns. A single cell with a 200 μm wide channel delivered a maximum power density of 33 mW cm^{-2} when using 2 M methanol feed at 80 °C.

Our group (CNR-ITAE, Messina, Italy) investigated two designs of flow-fields/current collectors for a passive DMFC monopolar three-cell stack (Figure 1.20) [196]. The first design (A) consisted of two plastic plates covered by thin gold film current collectors with a distribution of holes through which methanol (from a reservoir) and air (from the atmosphere) could diffuse into the electrodes. The second design (B) consisted of thin gold film deposited on the external borders of the fuel and oxidant apertures in the PCBs where electrodes were placed in contact. A 21 ml methanol reservoir with 3 small holes in the upper part to fill the containers and to release the produced CO_2, was attached to the anode side (Figure 1.21). The MEAs for the two stack designs (3 cells) were manufactured by assembling, simultaneously, three sets

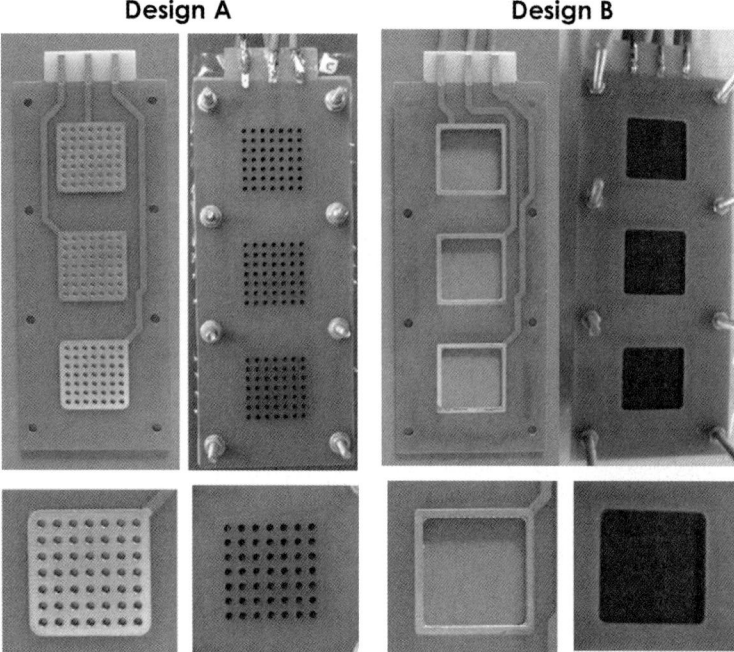

Figure 1.20 Pictures of two different monopolar plates for application in a DMFC three-cell stack operating under passive mode.

of anode and cathode pairs onto the membrane (Figure 1.21b); afterwards they were sandwiched between two PCBs. The geometrical area of each electrode was 4 cm² and the total area of the stack was 12 cm². The cells were connected in series externally through the electric circuit. The electrochemical characterization was carried out varying the catalyst loading and methanol concentration. A 4 mg cm^{-2} Pt loading provided the best electrochemical results in the presence of unsupported catalysts. This appeared to be the best compromise between electrode thickness and the amount of catalytic sites. Similar performances in terms of maximum power were recorded for the two designs whereas better mass transport characteristics were obtained with design B. On the contrary, OCV and stack voltage at low currents were higher for design A as a consequence of lower methanol crossover. Maximum power of 220–240 mW was obtained at ambient temperatures for the three-cell stack with 5 M methanol corresponding to a power density of about 20 mW cm^{-2}. An investigation of the discharge behavior of the two designs was carried out (Figure 1.22). A longer discharge time (17 h) with unique MeOH charge was recorded with design B at 250 mA compared with design A (5 h). This was attributed to easier CO_2 removal from the anode and better mass transport properties. In fact, in design A, CO_2 did not escape easily from the anode, which hindered methanol diffusion to the catalytic sites by natural convection. When the small stack based on A design was mechanically agitated, the effect of this forced convection increased the discharge time.

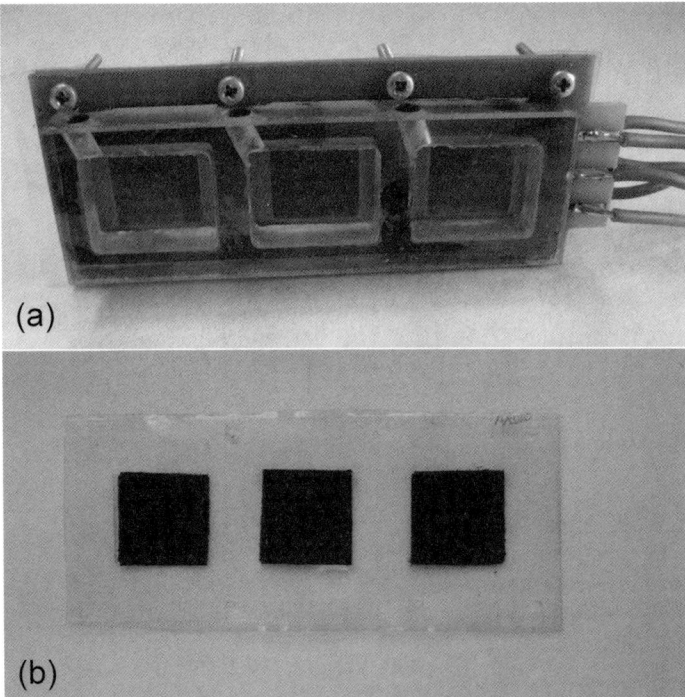

Figure 1.21 Pictures of the DMFC design B used for a three-cell stack (a) and MEA formed by a single membrane and three couples of electrodes (b).

A recent European project called Morepower addressed the development of a low cost, low temperature (30–60 °C) portable DMFC device of compact construction and modular design. The project was coordinated by GKSS (Germany) and included, as partners, Solvay, Johnson Matthey, CNR-ITAE, CRF, Polito, IMM and NedStack. The electrical characteristics of the device were 40 A, 12.5 V (total power 500 W). Single cell performance approached 0.2 A cm^{-2} at about 0.5 V/cell at 60 °C and atmospheric pressure [21]. Several new membranes were investigated in this project. One of the most promising was a low-cost proton exchange membrane produced by Solvay using a radiochemical grafting technology (Morgane CRA type membrane) which showed a suitable compromise in terms of reduced methanol crossover and suitable ionic conductivity [111]. Inorganic filler-modified S-PEEK membranes were also developed in the same project by GKSS (Germany) to reduce the permeability to alcohol while maintaining high proton conductivity [111].

1.4.2
Transportation

Though the application of fuel cells in transportation has drawn great enthusiasm and stimulated interest since the late 1970s, it is still considered a formidable venture

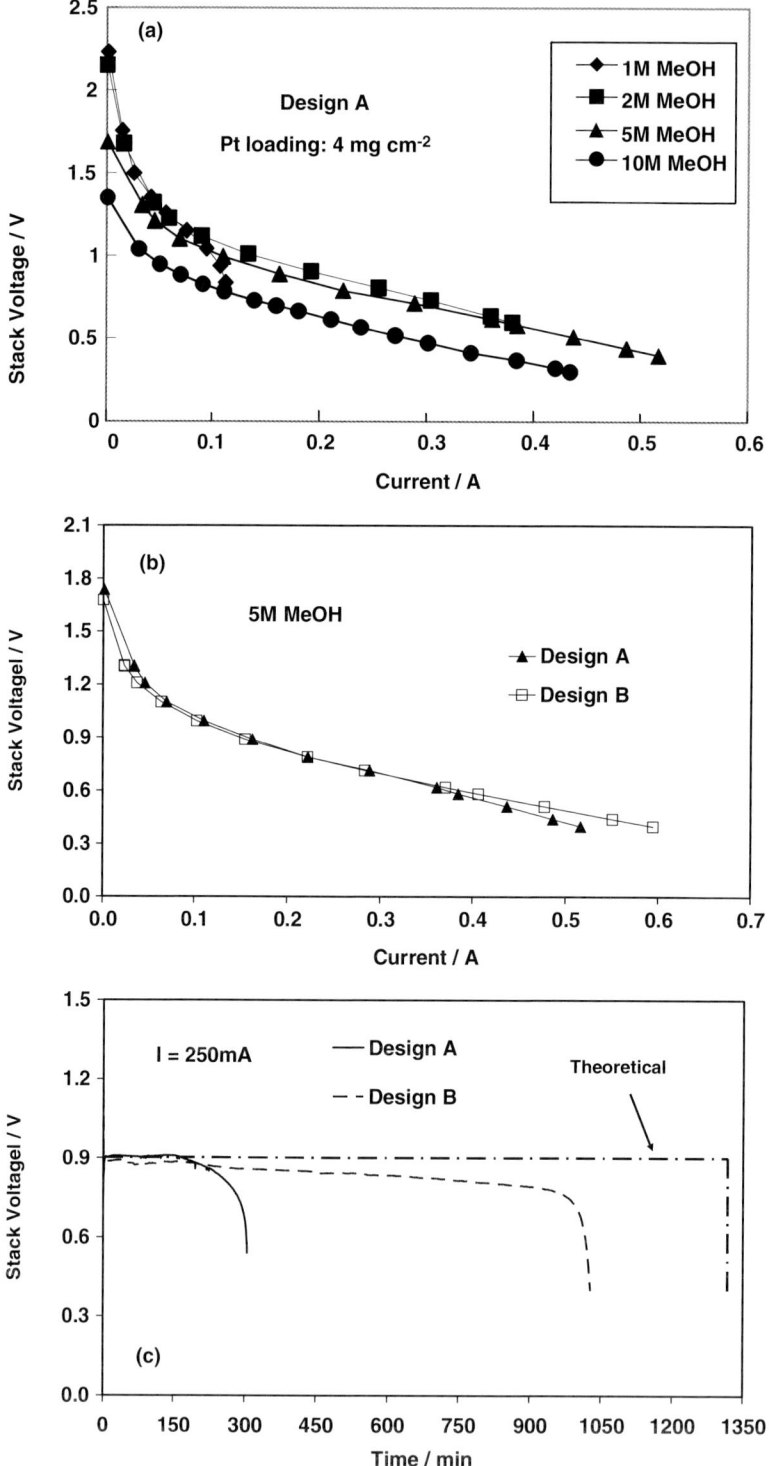

if fuel cell powered vehicles are to compete with the conventional internal combustion and diesel engine powered vehicles. This is not surprising since fuel cell development is still in its infancy, compared with the highly advanced IC or diesel engine technology which has taken over 100 years to reach high levels of performance with respect to operating characteristics (start-up time, acceleration, lifetime, considerable reduction in level of environmental pollutants, etc.). The impetus for developing battery and fuel cell-powered vehicles derived from the energy crisis in 1973; in the late 1980s and 1990s, environmental legislation to reduce greenhouse gas emissions provided further stimulation. The United States Partnership for new Generation of Vehicles Program was implemented to make 'Quantum Jumps' in the performance of automobiles, such as (i) tripling efficiency of fuel consumption, (ii) reaching a range of 500 km between refueling, and (iii) ultra low or zero emissions of pollutants, while remaining cost competitive with the current automobile technology [2, 4]. Similar objectives have been addressed in European Research Programs FP5 and FP6. The only types of vehicle that have the potential of reaching these goals are IC or diesel engine/battery and fuel cell/battery hybrid vehicles. The former type of power plants are more advanced than the latter and, in fact, Toyota and Honda have commercialized IC engine/battery hybrid vehicles in the last few years. Other companies have also started commercialization of diesel engine/battery hybrids [2, 4]. Nowadays, however, due to the considerable progress made in this field, DMFCs appear much more ready for application in electrotraction systems. With the development of highly active catalysts and appropriate ionomeric membranes, these systems have been successfully operated at temperatures close to or above 100 °C, allowing the achievement of interesting performances [104, 170]. In particular, it was shown that the overall efficiency of recent DMFC devices is comparable or superior to the combination of reformer-H_2:air fuel cells [171]. These aspects, together with the intrinsic advantages of methanol fuel cells with respect to hydrogen-consuming devices, which are due mainly to the liquid fuel feed and the absence of a cumbersome reformer, would claim for a close demonstration of DMFCs in electric vehicles. Yet, DMFC devices may be employed in a fuel cell vehicle if they fulfill specific requirements in terms of power density, durability, cost and system efficiency. Accordingly, more active catalysts need to be developed together with high temperature and crossover resilient membranes. In addition, a great deal of attention should be devoted to bipolar plates and flow-fields both in terms of design and materials.

Practically all worldwide activities on fuel cell/battery hybrid vehicles (Daimler/Chrysler/Ballard, Ford, Toyota, General Motors/Opel, Honda, Volkswagen, Fiat) are essentially on PEMFC or PEMFC/battery hybrid vehicles [2, 4]. In several demonstration vehicles, hydrogen was the fuel carried on board, mostly as a compressed gas

Figure 1.22 Polarization curves for the design A stack with a Pt loading of 4 mg cm^{-2} on each electrode at different methanol concentrations (a), comparison between the polarization curves obtained with the two different designs with a Pt loading of 4 mg cm^{-2} on each electrode and 5 M methanol solution (b), and chronopotentiometric results at 250 mA obtained with the two designs using a Pt loading of 4 mg cm^{-2} and 5 M methanol solution (c).

or as a metal hydride. However, in order to meet the technical targets of the vehicle and to minimize problems caused by changes needed in the infrastructure and fuel distribution network, emphasis has been on carrying conventional gasoline fuel or methanol on board and processing it into hydrogen. However, due to (i) the efficiency losses in fuel processing, (ii) the significant weight of the fuel processing system and (iii) the progress made in DMFC technology with respect to efficiency, specific power and power density, there has been an increasing interest in developing DMFCs in recent years; a 3 kW DMFC in a one-passenger vehicle prototype was demonstrated by Daimler-Chrysler/Ballard [2, 4]. Though DMFC technology is quite promising, major breakthroughs are still needed if it is to compete with PEMFC technology even though the latter has the burden of carrying a heavy fuel processor to produce hydrogen from gasoline or methanol or compressed hydrogen fuel [2, 4].

DMFC technology offers a solution for transportation applications in transition towards a zero emission future. Using methanol as a fuel circumvents one of the major hurdles plaguing PEMFC technology, that is the development of an inexpensive and safe hydrogen infrastructure to replace the gasoline/diesel fuel distribution network. It is well established that the infrastructure for methanol distribution and storage can be easily adapted from the current gasoline intensive infrastructure. Another drawback in using PEMFC technology is the need to store hydrogen (at very high pressures) or carry a bulky fuel processor to convert the liquid fuel into hydrogen on board the vehicle. Methanol is an attractive fuel because it is a liquid under atmospheric conditions and its energy density is about half of that of gasoline. Despite the compelling advantages of using DMFCs in transportation applications, major obstacles to their introduction remain. These barriers include the high cost of materials used in fabricating DMFCs (especially the high cost of platinum electrocatalysts), the crossover of methanol through the electrolyte membrane from the anode to the cathode, and the lower efficiency and power density performance of DMFCs in comparison to PEMFCs.

Despite these obstacles a number of institutions (particularly in the last ten years) have become actively engaged in the development of DMFCs for transportation applications. The most remarkable results achieved in this field are summarized in Table 1.3. These institutions have directed their resources toward improving every facet of the DMFC in a quest for competitive balance with PEMFCs, as stated below. Ballard Power Systems Inc. (BPSI, Canada) in collaboration with Daimler–Chrysler (Germany) recently reported the development of a 3 kW DMFC system that is at a very preliminary stage in comparison to Ballard's PEMFC products [197]. Daimler–Chrysler (Germany) demonstrated this system for transportation application in a small one-person vehicle at its Stuttgart Innovation Symposium in November, 2000. The DMFC go-cart weighed approximately 100 kg, required an 18 V/1 Ah battery system for starting the electric motor on its rear wheels, and had a range of 15 km and a top speed of 35 km/h. The stack used 0.5 l methanol (the concentration of methanol was unclear) as fuel and operated at approximately 100 °C. In January, 2001 Ballard revealed that they had built and operated a 6 kW stack (60 V) based on the same stack design as the prototype shown in Stuttgart. No details are available with respect to the

Table 1.3 DMFC prototypes for stationary, APU and automotive applications.

Single Cell/Stack Developer	Power/Cell Power density	Temperature (°C)	Oxidant	Methanol Concentration (M)	Anode Catalyst	Membrane Electrolyte	Cathode catalyst	Number of cells/ Surface area (cm^2)
Ballard Power Systems, Inc.	3 kW	100	Air	1	Pt/Ru	Nafion	Pt	—
IRD Fuel Cell A/s	100 mW cm^{-2}	90–100	1.5 atm air	—	Pt/Ru	Nafion	Pt	4/154 cm^2 bipolar
Thales, CNR-ITAE, Nuvera FCs	140 mW cm^{-2}	110	3 atm air	1	Pt/Ru	Nafion	Pt	5/225 cm^2 bipolar
Siemens Ag	250 mW cm^{-2}/ 90 mW cm^{-2}	110/80	3 atm O$_2$ (1.5 atm air)	0.5 (0.5)	Pt/Ru	Nafion 117	Pt-black	3 cm^2 per cell
Los Alamos National Labs	1 kW/-	100	3 atm air	0.75	Pt/Ru	Nafion 117	Pt	30/45 cm^2 bipolar
Thales, CRF-Fiat, CNR-ITAE, Solvay (DREAMCAR Project)	5 kW/160 mW cm^{-2}	130	3 atm air	1–2	85% PtRu/C	Hyflon	60%Pt/C	100/300 cm^2 bipolar

stack design and performance of the DMFC power source. However, the patent literature indicates fabrication techniques for producing DMFC electrodes [198]. The anode was prepared by first oxidizing the carbon substrate (carbon fiber paper or carbon fiber non-woven) via electrochemical methods in an acidic aqueous solution (0.5 M sulfuric acid) prior to the incorporation of the proton-conducting ionomer. The second step involves the impregnation of a proton-conducting ionomer such as a poly(perfluorosulfonic acid) into the carbon substrate. The anode preparation is completed by applying aqueous electrocatalyst ink to the carbon substrate without extensive penetration in the substrate. This method ensures that less electrocatalyst is used and that the catalyst is applied to the periphery of the electrode where it will be used more efficiently. The performance enhancements associated with the treatment of the carbonaceous substrate may be related to the increase in the wettability of the carbonaceous substrate. This may result in more intimate contact between the ionomer coating and the electrocatalyst, thereby improving proton access to the catalyst. Another theory concludes that the presence of the acidic groups on the carbon substrate itself may improve proton conductivity, or the surface active acidic groups may affect the reaction kinetics at the electrocatalyst sites. The assembly of the MEA and single cell was carried out via conventional methods, that is, hot pressing the anode and cathode to a solid polymer membrane electrolyte.

IRD Fuel Cell A/S (Denmark) has developed DMFCs primarily for transportation applications (0.7 kW) [199]. The stack was constructed with separate water and fuel circuits and the bipolar flow plates are made of a special graphite/carbon polymer material for corrosion reasons. The MEAs had an active cell area of 154 cm^2. The air pressure was 1.5 bar at the cathode. A nominal cell voltage of 0.5 V was observed for IRDs stack at a current density at 0.2 A/cm^2 and electric power was generated at 15 W per cell. More recently, IRD has developed a 3 kW DMFC stack.

A consortium composed of Thales-Thompson (France), Nuvera Fuel Cells (Italy), LCR (France) and Institute CNR-ITAE (Italy) has developed a five-cell 150 W stainless steel based air fed DMFC stack in the framework of the Nemecel project with the financial support of the European Union Joule Program [200]. Bipolar plates were used in the stack's design and MEAs were fabricated using Nafion as the solid polymer electrolyte and high surface area carbon supported Pt-Ru and Pt electrocatalysts for methanol oxidation and oxygen reduction, respectively. The electrode area was 225 cm^2 and the stack was designed to operate at 110 °C, using 1 M methanol and 3 atm air achieving an average power density of 140 mW/cm^2.

Siemens AG in Germany, in conjunction with IRF A/S in Denmark and Johnson Matthey Technology Center in the United Kingdom developed a DMFC stack with an electrode area of 550 cm^2 under the auspices of the European Union Joule Program [201–204]. The projected cell performance was a potential of 0.5 V at a current density of 100 mA/cm^2 with air pressure at 1.5 atm and the desirable stoichiometric flow rate. A 3-cell stack was demonstrated operating at a temperature of 110 °C and a pressure of 1.5 atm using 0.75 M methanol; this stack exhibited a performance level of 175 mA/cm^2 at 0.5 V per cell, and at 200 mA/cm^2 the cell potential was 0.48 V.

These performances were obtained at a high stoichiometric air flow rate (factor of 10) but in order to reduce auxiliary power requirements, one of the goals at Siemens was to improve the design to lower the air stoichiometric flow to the desired value of a factor of about two. A 0.85 kW air-fed stack composed of 16 cells and operating at 105 °C was demonstrated successively with a maximum power density of 100 mW cm^{-2}.

Los Alamos National Laboratory (LANL) is also actively pursuing the design and development of DMFC cell stacks for electric vehicle applications. According to the latest available information, a five-cell short stack with an active electrode area of 45 cm^2 per cell has been demonstrated [104, 170, 205]. The cells operated at 100 °C, an air pressure of 3 atm and a methanol concentration of 0.75 M. The maximum power of this stack was 50 W, which corresponds to a power density of 1 kW/l. At about 80% of the peak power, the efficiency of the cell stack with respect to the consumption of methanol was 37%.

Among the recent European community projects dealing with the development of DMFCs for automotive and APU applications, the Dreamcar project (ERK6-CT-2000-00315) that was carried out in the framework of the FP5 EC program should be mentioned. Dreamcar was the acronym of Direct Methanol Fuel Cell System for Car Applications; the project was coordinated by Thales Engineering & Consulting (France) and CRF- FIAT (Italy) and included, as partners, CNR-ITAE (Italy), Solvay (Belgium) and TAU-Ramot (Israel) [162]. The main objective of the project was to design, manufacture and test a 5 kW stack at high temperatures (up to 140 °C). There were three main research topics in the Dreamcar project: higher operating temperatures (up to 140 °C) to enhance the electrochemical reactions; development of new fluorinated (improvement of the membranes developed in the framework of a previous project, Nemecel JOE3-CT-0063) and hybrid inorganic-organic membranes; development of new carbon supported Pt-alloy catalysts to increase the efficiency of the electrodes and power density [162].

The Solvay Solexis Hyflon membrane was selected for the final stack. In order to allow stack operation at high temperatures with the Hyflon membrane, the operating pressure was 3 bar abs. The performance of the MEAs was first investigated in a single cell configuration based on the same materials of the final stack. In the framework of the same project, a nanoporous proton conducting membrane (NP-PCM) that showed superior performance in the presence of a liquid TFMSA acid electrolyte was also developed [103]. Yet, the use of an acidic liquid electrolyte, necessary to make the NP-PCM membrane conductive, was considered incompatible with the materials used in the construction and testing of the final stack (severe problems with corrosion and fluid management) [162].

The final stack (Figure 1.23) consisted of 100 cells of 300 cm^2 and provided an output electrical power of about 5 kW. The specific power output was 110 W/l The average single cell performance in the final stack was about 160 mW cm^{-2} compared with 300 mW cm^{-2} that was almost achieved in the single cell with the same membrane/electrode materials [162]. The main drawbacks concerned methanol crossover and heat management since significant heat/energy dissipated during operation at 130/140 °C at 3–4 bar using an external radiator [162].

Figure 1.23 A 5 kW DMFC stack developed in the framework of the Dreamcar project.

Technology that is presently considered promising for electro-traction consists of a hybrid system using both pure hydrogen-fed PEMFCs and advanced Li-batteries [206]. To make DMFCs competitive with regard to this technology, it is essential to increase the power density, decrease methanol crossover and reduce costs. Regarding the electrolyte, an appropriate membrane operating in a range that varies from sub-zero to 130 °C at ambient pressure is required. The same requirements apply to membranes for PEMFCs. In general, high temperature stack operation would simplify heat and water management.

1.4.3
Technology Development

1.4.3.1 DMFC Technology

MEAs are usually considered the most important components of a DMFC power source. They contain backing layers, gas diffusion layers, catalytic layers and membranes. However, a significant role is also played by the flow field/current collector, reactant manifold and the stack's housing. A stack module is usually formed by a connected series of cells (e.g., through bipolar plates). Several modules can be connected to each other in series or in parallel depending on the required electrical characteristics of the power source. Furthermore, several auxiliaries are necessary for thermal and water management, start-up, shut-down and normal operation. These include compressor/blowers, fuel tank and liquid pumps, methanol concentration sensors, gas/liquid separation devices, eventually catalytic burner, and

DC/DC (step-up) and DC/AC converters. All the above components form a DMFC system and they are the subject of development and integration studies.

1.4.3.2 Catalyst Preparation

The synthesis of a highly dispersed electrocatalyst phase in conjunction with a high metal loading on a carbon support is one of the present goals in DMFCs [2]. One of the main requirements for an optimal electrocatalyst is its high dispersion. The mass activity (A g^{-1}) of the catalyst for an electrochemical reaction is directly related to the degree of dispersion since the reaction rate is generally proportional to the active surface area. The main routes for the synthesis of Pt-Ru/carbon electrocatalysts include impregnation, colloidal procedures, self-assembling methods, decoration and so on [2].

1.4.3.3 Electrode Manufacturing and Membrane Electrode Assemblies Membrane Electrode Assembly (MEAs)

The performance of a DMFC is also strongly affected by the fabrication procedure of the MEA. Conventional technology that was used two decades ago consisted of the preparation of gas-diffusion electrodes having suitable polytetrafluoroethylene (PTFE) content in both diffusion and catalyst layers. Nafion ionomer was spread onto the electrocatalyst layer, followed by the preparation of membrane-electrode assembly by a hot-pressing procedure [19]. A disadvantage of this procedure is the poor electrochemically active area between electrocatalyst particles and ionomer, thus decreasing the catalyst use [207].

One of the approaches that has recently been used, especially for PEMFCs, concerns the direct deposition of the catalyst onto the membrane to form a catalyst coated membrane (CCM) [208]. The diffusion and backing layers are added subsequently for example, during cell and stack assembling. In this configuration, there is an intimate contact between the catalytic layer and the membrane whereas the diffusion layer is put in contact with the catalytic layer only.

Regarding the cathode operation, while oxygen reacts to produce water, nitrogen contained in the air stream remains entrapped in the pores of the electrode; the entrapped nitrogen is a diffusion barrier for the incoming oxygen, and it results in mass transport overpotential with, consequently, performance losses even at intermediate current densities. Furthermore, although it is known that oxygen permeability through the ionomer is high at high current densities, transport of this gas to the reaction sites is retarded by flooding of the electrocatalyst layer [2]. Due to this flooding of the active layer, the ionomer swells until it is saturated with water, thus increasing the hydrophilicity of the layer. Such drawbacks have been conveniently reduced in air feed-SPE fuel cells by using thin film electrodes. These are characterized by low electrocatalyst loadings (0.05–0.1 mg cm^{-2}) [208]. Due to the lower performance of the oxygen electrode with such low Pt loadings in DMFCs, alternative solutions should be investigated. Gas channels allowing a fast transport of the reaction gas and easy removal of the excess N_2 can be realized in the cathode layer by means of PTFE-carbon composite ducts [209]. This configuration does not affect electrocatalyst use or the continuity of ionomer in the catalyst layer, thus improving the mass-transport properties of the electrode. Another approach is to use pore formers such as $(NH_4)_2CO_3$ to increase the porosity in the active layer of the cathode [210, 211].

1.4.3.4 Stack Hardware and Design

The architecture of a DMFC stack for transportation and stationary applications, including the remote and distributed generation of electrical energy, is essentially similar to that of a PEFC stack for the same applications [5, 97, 98]. In contrast, a large variety of approaches and designs has been adopted for portable fuel cell stacks [4, 5]. Requirements for stacks vary depending on the applications. Compact size, fast start-up procedure and high performance are required for transportation applications [7]. Easy handling, miniaturization and rapid fuel refilling are especially important for portable applications [4, 5].

The conventional PEMFC stack architecture [97, 98] for transportation and stationary applications is based on bipolar plates connecting the various cells (MEAs) in series. There are also two end plates enabling current collection, a manifold, and appropriate gaskets which together with the flow-fields in the bipolar plates allow distribution of the reactants over the various cells. The flow fields are often based on flow channels machined into graphite (generally composite graphite is used) or consist of corrosion resistant alloy bipolar plates; a flow field can also be a corrosion resistant metal foam or a stamped flow pattern in a metal plate. The machined graphite flow field can be a simple design of dots, parallel channels, serpentine or interdigitated design [2]. The flow configuration may be cross-flow, co-flow or counter-flow. All these aspects significantly influence mass transport and thermal management by favoring diffusion or forced convection of the reactants to the catalytic sites and heat removal. In DMFC stacks cooling cells are not strictly required because efficient heat removal can be obtained either by increasing the recirculation rate of the liquid mixture of water and methanol at the anode or using an external radiator.

Significant progress has been made by improving the characteristics of the electrode backing layer in terms of composition and thickness to reduce mass transport limitations. Some investigations have focused on the design of reactant flow fields [11]. The most widely employed flow field in advanced fuel cells is based on the serpentine configuration. The reactant molecules have access to the electrocatalytic sites through diffusion across the so-called diffusion layer, that is, the backing layer, made of carbon cloth and carbon black, hydrophobized by the appropriate addition of PTFE.

A different approach to the flow of reactants and products within the electrode structure, that is, an interdigitated design, was proposed by Nguyen [212] and Wilson et al. [205] for H_2-O_2 solid polymer electrolyte fuel cells (SPEFCs). In practice, the reactant gases are forced to enter and exit the electrode pores under a gradient pressure achieved by making the inlet and outlet channels dead-ended. As pointed out by Nguyen [212], the flow through the electrode in an interdigitated design is no longer governed by diffusion but becomes convective in nature. The forced-flow-through characteristics created by the interdigitated flow fields in SPEFCs have been also investigated for DMFCs [11]. In general, it has been shown that enhanced mass transfer characteristics are achieved with the interdigitated flow field in DMFCs but these beneficial effects are especially observed only at high current densities, which correspond to low values of cell potential. At high cell potentials, that is, under practical operating conditions (above 0.5 V), higher efficiency for fuel use is obtained with the classical serpentine flow fields due to the lower methanol crossover.

In fuel cell stacks of significant size, graphite bipolar plates are being replaced with the more economic carbon based composite materials or by metallic foams [213, 214]. With composite materials, the same design of graphite plates may be reproduced whereas metallic foam operates, conceptually, under conditions similar to serpentine flow fields in that the reactant distribution over the electrocatalyst layer is controlled by diffusion. Alternative stack designs have been investigated by Scott and co-workers [215]. These authors analyzed the possibility of using more open structures at the anode (e.g., dots or open channels) to favor the diffusion of methanol. In other cases, the parallel flow channel pattern has been preferred due to an optimal combination of simplicity of design and suitable performance. Graphite or carbon composite based bipolar plates exhibit minimal corrosion. In the case of stainless-steel or metallic alloy-based materials, an appropriate evaluation of the chemical and electrochemical stability in the presence of hot methanol/water mixtures is necessary. In some cases surface treatments or special alloys are required to minimize corrosion.

DMFC stacks for portable applications may have different architectures [4, 5] especially if the power output is smaller than 50–100 W and passive mode operation is required. Several configurations have been proposed for the passive DMFC stacks; the most common are the bi-cell and monopolar-type [5]. In the bi-cell type, the methanol tank is allocated in between two anodes which belong to two different cells and the cathodes of these two cells are exposed to air [5]. Bi-cell units are grouped in a stack by leaving a gap between two cathodes belonging to two different bi-cells; the series connection of the various bi-cells is made externally. In the monopolar configuration all electrodes of the same type, for example, all anodes, are allocated on one face of the membrane and the cathodes on the other face. Each couple of electrodes forms a cell; the membrane is the same for all the cells. Series connection between two cells is created by an electric conductor passing through the membrane or by an external circuit [4, 5].

A planar architecture is often used for μDMFC stacks [216]. For example, a catalyzed membrane integrated on a silicon or polymeric matrix through micromachining processes has recently emerged as a possible way to fabricate miniaturized DMFCs [217, 218]. Thanks to integrated-circuit (IC) technology [219], micro-channel patterns of μDMFCs bipolar plates, into which reactants are fed, can be featured on an Si or polymeric matrix with high resolution and good repeatability. Basically, the μDMFC has a conventional single cell structure in which the MEA is sandwiched between two current collectors, made of gold, with fuel/air channels. These designs take advantage of the full wafer-level process capability. Alternatively, micro-channels can be created on a polymeric substrate, such as polycarbonate, by mechanical erosion with a numerical control mechanical device [193].

Micropumps can be used for fuel delivery in μDMFC stacks. For the passive mode operation, several approaches can be used for the methanol feed to the anode. These have been reviewed recently by Qian et al. [5]. Such approaches are based on natural circulation [220], capillary action [221] or self-pressurization using a controlled three-way valve [222]. It is appropriate to use a non-diluted fuel tank and to control the methanol and water feeds by valves, metering, orifices or pumps. Water should be recovered from the cathode, for example, by favoring back-diffusion through the membrane from the cathode.

1.4.3.5 DMFC Systems

The DMFC stack plant is generally designed on the basis of the power output level and the desired application. An interesting DMFC system design was proposed in a recent European project called Morepower [21]. The project investigated the development of a low-cost, low temperature, portable DMFC system of compact construction and modular design with nominal power 250 W for the potential markets of weather stations, medical devices, signal units, gas sensors and security cameras. The system was designed by the Institut fur Microtechnik of Mainz (Germany) and the modeling was carried out by Specchia *et al.* at the Politecnico of Turin (Italy) to evaluate heat, mass fluxes and pressure drops, for the integration and optimization of the DMFC components in a portable Auxiliary Power Unit [223]. The system design and components are reported in Figure 1.24 [224]. These consist of the DMFC stack, the radiator (E-201) to cool the fuel solution downstream the DMFC anode, the gas–liquid separator (S-201, an atmospheric adiabatic flash unit) to dump up the produced CO_2, the catalytic burner (R-401) to burn the residual MeOH vapor before

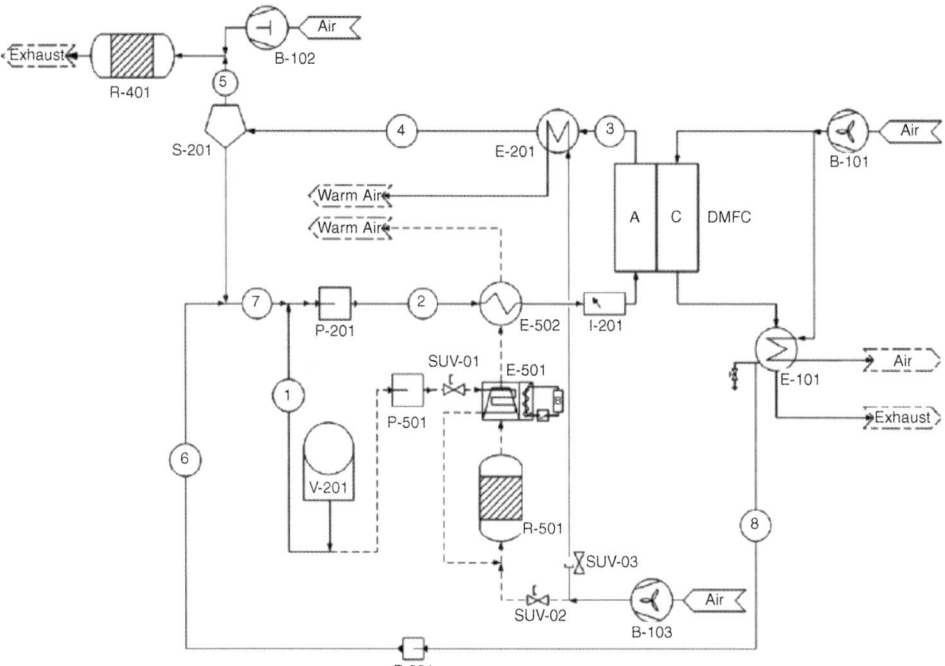

Figure 1.24 DMFC process scheme developed in the framework of the Morepower project [223].

releasing the anode exhausts in the atmosphere, the pump (P-201) to feed the fuel solution to the DMFC anode, the MeOH cartridge (V-201) to feed fresh MeOH into the system, the water condenser (E-101) to recover and make-up the water lost during operation, and the blower (B-101) to feed the fresh air necessary for the cathode reactions. The addition of fresh feed solution from the MeOH cartridge (V-201) to the exhaust solution is controlled via an MeOH sensor (I-201) [224, 225]; the controlled composition feed solution is then pumped into the DMFC where overall electrochemical reactions between the fuel and air produce power and heat. All the systems and sub-components necessary for the start-up are also present in the DMFC system. A small fraction of pure MeOH, taken directly from the MeOH cartridge (V-201) via a dedicated pump (P-501), is fed to an evaporator (E-501, electrically heated during the initial phase of the start-up procedure). The MeOH vapor obtained is then burned into a burner (R-501) with fresh air (B-103); the flue gas produced is used to heat-up the solution to be fed to the DMFC in the start-up heat exchanger (E-502).

The complexity of this design is determined by the characteristics required in terms of system control and rapid start-up and shut-down. A simpler system can be designed if a self-start up is preferred to rapid start-up; the burner (R-501) with its associated auxiliaries that is, evaporator (E-501), pump (P-501), blower (B-103) and relative valves in Figure 1.24 can be removed from the scheme. A concentrated methanol solution, for example, 10 M, remains liquid even at several degrees below zero. In terms of CO_2 escape, the use of a highly selective hydrophobic membrane can allow the removal of the catalytic burner (R-401) with the associated air supply (B-102) in Figure 1.24; furthermore, the radiator (E-201) may be more compact if no loss of water/methanol vapor occurs through the selective membrane even at high temperatures. In such a case, there is no need to cool down the unreacted fuel mixture significantly. Accordingly, it would not be necessary to spend much energy in the E-502 pre-heater. The heat released from the stack should be properly used to heat-up the methanol solution fed to the anode. The device necessary for recovering water from the cathode condenser to the anode may be quite compact if part of the water permeates or back-diffuses from a highly hydrophobic cathode to the anode through a membrane containing proper hydrophilic channels.

A simple passive methanol fuel cell stack usually does not need auxiliaries; on the other hand, miniaturized DMFCs may require some auxiliaries such as micropumps and so on. Miniaturized systems can also be quite complex. DMFC stacks and systems for portable uses have recently been reviewed [5]. Of course, a simplification of the systems allows a reduction in production costs. In some cases, the proper development of materials for MEAs and auxiliaries may help to simplify DMFC systems.

1.5
Perspectives of Direct Methanol Fuel Cells and Techno-Economical Challenges

The most challenging problem in the development of DMFCs has been, and still is, that significant enhancement of electrocatalytic activity for the 6-electron transfer electro-oxidation of methanol is needed. On the other hand, research in this area has encouraged many scientists and engineers to use highly sophisticated electrochemical surface science and material science techniques to unravel the mysteries of the reaction path, rate determining steps and physicochemical characteristics (electronic and geometric factors, adsorption/desorption energies and electrocatalyst/support interaction) which influence the activities of the various types of electrocatalysts. The sluggishness of the reaction, especially in the presence of protonic electrolytes, is caused by a strong chemical adsorption of CO-type species on an electrocatalyst subsequent to the dissociative adsorption of methanol (Pt is the best known electrocatalyst for this step). A neighboring chemisorbed labile OH species is vital for the electro-oxidation of the strongly adsorbed CO species. To date, the Pt-Ru electrocatalyst (50:50 at. wt%) has shown the best results. There has been little success with alternatives to Pt and its alloys in proton conductive electrolytes; those tested include transition metal alloys, oxides, and tungsten bronzes. One achievement has been using carbon-supported electrocatalysts, which help reducing the Pt loading by a factor of about two to four.

The reaction rates are higher in alkaline environments with respect to protonic electrolytes. This fact and the lower corrosion constraints in alkaline media allow the replacement of Pt with non-precious metal catalysts for example, Ni. Alternatively, PtRu can be used in an alkaline electrolyte to take advantage of the lower overpotentials. However, no significant enhancement in terms of power density has been achieved because this kinetic advantage is counteracted by the carbonation drawback and reduced ionic conductivity unless concentrated alkaline solutions are used.

The performance of the ORR on a platinum electrocatalyst is affected by the crossover of methanol from the anode to the cathode through the ion exchange membrane. The open circuit potential is reduced by about 200 mV due to the competitive adsorption of dissociated methanol and oxygen species. At present, for Pt alloys there is no clear evidence of catalytic enhancement in oxygen reduction when methanol crossover occurs. Non-platinum electrocatalysts, such as heat-treated phthalocyanines and porphyrins, as well as transition metal chalcogenides, have some chance of methanol tolerance but have considerably lower activity than platinum and also raise questions of stability. The short term prospect of replacing platinum as an electrocatalyst is very slim but the greater challenge is to reduce the noble metal loading in both electrodes by a factor of about 10 in order to reduce its cost to about \$10/kW.

If an anion exchange membrane is used instead of a protonic electrolyte, Pt-based cathode electrocatalysts can be replaced by silver or MnO_2, which are much less expensive and methanol tolerant. Although the oxygen reduction in alkaline media is faster than an acidic electrolyte, the performance enhancement achieved with anion exchange membranes is quite limited due to both the absence of a suitable ionomer to extend the triple-phase boundary in the electrode bulk and the low anionic conductivity.

The perfluorosulfonic acid polymer electrolyte in the DMFC is an equally expensive material (about \$300/kW, based on state-of-the-art performance). There has been a lot of research on alternative proton conducting membranes which allow CO_2

rejection (sulfonated polyetherketone, polyether sulfone, radiation grafted polystyrene, zeolites, electrolytes doped with heteropolyacids and sulfonated polybenzimidazole), but it is still a challenge to attain sufficiently high specific conductivity and stability in the DMFC environment. Nafion-based composite membranes with silicon oxide and zirconium hydrogen phosphate have enhanced performance in operation up to about 150 °C (lower activation and ohmic overpotentials); these can also operate suitably under ambient conditions with reduced crossover due to an increase of the tortuosity factor.

Alternatively, new emphasis has recently been placed on anion exchange membranes. Both anodic and cathodic reaction rates are enhanced in alkaline media. Yet, the kinetic advantage is counteracted by a thermodynamic loss due to the presence of a pH gradient between the anode and the cathode. This is caused by a carbonation process occurring at the anode. This drawback can be overcome by recirculating KOH or carbonate solution through the device, but several technical problems arise under these conditions, that is, precipitation of carbonate on the electrode pores, the need to frequently regenerate the electrolyte and so on. Other drawbacks of anion exchange membranes concern low anionic conductivity (about five times lower than Nafion at low temperatures), higher activation energy for ion conduction than Nafion, no proper ionomer solution to enable an extension of the three-phase reaction zone in the electrode bulk, and reduced stability at high temperatures. Some of these problems can be solved by improving the characteristics of the anionic polymer electrolyte. Regarding methanol crossover, there is no effect from electro-osmotic drag with anion exchange membranes. Yet, as is well known, most of the methanol crossover is due to the concentration gradient between the anode and the cathode and the hydrophilic properties of the present membranes.

A critical area to improve overall cell performance in is the fabrication of MEAs. Progress in the preparation of high performance MEAs has been made by preparing thin electrocatalyst layers (about 10 μm thick) composed of the electrocatalyst and ionomer in the electrode substrate or directly deposited onto the membrane (CCM). Problems caused by the barrier layer effects of nitrogen on oxygen access to the catalytically active sites and electrode flooding need further investigation. Possible solutions to these problems are heat treatments of the recast Nafion gel in the electrocatalytic layer to make it hydrophobic or the use of pore formers to increase porosity.

Direct methanol single cell development in the last decade has achieved very interesting results. Maximum power densities of about 500 mW cm^{-2} and 300 mW cm^{-2} under oxygen and air feed operation, respectively, and 200 mW cm^{-2} at a cell potential of 0.5 V have been reported for cells operating at temperatures close to or above 100 °C under pressurized conditions, with Pt loadings of 1–2 mg cm^{-2}. At ambient temperatures in passive mode operation, the power density ranges between 10 and 40 mW cm^{-2}.

The development of DMFC stacks for both transportation and portable applications has gained momentum in recent years. The rated power output of the DMFC stack varies from a few watts in the case of portable power sources up to a few kW for remote power generator and hybrid battery-fuel cell vehicles. The best results achieved with DMFC stacks for electrotraction are 1 kW/l power density with an overall efficiency of 37% at 0.5 V/cell. These performances make the DMFCs quite competitive with respect to the reformer-H_2/air SPE fuel cell, especially if one considers the complexity of the whole system; yet, the Pt loadings are still high in DMFCs (around 1–2 mg cm^{-2}). Reducing the loading of noble metals or using cheap non-noble metal catalysts is actually one of the breakthroughs which may allow an increase in DMFC competitiveness on the power source market.

In the short-term the high energy density of DMFCs and recent advances in the technology of miniaturized fuel cells make these systems attractive in terms of replacing the current Li-based batteries in cellular phones, lap top computers and other portable systems. This field appears the most promising for the near-term and a successful use of such systems is envisaged; the progress made in manufacturing DMFCs for portable systems may also stimulate new concepts and designs which may aid further development of these systems for electrotraction.

Table 1.4 summarizes the main drawbacks of DMFCs together with some potential solutions. Unfortunately, several proposed approaches create new drawbacks. For

Table 1.4 Drawbacks and potential solutions of DMFC devices.

Drawback	Potential solution	Present approach
Low power density	Enhance oxidation kinetics	-Multifunctional catalysts
		-Increase the operating temperature and pH
	Improve electrode performance	-Highly dispersed catalysts
		-Thin film electrodes
		-Optimization of the MEA
Fuel crossover	Membranes impermeable to methanol	-Anion exchange membranes
		-Composite membranes
		-Polyarylsulfonic membranes
		-Polyvinyl alcohol treated membranes
	Methanol tolerant oxygen reduction catalysts	-Chevrel-phase type ($Mo_4Ru_2Se_8$), transition metal sulfides ($Mo_xRu_yS_z$, $Mo_xRh_yS_z$) or other transition metal chalcogenides
		-Pt-alloys
High cost	Reduce noble metal loading	-Non-noble metal catalysts (anode and cathode) in conjunction with alkaline electrolytes
		-Oxide catalysts
		-Cathode catalysts based on iron or cobalt organic macrocycles (phenylporphyrins, phthalocyanines)
		-Cobalt polypyrrole-carbon composite catalysts (Co-PPY-C)
		-Decoration (anode catalyst)
	Membranes alternative to Nafion	-Anion exchange membranes
		-Grafted membranes
		-S-PEEK, SPSf, and so on

example, an increase in operating temperature to enhance the reaction kinetics causes membrane dehydration with most conventional membranes. This results in a significant increase of ohmic constraints. On the other hand, the membranes which allow high temperature operation, such as phosphoric acid doped polybenzoimidazole, do not appear appropriate in terms of suitable conductivity under low temperature operation, a pre-requisite for portable power sources. The use of non-noble metal catalysts is presently possible in DMFCs using alkaline electrolytes. However, low conductivity, carbonation and thermodynamic constraints limit the practical applications of this approach. The methanol tolerant cathode catalysts such as Chevrel-phase type or transition metal chalcogenides, do not allow oxidation to CO_2 of the methanol permeated through the membrane; the cathode outlets thus, contains traces of unreacted methanol that cannot be released in the atmosphere. This requires modification of the system and/or cell concept. Reduction of the catalyst layer thickness to reduce mass transport constraints can be achieved by increasing the concentration of the active phase on the support. However, this approach reduces catalyst use. The use of highly hydrophobic cathodes that favor oxygen transport and reduce flooding by water permeated through the membrane or formed by the reaction increases the resistance and reduces the interface between catalyst and ionomer (triple phase boundary).

These examples show that there are no unequivocal or radical solutions and a compromise is often necessary to enhance device characteristics. Furthermore, it also appears that materials should be tailored to specific applications. A chemical and dimensional stable electrolyte with high conductivity in a wide temperature range would be more appropriate if the conduction mechanism is not assisted by water. Methanol crossover is often associated with water permeation; these effects cause cathode poisoning and flooding. From a practical point of view, a carbon dioxide rejecting electrolyte appears more appropriate but new efforts should address the development of multifunctional catalysts with reduced noble metal loadings. Significant progress in the development of materials would be also beneficial to reduce system complexity.

The applications of DMFCs in portable power sources cover the spectrum of cellular phones, personal organizers, laptop computers, military back power packs, and so on. The infusion of semiconductor technology into the development of micro and mini fuel cells by leading organizations such as LANL, JPL, Motorola, has created an awareness of DMFCs replacing the most advanced type of rechargeable batteries, that is, lithium ion. For several of these applications, a DMFC working at room temperature and ambient pressure with an efficiency of only about 20% may perform strikingly better than lithium ion batteries with respect to operating hours between refueling/recharging because of the high energy density of methanol. Further, refueling in the case of DMFCs is instantaneous whereas it requires about 3–5 hrs for lithium ion batteries. There is still the challenge of reducing the weight and volume of DMFCs to a level competitive with lithium ion batteries, necessary, for instance, for cellular phone and laptop applications. What is most attractive about the portable power applications, compared with transportation and stationary applications, is that the cost per kW or cost per kWh could be higher by a factor of 10 to 100 without compromising the market application. For this application, there is hardly any competition for lithium ion and DMFCs from any other type of power source.

The present analysis indicates that the targets for each application may be achieved through thoughtful development of materials, innovative device design, and through an appropriate choice of operating conditions.

Acknowledgments

We acknowledge the financial support for DMFC activity from the European Community through the Nemecel (EU Joule), Dreamcar (EU FP5) and Morepower (EU FP6) projects, from Regione Piemonte through the Microcell project and from Pirelli Labs., Solvay-Solexis, De Nora and Nuvera through several contracts. In particular, we would thank the project leaders of these contracts, M. Dupont, M. Straumann, H. Hutchinson, G. Bollito, S. Nunes, G. Saracco, S. Specchia, P. Caracino, A. Tavares, A. Ghielmi, R. Ornelas and E. Ramunni.

We express our gratitude to our colleagues that have collaborated in DMFC activity at CNR-ITAE, in particular, A.K. Shukla, H. Kim, S. Srinivasan, C. Yang, R. Dillon, K. M. El-Khatib, Z. Poltarzewski, A.M. Castro Luna, G. Garcia, L.G. Arriaga, I. Nicotera and P. L. Antonucci.

We are indebted to our collaborators C. D'Urso, A. Stassi, A. Di Blasi, S. Siracusano, T. Denaro, F.V. Matera, E. Modica, G. Monforte and P. Cretì for their invaluable contribution.

References

1 Wasmus, S. and Kuver, A. (1999) *J. Electroanal. Chemistry*, **461**, 14–31.
2 Arico, A.S., Srinivasan, S. and Antonucci, V. (2001) *Fuel Cells*, **1** (2), 133–161.
3 Lamy, H., Léger, J.-M. and Srinivasan, S. (2000) Direct methanol fuel cells—from a 20th century electrochemists' dream to a 21st century emerging technology, in *Modern Aspects of Electrochemistry*, **34** (eds J.O'.M. Bockris and B.E. Conway), Plenum Press, New York, Ch 3, p. 53.
4 Dillon, R., Srinivasan, S., Aricò, A.S. and Antonucci, V. (2004) *J. Power Sources*, **127** (1–2), 112–126.
5 Qian, W., Wilkinson, D.P., Shen, J., Wang, H. and Zhang, J. (2006) *J. Power Sources*, **154** (1), 202–213.
6 Bockris, J.O'.M. and Srinivasan, S. (1969) *Fuel Cells: Their Electrochemistry*, McGraw-Hill Book Company, New York.
7 Shukla, A.K., Aricò, A.S. and Antonucci, V. (2000) *Renew. Sust. Energ. Rev.*, **5**, 137–155.
8 McNicol, B.D., Rand, D.A.J. and Williams, K.R. (1999) *J. Power Sources*, **83** (1–2), 15–31.
9 Ren, X., Wilson, M.S. and Gottesfeld, S. (1996) *J. Electrochem. Soc.*, **143** (1), L12–L15.
10 Aricò, A.S., Creti, P., Kim, H., Mantegna, R., Giordano, N. and Antonucci, V. (1996) *J. Electrochem. Soc.*, **143**, 3950–3959.
11 Aricò, A.S., Cretì, P., Baglio, V., Modica, E. and Antonucci, V. (2000) *J. Power Sources*, **91** (2), 202–209.
12 Moore, R.M., Gottesfeld, S. and Zelenay, P. (1999) Proton Conducting Membrane Fuel Cells – Second International Symposium, Electrochemical Society, Pennington, New Jersey, Proceedings

98-27 (eds S. Gottesfeld and T.F. Fuller), pp. 365–379.
13 Narayanan, S.R., Chun, W., Valdez, T.I., Jeffries-Nakamura, B., Frank, H., Surampudi, S., Halpert, G., Kosek, J., Cropley, C., LaConti, A.B., Smart, M., Wang, Q., Surya Prakash, G. and Olah, G.A. (1996) Program and Abstracts, 1996 Fuel Cell Seminar, pp. 525–528.
14 Baldauf, M. and Preidel, W. (1999) *J. Power Sources*, **84** (2), 161–166.
15 Shukla, A.K., Christensen, P.A., Hamnett, A. and Hogarth, M.P. (1995) *J. Power Sources*, **55** (1), 87–91.
16 Scott, K., Taama, W.M., Argyropoulos, P. and Sundmacher, K. (1999) *J. Power Sources*, **83** (1–2), 204–216.
17 Jung, D.H., Lee, C.H., Kim, C.S. and Shin, D.R. (1998) *J. Power Sources*, **71** (1–2), 169–173.
18 Ravikumar, M.K. and Shukla, A.K. (1996) *J. Electrochem. Soc.*, **143** (8), 2601–2606.
19 Srinivasan, S., Mosdale, R., Stevens, P. and Yang, C. (1999) *Annu. Rev. Energy Environ.*, **24**, 281–285.
20 Aricò, A.S., Antonucci, V., Alderucci, V., Modica, E. and Giordano, N. (1993) *J. Appl. Electrochem.*, **23** (11), 1107–1116.
21 EU funded project MOREPOWER (compact direct methanol fuel cells for portable applications), project nr. SES6-CT-2003-502652 (2004).
22 Carlstrom, C., Craft, J., Fannon, M., Manning, M., Marvin, R., Modi, A., Reichard, J., Scartozzi, P., Dolan, G. and Sievers, B. (2006) DMFC Prototype Demonstration for Consumer Electronic Applications in U.S. Department of Energy Hydrogen Program: Hydrogen, Fuel Cells & Infrastructure, Technologies, Annual Review, May 16–19, Arlington, Virginia.
23 Apanel, G. and Johnson, E. (2004) *Fuel Cells Bull.*, **11**, 12–17.
24 Kordesch, K. and Simader, G. (1996) *Fuel Cells and their Applications*, Wiley-VCH Verlag GmbH, Weinheim.
25 Pavela, T.O. (1954) *Ann. Acad. Sci. Fennicae AII*, **59**, 7–11.
26 Justi, E.W. and Winsel, A.W. (1955) Brit. Patent 821, 688.
27 Cathro, K.J. (1969) *J. Electrochem. Soc.*, **116** (11), 1608–1611.
28 Janssen, M.M.P. and Moolhuysen, J. (1976) *Electrochim. Acta*, **21** (11), 861–868.
29 Watanabe, M. and Motoo, S. (1975) *J. Electroanal. Chem.*, **60** (3), 275–283.
30 Bagotzky, V.S. and Vassiliev, Yu.B. (1967) *Electrochim. Acta*, **12** (9), 1323–1343.
31 Shibata, M. and Motoo, S. (1985) *J. Electroanal. Chem.*, **194** (2), 261–274.
32 McNicol, B.D. and Short, R.T. (1977) *J. Electroanal. Chem.*, **81** (2), 249–260.
33 Parsons, R. and VanderNoot, T. (1998) *J. Electroanal.Chem.*, **257** (1–2), 9–45.
34 Aramata, A., Kodera, T. and Masuda, M. (1988) *J. Appl. Electrochem.*, **18** (4), 577–582.
35 Beden, B., Kardigan, F., Lamy, C. and Leger, J.M. (1981) *J. Electroanal. Chem.*, **127** (1–3), 75–85.
36 Chandrasekaran, K., Wass, J.C. and Bockris, J. O'.M. (1990) *J. Electrochem. Soc.*, **137** (2), 518–524.
37 Christensen, P.A., Hamnett, A. and Troughton, G.L. (1993) *J. Electroanal. Chem.*, **362** (1–2), 207–218.
38 Ticianelli, E., Berry, J.G., Paffet, M.T. and Gottesfeld, S. (1977) *J. Electroanal. Chem.*, **81** (2), 229–238.
39 Goodenough, J.B., Hamnett, A., Kennedy, B.J., Manoharan, R. and Weeks, S. (1988) *J. Electroanal. Chem.*, **240** (1–2), 133–145.
40 Ravikumar, M.K. and Shukla, A.K. (1996) *J. Electrochem. Soc.*, **143** (8), 2601–2606.
41 Mc Breen, J. and Mukerjee, S. (1995) *J. Electrochem. Soc.*, **142** (10), 3399–3404.
42 Markovic', N., Widelov, A., Ross, P.N., Monteiro, O.R. and Brown, I.G. (1997) *Catal. Lett.*, **43** (1–2), 161–166.
43 Piela, P., Eickes, C., Brosha, E., Garzon, F. and Zelenay, P. (2004) *J. Electrochem. Soc.*, **151**, A2053–A2059.
44 Gurau, B., Viswanathan, R., Liu, R.X., Lafrenz, T.J., Ley, K.L., Smotkin, E.S., Reddington, E., Sapienza, E.A., Chan, B.C., Mallouk, T.E. and Sarangapani, S.

(1998) *J. Phys. Chem. B*, **102** (49), 9997–10003.
45 Fritts, S.D. and Sen, R.K. (1988) Assessment of Methanol Electrooxidation for Direct Methanol-Air Fuel Cells. *DOE Report*, Contract DE AC0676 RLO 1830 by Battelle.
46 Cameron, D.S., Hards, G.A., Harrison, B. and Potter, R.J. (1987) *Platinum Metals Rev.*, **31** (4), 173–181.
47 Beden, B., Kadirgan, F., Lamy, C. and Leger, J.-M. (1981) *J. Electroanal. Chem.*, **127** (1–3), 75–85.
48 Hamnett, A. and Kennedy, B.J. (1988) *Electrochimica Acta*, **33** (11), 1613–1618.
49 Miyazaki, E. (1980) *J. Catal.*, **65**, 84–92.
50 Goodenough, J.B., Hamnett, A., Kennedy, B.J., Manoharam, R. and Weeks, S.A. (1990) *Electrochimica Acta*, **35** (1), 199–207.
51 Hamnett, A. (1997) *Catal. Today*, **38** (4), 445–457.
52 Hogarth, M.P. and Hards, G.A. (1996) *Platinum Metal Rev.*, **40**, 150–155.
53 Wang, K., Gasteiger, H.A., Markovic, N.M. and Ross, P.N. Jr (1996) *Electrochim. Acta*, **41** (16), 2587–2593.
54 Frelink, T., Visscher, W. and Van Veen, J.A.R. (1995) *Surf. Sci.*, **335** (1–2), 353–360.
55 Frelink, T., Visscher, W. and Van Veen, J.A.R. (1996) *Langmuir*, **12** (15), 3702–3708.
56 Iwasita, T., Hoster, H., John-Anacker, A., Lin, W.F. and Vielstich, W. (2000) *Langmuir*, **16** (2), 522–529.
57 Tong, Y.Y., Rice, C., Wieckowski, A. and Oldfield, E. (2000) *J. Am. Chem. Soc.*, **122** (6), 1123–1129.
58 Gasteiger, H.A., Markovic, N., Ross, P.N. and Cairns, E.J. Jr (1994) *J. Phys. Chem.*, **98** (2), 617–625.
59 Friedrich, A.K., Geyzers, K.P., Linke, U., Stimming, U. and Stumper, J. (1996) *J. Electroanal. Chem.*, **402** (1–2), 123–128.
60 Schmidt, T.J., Noeske, M., Gasteiger, H.A., Behm, R.J., Britz, P., Brijoux, W. and Bonnemann, H. (1997) *Langmuir*, **13** (10), 2591–2595.
61 Dinh, H.N., Ren, X., Garzon, F.H., Zelenay, P. and Gottesfeld, S. (2000) *J. Electroanal. Chem.*, **491** (1–2), 222–233.
62 Schmidt, T.J., Gasteiger, H.A. and Behm, R.J. (1998) *Electrochem. Commun.*, **1**, 1–5.
63 Takasu, Y., Fujiwara, T., Murakami, Y., Sasaki, K., Oguri, M., Asaki, T. and Sugimoto, W. (2000) *J. Electrochem. Soc.*, **147** (12), 4421–4427.
64 Pletcher, D. and Solis, V. (1982) *Electrochim. Acta*, **27** (6), 775–782.
65 Chrzanowski, W. and Wieckowski, A. (1998) *Langmuir*, **14** (8), 1967–1970.
66 Kim, H. and Wieckowski, A. (1999) Third International Symp. On New Materials for Fuel Cells and Modern Battery Systems, Montréal, Canada (ed. O. Savadogo), pp. 125–126.
67 Liu, L., Pu, C., Viswanathan, R., Fan, Q., Liu, R. and Smotkin, E.S. (1998) *Electrochim. Acta*, **43** (241), 3657–3663.
68 Aricò, A.S., Shukla, A.K., El-Khatib, K.M., Creti, P. and Antonucci, V. (1999) *J. Appl. Electrochem.*, **29** (6), 671–676.
69 Gasteiger, H.A., Markovic, N., Ross, P.N. Jr and Cairns, E.J. (1994) *J. Electrochem. Soc.*, **141** (7), 1795–1803.
70 Aricò F A.S., Baglio, V., Cretì F P. and Antonucci, V. (2000) Fuel Cell Seminar Abstracts, October 30-November, Portland, Oregon, pp. 75–78.
71 Iwasita, T., Nart, F.C. and Vielstich, W. (1990) *Ber. Bunsenges Phys. Chem.*, **94** (9), 1030–1034.
72 Rolison, D.R., Hagans, P.L., Swider, K.L. and Long, J.W. (1999) *Langmuir*, **15** (3), 774–779.
73 Aricò, A.S., Monforte, G., Modica, E., Antonucci, P.L. and Antonucci, V. (2000) *Electrochem. Commun.*, **2** (7), 466–470.
74 Hamnett, A., Kennedy, B.J. and Wagner, F.E. (1990) *J. Catal.*, **124** (1), 30–40.
75 Aricò, A.S., Cretì, P., Modica, E., Monforte, G., Baglio, V. and Antonucci, V. (2000) *Electrochim. Acta*, **45** (25–26), 4319–4328.
76 Aricò, A.S., Baglio, V., Di Blasi, A., Modica, E., Antonucci, P.L. and

Antonucci, V. (2003) *J. Electroanal. Chem.*, **557**, 167–176.

77 Maillard, F., Lu, G.-Q., Wieckowski, A. and Stimming, U. (2005) *J. Phys. Chem. B*, **109** (34), 16230–16243.

78 Brankovic, S.R., Wang, J.X. and Adzic, R.R. (2001) *Electrochem. Solid-State Lett.*, **4** (12), A217–A220.

79 Arico,' A.S., Baglio, V., Di Blasi, A., Modica, E., Monforte, G. and Antonucci, V. (2005) *J. Electroanal. Chem.*, **576**, 161–169.

80 Aricò, A.S., Baglio, V., Modica, E., Di Blasi, A. and Antonucci, V. (2004) *Electrochem. Commun.*, **6**, 164–169.

81 Hays, C.C., Manoharan, R. and Goodenough, J.B. (1993) *J. Power Sources*, **45** (3), 291–301.

82 Niedrach, L.W. and Zeliger, H.I. (1969) *J. Electrochem. Soc.*, **116** (1), 152–153.

83 Okamoto, H., Kawamura, G., Ishikawa, A. and Kudo, T. (1987) *J. Electrochem. Soc.*, **134** (7), 1653–1658.

84 Bashyam, R. and Zelenay, P. (2006) *Nature*, **443** (7107), 63–66.

85 Poirier, J.A. (1994) *J. Electrochem. Soc.*, **141** (2), 425–430.

86 Giordano, N., Passalacqua, E., Pino, L., Aricò, A.S., Antonucci, V., Vivaldi, M. and Kinoshita, K. (1991) *Electrochim. Acta*, **36** (13), 1979–1984.

87 Watanabe, M., Saegusa, S. and Stonehart, P. (1989) *J. Electroanal. Chem.*, **271** (1–2), 213–220.

88 Ren, X., Zelenay, P., Thomas, S., Davey, J. and Gottesfeld, S. (2000) *J. Power Sources*, **86** (1), 111–116.

89 Garcia, G., Baglio, V., Stassi, A., Pastor, E., Antonucci, V. and Aricò, A.S. (2007) *J. Solid State Electr.*, **11**, 1229–1238.

90 Wang, Y., Li, L., Hu, L., Zhuang, L., Lu, J. and Xu, D. (2003) *Electrochem. Commun.*, **5** (8), 662–666.

91 Slade, R.C.T. and Varcoe, J.R. (2005) *Solid State Ionics*, **176** (5–6), 585–597.

92 Yu, E.H. and Scott, K. (2004) *Electrochem. Commun.*, **6** (4), 361–365.

93 Wynn, J.E. (1960) *Proc. Ann. Power Sources Conf.*, **14**, 52–57.

94 Murray, J.N. and Grimes, P.G. (1963) *Fuel Cells, CEP Technical Manual*, American Institute of Chemical Engineers, **57-65**.

95 Hunger, H.F. (1960) *Proc. Ann. Power Sources Conf.*, **14**, 55–59.

96 Varcoe, J.R. and Slade, R.C.T. (2004) *Fuel Cells*, **5** (2), 187–200.

97 Srinivasan, S., Dillon, R., Krishnan, L., Arico, A.S., Antonucci, V., Bocarsly, A.B., Lee, W.J., Hsueh, K.-L., Lai, C.-C. and Peng, A. (2003) in Proceedings of the 1st International Fuel Cell Science, Engineering and Technology Conference, April 21-23, 2003, Rochester, New York, 529–536.

98 Costamagna, P. and Srinivasan, S. (2001) *J. Power Sources*, **102** (1–2), 242–252.

99 Pivovar, B.S. et al. (2007) Sulfonated Poly (aryl ether)-type Polymers as Proton Exchange Membranes: Synthesis and Performance in *Membranes for Energy Conversion*, **2** (eds K.-V. Peinemann and S.P. Nunes), Wiley-VCH Verlag GmbH, Weinheim, pp. 1–46.

100 Aricò, A.S., Baglio, V., Di Blasi, A., Antonucci, V., Cirillo, L., Ghielmi, A. and Arcella, V. (2006) *Desalination*, **199**, 271–273.

101 Neburchilov, V., Martin, J., Wang, H. and Zhang, J. (2007) *J. Power Sources*, **169** (2), 221–238.

102 Peled, E., Duvdevani, T., Aharon, A. and Melman, A. (2000) *Electrochem. Solid-State Lett.*, **3** (12), 525–528.

103 Peled, E., Livshits, V., Rakhman, M., Aharon, A., Duvdevani, T., Philosoph, M. and Feigling, T. (2004) *Electrochem. Solid-State Lett.*, **7**, A507–A510.

104 Aricò, A.S., Antonucci, P.L., Modica, E., Baglio, V., Kim, H. and Antonucci, V. (2002) *Electrochim. Acta*, **47**, 3723–3732.

105 Wang, J., Wasmus, S. and Savinell, R.F. (1995) *J. Electrochem. Soc.*, **142**, 4218–4223.

106 Aricò, A.S., Creti, P., Antonucci, P.L. and Antonucci, V. (1998) *Electrochem. Solid-State Lett.*, **1**, 66–68.

107 Aricò, A.S., Baglio, V., Di Blasi, A., Cretì, P., Antonucci, P.L. and Antonucci, V. (2003) *Solid State Ionics*, **161**, 251–265.

108 Ruffmann, B., Silva, H., Schulte, B. and Nunes, S. (2003) *Solid State Ionics*, **162–163**, 269–275.

109 Aricò, A.S., Baglio, V., Di Blasi, A. and Antonucci, V. (2003) *Electrochem. Commun.*, **5**, 862–866.

110 Lufrano, F., Baglio, V., Staiti, P., Arico,' A.S. and Antonucci, V. (2006) *Desalination*, **199** (1–3), 283–285.

111 Antonucci, V., Aricò, A.S., Baglio, V., Brunea, J., Buder, I., Cabello, N., Hogarth, M., Martin, R. and Nunes, S. (2006) *Desalination*, **200**, 653–655.

112 Miyatake, K. and Watanabe, M. (2007) Polyimide Ionomer Membranes for PEFCs and DMFCs in *Membranes for Energy Conversion*, **2** (eds K.V. Peinemann and S.P. Nunes), Wiley-VCH Verlag GmbH, Weinheim, pp. 47–60.

113 Mauritz, K.A. and Moore, R.B. (2004) *Chem. Rev.*, **104**, 4535–4585.

114 Kreuer, K.D., Paddison, S.J., Spohr, E. and Schuster, M. (2004) *Chem. Rev.*, **104**, 4637–4678.

115 Savadogo, O. (2004) *J. Power Sources*, **127**, 135–161.

116 Savadogo, O. (1998) *J. New Mater. Electrochem. Sys.*, **1**, 47–66.

117 Cleghorn, S., Ren, X., Thomas, S. and Gottesfeld, S. (1997) Book of Extended Abstracts, ISE-ECS Joint Symposium, Paris, Sept., Abstract n. 182, pp. 218–219.

118 Nolte, R., Ledjeff, K., Bauer, M. and Mulhaupt, R. (1993) *J. Membr. Sci.*, **83**, 211–220.

119 Kerres, J., Ullrich, A., Meier, F. and Haring, T. (1999) *Solid State Ionics*, **125**, 243–249.

120 Alberti, G. and Casciola, M. (2003) *Annu. Rev. Mater. Res.*, **33**, 129–154.

121 Kreuer, K.D. (2001) *J. Membr. Sci.*, **185**, 29–39.

122 Li, Q. and Jensen, J.O. (2007) Membranes for High Temperature PEMFC Based on Acid-Doped Polybenzimidazoles in *Membranes for Energy Conversion*, **2** (eds K.-V. Peinemann and S.P. Nunes), Wiley-VCH Verlag GmbH, Weinheim, pp. 61–96.

123 Aricò, A.S., Baglio, V., Cretì, P., Di Blasi, A., Antonucci, V., Brunea, J., Chapotot, A. et al. (2003) *J. Power Sources*, **123** (2), 107–115.

124 Watanabe, M., Uchida, H., Seki, Y., Emori, M. and Stonehart, P. (1996) *J. Electrochem. Soc.*, **143**, 3847–3852.

125 Aricò, A.S., Baglio, V. and Antonucci, V. (2007) Composite Membrane for High Temperature Direct Methanol Fuel Cells in *Membranes for Energy Conversion*, **2** (eds K.-V. Peinemann and S.P. Nunes), Wiley-VCH Verlag GmbH, Weinheim, pp. 123–168.

126 Ogumi, Z., Matsuoka, K., Chiba, S., Matsuoka, M., Iriyama, Y., Abe, T. and Inaba, M. (2002) *Electrochemistry*, **70** (12), 980–983.

127 Ren, X., Sringer, T.E. and Gottesfeld, S. (1999) Proton Conducting Membrane Fuel Cells – Second International Symposium, Electrochemical Society, Pennington, New Jersey, Proceedings 98–27 (eds S. Gottesfeld and T.F. Fuller), pp. 341–357.

128 Ichinose, O., Kawaguchi, M. and Furuya, F. (2004) *J Appl Electrochem.*, **34** (1), 55–59.

129 Alonso-Vante, N. and Tributsch, H. (1986) *Nature (London)*, **323**, 431–432.

130 Sun, J.R., Wang, J.T. and Savinell, R.F. (1998) *J. Appl. Electrochem.*, **28** (10), 1087–1093.

131 Imaizunii, S., Shimanoe, K., Teraoka, Y. and Yamazoe, N. (2005) *Electrochem. Solid-State Lett.*, **8** (6), A270–A272.

132 Paffet, M.T., Beery, G.J. and Gottesfeld, S. (1988) *J. Electrochem. Soc.*, **135** (6), 1431–1436.

133 Watanabe, M., Tsurumi, K., Mizukami, T., Nakamura, T. and Stonehart, P. (1994) *J. Electrochem. Soc.*, **141** (10), 2659–2668.

134 Freund, A., Lang, J., Lehman, T. and Starz, K.A. (1996) *Cat. Today*, **27** (1–2), 279–283.

135 Jalan, V. and Taylor, E.J. (1983) *J. Electrochem. Soc.*, **130**, 2299–2301.

136 Gottesfeld, S., Paffett, M.T. and Redondo, A. (1986) *J. Electroanal. Chem.*, **205**, 163–168.

137 Beard, B.C. and Ross, P.N. (1990) *J. Electrochem. Soc.*, **137**, 3368–3374.

138 Kim, K.T., Hwang, J.T., Kim, Y.G. and Chung, J.S. (1993) *J. Electrochem. Soc.*, **140** (1), 31–36.

139 Mukerjee, S., Srinivasan, S., Soriaga, M.P. and McBreen, J. (1995) *J. Electrochem. Soc.*, **142** (5), 1409–1422.

140 Neergat, M., Shukla, A.K. and Gandhi, K.S. (2001) *J. Appl. Electrochem.*, **31** (4), 373–378.

141 Takasu, Y., Iwazaki, T., Sugimoto, W. and Murakami, Y. (2000) *Electrochem. Commun.*, **2** (9), 671–674.

142 Aricò, A.S., Shukla, A.K., Kim, H., Park, S., Min, M. and Antonucci, V. (2001) *Appl. Surf. Sci.*, **172** (1–2), 33–40.

143 Baglio, V., Di Blasi, A., D'Urso, C., Antonucci, V., Aricò, A.S., Ornelas, R., Morales-Acosta, D., Ledesma-Garcia, J., Godinez, L.A., Arriaga, L.G. and Alvarez-Contreras, L. (2008) *J. Electrochem. Soc.*, **155**, B829–B833.

144 Kinoshita, K. (1992) *Electrochemical Oxygen Technology*, John Wiley & Sons, Inc., New York.

145 Antolini, E., Lopez, T. and Gonzalez, E.R. (2008) *J. Alloys and Compounds*, **461**, 253–262.

146 Dohle, H., Divisek, J., Mergel, J., Oetjen, H.F., Zingler, C. and Stolten, D. (2000) Fuel Cell Seminar Abstracts, October 30–November 2, Portland, Oregon, pp. 126–129.

147 Paulus, U.A., Wokaun, A., Scherer, G.G., Schmidt, T.J., Stamenkovic, V., Radmilovic, V., Markovic, N.M. and Ross, P.N. (2002) *J. Phys. Chem. B*, **106** (16), 4181–4191.

148 Koffi, R.C., Coutanceau, C., Garnier, E., Leger, J.-M. and Lamy, C. (2005) *Electrochim. Acta*, **50** (20), 4117–4127.

149 Xiong, L. and Manthiram, A. (2005) *J. Electrochem. Soc.*, **152** (4), A697–A703.

150 Uchida, H., Ozuka, H. and Watanabe, M. (2002) *Electrochim. Acta*, **47** (22–23), 3629–3636.

151 Toda, T., Igarashi, H. and Watanabe, M. (1999) *J. Electroanal. Chem.*, **460** (1–2), 258–262.

152 Shukla, A.K., Raman, R.K., Choudhury, N.A., Priolkar, K.R., Sarode, P.R., Emura, S. and Kumashiro, R. (2004) *J. Electroanal. Chem.*, **563** (2), 181–190.

153 Zinola, C.F., Castro Luna, A.M., Triaca, W.E. and Arvia, A.J. (1994) *J. Applied Electrochem.*, **24** (2), 119–125.

154 Baglio, V., Stassi, A., Di Blasi, A., D'Urso, C., Antonucci, V. and Aricò, A.S. (2007) *Electrochim. Acta*, **53**, 1360–1364.

155 Stassi, A., D'Urso, C., Baglio, V., Di Blasi, A., Antonucci, V., Aricò, A.S., Castro Luna, A.M., Bonesi, A. and Triaca, W.E. (2006) *J. Appl. Electrochem.*, **36** (10), 1143–1149.

156 Jasinski, R. (1965) *J. Electrochem. Soc.*, **112**, 526–530.

157 Franke, R., Ohms, D. and Wiesener, K. (1989) *J. Electroanal. Chem.*, **260** (1), 63–73.

158 Faubert, G., Lalande, G., Coté, R., Guay, D., Dodelet, J.P., Weng, L.T., Bertrand, P. and Dénés, G. (1996) *Electrochim. Acta*, **41** (10), 1689–1701.

159 Elzing, A., Van der Putten, A., Visscher, W. and Barendrecht, E. (1987) *J. Electroanal. Chem.*, **233** (1–2), 113–123.

160 Reeve, R.W., Christensen, P.A., Hamnett, A., Haydock, S.A. and Roy, S.C. (1998) *J. Electrochem. Soc.*, **145** (10), 3463–3471.

161 Schmidt, T.J., Paulus, U.A., Gasteiger, H.A., Alonso-Vante, N. and Behm, R.J. (2000) *J. Electrochem. Soc.*, **147** (7), 2620–2624.

162 Final report of the EU funded project DREAMCAR (direct methanol fuel cell system for car applications), project nr. ERK6-CT-2000-00315 (2005).

163 Hockaday, R.G., DeJohn, M., Navas, C., Turner, P.S., Vaz, H.L. and Vazul, L.L. (2000) Proceedings of the Fuel Cell Seminar, Portland, Oregon, USA, 30 October–2 November, pp. 791–794.

164 Kelley, S.C., Deluga, G.A. and Smyrl, W.H. (2000) *Electrochem. Solid-State Lett.*, **3** (9), 407–409.

165 Narayanan, S.R., Valdez, T.I. and Clara, F. (2000) Proceedings of the Fuel Cell Seminar, Portland, OR, USA, 30 October–2 November, pp. 795–798.

166 Jung, D.-H., Jo, Y.-H., Jung, J.-H., Cho, S.-Y., Kim, C.-S. and Shin, D.-R. (2000) Proceedings of the Fuel Cell Seminar, Portland, OR, USA 30 October–2 November, pp. 420–423.

167 Bostaph, J., Koripella, R., Fisher, A., Zindel, D. and Hallmark, J. (2001) Proceedings of the 199th Meeting on Direct Methanol Fuel Cell Electrochemical Society, Washington, DC, USA 25–29 March.

168 Hockaday, R.G. (1998) US Patent No. 5,759,712.

169 Witham, C.K., Chun, W., Valdez, T.I. and Narayanan, S.R. (2000) *Electrochem. Solid-State Lett.*, **3**, 11, 497–500.

170 Gottesfeld, S., Ren, X., Zelenay, P., Dinh, H., Guyon, F. and Davey, J. (2000) Proceedings of the Fuel Cell Seminar, Portland, OR, USA, 30 October–2 November, pp. 799–802.

171 Dohle, H., Mergel, J., Scharmaan, H. and Schmitz, H. (2001) Proceedings of the 199th Meeting Direct Methanol Fuel Cell Symposium, Electrochemical Society, Washington, DC, USA 25–29 March.

172 Chang, H. (2001) The Knowledge Foundation's Third Annual International Symposium on Small Fuel Cells and Battery Technologies for Portable Power Applications, Washington, DC, USA, 22–24 April.

173 Chang, H., Kim, J.R., Cho, J.H., Kim, H.K. and Choi, K.H. (2002) *Solid State Ionics*, **148**, 601–605.

174 Jung, D.H., Jo, Y.-H.-., Jung, J.-H., Cho, S.-H., Kim, C.-S. and Shin, D.-R. (2000) Proceedings Fuel Cell Seminar, Portland, OR, USA, 30 October–2 November, pp. 420–423.

175 Lifton, R.F. (2001) The Knowledge Foundation's Third Annual International Symposium on Small Fuel Cells and Battery Technologies for Portable Power Applications, Washington, DC, USA, 22–24 April.

176 Martin, J.J., Qian, W., Wang, H., Neburchilov, V., Zhang, J., Wilkinson, D.P. and Chang, Z. (2007) *J. Power Sources*, **164** (1), 287–292.

177 Guo, Z. and Faghri, H. (2006) *J. Power Sources*, **160**, 1183–1187.

178 Liu, J.G., Zhao, T.S., Chen, R. and Wong, C.W. (2005) *Electrochem. Comm.*, **7**, 288–293.

179 Chen, R. and Zhao, T.S. (2007) *J. Power Sources*, **167** (2), 455–460.

180 Chen, R. and Zhao, T.S. (2007) *Electrochem. Comm.*, **9** (4), 718–724.

181 Chen, R. and Zhao, T.S. (2007) *Electrochim. Acta*, **52** (13), 4317–4324.

182 Zhang, H.F. and Hsing, I.-M. (2007) *J. Power Sources*, **167** (2), 450–454.

183 Abdelkareem, M.A. and Nakagawa, N. (2007) *J. Power Sources*, **165** (2), 685–691.

184 Chu, D. and Jiang, R. (2006) *Electrochim. Acta*, **51** (26), 5829–5835.

185 Yang, H., Zhang, T.S. and Ye, Q. (2005) *J. Power Sources*, **139** (1–2), 79–90.

186 Di Blasi, A., Baglio, V., Denaro, T., Antonucci, E. and Aricò, A.S. (2008) *J. New Mater. Electrochem. Syst.*, **11**, 165–174.

187 Lu, G.Q. and Wang, C.Y. (2004) *J. Power Sources*, **134** (1), 33–40.

188 Lai, Q.-Z., Yin, G.-P., Zhang, J., Wang, Z.-B., Cai, K.-D. and Liu, P. (2008) *J. Power Sources*, **175** (1), 458–463.

189 Blum, A., Duvdevani, T., Philosoph, M., Rudoy, N. and Peled, E. (2003) *J. Power Sources*, **117** (1–2), 22–25.

190 Kim, D., Cho, E.A., Hong, S.-A., Oh, I.-H. and Ha, H.-Y. (2004) *J. Power Sources*, **130** (1–2), 172–177.

191 Pan, Y.H. (2006) *J. Power Sources*, **161** (1), 282–289.

192 Lu, G.Q., Wang, C.Y., Yen, T.J. and Zhang, X. (2004) *Electrochim. Acta*, **49** (5), 821–828.

193 Motokawa, S., Mohamedi, M., Momma, T., Shoji, S. and Osaka, T. (2004) *Electrochem. Comm.*, **6** (6), 562–565.

194 Sabate, N., Esquivel, J.P., Santander, J., Torres, N., Gracia, I., Ivanov, P., Fonseca, L., Figueras, E. and Canè, C. (2008) *J. New Mater. Electrochem. Syst.*., **11** (2), 143–146.

195 Lim, S.W., Kim, S.W., Kim, J., Ahn, J.E., Han, H.S. and Shul, Y.G. (2006) *J. Power Sources*, **161** (1), 27–33.

196 Baglio, V., Stassi, A., Matera, F.V., Di Blasi, A., Antonucci, V. and Aricò, A.S. (2008) *J. Power Sources*, **180** (2), 797–802.

197 Harris, D. (2000) Ballard Power Systems Inc. News Release, 9 November.

198 Zhang, J., Colbow, K.M. and Wilkinson, D.P. (2001) US Patent No. 6,187,467.

199 http://www.ird.dk/product.htm/.

200 Buttin, D., Dupont, M., Straumann, M., Gille, R., Dubois, J.C., Ornelas, R., Fleba, G.P., Ramunni, E., Antonucci, V., Arico, V., Creti, P., Modica, E., Pham-Thi, M. and Ganne, J.P. (2001) *J. Appl. Electrochem.*, **31** (3), 275–279.

201 Baldauf, M. and Preidel, W. (1999) *J. Power Sources*, **84** (2), 161–166.

202 Baldauf, M., Frank, M., Kaltschmidt, R., Lager, W., Luft, G., Poppinger, M., Preidel, W., Seeg, H., Tegeder, V., Odgaard, M., Ohlenschloeger, U., Yde- Andersen, S., Lindic, M.H., Drews, T., Nguyen, C.V., Tachet, I., Skou, E., Engell, J., Lundsgaard, J., Hogarth, M.P., Hards, G.A., Kelsall, I., Theobald, B.R.C., Smith, E., Thompsett, D., Gunner, A. and Walsby, N. (1996–1999) Publishable Final Report of The European Commission Joule III Programme.

203 Baldauf, M. and Preidel, W. (1999) Book of Abstracts, Proceedings of the Third International Symposium on Electrocatalysis: Workshop, Electrocatalysis in direct and indirect methanol PEM fuel cells, Portoroz, Slovenia, 12–14 September.

204 Baldauf, M. and Preidel, W. (2001) *J. Appl. Electrochem.*, **31** (7), 781–786.

205 Wilson, M.S., Sringer, T.E., Davey, J.R. and Gottesfeld, S. (1995) Proton Conducting Membrane Fuel Cells I, Electrochemical Society, Pennington, NJ, Proceedings 95–23 (eds S. Gottesfeld, G. Halpert and A. Langrebe), p. 115.

206 http://world.honda.com/news/.

207 Aricò, A.S., Creti, P., Giordano, N., Antonucci, V., Antonucci, P.L. and Chuvilin, A. (1996) *J. Appl. Electrochem.*, **26** (9), 959–967.

208 Wilson, M.S. and Gottesfeld, S. (1992) *J. Appl. Electrochem.*, **22** (1), 1–7.

209 Shukla, A.K., Stevens, P., Hamnett, A. and Goodenough, J.B. (1989) *J. Appl. Electrochem.*, **19** (3), 383–386.

210 Fisher, A., Jindra, J. and Wendt, H. (1998) *J. Appl. Electrochem.*, **28** (3), 277–282.

211 Gamburzev, S., Boyer, C. and Appleby, A.J. (1999) Proton Conducting Membrane Fuel Cells II -Second International Symposium, Electrochemical Society, Pennington, New Jersey, Proceedings 98–27 (eds S. Gottesfeld and T.F. Fuller), pp. 23–29.

212 Nguyen, T.V. (1996) *J. Electrochem. Soc.*, **143** (5), L103–L105.

213 Busick, D.N. and Wilson, M. (1999) Proton Conducting Membrane Fuel Cells – Second International Symposium, Electrochemical Society, Pennington, New Jersey, Proceedings 98–27 (eds S. Gottesfeld and T.F. Fuller), pp. 435–445.

214 Zawodzinski, C., Wilson, M.S. and Gottesfeld, S. (1999) Proton Conducting Membrane Fuel Cells – Second International Symposium, Electrochemical Society, Pennington, New Jersey, Proceedings 98–27 (eds S. Gottesfeld and T.F. Fuller), pp. 446–456.

215 Scott, K., Taama, W.M. and Argyropoulos, P. (1998) *J. Applied Electrochem.*, **28** (12), 1389–1397.

216 Min, K.B., Tanaka, S. and Esashi, M. (2002) *Electrochemistry*, **70** (12), 924–927.

217 D'Arrigo, G., Spinella, C., Arena, G. and Lorenti, S. (2003) *Mater. Sci. Eng. C*, **23**, 13–18.

218 Yao, S.-C., Tang, X., Hsieh, C.-C., Alyousef, Y., Vladimer, M., Fedder, G.K.

and Amon, C.H. (2006) *Energy*, **31** (5), 636–649.
219 D'Arrigo, S., Coffa, C. and Spinella, C. (2006) *Sensors Actuators*, **99**, 112–118.
220 Ye, Q. and Zhao, T.S. (2005) *J. Power Sources*, **147** (1–2), 196–202.
221 Xie, C., Bostaph, J. and Pavio, J. (2004) *J. Power Sources*, **136** (2), 55–65.
222 Zhang, J., Colbow, K.M., Lee, A.N.L. and Lin, B. (2004) US Patent Pub. No. 20040131898.
223 Icardi, U.A., Specchia, S., Fontana, G.J.R., Saracco, G. and Specchia, V. (2008) *J. Power Sources*, **176** (2), 460–467.
224 Specchia, S., Icardi, U.A., Specchia, V. and Saracco, G. (2005) *Int. J. Chem. React. Eng.*, **3**, A24–A27.
225 Sgroi, M., Bollito, G., Innocenti, G., Saracco, G., Specchia, S. and Icardi, U.A. (2007) *J. Fuel Cell Sci. Technol.*, **4** (3), 345–349.

2
Nanostructured Electrocatalyst Synthesis: Fundamental and Methods

Nitin C. Bagkar, Hao Ming Chen, Harshala Parab, and Ru-Shi Liu

2.1
Introduction

Increasing demand for clean and sustainable energy sources as an alternative to environmental unfriendly fossil fuels, concerned with the economic development of the society and better quality of human life, has drawn great attention from researchers. This necessitates the development of highly efficient, cost-effective and environmental friendly options for electrochemical energy conversion and storage devices for a diverse range of consumer devices [1, 2]. Fuels cells are found to be the promising candidates in this regard. Fuel cell history can be traced back to the nineteenth century when it was invented in 1839 by William Grove, while different types of fuel cells were demonstrated in the mid-twentieth century [3]. Since then the development and use of fuel cells for various applications have come a long way.

In principle, an electrochemical reaction of oxygen with fuels to form water leads to the generation of electricity in a fuel cell [4, 5]. The main types of fuel cell include alkaline, direct methanol, molten carbonate, phosphoric acid, proton exchange membrane and solid oxide fuel cells [6]. Among these, the direct methanol fuel cell (DMFC) is considered to be a more promising fuel cell system in terms of commercialization in the short term, because the fuel used is liquid methanol, which is easy to handle, store and transport, compared with the fuels used in other types of fuel cells. DMFC has various advantages over the conventional power sources such as high efficiency, high energy density, silent operation/low noise, low heat transmission due to low working temperature, no use of oil or gas, stability and low or zero emissions [7–13]. Although much progress has been made in research and development of DMFC, some factors such as cost, power density and durability still pose a great challenge for the researchers in commercializing this kind of fuel cell. Fuel cell materials, such as electrocatalysts, electrolyte membranes and catalyst supports play a critical role in the performance of fuel cells and are directly related to the major challenges in DMFCs, which are methanol crossover and sluggish anode kinetics. To overcome these challenges, it is critical to develop new fuel cell materials, especially electrocatalysts. Electrocatalysts based on platinum play a key role in the

Electrocatalysis of Direct Methanol Fuel Cells. Edited by Hansan Liu and Jiujun Zhang
Copyright © 2009 WILEY-VCH Verlag GmbH & Co. KGaA, Weinheim
ISBN: 978-3-527-32377-7

electrocatalysis of fuel cells and are widely used in fuel cells today. The excellent catalytic activity of platinum for methanol oxidation at room temperature has drawn considerable interest in DMFCs [14–17]. However, the main drawbacks associated with the use of Pt as an electrocatalyst are the cost of platinum metal, low tolerance for methanol oxidation, Pt dissolution, and the poisoning of Pt by CO, which is an intermediate of methanol electro-oxidation. A number of strategies have been devoted to solving these problems, including alloying and use of a support. When pure Pt metal is used as a catalyst, the intermediate of methanol electro-oxidation, CO, becomes adsorbed on the surface of Pt, thus blocking the reactive sites on the surface and leading to low catalytic efficiency. Introducing another metal like ruthenium, which alloys with Pt, promotes the adsorption of the oxygen-containing species at lower potentials and facilitates the oxidation of CO to CO_2; thus reducing the effect of Pt poisoning by a multifunctional mechanism. Various Pt based alloys, such as PtRu, PtMo, PtW, PtSn, PtOs, have been investigated to enhance the efficiency of methanol oxidation in DMFC. Among them, PtRu alloy has been found to have better performance because of its suitable structural composition, alloying degree, particle size and morphology. Whilst the alloying strategy greatly reduces the poisoning of Pt catalyst due to the introduction of other metals, use of carbon in different forms as a support material is also becoming popular. The supporting strategy aims at getting a stable dispersion of the electrocatalyst and reducing the loading of Pt. The development of non-noble metal catalysts is another approach to reduce the cost of electrocatalyst, which is at its initial stages of research and needs to be explored further.

This chapter reviews recent progress in the synthesis of nanostructured Pt-based catalysts for DMFCs. Some fundamental understanding of the effect of structure parameters (such as particle size and crystal facet) on catalytic activity is briefly introduced. The synthetic approaches of conventional carbon-supported Pt-based catalysts and novel unsupported catalyst nanostructures are summarized. The focus of this chapter is to describe the versatile synthetic strategies of novel catalyst nanostructures as unsupported catalyst in DMFCs. Various novel catalyst nanostructures, such as nanospheres, nanotubes, nanofibers, nanochannels, core-shell structure, hollow structure, and so on, are presented, following the description of the synthesis methods, including template-based synthesis, galvanic replacement synthesis and electrochemical synthesis.

2.2
Fundamental Understanding of the Structure–Activity Relationship

Electrocatalysis in DMFC involves simultaneous occurrence of the methanol oxidation reaction (MOR) at the anode and oxygen reduction reaction (ORR) at the cathode [18–21]. The MOR reaction catalyzed by PtRu catalyst is presumed to occur through two mechanisms, namely the bifunctional effect and the ligand effect. The bifunctional effect involves dehydrogenation of the adsorbed methanol molecule to form poisonous intermediate CO, which further becomes adsorbed on the Pt surface

(Equation 2.1). The Ru atoms provide sites for water adsorption and dissociation, thus forming oxygen-containing species on Ru sites, which promotes the oxidative removal of CO_{ads} (Equation 2.2) [21, 22].

$$CH_3OH + Pt \rightarrow Pt - CO_{ads} + 4H^+ + 4e^- \quad (2.1)$$

$$Pt - CO_{ads} + Ru - OH_{ads} \rightarrow Pt + Ru + CO_2 + H^+ + e^- \quad (2.2)$$

The ligand effect assumes that the change of Pt electronic properties induced by the presence of Ru makes Pt atoms more susceptible for OH adsorption or even for dissociative adsorption of methanol [23–25]. It was observed that the threshold potential, where the decomposition of methanol proceeds by forming products other than CO_{ad}, depends on the catalyst surface and electronic structure. Hence, a fundamental understanding of the effects of structural parameters on the catalytic activity is necessary to design high performance DMFC catalysts. Understanding of the structure–activity relationship is necessary to reveal the basic nature of the catalysis.

The catalytic property was found to be affected by the size-dependent changes in the local electronic structure of the surface (electronic effects) and availability of sites presenting special geometric arrangements (geometric effects). For example, the surface–adsorbate interaction is affected by a change in the spatial configuration of surface atoms in small metal clusters, which enables chemisorption geometries that are not typically accessible on metal surfaces. The following sections deal with the special catalytic behavior of nanosized metal particles, classified under the category of 'electronic effects' or 'geometric effects.'

2.2.1
Particle Size Effect

Catalysis requires the use of nanoparticles of noble metal possessing high electroactive surface area that must be controlled to maximize their activity. The studies related to the effect of size variation on the efficiency of methanol electro-oxidation suggest that Pt nanoparticles in the size range of 3–10 nm show maximum mass specific activity [26, 27]. It was also found that the particles with sizes above or below this size range show loss of activity. For smaller nanoparticles the loss can be attributed to morphological considerations whereas for bigger nanoparticles, it can be accounted for by the desorption kinetics of the fuel. The ambiguous results obtained due to the large variation in size distribution of electrocatalyst materials necessitate employment of particles with narrow size distribution for efficient catalytic properties. The CO diffusion rate and interaction between CO_{ads} and OH_{ads} are highly size dependent [28, 29]. The strong adsorption of CO and OH during methanol oxidation is capable of inducing reconstruction and morphological changes on smaller particles. The observed particle size effect on the methanol electro-oxidation is reflected in terms of increasing strength of both Pt–CO and Pt–OH bonds with decreasing particle size. Strong dependence of the efficiency of methanol electro-oxidation on particle size distribution was observed [30], which can be realized by considering multitudes of parameters such as surface morphology of catalyst and the interactions among the oxidation products with different sized nanoparticles. The overall catalytic activity of the catalyst can thus be determined by the interplay of electronic and surface geometric effects.

Recently a spontaneous collapse in the crystalline structure of Pt nanoparticles with sizes below 1 nm has been observed [31] which resulted in reduced catalytic activity towards hydrogen oxidation reaction due to induced quantum size effects. The molecular dynamics simulations and density functional calculations suggest that Pt nanoparticles are converted from crystalline to amorphous structure, thus providing valuable estimates about the lower Pt use limit of 1 nm for catalytic properties. Density functional theory (DFT) calculations also suggest the upward shift in the d band center with respect to E_F and increase in density of states at the Fermi level, on account of narrowing of the d band caused by size reduction. Such electronic structure modification leads to an enhanced interaction between the surface and adsorbed molecules, such as CO, during methanol oxidation, thereby, lowering the catalytic efficiency with size reduction.

The computational studies of size-dependent catalytic properties of Pt nanoparticles by molecular dynamics provided valuable insight into the structure–property relationship with size. The molecular dynamics simulations using effective medium theory on four virtual Pt cubic particles with different diameters (1, 1.5, 2, and 2.5 nm) suggest different atom-packing structure as shown in Figure 2.1a. It was revealed that the ordered structure with well-defined (100) and (111) facets, was retained during subsequent heating and annealing for bigger nanoparticles. However, kink structure started appearing on the surface as particle diameter is reduced, which is associated with the reduction in the size of (100) and (111) facets, as shown in Figure 2.1b. These studies show that, the crystalline structure becomes amorphous as the particle size is reduced to 1 nm. Thus, it is clear from the above discussion that the size of Pt nanoparticles, in particular a specific size range, is a key factor in delivering higher catalytic performance for the electrocatalyst and the particles with sizes beyond this range are not as effective as was thought. Apart from the size, the shape of the nanoparticles is also very effective in determining the catalytic activity of the Pt nanoparticles, which is discussed in the following section.

2.2.2
Crystal Facet Effect

Recent studies have demonstrated that the catalytic activity is largely dependent on the shape of nanoparticles rather than size [32]. The structural sensitivity of methanol oxidation on low surface area of bulk platinum is very well known [33]. It was observed that the Pt (111) surface has the lowest activity and onset potential for methanol oxidation, but the highest CO tolerance in comparison with (100) and (110) miller index planes. The ORR activity of Pt (100) is higher than Pt (111) due to the different adsorption rates of sulfate ions on the facets as shown in Figure 2.2a [34]. The electrocatalytic activities of octahedral/tetrahedral Pt nanoparticles of same size

2.2 Fundamental Understanding of the Structure–Activity Relationship | 83

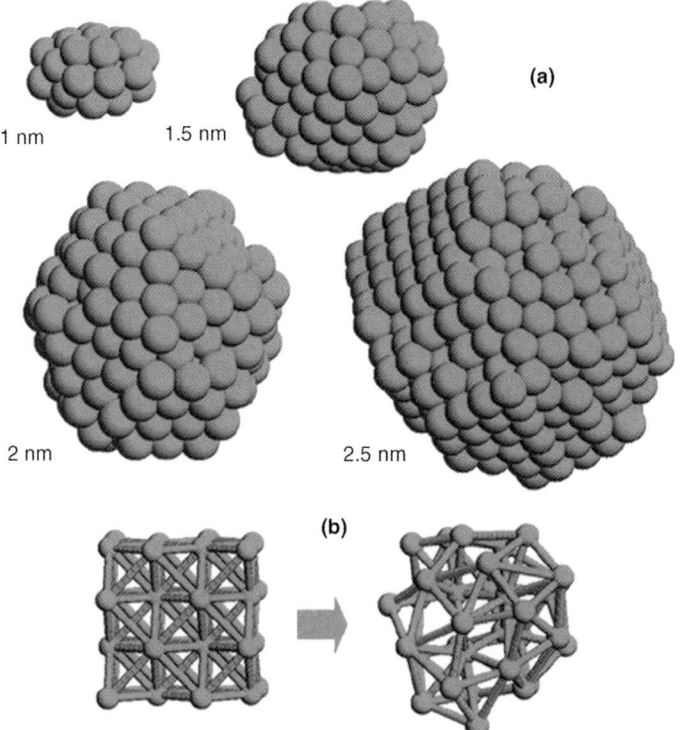

Figure 2.1 Molecular dynamic simulations of (a) four Pt cubic particles with different diameters, which were heated in vacuum to 2000 K and then slowly cooled down to 300 K, resulting in different atom-packing structures as a function of the particle size, and (b) a spontaneous quick collapse in the crystalline structure of a virtual Pt cubic particle of 1 nm diameter thermostatically kept at 300 K in vacuum. Reproduced with permission from Ref. [31]. Copyright © 2007 American Chemical Society.

(∼10 nm), but different shapes suggest an unexpected enhancement on Pt (111) nanofacets [35]. The enhancement is related to about threefold increase in transient intrinsic activity and 10-fold increase in CO tolerance steady-state activity compared with commercial Pt black. Generally, the step density and defects enhance the selectivity and overall rate of methanol oxidation. It was observed that bulk CO oxidation is also found to be sensitive to the step density which increases in the sequence Pt(111) < Pt(554) < Pt(553) [36]. The formation of oxygen-containing species needed for CO oxidation is enhanced at the surface with a higher step density resulting in lower onset potential for higher steps. It becomes obvious from Figure 2.2b that the overpotential for CO oxidation is lowered in the sequence Pt (111) < Pt(554) < Pt(553) and peak potential difference between Pt (553) and Pt (111) was found to be as high as 0.17 V.

The catalytic activity of platinum nanocubes and polyhedral Pt nanoparticles suggests higher current density during ORR for 7 nm Pt nanocubes which was found to be four times that of 3 nm polyhedral or 5 nm truncated cubic Pt

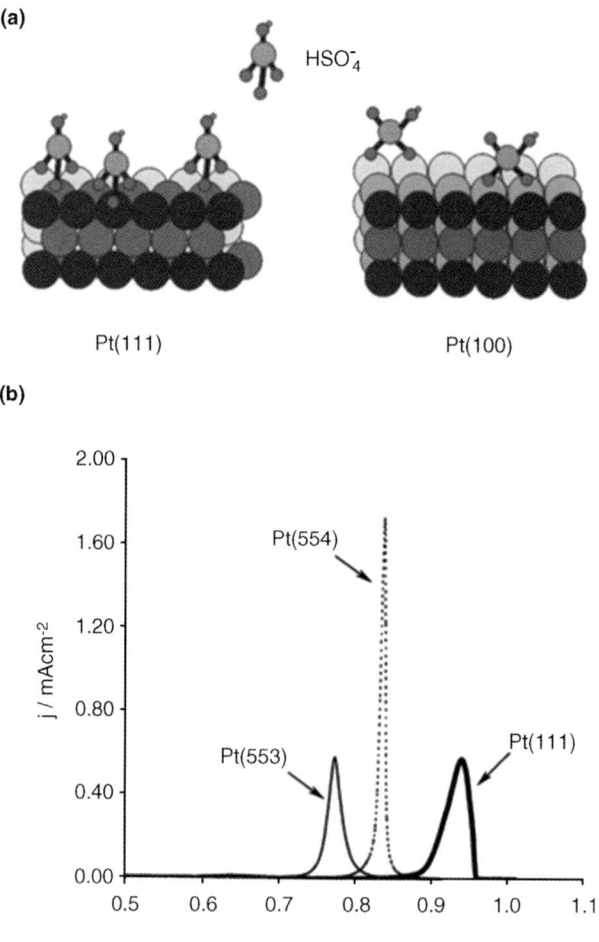

Figure 2.2 (a) Adsorption of sulfate anion on Pt (111) and Pt (100) surface, and (b) oxidation of saturated CO_{ad} layers on Pt (553) (thin solid line), Pt (554) (dashed line) and Pt (111) (thick solid line). Reproduced with permission from Ref. [34]. Copyright © 2008 Wiley-VCH and reproduced with permission from Ref. [36]. Copyright © 2000 Elsevier.

nanoparticles [34]. In the search for novel nanostructures of Pt, tetrahexahedral Pt nanocrystals with high-index facets such as {730}, {210}, and {520} have been synthesized, which showed high catalytic activity for electro-oxidation of small organic fuels such as formic acid and ethanol due to large density of atomic steps and kinks, which can serve as active sites for breaking chemical bonds [37]. All these studies suggest a stronger dependence of the shape of catalyst particles on its performance than that of its size. The catalytic activity can be further enhanced by synthesizing catalyst structures with high-index surface facets. The above studies are very helpful for researchers for the design and development of novel catalyst nanostructures with maximization of high-index surfaces and abundant corner and edge sites during crystal growth. It is very important to understand the fundamentals of crystal growth in order to develop multifaceted nanostructures. The essence of shape control lies in understanding the process of crystallization which largely depends on the fundamental steps of nucleation and growth. The simultaneous control over the factors affecting nucleation and growth kinetics, can lead to generation of uniform and monodispersed nanomaterials with desired shapes. However it is difficult to control the crystallization to obtain a perfect crystal. The nucleation results from aggregation of atoms or molecules to form nuclei which can grow further to give desired nanostructures. The most common architecture of these nanocrystals is isotropic particles, ranging from spherical to highly-faceted particles, such as cubic and octahedral. The nuclei generally act as seeds during the growth process and under kinetic as well as thermodynamic considerations, can give rise to variety of shapes including one-dimensional (1-D) anisotropic nanoparticles such as rods and wires, and two-dimensional (2-D) nanoparticles such as nanodiscs, plates, multipods and nanostars. The Figure 2.3 represents the conventional shapes of face-centered cubic (fcc) metals, which are enclosed by {111} and {100} facets, with a low proportion of corner and edge sites [38]. The shapes of fcc metals are decided by the surface energies of different crystallographic facets under thermodynamic considerations. The surface energy calculations of fcc metals revealed that the surface energies increase in the order $\gamma\{111\} < \gamma\{100\} < \gamma\{110\} < \gamma\{hkl\}$, where {hkl} represents high-index facets, with at least one h, k and l equal to 2 or more [39]. This alteration in surface energies of various facets during nanocrystal growth decides the final shape of the nanocrystal. The facets with high surface energies are usually eliminated from the crystal surface since they grow much faster than others during the crystal growth. As a result, low-index facets are enlarged at the expense of high-index facets and crystals take the shapes, as shown in Figure 2.3. From the above discussion, it is clear that the shape control of the Pt nanoparticles plays a key role in determining its catalytic performance which necessitates the development of novel synthetic strategies for designing an ideal electrocatalyst. Various approaches for the synthesis of electrocatalysts focusing on size and shape variation are outlined in the following sections.

2.3
Synthetic Methods of Conventional Carbon-Supported Catalysts

The current research in electrocatalyst development is focused on new avenues for synthesizing cost-effective and highly efficient catalysts to overcome the cost barrier imposed by the use of noble catalyst in DMFC for commercialization of fuel cells. There has been great interest in development of novel Pt nanostructures with high surface area exhibiting higher catalytic performance and use efficiency with low Pt content. An ideal electrocatalyst should be multifunctional, possessing surface reactivity, electronic conductivity and able to provide facile mass transport of

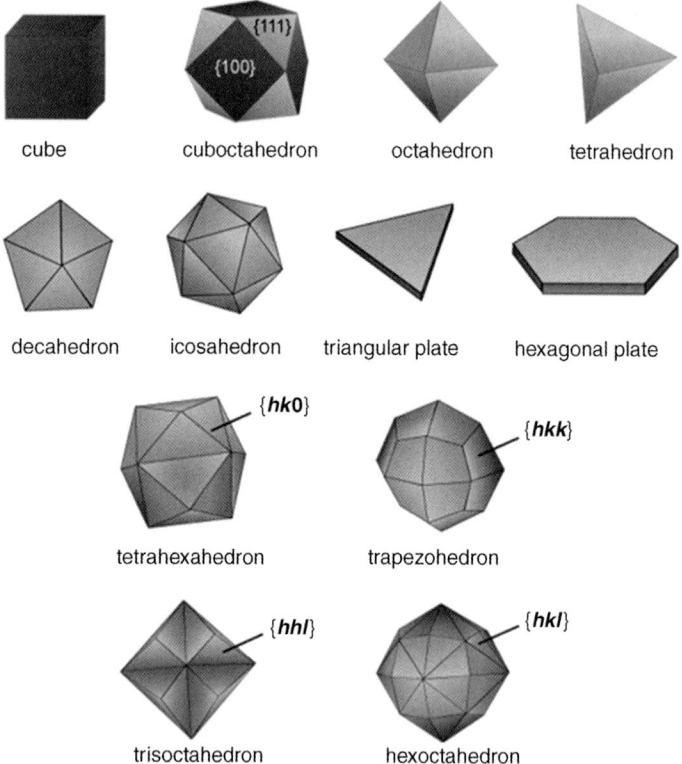

Figure 2.3 Conventional shapes of face-centered cubic (fcc) metals whose surface is enclosed by {100} and/or {111} facets. Black and gray colors represent the {100} and {111} facets, respectively. Unconventional shapes of fcc metals whose surface is enclosed by high-index facets. The Miller indices {hkl} obey the order h>k>l. Reproduced with permission from Ref. [38]. Copyright © 2007 Wiley-VCH.

molecules for enhanced molecular conversion. Figure 2.4 shows the strategies used for cost reduction and higher usage efficiencies to achieve higher catalytic activities. A cost-effective electrocatalyst can be obtained by reducing Pt loading without compromising its catalytic activity, which can be realized by using single crystal Pt layer, hollow nanostrustures of Pt and core shell assemblies with non-noble metal as a core and thin Pt layer as a shell. Platinum use can be improved by using nanotubular morphology, porous framework and novel nanostructured support materials with nanochannels. The novel multiple junction nanostructures can be created by a simple approach of improvement in the morphology of traditional catalyst [40]. Using abundant non precious metals, novel approaches by lowering the use of noble metal without sacrificing its catalytic activities can be developed. The use of support materials during synthesis of electrocatalyst is another way of reducing the cost of electrode materials and obtaining a stable dispersion of electrode material with

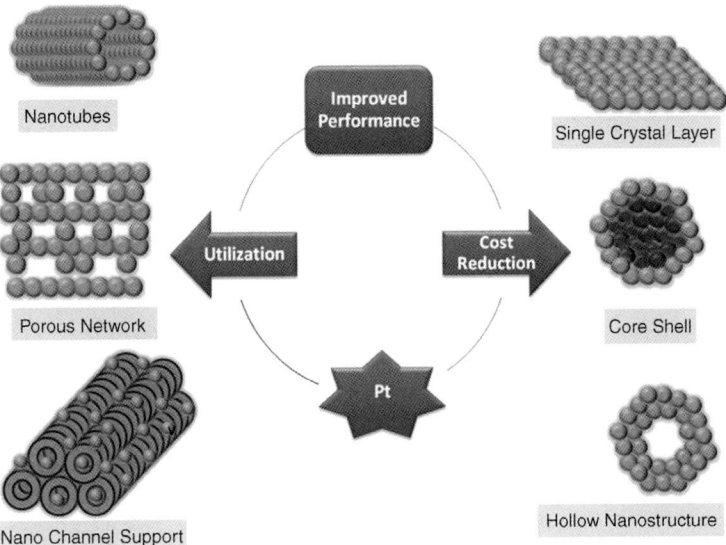

Figure 2.4 Schematic of strategies for designing an ideal platinum based electrocatalyst.

improved performance. The following sections emphasize on the recent advances in the conventional carbon supported electrocatalysts.

2.3.1 Impregnation Method

The conventional impregnation method has been widely adopted as an effective method for the preparation of high surface area due to its simplicity in preparing Pt/C and PtRu/C carbon supported electrocatalysts [41–60]. Conventional preparation techniques based on wet impregnation involve impregnation of the support with noble metal precursor salts that are subsequently reduced by reducing agents or a gaseous reducing environment. Generally, high surface area carbon black such as Vulcan XC-72R carbon black (surface area = $250\,m^2\,g^{-1}$) is used as a catalyst support in which metal precursors are mixed in aqueous solution to form homogeneous mixtures. The reduction of the metal precursors is carried out by using various reducing agents such as $Na_2S_2O_3$, $NaBH_4$, $Na_4S_2O_5$, N_2H_4 or formic acid. The concentration of reducing agent was also found to affect the dispersion and the surface composition of the prepared PtRu particles. For example, optimum molar ratio of $NaBH_4$ to metal ions was found to be between 5 and 15 for methanol electro oxidation. Catalysts with different metal loadings can be prepared by varying the concentration of metal precursors until high dispersion is obtained. Attempt to increase the metal loading resulted in aggregation of metal nanoparticles thereby lowering the methanol oxidation activities. Until now, it has been a challenge to prepare Pt catalysts with high metal loadings (more than 60 wt%) and small particle sizes (1–2 nm) using the conventional impregnation method. A significant increase in particle size was observed when the metal content in a commercial E-TEK Pt/Vulcan catalyst is increased from 10 to 60 wt%. For example, the particle size for a 10 wt% Pt catalyst was found to be 2.0 nm, which increased to 3.2 nm and 8.8 nm for a 30 and a 60 wt% Pt catalyst, respectively.

Structural and electrocatalytic properties of Pt/C and PtRu/C vary with the type of precursor employed in the impregnation method, and methanol electro-oxidation activity is sensitive to the heating temperature [60]. The choice of precursor has a large structural influence on the aggregation behavior of the Pt metal nanoparticles. Various precursors produced nanoparticles with different metal loadings and uniform dispersion on the carbon supports. For example, chlorine-containing precursors, such as H_2PtCl_6 and $RuCl_3$, were unable to give high dispersions for high metal loadings using the impregnation method [58]. This led to the use of non-chlorine-containing precursors such as $Pt(NH_3)_2(NO_2)_2$, $RuNO(NO_3)x$, $Pt(NH_3)_4(OH)_2$, Pt$(C_8H_{12})(CH_3)_2$, and $Ru_3(CO)_{12}$, which resulted in higher dispersion rates for PtRu/C catalysts. It was demonstrated that under appropriate conditions highly dispersed PtRu/C with metal loadings as high as 60 wt% can still be obtained by the impregnation method, even using chlorine-containing precursors with tunable particle size and alloyed degree of PtRu/C catalysts.

In simple terms, catalysts prepared using the conventional impregnation method show a large average size of the metal particles, and a broad particle size distribution which affects the electrocatalytic activity of these catalysts for methanol oxidation. Another major problem associated with the conventional impregnation process is that the catalyst does not show high activity even at higher metal loadings, because it is difficult for the reactant to access nanoparticles trapped inside the micropores of Vulcan XC-72 which show practically low electrochemical activity. Some of these drawbacks are avoided by using a modified impregnation method involving an additional precipitation step by changing the pH of the solution before a reduction step, which prevents nanoparticles from aggregating [44]. However, the control over size variation and monodispersity is still difficult to achieve using the impregnation method.

2.3.2
Colloidal Method

The colloidal method has been widely adopted for synthesizing PtRu nanoparticles with precise size control [61–75]. The colloidal methods include preparation of PtRu-containing colloids which results in highly dispersed PtRu catalysts by subsequent reduction generally followed by their deposition onto the carbon support. The stabilization of metallic colloids is a crucial aspect which necessitates the use of a stabilizing or protective agent to prevent the aggregation of the colloids. Various examples of stabilizing agents include polymers, block copolymers, solvents, long-chain alcohols, surfactants and organometallics. The nature of the protective agents (hydrophilic or hydrophobic) renders the hydrophilicity or hydrophobicity to the stabilized metal colloid. The relative rates of nucleation and particle growth determines the size of resulting metal colloids. A narrow size distribution is usually achieved either by electrostatic stabilization between the nanoparticles or by steric hindrance of organic molecules on the metallic surface of nanoparticles.

Although highly dispersed PtRu catalysts can be obtained, the colloidal method is more complex, relatively expensive with lower efficiency, and results in undesirable loss of noble metals through repeated filtering and washing. The major drawback of the method is the removal of the stabilizer after reduction which affects the electrochemical performance of the catalyst. Hence an alternative route of preparing colloidal nanoparticles without the use of protecting agents would be preferred; this can be achieved by appropriate combination of precursor, solvents and reducing agent. In this regard, colloidal Pt nanoparticles were synthesized using ethylene glycol, which serves as both a solvent and the protecting agent [74]. The glycol colloidal process is very attractive for large-scale synthesis of metal nanoparticles.

2.3.2.1 Polyol Synthesis

Polyol synthesis has been widely employed for the preparation of metal colloids by using the temperature-dependent reducing ability of liquid polyol or diol in the presence polyvinylpyrrolidone (PVP) as a stabilizing agent [76–87]. Generally, ethylene glycol (EG) is used as both solvent to dissolve metal precursor and reducing agent. The highly temperature-dependent reducing power of polyol proved to be a convenient, versatile and low-cost route for the synthesis of anisotropic metal nanostructures with narrow size distribution. The primary reaction of this process involves the reduction of an inorganic salt (the precursor) by polyol at an elevated temperature (boiling point of solvent). It is observed that at higher reduction rates, the growth process follows the thermodynamically favored shapes (quasi-spherical shapes), whereas at slower reduction rates, the nucleation and growth will be kinetically controlled and shape of the final product deviates from thermodynamically favored shapes. The different stages of growth process consist of the formation of metal clusters, followed by the nucleation of seeds, and subsequent growth of seeds into different morphologies. Recently, a precise shape control approach in metallic nanostructures has been demonstrated using the modified polyol process [88]. Careful control in the reduction kinetics, particularly at the seeding stage, is the key to the formation of these highly anisotropic structures. In the following section, we will discuss recent examples for the synthesis of Pt nanostructures and the parameters responsible for shape control of Pt nanostructure using a modified polyol process.

Platinum nanostructure In a recent study, the anisotropic growth of Pt was induced by addition of a trace amount of iron species (Fe [II] or Fe [III]) to a polyol process that resulted in alteration of the growth kinetics of Pt nanostructures [89]. The formation of single crystalline nanowires of uniform diameters with relatively high aspect ratio was thought to proceed by lowering of the super-saturation rate due to the addition of a small amount of iron species, which leads to highly anisotropic growth on the platinum agglomerates formed in the initial stages of the product as shown in Figure 2.5a–f. The aggregation resulted in the formation of urchin-like morphology as shown in Figure 2.5e and g. Platinum nanorods and hierarchically structured

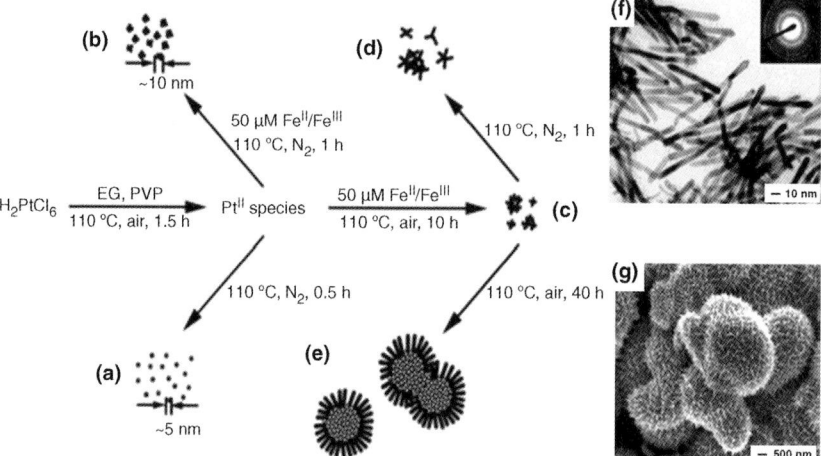

Figure 2.5 Schematic illustration of the formation of platinum nanoparticles (a and b), single-crystal Pt nanowires and urchin-like morphology (c–e) through an iron mediated polyol process. SEM images of hierarchically structured Pt agglomerates that were obtained as the final product of an iron-mediated polyol process (f and g). Reproduced with permission from Ref. [89] Copyright © 2004 American Chemical Society and reproduced with permission from Ref. [90]. Copyright © 2005 Wiley-VCH.

urchin-like Pt agglomerates were obtained as the final product of an iron-mediated polyol process (Figure 2.5f and g).

The reduction kinetics for the generation of Pt nanostructures was further manipulated to obtain four different morphologies: spheres, star-shaped particles, branched multipods, and nanowires [90]. The final morphology of nanostructures was found to depend on the way in which both the iron species and oxygen (or air) were supplied to the reaction system. The nitrogen atmosphere also played an important role in controlling morphologies in which a star-shaped growth of Pt heterostructures was observed.

A highly effective synthesis of Pt multipods (Figure 2.6a–d) was demonstrated from platinum 2,4-pentanedionate in organic solvents [91]. The anisotropic growth of Pt nanostructures was induced by introduction of trace amount of silver acetylacetonate, which followed kinetically controlled growth mechanism. A number of morphologies of Pt multipods including I- and V-shaped bipods, various types of tripods, and planar/three-dimensional (3-D) tetrapods were obtained (Figure 2.6e–h). The competitive growth between (111) and (100) planes leads to the formation of multipods in which the growth of branches was observed along (110) directions. The formation of silver seeds played an important role in directing the preferential growth along a particular crystallographic direction.

The effect of oxidation states of species in the active component of catalysts on electrocatalytic activities was investigated [92]. A diagram of the bichannel apparatus for catalyst preparation under an inert atmosphere using the modified polyol process is shown in the Figure 2.7a. The metallic species in Pt/C and PtRu/C, which have been

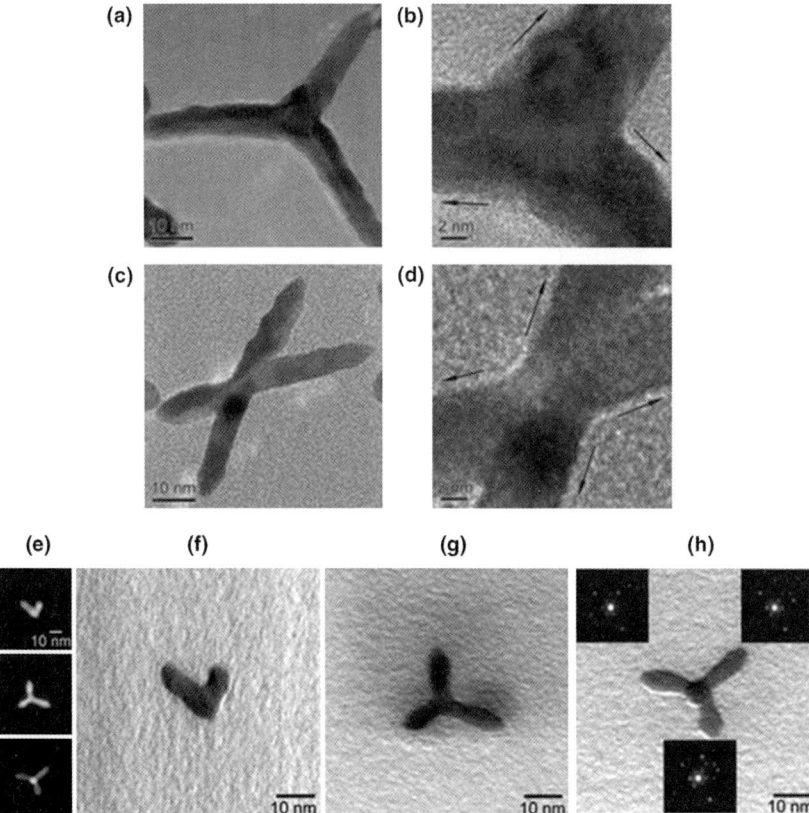

Figure 2.6 TEM and HRTEM images of two types of four-armed Pt nanostructures: (a and b) tetrahedral; (c and d) square-planar, and (e) dark and (f–h) bright field UHV-STEM images of Pt multipods synthesized at 180 °C. The insets show the nanodiffraction patterns. Reproduced with permission from Ref. [90] Copyright © 2005 Wiley-VCH and reproduced with permission from Ref. [91]. Copyright © 2005 American Chemical Society.

found to be responsible for high methanol oxidation activities, was controlled by adjusting the reducing atmosphere (nitrogen or air) in bichannel apparatus. The comparison of redox potential, shown in Table 2.1, suggests that the presence of oxygen alters the redox behavior of metal nanoparticles in alkaline EG solutions. The effect of incomplete reduction of Pt and Ru due to reoxidation of catalyst in the presence of oxygen was reduced by introducing high purity N_2 in bi-channel apparatus. As shown in Figure 2.7b, the electro-oxidation of methanol for PtRu/C catalyst prepared under inert atmosphere suggests superior behavior to the catalyst prepared under air atmosphere, as there is a smaller amount of oxygen species in the former catalyst.

Microwave polyol synthesis In the past few years, a microwave (MW)-assisted polyol synthesis method has attracted researchers, and has a number of advantages over the

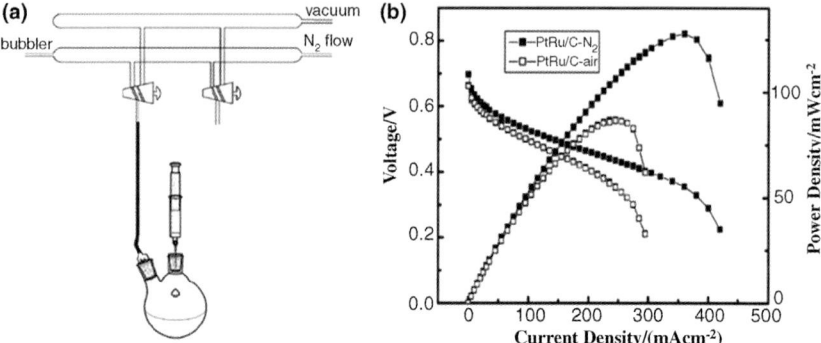

Figure 2.7 (a) Schematic representation of bichannel equipment for catalyst preparation under inert atmosphere, and (b) polarization and power density curves of PtRu/C-N2 and PtRu/C-air as anode catalysts in methanol oxidation. Reproduced with permission from Ref. [92]. Copyright © 2007 American Chemical Society.

conventional heating process [93–96], such as generation of nanostructures with smaller sizes, narrow size distributions, and a higher degree of crystallization. In addition, the MW-assisted polyol method has found to be promising for the synthesis of well-dispersed catalyst nanoparticles on carbon support since carbon is good MW absorber facilitating the creation of hot spots, which results in fast reduction and nucleation of metals on the surfaces of carbon support. Moreover, the catalyst nanoparticles adhere strongly to carbon nanofibers (CNF) at high temperature and leaching of catalysts from CNF surfaces is greatly suppressed under MW heating. A typical MW–polyol apparatus is shown in Figure 2.8a in which metal precursors in the presence of stabilizing agents are heated in microwave oven [97]. For a polar molecule such as water, energy in the form of heat is given out by molecular friction in microwave energy range, which is shown schematically in Figure 2.8b. The polar molecules try to orient themselves in the direction of electric field and when the field direction is changed alternatively, reorientation of solvent molecules in the direction of field results in loss of heat.

Table 2.1 Standard electrical potentials of redox couples in the preparation of Pt/C and PtRu/C catalysts.

Redox Couple	Reduction potential
O_2/OH^-	0.401
PtO/Pt	0.1526
PtO_2/Pt	0.1853
PtO_2/PtO	0.218
$Ru(OH)_3/Ru$	−0.0894
RuO_2/Ru	−0.024
$RuO_2/Ru(OH)_3$	0.109

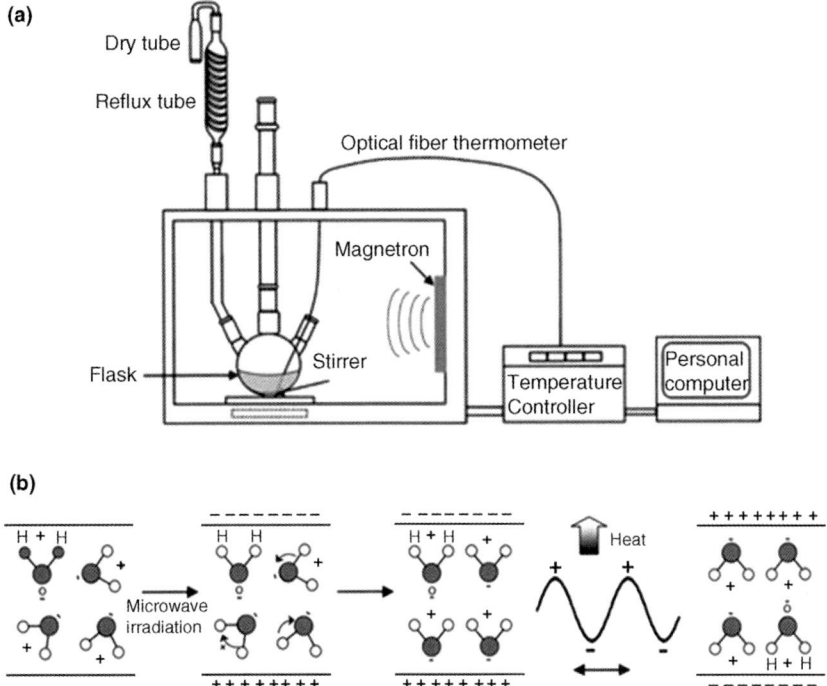

Figure 2.8 (a) Apparatus used for the microwave-assisted synthesis of metallic nanostructures, and (b) heating mechanism of H_2O by microwave irradiation. Reproduced with permission from Ref. [97]. Copyright © 2005 Wiley-VCH.

The MW–polyol method has been applied for the preparation of well-dispersed Pt nanoparticles supported on Vulcan XC-72 carbon black (CB), carbon nanotubes (CNT), CNFs and bimetallic nanoparticles [98]. Recently, PtRu alloy nanoparticles have been synthesized using microwave radiation within a few minutes on herringbone, platelate and tubular CNFs exhibiting different stacking structures of graphene sheets (Figure 2.9). The electrochemical activity during methanol oxidation for PtRu/CNF catalyst was found to be 70–200% higher compared with Pt/C catalyst. The highest activity was observed for PtRu alloy nanoparticles on the platelet CNF owing to the strong trapping of alloy nanoparticles on the flat edges of platelet CNF compared with others.

2.3.3
Microemulsion Method

A microemulsion consisting of an optically transparent dispersion of two immiscible liquids was first used to prepare monodispersed nanosized particles by S. Friberg and the late F. Gault in the mid-1970s. Since then, functional nanoparticles have been synthesized in different microemulsion systems [99–108]. The reverse

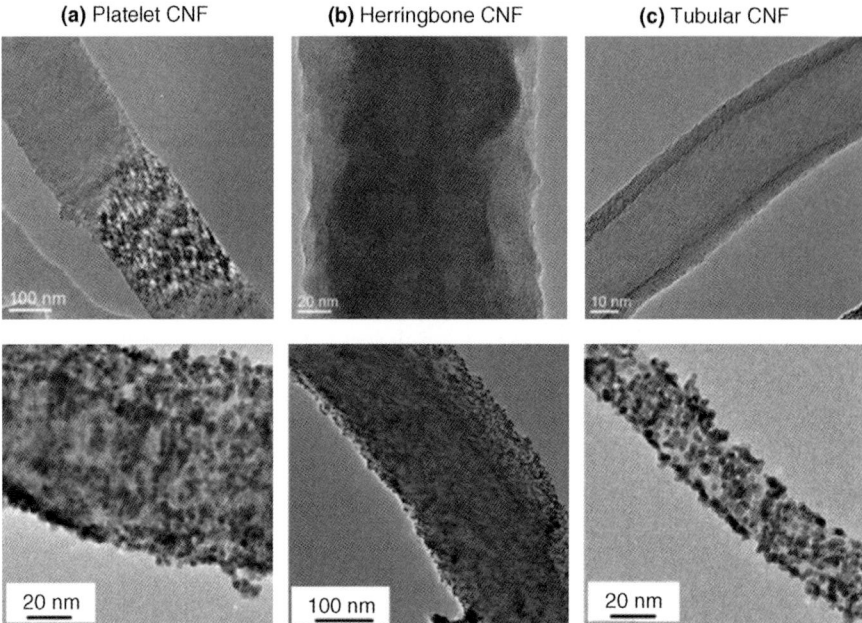

Figure 2.9 HRTEM images (a) platelet, (b) herringbone, and (c) tubular CNFs before and after PtRu nanoparticles deposition. Reproduced with permission from Ref. [98]. Copyright © 2007 American Chemical Society.

microemulsion is the water-in-oil microemulsion in which the dispersed liquid phase is water and the oil phase forms the continuous medium. Chemical reduction is carried out in a confined space of a tiny drop of precursor acting as a nanoscale reactor within the microemulsion which is surrounded by surfactant molecules. The diffusion of reducing agent in the water core leads to the nucleation of nanoparticles and an exchange among different water cores results in the growth of nanoparticles. The protective adsorption of surfactant molecules restricts further growth which results in the stabilization of nanoparticles. The surfactant-protected monodisperse nanoparticles can be used for deposition on a carbon support. The reducing agents generally used in the microemulsions are hydrazine (N_2H_4), sodium borohydride ($NaBH_4$) and hydrogen gas. The size variation in water-in-oil microemulsion is achieved by controlling the water–surfactant molar ratio (R = [H_2O]/[surfactant]). The role of the surfactant is to stabilize the nanoparticles against agglomeration and it can be anionic, cationic or nonionic in nature. Typical examples of microemulsions include anionic aerosol-OT (AOT)/heptane/water, cationic cetyltrimethylammonium bromide (CTAB)/hexanol/water and non ionic penta(ethylene glycol)-dodecylether (PEGDE)/hexane/water.

The advantage of the microemulsion method for catalytic nanoparticles includes better control over nanosize range with narrow size distribution and composition compared with the conventional impregnation and colloidal methods. Furthermore, by using the water-in-oil microemulsion route, it is possible to control the particle size by varying the water-to-surfactant molar ratio. Another major advantage of this method is the possible synthesis of a bimetallic electrocatalyst on carbon support. The bimetallic nanoparticles can be formed at room temperature with a high degree of alloying while conventional preparation methods usually require high temperatures. The bimetallic PtRu/C catalysts can be prepared either by one-step microemulsion method or by mixing individual microemulsion of Pt and Ru in two-step microemulsion method. Pojas *et al.* demonstrated that the particle size distribution is independent of both the total amount of metal precursor and the preparation route. The degree of alloying was found to be higher for the two-step microemulsion method. It has also been observed that pH of the environment strongly affects the electrosteric environment of surfactant. Despite control over size distribution, the microemulsion method cannot be useful for shape variation of nanoparticles. It uses expensive surfactant molecules with extra washing steps and may not be suitable for a large scale synthesis.

2.4
Synthetic Methods of Novel Unsupported Pt Nanostructures

The morphological control of electrocatalysts at the nanoscale is challenging and cannot be achieved by conventional synthesis methods. Hence the development of novel methods is envisaged to achieve control over the size and shape of electrocatalysts. By introducing various morphologies of Pt with higher surface area, low amount of metal loading and interesting chemical/catalytic properties at the nanoscale, it may be possible to improve the overall performance of catalyst material with reduced cost. The following sections emphasize the development of novel Pt nanostructures related to the above mentioned strategies, found in recent literature.

2.4.1
Template-Based Synthesis

Template-based synthesis of inorganic nanocrystals is a popular way of generating nanostructures wherein a template serves as a scaffold for the synthesis of nanostructures with the desired size and shape [109]. The synthesis involves the fabrication of desired material within the pores or channels of a nanoporous template. Electrochemical and electroless depositions, chemical polymerization, sol–gel deposition, and chemical vapor deposition have been presented as major template synthetic strategies. To date a variety of templates have been successfully demonstrated for the formation of nanomaterials including self-assembled organic surfactants/block copolymers, channels in porous materials, step edges on solid substrates, and the existing nanostructures [110–119]. Generally, these templates are classified into two types namely, hard templates and soft templates, where the hard templates are involved physically and need to be removed selectively after the synthesis of desired nanostructures. However the soft templates are normally involved in chemical synthesis procedures and it is possible to obtain the nanostructures in a pure form with soft templates. The popularity of the template-based method lies in its simplicity, cost-effectiveness and high efficiency. Template-based synthesis is considered as quite useful, since it allows the design and synthesis of a wide range of nanostructures with specified geometry and surface characteristics. Here we describe some of the recent examples of synthesis of Pt nanostructures by the template-based method.

2.4.1.1 Soft Template

In the past decade, the efficiency of soft templates in controlling the shape, size and crystallinity of inorganic nanoparticles has been studied in detail [120]. The variety of soft templates include surfactant micelles, reverse micelles, multilamellar vesicles, unilamellar liposomes, peptide tubes, liposomal aggregates and microemulsions. These nanoreactors are normally used to synthesize inorganic nanocrystals owing to their property of formation of ordered self-assembly when the concentration reaches a certain value, the critical micelle concentration (CMC), in order to decrease the system energy. These soft templates provide confined spaces for the growth of nanoparticles by acting as a scaffold. The synthesis of nanoparticles using soft templates highly depends upon the stability of the template structure which is dynamic in solution and generally changes with change in the composition during the reaction. Thus the stability of the template with respect to the reaction conditions is a major concern for the generation of nanomaterials using soft templates. Examples of soft templates for synthesis of Pt nanostructures include wormlike micelles for the growth of polycrystalline Pt nanowire, and liposomes for the growth of 2-D and 3-D Pt nanostructures [121–127].

Porous platinum nanoballs Porous Pt nanoballs were synthesized using soft templates from giant hexagonal liquid crystals formed by a quaternary system as nanoreactors [128]. The translucent and birefringent mesophase with cetyltrimethylammonium bromide (CTAB), as surfactant, and tetraamineplatinum(II), $Pt(NH_3)_4Cl_2$, as salt acts as a nanoreactor when slow reduction of platinum was carried out by passing γ rays. Growth kinetics studies revealed that the aggregation of Pt nanorods (2.8 nm diameter) results in the formation of a 3-D interconnected network of porous nanoballs of Pt templated by anisotropic hexagonal structure of soft template.

2.4.1.2 Hard Template

The synthesis of nanoparticles using hard templates involves a hard template as a central structure within which a network forms in such a way that the removal of template generates the nanostructures of desired shapes and sizes related to those of the templates. Examples of hard templates commonly used for the synthesis of the inorganic nanostructures include, mesoporous silica, porous alumina, track etched membranes, carbon nanotubes and so on [125–128]. The nanostructures are generally synthesized within the pores of these hard templates and the morphology of the nanostructures is dependent upon that of the templates, for example, if the template consists of cylindrical pores with uniform diameter, then monodispersed nanocylinders of the desired materials with solid or hollow nature can be synthesized within the voids of the template material depending on the experimental conditions. The use of hard templates enables us to generate monodispersed nanostructures with control over shape and size in a reproducible manner. However, the main disadvantage of this method lies in the removal of the template after the synthesis of nanostructures. In this section we describe some of the recent examples of synthesis of Pt nanostructures using hard templates and their catalytic activity with respect to application for DMFC.

Pt-Ru alloy nanoparticles Mesoporous silica has been used as a template for synthesis of highly stable Pt-Ru alloyed nanoparticles by pyrolysis of carbon, Pt and Ru precursors. The silica template was subsequently removed by etching with HF [129]. This method produced well dispersed bifunctional PtRu nanoparticles with an average size of 2–3 nm on the surface of carbon rods. The electrochemical behavior of PtRu electrocatalyst towards methanol oxidation showed higher mass current activity, and long-term stability, which can be attributed to higher surface area and higher diffusion kinetics due to uniform pore structure. A high tolerance to CO poisoning was also observed which can be due to the presence of oxygen species in Ru oxides on the surface of PtRu nanoparticles.

Sandwiched Ru/C nanocomposite Another recent alternative approach includes synthesis of a novel nanostructured bimetallic electrocatalyst (Pt/RuC) with Pt nanoparticles supported on the sandwiched RuC nanocomposite [130]. Ordered mesoporous silica SBA-15 was used as the template to prepare RuC composite with Ru nanocrystals sandwiched in the pore walls of porous carbons via chemical vapor deposition. It was observed that Pt nanoparticles were highly dispersed in the pore channels of RuC composite with unalloyed Pt and Ru metal phases (Figure 2.10). The catalytic activity and stability of Pt/RuC for methanol electro-oxidation was better than that of bimetallic PtRu/OMC and the current densities were found to be 132, 98 and 80 mA mg^{-1} for Pt/RuC, PtRu/OMC and Pt/OMC, respectively. Its CO tolerance performance was comparable to that of PtRu/OMC but largely improved compared with monometallic Pt/OMC. The improved performance could be interpreted by the bifunctional mechanism, the synergetic interaction between Pt and Ru, and between Ru and carbon interface coupled with the unique sandwiched structure of the composite.

Gold nanoporous structures A novel strategy of fabricating gold nanotubes with rough nanoporous walls obtained by silver dissolution using anodic aluminum oxide (*AAO*) as a hard template has been demonstrated [131]. Figure 2.11, depicts the synthetic procedure for the preparation of smooth as well as rough nanoporous walls. In this method, the electro-deposition of gold silver alloys followed by dealloying and polymer dissolution resulted in the formation of nanoporous walls. An ultrathin Pt layer was then electrodeposited on Au nanotubes without changing the nanoporous structure. The wall structure played an important role in imparting electrocatalytic activities towards methanol oxidation. The catalytic activity of nanoporous Pt tubes

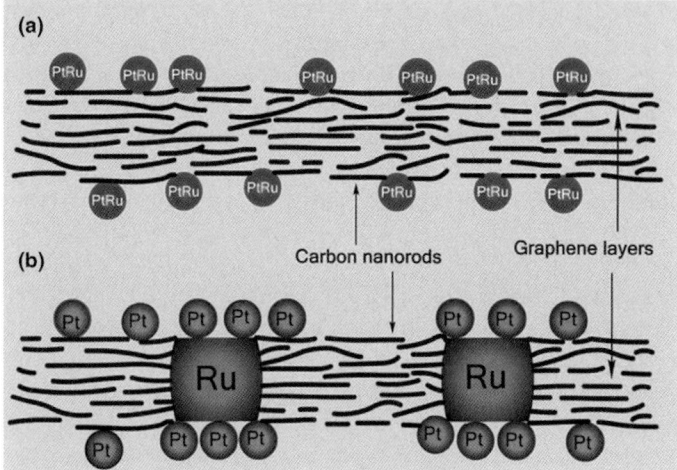

Figure 2.10 (a) Schematic illustration of two bimetallic catalysts: PtRu/OMC catalyst, and (b) Pt/RuC catalyst with Pt nanoparticles supported on sandwiched RuC nanocomposite. Reproduced with permission from Ref. [130]. Copyright © 2008 American Chemical Society.

was found to be higher than the commercially available Pt/C nanoparticles, which can be attributed to the higher CO poisoning tolerance and improved charge transfer kinetics. The current densities were found to be 2.8, 0.8 and 0.005 mA cm^{-2} for nanotubes with porous, smooth walls and flat Pt substrate, respectively. The

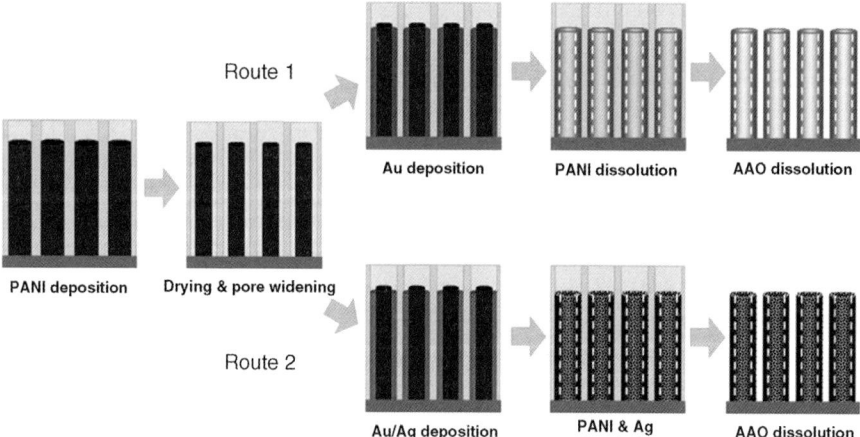

Figure 2.11 Schematic representation of the experimental procedure for the synthesis of nanotubes with smooth walls (route 1) and with nanoporous walls (route 2). Reproduced with permission from Ref. [131]. Copyright © 2008 American Chemical Society.

increased surface area due to exposure of inner as well as outer surfaces to reaction medium also contributed to the increased catalytic activity in comparison with nanotubes with smooth walls.

2.4.2
Galvanic Replacement Reaction

The use of hollow nanostructures as catalysts is an effective alternative for cost reduction with less metal content compared with other compact solid structures, reducing the loading amount when it is used as a catalyst. The galvanic replacement reaction has been exploited as a powerful means of preparing hollow metal nanostructures [132–140]. In galvanic displacement, metal ions from solution are reduced to the metal causing subsequent oxidation of the substrate which results in dissolution of exposed surface atoms. This simple method deposits metals selectively onto the oxidizable metal supports and does not require an external voltage source or a reducing agent in solution like electrodeposition and electroless plating. Noble metals that are of interest in catalysis (Au and Pt) can be deposited on the less noble metal such as Ag. The advantage of this method is the lower temperature used for the reaction since other state-of-the-art methods used a higher temperature for catalyst preparation. The support can be dissolved using chemical etching after galvanic displacement which forms the hollow metal nanostructures. The unique structural properties of these hollow structures lead to some interesting results in terms of their catalytic activities, which are described in the following sections with some recent examples. The work carried out in our laboratory on the development of the hollow Pt based electrocatalyst is also discussed in the later section.

2.4.2.1 Platinum Nanospheres
A large-scale synthesis of Pt hollow nanospheres with an average diameter of 24 nm was demonstrated by galvanic replacement reaction between cobalt particles and Pt solution [141]. In this procedure, Co nanoparticles were synthesized first using citric acid as a capping agent followed by its oxidation after addition of Pt salt during galvanic displacement reaction. This resulted in the nucleation of Pt atoms forming a thin shell around Co nanoparticles. The porous shell of Pt nanospheres with 3 nm thickness revealed enhanced electrocatalytic activity in methanol oxidation. The improved mass current activity of Pt nanospheres can be attributed to high surface area and increased platinum use.

2.4.2.2 Platinum Nanotubes
A large scale synthesis of Pt nanofiber/nanotubes junction structure by galvanic displacement of Ag from silver nanowires has been demonstrated [142]. The simple mixing of silver nanowires with Pt solution in the presence of CTAB results in the formation of a Pt nanotubes and nanofibers junction, which can be separated by sonication. The presence of CTAB is crucial for the anisotropic growth, which kinetically controls the morphology of crystal nuclei by preferential adsorption. The method is very simple and can produce large quantities of Pt nanotubes with a wall

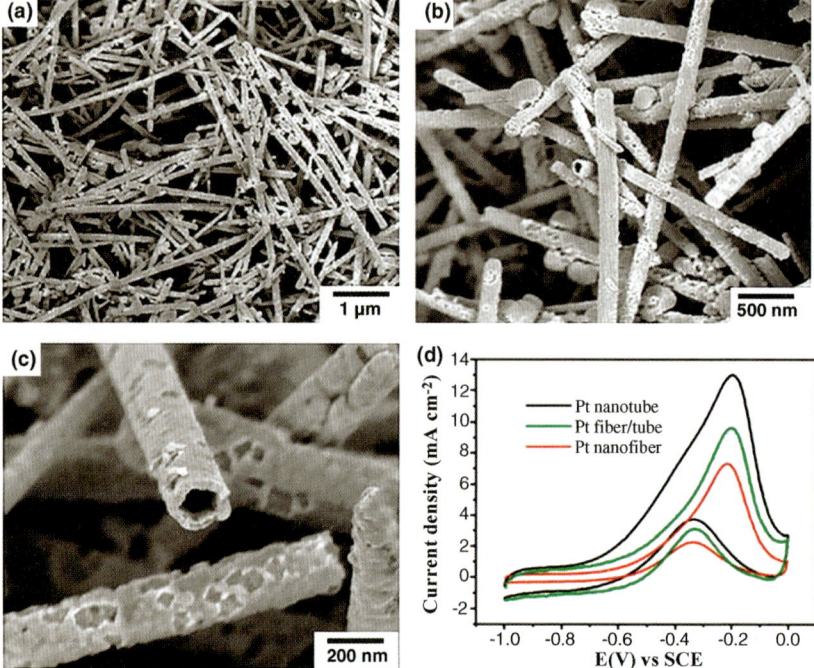

Figure 2.12 (a–c) SEM and FE-SEM images of the platinum nanotubes, and (d) cyclic voltammetry curves of the Pt nano products during methanol oxidation. Reproduced with permission from Ref. [142]. Copyright © 2008 American Chemical Society.

thickness between 20 and 30 nm as shown in Figure 2.12a–c. The electrochemical properties of Pt nanotubes over methanol oxidation revealed higher mass current density of 13 mA cm^{-2} compared with 9 and 7 mA cm^{-2} for Pt nanotube/nanofiber junction and nanofibers, respectively (Figure 2.12d). The superior performance of Pt nanotubes can be attributed to higher surface area, increased Pt use and improved mass transport kinetics since anisotropic morphology was found to improve mass transport and catalyst use.

The same method was used to prepare supportless PtPd nanotubes in order to obtain higher mass activity offered by PtPd alloys [143]. A galvanic displacement of silver from silver nanowires by simple mixing with Pt and palladium precursor solutions results in the formation of well defined PtPd nanotubes as shown in Figure 2.13a and b. The specific activity of PtPd nanotubes in oxygen reduction reaction was found to be 5.6 times higher than that of Pt/C electrocatalyst (Figure 2.13c). The improved ORR activity can be attributed to the variation in bond length of Pt fcc structure on account of alloying with Pd. In spite of higher durability, these catalysts suffer from low surface area and low use, hence it is necessary to optimize the dimensions of the PtNTs and PtPdNTs for future use in fuel cells.

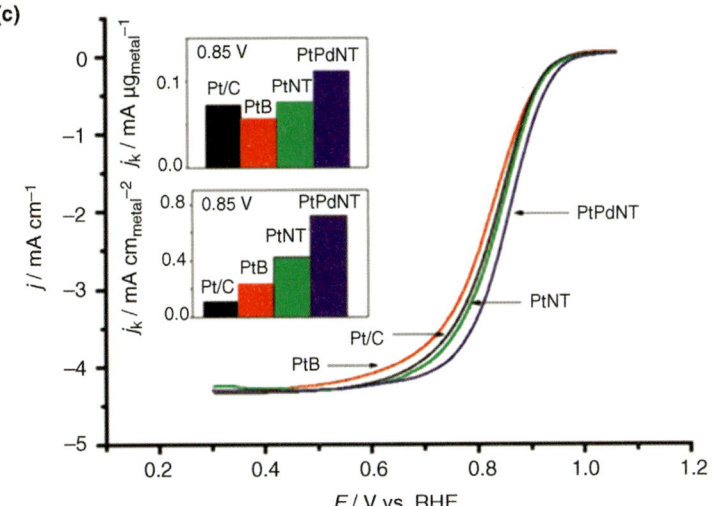

Figure 2.13 (a) SEM image of PtPdNTs, (b) TEM image and electron diffraction pattern (inset) of PtPdNTs, and (c) ORR curves of Pt/C, platinum black (PtB), PtNTs, and Pd PtNTs in O_2-saturated 0.5 m H_2SO_4 solution at room temperature. Inset: Mass activity (top) and specific activity (bottom) for the four catalysts at 0.85 V. Reproduced with permission from Ref. [143]. Copyright © 2007 Wiley-VCH.

2.4.2.3 Nanoporous Platinum

Recently the synthesis of porous Pt nanoparticles has been reported using diphenyl ether (DPE) as a medium, platinum acetylacetonate (Pt(acac)$_2$) as the precursor and 1,2-hexadecanediol (HDD) as the reducing reagent [144]. Hexadecylamine (HDA) functioned as both surfactant and reaction solvent, and 1-adamantanecarboxylic acid (ACA) was used to prevent free surface sites from coordinating with other capping ligands. The continuous formation of seeds and rapid autocatalytic growth has been discussed for the formation of metal nanoparticles including platinum. The amount of Pt precursor was found to be responsible for the flower-like porous nanoparticles. Methanol electro-oxidation and chronoamperometric analysis revealed that the mass current density of porous Pt nanoparticles were 60% higher than that of E-Tek catalysts.

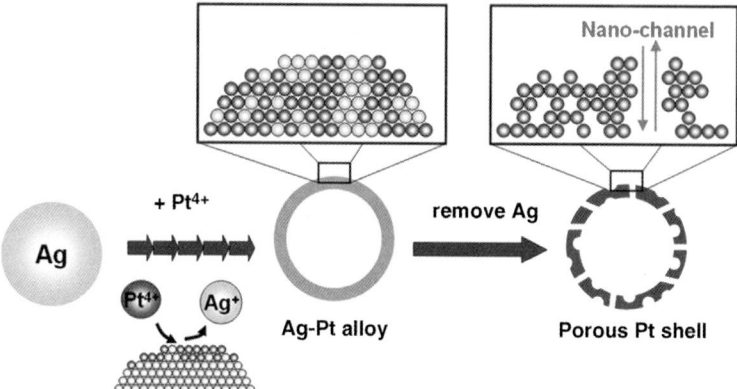

Figure 2.14 Synthetic routes of porous hollow Pt nanospheres with Ag cores and Ag-Pt hollow nanospheres. Reproduced with permission from Ref. [145]. Copyright © 2008 American Chemical Society.

2.4.2.4 Hollow Platinum Nanochannels

Recently, our group has reported the fabrication of Pt hollow spheres with nano-channels at room temperature using a modified galvanic replacement reaction and their applications as electrocatalysts [141, 145, 146]. Initially silver nanoparticles were prepared using PVP at 60 °C followed by the replacement of the silver by Pt at room temperature by dropwise addition of H_2PtCl_6 to the solution of silver nanoparticles as shown schematically in Figure 2.14. The mixture continued to undergo the reaction to yield nanoparticles with Ag-Pt alloy shells followed by the chemical etching treatment of newly synthesized nano shells using 1 M NH_4OH. The excess chloride ions were removed by centrifugation and the silver atoms from Ag-Pt alloy shells were dissolved using 1M HNO_3. Repetitive centrifugation helped to remove excess solvents and vigorous magnetic stirring maintained a stable dispersion of nanostructures. The ORR activity of as-synthesized hollow Pt nanospheres was determined as shown in Figure 2.15a and b. There was an enhancement in the current density of hollow Pt shells with nano-channels compared with the commercial Pt catalyst. Earlier reports on bimetallic alloy catalysts suggest that the enhanced ORR activity could be attributed to the modification of electronic structure of Pt (5d orbital vacancies). To elucidate the d orbital vacancies of Pt nanoparticles, we determined the oxidation state of commercial Pt and porous Pt hollow nanospheres by x-ray absorption near edge spectroscopy (XANES). As shown in Figure 2.15d, for porous Pt hollow nanospheres, the area of L_3 threshold resonance line, which is proportional to the number of vacant d-electron states [147], was greater than that of the commercial Pt nanoparticles indicating that the concentration of vacant d-electron states associated with a porous hollow structure is greater than that of commercial Pt nanoparticles. An increased 5d vacancy of Pt is believed to increase the interaction of O_2 and Pt, thereby enhancing the catalytic activity of porous Pt hollow nanospheres [148]. The increase in d-band vacancy may result from the presence of foreign atoms (silver)

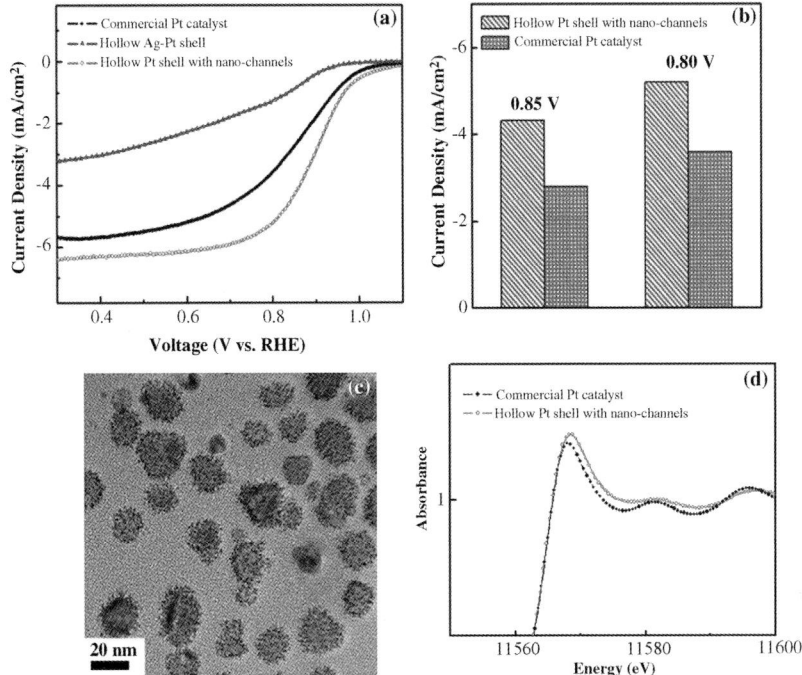

Figure 2.15 (a) Polarization curve for the ORR on the commercial Pt catalyst, hollow Ag-Pt shell, and the hollow Pt shell with nano-channels in 0.1M HClO$_4$. (b) Comparison of mass activity for commercial Pt and hollow Pt shell with nano-channels at 0.85 and 0.8 V, (c) TEM micrographs of the hollow Pt shell with nano-channels and (d) XANES spectra of the commercial Pt and hollow Pt shell with nano-channels at Pt L$_3$-edge. Reproduced with permission from Ref. [145]. Copyright © 2008 American Chemical Society.

and/or the atomic scale surface roughness of Pt [149]. Consequently, an enhanced atomic scale surface roughness and low coordination of some atoms may contribute to the observed activity.

2.4.2.5 Bimetallic Clusters

It has been observed that the ORR activity on Pt(111) substrates is inhibited due to the formation of a surface oxide layer. Zhang and co-workers have demonstrated that the coating of gold clusters on Pt surface stabilizes the Pt oxygen-reduction electrocatalysts [150]. The modification of Pt nanoparticles with gold clusters was carried out by galvanic displacement of Cu monolayer on Pt (111) substrate by Au atoms. The Pt surface was precoated with Cu monolayer by underpotential deposition. Au clusters, shown in Figure 2.16a and b, formed after repeated cycling are two to three monolayers thick, 2 to 3 nm in diameter and covered about 30 to 40% of the Pt surface. The presence of Au clusters was found to stabilize Pt (111) substrates against oxidation and resulted in improved ORR activity of Pt nanoparticles under oxidizing conditions as shown in Figure 2.16c–f.

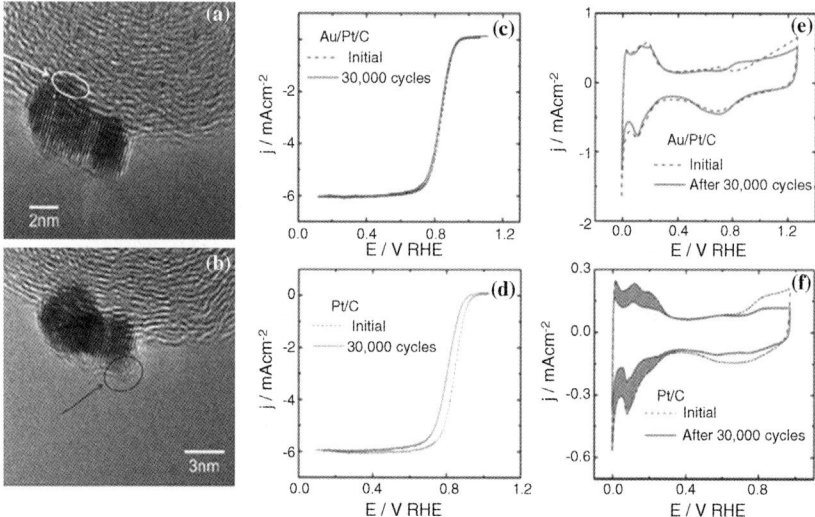

Figure 2.16 (a and b) TEM of a Au/Pt/C catalyst made by displacement of a Cu monolayer by Au showing atomic rows with spacings that are consistent with the Pt(111) single-crystal structure. A different structure in the areas indicated by the arrows is ascribed to the Au clusters, and polarization curves for the O_2 reduction reaction on Au/Pt/C (c) and Pt/C (e) catalysts on a rotating disk electrode, before and after 30 000 potential cycles. Cyclic voltammetry curves for Au/Pt/C (d) and Pt/C (f) catalysts before and after 30 000 cycles, respectively. The shaded area in (d) indicates the lost Pt area. Reproduced with permission from Ref. [150]. Copyright © 2007 American Association for the Advancement of Science.

2.4.3
Electrochemical Synthesis

The electrochemical approach for the synthesis of inorganic nanocrystals has been explored in detail. The electrochemical reduction route presents a simple, quick and economical way for the preparation nanostructured electrocatalyst. By changing the synthetic parameters, such as the concentration of metal precursors, current densities, and deposition time, the sizes of the nanoparticles and surface morphologies of the deposits can be well controlled. The main advantage of this method is that, it allows controlled, patterned and faceted crystal growth of nanostructures. Other advantages include reproducibility, simplicity and monodispersity of nanoparticles. There are many reports demonstrating generation of the faceted Pt nanostructures using electrochemical synthesis, a few of which are discussed in this section [151–157].

2.4.3.1 Spherical Platinum Clusters

A novel pulse electrodeposition method to synthesize spherical Pt nanoclusters by mixing a cocatalyst such as SnO_2 was demonstrated [158] (Figure 2.17a). The pulse electrodeposition was found to be advantageous in producing uniform metal nanoclusters compared with direct current deposition which produces larger metal

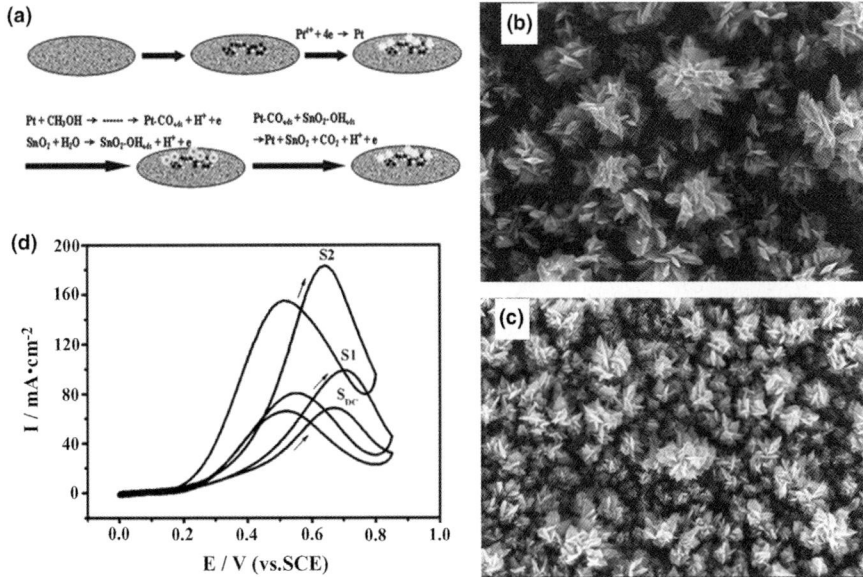

Figure 2.17 (a) Schematic diagram of structural designing of PtSnO2/C and the process from electrodepostion to electrooxide, (b) Scanning electron micrographs of Pt/C (S1), (c) PtSnO2/C (S2) by pulse electrodeposition, and (d) cyclic voltammetry of SDC, S1, and S2 in 0.5 M H2SO4 + 1.0 M CH3OH solution. Reproduced with permission from Ref. [158]. Copyright © 2008 American Chemical Society.

nanoparticles (Figure 2.17b). Electrodeposition in the presence of SnO_2 produced smaller spherical Pt nanoclusters facilitating uniform dispersion on a carbon support, as shown in Figure 2.17c. The current densities of electrodepositied platinum, Pt/C and PtSnO$_2$/C during methanol oxidation were found to be 68.47, 98.92, and 183.89 mA cm^{-2}, respectively. The increase in electrocatalytic activity of spherical nanoclusters catalyst (PtSnO$_2$/C) over methanol oxidation (Figure 2.17d) and oxygen reduction in the presence of SnO$_2$ was observed, which can be attributed to larger active surface area of Pt nanocluster and promotion of CO oxidation due to oxygen provided by SnO$_2$.

2.4.3.2 Nanoporous Platinum

The electrochemical lithiation of submicrometer PtO$_2$ has been employed to synthesize nanoporous Pt [159]. This results in the formation of a M/Li$_2$O nanocomposite with nanoscale metal clusters embedded in a Li$_2$O matrix. In an electrochemical lithiation process, 4Li can be inserted into the starting material of PtO$_2$, resulting in the formation of Pt/Li$_2$O nanocomposite. The Li extraction from these M/Li$_2$O nanocomposite can lead to a porous network of metal nanoparticles as shown in Figure 2.18. The synthesis of nanoporous Pt was carried out from PtO$_2$ by electrochemical

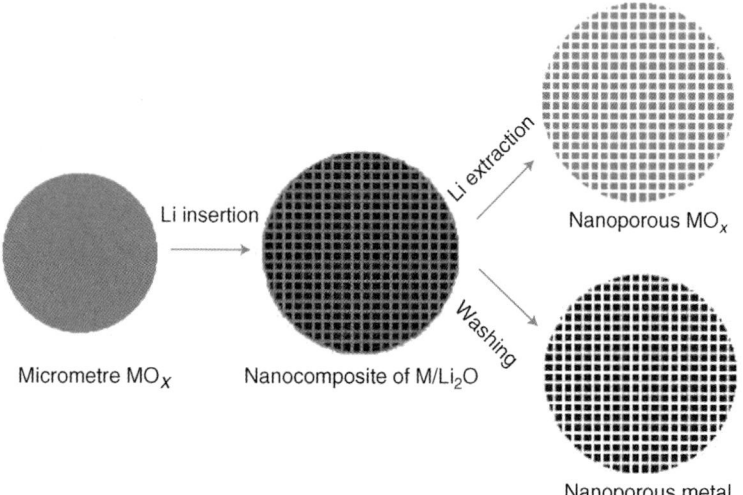

Figure 2.18 Schematic of the template-free electrochemical lithiation/delithiation synthesis of nanoporous structures. Reproduced with permission from Ref. [159]. Copyright © 2006 Nature Publishing Group.

lithiation followed by dissolving the Li$_2$O in acidic solution or water. The presence of various pore sizes (2–20 nm) was observed in the porous Pt nanostructures as shown in Figure 2.19a and b. The mass current density per unit mass of Pt was found to be 186 mA mg^{-1}, which decreased to 160 mA mg^{-1} after 100 scan cycles. This is the highest catalytic activity reported for pure Pt mixed with carbon support. As-prepared porous Pt showed enhanced electrocatalytic activities towards methanol oxidation, which can be due to the high surface area (Figure 2.19c and d), presence of porous structure and pronounced stability of nanoporous platinum.

2.4.3.3 Tetrahexahedral Platinum Nanostructures

A novel top down approach for the synthesis of highly faceted tetrahexadral Pt nanocrystals (with 24 faces) using an electrochemical method has been demonstrated (Figure 2.20a) [37, 160]. Tetrahexahedral (THH) Pt nanocrystals (NCs) with an abundance of low-coordination, high-reactivity atomic edges sites were grown on glassy carbon surface at the expense of Pt nanospheres by applying square wave pulse sequence alternated between reducing and oxidizing potentials at a rate of 10 Hz (Figure 2.20b–f). The alternating potential played a key role by controlling electrochemical reaction on the surface of nanoparticles which affects the adsorption reaction and metal deposition rates. The Pt tetrahexahedra were thermally stable, without morphological changes even at temperatures of 800 °C. The stability of {730} face of THH Pt nanocrystals was enhanced during positive potential steps as a result of formation of a monolayer of PtO and PtOH on the high index faces. The {730} face was preserved by deposition of fresh platinum in the place of atoms at {111} face as the electrode potential was made negative gradually, which resulted

2.4 Synthetic Methods of Novel Unsupported Pt Nanostructures | 107

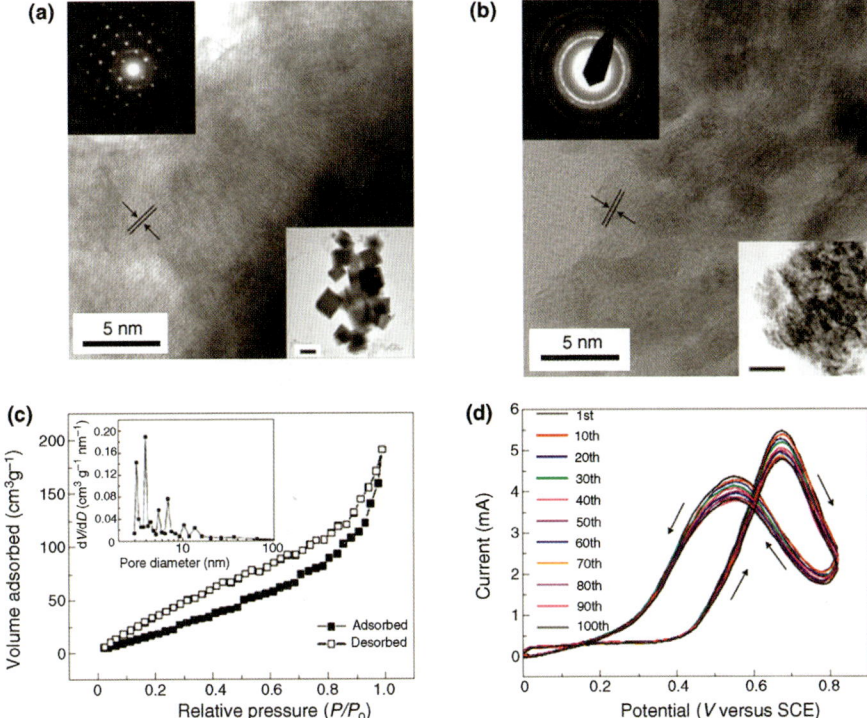

Figure 2.19 HRTEM images of (a) the initial situation, the top left and bottom right insets show the SAED pattern (scale bar, 200 nm); (b) discharged to 1.2 V, the top left and bottom right insets show the SAED pattern (scale bar, 30 nm). (c) Nitrogen adsorption/desorption isotherms after washing. Inset: The pore size distribution plot; (d) cyclic voltammetry of the nanoporous Pt electrode cycled at a scan rate of 20 mV s^{-1} in 1.0 M methanol in 0.5 M H$_2$SO$_4$ solution. Reproduced with permission from Ref. [159]. Copyright © 2006 Nature Publishing Group.

in the removal of oxide/hydroxide surface. The diffusion of oxygen atoms beneath the lattice of {111} faces creates unstable and disordered surface structure. The shape factor observed in the diffraction patterns revealed that Pt nanocrystals were capped by {730} faces, which contained a relatively high density of atomic step edges (Figure 2.21). It was found that about 43% of the total number of surface atoms resides along steps, which can be compared with 6%, 13%, and 35% for 5 nm diameter Pt cubes, spheres, and tetrahedral particles, respectively. The enhanced catalytic activity over ethanol and formate oxidation could be attributed to the existence of high density of steps and kinks at the surface of Pt nanocrystals. The fuel oxidation rate was found to be 400% for formate and 200% for ethanol compared with spherical nanoparticles.

Thus it is clear from the above discussion that the catalytic activity of Pt is highly dependent on its surface morphology which in turn is related to the synthetic approach for generation of Pt nanostructures. The anisotropic and hollow nanostructres show higher catalytic activity than the spherical nanostructures. Thus the

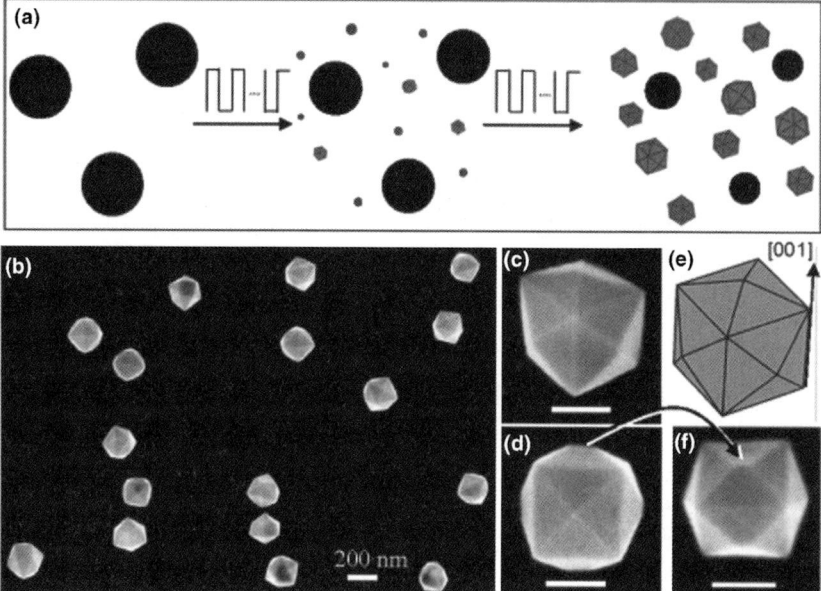

Figure 2.20 (a) Schematic of electrochemical preparation of THH Pt NCs from nanospheres; (b) low-magnification SEM image of THH Pt NCs with growth time of 60 min; (c and d) high-magnification SEM images of Pt THH viewed down along different orientations, showing the shape of the THH; (e) geometrical model of an ideal THH; (f) high-magnification SEM image of a THH Pt NC, showing the imperfect vertices as a result of unequal size of the neighboring facets. Reproduced with permission from Ref. [37]. Copyright © 2007 American Association for the Advancement of Science.

design of a cost-effective electrocatalyst with high efficiency is crucial for the commercialization of fuel cells.

2.5
Conclusions

Although a great amount of research work has been carried out in the area of development of novel, cost-effective, highly active, reliable/durable and stable fuel cell catalysts, a real breakthrough in terms of the commercialization of fuel cells is yet to be achieved, which remains a major challenge for fuel cell technology and commercialization. A theoretical approach for the design of catalysts provides a better understanding of the performance of catalysts as well as their structure–activity relationship. The recent advancements in the use of different morphologies of Pt nanostructures for cost reduction and improved performance of fuel cells may lead to a breakthrough in this field, which holds the key to new generations of clean energy devices. It was observed that the design of electrocatalyst by alloying technique or by using different morphologies of Pt metal greatly enhances the catalytic activity of the

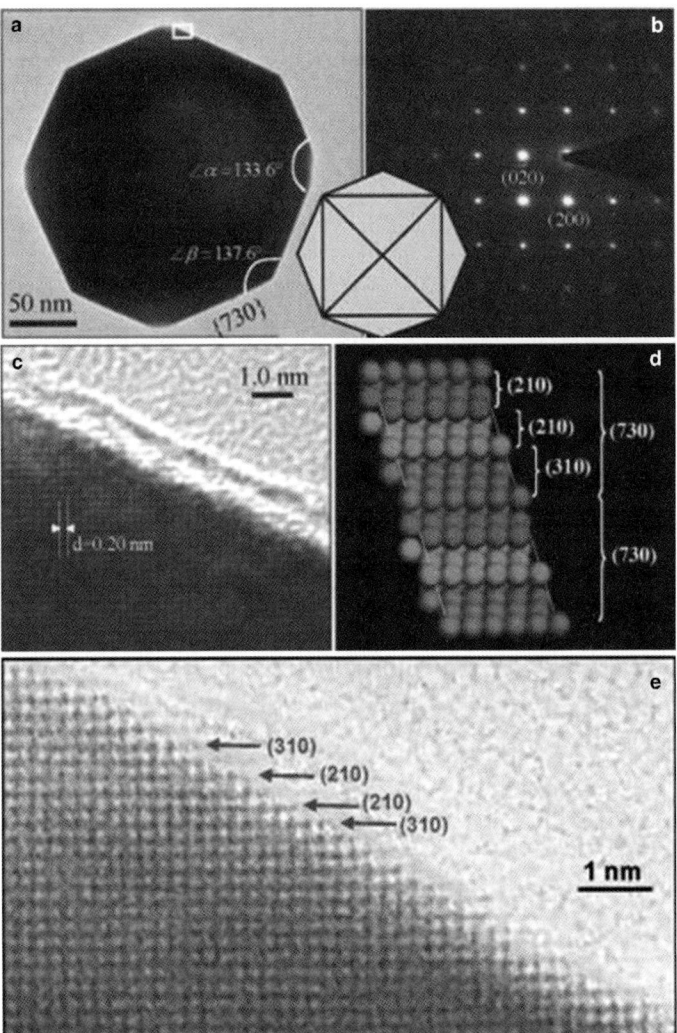

Figure 2.21 (a) TEM image of THH Pt NC recorded along [001] direction; (b) corresponding SAED pattern with square symmetry, showing the single-crystal structure of the THH Pt NC; (c) high-resolution TEM image recorded from the boxed area marked in (a); (d) atomic model of Pt (730) plane with a high density of stepped surface atoms. The (730) surface is made of (210) and (310) subfacets; (e) HRTEM image recorded from another THH Pt NC to reveal surface atomic steps in the areas made of (210) and (310) subfacets. The image reveals the surface atomic steps. Reproduced with permission from Ref. [37]. Copyright © 2007 American Association for the Advancement of Science.

materials with reduction in the cost and CO poisoning. The idea of development of non-noble catalyst for DMFC needs to be explored in detail in order to reduce the cost and improve the performance of the cell. Thus optimization of different approaches including various novel synthesis methods, support strategies, alloying techniques, are very much necessary for reduction of cost and improvement in the performance of DMFC with respect to its commercialization.

Acknowledgments

The authors would like to thank the National Science Council of the Republic of China (contract numbers: NSC 97–2113-M-002–012-MY3, NSC 97–2120-M-002–013 and NSC 97–2120-M-001–006) and the Industrial Technology Research Institute for financially supporting this research.

References

1 Arico, A.S., Bruce, P., Scrosati, B., Tarascon, J.M. and Van Schalkwijk, W. (2005) *Nat. Mater.*, **4**, 366.
2 Maier, J. (2005) *Nat. Mater.*, **4**, 805.
3 http://www.fctec.com/fctec_history.asp.
4 http://www.fctec.com/fctec_basics.asp.
5 http://en.wikipedia.org/wiki/DMFC.
6 http://www.fctec.com/fctec_types.asp.
7 Wasmus, S. and Kuver, A. (1999) *J. Electroanal. Chem.*, **461**, 14.
8 Arico, A.S., Srinivasan, S. and Antonucci, V. (2001) *Fuel Cells*, **2**, 133.
9 Reddington, E., Sapienza, A., Gurau, B., Viswanathan, R., Sarangapani, S., Smotkin, E.S. and Mallouk, T.E. (1998) *Science*, **280**, 1735.
10 Kim, J.G. (2001) *J. Ind. Eng. Chem.*, **7**, 58.
11 McGrath, K.M., Surya Prakash, G.K. and Olah, G.A. (2004) *J. Ind. Eng. Chem.*, **7**, 1063.
12 Kim, H.J. and Lim, T.H. (2004) *J. Ind. Eng. Chem.*, **7**, 1081.
13 Kim, M.S., Lee, S.J., Kang, J.U. and Bae, K.J. (2005) *J. Ind. Eng. Chem.*, **11**, 187.
14 Stamenkovic, V.R., Mun, B.S., Arenz, M., Mayrhofer, K.J.J., Lucas, C.A., Wang, G., Ross, P.N. and Markovic, N.M. (2007) *Nat. Mater.*, **6**, 241.
15 Liu, H., Song, C., Zhang, L., Zhang, J., Wang, H. and Wilkinson, D.P. (2006) *J. Power Sources*, **155**, 95.
16 Guo, Y.-G., Hu, J.-S. and Wan, L.-J. (2008) *Adv. Mater.*, **20**, 2878.
17 Park, K.-W. and Sung, Y.-E. (2006) *J. Ind. Eng. Chem.*, **12**, 165.
18 Yajima, T., Wakabayashi, N., Uchida, H. and Watanabe, M. (2003) *Chem. Commun.*, 828.
19 Roth, C., Benker, N., Buhrmester, T., Mazurek, M., Loster, M., Fuess, H., Koningsberger, D.C. and Ramaker, D.E. (2005) *J. Am. Chem. Soc.*, **127**, 14607.
20 Watanabe, M. and Motoo, S. (1975) *J. Electroanal. Chem.*, **60**, 267.
21 Markovic, N.M., Gasteiger, H.A., Ross, P.N., Jiang, X., Villegas, I. and Weaver, M.J. (1995) *Electrochim. Acta*, **40**, 91.
22 Goikovic, S.L., Vidakovic, T.R. and Durovic, D.R. (2003) *Electrochim. Acta*, **48**, 3607.
23 Christensen, P.A., Hamnett, A. and Troughton, G.L. (1993) *J. Electroanal. Chem.*, **362**, 207.
24 Iwasita, T. (2002) *Electrochim. Acta*, **47**, 3663.
25 Wang, D., Lu, J. and Zhuang, L. (2008) *Chem. Phys. Chem.*, DOI: 10.1002/cphc. 200800282.
26 Kinoshita, K. (1990) *J. Electrochem. Soc.*, **137**, 845.
27 Mukerjee, S. and McBreen, J. (1998) *J. Electroanal. Chem.*, **448**, 163.
28 Maillard, F., Savinova, E.R. and Stimming, U. (2007) *J. Electroanal. Chem.*, **599**, 221.
29 Lai, S.C.S., Lebedeva, N.P., Housmans, T.H.M. and Koper, M.T.M. (2007) *Top. Catal.*, **46**, 320.

30 Bergamaski, K., Pinheiro, A.L.N., Teixeira-Neto, E. and Nart, F.C. (2006) *J. Phys. Chem. B*, **110**, 19271.

31 Sun, Y., Zhuang, L., Lu, J., Hong, X. and Liu, P. (2007) *J. Am. Chem. Soc.*, **129**, 15465.

32 Bratlie, K.M., Lee, H., Komvopoulos, K., Yang, P. and Somorjai, G.A. (2007) *Nano Lett.*, **7**, 3097.

33 Housmans, T.H.M., Wonders, A.H. and Koper, M.T.M. (2006) *J. Phys. Chem. B*, **110**, 10021.

34 Wang, C., Daimon, H., Onodera, T., Koda, T. and Sun, S. (2008) *Angew. Chem. Int. Ed.*, **47**, 3588.

35 Susut, C., Chapman, G.B., Samjeské, G., Osawa, M. and Tong, Y.Y. (2008) *Phys. Chem. Chem. Phys.*, **10**, 3712.

36 Lebedeva, N.P., Koper, M.T.M., Herrero, E., Feliu, J.M. and van Santen, R.A. (2000) *J. Electroanal. Chem.*, **487**, 37.

37 Tian, N., Zhou, Z.-Y., Sun, S.- G., Ding, Y. and Wang, Z.L. (2007) *Science*, **316**, 732.

38 Xiong, Y., Wiley, B.J. and Xia, Y. (2007) *Angew Chem. Int. Ed.*, **46**, 7157.

39 Wang, Z.L. (2000) *J. Phys. Chem. B*, **104**, 1153.

40 Rolison, D.R. (2003) *Science*, **299**, 1698.

41 Fujiwara, N., Shiozaki, Y., Tanimitsu, T., Yasuda, K. and Miyazaki, Y. (2002) *Electrochemistry*, **70**, 988.

42 Qiao, H., Kunimatsu, M., Fujiwara, N. and Okada, T. (2005) *Electrochem. Solid-State Lett.*, **8**, A175–A178.

43 Ramkumar, R., Dheenadayalan, S. and Pattabiraman, R. (1997) *J. Power Sources*, **69**, 75.

44 Goodenough, J.B., Hamnett, A. and Kemmedy, B.J. (1990) *Electrochim. Acta*, **35**, 199.

45 Cui, Z., Liu, C., Liao, J. and Xing, Wei. (2008) *Electrochim. Acta*, **53**, 7807–7811.

46 Roman-Martinez, C., Cazorla-Amoros, D., Yamashita, H., de Miguel, S. and Scelza, O.A. (2000) *Langmuir*, **16**, 1123.

47 Vigier, F., Coutanceau, C., Vigier, F., Coutanceau, C., Perrard, A., Belgsir, E.M. and Lamy, C. (2004) *J. Appl. Electroanal.*, **34**, 439.

48 Coutanceau, C., Brimaud, S., Lamy, C., Leger, J.-M., Dubau, L., Rousseau, S. and Vigier, F. (2008) *Electrochim. Acta*, **53**, 6865.

49 Zhang, Y., Valiente, A.M., Ramos, I.R., Xin, Q. and Ruiz, A.G. (2004) *Catal. Today*, **93–95**, 619.

50 Fujiwara, N., Yasuda, K., Ioroi, T., Siroma, Z. and Miyazaki, Y. (2002) *Electrochim. Acta*, **47**, 4079.

51 Steigerwalt, E.S., Deluga, G.A. and Lukehart, C.M. (2002) *J. Phys. Chem. B*, **106**, 760.

52 Boxall, D.L., Deluga, G.A., Kenik, E.A., King, W.D. and Lukehart, C.M. (2001) *Chem. Mater.*, **13**, 891.

53 Dickinson, A.J., Carrette, L.P.L., Collins, J.A., Friedrich, K.A. and Stimming, U. (2002) *Electrochim. Acta*, **47**, 3733.

54 Friedrich, K.A., Geyzers, L.P., Dickinson, A.J. and Stimming, U. (2003) *J. Electroanal. Chem.*, **524/525**, 261.

55 Kawaguchi, T., Sugimoto, W., Murakami, Y. and Takasu, Y. (2004) *Electrochem. Commun.*, **6**, 480.

56 Kawaguchi, T., Sugimoto, W., Murakami, Y. and Takasu, Y. (2005) *J. Catal.*, **229**, 176.

57 Nashner, M.S., Frenkel, A.I., Somerville, D., Hills, C.W., Shapley, J.R. and Nuzzo, R.G. (1998) *J. Am. Chem. Soc.*, **120**, 8093.

58 Takasu, Y., Fujiwara, T., Murakami, Y., Sasaki, K., Oguri, M., Asaki, T. and Sugimoto, W. (2000) *J. Electrochem. Soc.*, **147**, 4421.

59 Steigerwalt, E.S., Deluga, G.A., Cliffel, D.E. and Lukehart, C.M. (2001) *J. Phys. Chem. B*, **105**, 8097.

60 Takasu, Y., Fujiwara, T., Murakami, Y., Sasaki, K., Oguri, M., Asaki, T. and Sugimoto, W. (2000) *J. Electrochem. Soc.*, **147**, 4421.

61 Bonnemann, H. and Nagabhushana, K.S. (2004) *J. New. Mater. Electrochem. Syst.*, **7**, 93.

62 Bonnemann, H., Brijoux, W., Brinkmann, R., Dinjus, E., Joussen, T. and Korall, B. (1991) *Angew. Chem.*, **103**, 1344–1346.

63 Watanabe, M., Uchida, M. and Motoo, S. (1987) *J. Electroanal. Chem.*, **229**, 395.

64 Luna, A.M.C., Camara, G.A., Paganin, V.A., Ticianelli, E.A. and Gonzalez, E.R. (2000) *Electrochem. Commun.*, **2**, 222.

65 Schmidt, T.J., Noeske, M., Gasteiger, H.A., Behm, R.J., Britz, P., Brijoux, W. and Bonnemann, H. (1998) *J. Electrochem. Soc.*, **145**, 925.

66 Radmilovic, V., Gasteiger, H.A. and Ross, P.N. Jr (1995) *J. Catal.*, **154**, 98.

67 Xue, X., Lu, T., Liu, C. and Xing, W. (2005) *Chem. Commun.*, **12**, 1601.

68 Liu, Z., Lee, J.Y., Han, M., Chen, W. and Gan, L.M. (2002) *J. Mater. Chem.*, **12**, 2453.

69 Schmidt, T.J., Gasteiger, H.A. and Behm, R.J. (1999) *Electrochem. Commun.*, **1**, 1.

70 Bonnemann, H., Brinkmann, R., Kinge, S., Ely, T.O. and Armand, M. (2004) *Fuel Cells*, **4**, 289.

71 Wang, X. and Hsing, I. (2002) *Electrochem. Acta*, **47**, 2897.

72 Liu, Z., Ling, X., Lee, J., Su, X. and Gan, L.M. (2003) *J. Mater. Chem.*, **13**, 3049.

73 Paulus, U.A., Endruschat, U., Feldmeyer, G.J., Schmidt, T.J., Bonnemann, H. and Behm, R.J. (2000) *J. Catal.*, **195**, 383.

74 Wang, Y., Ren, J., Deng, K., Gui, L. and Tang, Y. (2000) *Chem. Mater.*, **12**, 1622.

75 Ohde, H., Hunt, F. and Wai, C.M. (2001) *Chem. Mater.*, **13**, 4130.

76 Rioux, R.M., Song, H., Grass, M., Habas, S., Niesz, K., Hoefelmeyer, J.D., Yang, P. and Somorjai, G.A. (2006) *Top. Catal.*, **39**, 167.

77 Herricks, T., Chen, J.Y. and Xia, Y.N. (2004) *Nano. Lett.*, **4**, 2367.

78 Lee, E. P., Chen, J. Y., Yin, Y. D., Campbell, C. T. and Xia, Y.N. (2006) *Adv. Mater*, **18**, 3271.

79 Chen, J.Y., Herricks, T. and Xia, Y.N. (2005) *Angew. Chem. Int. Ed.*, **44**, 2589.

80 Fu, X.Y., Wang, Y., Wu, N.Z., Gui, L.L. and Tang, Y.Q. (2003) *J. Mater. Chem.*, **13**, 1192.

81 Fievet, F., Lagier, J.P. and Figlarz, M. (1989) *MRS Bull.*, **14**, 29.

82 Wiley, B., Sun, Y., Chen, J., Cang, H., Li, Z.- Y., Li, X. and Xia, Y. (2005) *MRS Bull.*, **30**, 356.

83 Zhang, W., Chen, P., Gao, Q., Zhang, Y. and Tang, Y. (2008) *Chem. Mater.*, **20**, 1699.

84 Yang, Y., Matsubara, S., Xiong, L., Hayakawa, T. and Nogami, M. (2007) *J. Phys. Chem. C*, **111**, 9095.

85 Kim, F., Connor, S., Song, H., Kuykendall, T. and Yang, P. (2004) *Angew. Chem. Int. Ed.*, **43**, 3673.

86 Xiong, Y. and Xia, Y. (2007) *Adv. Mater.*, **19**, 3385.

87 Han, S.B., Song, Y.J. and Lee, J.M. (2008) *Electrochem. Commun.*, **10**, 1044.

88 Susut, C., Nguyen, T.D., Chapman, G.B. and Tong, Y.Y. (2008) *Electrochim. Acta*, **53**, 6135.

89 Chen, J., Herricks, T., Geissler, M. and Xia, Y. (2004) *J. Am. Chem. Soc.*, **126**, 10854.

90 Chen, J., Herricks, T. and Xia, Y. (2005) *Angew. Chem. Int. Ed.*, **44**, 2589.

91 Teng, X. and Yang, H. (2005) *Nano Lett.*, **5**, 885.

92 Li, H., Sun, G., Gao, Y., Jiang, Q., Jia, Z. and Xin, Q. (2007) *J. Phys. Chem. C*, **111**, 15192.

93 Chen, W.X., Lee, J.Y. and Liu, J.L. (2002) *Chem. Commun.*, 2588.

94 Liu, Z., Ling, X.Y., Lee, J.Y., Su, X. and Gan, L.M. (2003) *J. Mater. Chem.*, **13**, 3049.

95 Liu, Z., Ling, X.Y., Su, X. and Lee, J.Y. (2004) *J. Phys. Chem. B*, **108**, 8234.

96 Liu, Z., Lee, J.Y., Chen, W.X., Han, M. and Gan, L.M. (2004) *Langmuir*, **20**, 181.

97 Tsuji, M., Hashimoto, M., Nishizawa, Y., Kubokawa, M. and Tsuji, T. (2005) *Chem. Eur. J.*, **11**, 440.

98 Tsuji, M., Kubokawa, M., Yano, R., Miyamae, N., Tsuji, T., Jun, M.- S., Hong, S., Lim, S., Yoon, S.- H. and Mochida, I. (2007) *Langmuir*, **23**, 387.

99 Liu, Z., Lee, J.Y., Han, M., Chen, W. and Gan, L.M. (2002) *J. Mater. Chem.*, **12**, 2453.

100 Zhang, X. and Chan, K. (2003) *Chem. Mater.*, **15**, 451.

101 Liu, Y., Qiu, X., Chen, Z. and Zhu, W. (2002) *Electrochem. Commun.*, **4**, 550.

102 Rojas, S., Garcia-Garcia, F.J., Jaras, S., Martinez-Huerta, M.V., Fierro, J.L.G. and Boutonnet, M. (2005) *App. Catal. A: Gen.*, **285**, 24.
103 Wu, M.-L., Chen, D.-H. and Huang, T.-C. (2001) *Chem. Mater.*, **13**, 599.
104 Boutonnet, M., Lögdberg, S. and Svensson, E.E. (2008) *Curr. Opin. Colloid Interface Sci.*, **13**, 270.
105 Bommarius, A.S., Holzarth, J.F., Wang, D.I.C. and Hatton, T.A. (1990) *J. Phys. Chem.*, **94**, 7232.
106 Xiong, L. and Manthiram, A. (2005) *Solid State Ionics*, **176**, 385.
107 Solla-Gullon, J., Vidal-Iglesias, F.J., Montiel, V. and Aldaz, A. (2004) *Electrochim. Acta*, **49**, 5079.
108 Wang, J., Ee, Lee See, Ng, S.C., Chew, C.H. and Gan, L.M. (1997) *Mater. Lett.*, **30**, 119.
109 Chang, H., Joo, S.H. and Pak, C. (2007) *J. Mater. Chem.*, **17**, 3078.
110 Napolskii, K.S., Barczuk, P.J., Vassiliev, S.Y., Veresov, A.G., Tsirlina, G.A. and Kulesza, P.J. (2007) *Electrochim. Acta*, **52**, 7910.
111 Yuan, J.H., Wang, K. and Xia, X.H. (2005) *Adv. Fun. Mater.*, **15**, 803.
112 Piao, Y., Lim, H., Chang, J.Y., Lee, W.Y. and Kim, H. (2005) *Electrochim. Acta*, **50**, 2997.
113 Kyotani, T., Tsai, L.F. and Tomita, A. (1997) *Chem. Commun.*, 701.
114 Fukuoka, A., Higashimoto, N., Sakamoto, Y., Sasaki, M., Sugimoto, N., Inagaki, S., Fukushima, Y. and Ichikawa, M. (2001) *Catal. Today*, **66**, 23.
115 Wang, D.H., Kou, R., Gil, M.P., Jakobson, H.P., Tang, J., Yu, D.H. and Lu, Y.F. (2005) *J. Nanosci. Nanotechnol.*, **5**, 1904.
116 Fukuoka, A., Higuchi, T., Ohtake, T., Oshio, T., Kimura, J., Sakamoto, Y., Shimomura, N., Inagaki, S. and Ichikawa, M. (2006) *Chem. Mater.*, **18**, 337.
117 Ramirez, E., Erades, L., Philippot, K., Lecante, P. and Chaudret, B. (2007) *Adv. Funct. Mater.*, **17**, 2219.
118 Song, Y., Garcia, R.M., Dorin, R.M., Wang, H.R., Qiu, Y., Coker, E.N., Steen, W.A., Miller, J.E. and Shelnutt, J.A. (2007) *Nano Lett.*, **7**, 3650.
119 Krishnaswamy, R., Remita, H., Imperor-Clerc, M., Even, C., Davidson, P. and Pansu, B. (2006) *Chem. Phys.*, **7**, 1510.
120 Pileni, M.P. (2003) *Nature Mater.*, **2**, 145.
121 Song, Y.J., Garcia, R.M., Dorin, R.M., Wang, H.R., Qiu, Y. and Shelnutt, J.A. (2006) *Angew. Chem. Int. Ed.*, **45**, 8126.
122 Wang, H.R., Song, Y.J., Medforth, C.J. and Shelnutt, J.A. (2006) *J. Am. Chem. Soc.*, **128**, 9284.
123 Song, Y.J., Garcia, R.M., Dorin, R.M., Wang, H.R., Qiu, Y., Coker, E.N., Steen, W.A., Miller, J.E. and Shelnutt, J.A. (2007) *Nano Lett.*, **7**, 3650.
124 Song, Y.J., Steen, W.A., Pena, D., Jiang, Y.B., Medforth, C.J., Huo, Q.S., Pincus, J.L., Qiu, Y., Sasaki, D.Y., Miller, J.E. and Shelnuttt, J.A. (2006) *Chem. Mater.*, **18**, 2335.
125 Song, Y.J., Yang, Y., Medforth, C.J., Pereira, E., Singh, A.K., Xu, H.F., Jiang, Y.B., Brinker, C.J., van Swol, F. and Shelnutt, J.A. (2004) *J. Am. Chem. Soc.*, **126**, 635.
126 Krishnaswamy, R., Remita, H., Imperor-Clerc, M., Even, C., Davidson, P. and Pansu, B. (2006) *Chem. Phys.*, **7**, 1510.
127 Song, Y.J., Challa, S.R., Medforth, C.J., Qiu, Y., Watt, R.K., Pena, D., Miller, J.E., van Swol, F. and Shelnutt, J.A. (2004) *Chem. Comm.*, 1044.
128 Surendran, G., Ramos, L., Pansu, B., Prouzet, E., Beaunier, P., Audonnet, F. and Remita, H. (2007) *Chem. Mater.*, **19**, 5045.
129 Liu, S.H., Yu, W.Y., Chen, C.H., Lo, A.Y., Hwang, B.J., Chien, S.-H. and Liu, S.B. (2008) *Chem. Mater.*, **20**, 1622.
130 Zeng, J., Su, F., Han, Y.-F., Tian, Z., Poh, C.K., Liu, Z., Lin, J., Lee, J.Y. and Zhao, X.S. (2008) *J. Phys. Chem. C*, **112**, 15908.
131 Shin, T.-Y., Yoo, S.-H. and Park, S. (2008) *Chem. Mater*, **20**, 5682.
132 Kim, S.J., Ah, C.S. and Jang, D.J. (2007) *Adv. Mater.*, **19**, 1064.
133 Chen, H.M., Liu, R.-S., Asakura, K., Jang, L.-Y. and Lee, J.-F. (2007) *J. Phys. Chem. C*, **111**, 18550.

134 Sun, S. and Xia, Y. (2003) *Nano Lett.*, **3**, 1569.

135 Yang, J., Lee, J.Y., Too, H.-P. and Valiyaveettil, S. (2006) *J. Phys. Chem. B*, **110**, 125.

136 Sun, S., Wiley, B., Li, Z.-Y. and Xia, Y. (2004) *J. Am. Chem. Soc.*, **126**, 9399.

137 Chen, H.M., Liu, R.-S., Asakura, K., Lee, J.-F., Jang, L.-Y. and Hu, S.-F. (2006) *J. Phys. Chem. B*, **110**, 19162.

138 Sun, S. and Xia, Y. (2004) *J. Am. Chem. Soc.*, **126**, 3892.

139 Chen, H.M., Chia, F.H., Liu, R.-S., Asakura, K., Lee, J.-F. and Jang, L.-Y. (2007) *J. Phys. Chem. C*, **111**, 5909.

140 Lee, W.-R., Kim, M.G., Choi, J.-R., Park, J.-I., Ko, S.J., Oh, S.J. and Cheon, J. (2005) *J. Am. Chem. Soc.*, **127**, 16090.

141 Liang, H.-P., Zhang, H.-M., Hu, J.-S., Guo, Y.-G., Wan, L.-J. and Bai, C.-L. (2004) *Angew. Chem. Int. Ed.*, **43**, 1540.

142 Bi, Y. and Lu, G. (2008) *Chem. Mater.*, **20**, 1224.

143 Chen, Z., Waje, M., Li, W. and Yan, Y. (2007) *Angew. Chem. Int. Ed.*, **46**, 4060.

144 Teng, X., Liang, X., Maksimuk, S. and Yang, H. (2006) *Small*, **2**, 249.

145 Chen, H.M., Liu, R.-S., Lo, M.-Y., Chang, S.-C., Tsai, L.-D., Peng, Y.-M. and Lee, J.-F. (2008) *J. Phys. Chem. C*, **112**, 7522.

146 Chen, J., Wiley, B., McLellan, J., Xiong, Y., Li, Z.-Y. and Xia, Y. (2005) *Nano Lett.*, **5**, 2058.

147 Mukerjee, S., Srinivasan, S. and Soriaga, M.P. (1995) *J. Electrochem. Soc.*, **142**, 1409.

148 Lytle, F.W., Wei, P.S.P., Greegor, R.B., Via, G.H. and Sinfelt, J.H. (1979) *J. Chem. Phys.*, **70**, 4849.

149 Ankudinov, A.L., Rehr, J.J., Low, J.J. and Bare, S.R. (2002) *J. Chem. Phys.*, **116**, 1911.

150 Zhang, J., Sasaki, K., Sutter, E. and Adzic, R.R. (2007) *Science*, **315**, 220.

151 Jia, J., Cao, L. and Wang, Z. (2008) *Langmuir*, **24**, 5932.

152 Tian, N., Zhou, Z.-Y., Sun, S.-G., Cui, L., Ren, B. and Tian, Z.-Q. (2006) *Chem. Commun.*, 4090.

153 Subhramannia, M., Ramaiyan, K. and Pillai, V.K. (2008) *Langmuir*, **24**, 3576.

154 Zoval, J.V., Lee, J., Gorer, S. and Penner, R.M. (1998) *J. Phys. Chem. B*, **102**, 1166.

155 Sieben, J.M., Duarte, M.M.E. and Mayer, C.E. (2008) *J. Appl. Electrochem.*, **38**, 483.

156 Loffler, M.-S., Gross, B., Natter, H., Hempelmann, R., Krajewski, Th. and Divisek, J. (2001) *Phys. Chem. Chem. Phys.*, **3**, 333.

157 Chen, X., Li, N., Eckhard, K., Stoica, L., Xia, W., Assmann, J., Muhler, M. and Schuhmann, W. (2007) *Electrochem. Commun.*, **9**, 1348.

158 Ye, F., Li, J., Wang, T., Liu, Y., Wei, H., Li, J. and Wang, X. (2008) *J. Phys. Chem. C*, **112**, 12894.

159 Hu, Y.-S., Guo, Y.-G., Sigle, W., Hore, S., Balaya, P. and Maier, J. (2006) *Nat. Mat.*, **5**, 713.

160 Ding, Y., Gao, Y., Wang, Z.L., Tian, N., Zhou, Z.-Y. and Sun, S.-G. (2007) *App. Phys. Lett.*, **91**, 121901.

3
Electrocatalyst Characterization and Activity Validation – Fundamentals and Methods

Loka Subramanyam Sarma, Fadlilatul Taufany, and Bing-Joe Hwang

3.1
Introduction

Low-temperature solid polymer electrolyte fuel cells such as direct methanol fuel cells (DMFCs) are gaining more and more attention from both scientific and technological perspectives due to their promising applications for fuel cell vehicles, stationary applications, and portable power sources [1, 2]. DMFC simplify the fuel cell system by operating without any external bulky fuel-reforming system and therefore would result in its rapid commercialization. Despite the significant progress, DMFCs still suffer from many obstacles such as low power density, which has been attributed to the poor kinetics of both anode [3], and cathode [4], high flux of water/methanol across membranes [5], and mixed potential at cathode [6]. These phenomena lead to high overpotentials at both the anode and the cathode and, hence, reduction in cell voltage. Other important constraints are the high cost of noble metal catalysts and perfluorosulfonic membranes and higher production costs of the various components of the device.

During the past two decades many significant research efforts have been made worldwide on the development of electrocatalyst materials, proton exchange membranes, and modeling and design of cell components and complete stacks for DMFC [7–9], and the performance achieved for the state-of-the-art membrane and electrode assemblies in DMFCs can compete favorably with hydrogen-fueled and on-board reforming fuel cell systems [10, 11]. Platinum is the best material for the adsorption and dehydrogenation of methanol but, unfortunately, formation of intermediate species such as CO [12], formic acid and formaldehyde poison the platinum anode and impede the catalytic performance for methanol oxidation. The alloying of Pt with promoter metals such as Ru [13–17], Sn [18–22], Mo [23–25], W [26, 27], and Os [28, 29] has been explored as a promising route to minimize the effect of poisonous intermediates. The superior catalytic efficiency of bimetallic Pt-M alloys was explained by the so-called bifunctional mechanism [13, 14, 30–32] and the ligand (electronic) mechanism [17, 33–35]. In the bifunctional mechanism the

Electrocatalysis of Direct Methanol Fuel Cells. Edited by Hansan Liu and Jiujun Zhang
Copyright © 2009 WILEY-VCH Verlag GmbH & Co. KGaA, Weinheim
ISBN: 978-3-527-32377-7

oxophilic promoter metal is thought to provide sites for the water dissociation reaction thereby increasing the methanol oxidation rate through bimolecular surface reaction. On promoter metal like Ru electrochemical dissociation of water is much facilitated at more negative potentials compared with pure Pt sites. In the ligand effects the promoter metal can alter the electronic state of the Pt by contributing d-electron density, and facilitating weakening of the Pt–CO bond and thus easy removal of CO is expected.

It is well known that the overpotential at the cathode in polymer electrolyte fuel cells (PEMFCs) is about 0.2 V due to the slow reaction kinetics of ORR on Pt catalysts. However, in the case of DMFCs, the methanol crossed over from the anode poisons the cathode and results in severe potential loss of about 0.1 V even under open-circuit conditions [36]. The development of new materials with catalytic properties to perform oxygen reduction with significant reduction in cathodic overpotential sis presently a challenging task of great technological importance. Several research works have focused on exploring methanol tolerant catalysts, such as ruthenium-based chalcogenides, metal carbides and oxides, metal phthalocyanines, and porphyrins [37–39]. Several bimetallic systems have also been extensively investigated in order to improve the ORR efficiency. Due to the structural changes caused by the alloying of second metal, some Pt-based bimetallic systems are reported to perform well in ORR [40]. In particular, replacement of some platinum with less expensive transition metals such as Ti, V, Cr, Co, and Ni is reported to yield higher electrocatalytic activities toward the ORR than platinum alone in DMFCs [41–43]. The superior catalytic activity of Pt-based alloy catalysts was attributed to several factors such as shortening of the Pt–Pt interatomic distance caused by alloying, changes in the electron density in the Pt 5d orbitals, the number of Pt nearest neighbors, and the presence of surface oxide layers as investigated from spectroscopic studies [44, 45].

Electrocatalysts advocated for methanol oxidation at the anode and oxygen reduction at the cathode in DMFCs are required to possess well-controlled structure, dispersion, and compositional homogeneity [46–49]. The electrocatalytic activities of both anode and cathode catalysts are generally dependent on numerous factors such as particle size and particle size distribution [50–54], morphology of the catalyst, catalyst composition and in particular its surface composition [55, 56], oxidation state of Pt and second metal, and microstructure of the electrocatalysts [49, 57, 58]. Surface manipulation strategies for nanosized electrocatalysts have been employed to increase their catalytic efficiencies towards MOR and ORR; rigorous characterization techniques which can provide information about nanoscale properties are required. For example, parameters such as particle size and variation in surface composition have strong influence on catalytic efficiency. Further, if the nanoparticles comprise two or more metals, both the composition and the actual distribution will determine the resulting catalytic properties. Thus surface structure, atomic distribution and composition are dominant nanostructure properties that require both control and careful characterization.

This chapter will provide an overview of characterization techniques for elucidating nanoscale characteristics of electrocatalysts and activity validation methodologies. We also highlight our own efforts into the development of XAS methodologies to determine the structure, atomic distribution, alloying extent and surface composition of electrocatalysts. Before discussing these issues in detail we will discuss briefly the role of anode and cathode electrocatalysts and the two principle reactions involved in DMFCs: methanol oxidation reaction (MOR) and oxygen reduction reaction (ORR).

3.2
Direct Methanol Fuel Cells – Role of Electrocatalysts

A DMFC is an electrochemical device that converts the chemical energy of a reaction directly into electrical energy. The principle and schematic diagram of a DMFC is shown in Figure 3.1. In a typical DMFC, methanol and water molecules are simultaneously electro-oxidized at anode to produce CO_2, electrons and protons through the MOR:

$$CH_3OH + H_2O \rightarrow CO_2 + 6H^+ + 6e^- \quad [E_a = 0.02 \text{ V vs. SHE}] \quad (3.1)$$

Protons generated at the anode pass through the proton exchange membrane to cathode and combine with the electrons and the oxidant air or oxygen simultaneously

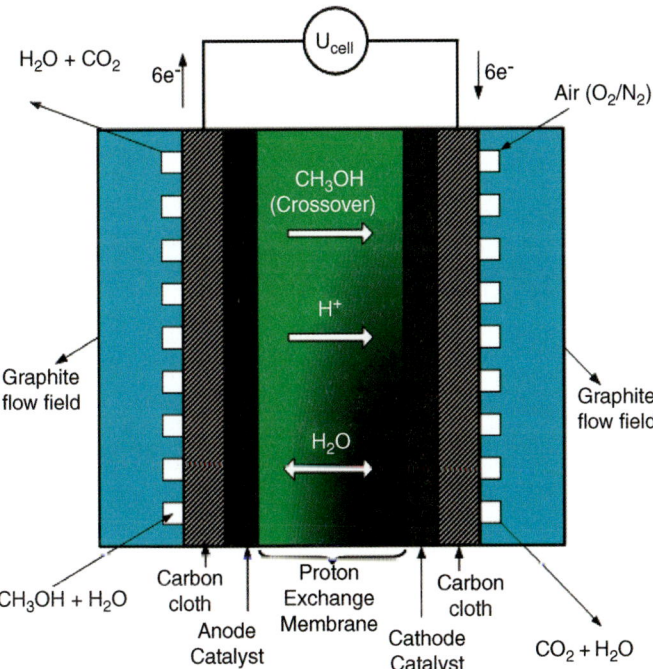

Figure 3.1 Schematic drawing of an operating DMFC single cell depicting H^+, H_2O, and CH_3OH transport through the proton exchange membrane (PEM).

reducing to water as (ORR):

$$3/2O_2 + 6H^+ + 6e^- \rightarrow 3H_2O \ [E_c = 1.23 \text{ V vs. SHE}] \tag{3.2}$$

Reactions (3.1) and (3.2) can be combined to give the overall reaction (Equation 3.3):

$$CH_3OH + 3/2O_2 \rightarrow CO_2 + 2H_2O \ [E_{cell} = 1.21 \text{ V}] \tag{3.3}$$

The free energy (ΔG) of overall reaction at 25 °C and 1 atm is -686 kJ mol^{-1} (for CH_3OH) [59].

3.2.1
Role of Anode Electrocatalysts – Methanol Oxidation Reaction (MOR)

The oxidation of methanol and water molecules is the principle reaction occurring at the anode of DMFCs. Pt is the best material for the adsorption and dehydrogenation of methanol. However, formation of intermediate species such as CO [12], formic acid and formaldehyde poison the platinum anode and impede the catalytic performance for methanol oxidation. Among them poisoning from CO species is severe on a Pt electrode surface during the electro-oxidation of methanol. The detailed methanol oxidation mechanism for the oxidation of methanol at Pt electrodes, involving the adsorption of CH_3OH and its successive dehydrogenation, yielding linearly bonded CO, has been proposed by Beden *et al.* [60] The methanol is adsorbed on the Pt surface (Equation 3.4) and then undergoes a sequence of dehydrogenation steps to yield linearly bonded CO species (Equations 3.5a to 3.5d), as represented below:

$$Pt + CH_3OH \rightarrow Pt(CH_3OH)_{ads} \tag{3.4}$$

$$Pt(CH_3OH)_{ads} \rightarrow Pt(CH_3O)_{ads} + H^+ + e^- \tag{3.5a}$$

$$Pt(CH_3O)_{ads} \rightarrow Pt(CH_2O)_{ads} + H^+ + e^- \tag{3.5b}$$

$$Pt(CH_2O)_{ads} \rightarrow Pt(CHO)_{ads} + H^+ + e^- \tag{3.5c}$$

$$Pt(CHO)_{ads} \rightarrow Pt(CO)_{ads} + H^+ + e^- \tag{3.5d}$$

On the pure Pt surface water dissociation will occur only at higher anodic overpotentials to form $Pt-(OH)_{ads}$ species (Equation 3.6):

$$Pt + H_2O \rightarrow Pt-(OH)_{ads} + H^+ + e^- \tag{3.6}$$

The last step is the reaction of Pt–OH species with Pt-adsorbed CO to give carbon dioxide (Equation 3.7):

$$Pt(CO)_{ads} + Pt-(OH)_{ads} \rightarrow Pt + M + CO_2 + H^+ + e^- \tag{3.7}$$

At potentials of technical interest for DMFCs (<0.6 V vs. RHE) the dissociation of water on Pt surface is the rate determining step [61].

However, water dissociation will occur at less positive potentials on promoter metal 'M' like Ru (0.2 V vs. RHE) and Ru^0 transfers oxygen more effectively than Pt^0 (Equations 3.8 to 3.10): [31, 32]

$$M + H_2O \rightarrow M-(H_2O)_{ads} \quad (3.8)$$

$$M-(H_2O)_{ads} \rightarrow M-(OH)_{ads} + H^+ + e^- \quad (3.9)$$

$$Pt(CO)_{ads} + M-(OH)_{ads} \rightarrow Pt + M + CO_2 + H^+ + e^- \quad (3.10)$$

During methanol oxidation the efficient catalyst must allow a complete oxidation to CO_2. Currently, carbon-supported Pt-Ru catalysts have been shown to be promising among the candidates available for electrochemical oxidation of methanol at anodes of DMFCs [13, 14]. A Pt/Ru atomic ratio of 1 : 1 was reported as the preferred atomic composition [62, 63].

The enhanced electrocatalytic activities enabled by promoter metals such as Ru, Sn, Mo or Os in mixed Pt–M catalysts have been explained by the so-called 'bifunctional' mechanism [56, 64] and the 'ligand' (electronic) mechanism [17, 33–35]. Lu and Masel have noticed that the bifunctional effect has a larger effect on CO removal than the ligand effect when they used 0.25 monolayers of Ru deposited on Pt (110) surface in ultra-high-vacuum (UHV) conditions. They found that out of the 4–6 kcal/mol (170–260 meV) reduction in the potential for CO removal, only about 1 kcal/mol (40 meV) is associated with the ligand effect, whereas 3–5 kcal/mol (130 to 220 meV) is associated with the bifunctional mechanism [65]. With the advent of computational methods more fundamental insights into the CO oxidation mechanism can be obtained. Ishikawa et al. have studied the adsorption of CO on Pt–M metal surfaces (M = Ru, Sn or Ge) with the relativistic density-functional self-consistent field X_α method. From their results it was found that the presence of M atoms weakens the Pt–C bond, and generally slightly lowers the CO stretching frequency of adsorbed CO. Substitution of Ru and Sn for platinum is found to alter the dissociation energy of H_2O. The results indicate that the promoting effect of alloying atoms involves both modification of Pt–CO binding and water activation [66].

3.2.2
Role of Cathode Electrocatalysts – Oxygen Reduction Reaction (ORR)

The electroreduction of O_2 is a multielectron reaction and, depending on the experimental conditions, it is known to take place through two different reaction pathways: the direct four-electron pathway (Equation 3.11) in which O_2 is reduced directly to water and the two-electron pathway (Equation 3.12) in which O_2 is reduced to peroxide followed by the decomposition or further reduction to water [67–69].

$$O_2 + 4H^+ + 4e^- \rightarrow 2H_2O \; (E^0 = 1.23 \text{ V vs. SHE}) \quad (3.11)$$

$$O_2 + 2H^+ + 2e^- \rightarrow H_2O_2 \; (E^0 = 0.67 \text{ V vs. SHE}) \quad (3.12)$$

Platinum has long been the best electrocatalyst for ORR, and oxygen reduction over platinum and other Pt bimetallic surfaces has been reviewed recently [70–72]. As discussed in these reviews, over the three low-index (100), (111), and (110) platinum surfaces in aqueous perchloric and sulfuric acid, the complete four-electron reduction to water can be observed. Two-electron reduction to hydrogen peroxide occurs on the Pt (111) surface in the hydrogen adsorption region which is evidence of surface site blocking by $H_{(ads)}$. Equally OH(ads), which begins forming from water decomposition at about 0.6 V on platinum electrodes, is also believed to inhibit oxygen reduction by blocking surface sites, and it may contribute to the need to operate oxygen electrodes at several millivolts relative to the 1.23 V reversible potential on the standard hydrogen scale.

On the Pt electrode surface the adsorption and reduction of O_2 via a four-electron reduction pathway follows the steps (Equations 3.13a–3.13d):

$$Pt-O_2 + H^+ + e^- \rightarrow Pt-OOH \tag{3.13a}$$

$$Pt-OOH + H^+ + e^- \rightarrow Pt-O + H_2O \tag{3.13b}$$

$$Pt-O + H^+ + e^- \rightarrow Pt-OH \tag{3.13c}$$

$$Pt-OH + H^+ + e^- \rightarrow Pt-OH_2 \tag{3.13d}$$

Sidik and Anderson have applied density functional theory to study the four-electron reduction mechanism of O_2 on platinum in aqueous acidic electrolytes [73]. Their studies suggest that the first electron transfer step (Equation 3.13a) is the r.d.s.. They have calculated the activation energy for the r.d.s. which is of about 0.60 eV at 1.23 V and close to the experimental value of 0.44 eV on Pt (111) in H_2SO_4.

In the two-electron reduction pathway the O_2 reduction follows the steps (Equations 3.14a and 3.14b):

$$Pt-O_2 + H^+ + e^- \rightarrow Pt-OOH \tag{3.14a}$$

$$Pt-OOH + H^+ + e^- \rightarrow Pt-(OHOH)_{ads} \tag{3.14b}$$

The adsorbed peroxide in Equation 3.14b can be electrochemically reduced to water (3.14c) and (3.14d):

$$Pt-(OHOH)_{ads} + H^+ + e^- \rightarrow Pt-(OH) + H_2O \tag{3.14c}$$

$$Pt-(OH) + H^+ + e^- \rightarrow Pt-OH_2 \tag{3.14d}$$

Or it may be catalytically decomposed on the electrode surface, or can be desorbed into the bulk of the solution. Although a number of important problems pertaining to the interpretation of the reaction pathway for the ORR on Pt (hkl) have not yet been resolved, recent studies of Marković et al. suggested that a 'series' pathway via an $(H_2O_2)_{ad}$ intermediate may be operative on Pt and Pt bimetallic surfaces [74–76].

Because of the high cost of platinum and the slow kinetics of the oxygen reduction, many efforts have been directed on finding alternative catalysts [44, 77].

Studies involving partial substitution of Pt with other transition metals like Co, Cr and Ni have indicated that these alloyed materials showed better performance towards ORR [45, 78–80]. Even though significant progress has been achieved in improving the ORR kinetics, mixed potential at the cathode due to the methanol crossover from anode makes the overpotential of this reaction at the desired current densities (e.g., 500 mA cm^{-2}) about 420–450 mV in DMFCs.

3.3
Characterization Techniques for Anode and Cathode Catalysts

3.3.1
Fundamental Aspects

Electrochemical energy conversion for technical applications relies largely on the high catalytic reactivity of electrocatalysts [81]. The optimum catalytic activity is strongly dependent on various factors such as atomic distribution of catalytic metal sites in catalyst matrix [82–85], as well as the surface structure and composition, particle size and particle size distribution of the catalysts [17, 86]. In order to select the proper electrocatalyst material for DMFC applications, characterization methods play an important role in fuel cell research. Ideally, the characteristics of the electrocatalyst material should be used as a selection criterion and they should allow researchers to forecast the corresponding DMFC performance. Several experimental techniques, including microscopy, diffraction, and numerous spectroscopies have been successfully applied to deduce the structural aspects of electrocatalysts. Size, size distribution, morphology, segregation, alloying extent, atomic distribution, degree of alloying, and surface composition are some of the key properties given considerable importance in the characterization of bimetallic alloy or core-shell structured electrocatalysts. The commonly available X-ray diffraction (XRD) technique to understand the alloy structure of bimetallic nanoparticles (bi-MNPs) may not give real structural information. It does not identify the short-range ordering (local environment) and only provides information about the long-range ordering and periodicities preferably on single crystals or polycrystals and hence conclusions about the alloy structure of nanosized particles cannot be simply drawn from XRD. However, if the lattice spacings of the two metals are distinct, XRD can offer some information on the degree of mixing [87, 88]. Electron microscopy techniques like transmission electron microscopy (TEM) [89], microscopic techniques like TEM [90] and high resolution electron microscopy (HRTEM) [91], will provide only qualitative understanding of the structure of electrocatalysts. For example the HRTEM method enables one to distinguish core/shell structure on the basis of observing different lattice spacing and crystal orientation. However, if the core and shell components comprise similar crystal orientation and lattice spacing then it would be very difficult to identify the core-shell structured nanoparticle with HRTEM. TEM offers better understanding of the core-shell structure if the core and shell region comprise metal atoms with a high contrast. Scanning TEM (STEM) which relies on the scanning of electron beam across the sample combined with high angle annular dark field (HAADF) or Z-contrast imaging technique (as the HAADF image contrast is proportional to Z^{α}, where Z is the atomic number and α is in the range 1.5–2) offers possibilities to identify the internal structure of the nanoparticle [92]. Surface alloying and segregation phenomena of electrocatalysts can be reasonably obtained from the surface sensitive techniques such as X-ray photoelectron spectroscopy (XPS), Auger electron spectroscopy (AES) [93], and infrared spectroscopy (IR).

Another parameter of great importance in assessing the suitability of core-shell or alloy structured electrocatalysts towards DMFC applications is the surface composition. Surface sensitive techniques such as electron spectroscopy for chemical analysis (ESCA), secondary ion mass spectrometry (SIMS), ion scattering spectroscopy (ISS), XPS, and AES have enabled investigators to obtain reasonably good estimates of surface compositions of bulk or thin film alloys [94, 95]. However, these techniques have severe limitations when analyzing the surface composition of bi-MNPs. For example, due to the escape depth of the Auger electrons and photoelectrons, the application of AES and XPS to particles with diameters of about or less than 30 Å is restricted [96, 97]. Although SIMS offers depth information in the range of 10 to 20 Å it does not provide the spatial resolution of electron beam techniques such as scanning electron microscopy (SEM), TEM, and AES.

One promising method for probing the internal structures of core-shell and alloy type bi-MNPs and species adsorbed on them is X-ray absorption spectroscopy (XAS), which consists of both X-ray absorption near-edge structure (XANES) and extended X-ray absorption fine structure (EXAFS) regions. XAS studies in the XANES region (conventionally from below the edge up to $\sim$30–50 eV) can reveal the oxidation state and d-band occupancy of a specific atom in nanoparticles as well as the size of nanoparticles [98, 99]. EXAFS (above $\sim$30–50 eV) provides a powerful tool for the analysis of local atomic structure, giving accurate information about the average local atomic environment [100, 101]. EXAFS is particularly sensitive to interatomic distances and local disorder and has been successfully used to resolve subtle nanoparticle structural details [102]. XAS has been proved as a powerful technique for the characterization of bimetallic electrocatalysts [103–105].

A large number of experimental techniques have been explored for the atomic characterization of electrocatalyst nanostructures and significant advances are made on understanding their unique structures and properties. The capabilities of diffraction, microscopy and several spectroscopic techniques in realizing the physical properties and atomistic structures of electrocatalysts are reviewed in the following section.

3.3.2
Evaluation of Catalyst Crystallite Size, Elemental Composition, Morphology, Dispersion

3.3.2.1 X-ray Diffraction (XRD)
XRD has been particularly widely employed to study the supported and unsupported electrocatalyst nanoparticles to realize information on structure, crystallinity, lattice spacing and grain size. In practice the average particle size of PtRu based

electrocatalysts was calculated from the broadening of (220) peak using the Scherrer equation (Equation 3.15) [106].

$$d = 0.94 \lambda_{k\alpha 1}/B_{(2\theta)} \cos\theta_B \qquad (3.15)$$

where d is the average particle diameter, $\lambda_{k\alpha 1}$ is the wavelength of X-ray radiation, θ_B is the angle of the (220) peak, and $B_{(2\theta)}$ is the width in radians of the diffraction peak at half-height. The determination of average particle diameter using (220) peak broadening in $2\theta \sim 67°$ is particularly reliable for carbon-supported PtRu electrocatalysts since in this region there are no reflection signals associated with the carbon support. By careful XRD measurements several authors have followed the changes in the lattice constant caused by alloying in electrocatalysts in order to obtain information about the alloying degree [107–110]. Antolini and co-workers have proposed an equation for determining the alloying degree of a PtRu catalyst which is defined as the Ru atomic fraction (x_{Ru}) through Equation 3.16 [111, 112].

$$a = a_0 - 0.124\, x_{Ru} \qquad (3.16)$$

where a_0 is the lattice constant of pure Pt. For unsupported pure Pt, a_0 has the value of 0.39231 nm whereas for supported pure Pt, $a_0 = 0.39155$ nm reported for Pt/C catalyst of E-Tek [111]. Radmilović and co-workers also proposed similar type of relationship for single-phase PtRu bulk alloys: $a = 0.39262 - 0.124\, x_{Ru}$ (or $a = 0.38013 + 0.1249\, x_{Pt}$). Antolini and Cardellini used the peak height ratio of the Pt [111] crystal face and the C[0015] reflection of the carbon in order to evaluate the thermal crystallization considered as a crystallinity degree index of PtRu/C nanoparticles [112]. In some cases XRD has been also used to calculate the surface area of the catalyst, if the shape of the catalyst particles is spherical, by using Equation 3.17:

$$S = 6000/\rho\, d \qquad (3.17)$$

where S is the surface area ($m^2\, g^{-1}$), d is the average particle size (nm), and ρ is the Pt density (21.4 g cm^{-3}). Information on chemical composition can be obtained first by constructing a calibration graph of lattice parameter vs. atomic fraction and later by quantifying a composition based on the measured lattice parameter. Gasteiger et al. carried out detailed studies on the correlation of the lattice parameter with the alloy composition in the binary Pt-Ru system [17], and found a linear dependence, according to the Vegard's law. These studies allowed the authors to establish the relationship between the electrode composition and electrocatalytic activity. However XRD is a bulk method, and reveals information on the bulk structure of the catalyst and its support and its application to the interpretation of nanosized particles is rather difficult. By using simulation calculations Debye function analysis (DFA) offers a convenient approach to determine the size distribution and structure of small clusters [113]. In contrast to the analysis restricted to the limited regions such as the integral intensities of single Bragg peaks, DFA provides detailed structural information in a range of dispersion (approx. > 40%). In a typical DFA analysis, the measured diffraction curves are fitted by a set of Debye functions for clusters with the 'magic' numbers $N = 13, 55, 147, \ldots$ (for cuboctahedra and icosahedra) and N = 54, 181 (for decahedra) and N = 13, 57, 154 (for hexagonal close-packed clusters) [114]. A histogram plotting the mass fraction of specific structural units present within the sample against their average size can then be constructed through the evaluation of the full pattern. By carrying out the systematic numerical simulations using Debye functions, Vogel et al. obtained the intrinsic structure including the average lattice constant and the size distribution of surfactant-stabilized Pt-Ru catalysts and silica-supported Pt-Ru colloids in the as-synthesized state and after several heat treatments under various atmospheres [114].

3.3.2.2 Transmission Electron Microscopy (TEM)

TEM in which the electrons pass through the sample, generally requires electrocatalyst particles to be dispersed onto an electron-transparent substrate such as a thin copper-coated microgrid. TEM is particularly useful because of the high contrast between the metal atoms (especially heavy metals) and gives the information about the size, size distribution, dispersion and even the morphology of various shapes of particles. In general, for fuel cell catalysts microscopic investigations are combined with other spectroscopic and diffraction techniques to obtain a comprehensive idea of the catalysts real structure. Radmilovic et al. reported detailed study of carbon-supported nanoparticles by TEM and XRD [106]. The focus of their work was on the characterization of a commercially-available carbon-supported Pt-Ru (1 : 1) catalyst in terms of both, particle size and completeness of alloy formation. In the authors' opinion, many of the difficulties of XRD can be addressed by TEM. Especially the lattice structure can be studied by HRTEM, including the presence of defects as dislocations, twins, and so on. HRTEM presents an interesting tool in catalyst characterization, as it can be used to determine the geometric shape of faceting planes, the presence of surface steps, the surface roughness, as well as the size and distribution of electrocatalyst nanoparticles. HRTEM offers resolution down to the Ångstrom level and enables information to be obtained on the structure (atomic packing) rather than just morphology of the nanoparticles. Zhang and Chan presented TEM images along with the selected area electron diffraction patterns of Pt-Ru nanoparticles synthesized by a two-microemulsion route in which the metal precursors and reducing agent formed two individual microemulsion systems [115]. Based on the presence of only diffractions from the face centered cubic (fcc) in the electron diffraction pattern of Pt-Ru nanoparticles, the authors have concluded the formation of binary Pt-Ru alloy with the fcc structure. Further evidence of the formation of fcc structured Pt-Ru binary alloy was given through the established linear relationship of the root of the sum of squares of the lattice coordinates versus the radius of the concentric rings with the lattices (111), (200), (220), (311), and (222). The authors have indicated that the calculated lattice cell constant through such a relationship is 3.862 Å which is in between those of platinum and ruthenium and is in agreement with that of a 1 : 1 PtRu alloy [116]. Once the particle size distributions are obtained through TEM images, the mean particle size d_m can be calculated with Equation 3.18 [117]:

$$d_m = \Sigma n_i d_i / \Sigma n_i \tag{3.18}$$

where n_i is the number of particles with diameter d_i. It is also possible to estimate the dispersion (ratio of surface atoms to total number of atoms) of the spherically shaped PtRu clusters through the information of cluster composition and particle size distribution [118].

3.3.2.3 Scanning Electron Microscopy (SEM)

In SEM, the surface of the sample is scanned in a raster pattern with a beam of energetic electrons. The SEM image is produced due to secondary electrons emitted by the sample surface following excitation by the primary electron beam [119, 120]. Bi and Lu used SEM to follow the growth process and morphological control of platinum nanostructure, nanofiber, and nanotube junction structures [121]. These Pt nanostructures with various anisotropies were obtained by the galvanic replacement reaction between Ag nanowires and platinum salt solution in the presence of CTAB solution. From SEM observations, the authors have found that the platinum nanostructures growth follows three steps: at first platinum nanoparticles will grow on the surface of Ag nanowire; then Ag/Pt composite nanowires will be formed; finally the Pt nanofibers and nanotubes will grow. From the field-emission SEM images the authors are able to determine the length, inner and outer diameter of platinum nanotubes. Besides this, the information of Pt nanotubes uniformity was conveniently obtained from SEM images. Kawaguchi et al. studied the process of particle growth of Pt, Ru, and binary PtRu supported on carbon as a function of pyrolysis time [122]. The catalyst nanoparticles were prepared by an impregnation-reductive pyrolysis method at various temperatures. The authors have discussed the particle growth behavior from the high-resolution SEM images. Although SEM images have lower resolution than TEM, SEM offers better three dimensional images of the electrocatalysts [123, 124].

3.3.2.4 Atomic Force Microscopy (AFM)

Atomic force microscopy (AFM) is a non-destructive method for investigating the microscopic surface topography of nanostructures. In this method, a probe scans the surface of a material with a sharp tip in order to clearly image the features of a sample and senses the small (approximately 1 nN) repulsive force between the probe tip and the surface. Rodríguez-Nieto et al. used AFM in order to obtain morphological and microscopic surface characterization of Pt–Ru electrodeposits produced on activated highly ordered pyrolytic graphite (HOPG) substrates [125]. The authors are able to deduce the surface roughness of Pt–Ru electrodeposits at the nm scale from AFM images. Schmidt et al. employed UHV-AFM to determine the particle size distribution and corresponding dispersion of PtRu nanoparticles from height measurements of the imaged PtRu nanoparticles [126].

3.3.2.5 Energy Dispersive X-ray Spectroscopy (EDS/EDX)

Energy Dispersive X-ray Spectroscopy (EDS) (also called energy-dispersive X-ray analysis (EDXA)) is an analytical technique that uses a scanning electron microscope for chemical microanalysis [127]. As the electron beam (typically 10–20 keV) displaces electrons in the sample, detection equipment converts the electrons scattered by the electron beam into a microscopic image. Since each element produces characteristic X-ray energies, information on the elemental composition of individual nanoparticles can be conveniently determined. Also, the amount of each element can be determined from the relative counts of the detected X-rays. EDS has been widely employed to determine the elemental composition and amount of Pt and Ru in PtRu nanoparticles [115, 128, 129].

3.3.3
Elucidation of Nanostructural Characteristics of Electrocatalysts – Alloying Extent or Atomic Distribution, Electronic Structural Features, State of Catalyst Species, Surface Composition, Segregation Phenomena

3.3.3.1 X-ray Absorption Spectroscopy

In order to realize the actual structure of electrocatalysts of either alloy or core-shell structure, especially when those bimetallic systems are in sub-nm to 2–3 nm size range, a combination of characterization techniques is required. In general, the X-ray absorption spectrum of a sample can be divided into two regions: the near edge region (XANES, 0–50 eV above the absorption edge) and the oscillatory part of the spectrum (, EXAFS, >50 eV above the absorption edge). The capability of tuning the X-ray energy to the absorbing edge of each participating metal in bimetallic systems makes EXAFS as an attractive technique to elucidate the local structure and can provide information on the environment about a particular atom. By analyzing the EXAFS spectrum of each metal in bimetallic nanomaterials concurrently, valuable structural and chemical information (e.g., interatomic distance, coordination number, oxidation state of chemical species) about the nanostructures can be conveniently obtained and this information can supplement the microscopy data. By XANES measurements information on oxidation states, valence band vacancies and adsorption geometries of molecules at the surface can be obtained. However, as the evaluation of the spectra is quite complex, due to multiple scattering processes, EXAFS analysis is generally preferred. Several researchers have successfully used EXAFS to study the bonding habits, geometry, and surface structures of many electrocatalyst nanoparticles (e.g., Pt/Ru, Pt/Co, Pt/Mo and etc.), from which the shape, size, and the short range order in atomic distributions occurring within the particles were reliably obtained [130–135]. Recently, Russell and Rose thoroughly reviewed the capabilities of XAS towards analyzing the structural aspects of low temperature fuel cell catalysts [136].

By collecting the XAS data at the absorption edges corresponding to each element in the bi-MNPs under investigation, the extent of intermixing (alloying extent), homogeneity (atomic distribution) of the bi-MNPs may be assessed [82]. In general one can assess the alloyed or core-shell structure of nanomaterials simply from the coordination numbers of participating elements in core-shell materials. For a homogeneous bimetallic system of A_{core}–B_{shell} cluster in which the core of the cluster is composed of N atoms of A (N_A) and the surface is made of N atoms of B (N_B), the total coordination number ($N_{AA} + N_{AB}$) for the 'A' atom will be greater than the total coordination for the 'B' atoms ($N_{BA} + N_{BB}$) [105, 137]. If bi-MNPs possess random alloyed structure, the ratios of coordination number of A and B coordination, N_{A-A}/N_{A-B} and N_{B-A}/N_{B-B}, should be consistent with the ratio of atomic fraction x_A/x_B.

The alloy or core-shell type structure of metallic nanostructues from XAS measurements can be better understood if we can obtain knowledge about the atomic distribution and alloying extent of the participating elements. This is particularly important since among the various structural aspects it is of most important to control the homogeneity, dispersion, and alloying extent as they have profound influence on the surface properties which affect the biocompatibility and stability of the bi-MNPs. Hence methods to get more insight into structural aspects are needed. Even though alloying is a well-known phenomenon, detailed studies on quantitative assessment of alloying extent in bi-MNPs have been lacking so far. In our research group, by deriving the structural parameters from XAS analysis a general methodology to estimate the alloying extent or atomic distribution in bi-MNPs was demonstrated [82].

By estimating the ratio of the coordination number (CN) of A around B and also the CN of B around A to the total CNs, one can conveniently estimate the alloying extent of A (J_A) and B (J_B) in A–B bi-MNPs. The parameters that are needed to derive the extent of alloying are represented as $P_{observed}$, $R_{observed}$, P_{random}, and R_{random}. The parameter $P_{observed}$ can be defined as a ratio of the scattering atoms 'B' CN around absorbing 'A' atoms (N_{A-B}) to the total CN of absorbing atoms (ΣN_{A-i}), ($P_{observed} = N_{A-B}/\Sigma N_{A-i}$). Similarly, $R_{observed}$ can be defined as a ratio of the scattering atoms 'A' CN around absorbing 'B' atoms (N_{B-A}) to the total CNs of absorbing atoms (ΣN_{B-i}), ($R_{observed} = N_{B-A}/\Sigma N_{B-i}$). Whereas, P_{random} and R_{random} can be taken as 0.5 for perfect alloyed bi-MNPs if the atomic ratio of 'A' and 'B' is 1 : 1. The J_A and J_B for 1 : 1 A − B bi-MNPss can then be estimated by using Equations 3.19 and 3.20 respectively.

$$J_A = (P_{observed}/P_{random}) \times 100\% \quad (3.19)$$

$$J_B = (R_{observed}/R_{random}) \times 100\% \quad (3.20)$$

Based on the ΣN_{A-i}, ΣN_{B-i}, J_A and J_B it is possible to predict the structural models of nanoparticles. For example if $\Sigma N_{A-i} > \Sigma N_{B-i}$ there appears core is rich in 'A' atoms and shell is rich in 'B' atoms. In this case if both $J_A < J_B$ then bi-MNPs structure is almost close to pure $A_{core} - B_{shell}$ (case 1, Figure 3.2). However if $J_A > J_B$ with a coordination parameter relationship $\Sigma N_{B-i} > \Sigma N_{A-i}$ then the bi-MNPs possess 'B' rich in core–'A' rich in shell structure (case 2, Figure 3.2). If $\Sigma N_{A-i} = \Sigma N_{B-i}$ and J_A & $J_B \approx 100\%$ then bi-MNPs adopt an alloy structure (case 3, Figure 3.2).

It is possible to construct the structural models emphasizing the atomic distribution in the bi-MNPs with the knowledge of the ΣN_{A-i}, ΣN_{B-i}, J_A and J_B values derived from XAS. With the help of extent of alloying values and structural parameters extracted from EXAFS it is possible to predict the structure models of Pt−Ru/C catalysts. We have calculated the alloying extent of Pt (J_{Pt}) and Ru (J_{Ru}) for commercial 30 wt.% Pt−Ru/C catalysts.

In the case of JM 30 catalyst the coordination numbers of Pt and Ru atoms around the Pt atom are found to be 5.6 ± 0.3 and 1.4 ± 0.1 respectively and the total

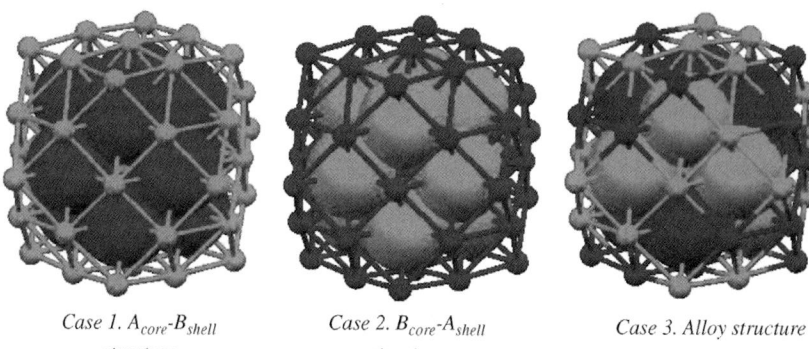

Case 1. A_{core}-B_{shell} structure   Case 2. B_{core}-A_{shell} structure   Case 3. Alloy structure

Figure 3.2 Structural models of bimetallic nanoparticles with core-shell and alloy structures deduced from XAS structural parameters. Key: Blue, A; Pink, B.

coordination number ΣN_{Pt-i} is 7.0. The coordination numbers of Ru and Pt atoms around the Ru atom are determined as 3.4 ± 0.2 and 2.2 ± 0.3 respectively and the total coordination number ΣN_{Ru-i} calculated as 5.6. From these values $P_{observed}$ and $R_{observed}$ determined as 0.20 and 0.39 respectively and J_{Pt} and J_{Ru} values are calculated as 40 and 78% respectively. For E-Tek 30 catalyst we have calculated the coordination numbers of Pt and Ru atoms around the Pt atom as 6.2 ± 0.3 and 0.9 ± 0.1 respectively and ΣN_{Pt-i} as 7.1; the coordination numbers of Ru and Pt atoms around the Ru atom are determined as 3.7 ± 0.2 and 1.2 ± 0.2 respectively and the ΣN_{Ru-i} as 4.9. The other two structural parameters $P_{observed}$ and $R_{observed}$ in the case of E-Tek 30 calculated as 0.13 and 0.24 respectively and the J_{Pt} and J_{Ru} values are calculated as 26 and 48% respectively. It is clear from the structural coordination parameter values of both the catalysts that $\Sigma N_{Pt-i} > \Sigma N_{Ru-i}$ and $J_{Ru} > J_{Pt}$ and indicates that the catalysts adopt a Pt rich in core and Ru rich in shell structure.

From the quantitative extent of alloying values we can see that in both the catalysts considerable amount of Ru is segregated on the shell layer but the extent of segregation of Ru is higher in E-Tek 30 when compared to the JM 30. The increased value of J_{Ru} in JM 30 catalyst indicated that most of the Ru is involved in alloying and hence less segregation of Ru in the shell whereas in the case of E-Tek 30 catalyst less extent of Ru is involved in the alloying and considerable extent of segregation of Ru can be expected in the shell region. The segregation of Ru in the case of E-Tek 30 in part may be responsible for its lower methanol oxidation activity compared to JM 30. Recent IR measurements on the Pt-Ru alloy particle electrodes indicates two modes of adsorbed CO vibrations related to both Pt and Ru domains present on the surface supports the surface segregation of Ru in commercial catalysts [138]. The XAS results support the Pt-rich core and Ru rich shell structure for commercial carbon-supported Pt-Ru catalysts. Increase in J_{Pt} and J_{Ru} values in JM 30 compared to E-Tek 30 indicates that the atomic distribution of Pt and Ru atoms are much facilitated. Increase in atomic distribution can be taken as a measure for enhanced homogeneity.

Lin et al. investigated the commercial and in-house prepared Pt-Ru catalysts by ex-situ EXAFS [139]. From the observed EXAFS parameters, the authors have proposed that the two catalysts differed in the degree of Pt-Ru alloying. The EXAFS data of the in-house prepared catalyst indicated the signatures of Pt–Ru bonds, whereas no significant contributions were found in the commercial catalyst. These observations led the authors to conclude that the catalyst synthesized in-house was at least partially alloyed, while the commercial catalyst system seemed to contain mixed phases of Pt and RuO_x. Greegor and Lytle demonstrated the feasibility of EXAFS technique for measuring size and shape of small metal particles [140]. This methodology relies on developing a two-region model for various geometrical shapes like spheres, cubes, and disks and calculating the EXAFS average coordination number for first, second, and third coordination spheres as a function of cluster size. Nuzzo and co-workers have also elaborated the modeling nanoparticle size and shape with EXAFS [102]. They considered two model particles characterized by a common average first-shell coordination number (a value of ~8 for a 92 atom hemispherical and a 55 atom spherical cubooctahedral cluster). The authors have emphasized that the geometry of these two clusters is significantly different with different sizes, shapes or lattice symmetries so each cluster can generate a unique sequence of average coordination numbers in the first few nearest-neighbor shells. Once such a sequence is obtained experimentally, then the corresponding cluster size, shape, and symmetry may be conveniently determined. Several authors have studied the effect of particle size on the XANES region of the XAS spectra for Pt/C catalysts [102, 141, 142]. In their potential-dependent XANES studies on Pt/C catalyst particles with a 3.7 and ≤1.0 nm diameter, Yoshitake et al. have observed that the white line intensity was increased for both particle sizes as the potential was increased [141]. In general the white line at the Pt L_{III}-edge is an absorption threshold resonance, attributed to electronic transitions from $2p_{3/2}$ to unoccupied states above the Fermi level and is sensitive to changes in electron occupancy in the valence orbitals of the absorber [143]. Hence, changes in the white line intensity have been directly related to the density of unoccupied d states and indicate the changes in the oxidation state of the Pt absorber. In general, if the white line intensity decreases, the density of unoccupied d states and the oxidation state of Pt are both lower. The lower white line intensity observed at negative potentials thus corresponds to a more metallic state. Mansour and co-workers [144] have proposed that by comparing the white line intensities of Pt L_3 and Pt L_2 edges of a sample with those of a reference metal foil one can determine the fractional d-electron occupancy (f_d) of the absorber atoms in the sample by Equation 3.21:

$$f_d = (\Delta A_3 + 1.11\, \Delta A_2)/(A_{3,r} + 1.11\, A_{2,r}) \tag{3.21}$$

where $A_{3,r}$ and $A_{2,r}$ represents the areas under the white line at the L_{III} edge and L_{II} edge, respectively of the reference foil spectrum (Equation 3.22).

$$\Delta A_x = A_{x,s} - A_{x,r} \tag{3.22}$$

with $x = 2$ or 3 and $A_{x,s}$, the area under the white line at the L_x edge of the sample spectrum. f_d can then be used to calculate the total number of unoccupied d-states per Pt atom in the samples (Equation 3.23):

$$(h_J)_{t,s} = (1.0 + f_d)\,(h_J)_{t,r} \tag{3.23}$$

where $(h_j)_{t,T}$, t = total for Pt has been shown to be 0.3 [145]. A large $(h_j)_{t,s}$ value thus indicates a smaller d-electron density and an increased d-band vacancy compared with those for bulk Pt.

Further, Mukerjee et al. [40] and Min and co-workers [146] studied detailed particle size effects in several binary anode and cathode electrocatalysts. Mukerjee et al. calculated the values for Pt/C particles with four different diameters at potentials corresponding to the hydrogen adsorption (0.0 V vs. RHE), the double layer (0.54 vs. RHE), and the oxide formation (0.84 V vs. RHE) regions. With the decreasing particle size authors have observed an increased widening of the white line. The authors observed an increase in Pt L_{III} white line intensity at 0.84 V vs. RHE due to the adsorption of OH species at higher potentials, whereas the broadening of the white line at 0 V vs. RHE is related to adsorbed hydrogen. It has been shown that with increasing particle size the d band vacancy decreases, indicating that the electronic effects due to adsorption of H and OH are more pronounced for smaller particles. The authors have proposed that as the adsorption strength of H, OH and CO is increased with decreasing particle size, below a certain size resulted in reduced methanol oxidation activities. It has been proposed that the intrinsic activity of Pt-based electrocatalysts for ORR in acidic solutions depends on both the shape, size of the particles, and the adsorption strength of oxygen intermediates [147, 148]. Min and coworkers carried out detailed investigations on the particle size and alloying effects in Pt-based Pt-Co, Pt-Ni and Pt-Cr catalysts [146]. From the understanding of the XANES region of the spectra, the authors observed a decrease of the d band vacancy with increasing particle size which is in agreement with Mukerjee et al. This observation suggests lowered adsorption strength of adsorbed oxygen species, thus facilitating the ORR reaction at larger particles.

Nashner and co-workers reported XAS characterization of carbon-supported Pt-Ru nanoparticles with exceptionally narrow size and compositional distributions synthesized from the molecular cluster precursor $PtRu_5C(CO)_{16}$ [134]. The authors have deduced structural variations in the Pt-Ru nanoparticles exposed to different gaseous atmospheres such as hydrogen and oxygen on the basis of ex-situ EXAFS measurements in combination with transmission electron microscopy. In the case of Pt-Ru nanoparticles exposed to H_2 atmosphere, the authors found that the ratio of Ru–Pt bonds to Ru–M (N_{Ru-Pt}/N_{Ru-M}) as well as Pt–Ru bonds to Pt–M (N_{Pt-Ru}/N_{Pt-M}) obtained from experimental EXAFS data is always lower than the statistically predicted ratios indicating stronger weighting of the homometallic coordination in nanoparticles. The authors proposed that Pt shows a pronounced preference for segregation to the particle surfaces based on the fact that $N_{Pt-Ru}/N_{Pt-M} > N_{Ru-Pt}/N_{Ru-M}$. Upon chemisorption of oxygen, the authors found increase in the disorder in the first shell metal bond lengths accompanied by the average bonding of two oxygens to both Pt and Ru with bond distances similar to those found in structures with binding oxygen atoms. In another interesting study, Nuzzo and co-workers used XAS to follow core-shell inversion in Pt-Ru nanoparticles during hydrogen treatment at various temperatures [149]. Based on XAS structural parameters, the authors found that the incipient Pt-Ru nanoparticles initially formed a disordered structure at 473 K in which Pt is found preferentially at the core of condensing particle.

After exposure to high temperature treatment to 673 K, the nanoparticle undergoes a core-shell inversion takes place leading to preferential migration of Pt to the equilibrated bi-MNP.

Very recently, by using XAS we deduced that the heat-induced changes in the surface population of Pt and Ru in PtRu/C catalyst nanoparticles and the corresponding changes were correlated to the electrocatalytic activity [150]. In this study, the thermal-treatment procedure was designed in such a way that the particle size of initial nanoparticles was not altered upon thermal-treatment but can change only the surface population of Pt and Ru allowing us to deduce the structural information independent of particle size effect. We used XAS to deduce the structural parameters that can provide information on atomic distribution (or) alloying extent as well as surface population of Pt and Ru in PtRu/C nanoparticles. The PtRu/C catalyst sample obtained from Johnson Matthey was subjected to the heat treatment in two environments. At first the as-received catalyst was reduced in 2% H_2 and 98% Ar gas mixture at 300 °C for 4 h (PtRu/C as-reduced). Later this sample was subjected to either oxygen (PtRu/C-O_2-300) or hydrogen thermal treatment (PtRu/C-H_2-350). XAS results reveal that when the as-reduced PtRu/C catalyst was exposed to the O_2 thermal-treatment strategy, considerable amount of Ru was moved to the catalyst surface. In contrast, H_2 thermal-treatment strategy led to the higher population of Pt on the PtRu/C surface. Characterization of the heat-treated PtRu/C samples by XRD and TEM reveal that there is no significant change in the particle size of thermal-treated samples when compared to the as-received PtRu/C sample. Both XAS and electrochemical CO_{ads} stripping voltammetry results suggested that the PtRu/C-H_2-350 sample exhibit significant enhancement in reactivity toward CO-oxidation as a result of the increased surface population of the Pt when compared with the PtRu/C-O_2-300 and PtRu/C as-reduced samples.

Surface and core composition in bimetallic nanoparticles – an XAS methodology
Recently, we developed a general methodology based on EXAFS techniques to quantitatively determine the surface and core composition of bi-MNPs by combining modeling and experimental approaches [151]. The advantage of this method is that it can be used to extract the surface and core composition of various bi-MNPs in reaction conditions, even in the liquid phase, and can be also extended to ternary nanoparticles. The key to our methodology is to use geometric arguments to relate inaccessible surface and core compositions to quantities that can be measured robustly by XAS. By knowing only the basic crystal structure (fcc, bcc, etc.) and overall nanoparticle shape (cubooctahedron, icosahedron, truncated octahedron etc.) it is possible to write the equations for surface and core composition of bi-MNPs in terms of measurable quantities. A system of AB bi-MNPs with an fcc structure is employed to demonstrate the feasibility of the developed methodology. Based on the cluster model, the dependence of the total number of atoms (n^t), the number of surface atoms (n^s), and the total average surface coordination number ($\underline{N}^s$) on the total average coordination number ($\underline{N}$) for the fcc cluster of icosahedron, cubooctahedron, and truncated octahedron shapes can be evaluated. The equations for surface and core composition of a system of AB bi-MNPs with an fcc structure can be given by

Equations 3.24 to 3.27:

$$X_A^s = \frac{x(\Sigma N_{A-i}-12)}{x(\Sigma N_{A-i}-12)+(1-x)(\Sigma N_{B-i}-12)} \quad (3.24)$$

$$X_B^s = \frac{(1-x)(\Sigma N_{B-i}-12)}{(1-x)(\Sigma N_{B-i}-12)+x(\Sigma N_{A-i}-12)} \quad (3.25)$$

$$X_A^c = \frac{x(\underline{N}^s-\Sigma N_{A-i})}{x(\underline{N}^s-\Sigma N_{A-i})+(1-x)(\underline{N}^s-\Sigma N_{B-i})} \quad (3.26)$$

$$X_B^c = \frac{(1-x)(\underline{N}^s-\Sigma N_{B-i})}{(1-x)(\underline{N}^s-\Sigma N_{B-i})+x(\underline{N}^s-\Sigma N_{A-i})} \quad (3.27)$$

where ΣN_{A-i} and ΣN_{B-i} represent the average coordination numbers of A and B. We can determine the total average coordination numbers of A and B, ΣN_{A-i}, and ΣN_{B-i}, from EXAFS measurements and the bulk composition (x) from the edge jump in XANES measurements. Once ΣN_{A-i} and ΣN_{B-i} are determined, the surface composition of the nanoparticles can be obtained by Equations 3.24 and 3.25. Meanwhile, the average coordination number of the cluster $\underline{N}$ can be calculated by Equation 3.28:

$$\underline{N} = x\Sigma N_{A-i} + (1-x)\Sigma N_{B-i} \quad (3.28)$$

The core composition of A and B atoms in the fcc nanoparticles with various shapes can be estimated by Equations 3.26 and 3.27, and $\underline{N}^S$ needs to be obtained from the experimental total average coordination number ($\underline{N}$) associated with the fcc model calculation considering various shapes. In general, the surface composition depends only on the structure of the nanoparticles, while core composition relies not only on the structure, but also on the shape of the nanoparticles.

3.3.3.2 X-ray Photoelectron Spectroscopy (XPS)

XPS analyses are a common supplement in the characterization of fuel cell electrocatalysts. XPS works based on the photoelectric effect. When the energy of X-ray radiation is sufficient to overcome the binding energy of the electron of the analyte atom, molecule, or solid/surface, then the electrons will be ejected. In general, both valence and core electrons can be ejected by X-ray radiation. The composition of material can be determined by using the peak areas under the binding energy curves of the core electron which is characteristic of each element present in the sample. Further, information on chemical bonding can also be conveniently obtained through XPS since the peak shape and binding energy are sensitive to the oxidation and chemical state of the emitting atom [152, 153]. XPS is also particularly useful to realize the particle size effects in fuel cell catalysts. In their studies Kao et al. observed a 0.3 eV increase of the Pt 4f binding energy when compared to bulk Pt systems [154]. Eberhard et al. also found a continuous decrease of the Pt 4f binding energy with decreasing particle size [155]. Zhang and Chan present XPS analyses of PtRu nanoparticles prepared in water-in-oil reverse microemulsion [115]. The Pt $4f_{7/2}$

and Pt $4f_{5/2}$ lines appeared at 71.30 eV and 74.57 eV, respectively were attributed to metallic Pt^0. The peaks appeared at 72.49 eV and 75.88 eV were assigned to Pt^{II} in PtO and $Pt(OH)_2$, respectively. Based on the relative height of the peaks the authors suggested that metallic Pt^0 is the predominant species in the nanoparticles. The authors observed three components with binding energies of 461.32, 463.41, and 465.72 eV in the corresponding Ru $3p_{3/2}$ spectrum corresponding to the Ru^0 metal, Ru^{IV} (e.g., RuO_2), and Ru^{VI} (in RuO_3), respectively. From these results, the authors concluded that the surface of nanoparticles contains metal and Ru oxides species. Although XPS is suitable for realizing the chemical state and bonding in electrocatalysts its application towards the determination of surface composition is limited. For particle sizes <3 nm not less than half of the atoms in the cluster belong to the surface. Hence, the surface specific XPS with the escaping depth of an electron of about ($\lambda \sim 3$ nm) becomes a bulk method for small particles [156, 157].

3.3.3.3 Electrochemical-Nuclear Magnetic Resonance Spectroscopy (EC-NMR)

Electrochemical nuclear magnetic resonance (EC-NMR) which combines both solid state NMR and electrochemistry has emerged as a powerful technique to elucidate the electronic properties of metal surfaces [158]. In particular, EC-NMR provides an electronic level description based on the Fermi level local density of states (E_f-LDOS) [159]. Wieckowski's group carried out detailed EC-NMR studies in order to explore the structure of electrocatalyst nanoparticles, estimation of various E_f-LDOS that are involved in construction of the metal-adsorbate bonds, and diffusional behavior of CO on PtRu bimetallic catalysts as well as an interesting relationship between electrochemical current generation and the E_f-LDOS of CO on Pt [160, 161]. ^{13}C and ^{195}Pt are particularly useful nuclei for investigating electrochemical interfaces. Quantitative information about the E_f-LDOS of both 5σ and $2\pi^*$-orbitals of the chemisorbed CO on Pt nanoparticles can be conveniently achieved by the ^{13}C EC-NMR. This analysis is particularly based on metal and ligand Knight shifts and spin-lattice relaxation rates and it is important since the variation of these E_f-LDOS reflects the changes in Pt–CO chemisorption bonds. Similarly, from the ^{195}Pt EC-NMR the 6s and 5d E_f-LDOS of Pt surfaces can reasonably be obtained. The electronic alterations of the metal surfaces can be realized through variations in E_f-LDOS. In an elegant work, Wieckowski and co-workers carried out thorough ^{195}Pt EC-NMR measurements on commercial Pt–Ru alloy nanoparticles and ^{13}C EC-NMR for CO chemisorbed on these catalysts [162]. The authors showed ^{195}Pt EC-NMR spectra of a Pt-black sample (with an average particle diameter of 2.8 nm) and Pt-Ru nanoparticles (with an average particle diameter of 2–3 nm). The authors found that for the Pt-black sample, the Pt NMR spectrum extends from 1.095 to 1.14 G/kHz, whereas for Pt–Ru nanoparticles, a much narrower NMR signal extending only from 1.095 to 1.115 G/kHz was found. Based on the observation that the whole spectrum is shifted toward lower Knight shifts, the authors reached the conclusion that there are no Pt atoms whose electronic properties resemble those of bulk Pt. The authors have suggested that if the nanoparticle retains a homogeneous composition then the corresponding NMR spectra of bimetallic catalysts can be expected to show broad, layer like structures. For example, the ^{195}Pt NMR spectrum of a 2.4 nm sized Pt-Pd bimetallic catalyst extended from 1.09 to 1.13 G/kHz [163]. In contrast, the presence of a relatively narrow peak can be found if there is a surface segregation of one component in bimetallic catalysts. Based on the fact that the ^{195}Pt NMR spectrum of Pt-Ru nanoparticles exhibited a relatively narrow peak centered at about 1.104 G/kHz, the authors have suggested that there is a major surface enrichment of Pt atoms in the Pt–Ru alloy nanoparticles. From the spin-lattice relaxation measurements the authors have found significant reduction in E_f-LDOS at Pt sites and also on the C-sites of adsorbed CO due to Ru addition, indicating a decrease in the total DOS at E_F for the Pt atoms. Thus EC-NMR is useful to evaluate the electronic effects in bimetallic electrocatalysts and for investigating electrochemical interfaces.

3.3.3.4 Low-Energy Electron Diffraction (LEED)

Low-energy electron diffraction (LEED) is the principal technique used for the determination of surface structures. It can provide information on size, symmetry, and atomic positions of ordered alloys. The LEED experiment uses a beam of electrons of a well-defined low energy (typically in the range 10–200 eV) incident normally on the sample. Usually the electron gun emitting the primary beam is mounted in the center of a hemispherical phosphorescent screen, on which backscattered electrons appear as bright spots. Overall LEED can yield a two dimensional reciprocal lattice of the imaged surface. However, one requisite is that the sample itself must be a single crystal with a well-ordered surface structure in order to generate a back-scattered electron diffraction pattern. Lin *et al.* reported the surface structural characterization and the coverage determination of Ru-modified Pt(111) electrode by *ex situ* LEED [164]. Based on the LEED patterns, the authors have found that the electrodeposition of Ru on Pt (111) forms a monatomic commensurate (1×1) layer with Pt (111) at low coverages ($\theta_{Ru} = 0.25$). Gasteiger *et al.* investigated the surface composition and structure of Pt$_3$Sn single-crystals by LEED [165]. From the combination of LEED patters, LEIS and AES results the authors have found that the annealed Pt$_3$Sn(111) surface and the sputtered but not-annealed Pt$_3$Sn(110) surfaces have the same nominal surface composition 20–25 wt.% Sn, but different local structures.

3.3.3.5 Auger Electron Spectroscopy (AES)

AES is a powerful tool for determining the composition of the top few layers of a surface. In AES the sample of interest is irradiated with a high energy (2–10 keV) primary electron beam. This bombardment results in the emission of backscattered, secondary, and Auger electrons that can be detected and analyzed. The backscattered and the secondary electrons are used for imaging purposes similar to that in SEM. The Auger electrons are emitted at discrete energies that are characteristic of the elements present on the sample surface. When analyzed as a function of energy, the peak positions are used to identify the elements and the chemical states present. AES is widely employed on electrocatalysts to realize the surface structure. Stamenkovic *et al.* investigated the surface structure of PtM (M = Co, Ni, Fe) poly crystalline alloys with the combination of AES, low energy ion scattering (LEIS) and ultraviolet photoemission spectroscopy (UPS) [166]. By careful modeling of emission from several subsurface layers with dynamical scattering of the outgoing Auger electron, the authors have observed that the Co_{775}/Pt_{237} AES peak ratio for Pt_3Co sample is different on sputtered and annealed surfaces indicating that the concentration profile of Pt and Co atoms in the surface region may depend on the respective UHV treatment of the alloy sample. From the combination of spectroscopic results, the authors found that in the case of annealed Pt_3Co sample at 1000 K, a complete Pt-skin surface was formed and due to compete segregation of Pt atoms the surface composition of Pt is calculated was 100 at%. In the case of ion sputtered Pt_3Co sample, surface composition corresponds to the ratio of alloying elements in the bulk that is, 75 at% Pt and 25% Co. In another interesting study, Tremiliosi-Filho et al. used AES as a primary characterization technique to investigate the ruthenium coverage on Pt (111) surface [167]. The authors have calculated the amount of Ru monolayers formed on Pt(111) surface from the intensity of AES peak observed at 274 eV as a result of Pt(111) exposure to $RuCl_3$ solution. With the increasing concentration of $RuCl_3$ solution the intensity of AES peak was found to be increased and corresponding amount of Ru monolayer coverage on Pt(111) was calculated to be higher.

3.3.3.6 Low-Energy Ion Scattering (LEIS)

Stamenkovic et al. used LEIS to estimate the surface composition of sputtered and annealed samples of Pt_3Co alloys [166]. By using the LEIS spectra, the scattering peaks of the Pt and Co atoms were calculated from the classical equation for elastic collisions and the true surface composition of the outermost atomic layer was estimated. From the LEIS analyses the authors found that the surface composition of the outermost atomic layer of Pt_3Co alloy sample after mild sputtering was 75 at% of Pt and 25 at% of Co corresponds to the bulk ratio of alloying components and after annealing the topmost surface layer consisted of only Pt atoms. This study is particularly important since the surface electrochemistry of Pt alloyed by the 3d transition metals (Pt-M alloys, M = Co, Ni, Fe, etc.) is a major contributing factor in enhancing the catalytic activity when compared with Pt for the electrochemical oxygen reduction reaction and methods to estimate the surface is of great interest. Gasteiger et al. also used LEIS to assess the surface composition of sputtered and annealed samples of Pt-Ru alloys [17]. The authors found that Pt–Ru alloy surfaces annealed in UHV conditions showed a strong enrichment of Pt to the surface, whereas the surface composition of sputtered-cleaned alloys was nearly identical to their bulk composition.

3.3.3.7 Temperature Programmed Reduction (TPR)

Temperature programmed reduction (TPR) has been successfully explored to evaluate the surface composition of bimetallic PtRu/C catalysts [168–170]. In their work, the authors suggested that upon calcination surface platinum (Pt^s) on reduced Pt crystallites is oxidized to Pt^sO and Pt^sO_2 as shown in Equations 3.29 and 3.30:

$$2\,Pt^s + O_2 \rightarrow 2\,Pt^sO \tag{3.29}$$

$$Pt^s + O_2 \rightarrow Pt^sO_2 \tag{3.30}$$

After calcinations, the state of Pt^sO_x can easily be characterized with the TPR technique by reducing the calcined catalysts by flowing H_2 as shown in Equation 3.31:

$$MO_x + x\,H_2 \rightarrow x\,H_2O + M \tag{3.31}$$

Similarly, the state of Ru also can be characterized by calcinations followed by the reduction. The authors found that oxygen chemisorbed on Ru exhibited a higher reduction temperature ($T_r = 300$ K) than that chemisorbed on Pt ($T_r = 250$ K). In case of bimetallic PtRu alloy nanoparticles the authors experiments suggested that T_r varies with Pt/Ru surface composition [168]. For example a Pt-rich surface displays a lower T_r (~ 300 K) when compared to Ru-rich surface ($T_r = 320$ K). Based on these observations authors evaluated the surface enrichment in bimetallic PtRu catalysts.

3.4
Evaluation of Electrocatalyst Activity, Electrochemical Active Surface Area, Catalyst – Adsorbate Interactions, and Activity Validation Techniques

Characterizing and evaluating the electrocatalyst activity in DMFCs is an extremely important aspect towards performance optimization of electrocatalysts and requires several electrochemical methods. For instance, cyclic voltammetry (CV) (potential cycling), linear sweep voltammetry (LSV), rotating disk electrode (RDE), and rotating ring-disk electrode (RRDE), CO-stripping voltammetry are methods that are widely employed to get structural as well as electrocatalytic activity information about electrocatalysts. An electrochemical reaction generally involves a sequence of steps. It may start by the transport and adsorption of the reactants to the surface of the electrode/catalyst, charge transfer related to either oxidation or reduction on the surface of electrode/catalyst, and terminated by the transport of product(s) from this contacting surface of electrode/catalyst.

In this section, we attempt to cover several electrochemical techniques commonly practiced in DMFCs fields, with emphasis on applied aspects. We also consider the electrochemical cell and its instrumentation in order to obtain valid results of the characterization and evaluation for electrocatalyst activity, and these will be presented where appropriate but not described in detail.

3.4.1
Cyclic Voltammetry (CV)

Cyclic voltammetry is a type of potentiodynamic electrochemical measurement and generally used to study the basic characteristics of the studied system mainly regarding the mechanism of electrode reactions and their kinetic parameters. It offers a rapid determination of redox potentials of electroactive species. CV is characterized by the smooth increase of a working electrode potential from one potential limit to the other and back. In that case, the current at the working electrode is plotted versus the applied potential to give the cyclic voltammogram trace [171–176].

A catalyst is usually prepared in the form of slurry before it is applied on the surface of the electrode. A catalyst is first well-mixed with other components: solvent and additives, through ultrasonication. A solvent could be typically of water and short chain alcohols such as ethanol. Naturally, the catalyst particles adhere to the surface of the working electrode. However, in order to increase the adhesion, a typical additive such as perfluorinated ionomers of DuPont's Nafion was used. Nafion can also be used as a binding material between each catalyst particle, which may lead to higher catalyst use. Furthermore, the dispersion of the catalyst particles is also crucial regarding optimum catalyst use. Therefore, it is important to discover the optimum loading of the metal under characterization [80, 150, 173, 175]. To the best of our knowledge, we found that an optimum Pt loading for the working electrode preparation was 0.22 mg-Pt/cm^2. The working electrode was made of unsupported or supported Pt-based catalysts immobilized on a glassy carbon electrode (GCE) surface (0.1964 cm^2 area). The procedure for electrode fabrication involved three steps: first, the preparation of a clear suspension by sonicating a known amount of catalyst powder dispersed in 0.5% Nafion; second, placing an aliquot of the suspension (7 μL of 6.2 μg-Pt mL^{-1} of the catalyst) on the GCE disc; third, air-drying about 5 min at room temperature and then at 80 °C to yield a uniform thin film of the catalyst [80, 150, 173, 175].

3.4.1.1 Applications

In this particular study related to the DMFCs, CV can be applied as a powerful tool to characterize and evaluate intrinsic properties of electrocatalyts, such as:

- State of the Pt-based catalysts in the absence of methanol in acidic electrolyte.
- Qualitative indicator for the activity of Pt-based catalysts towards electro-reduction of oxygen.
- Qualitative indicator for the activity of Pt-based catalysts towards electro-oxidation of methanol.
- Reversibility of hydrogen adsorption/desorption reaction and roughness factor of Pt-based catalysts.
- Electrochemical active surface area (ECASA) of Pt-based catalysts through the hydrogen adsorption/desorption peak area, CO stripping, and underpotential deposition of copper (Cu-UPD) methods.
- Surface composition of Pt-based catalysts through the Cu-UPD method.

3.4.1.2 State of the Pt-Based Catalysts

It is well known that unsupported or Pt supported on carbon was mainly chosen as the catalyst in PEMFCs and DMFCs rather than other monometallic catalysts for the following reasons: (a) high exchange current densities for anodic and cathodic reactions (b) reasonable Tafel slope at all potentials, and (c) ability to oxidize CO impurity on the anode side of PEMFCs. However, several studies on a bimetallic system have also been extensively explored in order to improve the performance

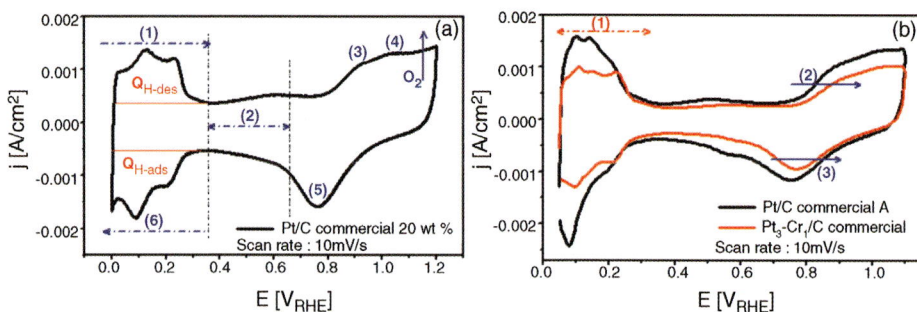

Figure 3.3 Cyclic voltammetry of Pt-based catalysts obtained in 0.5 M sulfuric acid in the absence of methanol solution at ambient temperature of $25 \pm 1\,°C$ (a) Pt/C commercial 20 wt.% (b) commercial Pt_3-Cr_1/C and Pt/C catalysts.

of either of the anodic and cathodic reactions. Figure 3.3 shows the cyclic voltammograms of Pt-based catalysts obtained in 0.5 M sulfuric acid in the absence of methanol solution. In the absence of redox active species in the potential scan range, the observed current-potential behavior is exclusively determined by the formation and dissolution of hydrogen ad-layers and oxygen ad-layers on the surface of catalyst, as shown in the Figure 3.3(a) for the Pt/C catalyst.

In Figure 3.3(a), the different regions from 1 to 6 are named to describe different processes occurring at the surface of catalyst and can be described as follows: region 1 extending from 0 to ~ 0.35 V vs. RHE represents the desorption process of adsorbed hydrogen (H_{ads}) from the surface of Pt catalyst. The integrated anodic charge under hydrogen desorption area in Coulombs is denoted as $Q_{H\text{-}des}$. The hydrogen desorption exhibits a few pairs of well-defined peaks due to the presence of different Pt crystalline facets. The double layer (DL) charging and discharging are shown in region 2, ranging from $0.35 \sim 0.65$ V vs. RHE. In the anodic direction higher than $0.65\ V_{RHE}$, one can observe 2 peaks in regions 3 and 4. The consecutive reactions which occur at regions 3 and 4 are the formation of OH and O on the surface of Pt catalyst, respectively. Pt catalyzes water dissociation and finally the Pt surface was covered with an oxide layer. Beyond region 4, the oxygen gas evolution starts to dominate so a reverse scan would immediately occur. When reversing the scan direction, the reduction of oxide layer is observed as shown in region 5. Compared to the peak of oxide formation, the peak of oxide reduction in region 5 is shifted to lower potentials. During the cathodic scan from $0.35\ V_{RHE}$ down, the underpotential deposition of hydrogen atoms occurs from the reduction of the protons. The integrated cathodic charge in Coulombs of the adsorption of hydrogen is denoted as $Q_{H\text{-}ads}$. The hydrogen adsorption exhibits a few pairs of well-defined peaks for different Pt crystalline facets, which is similar to the observation in the hydrogen desorption. Therefore, the ratio of $Q_{H\text{-}des}/Q_{H\text{-}ads}$ can give us an information of the reversibility of hydrogen adsorption/desorption ($H_{ads/des}$) reaction. Furthermore, it is also possible to obtain the roughness factor (RF), particle size, and electrochemical active surface area (ECASA) of Pt catalyst.

3.4.1.3 Qualitative Indicator for the Activity of Pt-Based Catalysts Towards ORR

From the CV of Pt/C catalyst in Figure 3.3(a), the $H_{ads/des}$ peaks on Pt are clearly seen, indicating the presence of polycrystalline Pt. It is quite different in the case of Pt_3-Cr_1/C catalyst, where the well-defined $H_{ads/des}$ peaks are not clearly shown in Figure 3.3(b). Furthermore, the $H_{ads/des}$ peaks of Pt_3-Cr_1/C catalyst are suppressed. This phenomenon suggests that high dispersion of the catalysts with the disordered surface structure is obtained and that their surfaces remain stable in the potential interval explored. In addition, all the Pt/C and Pt-Cr catalysts have a similar double-layer behavior, indicating that both catalysts have a similar double layer capacitance. Furthermore, it can be found that the onset of the oxide formation and the peak potential of the oxide reduction (indicated with blue arrow in the Figure 3.3(b)) of Pt-Cr catalyst are also shifted to more positive potentials, indicating that the alloying of Pt with Cr inhibits the chemisorption of OH on the Pt sites at high potentials (above 0.8 V). This may have a beneficial effect on the oxygen adsorption at low overpotential and thus may lead to an enhancement of the ORR kinetics [54, 177].

3.4.1.4 Qualitative Indicator for the Activity of Pt-Based Catalysts Towards MOR

Figure 3.4a shows that both of carbon supported Pt commercial A and B samples produce 2 main peaks in the forward and backward scan, which reflects the electro-oxidation of methanol. The peak current obtained at the forward and backward scan are denoted as i_f and i_b, respectively. It is accepted that the ratio of i_f to i_b can be used to describe the catalyst tolerance to carbonaceous species accumulation [178]. A higher value of i_f/i_b indicates better oxidation of methanol to carbon dioxide during the anodic scan and less accumulation of carbonaceous residues on the catalyst surface.

As shown in the inset of Figure 3.4a, one may find that Pt/C comm.-A exhibits higher i_f/i_b than that of Pt/C comm.-B. Moreover, the onset potential of Pt/C comm.-A is more negative than that of Pt/C comm.-B. This information strongly suggests that Pt/C comm.-A will tend to give a better performance towards methanol

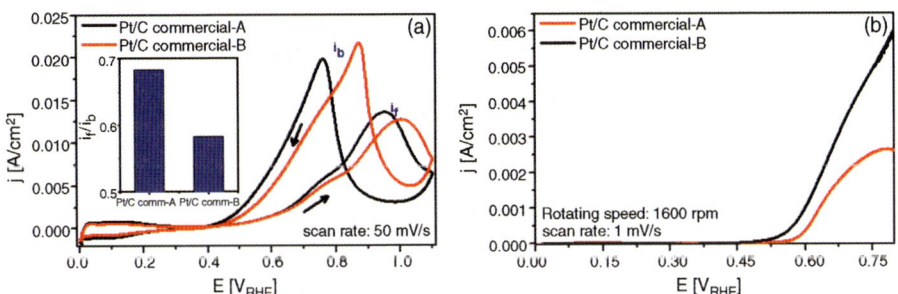

Figure 3.4 The feature of commercial Pt/C catalysts obtained in 0.5 M sulfuric acid in the presence of 5 wt.% methanol solution at ambient temperature of $25 \pm 1\,°C$ (a) Cyclic voltammetry, and inset: ratio of i_f/i_b (b) Linear sweep voltammetry at rotating speed of 1600 rpm.

oxidation. Indeed, Pt/C comm.-A is truly exhibiting a better performance of MOR, as investigated by other electrochemical techniques like LSV as shown in Figure 3.4b.

3.4.1.5 Reversibility of $H_{ads/des}$ Reaction and Roughness Factor (RF) of Pt-Based Catalysts

All measurements of the above mentioned parameters are based on $H_{ads/des}$ charges. Therefore, while investigating the intrinsic parameters of the catalyst, one must carefully consider faradaic current separation contributed from the double layer charging and discharging and other possible competing side reactions. In general, a capacitance from the double layer region is manifested by carbon support in the catalyst, although Pt or a second metal in Pt-based bimetallic catalysts also contribute to some double layer capacitance. Furthermore, one should note that the poisoning effect from the adsorbed impurities would mask all the details in $H_{ads/des}$ region and might even be manifested as bumps or peaks in the oxide formation region.

The coulombic charge for the hydrogen desorption ($Q_{H\text{-des}}$) and adsorption ($Q_{H\text{-ads}}$) can be obtained by dividing the area under the hydrogen desorption and adsorption regions, respectively with the scan rate.

The mean value between $Q_{H\text{-des}}$ and $Q_{H\text{-ads}}$ is denoted as $Q_{H\text{-upd}}$ and can be used to evaluate the electrochemical active surface area (ECASA) [179].

$$Q_{H\text{-upd}} = [Q_{H\text{-des}} + Q_{H\text{-ads}}]/2 \tag{3.32}$$

As mentioned earlier, the ratio of $Q_{H\text{-des}}/Q_{H\text{-ads}}$ can give the information of the reversibility of the $H_{ads/des}$ reaction. Furthermore, a catalyst parameter such as roughness factor (RF) can be obtained from the value of $Q_{H\text{-upd}}$. RF is defined as the ratio between the active surface area (EAA) and geometric area of the electrode (A_g) as shown in the following equation [180]:

$$RF = \frac{EAA}{A_g} = \frac{[Q_{H-upd}(C)/210(\mu C/cm^2-Pt)] \times 10^6}{A_g(cm^2)} \tag{3.33}$$

where $210\,\mu C/cm^2$ represents a charge required to oxidize a monolayer of hydrogen adsorbed on Pt [181–183]. It is reasonable to use a charge of $210\,\mu C/cm^2$ as a reference in Equation 3.33, because it is established based on the polycrystalline surface of Pt which possesses an average atom surface density of 1.3×10^{15} per cm^2 and assuming that each Pt site is covered by one hydrogen atom. The polycrystalline surface of Pt is composed of three dominant basal planes, which are Pt(100), Pt(110), and Pt(111) planes.

3.4.1.6 Electrochemical Active Surface Area (ECASA) and Particle Size of Pt-Based Catalysts

Typically, two electrochemical based methods are used to estimate the ECASA. The first method relies on the hydrogen adsorption/desorption ($H_{ads/des}$) charges [179–187]. In the earlier discussion, we have mentioned that Pozio et al. [179] have used a mean value of $Q_{H\text{-upd}}$ to estimate ECASA If the amount of Pt used within the catalyst layer (w_{Pt}) is known, then ECASA can be calculated by using Equation 3.34:

$$ECASA_{H\text{-upd}}(m^2/g) = \frac{Q_{H\text{-upd}}(C)}{210(\mu C/cm^2).w_{Pt}(g)} \times 100 \tag{3.34}$$

It is clear that from the Equation 3.34, that one may change a wide range of loading (w_{Pt}) to obtain ECASA. This is only true if all of the platinum loading on the electrode was totally electrochemical active. However, the fact is that as the platinum loading increases, the electrode thickness grows. As a consequence some platinum particles are blocked on the carbon support and not exposed to the electrolyte solution. The $H_{ads/des}$ mechanism is dominated by ionic (H^+) conduction, where it very much depends on the Nafion layer resistance [184, 185]. An increase in platinum loading will be followed with an increase in catalyst layer thickness on the electrode and also its resistance. Therefore, it is a true phenomenon that by increasing platinum loading, we may observe a decrease in ECASA due to the obvious inaccessibility of the deepest region of the catalyst layer. In contrast, when the platinum loading is too small, it becomes very difficult to obtain a precise loading (w_{Pt}) and therefore the uncertainty of the obtained ECASA is very high. From the statistical point of view, the standard deviation of ECASA (Δ ECASA) obtained from the Equation 3.35 suggests that too small a platinum loading should be avoided during the experiments [179].

$$\Delta ECASA = \left| \frac{\Delta Q_{H\text{-upd}}}{210(\mu C/cm^2) \cdot w_{Pt}} \right| + \left| \frac{\Delta w_{Pt} \cdot Q_{H-upd}}{210(\mu C/cm^2) \cdot w_{Pt}^2} \right| \tag{3.35}$$

ECASA obtained from $H_{ads/des}$ charges may not be suitable for adoption in the particular bimetallic catalyst such as Pt-Ru. Pt-Ru catalyst commonly showed an unclear 'double-layer' region and overlapping of hydrogen and oxygen adsorption regions [188]. This condition may prevent any straightforward application of $H_{ads/des}$ charges and hinder the determination of exact value of its ECASA and other parameters. This technique also showed the same overlapping of hydrogen and oxygen regions for other transition metal, such as Ni, Fe, Os, and so on.

The $H_{ads/des}$ method has been applied also to fine powders. In the case of supported metals, the H atoms deposited on the metallic particles may diffuse along the surface to regions where the support is uncovered popularly known as 'spillover effect' [189, 190]. This spillover effect may render the results of $H_{ads/des}$ rather ambiguous, thus invalidating the quantitative significance of the measured $Q_{H\text{-upd}}$.

3.4.2
Adsorptive CO-Stripping Voltammetry (CO_{ads}-SV)

The second electrochemical method for the estimation of ECASA is CO stripping voltammetry [179, 180, 186, 191–197]. Figure 3.5(a) shows two cyclic voltammograms obtained on the bimetallic home-made PtRu catalyst with a CO adsorbed ad-layer. CO can be strongly adsorbed onto the surface of Pt to form a monolayer. However, this

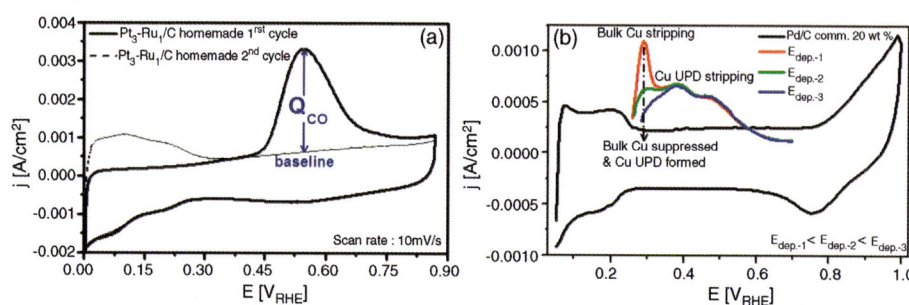

Figure 3.5 (a) CO stripping voltammograms for the home-made Pt_3-Ru_1/C catalysts in N_2-saturated 0.5 M sulfuric acid electrolyte at a scan rate of 10 mV/s, and 25 ± 1 °C; (b) Cu deposition in N_2-saturated 0.5 M sulfuric acid and in presence of 32 mM $CuSO_4$ on a commercial carbon supported Pd (20 wt.%)/GC electrode under various deposition potentials for 120 s at 30 ± 1 °C. Scan rate for Pd/C in the absence of $CuSO_4$ solution was 20 mV/s while the Cu deposition was scanned at 10 mV/s. Note: $E_{dep\text{-}1} < E_{dep\text{-}2} < E_{dep\text{-}3}$.

irreversible monolayer of CO will be removed quickly and completely by electrochemical oxidation at sufficiently higher potentials in the first cycle of CV.

During the first forward step the peaks characteristic of hydrogen/desorption ($H_{ads/des}$) were suppressed due to the presence of adsorbed CO. Moreover, the observed peaks at 0.45–0.50 V correspond to the oxidative stripping of the adsorbed CO layer. On the second sweep after electro-oxidation of CO, the voltammograms return to those observed in the absence of adsorbed CO. The calculated charge under the CO oxidation peak (Q_{CO}) is related to the following oxidation process that is, conversion of CO to CO_2:

$$CO + H_2O \rightarrow CO_2 + 2H^+ + 2e^- \tag{3.36}$$

CO oxidation to CO_2 involves two electrons as shown in Equation 3.36. Therefore, if one CO molecule bonded with one Pt atom in a linear adsorption configuration (Pt-CO_{ad}), then a charge required to oxidize a monolayer of CO adsorbed on Pt is equal to 420 µC/cm². ECASA can then be calculated by using Equation 3.37:

$$ECASA_{CO}(m^2/g) = \frac{Q_{CO}(C)}{420(\mu C/cm^2) \cdot w_{Pt}(g)} \times 100 \tag{3.37}$$

However, a bridge adsorption configuration (2Pt-CO_{ad}) may happen if 1 CO molecule occupies 2 Pt atoms. In that case, a charge required to oxidize a monolayer CO adsorbed on Pt is equal to 210 µC/cm². Those two CO adsorption configurations are strongly influenced by the applied potential. A linear adsorption may dominate if the CO adsorption occurs at a potential close to 0 V, as shown in a recent study [198].

Pozio et al. [179] has suggested that ECASA calculated by means of CO adsorption seems not dependent of the platinum loading. It may due to the stripping mechanism dominated by the electronic conduction of the oxidation reaction with hydroxyl group from water. The electronic conduction would not be influenced by the Nafion layer resistance in the catalyst. In the particular study of bimetallic PtRu, the CO stripping method basically gives a separate determination of platinum and second metal (Ru) in Pt-based alloy catalysts to bring a true ECASA, as desorption of CO from 'Pt' and 'Ru' usually takes place at different potentials [194–197].

3.4.3
Underpotential Deposition (UPD)

Up to now, we have described two typical electrochemical methods which are used for the estimation of ECASA. However, other studies have reported an alternative electrochemical method based on foreign metal ad-atoms, such as copper (Cu) [180, 192, 199–204], silver (Ag) [205, 206], lead (Pb) [207], antimony (Sb) [208], and so on. This technique, so called underpotential deposition (UPD) of metal is extremely useful for carbon supported catalysts as the ad-atoms can only be formed on metallic portion in the complex surface. In that case, an advantage of this method over $H_{ads/des}$ methods is that no spillover effect. Indeed, in a very recent study, Green and Kucernak [202, 203] showed the viability of underpotential deposition of copper (Cu-UPD) to characterize supported and unsupported Pt, Ru and PtRu catalysts. In their particular study of Pt, Ru and PtRu catalysts, the choice of Cu ad-atoms as a probe was motivated by the close atomic radii of Cu (0.128 nm), Ru (0.134 nm), and Pt (0.138 nm) and a suitable potential region of Cu desorption at relatively low potential. This privilege at a certain extent can introduce a correction for the double layer charging and oxygen adsorption. Furthermore, they also showed that it is also possible to determine the surface composition of PtRu catalyst via Cu-UPD method due to the difference in adsorption energies for Cu on either Pt or Ru. The UPD studies have been performed using a variety of metals include Cu, Pb, Sn, and Fe deposited on particular Pt catalyst [209].

The phenomenon of UPD itself refers to the deposition of metal on a foreign metal substrate at the potential more positive than that predicted by the Nernst equation for a bulk deposition. This implies that the depositing ad-atoms are more strongly to the foreign metal electrode [210–213]. In case of Cu-UPD, the metal deposition processes at the electrode surface can be represented by Equation 3.38:

$$Cu^{2+} + 2e^- \rightarrow Cu \quad (420\,\mu C/cm^2) \tag{3.38}$$

The Nernst equation was applied in order to predict the equilibrium potential (E_{eq}) at which the deposition and dissolution of the bulk-metal phase happens (Equation 3.39):

$$E_{eq} = E^0 + \frac{RT}{2F}\ln\frac{a_{Cu^{2+}}}{a_{Cu}} \cong 0.34\,V_{RHE} \quad \text{for Cu} \tag{3.39}$$

where E^0 is the standard potential and a is the activity. R, T and F are the molar gas constant, temperature, and Faraday constant, respectively. The formation of the first monolayer is inferred from the pronounced current peaks at the potential $E > E_{eq}$, while the bulk deposition occurs at $E < E_{eq}$. Taking an example of Cu on Au system, UPD Cu on Au occurs at ~0.46 V while bulk deposition occurs at ~0.25 V.

The cyclic voltammogram of platinum in the solution of H_2SO_4 and $CuSO_4$ at wide scan range of 0–1.5 V commonly exhibited four main peaks related to different mechanisms. The CV is dominated by the deposition and stripping of both bulk and underpotential-deposited copper. In a forward anodic scan, a distinct sharpened peak represents the bulk copper stripping, while in the backward cathodic scan, the bulk copper deposition occurs at almost a similar potential as bulk copper stripping. In detailed observation, the bulk copper deposition occurs at around 0.25 V, and happens at slightly lower potential than its bulk copper stripping. At the platinum double-layer region, we can observe a several peaks regarding with the UPD processes of copper on platinum, indicating the existence of copper sites with different adsorption energies. In a forward anodic scan, the UPD of copper stripping occurs at more positive potentials compared with that of bulk copper striping. Furthermore, the UPD of copper deposition is also shifted to more positive potentials compared with the deposition of bulk copper. In addition, the hydrogen adsorption region is suppressed due to the presence of copper, and the oxide reduction is distorted due to the onset of copper UPD [202].

Green and Kucernak used the plot of the ratio of copper stripping charge to hydrogen charge (Q_{Cu-UD}/Q_{H-des}) as a function of deposition potential on platinum electrode in order to obtain an optimum condition for the formation of well-ordered mono-layer of UPD copper without the possibility of three-dimensional growth of bulk copper [202]. The charge of 420 µC/cm^2 in the copper metal deposition reaction (Equation 3.38) is used in the calculation of Q_{Cu-UPD}, while Q_{Cu-UPD} itself is obtained after subtracting the total measured UPD copper stripping charges with the platinum background. The ratio of Q_{Cu-UPD}/Q_{H-des} is expected to be 2, where copper atoms adsorbs on platinum surface at the same sites with hydrogen to form a completely UPD copper layer. Their results showed that $Q_{Cu-UPD}/Q_{H-des} \approx 2$ was achieved when the deposition potential (E_{dep}) is in the range of 0.25–0.3 V. When E_{dep} is below 0.25 V, deposition of bulk copper occurs, while at potential higher than 0.3 V incomplete formation of UPD layer occurs. This technique can also be widely applied in finding an optimum deposition time of UPD copper.

In our recent results shown in Figure 3.5(b), we also found a similar trend, where the UPD of Cu occurs at the potential more positive than that of the bulk Cu deposition. The bulk copper deposition on commercial Pd/C catalysts occurs at the deposition potential of $E_{dep.-1}$. In detailed observation, an increase in deposition potentials at $E_{dep.-2}$ and $E_{dep.-3}$ is followed by a decrease in current density corresponding to bulk copper stripping while the UPD copper starts to grow, as shown with the black arrow.

In a particular study of Pt-based catalysts, the determination of *ECASA* from the Cu-UPD method should consider several important aspects as follows: (i) applied deposition potential and time for the formation of UPD copper; (ii) applied potential scan range for both Pt background and UPD of copper stripping; and (iii) a correction for double layer charging, oxygen adsorption, and further possibility of adsorbed anion. In earlier discussion, we have mentioned that an optimum condition for the formation of well-ordered mono-layer of UPD copper can be achieved when the ratio of Q_{Cu-UD}/Q_{H-des} is nearly equal to 2, as suggested by Green and Kucernak [202]. Thus, for the first important aspect, the plot between Q_{Cu-UD}/Q_{H-des} vs. potential and/or time deposition should be made in order to find in which potential

3.4 Evaluation of Electrocatalyst Activity, Electrochemical Active Surface Area

and/or time deposition region a value of $Q_{Cu\text{-}UD}/Q_{H\text{-}des} \approx 2$ occurred. In the second aspect, the CV for a bare Pt (as background) should be scanned from 0 to 0.85 V. The end potential in the forward scan was chosen to be 0.85 V, at the point where the oxide growth commonly starts on platinum as can be seen in Figure 3.3(a), and thus only a very small oxide reduction peak was expected to be observed at the backward scan. The linear potential scan for the Cu UPD stripping should be also scanned until the end potential of 0.85 V is reached. In case of PtRu catalyst, the linear potential scan for the Cu UPD stripping could be started from 0.3 V until 0.85 V. The applied potential of 0.3 V was chosen as the starting point of the scan due to a completion of mono-layer of UPD copper, while 0.85 V as the end potential was due to a completion of oxidative removal of UPD copper layer and also a suppression of oxide adsorption/desorption. Once the oxide adsorption/desorption mechanism was suppressed, we may consider it as an advantage for the calculation of $Q_{Cu\text{-}UPD}$ as a very small correction would be expected, and will be discussed in the following third aspect. In the third aspect, the charge of $Q_{Cu\text{-}UPD}$ should be a result after subtracting the total measured charge from experiment (Q_{exp}) with the charge due to charging of double-layer capacitance (Q_{DL}), the charge due to the growth of any oxide and (or) oxygenated species (Q_{ox}), and the charge due to adsorption/desorption of any adsorbed anion (Q_{anion}) as shown in Equation 3.40 [203]:

$$Q_{Cu\text{-}UPD} = Q_{exp} - Q_{DL} - Q_{ox} - Q_{anion} \quad (3.40)$$

Therefore, it is important to include the CV for bare platinum as a background for the matter of correction for $Q_{Cu\text{-}UPD}$. ECASA can be calculated by assuming that a charge required to oxidize a monolayer of Cu adsorbed on each metal surface is equal to 420 µC/cm² as shown in Equation 3.41 [203]:

$$ECASA_{Cu\text{-}UPD}(m^2/g) = \frac{Q_{Cu\text{-}UPD}(C)}{420\,(\mu C/cm^2) \cdot w_{Pt}(g)} \times 100 \quad (3.41)$$

Once the ECASA was obtained either from H-UPD, CO stripping, and Cu-UPD methods, it was possible to determine the average particle size (d) by assuming the shape of the catalyst particle to be spherical, as shown in Equation 3.42:

$$d(nm) = \frac{6000}{ECASA\,(m^2/g) \cdot \rho(g/cm^3)} \quad (3.42)$$

where ρ is the average particle density, and it was 21.4 g/cm³ for Pt. However for the bimetallic system the average particle density could be obtained from Equation 3.43:

$$\rho_{Pt\text{-}M} = x_{Pt}^{bulk}\rho_{Pt} + x_{M}^{bulk}\rho_{M} \quad (3.43)$$

where x_{Pt}^{bulk} and x_{M}^{bulk} are the bulk composition of Pt and second metal 'M' respectively, while ρ_M is the average particle density of second metal 'M'.

3.4.3.1 Surface Composition of Pt-Based Catalysts Through the Cu-UPD method

It is of interest to describe the feature for the stripping of Cu UPD layer formed on a properly dispersed bimetallic PtRu catalyst. Commonly, a peak at low potential around 0.42 V is accompanied by a shoulder that continues to much higher potential. The first peak at around 0.42 V was due to the oxidative removal of adsorbed Cu UPD layer while the shoulder was removal from the Pt sites. This information strongly supports that the different adsorption energies for Cu on either Pt or Ru could be used to determine the Ru metal surface content of PtRu catalyst. Further the deconvolution of linear anodic stripping of Cu UPD can be made to separate the charges contributed by Ru and Pt sites, by assuming that the feature of these Cu UPD has a Gaussian line shape. The Ru coverage on the surface of bimetallic PtRu catalyst (x_{Ru}^s) can be quantified from the ratio of charge in first peak contributed by Ru (Q_{Ru}^{Cu-UPD}) to Q_{Cu-UPD} as shown in Equation 3.44 [203]:

$$x_{Ru}^s = Q_{Ru}^{Cu\text{-}UPD}/Q_{Cu\text{-}UPD} \tag{3.44}$$

3.4.4
Rotating Disk Electrode (RDE)

It is possible to increase the mass transport by introducing a forced convection in which the analyte solution flows relative to a working electrode. A popular method for creating such a relative movement is by rotating a working electrode. Such a method is called a rotating disk electrode (RDE). Therefore, RDE could also be called as a hydrodynamic working electrode, where the steady-state current is determined by solution flow rather than diffusion [171, 174].

In a common 3-electrode cell, when an working electrode spins, the reactant is dragged to the surface of the working electrode, and the resulting centrifugal force makes the product fling away from the surface of the working electrode. However, one should notice that there is still a stagnant reaction layer covering the surface of the working electrode, and the reactant transports through this layer by diffusion. There is a strong dependency of thickness of diffusion layer (δ) with applied rotation rate (ω) of working electrode, as shown in Equation 3.46. An increase in rotation rate will lead to a thinner diffusion layer. A rotation rate in the range between 5–10000 rpm can control the flow of the reactant through the surface of working electrode follows a laminar pattern. Figure 3.6 shows a schematic diagram of RDE setup. The working electrode was connected to the electrode rotator with very fine control of rotation rate.

The flow pattern under the RDE was obtained by numerically solving the Navier–Stokes equation and continuity equation under the following conditions: (i) the radius of the disk on the working electrode is large enough compared to that of boundary layer thickness, and thus the small distortion of flow pattern at the center and edge can be neglected; (ii) the roughness of the disk surface of the working electrode is small enough compared to that of the boundary layer thickness; (iii) the radius of the electrochemical cell is large enough compared to that of the disk on the working electrode, and thus the reflection of the flow at the vessel of electrochemical cell's wall does not affect the flow pattern under the disk; and (iv) The rotating speed is larger enough than that of the lower critical value in which the effect of natural

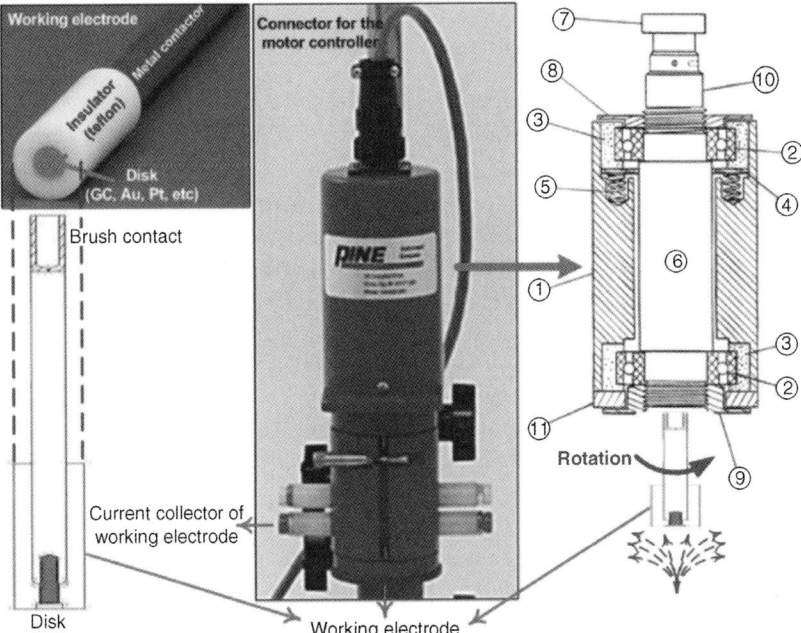

Figure 3.6 The RDE setup of Pine instrument and its schematic diagram. (1) rotator body (2) bearing: 2 (3) rubber housing: 2 (4) washer (5) spring: 4 (6) spindle for electrode (7) pulley (8) top lock nut (9) bottom lock nut (10) bushing/brush contact, and (11) retainer.

convection is negligible. Furthermore, the rotating speed is also smaller than that of the higher critical value in which the flow gets to be turbulent.

The solution of the Navier–Stokes equation and continuity equation under those mentioned conditions may bring two possibilities of the net mass transport rate of a reactant to the surface of the working electrode [171, 174]. The first net mass transport is the convection which can control the thickness of the diffusion layer, while the second one is the diffusion which can control the reactant through the diffusion layer. The flux normal to the electrode surface due to diffusion is given by $D(\partial^2 C/\partial x^2)$, and that due to convection is given by $v_x(\partial C/\partial x)$, where D is the diffusion coefficient of the species, C is the bulk concentration of the species, and v_x is the solution velocity in the x direction, which is normal to the electrode surface [171, 174].

The RDE is becoming one of the most powerful methods for studying both diffusion in electrolytic solutions and the kinetics of moderately fast electrode reaction, because the hydrodynamics and the mass-transfer characteristics are well understood and the current density on the disk electrode is supposed to be uniform.

V.G. Levich [214] solved the family of equations and provided an empirical relationship between diffusion limiting current (i_d) and rotation rate (ω) as shown in Equation 3.46. In particular application in fuel cells, the empirical relationship which is given by Levich was also used in a LSV experiment performed on a RDE to study the intrinsic kinetics of the catalyst [186, 194, 215–225]. However, it is more appropriate to continue the discussion in detail in Section 3.4.6.

3.4.5
Rotating Ring-Disk Electrode (RRDE) Method

The rotating ring-disk electrode (RRDE) is very similar to a RDE, where both RDE and RRDE use a similar hydrodynamic approach. However, the main difference between a RRDE and RDE is the addition of a second working electrode in the form of a ring around the central disk of the first working electrode. The structure of the RRDE itself consists of mainly two electrodes which are the disk and ring electrodes, separated by an insulator material such as Teflon, and connected to the potentiostat through different leads as shown in Figure 3.7.

Since RRDE experiments involve the examination of two potentials contributed by the disk electrode (E_D) and ring electrode (E_R), the representation of the results involves more information than that of a single working electrode in RDE experiments. Thus, RRDE experiments are usually carried out with a bipotentiostat which are capable of controlling 4 electrodes and allowing an adjustment of E_D and E_R separately [174].

The disk is held at a potential where the reaction of interest takes place, and a current-potential curve is then recorded at the ring. This allows the identification of intermediates and/or products. A current–potential curve is recorded at the disk while the ring potential is held at a constant value where the intermediates or products are reduced or oxidized. This allows the identification of the exact potential range over which they are formed. Alternatively, the disk is held at a potential where intermediates or potentials are formed and the ring is maintained at a potential at which they undergo electron transfer. This allows quantitative kinetic measurements to be performed [171, 174].

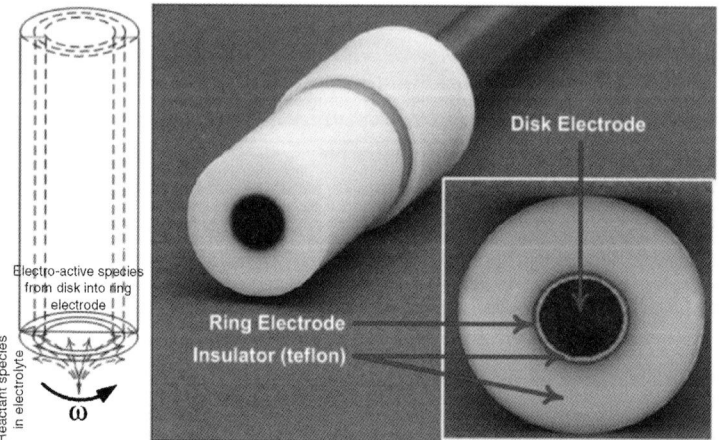

Figure 3.7 The rotating ring disk electrode: electrode/solution interface at rotation rate of ω.

The essentials of RRDE operation are as for the RDE with the addition of one extremely important parameter: the collection efficiency (N) could be defined for a reversible reaction (Equation 3.45) [171, 174]:

$$O + e^- \rightarrow R \text{ as } N = -\frac{I_R}{I_D} \qquad (3.45)$$

where I_R and I_D are the ring current and disk current, respectively, and the negative sign indicates that the currents are opposed. The collection efficiency depends only on the geometry the disk (radius of r_1), and the ring electrode (inner and outer radius of r_2 and r_3, respectively). However, for a particular practice in a RRDE cell system, N is also determined by using a standard redox couple, such as $Fe(CN)_6^{3-}$ to $Fe(CN)_6^{4-}$ [226]. N is calculated by the ratio of two currents measured from those disk currents (reduction of $Fe(CN)_6^{3-}$ to $Fe(CN)_6^{4-}$) and ring currents (oxidation of the resulting $Fe(CN)_6^{4-}$).

An informative example of the application of the RRDE concerns the study of the reduction of O_2 at a Pt disk electrode and the detection of the intermediates and/or products at the Pt ring. The technological importance of oxygen reduction in devices such as fuel cells has led to an extensive investigation of the reduction mechanism and particularly of the role of hydrogen peroxide (H_2O_2) as an intermediate species [227–240]. One should note that an intermediate species of H_2O_2 is harmful to the fuel cell components. Again, it is more appropriate to continue our discussion in detail in Section 3.4.6.

3.4.6
Linear Sweep Voltammetry (LSV)

It is appropriate to continue our discussion on a linear sweep voltammetry (LSV) experiment performed on a RDE to study the intrinsic kinetics of the catalyst [186, 194, 215–225]. The knowledge of the velocities in radial and vertical direction, which can be obtained via the Navier–Stokes equation, allows the calculations of the mass transport to the disk surface through a diffusion layer with the thickness of δ according to Equation 3.46:

$$\delta = 1.61 \cdot v^{1/6} \cdot D^{1/3} \cdot \omega^{-1/2} \qquad (3.46)$$

where v represents the kinematic viscosity of the electrolyte. As we mentioned earlier, according to Equation 3.46 the thickness of the diffusion layer of a chosen system strongly depends on the rotation rate. The potential of the working electrode in the LSV experiment is scanned from a potential in which no reaction occurs to a potential where a reaction occurs. Furthermore, when the overpotential is high enough, the reaction rate will be determined by the diffusion mass transport of the reactant at a given electrode rotation rate. At this particular condition, a diffusion limiting current is achieved and can be described as a function of the diffusion layer thickness as shown in Equation 3.47, by assuming that the Fick's law can be applied.

$$i_d = nFAC\frac{D}{\delta} \qquad (3.47)$$

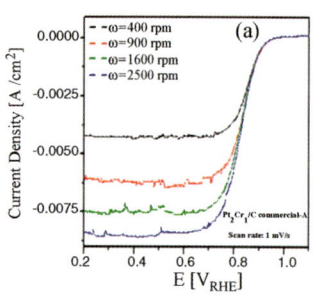 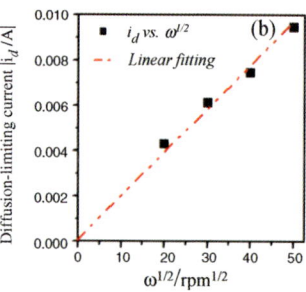 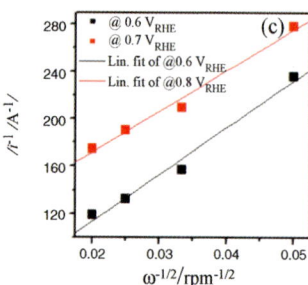

Figure 3.8 Evaluation of intrinsic kinetics activity towards oxygen reduction reaction for commercial carbon supported Pt_3-Cr_1 catalyst (a) LSV recorder at 1 mV/s under various rotation rate in oxygen saturated 0.5 M sulfuric acid at 25 °C (b) Plot of $|i_d|$ vs. $\omega^{1/2}$ (c) Koutecky–Levich plots at various potential.

n, A and F are the number of electrons involved, geometric electrode area, and Faraday constant, respectively. The combination of Equations 3.46 and 3.47 results in the *Levich equation* for the diffusion limited current on a RDE as shown in Equation 3.48. The diffusion limited current is for a given system only determined by the rotation rate with B being the *Levich constant* [171, 174]:

$$i_d = 0.620nFACD^{2/3}\nu^{-1/6}\omega^{1/2} = BC\omega^{1/2} \qquad (3.48)$$

It is clearly seen in Equation 3.48 that a linear relationship exists in the plotting of i_d versus $\omega^{1/2}$, and goes through the (0, 0) origin. Indeed, we also found a same linear relationship in part of our recent result shown in Figure 3.8(b). It is of great interest to point out that at the onset potential, the current is controlled mainly by reaction kinetics rather than mass transport, and is expressed in Equation 3.49 [172, 175].

$$i_k = nFk_\eta C \qquad (3.49)$$

where k_η is the rate constant and is the function of overpotential ($_\eta$). Furthermore, i_k is the current that would flow under the kinetic limitation if the mass transfer were efficient enough to keep the concentration at the electrode surface equal to the bulk value, regardless of the electrode reaction.

In an entire potential scan range, the overall current (i) in Equation 3.50 is described by Koutecky–Levich equation, as the partition of the overall current in a kinetically determined and a diffusion determined part as shown [171, 174, 214]:

$$\frac{1}{i} = \frac{1}{i_k} + \frac{1}{i_d} = \frac{1}{i_k} + \frac{1}{BC\omega^{1/2}} \qquad (3.50)$$

It is clearly seen in Equation 3.50 that a plot of i versus $\omega^{1/2}$ will be curved and tend toward the limit $i = i_k$ as $\omega^{1/2} \to \infty$ (Figure 3.9a). The plot of i^{-1} versus $\omega^{-1/2}$ as observed in Figures 3.8(c) and 3.9(b) will yield a straight line, where its slope can be

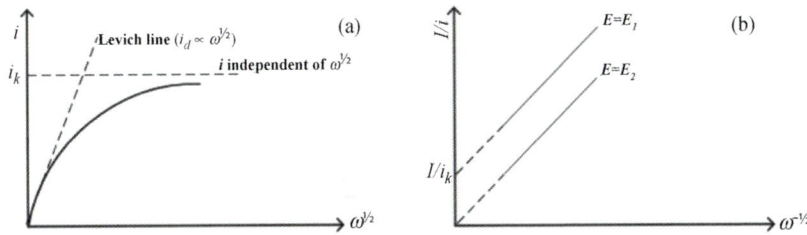

Figure 3.9 (a) Variation of i with $\omega^{1/2}$ in the RDE experiment under a constant E_D for the slow reaction on the electrode; (b) Koutecky–Levich plots at potential E_1, where the rate of electron transfer is sufficiently slow to act as limiting factor, and at E_2, where electron transfer is rapid.

used to determine *Levich constant* of B, from which the number of electrons involved in the reaction can be calculated using known values of concentration and the diffusivity of particular reactant in the medium under investigation. The intercept of the plot on the ordinate axis at $\omega^{1/2} = 0$ gives the values of i_k^{-1}, which can be used for further determination of the kinetic parameter k_n according to Equation 3.49.

The combination of LSV and RDE methods can be further used to obtain several intrinsic parameters of catalyst. These intrinsic parameters include kinetic parameter of Tafel slope, mass activity, and specific activity together defined the activity of catalyst. Figure 3.10 shows a representative sequence steps for evaluating the activity of carbon supported Pt and Pt-Co catalysts towards oxygen reduction reaction (ORR) [80].

The ORR on all those catalysts is diffusion-controlled when the potential is less than 0.7 V and is under a mixed control region of diffusion-kinetics from 0.7 to 0.85 V, as shown in the Figure 3.10(a). It is possible to take advantage of these observed phenomenon, to measure the value of diffusion limited current by linearizing the observed current of LSV under the region of 0–0.7 V. In the region of 0–0.7 V,

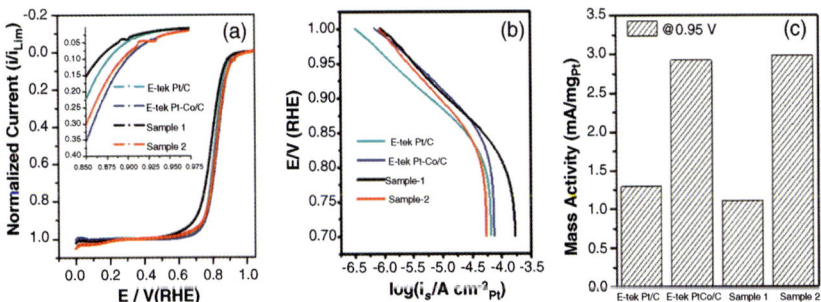

Figure 3.10 Evaluation of electro-activity towards oxygen reduction reaction for various carbon supported Pt and Pt-Co catalysts (a) LSV recorder at 1 mV/s under rotation rate of 2500 rpm in oxygen saturated 0.5 M sulfuric acid at 25 °C; (b) Tafel plots. Currents are per gram of Pt used in the electrode preparation (c) Mass activity measured at 0.95 V_{RHE} [80].

the observed current commonly shows almost a constant value, and thus one can easily linearize that observed current. The result of the linearization was known as the diffusion limited current and again denoted as i_d. To the best of our knowledge, we believe that it is more appropriate to use the normalized current in Figure 3.10(a) for the qualitative comparison of the ORR activity of the catalysts. The normalized current is obtained by simply dividing the measured currents with the obtained i_d. In the Tafel region (higher than 0.85 V) and the mixed potential region, the ORR activities show a significant difference in their magnitudes. It can be clearly visualized from the inset of Figure 3.10(a) that E-Tek Pt-Co/C and sample 2 show a drastically enhanced activity in raising a portion of the curves compared with that for E-Tek Pt/C. On the other hand, the sample 1 displays comparatively poor activity toward ORR.

To compare the specific activity (i_s) of the different Pt-Co/C catalysts for ORR, mass transfer-corrected Tafel plots of kinetic current densities based on the active area of platinum were evaluated. For such an evaluation, the prerequisite factor, namely, the chemically active surface area of Pt (*EAA*) for all catalysts was determined from Equation 3.51a:

$$EAA_{\text{H-UPD}}(\text{cm}^2) = \frac{Q_{\text{H-UPD}}(C)}{210(\mu C/\text{cm}^2)} \times 10^6 \quad \text{from H-UPD method} \quad (3.51a)$$

$$EAA_{\text{CO}}(\text{cm}^2) = \frac{Q_{\text{CO}}(C)}{420(\mu C/\text{cm}^2)} \times 10^6 \quad \text{from CO stripping method} \quad (3.51b)$$

$$EAA_{\text{Cu-UPD}}(\text{cm}^2) = \frac{Q_{\text{Cu-UPD}}(C)}{420(\mu C/\text{cm}^2)} \times 10^6 \quad \text{from Cu-UPD method} \quad (3.51c)$$

The mass transport correction for a RDE was obtained by rearranging the Koutecky–Levich equation (Equation 3.50), and shown in Equation 3.52:

$$i_k = \frac{i_d \cdot i}{i_d - i} \quad (3.52)$$

The specific activities (i_s) are obtained via the calculation of i_k using Equation 3.52 and the normalization with the Pt active area (*EAA*) of corresponding catalysts, as given in Equation 3.53:

$$\text{Specific activity } (i_s) = \frac{i_k(A)}{EAA(\text{cm}^2)} \quad (3.53)$$

Figure 3.10(b) showed the extraction of Tafel slope from the linear region of the plot about 1.0 to 0.90 V where the ORR is considered as an activation-controlled process. The Tafel plots are almost parallel, indicating that the reaction pathway and r.d.s. are the same on all the catalysts. The observed Tafel slope values for the various Pt-Co/C catalysts, in the range of 75–90 mV/decade, are in good agreement with other reported values for the Pt-Co/C catalysts [241]. Furthermore, it is possible to deduce the number of electrons involved in the reactions based on the obtained slope.

The mass activity as an index to assess the applicability of the catalyst toward ORR can be calculated from Equation 3.54:

$$\text{Mass activity (MA)} = \left[\frac{i_k(A)}{w_{Pt}(g)}\right]_{0.95\ V_{RHE}} \tag{3.54}$$

It is appropriate to continue our discussion on LSV experiments performed on a RRDE in this section. RRDE was extensively used for the investigation of the electrochemical partial reduction reaction of oxygen to hydrogen peroxide [227–240], with regard to chemical stability of fuel cell components, particularly in Nafion. The formation of H_2O_2 during the ORR is involving a two-electron process via the side reaction in Equation 3.55:

$$O_2 + 2H^+ + 2e^- \rightarrow H_2O_2 \tag{3.55}$$

It is believed that radicals such as •OH and •OOH are formed from H_2O_2, and may attack the ionomer membrane. It is reasonable that designed catalysts with the ability to bypass the formation of H_2O_2 under fuel cell operation are preferred in order to avoid such radicals attack, and thus the four-electron process during the ORR is expected.

Markovic et al. [239] reported a new experimental method for the fast assembly of rotating ring-platinum single crystal disk electrodes. They have successfully prepared a various extremely clean surface of disk electrodes with platinum low-index single crystal such as Pt(100), Pt(110), and Pt(111), and applied it in the RRDE method to present the results for the detection of H_2O_2 on the ring electrode during oxygen reduction. The lowest activity of Pt(111) compared to that of Pt(100) and Pt(110) in sulfuric acid is attributed to the highest structure sensitivity of (bi)sulfate anion adsorption and its inhibiting effect. The collection efficiency of N was measured using the well-known potassium ferrocyanide technique, and was further used to calculate the fluxes of H_2O_2. They found a very small fluxes (i.e., $I_R/N < 0.01 I_D$) of H_2O_2 at the ring at potential above 0 V_{SCE} for all surfaces of single crystal disks. This information is strongly implies that oxygen reduction proceeds almost entirely through the direct four-electron reaction pathway. However, below 0 V_{SCE}, the adsorbed hydrogen affects the partial reduction of oxygen to H_2O_2 in the order of Pt(111) < Pt(100) < Pt(110) where almost 100% H_2O_2 formation occurred on Pt(111). At this particular negative potential limit, the limiting current on Pt(111) corresponds to an exactly two-electron reduction of O_2. Furthermore, the activity for ORR on the H_{ads}-modified electrodes thus decreases in the order of Pt(110) < Pt(100) < Pt(111).

Several studies of the RRDE measurements during the ORR on supported Pt catalysts with the effect of presence of Nafion were extensively reported [235–238]. Claude et al. [238] reported a new method for the electrochemical characterization of composite electrode materials consists of a paste of catalytic material in Nafion. The resulting paste was further coated on a RRDE which partially mimic the working environment of PEMFC operation. However, the high loading of both Nafion and catalyst restricted the quantification of the H_2O_2 formation to a very low rotation rate (about 100 rpm) due to the instability of the collection efficiency at higher rotation rate. In contrast, Behm and co-workers [235] have recently developed a method of thin-film RRDE to allow much higher rotation rate in the particular study on the effect of the presence of Nafion.

In order to determine the lower limit of the Nafion film diffusion current, i_f, the measured current, i, is described by simply modifying the raw Koutecky–Levich equation as shown in Equation 3.56:

$$\frac{1}{i} = \frac{1}{i_k} + \frac{1}{i_d} + \frac{1}{i_f} = \frac{1}{i_k} + \frac{1}{BC\omega^{1/2}} + \frac{1}{nFAC_f D_f/\delta_f} \qquad (3.56)$$

where C_f and D_f are the concentration and diffusion coefficient of the species within the film, respectively, while δ_f is the thickness of the Nafion film. The term of $\frac{1}{i_k}$ in Equation 3.56 can be neglected, since the reaction kinetic is very fast at the limiting current, thus i_k becomes much larger compared to that of i_d or i_f, and Equation 3.56 can be simplified to Equation 3.57 [235]:

$$\frac{1}{i_{lim}} = \frac{1}{i_d} + \frac{1}{i_f} = \frac{1}{BC\omega^{1/2}} + \frac{1}{nFAC_f D_f/\delta_f} \qquad (3.57)$$

Based on Equation 3.57, it now clearly becomes an advantage that one can determine the corresponding Nafion films parameters of C_f and D_f from the intercept obtained on the plot of i_{lim}^{-1} versus $\omega^{-1/2}$.

In particular, for evaluating the amount of H_2O_2 produced during ORR one should consider the important experimental condition, that the potential range of disk electrode should be scanned from 1.0 to 0.4 V and the ring potential is kept constant at 1.2 or 1.4 V where the H_2O_2 can be quickly oxidized. The ratio of H_2O_2 formed from 1 molecule of O_2 can be calculated from Equation 3.58 [232–236]:

$$X_{H_2O_2} = \frac{2I_R/N}{I_D + I_R/N} \qquad (3.58)$$

When the disk potential lower than 0.6 V the ORR reached the diffusion-limited current [235]. On the other hand, when the disk potential higher than 0.6 V a negligible amount of H_2O_2 was observed at the ring electrode. It should be noted that a significant amount of H_2O_2 was detected at the ring electrode when the disk potential was in the H-UPD region (e.g., <0.3 V). When the disk potential is higher than 0.2 V the amount of H_2O_2 is less than 1%, and it increases to be higher than 5% at the disk potential lower than 0.1 V.

Finally, they suggested that an increase in the amount of H_2O_2 in the hydrogen adsorption region was believed to be due to the coverage of hydrogen atoms on the corresponding disk electrode, and thus inhibited the directly four-electron reduction of O_2. It seems that their explanation of the hydrogen-inhibition on the contribution of two-electron reduction of O_2 is found in good agreement with other reports [239]. Although the amount of H_2O_2 produced at the cathode is low when a PEMFC is under normal operation, a significant amount of H_2O_2 may form at the anode if some of oxygen is introduced into the anode (through diffusion or air bleed).

Inaba et al. [231] investigated the effect of agglomeration of Pt/C catalyst on H_2O_2 formation in oxygen reduction by means of RRDE technique. Various amounts of platinum loading from 20 wt.% Pt/C catalysts were loaded on glassy carbon disk electrode in order to give a distinction to the formed bulk and particle catalysts. Their results revealed that the series two-electron reduction pathway which leads to the formation of H_2O_2 does exist on carbon-supported Pt particles. In opposite, the series two-electron reduction pathway is negligible on a clean bulk Pt surface. However, in very recent study, Lin *et al.* [228] showed that after the correction of internal diffusion, the specific ORR currents are independent of the catalyst loading and are considered to represent the intrinsic ORR activity of the catalyst. Furthermore, they showed that for all differently prepared catalysts regardless of its structure showed a similar specific activity and Tafel slope (~120 mV/dec), and thus gave a similar reaction pathway of four-electron reduction of O_2 to form H_2O. It is of interest to point out the progress that have been achieved from the group of Stamenkovic' *et al.* [78, 233], where they successfully showed that the unique properties of monolayer thin films of Pt, so called 'Pt-skin', on particular Pt–Co and Pt–Ni bimetallic system could be essential in the future design of high activity catalysts for the application of oxygen reduction reaction. Furthermore, the reaction mechanism on the 'Pt-skin' catalysts follows the four-electron reduction of O_2.

3.5
Conclusions and Outlook

In conclusion, the material reviewed here highlights the importance of thorough characterization of electrocatalysts in terms of their nanoscale properties such as particle size, size distribution, alloying degree, atomic distribution and surface composition. The capabilities of several microscopic and spectroscopic techniques to realize the nanoscale properties of electrocatalysts are reviewed. In particular capabilities of XAS to realize the structure of electrocatalysts and in providing the atomic level information on alloying extent, atomic distribution, and surface composition – the properties which have strong influence on the corresponding activity of electrocatalysts are discussed in detail. The most striking advantage of XAS is that detailed electronic and structural information about electrocatalysts can be obtained even in their active chemical environments and structure–performance relationships can be properly understood. Although XAS is a bulk technique and provides average information of all the elements present in the material, due to the large number of atoms present on the surface in case of nanoscale entities one can obtain information about the surface properties. Such information is critical since electrocatalysts' surface chemistry predominantly determines their electrochemical properties. Significant progress has been made in developing suitable electrochemical techniques to validate the activity of electrocatalysts. Care must be taken regarding the stability of electrocatalysts under the potential range of operation in order to use the developed electrochemical techniques properly. Combination of the structural information obtained by the promising characterization techniques and activity information through excellent electrochemical validation techniques will certainly aid our current understanding of the structure–activity relationships and help us to explore promising electrocatalysts for DMFCs applications. It is equally important that the advances in experimental techniques will be combined with theoretical insights in order to achieve a better understanding of the structure–activity relationships.

Acknowledgments

The authors gratefully acknowledge the financial support from the National Science Council (NSC), and facilities from the National Synchrotron Radiation Research Center (NSRRC) and the National Taiwan University of Science and Technology (NTUST), Taiwan.

References

1 Wasmus, S. and Vielstich, W. (1993) *J. Appl. Electrochem.*, **23**, 120.
2 Ren, X., Wilson, M.S. and Gottesfeld, S. (1996) *J. Electrochem. Soc.*, **143**, L12.
3 Reddington, E., Sapienza, A., Gurau, B., Viswanathan, R., Sarangapani, S., Smotkin, E.S. and Mallouk, T.E. (1998) *Science*, **280**, 1735.
4 Ralph, T.R. and Hogarth, M.P. (2002) *Platinum Met. Rev.*, **46**, 3.
5 Kauranen, P.S. and Skou, E. (1996) *J. Electroanal. Chem.*, **408**, 189.
6 Aricò, A.S., Creti, P., Antonucci, P.L. and Antonucci, V. (1998) *Electrochem. Solid-State Lett.*, **1**, 66.
7 Carrette, L., Friedrich, K.A. and Stimming, U. (2001) *Fuel Cells*, **1**, 5.
8 Arico, A.S., Srinivasan, S. and Antonucci, V. (2001) *Fuel Cells*, **1**, 133.
9 Wasmus, S. and Küver, A. (1999) *J. Electroanal. Chem.*, **14**, 461.
10 Ren, X.M., Zelenay, P., Davey, J. and Gottesfeld, S. (2000) *J. Power Sources*, **86**, 111.
11 Baldauf, M. and Preidel, W. (2001) *J. Appl. Electrochem.*, **31**, 781.
12 Hamnett, A. (1999) *Interfacial Electrochemistry, Theory, Experiment and Applications* (ed. A. Wieckowski), Marcel Dekker, New York, Ch. 47.
13 Watanabe, M., Uchida, M. and Motoo, S. (1987) *J. Electroanal. Chem.*, **229**, 395.
14 Watanabe, M., Furuchi, Y. and Motoo, S. (1985) *J. Electroanal. Chem.*, **191**, 367.
15 Lizcano-Valbuena, W.H., Paganin, V.A. and Gonzalez, E.R. (2002) *Electrochim. Acta*, **47**, 3715.
16 Kelly, S.C., Deluga, C.A. and Smyrl, W.H. (1999) *Electrochem. Solid-State Lett.*, **3**, 407.
17 Gasteiger, H.A., Markovic, N., Ross, P.N. and Cairns, E.J. (1993) *J. Phys. Chem.*, **97**, 12020.
18 Anderson, A.B., Grantscharova, E. and Shiller, P. (1995) *J. Electrochem. Soc.*, **142**, 1880.
19 Bittins-Cattaneo, B. and Iwasita, T. (1987) *J. Electroanal. Chem.*, **238**, 151.
20 Cathro, K.J. (1969) *J. Electrochem. Soc.*, **116**, 1609.
21 Aricò, A.S., Poltarzewski, Z., Kim, H., Morana, A., Giordana, N. and Antonucci, V. (1995) *J. Power Sources*, **55**, 159.
22 Götz, M. and Wendt, H. (1998) *Electrochim. Acta*, **43**, 3637.
23 Shrosphire, J.A. (1965) *J. Electrochem. Soc.*, **112**, 465.
24 Grgur, B.N., Markovic, N.M. and Ross, P.N. (1999) *J. Electrochem. Soc.*, **146**, 1613.
25 Mukerjee, S., Lee, S.J., Ticianelli, E.A., McBreen, J., Grgur, B.N., Markovic, N.M.,

Ross, P.N., Giallombardo, J.R. and De Castro, E.S. (1999) *Electrochem. Solid-State Lett.*, **2**, 12.

26 Shukla, A.K., Ravikumar, M.K., Aricò, A.S., Candiano, G., Antonucci, V., Giordano, N. and Hamnett, A. (1995) *J. Appl. Electrochem.*, **25**, 528.

27 Hen, P.K. and Tseung, A.C.C. (1994) *J. Electrochem. Soc.*, **141**, 3082.

28 Zhu, Y. and Cabrera, C.R. (2001) *Electrochem. Solid-State Lett.*, **4**, A45.

29 Gurau, B., Viswanathan, R., Liu, R., Lafrenz, T.J., Ley, K.L., Smotkin, E.S., Reddington, E., Sapienza, A., Chan, B.C., Mallouk, T.E. and Sarangapani, S. (1998) *J. Phys. Chem. B*, **102**, 9997.

30 Ley, K.L., Liu, R., Pu, C., Fan, Q., Leyarovska, N., Segre, C. and Smotkin, E.S. (1997) *J. Electrochem. Soc.*, **144**, 1543.

31 Ticianelli, E., Beery, J.G., Paffett, M.T. and Gottesfeld, S. (1989) *J. Electrochem. Soc.*, **258**, 61.

32 Freelink, T., Visscher, W. and Van Veen, J.A.R. (1995) *Surf. Sci.*, **335**, 353.

33 Jiang, X., Permeter, J.E., Estrada, C.A. and Goodman, D.W. (1991) *Surf. Sci.*, **249**, 44.

34 Goodnough, J.B., Manoharan, R., Shukla, A.K. and Ramesh, K.V. (1989) *Chem. Mater.*, **1**, 391.

35 Goodnough, J.B., Harnett, A., Kennedy, B.-J., Manoharan, R. and Weeks, S.A. (1988) *J. Electroanal. Chem.*, **240**, 133.

36 Aricò, A.S., Srinivasan, S. and Antonucci, V. (2001) *Fuel Cells*, **2**, 1.

37 Reeve, R.W., Christensen, P.A., Hamnett, A., Haydock, S.A. and Roy, S.A. (1998) *J. Electrochem. Soc.*, **145**, 3463.

38 Tributsch, H., Bron, M., Hilgendorff, M., Schulenberg, H., Dorbant, I., Eyert, V., Bogdanoff, P. and Fiechter, S. (2001) *J. Appl. Electrochem.*, **31**, 739.

39 Trapp, V., Christinsen, P.A. and Hamnett, A. (1996) *J. Chem. Soc., Faraday Trans.*, **21**, 4311.

40 Mukerjee, S., Srinivasan, S., Soriaga, M.P. and McBreen, J. (1995) *J. Electrochem. Soc.*, **142**, 1409.

41 Shukla, A.K., Neergat, M., Parthasarathi, B., Jayaram, V. and Hegde, M.S. (2001) *J. Electroanal. Chem.*, **504**, 111.

42 Neergat, M., Shukla, A.K. and Gandhi, K.S. (2001) *J. Appl. Electrochem.*, **31**, 373.

43 Aricò, A.S., Shukla, A.K., Kim, H., Park, S., Min, M. and Antonucci, V. (2001) *Appl. Surf. Sci.*, **172**, 33.

44 Toda, T., Igarashi, H. and Watanabe, M. (1999) *J. Electroanal. Chem.*, **460**, 258.

45 Mukerjee, S., Srinivasan, S., Soriaga, M.P. and McBreen, J. (1995) *J. Phys. Chem.*, **99**, 4577.

46 Stonehart, P. (1992) *J. Appl. Electrochem.*, **22**, 995.

47 Maoka, T., Kitai, T., Segawa, N. and Ueno, M. (1996) *J. Appl. Electrochem.*, **26**, 1267.

48 Antolini, E. (2003) *Mater. Chem. Phys.*, **78**, 563.

49 Mukerjee, S. (1990) *J. Appl. Electrochem.*, **20**, 537.

50 Mukerjee, S. and McBreen, J. (1998) *J. Electroanal. Chem.*, **448**, 163.

51 Takasu, Y., Iwazaku, T., Sugimoto, W. and Murakami, Y. (2000) *Electrochem. Commun.*, **2**, 671.

52 Cherstiouk, O.V., Simonov, P.A. and Savinova, E.R. (2003) *Electrochim. Acta*, **48**, 3851.

53 Kinoshita, K.(ed.) (1992) *Electrochemical Oxygen Technology*, John Wiley & Sons, Inc., New York, pp. 46–48.

54 Yang, H., Alonso-Vante, N., Le'ger, J.M. and Lamy, C. (2004) *J. Phys. Chem. B*, **108**, 1938.

55 Herrero, E., Franaszczuk, K. and Wieckowski, A. (1993) *J. Electroanal. Chem.*, **361**, 269.

56 Davis, J.C., Hayden, B.E. and Pegg, D.J. (1998) *Electrochim. Acta*, **44**, 1181.

57 Arenz, M. Mayrhofer, K.J.J. Stamenkovic, V., Bilzanac, B.B., Tomoyuki, T., Ross, P.N. and Markovic F N.M. (2005) *J. Am. Chem. Soc.*, **127**, 6819.

58 Maillard, F., Eikerling, M., Cherstiouk, O.V., Schreier, S., Savinova, E. and Stimming, U. (2004) *Faraday Discuss.*, **125**, 357.

59 Bockris, J.O.M. and Srinivasan, S. (1969) *Fuel Cells: Their Electrochemistry*, McGraw-Hill Book Company, New York.

60 Beden, B., Leger, J.M. and Lamy, C. (1992) *Modern Aspects of Electrochemistry* (eds J.Ò.M. Bockris, B.E. Conway and R.E. White), Plenum Press, New York, p. 97.

61 Chandrasekaran, K., Wass, J.C. and Bockris, J.Ò.M. (1990) *J. Electrochem. Soc.*, **137**, 518.

62 Ianniello, R., Schmidt, V.M., Stimming, U., Stumper, J. and Wallam, A. (1994) *Electrochim. Acta*, **39**, 1863.

63 Gasteiger, H.A., Markovic, N., Ross, P.N. Jr and Cairns, E.J. (1994) *J. Phys. Chem.*, **98**, 617.

64 Watanabe, M. and Motoo, S. (1975) *J. Electroanal. Chem.*, **60**, 275.

65 Lu, C. and Masel, R.I. (2001) *J. Phys. Chem. B*, **105**, 9793.

66 Liao, M.S., Cabrera, C.R. and Ishikawa, Y. (2000) *Surf. Sci.*, **445**, 267.

67 Adzic, R.R. and Wang, J.X. (1998) *J. Phys. Chem. B*, **102**, 8988.

68 Shukla, A.K. and Raman, R.K. (2003) *Annu. Rev. Mater. Res.*, **33**, 155.

69 Damjanovic, A. and Brusic, V. (1967) *Electrochim. Acta*, **12**, 615.

70 Marković, N.M., Schmidt, T.J., Stamenković, V. and Ross, P.N. (2001) *Fuel Cells*, **1**, 105.

71 Adzic, R. (1998) *Electrocatalysis* (eds J. Lipkowski and P.N. Ross), John Wiley & Sons, Inc., New York, p. 197.

72 Marković, N.M. and Ross, P.N. (1999) *Interfacial Electrochemistry. Theory, Experiment and Applications* (ed. A. Wieckowski), Marcel Dekker, New York, p. 821.

73 Sidik, R.A. and Anderson, A.B. (2002) *J. Electroanal. Chem.*, **528**, 69.

74 Marković, N.M., Gasteiger, H.A., Grgur, B.N. and Ross, P.N. (1999) *J. Electroanal. Chem.*, **467**, 157.

75 Stamenković, V., Marković, N.M. and Ross, P.N. Jr (2000) *J. Electroanal. Chem.*, **500**, 44.

76 Grgur, B.N., Marković, N.M. and Ross, P.N. Jr (1997) *Can. J. Chem.*, **75**, 1465.

77 Drillet, J.F., Ee, A., Friedmann, J., Kotz, R., Schnyder, B. and Schmidt, V.M. (2002) *Electrochim. Acta*, **47**, 1983.

78 Stamenković, V., Schmidt, T.J., Ross, P.N. and Marković, N.M. (2002) *J. Phys. Chem. B*, **106**, 11970.

79 Mukerjee, S. and Srinivasan, S. (1993) *J. Electroanal. Chem.*, **357**, 201.

80 Hwang, B.-J., Kumar, S.M.S., Chen, C.-H. Monalisa, Cheng, M.-Y., Liu,D.-G. and Lee,J.-F. (2007) *J. Phys. Chem. C*, **111**, 15267.

81 Kordesch, K. and Simader, G. (1996) *Fuel Cells and Their Applications*, 1st edn, Wiley-VCH Verlag GmbH, Weinheim.

82 Hwang, B.-J., Sarma, L.S., Chen, J.M., Chen, C.-H., Shih, S.C., Wang, G.R., Liu, D.-G., Lee, J.-F. and Tang, M.T. (2005) *J. Am. Chem. Soc.*, **127**, 11140.

83 Chen, J.M., Sarma, L.S., Chen, C.-H., Cheng, M.-Y., Shih, S.C., Wang, G.R., Liu, D.-G., Lee, J.-F., Tang, M.T. and Hwang, B.-J. (2006) *J. Power Sources*, **159**, 29.

84 Bock, C., Paquet, C., Couillard, M., Botton, G.A. and MacDougall, B.R. (2004) *J. Am. Chem. Soc.*, **126**, 8028.

85 Iwasita, T., Hoster, H., John-Anacker, A., Lin, W.F. and Vielstich, W. (2000) *Langmuir*, **16**, 522.

86 Friedrich, K.A., Geyzers, K.P., Dickinson, A.J. and Stimming, U. (2002) *J. Electroanal. Chem.*, **524–525**, 261.

87 Xiong, L. and Manthiram, A. (2005) *J. Electrochem. Soc.*, **152**, A697.

88 Watanabe, M., Tsurumi, K., Mizukami, T., Nakamura, T. and Stonehart, P. (1994) *J. Electrochem. Soc.*, **141**, 2659.

89 Roth, C., Martz, N. and Fuess, H. (2001) *Phys. Chem. Chem. Phys.*, **3**, 315.

90 Liao, S., Holmes, K.A., Tsaprailis, H. and Birss, V.I. (2006) *J. Am. Chem. Soc.*, **128**, 3504.

91 Mu, Y., Liang, H., Hu, J., Jiang, L. and Wan, L. (2005) *J. Phys. Chem. B*, **109**, 22212.

92 Voyles, P.M., Muller, D.A., Grazul, J.L., Citrin, P.H. and Gossmann, H.-J.L. (2002) *Nature*, **416**, 826.

93 Gale, R.J.(ed.) (1988) *Spectroelectrochemistry: Theory and Practice*, Plenum Press, New York.
94 Somorjai, G.A. (1978) *Surf. Sci.*, **201**, 489.
95 Ross, P.N. (1991) *Electrochim. Acta*, **36**, 2053.
96 Powell, C.J. (1978) Quantitative surface analysis of materials, in *ASTM Special Publication 643* (ed. N.S. McIntyre,), American Society for Testing and Materials, Philadelphia, p. 5.
97 Briggs, D. and Seah, M.P. (1990) *Practical Surface Analysis*, vol. 1, John Wiley & Sons, Inc., New York.
98 Bazin, D. and Rehr, J.J. (2003) *J. Phys. Chem. B*, **107**, 12398.
99 Bazin, D., Sayers, D., Rehr, J.J. and Mottet, C. (1997) *J. Phys. Chem. B*, **101**, 5332.
100 Konnigsberger, D.C. and Prins, R.(eds) (1988) *X-Ray Absorption: Principles, Applications, Techniques of EXAFS, SEXAFS, and XANES*, John Wiley & Sons, Inc., New York, p. 362.
101 Iwasawa, Y. (ed.) (1996) *XAS for Catalysts and Surfaces*, World Scientific, Singapore, p. 113.
102 Frenkel, A.I., Hills, C.W. and Nuzzo, R.G. (2001) *J. Phys. Chem. B*, **105**, 12689.
103 Nitani, H., Nakagawa, T., Daimon, H., Kurobe, Y., Ono, T., Honda, Y., Koizumi, A., Seino, S. and Yamamoto, T. (2007) *Appl. Catal. A: Gen.*, **326**, 194.
104 Sinfelt, J.H., Via, G.H. and Lytle, F.W. (1984) *Catal. Rev. Sci. Eng.*, **26**, 81.
105 Bazin, D.C., Sayers, D.A. and Rehr, J.J. (1997) *J. Phys. Chem. B*, **101**, 11040.
106 Radmilović, V., Gasteiger, H.A. and Ross, P.N. (1995) *J. Catal.*, **154**, 98.
107 Wang, D., Zhuang, L. and Lu, J. (2007) *J. Phys. Chem. C*, **111**, 16416.
108 Aricò, A.S., Creti, P., Modica, E., Monforte, G., Baglio, V. and Antonucci, V. (2000) *Electrochim. Acta*, **45**, 4319.
109 Rolison, D., Hagans, P., Swider, K. and Long, J. (1999) *Langmuir*, **15**, 774.
110 Gurau, B., Viswanathan, R., Liu, R.X., Lafrenz, T.J., Ley, K.L., Smotkin, E.S., Reddington, E., Sapienza, A., Chan, B.C., Mallouk, T.E. and Sarangapani, S. (1998) *J. Phys. Chem. B*, **102**, 9997.
111 Antolini, E., Cardellini, F., Giorgi, L. and Passalacqua, E. (2000) *J. Mater. Sci. Lett.*, **19**, 2099.
112 Antolini, E. and Cardellini, F. (2001) *J. Alloys Compd.*, **315**, 118.
113 Gnutzmann, V. and Vogel, W. (1990) *J. Phys. Chem.*, **94**, 4991.
114 Vogel, W., Rosner, B. and Tesche, B. (1993) *J. Phys. Chem.*, **97**, 11611.
115 Zhang, X. and Chan, K.Y. (2003) *Chem. Mater.*, **15**, 451.
116 Bommarius, A.S., Holzarth, J.F., Wang, D.I.C. and Hatton, T.A. (1990) *J. Phys. Chem.*, **94**, 7232.
117 Dobrosz, I., Jiratova, K., Pitchon, V. and Rynkowski, J.M.J. (2005) *Mol. Catal. A: Chem.*, **234**, 187.
118 Bönnemann, H. and Braun, G.A. (1996) *Angew. Chem. Int. Ed. Engl.*, **35**, 1992.
119 Skoog, D.A., Holler, F.J. and Crouch, S.R. (2007) *Principles of Instrumental Analysis*, 6th edn., Thomson Brooks/Cole Publishing, Belmont, USA.
120 Ferrando, R., Jellinek, J. and Johnston, R.L. (2008) *Chem. Rev.*, **108**, 845.
121 Bi, Y. and Lu, G. (2008) *Chem. Mater*, **20**, 1224.
122 Kawaguchi, T., Sugimoto, W., Murakami, Y. and Takasu, Y.J. (2005) *Catal.*, **229**, 176.
123 Bauer, A., Gyenge, E.L. and Oloman, C.W. (2006) *Electrochim. Acta*, **51**, 5356.
124 Park, K.W., Sung, Y.E., Han, S., Yun, Y. and Hyeon, T.J. (2004) *Phys. Chem. B*, **108**, 939.
125 Rodríguez-Nieto, F.J., Morante-Catacora, T.Y. and Cabrera, C.R. (2004) *J. Electroanal. Chem.*, **571**, 15.
126 Schmidt, T.J., Noeske, M., Gasteiger, H.A., Behm, R.J., Britz, P. and Bönnemann, H. (1998) *J. Electrochem. Soc.*, **145**, 925.
127 Skoog, D.A. and Leary, J.J. (1992) *Principles of Instrumental Analysis*, 4th edn, Saunders, Fort Worth, TX, USA.
128 Lee, D., Hwang, S. and Lee, I. (2006) *J. Power Sources*, **160**, 155.

129 Chen, J.M., Sarma, L.S., Chen, C.-H., Cheng, M.-Y., Shih, S.C., Wang, G.R., Liu, D.-G., Lee, J.F., Tang, M.T. and Hwang, B.-J. (2006) *J. Power Sources*, **159**, 29.

130 Mukerjee, S. and Urian, R.C. (2002) *Electrochim. Acta*, **47**, 3219.

131 Harada, M., Toshima, N., Yoshida, K. and Isoda, S. (2005) *J. Colloid Interf. Sci.*, **283**, 64.

132 O'Grady, W.E., Hagans, P.L., Pandya, K.I. and Maricle, D.L. (2001) *Langmuir*, **17**, 3047.

133 Russell, A.E., Maniguet, S., Mathew, R.J., Yao, J., Roberts, M.A. and Thompsett, D. (2001) *J. Power Sources*, **96**, 226.

134 Nashner, M.S., Frenkel, A.I., Adler, D.L., Shapley, J.R. and Nuzzo, R.G. (1997) *J. Am. Chem. Soc.*, **119**, 7760.

135 Liu, D.-G., Lee, J.F. and Tang, M.T. (2005) *J. Mol. Catal. A: Chem.*, **240**, 197.

136 Russell, A.E. and Rose, A. (2004) *Chem. Rev.*, **104**, 4613.

137 Via, G.H. and Sinfelt, J.H. (1996) in *X-ray Absorption Fine Structure for Catalysts and Surfaces* (ed. Y. Iwasawa), World Scientific, London.

138 Park, S., Wieckowski, A. and Weaver, M.J. (2003) *J. Am. Chem. Soc.*, **125**, 2282.

139 Lin, S.D., Hsiao, T.C., Chang, J.R. and Lin, A.S. (1999) *J. Phys. Chem. B*, **103**, 97.

140 Greegor, R.B. and Lytle, F.W. (1980) *J. Catal.*, **63**, 476.

141 Yoshitake, H., Yamazaki, O. and Ota, K. (1994) *J. Electrochem. Soc.*, **141**, 2516.

142 Siepen, K., Bönnemann, H., Brijoux, W., Rothe, J. and Hormes, J. (2000) *Appl. Organometal. Chem.*, **14**, 549.

143 Horsely, J.A. (1982) *J. Chem. Phys.*, **76**, 1451.

144 Mansour, A.N., Cook, J.W. and Sayers, D.E. (1984) *J. Phys. Chem.*, **88**, 2330.

145 Brown, M., Peierls, R.E. and Stern, D.E. (1977) *Phys. Rev. B*, **15**, 738.

146 Min, M.-K., Cho, J., Cho, K. and Kim, H. (2000) *Electrochim. Acta*, **45**, 4211.

147 Markovic, N., Gasteiger, H. and Ross, P.N. (1997) *J. Electrochem. Soc.*, **144**, 1591.

148 Kinoshita, K. (1992) *Electrochemical Oxygen Technology*, John Wiley & Sons, Inc., New York.

149 Nashner, M.S., Frenkel, A.I., Somerville, D., Hills, C.W., Shapley, J.R. and Nuzzo, R.G. (1998) *J. Am. Chem. Soc.*, **120**, 8093.

150 Hwang, B.-J., Sarma, L.S., Wang, G.R., Chen, C.-H., Liu, D.-G., Sheu, H.S. and Lee, J.F. (2007) *Chem. Eur. J.*, **13**, 6255.

151 Hwang, B.-J., Chen, C.-H., Sarma, L.S., Al Andra, C.C., Lai, F.J., Chen, H.H., Hsaio, S.Y., Sheu, H.S., Liu, D.-G. and Lee, J.F. Unpublished results.

152 Park, K.W., Choi, J.H. and Sung, Y.E. (2003) *J. Phys. Chem. B*, **107**, 5851.

153 Reetz, M.T., Lopez, M., Grünert, W., Vogel, W. and Mahlendorf, F. (2003) *J. Phys. Chem. B*, **107**, 7414.

154 Kao, C.C., Tsai, S.C., Bahl, M.K., Chung, Y.W. and Lo, W.J. (1980) *Surf. Sci.*, **95**, 1.

155 Eberhardt, W., Fayet, P., Cox, D., Fu, Z., Kaldor, A., Sherwood, R. and Sondericker, D. (1990) *Phys. Rev. Lett.*, **64**, 780.

156 Roth, C., Martz, N. and Fuess, H. (2004) *J. New Mater. Electrochem. Syst.*, **7**, 117.

157 Phung, X., Groza, J., Stach, E.A., Williams, L.N. and Ritchey, S.B. (2003) *Mater. Eng.*, **A359**, 261.

158 Tong, Y.Y., Oldfield, E. and Wieckowski, A. (1998) *Anal. Chem.*, **70**, 518.

159 Babu, P.K., Tong, Y.Y., Kim, H.S. and Wieckowski, A. (2002) *J. Electroanal. Chem.*, **524–525**, 157.

160 Babu, P.K., Kim, H.S., Oldfield, E. and Wieckowski, A. (2003) *J. Phys. Chem. B*, **107**, 7595.

161 Lu, C., Rice, C., Masel, R.I., Babu, P.K., Waszczuk, P., Kim, H.S., Oldfield, E. and Wieckowski, A. (2002) *J. Phys. Chem. B*, **106**, 9581.

162 Babu, P.K., Kim, H.S., Oldfield, E. and Wieckowski, A. (2003) *J. Phys. Chem. B*, **107**, 7595.

163 Tong, Y.Y., Yonezawa, T., Toshima, N. and Van der Klink, J.J. (1996) *J. Phys. Chem.*, **B100**, 730.

164 Lin, W.F., Zei, M.S., Eiswirth, M., Ertl, G., Iwasita, T. and Vielstich, W. (1999) *J. Phys. Chem. B*, **103**, 6968.

165 Gasteiger, H.A., Marković, N.M. and Ross, P.N. Jr (1996) *Catal. Lett.*, **36**, 1.
166 Stamenkovic, V.R., Mun, B.S., Mayhofer, K.J.J., Ross, P.N. and Markovic, N.M. (2006) *J. Am. Chem. Soc.*, **128**, 8813.
167 Tremiliosi-Filho, G., Kim, H., Chrzanowski, W., Wieckowski, A., Grzybowska, B. and Kulesza, P. (1999) *J. Electroanal. Chem.*, **467**, 143.
168 Huang, S.Y., Chang, S.M. and Yeh, C.T. (2006) *J. Phys. Chem. B*, **110**, 234.
169 Huang, S.Y., Chang, S.M., Lin, C.L., Chen, C.-H. and Yeh, C.T. (2006) *J. Phys. Chem. B*, **110**, 23300.
170 Wang, K.W., Huang, S.Y. and Yieh, C.T. (2007) *J. Phys. Chem. C*, **111**, 5096.
171 Bard, A.J. and Faulkner, L.R. (2000) *Electrochemical Methods: Fundamentals and Applications*, 2nd edn, John Wiley & Sons, Inc., New York.
172 Zoski, C.G. (2006) *Handbook of Electrochemistry*, 1st edn, Elsevier.
173 Sawyer, D.T. and Roberts, J.L. Jr (1974) *Experimental Electrochemistry for Chemist*, John Wiley & Sons, Inc., New York.
174 Rieger, P.H. (1994) *Electrochemistry*, 2nd edn, Prentice-Hall.
175 Hwang, B.-J., Senthil Kumar, S.M., Chen, C.-H., Chang, R.-W., Liu, D.-G. and Lee, J.-F. (2008) *J. Phys. Chem. C*, **112**, 2370.
176 Sarangapani, S. and Luczak, F.J. (2006) *Experimental Methods in Low Temperature Fuel Cells*, Springer, US.
177 Koffi, R.C., Coutanceau, C., Garnier, E., Léger, J.M. and Lamy, C. (2005) *Electrochim. Acta*, **50**, 4117.
178 Liu, Z.L., Ling, X.Y., Su, X.D. and Lee, J.Y. (2004) *J. Phys. Chem. B*, **108**, 8234.
179 Pozio, A., De Francesco, M., Cemmi, A., Cardellini, F. and Giorgi, L. (2002) *J. Power Sources*, **105**, 13.
180 Watt-Smith, M.J., Friedrich, J.M., Rigby, S.P., Ralph, T.R. and Walsh, F.C. (2008) *J. Phys. D: Appl. Phys.*, **41**, 174004.
181 Bakotzky, V.S. and Vassilyev, Y.B. (1967) *Electrochim. Acta*, **12**, 1323.
182 Biegler, T., Rand, D.A.J. and Woods, R. (1971) *J. Electroanal. Chem.*, **29**, 269.
183 Woods, R. (1974) *J. Electroanal. Chem.*, **49**, 217.
184 Tamizhmani, G., Dodelet, J.P. and Guay, D. (1996) *J. Electrochem. Soc.*, **143**, 18.
185 Antolini, E., Giorgi, L., Pozio, A. and Passalacqua, E. (1999) *J. Power Source*, **77**, 136.
186 Schmidt, T.J., Gasteiger, H.A., Stab, G.D., Urban, P.M., Kolb, D.M. and Behm, R.J. (1998) *J. Electrochem. Soc.*, **145**, 2354.
187 Ticianelli, E.A., Beery, J.G. and Srinivasan, S. (1991) *J. Appl. Electrochem.*, **21**, 597.
188 Petrii, O.A. (2008) *J. Solid State Electrochem.*, **12**, 609.
189 Mitchell, P.C.H., Ramirez-Cuesta, A.J., Parker, S.F., Tomkinson, J. and Thompsett, D. (2003) *J. Phys. Chem. B*, **107**, 6838.
190 Mallát, T., Petró, J. and Sárkány, A. (1980) *React. Kinet. Catal. Lett.*, **13**, 33.
191 Vidaković, T., Christov, M. and Sundmacher, K. (2007) *Electrochimica Acta*, **52**, 5606.
192 Nagel, T., Bogolowski, N. and Baltruschat, H. (2006) *J. Appl. Electrochem.*, **36**, 1297.
193 Jusys, Z., Kaiser, J. and Behm, R.J. (2002) *Electrochim. Acta*, **47**, 3693.
194 Schmidt, T.J., Noeske, M., Gasteiger, H.A., Behm, R.J., Britz, P., Brijoux, W. and Bönnemann, H. (1997) *Langmuir*, **13**, 2591.
195 Gasteiger, H.A., Markovic, N., Ross, P.N. Jr and Cairns, E.J. (1994) *J. Electrochem. Soc.*, **141**, 1795.
196 Dinh, H.N., Ren, X., Garzon, F.H., Zelenay, P. and Gottesfeld, S. (2000) *J. Electroanal. Chem.*, **491**, 222.
197 Paseka, I. (2006) *J. Solid-State Electrochem.*, **11**, 52.
198 Rush, B.M., Reimer, J.A. and Cairns, E.J. (2001) *J. Electrochem. Soc.*, **148**, A137.
199 Thiel, K.-O., Hintze, M., Vollmer, A. and Donner, C. (2008) *J. Electroanal. Chem.*, **621**, 7.
200 Fernando Hernandez, F. and Baltruschat, H. (2007) *J Solid State Electrochem.*, **11**, 877.

201 Zhang, J., Vukmirovic, M.B., Sasaki, K., Uribe, F. and Adzic, R.R. (2005) *J. Serb. Chem. Soc.*, **70**, 513.
202 Green, C.L. and Kucernak, A. (2002) *J. Phys. Chem. B.*, **106**, 1036.
203 Green, C.L. and Kucernak, A. (2002) *J. Phys. Chem. B.*, **106**, 11446.
204 Bakos, I. and Szabó, S. (1990) *React. Kinet. Catal. Lett.*, **41**, 53.
205 Takakusagi, S., Kitamura, K. and Uosaki, K. (2008) *J. Phys. Chem. C*, **112**, 3073.
206 Mascaro, L.H., Santos, M.C., Machado, S.A.S. and Avaca, L.A. (1997) *J. Chem. Soc., Faraday Trans.*, **93**, 3999.
207 Uhm, S., Chung, S.T. and Lee, J. (2007) *Electrochem. Comm.*, **9**, 2027.
208 Lee, J.K., Jeon, H., Uhm, S. and Lee, J. (2008) *Electrochim. Acta*, **53**, 6089.
209 Lamy-Pitara, E. and Barbier, J. (1997) *Appl. Catal. A: General*, **149**, 49.
210 Kolb, D.M., Prazasnyski, M. and Gerischer, H. (1974) *J. Electroanal. Chem.*, **54**, 25.
211 Kolb, D.M. (1978) *Advances in Electrochemistry and Electrochemical Engineering*, vol. 11, John Wiley Interscience, New York, p. 125.
212 Aramata, A. (1997) *Underpotential Deposition on Single-Crystal Metals, Modern Aspects of Electrochemistry*, vol. 31, Plenum Press, New York, p. 181.
213 Sudha, V. and Sangaranarayanan, M.V. (2005) *J. Chem. Sci.*, **117**, 207.
214 Levich, V.G. (1962) *Physical Hydrodynamics*, Prentice-Hall, Inc., Englewood Cliffs, N.J.
215 Schmidt, T.J., Gasteiger, H.A. and Behm, R.J. (1999) *J. Electrochem. Soc.*, **146**, 1296.
216 Mayrhofer, K.J.J., Strmcnik, D., Blizanac, B.B., Stamenkovic, V., Arenz, M. and Markovic, N.M. (2008) *Electrochim. Acta*, **53**, 3181.
217 Wang, J.X., Markovic, N.M. and Adzic, R.R. (2004) *J. Phys. Chem. B.*, **108**, 4127.
218 Anderson, A.B., Roques, J., Mukerjee, S., Murthi, V.S., Markovic, N.M. and Stamenkovic, V. (2005) *J. Phys. Chem. B.*, **109**, 1198.
219 Liu, Z., Yu, C., Rusakova, I.A., Huang, D. and Strasser, P. (2008) *Top. Catal.*, **49**, 241.
220 Koh, S., Toney, M.F. and Strasser, P. (2007) *Electrochim. Acta*, **52**, 2765.
221 Dundar, F., Smirnova, A., Dong, X., Ata, A. and Sammes, N. (2006) *J. Fuel Cell Sci. and Tech.*, **3**, 428.
222 Baglio, V., Di Blasi, A., D'Urso, C., Antonucci, V., Aricò A.S., Ornelas, R., Morales-Acosta, D., Ledesma-Garcia, J., Godinez, L.A., Arriaga, L.G. and Alvarez-Contreras, L. (2008) *J. Electrochem. Soc.*, **155**, B829.
223 Li, H., Sun, G., Li, N., Sun, S., Su, D. and Xin, Q. (2007) *J. Phys. Chem. C*, **111**, 5605.
224 Li, W., Zhoua, W., Li, H., Zhoua, Z., Zhou, B., Suna, Q. and Xin, Q. (2004) *Electrochim. Acta*, **49**, 1045.
225 Ioroi, T. and Yasuda, K. (2005) *J. Electrochem. Soc.*, **152**, A1917.
226 Albery, W.J. and Hitchman, M.L. (1971) *Ring-Disc Electrodes*, Clarendon Press, Oxford.
227 Bonakdarpour, A., Dahn, T.R., Atanasoski, R.T., Debe, M.K. and Dahn, J.R. (2008) *Electrochem. Solid-State Lett.*, **11**, B208.
228 Shih, Y.-H., Sagar, G.V. and Lin, S.D. (2008) *J. Phys. Chem. C*, **112**, 123.
229 Wang, F. and Hu, S. (2006) *Electrochim. Acta*, **51**, 4228.
230 Santos, L.G.R.A., Oliveira, C.H.F., Moraes, I.R. and Ticianelli, E.A. (2006) *J. Electroanal. Chem.*, **596**, 141.
231 Inaba, M., Yamada, H., Tokunaga, J. and Tasaka, A. (2004) *Electrochem. Sol. State Lett.*, **7**, A474.
232 Schmidt, T.J. and Gasteiger, H.A. (2003) *Handbook of Fuel Cells: Fundamentals, Technology and Applications*, 1st edn, John Wiley & Sons, Ltd, Chichester, p. 316.
233 Stamenković, V., Schmidt, T.J., Ross, P.N. and Marković, N.M. (2003) *J. Electroanal. Chem.*, **554–555**, 191.
234 Ramaswamy, N., Hakim, N. and Mukerjee, S. (2008) *Electrochim. Acta*, **53**, 3279.

235 Paulus, U.A., Schmidt, T.J., Gasteiger, H.A. and Behm, R. (2001) *J. Electroanal. Chem.*, **495**, 134.
236 Schmidt, T.J., Paulus, U.A., Gasteiger, H.A., Alonso-Vante, N. and Behm, R.J. (2000) *J. Electrochem. Soc.*, **147**, 2620.
237 Antoine, O. and Durand, R. (2000) *J. Appl. Electrochem.*, **30**, 839.
238 Claude, E., Addou, T., Latour, J.M. and Aldebert, P. (1998) *J. Appl. Electrochem.*, **28**, 57.
239 Markovic, N.M., Gasteiger, H.A. and Ross, P.N. (1995) *J. Phys. Chem.*, **99**, 3411.
240 Geniès, L., Faure, R. and Durand, R. (1998) *Electrochim. Acta*, **44**, 1317.
241 Paulus, U.A., Wokaun, A., Scherer, G.G., Schmidt, T.J., Stamenkovic, V., Radmilovic, V., Markovic, N.M. and Ross, P.N. (2002) *J. Phys. Chem. B*, **106**, 4181.

4
Combinatorial and High Throughput Screening of DMFC Electrocatalysts

Rongzhong Jiang and Deryn Chu

4.1
Introduction

Portable power sources play an important role in our daily life due to the increasing needs of using various power consuming electronics such as notebook computers, cell phones and digital cameras. Currently, the most widely used portable power sources are batteries. The high cost and low energy density of batteries are far behind our demands for power requirement. Even for the state-of-the-art battery, such as lithium ion battery, the energy density is less than 180 Wh/Kg. Innovative power sources are urgently needed to be an alternative to batteries. In recent years, fuel cells have attracted much attention, as they can directly convert the chemical energy of fuel to electric energy through electrochemical reactions. A fuel cell consists of an anode, a cathode, and an electrolyte membrane between the anode and the cathode. The most common type of fuel cell is polymer electrolyte membrane (PEM) fuel cell, which uses hydrogen or alcohol as fuel. Direct methanol fuel cells (DMFCs) directly uses methanol as fuel. Because of the high theoretical fuel energy density (6081 Wh/kg), and the ease of storage and transport of the liquid fuel, DMFC has become one of the more promising energy conversion sources for portable applications. People began to study direct use of methanol to generate electricity by a fuel cell early in 1960s [1–4]. However, the research on DMFC was relatively slow during the first 30 years [5–10] due to two main technical barriers: slow catalytic kinetic rates for methanol oxidation at the anode and oxygen reduction at the cathode; as well as methanol crossover from the anode to the cathode through the electrolyte membrane. Methanol crossover causes not only fuel waste, but also cathode electrode depolarization. The application of Nafion perfluorosulfonic acid as a solid polymeric electrolyte membrane has largely blocked the methanol crossover in comparison to using liquid electrolyte [11–15].

Much effort has been made in seeking electrode catalysts. Various non-platinum catalysts, such as metallo-porphyrins and metallo-phthalocyanines [16–19] were investigated for catalytic reduction of oxygen. For example, Anson and Collman et al. [20, 21] have synthesized and studied dimeric cofacial cobalt porphyrins, which demonstrated the capability of catalyzing the oxygen 4-electron reduction to water. However, the catalytic activity of these transition metal macrocycles is not stable in acidic electrolytes. Although an approach of heat-treatment [22–24] was proposed to enhance the stability, these transition metal macrocycles have not actually been used as catalysts in a fuel cell because of uncompetitive catalytic activity and stability as compared to platinum-based catalysts. The catalytic activity of this type of catalyst is not only dependent on the transition metals, but also is highly dependent on the organic ligand. So far, only a small number of transition metal complexes have been tested for electrode catalysts, and various new organic ligands need to be synthesized.

The majority of work has been focused on noble metal catalysts for practical fuel cell applications. So far, the most widely used catalyst is still pure and supported platinum for catalytic reduction of oxygen. Various reports of research indicate that the addition of a secondary transition metal to the platinum to form an alloy will improve the catalytic activity for oxygen reduction, and also improve the tolerance to the poisoning of methanol that causes the cathode depolarization when the methanol crosses over from the anode to the cathode in a DMFC.

Pure and supported platinum was also used for catalytic oxidation of methanol at the early time. Because of the anode catalyst poisoning by the intermediate products (such as CO) either pure or supported platinum showed a poor performance for catalytic methanol oxidation. Significant improvement for catalytic methanol oxidation was obtained when a platinum-based bi-functional alloy, such as PtRu, was used [10]. The platinum catalyst's main functions are adsorption and consecutive dehydrogenation of methanol molecules, and the schemes are described in Equation 4.1. However, pure platinum is poor for water adsorption and water dehydrogenation. The electrode reaction (Equation 4.2) occurs at a potential higher than 0.4 V. Equation 4.3 shows that CO_2 forms at pure platinum catalysts. Apparently, a secondary transition metal is needed to form an alloy with platinum in order to reduce the activation potential for methanol oxidation. For example, when PtRu alloy is used, the Ru atoms will substitute Pt's function in Equation 4.2; and the reaction (4.4) happens at an electrode potential of 0.2–0.3 V, which is lower than the potential needed for the pure Pt catalyst. The function of Pt-Ru for catalytic methanol oxidation is commonly described in the term of the 'bi-functional mechanism' [25–27]. Therefore, the overall potential of methanol oxidation shown in Equation 4.5 can be significantly reduced.

$$CH_3OH + Pt = Pt(CO)_{ads} + 4H^+ + 4e^- \tag{4.1}$$

$$H_2O + Pt = Pt(OH)_{ads} + H^+ + e^- \;\; > 0.4 \text{ V vs. NHE} \tag{4.2}$$

$$Pt(CO)_{ads} + Pt(OH)_{ads} = 2Pt + CO_2 + H^+ + e^- \tag{4.3}$$

$$H_2O + Ru = Ru(OH)_{ads} + H^+ + e^- \;\; 0.2-0.3 \text{ V vs. NHE} \tag{4.4}$$

$$Pt(CO)_{ads} + Ru(OH)_{ads} = Pt + Ru + CO_2 + H^+ + e^- \tag{4.5}$$

Most recently, binary, ternary, and quaternary transition metal alloys are being investigated for catalytic oxidation of methanol and for catalytic reduction of oxygen. However, the amount of the work for development of catalysts increases tremendously as the number of metals in the alloy increases. Thousands or tens of thousands of transition metal compositions must be tested for their activity and stability.

In addition to research on multiple-elements and multiple-functions of catalysts, much attention has been paid to seeking the catalysts with smaller particle size and larger surface area. Nanotechnologies have been widely applied in catalyst development [27–29]; and carbon nanotubes are used as catalysts supports. There is much work to do, not only for the development of platinum-based multi-element transition metal alloys, non-platinum catalysts, but also nanosized, high surface area catalysts, and catalyst supporters. The traditional methods either for synthesis or for characterization are far behind our demands for highly efficient innovative catalysts.

In recent years, there is much interest in using combinatorial methods for high throughput synthesis and screening of electrode catalysts [30]. The combinatorial method was initially used to screen a library of biological structures [31], and development of catalytic antibodies [32]. Since Xiang *et al.* [33] introduced it for the discovery of superconductors in 1995; combinatorial method has been widely applied in fast synthesis and characterization of materials, chemicals, and catalysts. As shown in Figure 4.1, there is a great increase of annual publications for the research on catalysts with combinatorial method in the late 1990s, and thereafter, the annual publications are kept at a rate of about 100. By the end of 2007, the total number of publications for research on catalysts with combinatorial method has been 933. Among these, there are 40 publications for research on methanol oxidation catalysts, 29 on oxygen reduction catalysts, and 56 on fuel cell catalysts. The present article focuses on combinatorial and high throughput screening of DMFC catalysts, including the principle, methods and applications.

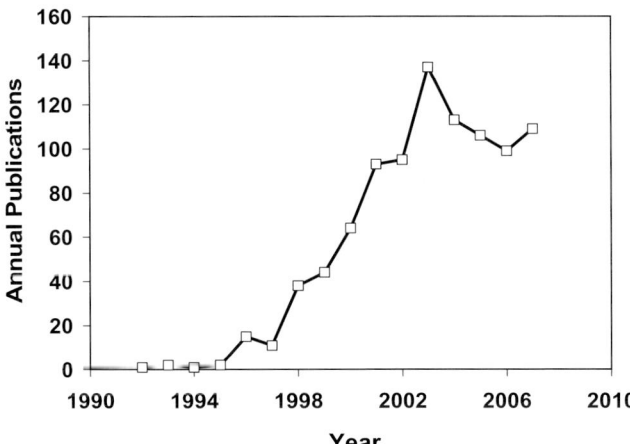

Figure 4.1 Annual publications on catalyst discovery with combinatorial method searched with ISI Web of Knowledge by March 2008.

4.2
Common Procedures for the Development of DMFC Catalysts

Traditional development of DMFC catalysts involves multiple steps and time consuming work. Figure 4.2 shows common procedures for the synthesis and characterization of DMFC catalysts. The first step is chemical synthesis. There are many technologies for the synthesis of DMFC catalysts, such as impregnation, co-precipitation, colloidal, and micro-emulsion methods. For the impregnation method, Pt and transition metal precursors are first adsorbed on a carbon support and subsequently the mixture is dried and reduced with a hydrogen stream at 200 to 700 °C in a tubular furnace. The high temperature of the hydrogen stream drives the reduction to completion. The disadvantage is that the high temperature decreases the active surface area because of the aggregation of the nanoparticles. For the co-precipitation method, the platinum halide and transition metal oxide are mixed with an excess of sodium nitrate, and heated to 500 °C for a few hours. Then the solidified

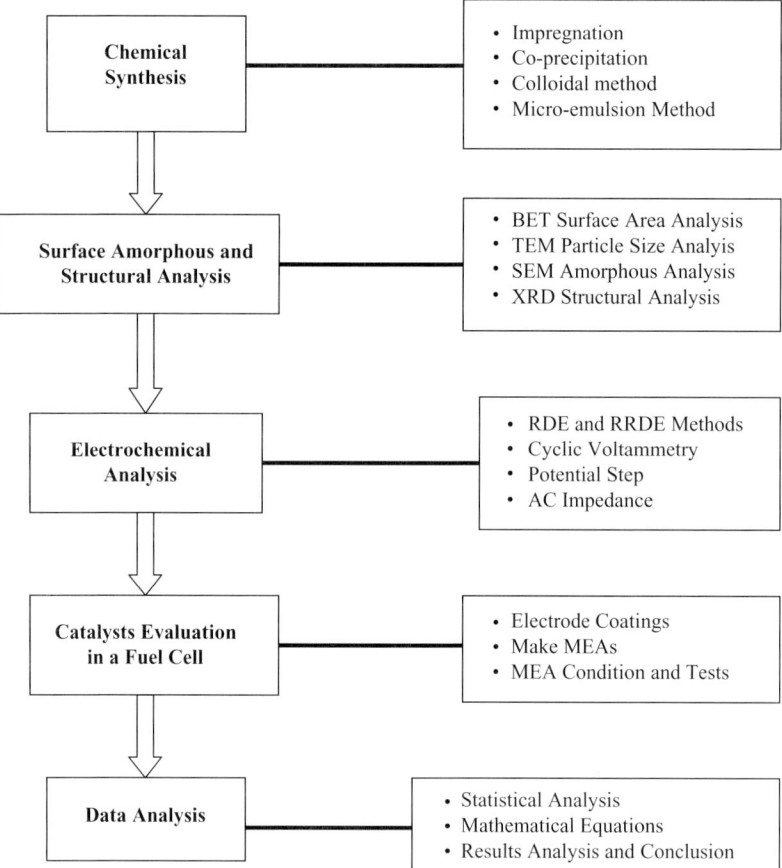

Figure 4.2 Common procedures of development of DMFC catalysts.

melt is washed with water to remove nitrate and chloride, and reduced by hydrogen gas between room temperature and 200 °C. For the colloidal method, surfactant or polymer is often used as stabilizers to protect the nanosized particles from recombination; and the reduction temperature is well below 200 °C. However, it has been discovered that the conditions of the colloidal method also create a space for the nanoparticle to grow [25], which is a main factor to limit the particle size for this method. In addition, the removing of stabilizer after reduction becomes a major concern. The micro-emulsion is a new method for synthesis of nanosized metal alloys developed in recent years, which also uses surfactant as a stabilizer, but does not leave space for nanoparticles to grow. A micro-emulsion consists of nanosized water droplets surrounded by an organic phase and stabilized by a surfactant. The dispersion of metal nanoparticles is dependent on the ratio of surfactant to metal precursor, and the pH environment.

After synthesis, the DMFC catalysts need to be characterized for surface area, chemical structure, composition, and catalytic activity. As shown in Figure 4.2, BET surface area analysis is used to measure the surface area of materials, transmission electron microscope (TEM) for the particle size, scanning electron microscope (SEM) for surface morphology, and X-ray diffraction (XRD) for structural analysis. Many electrochemical technologies are used to determine the catalytic activity and stability of the catalysts. The rotating disk electrode (RDE) and rotating ring disk electrode (RRDE) are used to measure kinetic rates and the number of electron transfers for oxygen reduction. Cyclic voltammetry (CV) and potential step are used to measure the catalytic activity for methanol oxidation. AC impedance is used to determine the reaction impedance for oxygen reduction and for methanol oxidation at the interface of the catalytic layer and electrolyte in an electrochemical cell, and also to measure the impedance of the electrolyte membrane. Complex instrument and equipment are required in both physical and electrochemical methods.

The final step in development of a DMFC catalyst is to make a membrane electrode assembly (MEA), and test how the catalyst performs in a single fuel cell. Furthermore, a large amount of experimental data must be analyzed after these characterizations. For traditional synthesis and characterization as shown in Figure 4.2, only a single experiment is carried out in a time period. There are a large number of experiments to do, either for synthesis or for characterization in the development of a DMFC catalyst.

4.3
General Methods for Combinatorial and High Throughput Screening

As opposed to the traditional methods, the combinatorial method requires a large number of catalyst samples to be synthesized or characterized in parallel within a small spatial area. The combinatorial method does not shorten the reaction time for synthesis of a single catalyst sample, nor speed up the time for characterization of it. However, it significantly saves the total time of synthesis and characterization for a large number of catalyst samples than doing them one by one. Unfortunately, there is no one combinatorial method that is generally suitable for all types of DMFC catalyst synthesis, nor for all types of characterization. According to the specific requirements, each type of catalyst synthesis or characterization needs to create its unique combinatorial method. General steps for carrying out combinatorial experiment are: (1) generate an array of catalyst samples; (2) design a screening method; (3) prepare equipment and instruments especially for the combinatorial measurements; (4) design experimental procedures and carry out experiment; and (5) data treatment and theoretical analysis.

The first step is generating an array of catalyst samples. Figure 4.3 shows a schematic drawing of arrays that are used for combinatorial synthesis and high

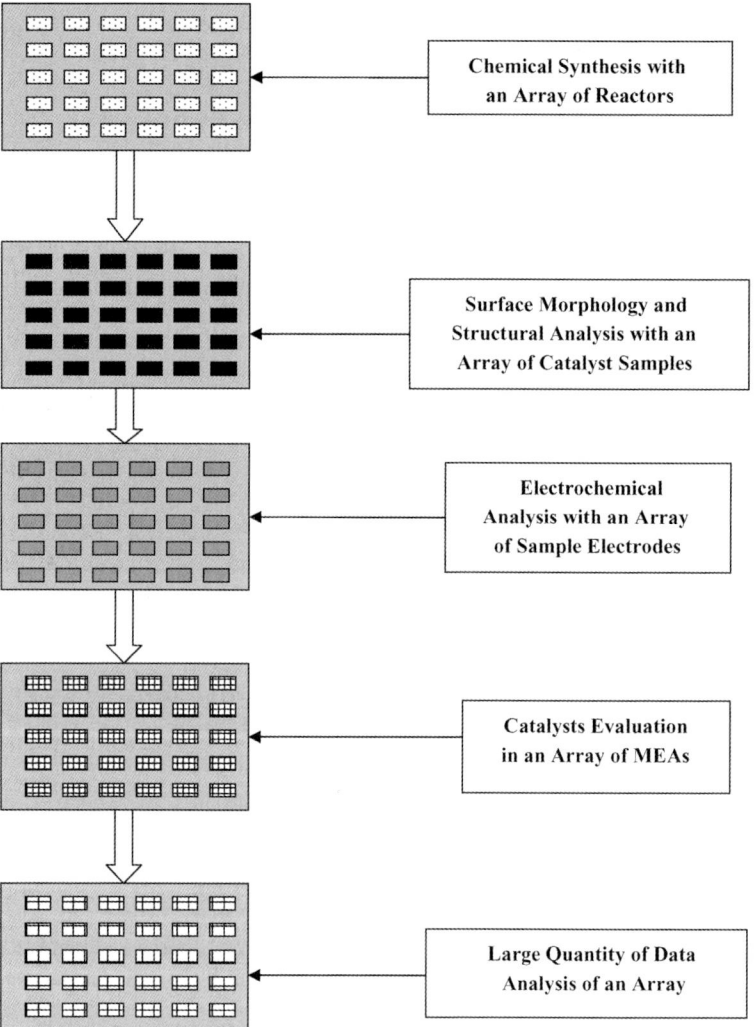

Figure 4.3 Schematic drawing of arrays for DMFC catalyst synthesis, characterization, and data analysis; each of them consists of 30 parallel and spatially separated samples.

throughput screening of DMFC catalysts. The members in the array must be spatially separated and be tested or scanned in parallel at about the same time within a small spatial area. The quantity of catalyst members in the array is a primary criterion for evaluation of the capacity of the combinatorial method. If the array contains more catalyst members, it has a greater capability for fast screening of DMFC catalysts. For chemical synthesis, the array consists of a number of small reactors, each of which has about the same function as that in traditional chemical synthesis. For carrying out combinatorial synthesis, each time we change only one experimental factor, and keep the others constant among various reaction and environmental conditions, such as temperature, pressure, gas flow rate, stirring status and composition of reactants. For example, if we synthesize a series of catalysts with different ratio of transition metal contents, we change only the chemical compositions in the reactor array, but keep the other operational conditions constant. How to change or maintain the environmental conditions for each of members in a reactor array is a great challenge for chemical and catalyst synthesis by the combinatorial method.

The second step for carrying out combinatorial experiments is designing a high throughput screening method. One type of screening method is based on measuring the physical properties of the targeted materials, where neither chemical nor electrochemical reaction occurs. This type of screening method includes BET analysis, SEM, TEM, and XRD measurements. Another type of screening method is based on measuring the chemical and electrochemical properties, where chemical or electrochemical reaction occurs, such as pH change, heat release, gas generation, and color change. The latter type of screening method includes measuring pH, temperature, current, voltage and impedance. The electrochemical methods, such as RDE, RRDE, potential step, AC impedance and CV belong to the second type of screening method. The efficiency of the combinatorial method is significantly dependent on how fast the screening method is. In most cases, the arrays for catalysts synthesis and screening are different. However, in some special conditions, an array can be used for both catalyst synthesis and catalyst screening.

4.4
Methods of Combinatorial Synthesis

4.4.1
Chemical Synthesis

Combinatorial chemical synthesis is generally conducted in an array of chemical reactors, where the reactants are added; and the temperature, pressure and environmental conditions are controlled. It is difficult to accurately control the environmental conditions for a large number of chemical reactors in a reactor array at the same time within a small spatial area. In order to realize effective combinatorial synthesis, some automated or robotic methods are designed to automatically generate reactor array and conduct parallel chemical synthesis. Many automated or robotic methods [34–36] were reported to control the environmental conditions of reactor arrays for chemical synthesis. Dellamorte *et al.* [34] described an instrument for investigation of monolith catalysts. The instrument consists of eight-member reactor array, in which the eight reactors were designed with separate thermocouples and radiant heaters, allowing for the independent measurement and control of each reactor temperature. Yamada *et al.* [35] developed an instrument for the preparation of heterogeneous catalysts by an impregnation method, which consists of a powder dispensing robot and an automated liquid handling machine equipped with ultrasonic and vortex mixer. The powder dispersion instrument connected with a robot that automatically prepares 56 tubes of powders at one time. Hoogenboom *et al.* [36] developed a robot system for combinatorial chemical synthesis. Figure 4.4 shows a picture of the machine and a schematic overview of the workplace of the synthesis robot. The picture also shows the modular approach of this robot with five positions for reactor arrays (bottom left of the scheme), two positions to place microtiter plates or custom racks (top left of the scheme; here a rack for 2 ml vial is programmed), one position for a large vial rack (here for 8 ml vials) and stock solution

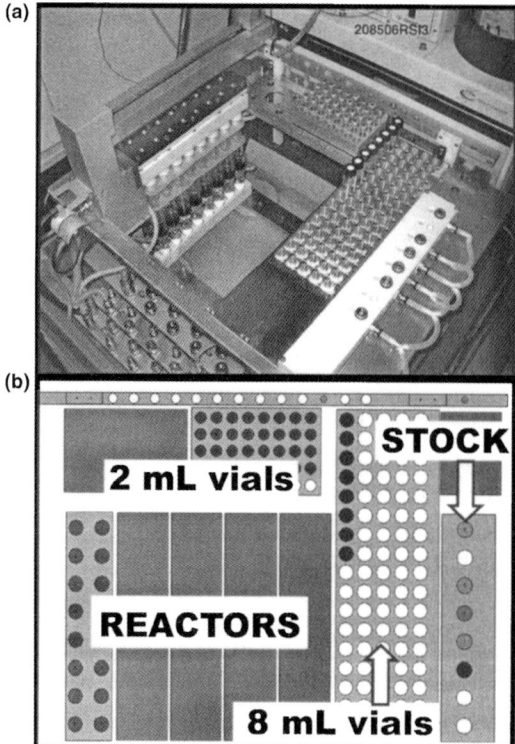

Figure 4.4 An automated parallel chemical synthesis platform. (a) picture of the ASW2000 synthesis robot; (b) a schematic overview of the workspace of the synthesis robot. (Reproduced with permission from [36]. Copyright © 2005 American Institute of Physics.)

rack (right of the scheme). Moreover, the synthesizer is equipped with a XYZ-liquid handling system, and agitation is achieved by a vortex movement of the reactors. In addition, the synthesizer is covered with a glove box to retain an inert atmosphere for oxygen and moisture sensitive reactions. Automated parallel synthesis can be performed in the arrays containing 80-member 13 ml reactors, 40-member 27 ml reactors, and 20-member 75 ml reactors. Although these robotic machines described above are designed for chemical or heterogeneous catalyst synthesis, they can also be used for synthesis of DMFC catalysts by appropriate modifications.

As opposed to the combinatorial synthesis described above, Reddington et al. [30] used a very simple method to synthesize multi-element metal alloys for DMFC catalysts. They used a commercially available inkjet to print metal salt inks onto a carbon paper, and subsequently reduced the salt inks to metals or alloys by borohydride to quickly generate an array of a large number of catalysts. After the array was dried, washed, and contacted electrically, these catalysts were screened for activity. The liability of the printing method for the synthesis of alloys is questionable because the printed inks may not be completely reduced and alloyed. However, even if the early work of combinatorial synthesis is not perfect, the method proposed by Reddington et al. [30] has been recognized as a pioneer work of combinatorial electrochemistry.

4.4.2
Electrochemical Synthesis

Electrochemical reactions are a type of chemical reaction that occur at the interface of an electrode and electrolyte. Chemicals, metals or catalysts can be synthesized through electrochemical reactions in an electrochemical cell. There were a number of reports of combinatorial synthesis using electrochemical methods [37–41]. Brandli et al. [37] generated a 256- member array of porous aluminum oxide with different pore size and pore density by an automated electrochemical synthesis. Beattie et al. [38] electrochemically synthesized Cu-Sn-Zn alloys by a combinatorial method. Jayaraman et al. [39, 40] developed an automated system for electrochemical synthesis and high throughput screening of methanol and CO oxidation catalysts for fuel cells. The system consists of three major subsystems: a motion system, a computer with a software interface, and an electrochemical subsystem. The electrochemical subsystem contains an array of 9×7 working electrodes uniformly arranged on a polypropylene block, and an electrode probe. The electrode probe contains a counter and a reference electrode, which is attached to an automated XYZ motion system. During electrochemical synthesis the deposition of Pt-WO_3 can be controlled by a potential pulse or current pulse with the programmed software. Erichsen et al. [41] developed and evaluated an electrochemical robotic system for combinatorial synthesis. This system uses standard microtiter plates as reaction wells for potentiostatic and galvanostatic electrochemical synthesis, and also for high throughput electroanalysis. This system allows electrochemical experiments under inert gas atmosphere, in aqueous or organic solvents, and for adding or removing solutions by means of integrated syringe pumps. Figure 4.5 shows a schematic

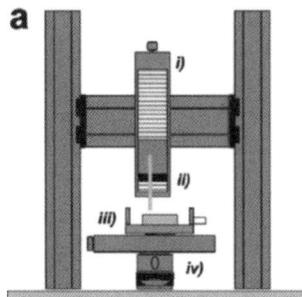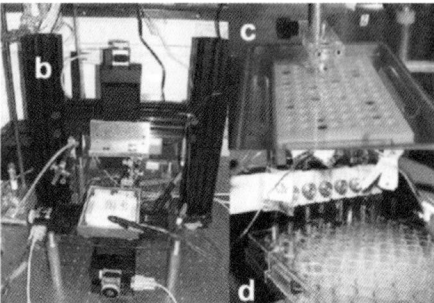

Figure 4.5 An automated parallel electrochemical synthesis system. (a) Schematic representation of the electrochemical robotic system: (i) Z-positioning table, (ii) electrode holder with working electrode, counter electrode and reference electrode, (iii) chamber with argon supply, (iv) x-y position table; (b) Over view; (c) Measurement using a three electrode setup; and (d) Eight-fold electrode holder equipped with eight electrode setups. (Reproduced with permission from [41]. Copyright © 2005 American Institute of Physics.)

drawing of the design and the photographs of the automated electrochemical system. The system contains a Z-position table; an electrode holder with working, counter and reference electrodes; a chamber with inert gas supply; and an x–y position table. The electrochemical synthesis can be consecutively conducted in the array of the reaction wells.

4.4.3
Physical Synthesis

Physical methods, such as physical vapor deposition, are commonly used to synthesize multiple element alloys and metal oxides. Pulsed laser deposition, sputtering, evaporation, and molecular beam epitaxy have all been used for material library fabrication [42–47]. Whitacre et al. [42] used co-sputtering to synthesize thin film alloy libraries for DMFC anode catalysts. They prepared Pt-Ru and Ni-Zr-Pt-Ru catalyst thin films (<10 nm thickness) on an array of 36 gold electrodes that were micro-fabricated on 12.5 × 12.5 cm glass substrate. Bonakdarpour et al. [43] used a magnetron sputtering-based deposition method to synthesize oxygen reduction electrocatalysts for polymer electrolyte membrane (PEM) fuel cells. They prepared $Pt_{1-x}M_x$ (M = Fe, Ni) alloys onto nanostructured whisker-like supporters, and studied the dissolution of Fe and Ni from the alloys under simulated fuel cell operating conditions. Cooper et al. [46, 47] designed a plasma sputtering system for deposition of thin-film combinatorial libraries. As shown in Figure 4.6, they used a rotating carousel to position a shadow mask between the targets and the substrate. Multilayer films are built up by deposition sequentially through various masks. They used post-deposition annealing treatment to promote inter-diffusion of the layered structure to form alloys. Either discrete or compositional gradient alloy libraries can be deposited with this system. They investigated Pt-Ru-W and Pt-Ru-Co ternary alloys for DMFC anode catalysts using plasma sputtering to prepare catalyst libraries.

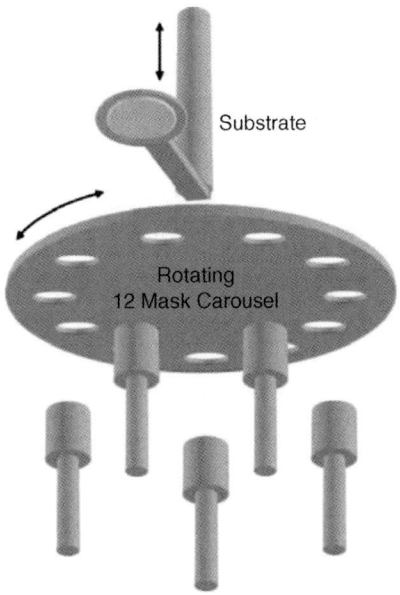

Figure 4.6 Basic sputter system chamber lay out showing the relative positions of the substrate holder, mask carousel, and sputter sources. (Reproduced with permission from [46]. Copyright © 2005 American Institute of Physics.)

In comparison with chemical and electrochemical synthesis, the method of physical vapor deposition is simpler, and can be used to make larger member catalyst libraries within a smaller spatial area on a substrate by deposition sequentially through various masks. The disadvantage of this method is that the resulting catalysts have much smaller surface area than that of a chemical method, so that the number of the active catalytic sites that can be contacted by the reactants and electrolyte is significantly decreased.

4.5
Electrode Arrays for High Throughput Screening

4.5.1
Direct Electrode Array and Automated Screening Method

An array of working electrodes can be generated directly by integration of a number of single electrodes on a same geometrical plane. Figure 4.7 shows an electrochemical cell that contains a direct working electrode array for combinatorial experiments. Apparently, this combinatorial electrochemical cell is different from that of the common 3-electrode electrochemical cell that contains a working electrode, a reference electrode and a counter electrode. In the combinatorial electrochemical

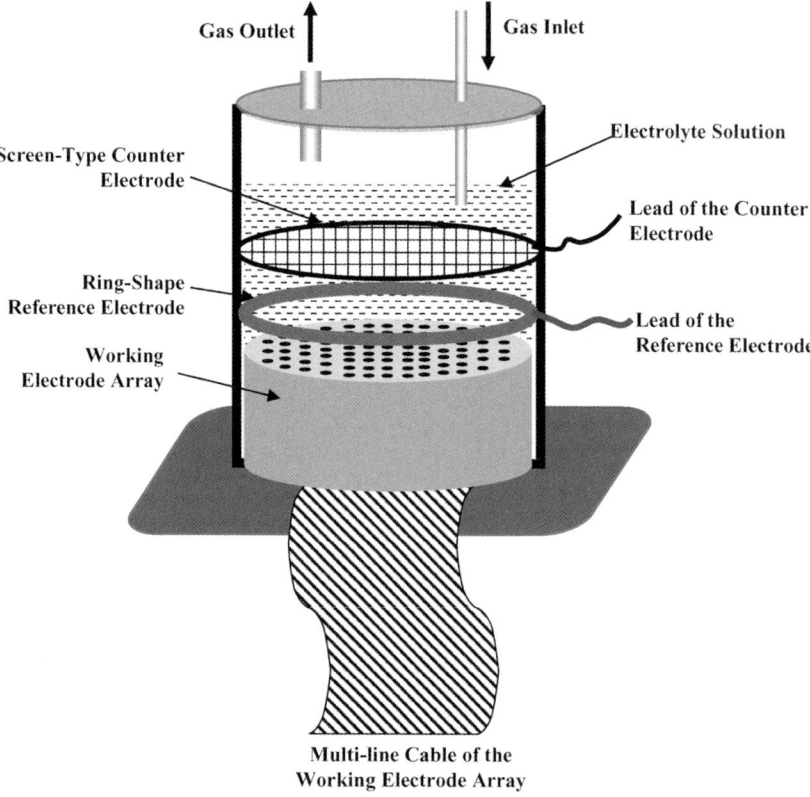

Figure 4.7 An electrochemical cell with direct working electrode array. The multi-line electric cable is used to lead each of the electrodes individually. These electrodes are separated by electric insulating material; but they share the same electrolyte solution.

cell, the working electrode is replaced by a working electrode array that contains a number of working electrodes. Each of these working electrodes in the array is sealed separately. The electric leads from these electrodes are led out by an electric cable with multiple channels. In this type of combinatorial electrochemical cell, all the working electrodes in the electrode array share the same counter electrode (screen-type) and the same reference electrode (ring-shape), which looks like a single electrochemical cell.

A special electrochemical instrument with multiple channels is needed to operate the combinatorial electrochemical cell, which causes a complexity for parallel electrochemical screening. In order to solve this problem, a device with a function of automatic electrode connection and disconnection is designed, and controlled by a computer with appropriate programs. The working electrodes in the electrode array can be connected individually, by group, or entirely. Sullivan et al. [48] reported an automated screening device in 1999 for the study of organo-sulfur mono-layers on gold electrode surfaces. In their electrochemical cell, the working electrodes in the direct electrode array can be automatically connected to a potentiostat by a robot controlling device.

The primary advantage of the direct electrode array is a capability of distinguishing small differences in current and voltage with high accuracy. Secondly, the common traditional electrochemical methods, such as CV, chronoamperometry, chronopotentiometry, potential-step and AC impedance can be directly used in the combinatorial electrochemical cell. Thirdly, all the electrodes in the array can be polished to refresh the surfaces to achieve reproducible experimental results. The disadvantages of the direct electrode array are the significantly increased complexity and cost associated with electrode processing and equipment design when the electrode numbers in the electrode array increases. For the requirements of combinatorial research some multiple channel devices and instruments have been developed [48–50] and improved in recent years.

4.5.2
Material Spot Array on a Single Electrode and Optical Screening Method

A number of targeting electrochemical materials can be studied simultaneously by forming a material spot array on a single electrode. Figure 4.8 shows a schematic view of a material spot array containing 82 material spots on a planar electrode surface. Each of the spots is used to deposit different electrochemical materials or metal alloys or compositions. For example, a number of DMFC catalysts can be coated as spots on a single planar electrode and electrochemically screened by a single electrochemical measurement. The electrode substrates for the planar working electrode are generally graphite, carbon or metals. Actually, such a system is very similar to the traditional single electrochemical cell, except that the working electrode is elaborately tailored into many small regions or spots to adopt the targeting electrochemical materials. In the material spot array all the spots are electrically shorted, which is very

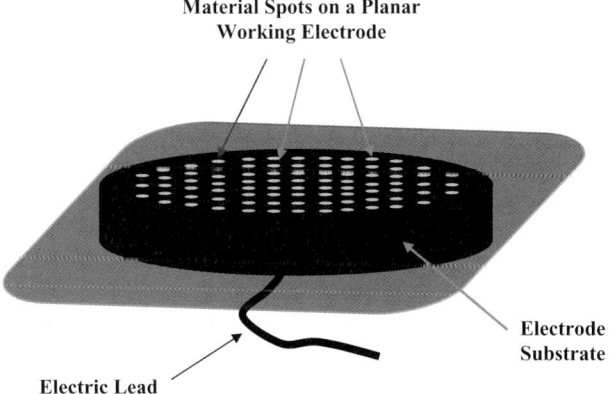

Figure 4.8 A material spot array containing 82 material spots that are coated on a planar electrode surface.

different from that of the direct electrode array where all the electrodes are electrically separated and led with individual wires. Reddington et al. [30], and then Choi et al. [51] reported that hundreds of combinations of binary, ternary, quaternary catalysts were coated on a planar carbon substrate to form material spot arrays on a single electrode surface by using a computer controlled printing technique. The traditional electrochemical method, such as potential-step, can also be used for indirect screening of these material spots. As long as a potential is applied between the working and the counter electrodes, all the spots in the array will have the same potential but different local current because they have different electrochemical activity. The most obvious advantage of this kind of array is that only one electrochemical measurement is needed for analysis of all of the material spots in the array, and that no special electrochemical instrument or device is needed. Furthermore, it is much easier to generate a large member array for a material spot array than for a direct electrode array. After creation of the material spots array, the second step is to find an appropriate screening method to identify the differences of the local current density on each of the material spots, which is a great challenge to this type of combinatorial method.

Reddington et al. [30] first used fluorescent proton indicators as an indirect screening method to identify the local current density on the material spots array. They electrochemically studied methanol oxidation on hundreds of catalyst combinations that were coated on a carbon electrode surface to form material spot arrays. The electrochemical and optical reactions are described by the Equations 4.6 and 4.7.

$$CH_3OH + H_2O \rightarrow CO_2 + 6H^+ + 6e^- \tag{4.6}$$

$$Indicator + H^+ \rightarrow Fluorescent\ Compound + Fluorescence \tag{4.7}$$

Here, quinine and Ni^{2+} complex of 3-pyridin-2yl-(4,5,6) triazolo-(1,5-a) pyridine (PTP) are used as indicators at the neutral and low pH, respectively. The more active electrocatalyst will generate higher current, and produce more protons while a potential difference is applied between the working electrode and the counter electrodes, and in turn, the protons stimulate the indicator to generate fluorescence on the spots. The higher concentration of protons results in brighter spots because the spots in the array are spatially isolated on the electrode surface.

The disadvantage of the optical screening is the dependence of proton generation or consumption in the electrochemical reaction. Because many electrochemical reactions are studied in concentrated base or acid, or do not involve significant proton change, the application of the optical screening method is greatly limited. The accuracy of the optical method is also an issue, which is dependent on visually comparing the strength of the fluorescence, which is more affected by errors and irreproducibility than the direct electrochemical measurements [48].

Most recently, a number of reports [48, 51–53] described the optical screening method with some improvements in accuracy using the same or different fluorescent indicators. Sullivan et al. [48] reported that a thin-layer cell with 28 members of material spots in an array was used to study the electrochemical reactions of 1,4-benzoquinone and alkanethiols using fluorescein as an indicator. However, the

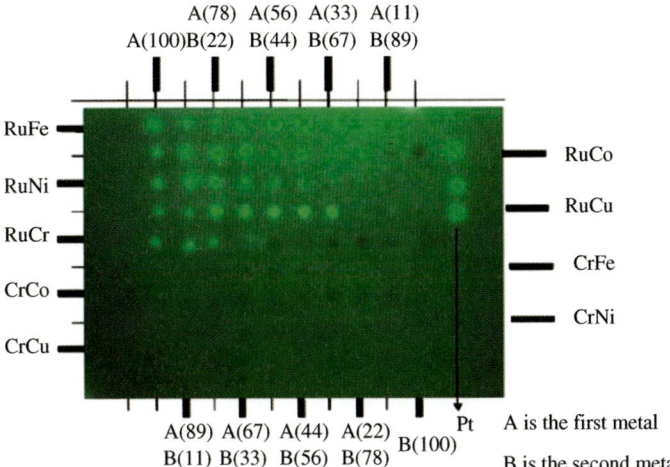

Figure 4.9 Fluorescence image of non-Pt binary array at 0.55 V vs. RHE in oxygen saturated electrolyte, which contained, 100 mM fluorescein sodium salt and 0.2 M $NaNO_3$. (Reproduced with permission from [54]. Copyright © (2006) Elsevier.)

accuracy of the optical screening method is improved only a little due to the minimization of the fluorescent contribution from the background by the thin-layer electrochemical cell.

Figure 4.9 shows an example of optical screening of material spots array on a single electrode by Liu *et al.* [54]. They reported that the catalytic activity of binary alloys for oxygen reduction can be identified by observing the fluorescence. They used a fluorescein as a pH indicator for detecting pH change of the electrolyte in the vicinity of the cathode caused by oxygen reduction.

4.5.3
Indirect Electrode Array on a Single Conductive Substrate and Electrolyte Probe Screening Methods

Jiang *et al.* [55, 56] reported a type of electrode array that is different from both direct electrode array and material spots array. They successfully screened metal alloy DMFC catalysts coated on the electrode array for combinatorial electrochemical analysis. Figure 4.10 shows an indirect electrode array containing 144 electrodes arranged on a single graphite substrate and the magnified side-view of two adjacent electrodes [56]. The surface of the graphite substrate is covered with a plastic coating except for certain areas reserved for loading of electrode materials. The plastic material is used to insulate only the electrode surfaces but not their internal contacts. The small round circles on the exposed graphite surfaces of the electrode array are coated with different combinations of catalysts, and then covered with microporous separators to load electrolyte. The electrodes in the working electrode array are electrically shorted to each other because they share the same electrically conductive

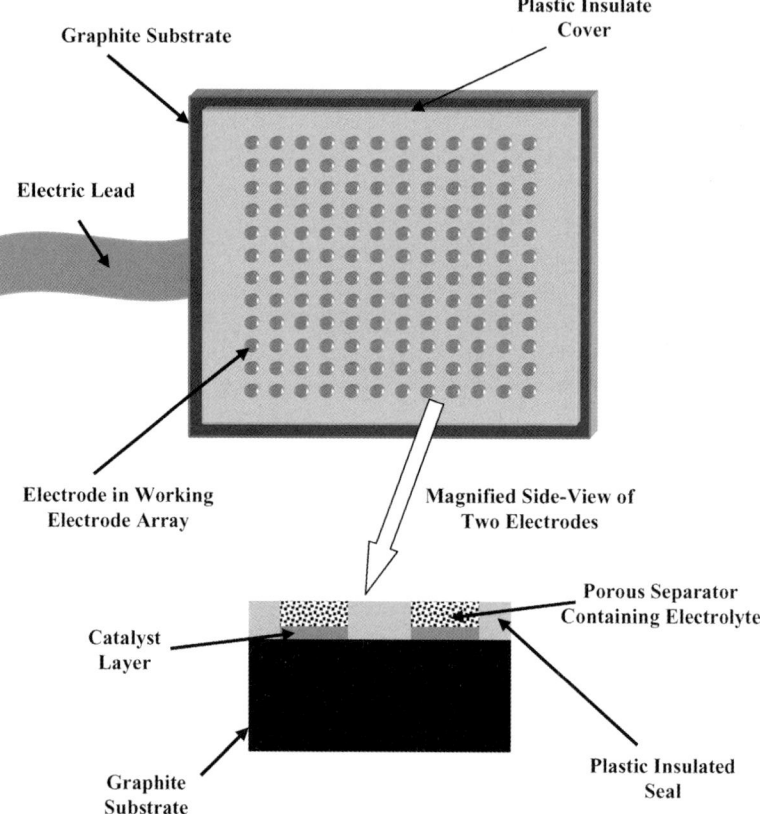

Figure 4.10 An indirect electrode array containing 144 electrodes arranged on a single graphite substrate and the magnified side-view of two adjacent electrodes. (Reproduced with permission from [56]. Copyright ©2003 Kluwer Academic/Plenum Publishers.)

substrate. Therefore, we define this type of electrode array as an indirect electrode array in order to identify its difference from the direct electrode array. On the other hand, the microporous separators on the small circles in the indirect electrode array are isolated from each other; these electrode surfaces are ionically insulated when the electrolyte is added into the porous separators. However, in the material spot array as shown in Figure 4.7, all the material spots are ionically shorted when electrolyte is added into the electrochemical cell.

The indirect electrode array can be screened with an electrolyte probe. The upper drawing in Figure 4.11 shows a design for the electrolyte probe that is formed by filling concentrated electrolyte in a flexible tube whose terminals are sealed by microporous material [55]. The ionic conductance can be adjusted by the thickness and length of the electrolyte probe, and by the concentration of the electrolyte. A tiny reference electrode can be inserted in the electrolyte tube where it is closest to the

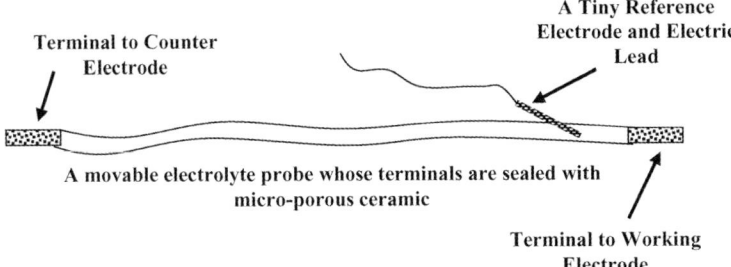

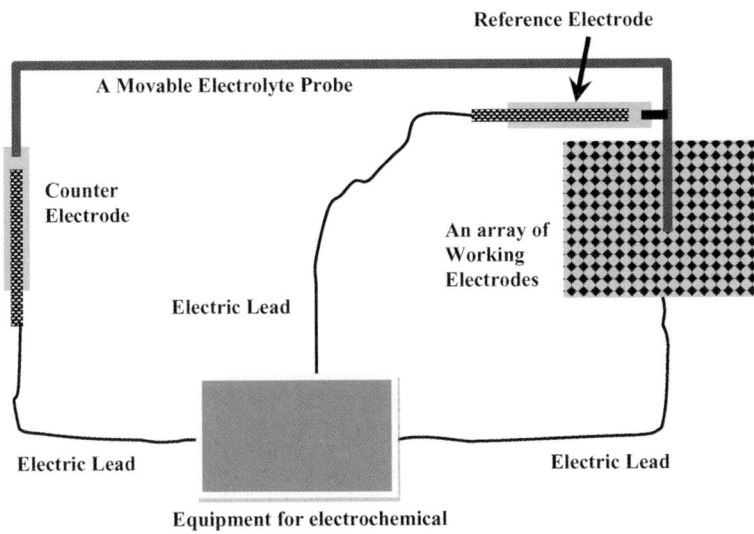

Figure 4.11 Combinatorial measurement with an indirect electrode array for working electrodes and a corresponding movable electrolyte probe for fast screening. (Reproduced with permission from [55]. Copyright © (2002) Elsevier.)

terminal toward the working electrodes. For studying two-electrode cells, such as batteries or fuel cells, the reference electrode is not necessary. All the working electrodes in the indirect electrode array have a common counter electrode that is located separately, a certain distance from the indirect electrode array.

The lower drawing in Figure 4.11 shows the principle of combinatorial measurement with the indirect electrode array and the movable electrolyte probe [55]. During the combinatorial analysis, an individual electrochemical cell is formed instantly, when the movable terminal of the electrolyte probe contacts one of the working electrodes in the electrode array. The electrochemical responses (for example, current, voltage, impedance, conductivity, etc.) recorded by the electrochemical instrument relate only to the working electrode spot that the electrode probe touches, but other working electrode spots, while ionically insulated, do not make any contribution. After completing one working electrode analysis, the terminal of the electrolyte probe is moved onto the next working electrode for the next analysis, and so on, until all the working electrodes are screened. The major difference of this electrochemical cell's configuration lies in the fact that, a traditional electrochemical cell has a working electrode always facing toward the counter electrode at a very short distance, while the counter electrode for the combinatorial analysis is physically separated from the indirect electrode array.

The screening method with an electrolyte probe has many advantages. Firstly, it needs only a common electrochemical instrument, and most of the traditional electrochemical methods, such as CV, chronoamperometry, chronopotentiometry, potential-step and AC impedance can be directly used. Secondly, it actually uses electrochemical cells as an array because each individual electrode in the electrode array can form a different electrochemical cell with the electrolyte probe and the counter electrode. Therefore, it can be used to study a wide range of electrochemical materials, such as catalysts, electrolytes and reactants, by simply arranging them in the indirect electrode array. For example, it can be used to study a group of parallel electrochemical reactions that occur in different electrolytes or have different reactant concentrations.

In addition to the electrolyte probe screening method, Jiang et al. reported a pen-shaped O_2-electrode-probe screening method [57] for high throughput analysis of anode electrode materials coated on an indirect electrode array. The O_2-electrode-probe consists of a large-area O_2 electrode and a cylindrical electrolyte sponge with a short cone tip for screening. The electrolyte sponge is saturated with electrolyte. The working electrode array is generated on a graphite plate which is similar to that shown in Figure 4.10. During screening, the wire of the electrode-probe is connected to a potentiostat, and the sponge tip is moved to touch the surface of an electrode in the indirect electrode array to form an electrochemical cell. After one electrode is screened, the sponge tip is moved to another electrode in the indirect electrode array.

4.6
Other Screening Methods for Catalyst Discovery

The method of high throughput screening is a primary important factor for combinatorial research. A screening method must be combined with an appropriate electrode or material array to carry out a combinatorial experiment. For the development of electrode catalysts, the combinations of screening methods and sample arrays can be classified as three types. The first type needs only one electrochemical measurement, such as the combination of optical screening and material spots array with a potential-step measurement. The second type needs consecutive electrochemical scanning, such as the combination of automated screening and direct electrode array, or the combination of electrolyte probe screening and indirect electrode array, to scan the electrodes in the array continuously. The third type does not need any electrochemical measurement, such as SEM, BET, or XRD for the measurement of material sample array. Unfortunately, the third type of screening method is rarely reported in combinatorial measurements due to some limitations of the instrumental design and some special requirements of sample preparations. The efficiency of a screening method is highly dependent on the speed of screening, the resolution of samples in the material array, and the accuracy of data acquisition. In addition to the automated screening, the optical screening, and the electrolyte-probe screening as described above, there are some other screening methods.

4.6.1
Infrared (IR) Thermography

Since Moates et al. [58] used IR thermography to screen libraries of heterogeneous catalyst formulations, IR thermography has attracted much interest to be used as a tool for high throughput screening [59–66]. The principle of IR thermographic imaging is based on the measurement of thermal energy emitted from an object. Unlike visible light, everything with a temperature above absolute zero emits heat. The higher the object's temperature, the greater the IR radiation emitted. The IR emission allows us to record what our eyes cannot see. Moates et al. [58] reported that a library of metal catalysts supported on gamma-alumina is screened for catalytic oxidation of hydrogen. The reaction heat liberated on the surface of the catalysts was detected by an IR thermography. Cypes and Brooks et al. [59, 60] primarily screened a 256-member catalyst library for CO oxidation in about one hour with an IR thermography. Then they used imaging reflection FTIR spectroscopy for the secondary screening, and finally they discovered and improved a series of novel RuCoCe catalyst compositions for CO oxidation. Yamada et al. [66] used IR thermography for the rapid evaluation of anode catalysts for PEM fuel cells. A series of Pt/C modified metal oxides were tested for the combustion of diluted CO, H_2 and a mixture of CO and H_2 in air. By adding the identified metal oxides into Pt/C, some potential CO-tolerance catalyst combinations were found. Figure 4.12 shows IR thermographs of passing H_2 and CO gases over the metal oxide-modified Pt/C catalyst array [66]. The array was made by placing each catalyst in an aluminum vessel, and then the vessels were arranged in a tube with a ZnSe window. The temperature increase at the modified Pt/C catalysts was detected by the IR thermography when they were exposed to the air that contained 2000 ppm H_2 and 1000 ppm CO (Figure 4.12b). However, the unmodified Pt/C catalysts showed no temperature increase because they were not active for hydrogen oxidation in the presence of CO (middle circles in Figure 4.12b). Both the modified and unmodified Pt/C catalysts are not active to CO alone (Figure 4.12a), as no temperature increase was detected.

4.6.2
Scanning Mass Spectrometry

Scanning mass spectrometry is of growing importance for the characterization of catalytically active surfaces [67–70]. It may possibly be applied for screening of fuel cell catalysts in the future. The instrument is capable of measuring catalytic activity spatially resolved by means of two concentric capillaries. Yaccato et al. [68] prepared

(a) CO oxidation

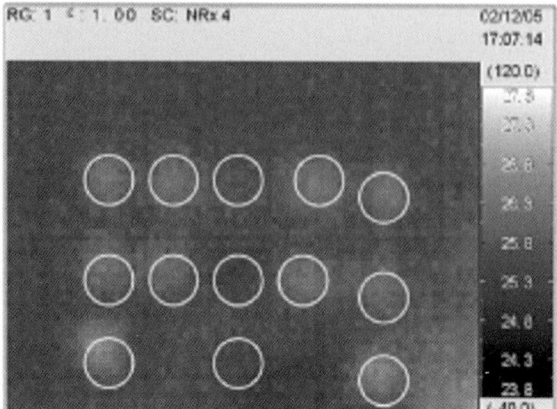

(b) H$_2$ oxidation with CO

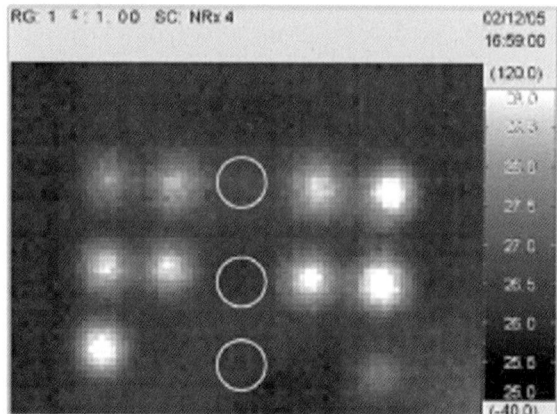

Figure 4.12 IR thermographs of (a) 1000 ppm CO in air passed over Pt/Cs with/without additives; and (b) 2000 ppm H$_2$ and 1000 ppm CO in air. (Reproduced with permission from [66]. Copyright © (2004) Elsevier.)

a library of catalyst samples by liquid dispensing techniques and screened them for catalytic activity by scanning mass spectrometers. This primary screening tool uses quadrupole mass spectrometry (QMS) for rapid serial detection. More than 500 potential catalysts could be screened in a single day. Eckhard et al. [69] reported that a calibrated QMS is combined with a positioning unit derived from a scanning electrochemical microscope (SECM) to sequentially address 25 catalysts that are placed in wells. Highly reproducible catalytic data are obtained under stagnant-point flow conditions by means of coupled gas feed and QMS capillaries. The reaction array can be heated and is fully sealed from the atmosphere by covering with a glass lid.

4.6 Other Screening Methods for Catalyst Discovery

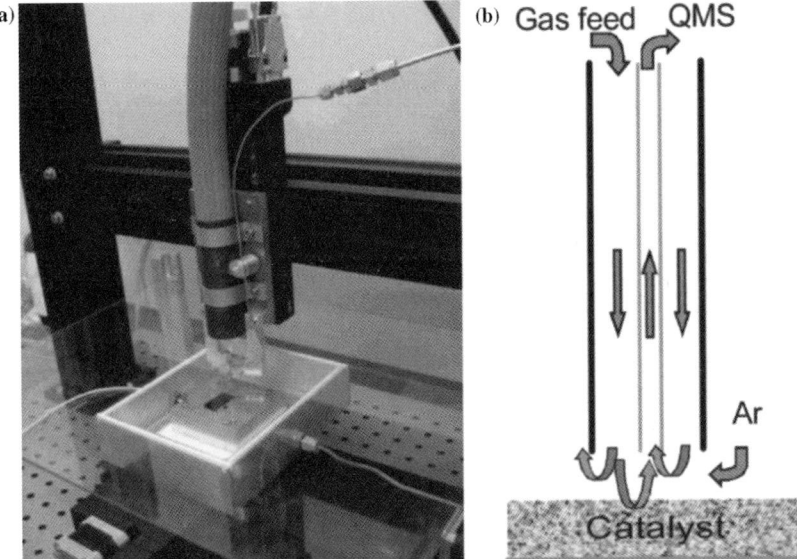

Figure 4.13 (a) Photograph of a scanning QMS system, where the sample is placed on a heating block underneath the scanned QMS capillary inside a chamber flooded with argon; (b) gas flow within the capillary bundle. (Reproduced with permission from [67]. Copyright © (2007) American Chemical Society.)

Figure 4.13 shows the working principle of the scanning mass spectrometer reported by Li *et al.* [67]. In the scanning system, the reactant mixture is fed through a stainless steel tube onto the catalytically active surface of the sample. A quartz capillary of the QMS was placed inside the gas feed tube to achieve a concentric bundle of capillaries with open ends at the same level. A reservoir containing the sample holder was purged with argon throughout the measurement via a tube. The positioning system precisely moves the concentric capillaries in the z-direction and the sample in the x and y directions. The step-motors can be driven either manually or automatically by the program to position the QMS capillary. It took 72 seconds to perform a measurement at one position over the sample surface, including moving the capillary bundle, stabilizing, and data acquisition.

4.6.3
Scanning Electrochemical Microscope (SECM)

SECM was first reported by Husser and Bard *et al.* in 1989 [71]. It is an electrochemical imaging technique for mapping the surface of a substrate when it is immersed in an electrolyte solution. SECM involves the measurement of current through an ultra-micro electrode (with a radius of a few nm to 25 µm) when it is moved in a solution in a vicinity of the substrate. In recent years, this technology has been used as a screening tool for high throughput characterization of electrode catalysts [72–75]. Lu et al. [72, 73] reported that SECM offers the capability for both qualitative and quantitative characterization of electrocatalytic activity of thin-film catalysts on planar substrate for fuel cells. They performed a rapid screening of electrocatalytic activity of ternary Pt-Ru-WC and Pt-Ru-Co thin film gradient material libraries towards hydrogen oxidation in the presence or absence of CO adsorption using this equipment; and identified potential electrocatalysts for PEM fuel cells. Black et al. [74] formed Pt-Ru thin-film libraries by sequential sputter deposition through masks onto Si wafers; and employed SECM to characterize the electrocatalytic activity of the thin film alloys for proton reduction to hydrogen in acidic water solution. Their results show that the most active alloy members have a Pt-Ru ratio about 50/50. The imaging of electrocatalytic activity for oxygen reduction in an acidic medium by SECM was first successfully achieved by Bard and co-workers [75] on discrete catalyst spots on glassy carbon substrate, where a modified SECM mode of 'tip generation–substrate collection (TG–SC)' was proposed as an alternative operation mode from the previous feedback mode [73].

Figure 4.14 shows the scheme of the electrochemical setup for SECM experiment using TG-SC mode to study oxygen reduction on thin-film catalyst substrate [73]. The electrochemical cell is composed of four electrodes with the tip as the oxygen generator and the thin-film catalysts substrate as the working electrode. The constant external positive current of the tip is controlled between the tip and the

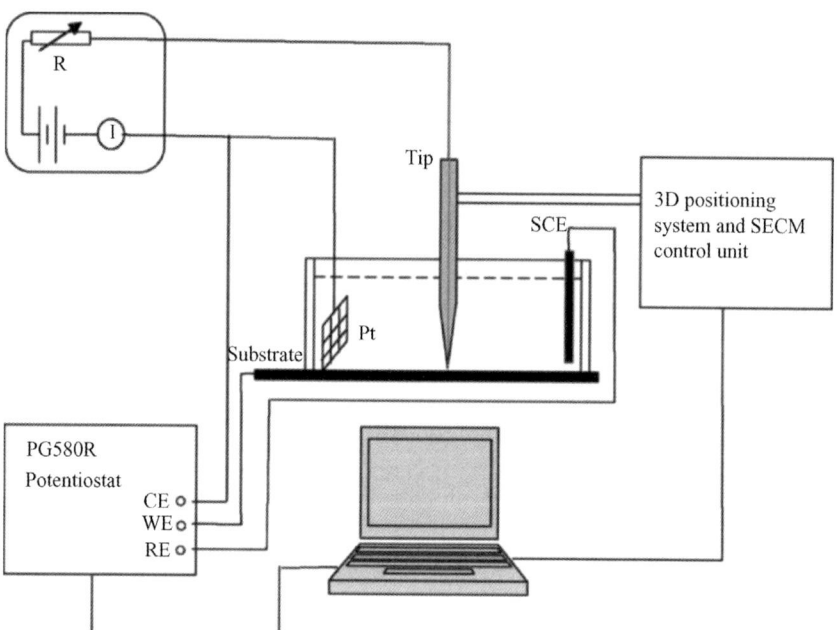

Figure 4.14 Schematic diagram of the electrochemical setup used for SECM experiments by modified TG–SC mode. (Reproduced with permission from [73]. Copyright © (2007) Elsevier.)

counter electrode. The tip current is measured in the accuracy of nA scale. The substrate was positioned at the bottom of the cell horizontally with the deposited thin-film surface facing upwards.

4.7
Combinatorial Methods for DMFC Evaluation and Data Analysis

The DMFC catalyst samples obtained from combinatorial synthesis can either be evaluated with a half electrochemical cell as shown in Figure 4.7, or directly evaluated in actual fuel cells. After a half cell evaluation, the catalysts must be tested in real fuel cells to further confirm the catalytic activity and stability. However, it is more difficult for high throughput screening of catalyst samples in practical fuel cells than that in a half cell, because it is not easy to generate a large member fuel cell array. There are a few reports of generating small fuel cell arrays [76–81] for high throughput screening of fuel cell catalysts.

4.7.1
Micro Fuel Cell Array

Liu and Smotkin et al. [76, 77] developed a small DMFC array containing 25 membrane electrode assemblies (MEAs). The MEAs consist of a common polymer electrolyte membrane, a common counter electrode, and an array of disk electrodes. The flow field on the counter electrode side was used for fuel or air flow. The disk electrodes were formed by inserting a bunch of graphite rods into a fuel cell holder in order to obtain good contact with the gas diffusion layers on the electrolyte membrane. The small fuel cell array can be used for very precise screening of libraries of DMFC catalysts. Ito et al. [78, 79] fabricated a micro DMFC array containing 10 single cells, which were arranged in the micro holes (0.5 mm in diameter) on a flexible polymer substrate. The DMFC array, in which the cells are linked in series, has open circuit voltage 5.6 V; and the average cell power density is 3 mW cm^{-2}. Jiang et al. [82] designed a combinatorial electrochemical cell-array for high throughput screening of micro fuel cells and metal/air batteries. The cell array contains a common air electrode and 16 micro anode electrodes for high throughput screening of both fuel cells (based on polymer electrolyte membrane) and metal/air batteries (based on liquid electrolyte). Catalysts for methanol oxidation can easily be coated on the anodes of the electrochemical cell array and screened by switching a graphite probe from one cell to the others. The electrochemical cell array was also used for determination of optimum experimental operating conditions. Figure 4.15 shows a longitudinal cross-section view of the disassembled combinatorial electrochemical cell-array [82]. Here, a dashed line is used to separate the anode-array part and the common air-electrode part. The area of the whole cell-array is $6.0 \times 6.0 \text{ cm}^2$, on which the 16-member anode-array takes $3.5 \times 3.5 \text{ cm}^2$. The common air-electrode supplies air to the cathodes of the cell-array, and catalyzes oxygen reduction by air-convection. From the top to the bottom of the air-electrode,

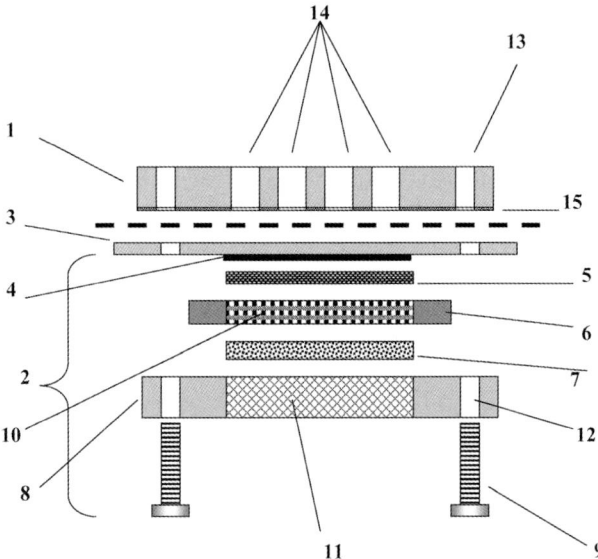

Figure 4.15 A longitudinal cross-section view of a disassembled combinatorial electrochemical cell-array. 1, Top anode end-plate with 16 cylindrical holes for anode material coating and screening; 2, Components of air electrode on lower part of the cell-array; 3, Electrolyte membrane (or battery separator); 4, Platinum catalyst coating for catalytic oxygen reduction; 5, Carbon cloth for gas diffusion electrode; 6, Cathode current collector with small holes for air-passage; 7, Air-filter; 8, Bottom cathode end-plate with small holes for air-passage; 9, bolt for assembling; 10, Area for air-passing on the cathode current collector; 11, Area for air-passing on cathode end-plate; 12, Hole for inserting bolt on cathode end-plate; 13, Hole for inserting bolt on anode end-plate; 14, Holes for anode material coating and screening, and 15, Teflon gasket for the electrode array on the bottom of the anode end-plate. (Reproduced with permission from [82]. Copyright © 2007 American Institute of Physics.)

there is a polymerelectrolyte membrane layer, 250 μm thick (or a battery separator layer 150 μm thick), a cathode catalyst layer (10–15 μm thick), gas diffusion layer (450 μm), cathode current collector layer (1.1 mm thick), air filter layer (2.0 mm thick), and cathode end-plate layer (8.5 mm thick). For fuel cell research, Nafion 117 was used as electrolyte membrane. For metal/air battery research, a micro porous polyestersulfone film was used as the separator. The gas diffusion layer was a piece of carbon cloth. A titanium sheet with small holes was used as the cathode current collector, which is inert to chemical and electrochemical reactions. The cathode end-plate was made of a square shape of organic glass that was processed with small holes on the center for air passage. The cathode catalyst was platinum black. The total thickness of the cathode was 12.5 mm. The top anode-array of the electrochemical cell-array is relatively simple with only two layers. The top layer of the cell was made of a plate of organic glass (8.5 mm thick), which was covered by a Teflon gasket (200 μm thick) on its bottom face in order to obtain the best sealing between the anodes and the membrane that is located on the top of the cathode side. The micro electrochemical cell array was successfully used for screening of mixed metal catalysts of DMFCs, and for studying of the effect of methanol concentration on discharge performance.

4.7.2
A Method for High Throughput Screening of DMFC Single Cells

Jiang et al. [83] reported a simple method for high throughput screening of DMFC single cells. They formed a 40-member array of DMFCs by electrical connection of the small fuel cells in series, one cell's negative terminal linked to another's positive terminal. Figure 4.16 shows a schematic view of the experimental steps for preparation of a single fuel cell and the formation of the fuel cell array [83]. The fuel cell array was characterized electrochemically under constant current discharge. Here, every single fuel cell in the array has the same current but different voltages. High throughput screening was realized by probing the voltages between the negative

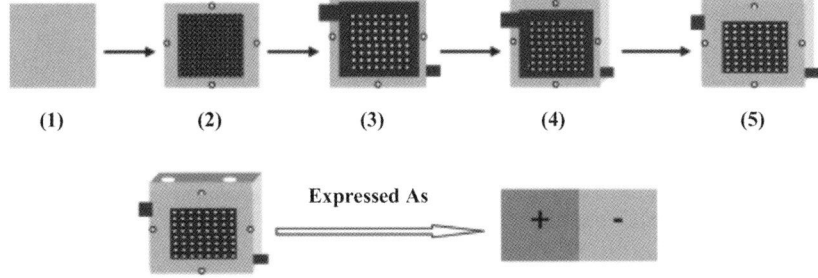

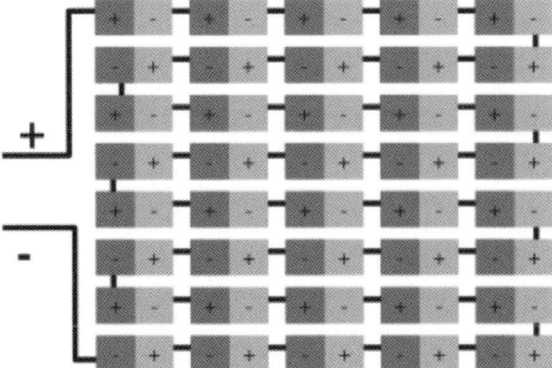

Figure 4.16 Schematic views of the steps of preparation of a single fuel cell (upper panel), and a fuel cell array (lower panel) containing 40 fuel cells that are generated by electrically connecting individual fuel cells in series for combinatorial measurements under constant current discharge. Steps of preparation a single fuel cell: (1) Start from a polymer electrolyte membrane; (2) Attach catalysts to the membrane; (3) Attach current collectors; (4) Attach fuel compartment and anode end-plate; (5) Attach cathode end-plate. (Reproduced with permission from [83]. Copyright © (2005) American Chemical Society.)

and the positive terminals of each single fuel cell with two electric tips linked with a multimeter. The screening of 40 fuel cells was completed in only a few minutes for one current-step. Since all of these fuel cells were measured under the same conditions (that is, the same current and temperature) and these measurements were finished in a short period of time, the experimental accuracy was improved significantly. The voltage–current curves of the DMFCs were obtained by changing the discharge current from low to high and measuring the cell voltages by several times of current-step.

4.7.3
Data Analysis of Combinatorial Results

Combinatorial and high throughput screening experiment generates a large quantity of data. In most cases, only a small part of data for the best catalyst members is analyzed for further scale-up experiments; and the majority of data from the poor catalyst members is not used. For example, after a current-step experiment, only a few of the fuel cells that have the highest discharge potentials are selected for the further scale-up research. However, sometimes we need to map or analyze all of the data in order to theoretically understand the catalytic activity and stability of these catalysts. Analysis of large quantity of experimental data creates a great challenge for combinatorial research. Figure 4.17 shows voltage–current curves obtained from a 40-member fuel cell array for methanol concentrations from 0.5 to 4.0 M [83]. There are 200 experimental curves that were obtained by multiple steps of constant current discharge and measuring the voltages of each cell. The total number of data points acquired was about 2000. Each of the voltage–current curves contains a complete description of an individual single fuel cell's performance under a specific experimental condition. The better performing fuel cells have the higher voltage if the discharge current is the same. As shown in Figure 4.17, these voltage–current curves of single fuel cells are distributed in a wide voltage range because of significant performance differences. These data need to be figured, mapped, statistically treated, analyzed, and explained. Sometimes, mathematical simulation and modeling are useful to theoretically understand the experimental results.

4.8
Challenge and Perspective

Nanosized, high surface area, multi-element transition metal alloys are being sought for highly active and long-term stable DMFC catalysts. Combinatorial and high throughput screening methods have provided a potential capability for fast identification of the best from innumerable known and unknown candidates. One of the challenges is combinatorial synthesis. The early work for combinatorial synthesis of binary, ternary and quaternary metal alloys is based on a method of simply printing metal salt inks onto a conductive substrate and subsequent borohydride reduction [30]. Such a method is far from the practically used synthetic

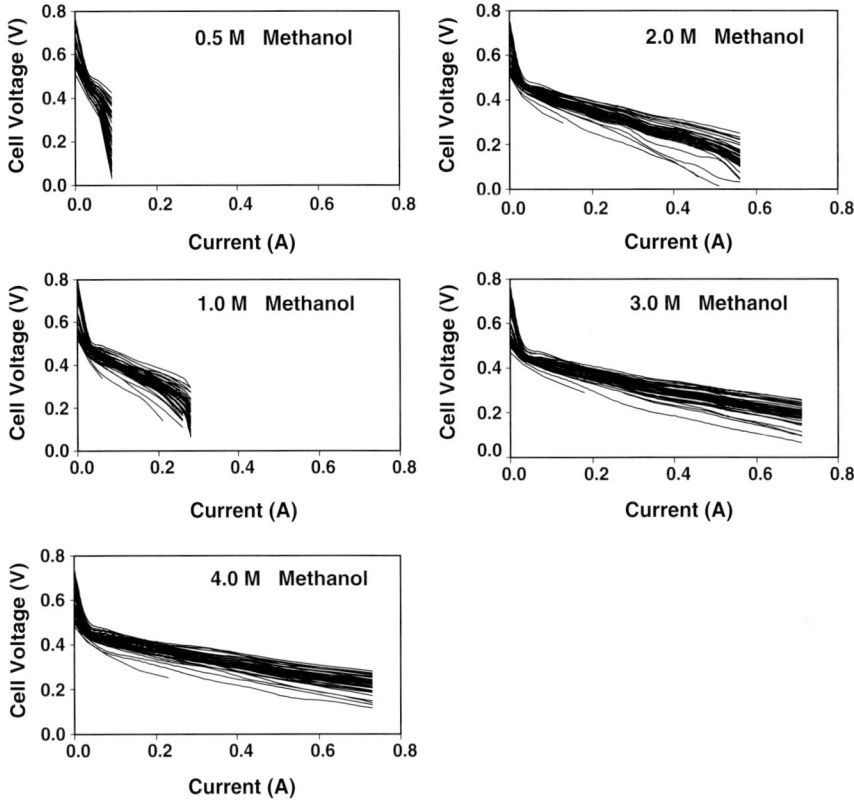

Figure 4.17 Voltage–current curves of a 40-member DMFC array with various fuel concentrations. (Reproduced with permission from [83]. Copyright © (2005) American Chemical Society.)

methods for making nanosized catalysts, and gives unreliable results even for the alloy compositions. The secondary method for combinatorial synthesis is to use small reactor arrays and robotic devices to make alloys, which is much closer to the real synthetic methods. The latter method is able to partially control the reaction conditions to form nanosized catalyst particles. The synthesized alloy catalysts using this method need to be coated on an array of electrodes for further combinatorial characterization. The complexity of such a robotic system and its noncommercial availability has limited its wide application. In the electrochemical method to synthesize alloys, the metals are directly deposited on the electrode surface tightly or loosely. The deposition is highly dependent on the redox potentials of the selected metal salts. In order to generate an array of metal alloy samples on a flat electrode surface, the electrode needs to be divided into a number of small regions that are spatially and ionically separated for deposition of metals through electrochemical reactions. A single or multiple-channel potentiostat is needed to carry out experiment by consecutive synthesis or parallel synthesis, respectively. The advantage of electrochemical synthesis is that the generated catalyst array can be electrochemically characterized without further coating processes. However, due to the complexity of this method for combinatorial synthesis, it is not popular. The physical vapor deposition method such as sputtering is a relative simple method to synthesize DMFC alloy catalysts. A large member library of binary, ternary and quaternary metal alloy catalysts can be generated easily by sequential deposition through various masks. A post-deposition annealing procedure is needed to promote inter-diffusion of the metal layers to form alloys. The annealing temperature is essential for alloy formation. The disadvantage of the PVD method is that the resulting catalysts have little surface area and lower activity in comparison with the nanosized catalysts that have the same compositions. On the other hand, the high cost of sputtering equipment makes it less popular for making practical DMFC catalysts. There is a great challenge in using the combinatorial method to synthesize nanosized catalyst particles. Although the impregnation, co-precipitation, colloidal, and micro-emulsion are well known traditional synthetic methods to generate nanomaterials, so far, these methods are not popular for applications in combinatorial synthesis. In order to widely use combinatorial methods for DMFC catalyst synthesis, we need to continue to seek innovative methods to generate large member catalyst libraries that are reliable, simple, low cost, and highly efficient for making nanosized, high surface area, highly active and long term stable catalysts.

The secondary challenge for combinatorial and high throughput screening is to find a fast, reliable, efficient and popular screening method. The optical screening method gives a good example for high throughput screening that needs only one electrochemical measurement to screen hundreds of catalyst compositions. Because the optical screening method needs an electrolyte condition near neutral, its wide application is limited. The automated screening and electrolyte probe screening methods can be applied to any electrolyte conditions. The disadvantage of these methods is a need for many electrochemical measurements by consecutive scanning of the catalyst samples in the electrode array. The IR thermography screening method is able to quickly probe the activity of catalytic reaction in the conditions for catalysts to directly contact the reactants, but it provides no information of the electrocatalytic activity. The microprobe scanning mass spectrometry is able to consecutively screen the products generated by the catalytic reactions and indirectly detect the catalytic activity. Because this method requires transport of minute gas samples from each library site to the detection system, it is difficult to implement. The SECM screening method involves the measurement of nanocurrents through an ultra-micro electrode tip when it is moved in a solution in a vicinity of a substrate. It has demonstrated the capability for fast screening of a catalyst thin-film library using SECM on the relatively smooth substrate that is prepared by sequential sputter depositions. As the SECM results are highly dependent on the topographic conditions of the substrate, it is not a popular method for screening of DMFC catalysts. Because of the complexity and experimental limitations of the present screening methods, further searching for simple, reliable, and efficient screening methods is still a primary work for combinatorial research on DMFC catalysts.

Acknowledgement

The author wishes to thank the U. S. Department of Army and the Army Materiel Command for their support to this work, and Dr. Cynthia Lundgren for reviewing the manuscript and helpful suggestions.

References

1. Boies, D.B. and Dravnieks, A. (1962) *J. Electrochem. Soc.*, **109**, C198.
2. Cairns, E.J. and Bartosik, D.C. (1964) *J. Electrochem. Soc.*, **111**, C78.
3. Cairns, E.J. and Bartosik, D.C. (1964) *J. Electrochem. Soc.*, **111**, 1205.
4. Fussnegg, K., Jahnke, H., Rhein, H. and Schnepf, G. (1968) *J. Electrochem. Soc.*, **115**, C225.
5. Cathro, K.J. (1971) *J. Electrochem. Soc.*, **118**, 1523.
6. Hampson, N.A., Willars, M.J. and Mcnicol, B.D. (1979) *J. Power Sources*, **4**, 191.
7. Abens, S. and Jacobs, R. (1983) *J. Electrochem. Soc.*, **130**, C331.
8. Goodenough, J.B., Hamnett, A., Kennedy, B.J. and Weeks, S.A. (1987) *Electrochim. Acta*, **32**, 1233.
9. Kanda, T., Kawashima, A., Asami, K. and Hashimoto, K. (1987) *J. Electrochem. Soc.*, **134**, C41.
10. Hamnett, N.A. and Kennedy, B.J. (1988) *Electrochim. Acta*, **33**, 1613.
11. Grot, W.G.F., Mehra, V., Munn, G.E. and Solenberger, J.C. (1975) *J. Electrochem. Soc.*, **122**, C101.
12. Hora, C.J. and Maloney, D.E. (1977) *J. Electrochem. Soc.*, **124**, C319.
13. Huba, F., Yeager, E.B. and Olah, G.A. (1979) *Electrochim. Acta*, **24**, 489.
14. Nakajima, H. (1991) *J. Chem. Technol. Biotechnol.*, **50**, 555.
15. Springer, T.E., Zawodzinski, T.A. and Gottesfeld, S. (1991) *J. Electrochem. Soc.*, **138**, 2334.
16. Scherson, D.A., Gupta, S.L., Fierro, C., Yeager, E.B., Kordesch, M.E., Eldridge, J., Hoffman, R.W. and Blue, J. (1983) *Electrochim. Acta*, **28** (9), 1205–1209.
17. Chang, C.K., Liu, H.Y. and Abdalmuhdi, I. (1984) *J. Am Chem. Soc.*, **106** (9), 2725–2726.
18. Jiang, R.Z. and Dong, S.J. (1990) *J. Phys. Chem.*, **94** (9), 7471–7476.
19. Jiang, R.Z. and Dong, S.J. (1988) *J. Electroanal. Chem.*, **246** (1), 101–117.
20. Collman, J.P., Marrocco, M., Denisevich, P., Koval, C. and Anson, F.C. (1979) *J. Electroanal. Chem.*, **101** (1), 117–122.
21. Collman, J.P., Denisevich, P., Konai, Y., Marrocco, M., Koval, C. and Anson, F.C. (1980) *J. Am. Chem. Soc.*, **102** (19), 6027–6036.
22. Vanveen, I.A.R., Colijn, H.A. and Vanbaar, J.F. (1988) *Electrochim. Acta*, **33** (6), 801–804.
23. Dong, S.J. and Jiang, R.Z. (1987) *Ber. Bunsenges. Phys. Chem.*, **91** (4), 479484.
24. Gojkovic, S.L., Gupta, S. and Savinell, R.F. (1999) *J. Electroanal. Chem.*, **462** (1), 63–72.
25. Tu, H.C., Wang, W.L., Wan, C.C. and Wang, Y.Y. (2006) *J. Phys. Chem. B*, **110**, 15988–15993.
26. Gurau, B., Viswanathan, R., Liu, R., Lafrenz, T.J., Ley, K.L., Smotkin, E.S., Reddington, E., Sapienza, A., Chan, B.C., Lallouk, T.E. and Sarangapani, S. (1998) *J. Phys. Chem. B*, **102**, 9997–10003.
27. Liu, H.S., Song, C.J., Zhang, L., Zhang, J.J., Wang, H.J. and Wilkinson, D.P. (2006) *J. Power Sources*, **155**, 95–110.
28. King, W.D., Corn, J.D., Murphy, O.J., Boxall, D.L., Kenik, E.A., Kwiatkowski, K.C., Stock, S.R. and Lukehart, C.M. (2003) *J. Phys. Chem. B.*, **107**, 5467–5474.
29. Lee, S.A., Park, K.W., Choi, J.H., Kwon, B.K. and Sung, Y.E. (2002) *J. Electrochem. Soc.*, **149** (10), A1299–A1304.

30 Reddington, E., Sapienza, A., Gurau, B., Viswanathan, R., Sarangapani, S., Smotkin, E.S. and Mallouk, T.E. (1998) *Science*, **280**, 1735.

31 Nisonoff, A., Hopper, J.E. and Spring, S.B. (1975) *The Antibody Molecule*, Academic Press, New York.

32 Pollack, S.J., Jacobs, J.W. and Schultz, P.G. (1986) *Science*, **234**, 1570.

33 Xiang, X.D., Su, X., Briceno, G., Lou, Y., Wang, K.A., Chang, H., Freedman, W.G.W., Chen, S.W. and Schultz, P.G. (1995) *Science*, **268**, 1738.

34 Dellamorte, J.C., Vijay, R., Snively, C.M., Barteau, M.A. and Lauterbach, J. (2007) *Rev. Sci. Instrum.*, **78**, 072211.

35 Yamada, Y., Akita, T., Ueda, A., Shioyama, H. and Kobayashi, T. (2005) *Rev. Sci. Instrum.*, **76**, 062226.

36 Hoogenboom, R. and Schubert, U. (2005) *Rev. Sci. Instrum.*, **76**, 062202.

37 Brandli, C., Jaramillo, T.F., Ivanovskaya, A. and McFarland, E.W. (2001) *Electrochim. Acta.*, **47**, 553–557.

38 Beattie, S.D. and Dahn, J.R. (2005) *J. Electrochem. Soc.*, **152**, C542–C548.

39 Jayaraman, S., Baeck, S.H., Jaramillo, T.F., Shwarstein, A.K. and MacFarland, E.W. (2005) *Rev. Sci. Instrum.*, **76**, 062227.

40 Jayaraman, S., Jaramillo, T.F., Baeck, S.H. and MacFarland, E.W. (2005) *J. Phys. Chem. B*, **109**, 22958–22966.

41 Erichsen, T., Reiter, S., Ryabova, V., Bonsen, E.M. and Schuhmann, W. (2005) *Rev. Sci. Instrum.*, **76**, 062204.

42 Whitacre, J.F., Valdez, T. and Narayanan, S.R. (2005) *J. Electrochem. Soc.*, **152**, A1780–A1789.

43 Bonakdarpour, A., Wenzel, J., Stevens, D.A., Sheng, S., Monchesky, T.L., Lobel, R., Atanasoski, R.T., Schmoekel, A.K., Vernstrom, G.D., Debe, M.K. and Dahn, J.R. (2005) *J. Electrochem. Soc.*, **152**, A61–A72.

44 Yang, R., Stevens, K., Bonakdarpour, A. and Dahn, J.R. (2007) *J. Electrochem. Soc.*, **154**, B893–B901.

45 He, T., Kreidler, E., Xiong, L., Luo, J. and Zhong, C.J. (2006) *J. Electrochem. Soc.*, **153**, A1637–A1643.

46 Cooper, J.S., Zhang, G. and McGinn, P.J. (2005) *Rev. Sci. Instrum.*, **76**, 062221.

47 Cooper, J.S. and McGinn, P.J. (2006) *J. Power Sources*, **163**, 330–338.

48 Sullivan, M.G., Utomo, H., Fagan, P.J. and Ward, M.D. (1999) *Anal. Chem.*, **71**, 4369–4375.

49 Zhou, L., Savvatelev, V., Booher, J., Kim, C.H. and Shinar, J. (2001) *Appl. Phys. Lett.*, **79**, 2282–2284.

50 Liu, H., Felten, C., Xue, Q., Zhang, B., Jeddrzejewski, P., Karger, B.L. and Foret, F. (2000) *Anal. Chem.*, **72**, 3303–3310.

51 Choi, W.C., Kim, J.D. and Woo, S.I. (2002) *Catal. Today.*, **74**, 235–240.

52 Chen, G., Delafuente, D.A., Sarangapani, S. and Mallouk, T.E. (2000) *Catal. Today*, **67**, 341–355.

53 Gurau, B., Visvanathan, R., Liu, R., Lafrenz, T.J., Ley, K.L., Smotkin, E.S., Reddington, E., Sapienza, A., Chan, B.C., Mallouk, T.E. and Sarangapani, S. (1988) *J. Phys. Chem. B*, **102**, 9997–10003.

54 Liu, J.H., Jeon, M.K. and Woo, S.I. (2006) *Appl. Surf. Sci.*, **252**, 2580–2587.

55 Jiang, R.Z. and Chu, D. (2002) *J. Electroanal. Chem.*, **527**, 137–142.

56 Jiang, R.Z. and Chu, D. (2003) In *High Throughput Analysis: A Tool for Combinatorial Materials Science* (eds A. Totyrailo Radislav and Amis Aric), Kluwer Academic/Plenum Publishers, New York, pp. 447–466.

57 Jiang, R.Z. and Chu, D. (2005) *Rev. Sci. Instrum.*, **76** (6), 062213.

58 Moates, F.C., Somani, M., Annamalai, J., Richardson, J.T., Luss, D. and Willson, R.C. (1996) *Ind. Eng. Chem. Res.*, **35** (12), 4801–4803.

59 Cypes, S., Hagemeyer, A., Hogan, Z., Lesik, A., Streukens, G., Volpe, A.F., Weinberg, W.H. and Yaccato, K. (2007) *Comb. Chem. High Throughput Screening*, **10** (1), 25–35.

60 Brooks, C., Cypes, S., Grasselli, R.K., Hagemeyer, A., Hogan, Z., Lesik, A., Streukens, G., Volpe, A.F., Turner, H.W., Weinberg, W.H. and Yaccato, K. (2006) *Top. Catal.*, **38** (1–3), 195–209.

61 Kubanek, P., Busch, O., Thomson, S., Schmidt, H.W. and Schuth, F. (2004) *J. Comb. Chem.*, **6** (3), 420–425.

62 Snively, C.M., Oskarsdottir, G. and Lauterbach, J. (2001) *Catal. Today*, **67** (4), 357–368.

63 Biniwale, R.B. and Ichikawa, M. (2007) *Chem. Eng. Sci.*, **62** (24), 7370–7377.

64 Biniwale, R.B., Yamashiro, H. and Ichikawa, M. (2005) *Catal. Lett.*, **102** (1–2), 23–31.

65 Kirsten, G. and Maier, W.F. (2004) *Appl. Surf. Sci.*, **223** (1–3), 87–101.

66 Yamada, Y., Ueda1 F A., Shioyama, H. and Kobayashi, T. (2004) *Appl. Surf. Sci.*, **223**, 220–223.

67 Li, N., Assmann, J., Schuhmann, W. and Muhler, M. (2007) *Anal. Chem.*, **79** (15), 5674–5681.

68 Yaccato, K., Carhart, R., Hagemeyer, A., Lesik, A., Strasser, P., Volpe, A.F., Turner, H., Weinberg, H., Grasselli, R.K. and Brooks, C. (2005) *Appl. Catal. A-Gen.*, **296** (1), 30–48.

69 Eckhard, K., Schluter, O., Hagen, V., Wehner, B., Erichsen, T., Schuhmann, W. and Muhler, M. (2005) *Appl. Catal. A-Gen.*, **281** (1–2), 115–120.

70 Li, N., Eckhard, K., Abmann, J., Hagen, V., Otto, H., Chen, X., Schuhmann, W. and Muhler, M. (2006) *Rev. Sci. Instrum.*, **77**, 084102.

71 Husser, O., Craston, D.H. and Bard, A.J. (1989) *J. Electrochem. Soc.*, **136** (11), 3222–3229.

72 Lu, G.J., Cooper, J.S. and McGinn, P.J. (2006) *J. Power Sources*, **161** (1), 106–114.

73 Lu, G.J., Cooper, J.S. and McGinn, P.J. (2007) *Electrochim. Acta*, **52** (16), 5172–5181.

74 Black, M., Cooper, J.S. and McGinn, P.J. (2005) *Meas. Sci. Technol.*, **16** (1), 174–182.

75 Fernandez, J.L. and Bard, A.J. (2003) *Anal. Chem.*, **75** (13), 2967–2974.

76 Smotkin, E.S., Jiang, J., Nayar, A. and Liu, R. (2006) *Appl. Surf. Sci.*, **252**, 2573–2579.

77 Liu, R.X. and Smotkin, E.S. (2002) *J. Electroanal. Chem.*, **535** (1–2), 49–55.

78 Ito, T., Kimura, K. and Kunimatsu, M. (2006) *Electrochem. Commun.*, **8**, 973–976.

79 Ito, T. and Kunimatsu, M. (2006) *Electrochem. Commun.*, **8**, 91–94.

80 Wan a, N., Maoa, Z., Wang, C. and Wang, G. (2007) *J. Power Sources*, **163**, 725–730.

81 Lim, S.W., Kim, S.W., Kim, H.J., Ahn, J.E., Han, H.S. and Shul, Y.G. (2006) *J. Power Sources*, **161**, 27–33.

82 Jiang, R.Z. (2007) *Rev. Sci. Instrum.*, **78**, 072209.

83 Jiang, R.Z., Rong, C. and Chu, D. (2005) *J. Comb. Chem.*, **7** (2), 272–278.

5
State-of-the-Art Electrocatalysts for Direct Methanol Fuel Cells
Hanwei Lei, Paolina Atanassova, Yipeng Sun, and Berislav Blizanac

5.1
Introduction

Direct methanol fuel cells (DMFCs) get special attention for portable electronics applications [1–19], but potentially could be used for transportation systems [20]. This is because methanol not only is a common chemical industry fuel, but also has very high energy density. Theoretically the energy density of methanol is about 6.0 Wh/g and 4.8 Wh/cc, which is almost one order of magnitude higher than that of the lithium ion batteries. Considering current DMFC working with 20–30% system efficiency at low temperature (<80 °C), a DMFC has the potential to offer 2–3 times the energy density of the Li-ion batteries in the future.

Contrary to the lack of infrastructure for hydrogen fuel, liquid methanol is easily obtained as a fuel for portable devices. Charging time is almost zero as just changing a methanol cartridge makes the DMFC very attractive as a power source for portable devices. In general, DMFC can provide high energy for longer run without limit of the charging time, but the power density is still not high enough to power up a device, especially at initial start-up. On the other hand, a Li-ion battery can deliver high power in a short period of time, which is critical at start-up or when there is a need for a peak power to follow up the load change immediately during the operation. It is expected that in the near future a hybrid system of DMFC and Li-ion battery could be a good solution for a portable device [18].

There are several key technical barriers to be overcome for DMFC to be commercialized as a power source for a wide range of applications. First, DMFC power density needs improvement. This requires better electrocatalysts, especially for the anode methanol oxidation, though the cathode also needs improvement. As is known, the anode overpotential of about 250 mV to oxidize the methanol is a big factor for the overall polarization loss at low temperatures <80 °C. Second, fuel use needs improvement for better system efficiency; this requires better methanol crossover (MCO) management. As demonstrated [20–23], MCO not only results in the waste of anode fuel, but also a decline in the system efficiency due to the

Electrocatalysis of Direct Methanol Fuel Cells. Edited by Hansan Liu and Jiujun Zhang
Copyright © 2009 WILEY-VCH Verlag GmbH & Co. KGaA, Weinheim
ISBN: 978-3-527-32377-7

formation of a mixed potential at the cathode side. The intermediate such as carbon monoxide due to the oxidation of the crossover methanol at the cathode side also causes the cathode catalyst to deactivate. Though there is significant progress in hydrocarbon (HC) membrane recently compared to Nafion, it hardly eliminates the MCO completely [3–5, 19].

Of course, the durability of DMFC needs big improvement to meet >5000 hours operation requirement for a portable device. This requires a highly stable electrocatalyst and very durable membrane electrolyte. It has been identified that the particle size growth and the ruthenium leaching out of PtRu anode catalyst which causes Ru crossover are the biggest problems for the durability of DMFC anode catalyst [24–27].

Currently, the cost of DMFC is still too high to compete with other power sources such as Ni-MH and Li-ion batteries. High Pt content of the electrocatalyst in the catalyst coated membrane (CCM) or membrane electrode assembly (MEA) is the key factor for the high material cost for a DMFC. In order to have a low Pt content MEA, developing a highly dispersed supported catalyst is a reasonable approach. For the longer term, a non-Pt catalyst will win if the performance is advanced significantly compared with Pt-based catalyst [28, 29].

This chapter will mainly discuss progress in the DMFC electrocatalyst area especially in advances with supported anodes. DMFC MEA performance and durability will also be discussed to a certain degree because it is so critical for DMFC commercialization and is difficult to separate from electrocatalyst development. Progress in DMFC cathode catalysts can be seen from recent publications [28–32] and will be less addressed in this chapter.

5.2
Electrocatalysis and Electrocatalysts for DMFC

5.2.1
Electrocatalysis for Methanol Oxidation

Thermodynamically, methanol should be oxidized easily because the methanol oxidation free energy $\Delta G = -686\,\text{kJ}\,\text{mol}^{-1}$ at 25 °C and 1 atm corresponds to $\Delta E = 1.18\,\text{V}$, similar to that of the hydrogen/air fuel cells [11, 33]. However, the methanol oxidation kinetics is very sluggish at low temperatures (<80 °C) which results in an overpotential of about 250 mV even for the best available anode Pt-Ru catalyst. So far intensive studies have been concentrated on the development of better catalysts [2, 15, 33, 34], typically platinum based alloys. The related mechanism study has been investigated in the past 30 years. So far, the 'bifunctional theory' has been widely accepted for methanol oxidation on Pt-based catalysts which can be expressed as Equations 5.1 to 5.4 [2, 15]:

$$\text{Pt-CH}_3\text{OH} \rightarrow \text{Pt-(CH}_3\text{OH)}_{\text{ads}} \tag{5.1}$$

$$\text{Pt-(CH}_3\text{OH)}_{\text{ads}} \rightarrow \text{Pt-CO}_{\text{ads}} + 4\text{H}^+ + 4\text{e}^- \tag{5.2}$$

$$M + H_2O \rightarrow M\text{-}(H_2O)_{ads} \qquad (5.3a)$$

$$M\text{-}(H_2O)_{ads} \rightarrow M\text{-}OH_{ads} + H^+ + e^- \qquad (5.3b)$$

$$Pt\text{-}CO_{ads} + M\text{-}(H_2O)_{ads} \rightarrow Pt + M + CO_2 + 2H^+ + 2e^- \qquad (5.4a)$$

$$Pt\text{-}CO_{ads} + M\text{-}OH_{ads} \rightarrow Pt + M + CO_2 + H^+ + e^- \qquad (5.4b)$$

Equation 5.2 involves C−H bond cleavage (or C−H activation), and Equations 5.3a and 5.3b relate to water activation. Here Pt and element M play complementary roles in the bifunctional mechanism. The rate-determining step has long been thought to be within steps (5.3a, 5.3b) and (5.4a, 5.4b), thus some oxophilic elements such as Pt-Ru [28,35–38], Pt-Sn [39, 40] and Pt-based ternary alloys [2, 41–43] have been investigated extensively.

A recent deuterium isotope analysis of methanol oxidation on Pt, Pt-Ru and Pt-Ru-Os-Ir electrodes sheds light on the mechanism [2]. It illustrates that there is a characteristic transition potential where the primary reaction in the rate-determining step changes from water activation to C−H bond activation. It is observed that a crossover potential is about 0.35 V for PtRu and PtRuOsIr electrodes which is within the DMFC potential window (0–0.4 V). This indicates that the catalyst discovery strategy involving Pt promoter metals must consider not only improving water activation, but also enhancing C−H bond activation.

Since methanol oxidation mechanism has been extensively discussed in Chapter 2 here we focus more on the synthesis of the practical DMFC anode catalysts and the related characterization and evaluation.

5.2.2
Electrocatalyst Development

It is generally accepted that Pt-Ru catalysts with a Pt : Ru atomic ratio of 1 : 1 are the best for methanol oxidation in DMFC due to overall activity and durability [9, 15], though there has been an argument that Pt-RuO$_x$H$_y$ phase [44] and Ru-rich or Pt-rich compositions at different DMFC operating conditions were more active [45]. While there will be more studies in the future to investigate new compositions, structures, particle size and distributions, it is worthwhile to summarize some important progress in DMFC anode catalyst development for commercialization.

Conventional electrocatalyst preparation methods and the interaction between electrocatalyst and support were systematically discussed by Kinoshita and Stonehart in 1977 [46], where they summarized seven methods for fuel cell catalyst preparation and their applications in binary alloy catalysts. These methods, applied to both Pt and Pt-Ru catalyst preparation, include: impregnation, ion-exchange and adsorption, the colloidal method, Adams' method, Raney's method, thermal decomposition and freeze-drying. Later, Kinoshita particularly discussed the applications of impregnation, ion-exchange, adsorption and organic macrocycles in the preparation of the carbon-supported electrocatalysts [47]. A very recent DMFC anode catalysis review by Liu et al. summarized the progress of DMFC PtRu anode synthesis in three categories: impregnation, colloidal, microemulsion methods [9]. It should be emphasized that a new catalyst manufacturing platform called spray conversion reaction (SCR) has been developed recently by Cabot [48–50] and applied to make commercial grade carbon-supported Pt and Pt-Ru catalysts for both H_2/air and direct methanol fuel cells [3–8]. The SCR process and related catalyst performance advancement will be discussed in detail in the following sections.

Table 5.1 summarizes the preparation methods for PtRu catalysts for DMFC applications. In fact, there is an overlap of applying methods when practical catalysts are synthesized, for example, borohydride reduction can be used in both the impregnation and colloidal methods [9, 46]; the Watanabe method can be regarded as a special case for colloidal technique. No matter what kind of methods or the combination of several techniques are used for catalyst preparation, the practical electrocatalysts for DMFC applications should possess most or all of the following features:

- *High intrinsic activity* for a higher current density at DMFC operating voltage
- *Excellent electrochemical stability* for better MEA durability
- *Easy scale-up* for high volume production with good reproducibility and low manufacturing cost
- *Low cost* with less Pt content on a carbon supported catalyst with high catalyst use
- *Good ink formulation* with desirable viscosity, particle size and hydrophilic/hydrophobic properties to fit MEA manufacturing process.

Besides catalyst preparation methods, the support materials significantly influence the catalyst performance. Vulcan XC-72 (product of Cabot corporation, BET, about 250 m^2/g) has been used as standard electrocatalyst support for a long time because of its good electrical conductivity. Other high surface area carbon support such as Ketjen Black (KB, Product of Akzo Nobel, BET is usually between 800–1400 m^2/g) has also demonstrated promise in fuel cell electrocatalysts [3–9]. Catalysts based on nanostructure carbons such as carbon nanotubes, nanofibers and nanowires have presented great performance in fuel cells [9, 67–70], though the cost of these materials needs significant reduction for practical adoption.

Catalyst composition plays an important role in catalytic activity advancement. In recent years, combinatorial chemistry has been carried out to discover new catalyst compositions for the methanol oxidation [42, 43, 63, 71–73]. It should be pointed out that the combinatorial method can be applied to screen or down-select catalyst compositions with certain throughput. However, the catalyst activity is very dependent upon the preparation methods under certain conditions and some compositions discovered by the combinatorial approach may not possess the optimal performance in the real fuel cell system. For example, the work in [42] illustrated that Pt-Ru-Os-Ir was a better catalyst than Pt-Ru in a combinatorial study based on the $NaBH_4$ reduction process. The $NaBH_4$ reduction method may be good to make Pt-Ru-Os-Ir but it is not the best method to make Pt-Ru catalyst. It is known that Pt-Ru catalyst powders can be made by the Watanabe method [35] with much higher catalyst

Table 5.1 Summary of Pt-Ru catalyst preparation methods.

Preparation method	Process	Catalyst features	Reference
Spray conversion reaction	Generate droplets to contain metal precursors and carbon support, then thermally decompose the metal precursors under controlled temperatures and pressure to form supported catalyst on carbon.	Supported catalyst with spherical agglomerates, high activity and stability. Crystalline size: 2–4 nm.	[48–50]
Borohydride reduction	Metal precursors with or without carbon supports are simply reduced by $NaBH_4$.	Supported and/or unsupported catalysts with crystalline size of 5–10 nm.	[42, 51, 52]
Watanabe method	Colloidal Pt and Ru oxides are prepared in aqueous solution with strict control of pH when add reaction agents, then they are reduced for PtRu by H_2 bubbling.	High surface area supported and/or unsupported catalysts with crystalline size of 2–3 nm.	[35]
Impregnation	Impregnated precursors with supports are reduced by a reductive agent such as, N_2H_4, HCOOH, HCHO, $Na_2S_2O_3$, H_2.	Supported catalysts with strong influence of support structures; Need careful control of preparation conditions to make nanosize with good distribution, crysytalline size: 2–6 nm.	[9, 46, 47, 53]
Colloidal method	Prepare the Pt-Ru containing colloids in the presence of carbon supports, then reduce the mixture by H_2, sodium citrate, HCHO, and so on.	Mainly for the supported catalysts; Catalysts usually have higher surface area than those by impregnation. Crystalline size: 2–4 nm.	[9, 46, 54, 55]
Ad-atom	Deposit Pt onto Ru or Ru onto Pt to make submonolayer catalysts with less Pt or Ru content.	High catalytic activity but with much less noble metal content compared with conventional PtRu catalysts.	[56–58]
Polyol	Pt and Ru precursors are dissolved in ethylene glycol, then mixed with carbon, the resulted suspension is dropwised into reactor filled with ethylene glycol under refluxing at 195 °C, at last immediately cool to room temperature.	Supported PtRu/C with a uniformly distribution of nanoparticles. Crystalline size: about 2.6 nm.	[59–61]
Plasma sputtering	Plasma sputtering is used to deposit Pt and Ru onto targeting substrates.	Deposit Pt-Ru directly onto GDL or membrane with loading of 40 μg/cm^2. Crystalline size: about 4 nm.	[62, 63]
Ball milling	Pt and Ru powders are high energy ball milled with dispersive agent MgH_2 and then MgH_2 is leached in HCl solution to make unsupported PtRu.	High surface area unsupported PtRu of 60–70 m^2/g but with some unalloyed metal phase.	[64–66]

activity due to very small Pt-Ru crystallite size. However, it is difficult to make a number of combinatorial catalysts rapidly by the Watanabe method for Pt-Ru-Os-Ir with a desirable alloy phase. The catalytic activity of actual fuel cell Pt-Ru powders made by the Watanabe method is higher than that of Pt-Ru-Os-Ir made by $NaBH_4$ reduction.

5.2.3
Spray Conversion Reaction Platform for Electrocatalyst Manufacturing

Cabot has developed a spray conversion reaction (SCR) platform for the discovery and low cost manufacture of high performance fuel cell electrocatalysts. The method was originally developed by Superior MicroPowders before it was acquired by Cabot in 2003. The details of the manufacturing process were described in [48–50].

Figure 5.1 shows a schematic of SCR processing. The process starts from the formation of an aqueous liquid which includes precursor chemicals for the desired final dispersed phases such as Pt, PtRu, PtCo together with a carbon support colloidal suspension. The liquid is converted into droplets containing the dissolved or suspended ingredients via a mechanical generator, the droplets are entrained in a heated gas stream, the solvent in the droplets evaporates and the precursors are thermally decomposed and deposit onto the surface of the carbon support. Finally the dry powders can be collected. For alloy catalysts such as PtRu or PtCo, a post-processing step will be followed in a reduced environment to make the alloys.

Because the reaction is processed at a relatively high temperature for a relatively short period of time, the particle size and size distribution can be maintained which is important for making supported electrode catalysts. When the precursors are carefully selected to avoid use of chloride, the resulting catalysts are less contaminated and there is no need to do follow-up cleaning.

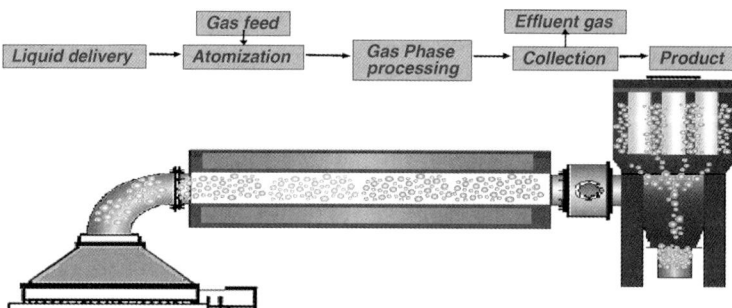

Figure 5.1 Schematic of SCR platform to make fuel cell electrocatalysts. Catalyst precursors are mixed with carbon dispersion and then sent to the reaction zone via a carrier gas in the form of droplets after the atomization through a spray nozzle. At controlled temperatures and pressures, precursors will be thermally decomposed and result in Pt or Pt-Ru or other Pt alloys deposited onto supported carbon for a supported electrocatalyst which is collected at the end.

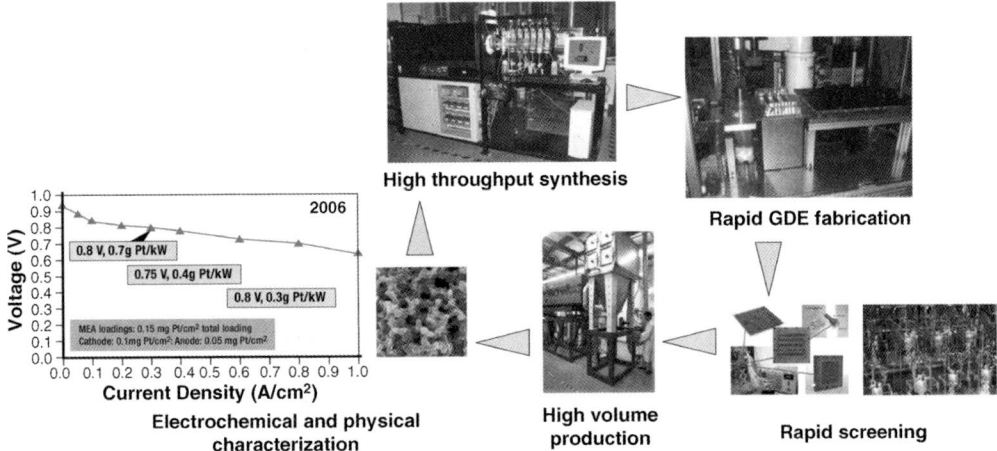

Figure 5.2 High throughput SCR electrocatalyst development platform. The SCR platform can be used to generate multiple samples with different compositions, the resulting powders can be deposited onto carbon paper or cloth to make gas diffusion electrode (GDE) for rapid screening in a liquid 3-electrode system or in a multi-channel combi-fuel cell in the form of MEA made by laminating GDE with Nafion membrane. The physical characterization and electrochemical evaluation results are analyzed further in order to decide if high volume production is of value.

On the other hand, because key parameters such as residence times, temperature distributions, inlet/outlet temperatures are well understood and modeled [48, 49], this SCR can be operated continuously to specified batch volumes, ensuring the scalability of this manufacturing platform and reproducibility of the materials characteristics. Cabot has developed several reaction units: the small one can make several tens of grams per hour for R&D of catalyst composition discovery and screening; the high capacity type can make kilograms of catalyst per hour for commercial products.

With control of the process conditions and carbon support and precursors, the catalyst structure, morphology and porosity of the catalysts can be designed. Besides making new catalyst to improve catalytic activity, the SCR platform has been expanded for a high throughput approach for a discovery of new alloy electrocatalysts, particularly for PEMFC cathodes and DMFC anodes [74–76]. As illustrated in Figure 5.2, this SCR catalyst platform allows for rapid investigation of complex alloy systems with high efficiency and high reproducibility.

5.3
DMFC Electrocatalyst Characterization and Evaluation

Because SCR electrocatalysts are very different from the conventional electrocatalysts, the following sections will be focused more on the characterization and evaluation of this type catalyst in DMFC.

5.3.1
Physical Characterization

The new catalysts made by SCR present unique morphology and excellent fuel cell performance as described in [3–8, 74–76]. Key physicochemical features of this type catalyst will be reviewed first.

Because the catalysts are formed by heated treatments of liquid droplets, the catalysts are in the form of soft spherical agglomerates comprised of aggregates of carbon particles. The active phase, which is PtRu nanoparticles, is well dispersed on the surface of carbon, and in the range of 2–4 nm.

Figure 5.3 shows the SEM and TEM structures of the supported electrocatalysts made by SCR platform. The porous aggregates with high surface area (for example, 60 wt% PtRu/C has a BET area of about 280 m^2/g) present spherical morphology with a diameter of about 5 μm. The pore was formed simultaneously with deposition of PtRu. The pore size inside the aggregates is about 20–100 nm. This pore size is large enough to allow Pt or Pt alloy (such as PtRu) to deposit inside to use the high surface area of the carbon support. This has been confirmed by cross-sectional TEM imaging of 60 wt% PtRu supported on KB as illustrated in Figure 5.4. Cross-sectional image of spherical agglomerates were obtained from a microtome sample, which is an ultrathin slice of epoxy resin embedded with Cabot catalysts. It becomes very clear that the spherical agglomerate is solid inside. The higher resolution images also show well-dispersed PtRu nanoparticles throughout the entire agglomerate.

Figure 5.3 Hierarchical structure of Cabot SCR carbon supported electrocatalysts.

5.3 DMFC Electrocatalyst Characterization and Evaluation | 205

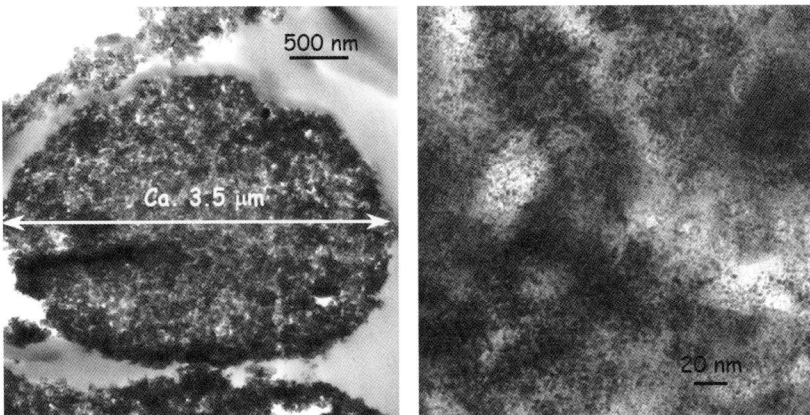

Figure 5.4 Cross-sectional TEM image of a 60 wt% PtRu/C catalyst powder aggregate made from SCR. It indicates uniform aggregate distribution with the agglomerate.

A further study of high resolution TEM illustrates that the crystallite size of PtRu is very small. An example is given in Figure 5.5 where a 60 wt% PtRu/KB crystallite size is on average 2–2.5 nm, and the catalysts are distributed uniformly on the carbon surface. Similar studies have been done for other PtRu catalyst compositions, after investigating over 30 TEM images and 10 000 particulates to identify the particle size,

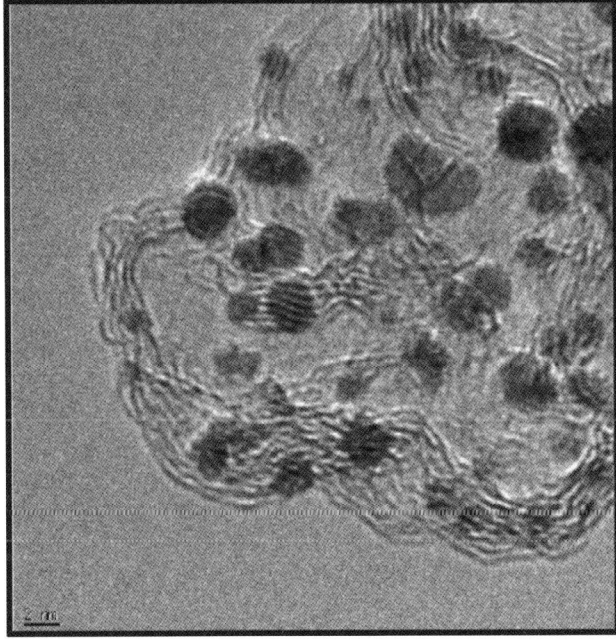

Figure 5.5 High resolution TEM of 60 wt% PtRu/KB made from SCR.

Table 5.2 High resolution TEM results of different SCR supported catalysts.

Catalysts	45 wt% PtRu/C	50 wt% PtRu/C	55 wt% PtRu/C	65 wt% PtRu/C	70 wt% PtRu/C
Characterized images	31	24	31	31	29
Identified PtRu particulates	14 355	12 866	14 844	13 030	11 963
Nano particle average diameter (nm)	2.35	2.46	2.40	2.42	2.52
Particulate diameter std. deviation (nm)	1.1	1.24	1.12	1.18	1.18

the statistical results have been summarized in Table 5.2 for PtRu/KB with weight loadings from 45% to 70%. It shows that all PtRu crystallite sizes are around 2.3–2.5 nm. The small crystalline size and high dispersion onto the carbon support will contribute high performance in the fuel cell which will be discussed in more detail in the following sections.

Besides the crystallite size, the alloy phase will be key for PtRu catalyst activity and stability. Pt and Ru can form 1 : 1 atomic ratio in an alloy with good high catalyst activity [9, 15]. The Pt lattice parameter is usually used to indicate the alloy phase formation [43]. X-ray diffraction (XRD) spectroscopy data are used to calculate lattice parameters of different PtRu alloys. The result of the SCR 60 wt% PtRu supported on KB catalyst is presented in Figure 5.6, together with several other commercially available catalysts such as Pt, PtRu black and 60 wt% PtRu/C. Taking the lattice parameter of the bulk PtRu with 1 : 1 atomic ratio (or 50 : 50) as a baseline of a true PtRu alloy, the SCR PtRu/C presents a 1 : 1 mole ratio Pt-Ru alloy. Commercially available PtRu/C made by other makers shows little Pt rich feature. Good alloy phase

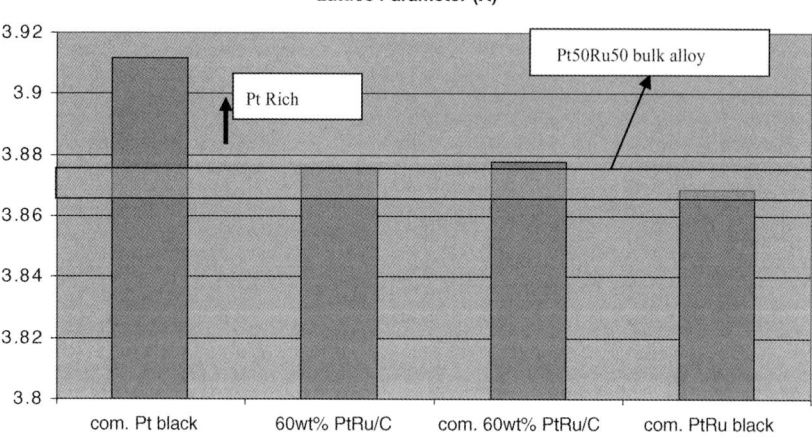

Figure 5.6 Comparison of XRD lattice parameter of SCR 60 wt% PtRu/KB against commercially available Pt black, PtRu black as well as 60 wt% PtRu/C made by other catalyst makers.

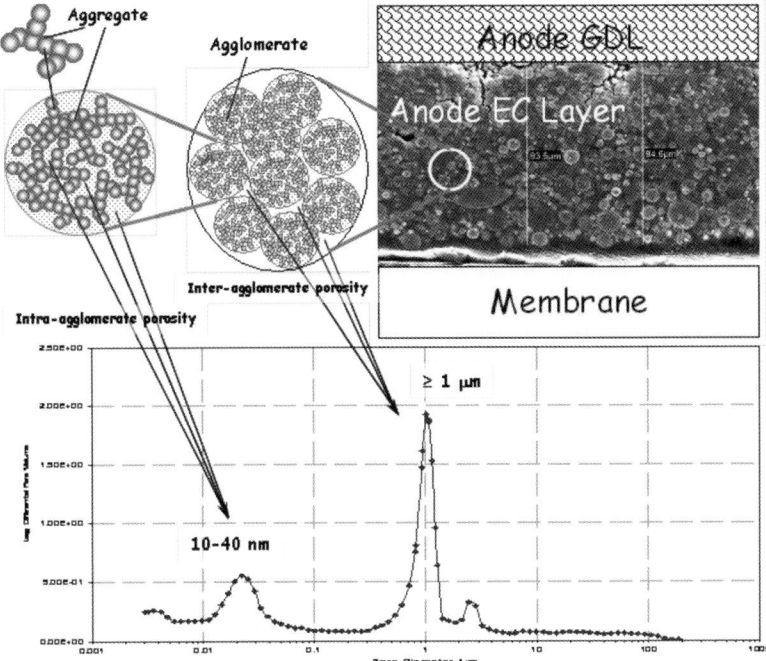

Figure 5.7 Pore size distribution of SCR 60 wt% PtRu/KB. The majority of the pores are inter pores between agglomerates with pore size about 1 μm and some in the range of 2–3 μm. The intra pores inside agglomerates are around 10–40 nm.

helps to improve the catalyst activity and stability. In Section 5.3.3 catalyst activity and stability will be discussed in detail.

The structure of the SCR electrocatalysts allows for the unique design of electrode layer structures and morphologies when engineered into a CCM or MEA because it affects mass transport phenomena and use of active sites. Figure 5.7 shows an example of SCR anode catalyst (60 wt% PtRu/KB EC600) pore size distribution by Hg porosimetry and the anode layer structures by cross-sectional SEM with this catalyst. The porosity of the anode PtRu/KB material is about 82%. There are mainly two types of pores: Type I is intra pores inside agglomerates with pore size of 10–40 nm. Type II has a majority of pores with size around 1 μm due to the inter voids between agglomerates (and some in the range of 2–4 μm). The two types influence methanol diffusion and crossover and CO evolution differently. The effect on MEA performance will be discussed in Section 5.4.1.

Particle size distribution and overall powder pore distribution were not paid too much attention in the past. Recent studies showed that these features influence the catalyst ink formulation and the layer structure. For example, the catalyst layer can be packed more densely after certain ink processing. The dense anode layer can help manage MCO to a certain degree, thus help improve the MEA performance. Figure 5.8 shows a regular PSD (particle size distribution) of an anode PtRu/KB

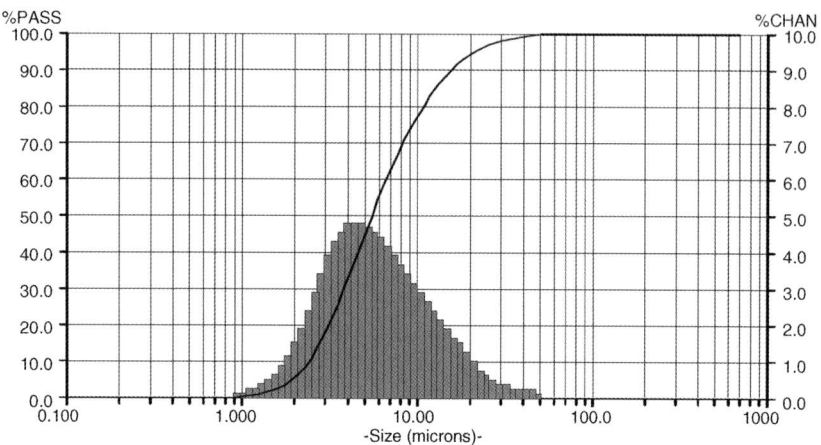

Figure 5.8 Particle size distribution of 60 wt% PtRu/KB with $d_{50} = 5.5\,\mu m$, $d_{90} = 15.6\,\mu m$, $d_{mv} = 7.3\,\mu m$.

made by SCR platform. Cabot has also developed a method to make smaller particle size catalysts with narrow PSD for a dense catalyst layer in a MEA [77].

5.3.2
Electrochemical Evaluation

5.3.2.1 Thin Film Rotating Disc Electrode (TFRDE) Catalyst Characterization

As discussed previously, Cabot's catalyst production technology results in catalyst powders with unique hierarchical and morphological properties. It is expected that this unique secondary catalyst structure will have an impact on MEA performance through the anode catalyst layer morphology effect on mass transport and intrinsic kinetic phenomena. The purpose of *ex-situ* catalyst screening by utilizing the thin film rotating disk electrode (TFRDE) experimental configuration is to determine intrinsic catalytic activity of supported catalysts for anode and cathode reactions in fuel cells. TFRDE technique has been developed to characterize the electrocatalyst activity for several years [78–80]. The main advantage of TFRDE methodology over testing in MEA configuration is the ability to measure the catalytic activity which is not influenced by catalyst use, preparation conditions (compression, temperature, etc.), or by the selection of different components of MEA, that is, gas diffusion media, type and the amount of ionomer in catalyst layer, membrane, flow field plates, and so on. The MEA test still represents the ultimate test of the catalyst performance, but those measurements in conjunction with TFRDE catalysts characterization enables optimization of complex MEA structure by addressing/optimizing the major performance loss contributions, that is, kinetic vs. mass transport limitations.

The TFRDE electrodes were made by depositing about 82 µg/cm² catalyst onto a glass carbon disk electrode followed by a coating of thin Nafion film (~0.2 µm) as described in related publications [78–80]. First, a comparison study of SCR 60 wt% supported anode catalyst versus other company's commercial 60 wt% PtRu/C and

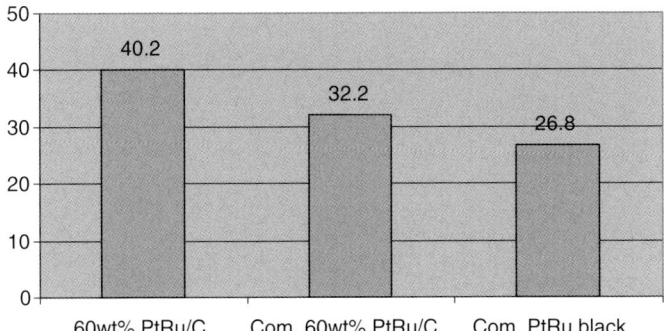

Figure 5.9 Comparison of methanol oxidation mass-specific activity for SCR 60 wt% PtRu/C anode and other company's commercial supported 60 wt% and un-supported PtRu-black catalyst in 0.5 M H_2SO_4 and 1 M methanol solution after 10 min of potential hold at 0.4 V (RHE).

unsupported PtRu black has been performed in 1 M methanol and 0.5 M H_2SO_4 solution. All electrodes were contacted with the solution at 0.075 V before the potential was stepped into the methanol oxidation region, for example, 0.4 V vs. RHE. The electrodes were held for additional 5 min at 0.075 V before stepping to 0.4 V. To simulate the steady-state conditions, the electrodes were held at 0.4 V for 10 min. The disk electrode for methanol electro-oxidation was used in stationary regime. The results of methanol electro-oxidation for the 3 electrode samples mentioned are summarized in Figure 5.9. It is obvious that SCR supported PtRu catalyst shows 25% higher mass activity than other company's commercial 60 wt% PtRu/C catalyst, it also has 50% higher activity than the commercial PtRu black catalyst for the methanol oxidation.

Another experiment has been investigated to find the dependence of methanol oxidation activity upon the catalyst metal loading in the range of 45 wt% to 75 wt% by TFRDE method via sweeping potential with 20 mV/sec from 0.075 V to 0.6 V in 0.5 M H_2SO_4 and 1 M methanol solution. Figure 5.10 summarizes the testing results, where Figure 5.10(a) shows the mass activity for different SCR anodes for methanol oxidation, an insert (a′) in Figure 5.10(a) summarizes the activity comparison at 0.4 V for different metal weight loading catalysts. Note that the mass activity here, by relatively fast potentiodynamic scan, is almost twice of that by potentiostatic mode after holding at 0.4 V for 10 min.

Figure 5.10(b) shows the result of surface specific activity with an insert of (b′) for the comparison of the activity of different weight loading catalysts at 0.4 V. In Figure 5.10(b), the electrochemical surface area was used to calculate the surface specific activity which was measured by CO stripping in 0.5 M H_2SO_4 separately. After finishing the surface area measurement, a certain amount of methanol was injected into the cell for a 1 M methanol oxidation study. Figure 5.10(a′) and (b′) indicate that the SCR anodes with Pt-Ru metal loading of 45–65 wt% show higher

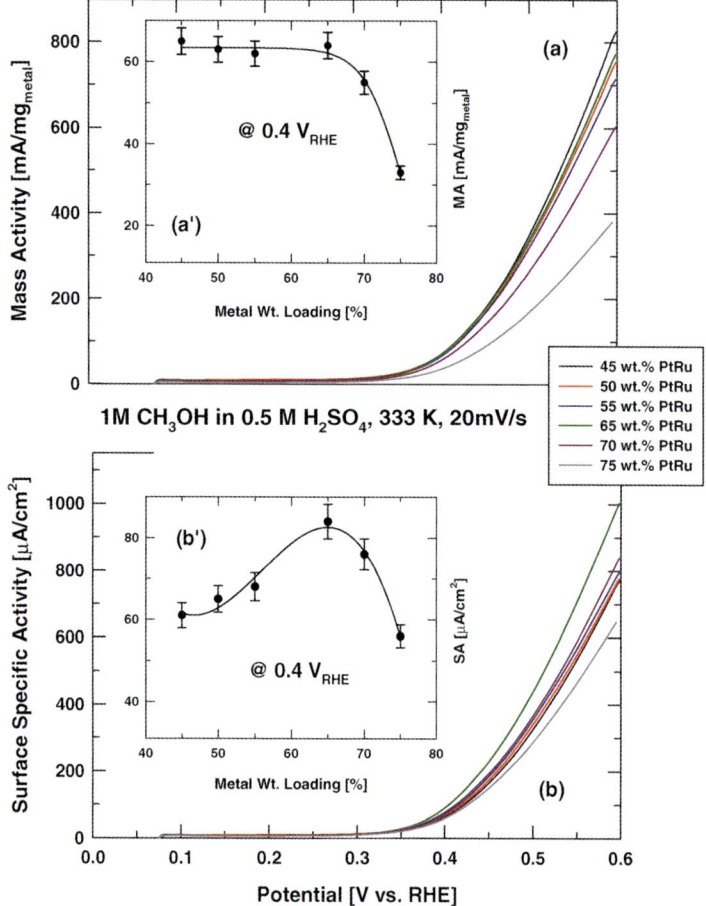

Figure 5.10 Methanol electro-oxidation on different metal loading SCR anode catalysts in 1 M methanol and 0.5 M H$_2$SO$_4$ solution. Here current densities for mass activity and surface specific activity are normalized based on catalyst mass and surface area separately.

mass activity than those of 70 wt% and 75 wt% ones, and 65 wt% PtRu/C catalyst shows the highest surface specific activity for the methanol oxidation. In the actual DMFC evaluation, we found 60 wt% PtRu/C showed the best MEA performance.

5.3.2.2 Fuel Cell Evaluation

Fuel cell evaluation is a critical step to validate the catalyst catalytic performance in a MEA under practical cell operating conditions. Because of the MCO effect, the measurement of anode polarization is important to understand the anode catalyst activity in a cell. Figure 5.11 shows the regular design for anode polarization measurement, where a power supply is used to polarize the anode to different potentials under various temperatures and methanol concentrations. The DMFC

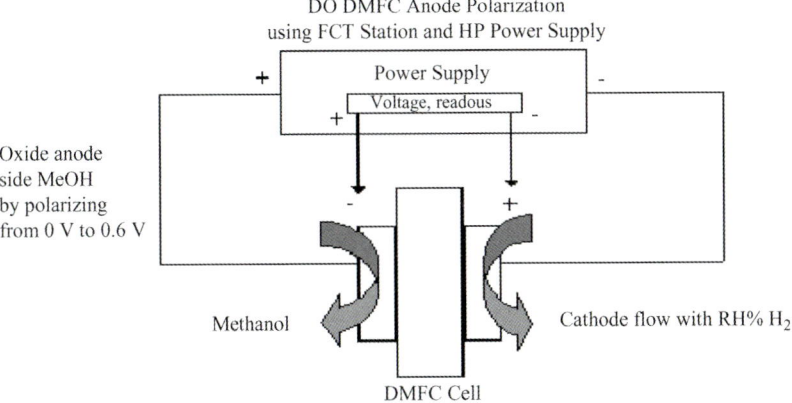

Figure 5.11 Schematic of methanol anode polarization measurement.

anode side connects to the positive end of the power supply to be polarized to different potentials such as from 0 to 0.6 V. The DMFC cathode, connecting with the negative end of the power supply, is supplied with 100% humidified hydrogen and works as a dynamic hydrogen electrode and the counter electrode as well.

Figure 5.12 shows the comparison study of SCR 60 wt% carbon supported PtRu/C anode versus commercially available PtRu/C made by other company. Figure 5.12(a) summarizes the anode polarization comparison study at 60 °C, and Figure 5.12(b) shows the full fuel cell MEA performance. The MEAs were made on Nafion 1135 membrane. Under the same testing conditions such as 3 ml/min 1 M methanol at 60 °C and the same catalyst metal loading, Cabot 60 wt% PtRu/C anode presents 240 mA/cm^2 at 0.4 V, about 20% higher than that of the commercial anode which is about 200 mA/cm^2. Similar advancement of Cabot catalyst is observed in full DMFC polarization measurement in Figure 5.12(b), where Cabot 60 wt% PtRu/C shows

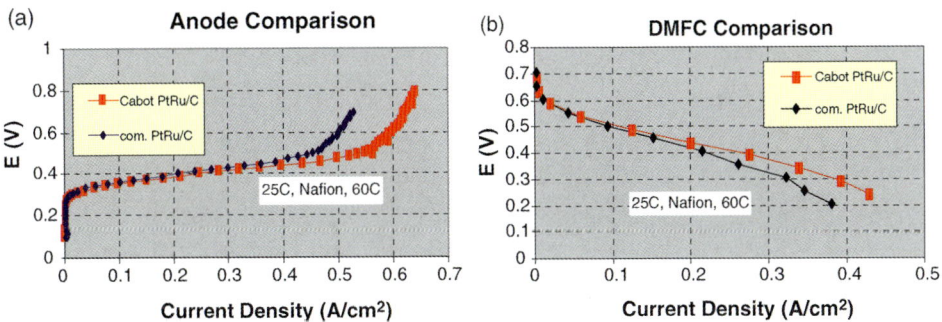

Figure 5.12 DMFC anode catalyst performance comparison of Cabot SCR 60 wt% PtRu/C vs. other maker's state-of-the-art 60 wt% PtRu/C (a) in anode polarization measurement; (b) in full cell polarization measurements at 1 M methanol solution at 60 °C. MEAs were made on Nafion 1135 membrane with same anode PtRu metal loading of 3 mg/cm^2.

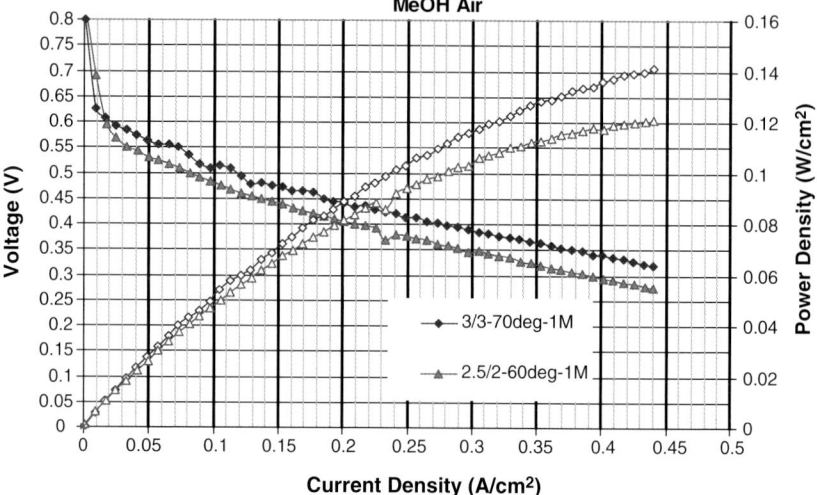

Figure 5.13 Performance of 60 wt% SCR anode catalyst in DMFC with Nafion 115 based MEA and 1 M methanol under different operating conditions: blue line, anode/cathode stoich = 3/3, at 70 °C; red line, anode/cathode stoich = 2.5/2, at 60 °C.

current density of 260 mA/cm^2 at 0.4 V, about 20% higher than 215 mA/cm^2 of the commercial PtRu/C.

Figure 5.13 shows another example of the SCR anode in a MEA based on Nafion 115 membrane with different stoichiometric flows at 60–70 °C for a portable DMFC application. With the anode loading of 3 mg/cm^2 and cathode loading of 1.5 mg/cm^2 (65 wt% Pt/C Cabot SCR cathode), it shows a current density of 275 mA/cm^2 at 0.4 V with stoichiometric ratio of 1 M methanol : dry air = 3 : 3 at 70 °C, corresponding to a power density of 110 mW/cm^2. When temperature changes to 60 °C with reduced stoichiometric flows of methanol : dry air of 2.5 : 2, the current density drops to 215 mA/cm^2, corresponding to 86 mW/cm^2 at 0.4 V. Both high power densities at 60–70 °C demonstrates good performance for portable DMFC applications.

5.3.3
Durability Study

The intent of durability testing of fuel cell catalysts and MEAs is to assess the performance decay rate or what percentage of the initial performance is lost over time. It may also help calculate after how many hours operation the beginning-of-life performance will degrade 20%. Recently the durability study of DMFC MEA and related electrocatalysts has been investigated more extensively due to the commercialization demand to use DMFCs for portable electronics [3–7, 9, 24–26]. Usually 1000 hours lifetime testing of MEA performance is required for materials screening for DMFC applications. For portable electronics product, DMFC should have 3000–5000 hours durability to enter the market. Figure 5.14 shows durability test

5.3 DMFC Electrocatalyst Characterization and Evaluation

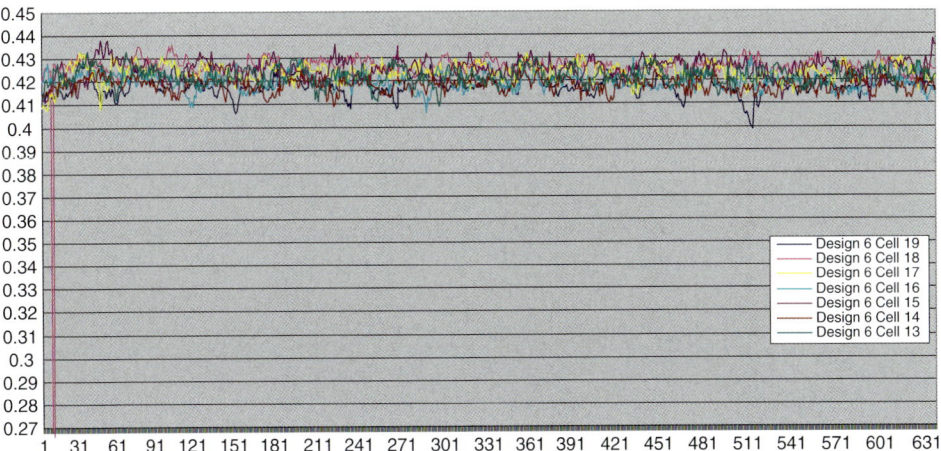

Figure 5.14 DMFC stack durability test for Nafion 115 based MEAs over 600 hours at 70 °C.

data for a stack with Cabot DMFC MEAs based on SCR anode and cathode electrodes. Here only 600 hours long term durability testing data was plotted, though the stack ran over 2000 hours, and the performance decay rate was observed to be about 20 μV/h per cell.

MEA durability must be based on excellent catalyst stability but the other MEA components such as membrane or GDLs could fail first during DMFC operation. For electrocatalyst, particle size growth and the leaching of the metal(s) from the alloy phase are two key factors for DMFC anode degradation. Besides using the electrochemical evaluation to monitor the performance change, spectroscopic methods are usually used to help identify the root cause of the electrocatalyst activity decay or MEA failure. We have used XRD to monitor the catalyst crystalline size change in a DMFC MEA after fuel cell operation for a length of time. Figure 5.15 shows an example of post mortem XRD analysis of the MEA samples from several DMFC stacks after operating for up to 3000 hours. The analysis showed that the SCR PtRu/C crystalline size (based on XRD) did not really change during 3000 hours operations. This demonstrates that during the long 3000 hours' operation, the SCR anode is very stable without crystallite size growth. As mentioned before, the crystallite size growth usually results in a decrease in electrochemical surface area and the catalyst deactivation. This was not observed for Cabot SCR carbon supported Pt-Ru catalysts.

Further investigation has been done to understand the catalyst stability via *ex-situ* Ru leaching test. The experiment was performed by soaking the anode catalyst in sulfuric solutions with and without methanol for several hours, following with a measurement of Ru content in the solution phase by ICP (induced coupling plasma) method. Figure 5.16 presents the Ru leaching experiment results of SCR anode, in comparison with other companies' commercial PtRu/C and PtRu black catalysts. In the case of acid only such as 1 M H_2SO_4, the PtRu catalysts are stable with less than 0.2% Ru coming out. However, in the presence of methanol in acid, soaked for

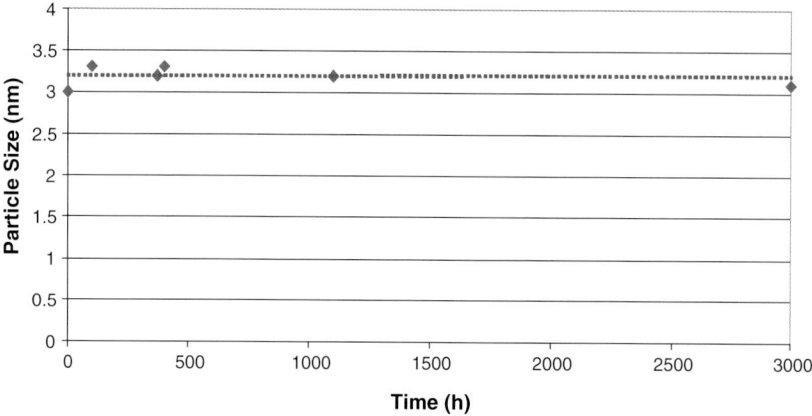

Figure 5.15 Change of catalyst crystallite size (by XRD) of Cabot SCR 60 wt% PtRu/C anode with time during 3000 hours operation in DMFC stacks. No particle size change was observed, indicating excellent catalyst stability.

2 hours at room temperature, Ru leaching occurs. The leaching amount of Ru for Cabot-supported PtRu catalyst is 3.8% in 1 M methanol and 1 M H_2SO_4, but it is about 1/3 of other companies' supported and/or PtRu black catalysts. When methanol concentration increases to 3 M in the mixed solution, the amount of Ru leaching for Cabot anode was 2.9%. (It is not clear why it decreased from 3.8% in 1 M to 2.9% in 3 M methanol, but could be related to ICP sensitivity). This is about $1/4$ of the other two catalysts which show an increased Ru leaching level. These experiments indicate

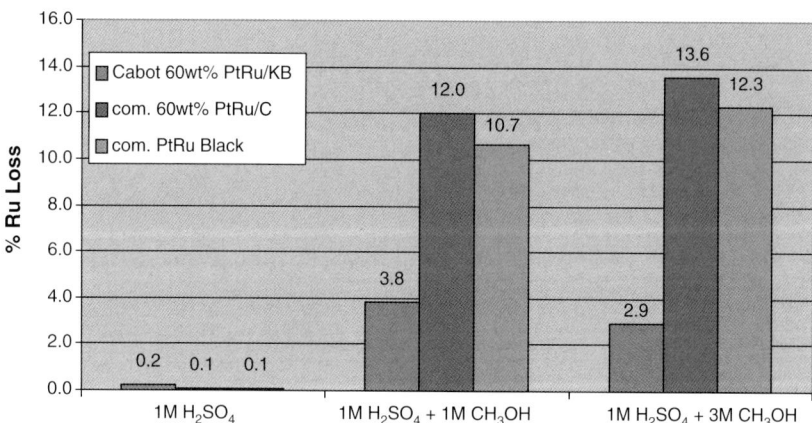

Figure 5.16 *Ex-situ* Ru leaching amount of PtRu catalysts in sulfuric solution with and without methanol at room temperature. The experiments were performed by soaking the catalysts in 1 M H_2SO_4 solution without and with 1 M and 3 M methanol for 2 hours separately, then analyzing the solutions by ICP to measure the Ru amount.

a significantly better chemical stability for the anode catalyst made by SCR platform in Cabot. The Pt content was monitored in 1 M H_2SO_4 in the presence of methanol for above samples and no significant amount of Pt was detected, which shows that the Pt phase is stable in such a solution. We believe that the less Ru leaching and stable PtRu crystalline size of Cabot SCR anode electrocatalyst result in good durability performance in DMFC MEA as shown in Figure 5.15.

5.4
DMFC Performance Advancement via MEA Design

5.4.1
Electrocatalyst Layer Design

High electrocatalyst activity is the prerequisite for high MEA performance. However, the MEA performance is also dependent upon the electrode layer design as well as the membrane electrolyte properties. Here we discuss briefly how the catalyst layer influences MEA performance and can be improved through the anode layer design.

The key for the anode layer structure is to develop a network to balance the transport processes such as the methanol diffusion to the catalyst sites, MCO through the membrane, proton conduction from the reactive sites to the membrane, electron release from the reaction sites to gas diffusion layer, which is connected to the electrical connectors, water transport through the membrane, and the release of the anode product carbon dioxide. For an active DMFC system, managing MCO is a key factor for cell performance improvement. Beside the membrane, MCO can be controlled to a certain degree by tailoring the layer structure (such as the morphology, size and distribution of the catalyst particles and pores) as well as the layer hydrophilicity. For a passive DMFC, depending on whether the system uses highly concentrated methanol or pure methanol at vapor mode as fuel, the anode layer design could be very different. Creating efficient oxygen mass transport channels and managing water flooding at the cathode layer also have to be designed carefully.

Catalyst ink formulation is critical to deposit an optimal catalyst layer for the MEA performance. Figure 5.17 shows a schematic of an ink development. Good ink requires a narrow catalyst particle size distribution, desirable viscosity and surface tension and good stability so that catalyst ink can be printed or deposited for an optimal catalyst layer. Because of the unique spherical structure, the SCR catalyst inks (containing Nafion ionomer) usually have a low viscosity of 7–30 cP with surface tension of about 30 dynes/cm, and they have good stability (>48 hours) for continuous printing of CCMs or GDEs for MEA manufacture.

Figure 5.18 shows the cross-sectional SEM of MEA catalyst layer structures developed by spray deposition onto Nafion membrane. These images confirm a good catalyst layer uniformity (with an anode loading of about 3 mg/cm^2 of SCR 60 wt% PtRu/C on the Nafion 115 membrane) as well as a clear spherical catalyst structure. A comparison of MEA performance of Cabot SCR 60 wt% PtRu/C anode (with anode 3 mg/cm^2 PtRu/C and cathode Pt/C 1.5 mg/cm^2) against state-of-the-art

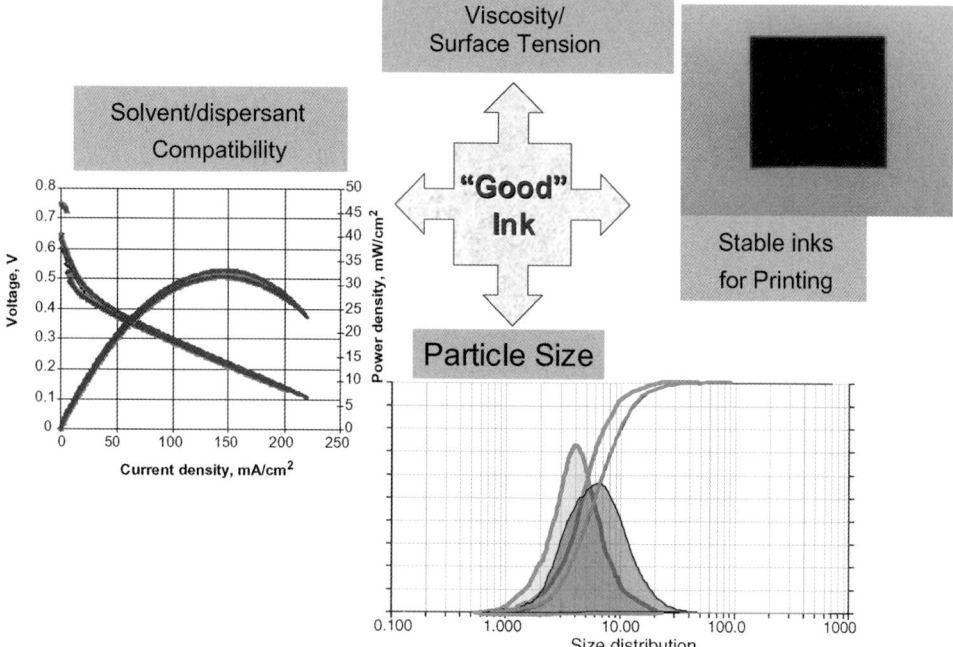

Figure 5.17 Ink development for depositing catalyst layer onto membrane or GDL. It is critical to have good control of particle size, ink viscosity and surface tension, compatibility with solvent and dispersant and good stability so that the catalyst ink can be deposited sustainably for MEA manufacturing.

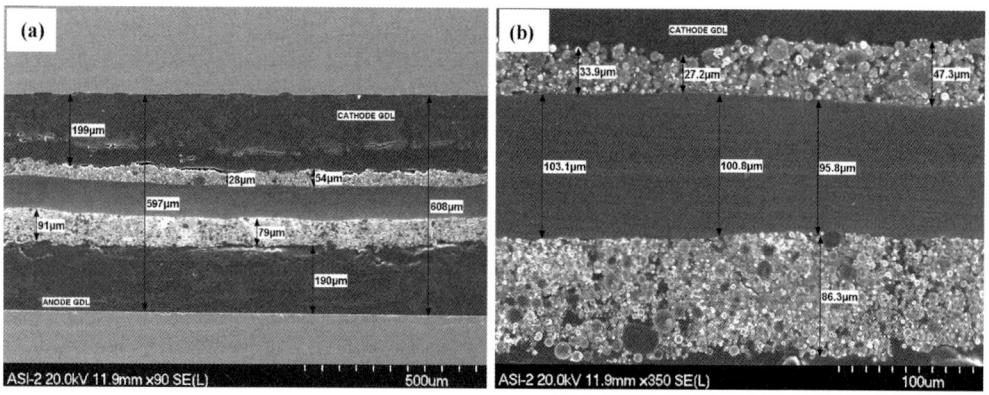

Figure 5.18 Cross-sectional SEM images of a DMFC MEA structure. (1) 5-layer MEA structure; (2) close-up of a 3-layer catalyst-coated membrane structure for the same MEA. The catalyst layer shows good uniformity with a clear spherical catalyst structure.

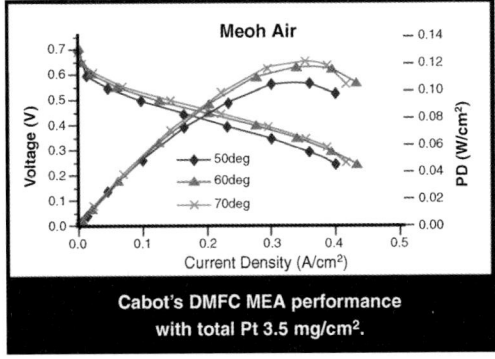
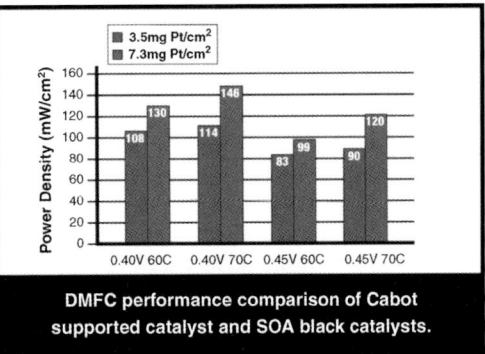

Cabot's DMFC MEA performance with total Pt 3.5 mg/cm². DMFC performance comparison of Cabot supported catalyst and SOA black catalysts.

Figure 5.19 Example of Cabot DMFC MEA with 3.5 mg/cm² total Pt loading of our supported catalysts and the performance comparison to that of high loading black catalysts.

(SOA) PtRu black anode (with anode 8 mg/cm² PtRu black and same cathode) is summarized in Figure 5.19. The SCR supported PtRu/C shows good performance though the anode loading is about 2.5 times lower than that of a black PtRu, demonstrating about 83% and 75% of relative MEA performance at 60 °C and 70 °C respectively against the one with high loading Pt-Ru black. This result indicates how optimized catalyst layers in a MEA can help address the cost and performance challenge issues for DMFC commercialization. On the other hand, for PtRu black catalyst, over 2 times Pt loading in MEA versus SCR carbon supported PtRu just resulted in very limited performance gain. This hints that unsupported PtRu black is not a cost-effective catalyst for DMFC. The doubled Pt normalized performance indicates a clear advantage of SCR supported electrocatalyst over the unsupported catalyst for DMFC.

5.4.2
Hydrocarbon Membrane for DMFC Performance Improvement

Though catalyst activity is a major factor for MEA performance, due to the negative impact of MCO on performance and fuel efficiency, the membrane plays a key role in managing the crossover. It is known that Nafion is a standard electrolyte for DMFC because of its excellent electrochemical stability and proton conductivity and good durability. However, due to the high methanol permeability, Nafion normally results in low methanol use in DMFC. Usually the water drag of Nafion is about 3, which can easily cause cathode flooding, resulting in making water management more challenging. The dimensional stability of Nafion needs further improvement because its severe swelling in water and high sensitivity to moisture sometimes cause problems during MEA manufacturing.

Recent progress in new membranes for DMFC application is encouraging [19]. Thanks to new chemistry structure, hydrocarbon (HC) membrane shows advantages over Nafion with reduced MCO (or increased methanol use), smaller water drag number and possibly improved dimensional stability. There is significant progress in

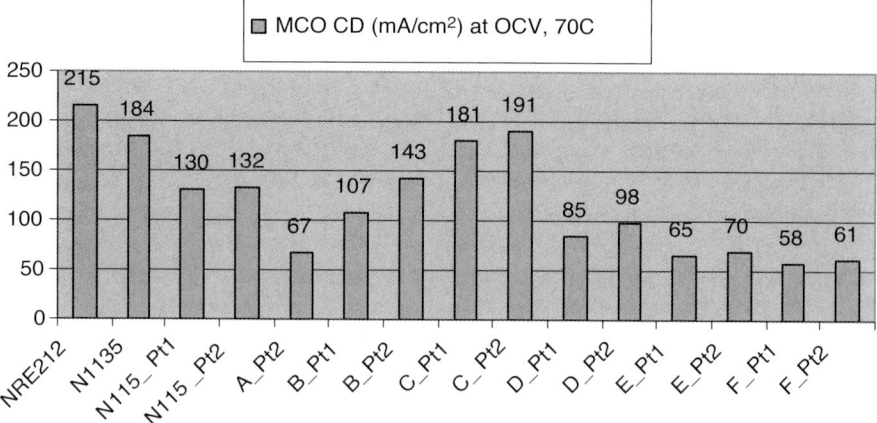

Figure 5.20 Comparison of MCO current at OCV of Nafion vs. HC membranes.

HC membrane based MEA development for DMFC [3–5]. As an example, Figure 5.20 summarizes a comparison study of MCO current at OCV of Nafion against several types of HC membranes. Here NRE212, N1135 and N115 are Nafion type membranes with thickness of 2, 3.5 and 5 mil separately. A–F are HC membranes with thickness of 25–50 μm made by different membrane developers. Pt 1 and Pt 2 refer to different anode layer structures with different Pt loading in the anode. It is clear that most HC membranes have lower MCO rates than Nafion. For example, type A has a MCO rate of 67 mA/cm² at OCV which is about half of the MCO rate of Nafion 115 membrane. The HC membrane thickness is about 30 μm, almost 1/4 of Nafion 115 thickness. If normalized, MCO rate with the membrane thickness, all HC membranes will present low MCO rate than Nafion.

The low MCO of HC can be translated into a performance improvement, not only resulting in power density increase, but also methanol use improvement. Figure 5.21 shows an example of HC-membrane-based MEA performance at 70 °C with total Pt loading of 3.5 mg/cm² in the MEA. It shows good power density such as 140 mW/cm² at 0.4 V and 110 mW/cm² at 0.45 V. The methanol use, that is, the percentage of methanol used for generating the electricity, is usually obtained in the mass balance measurement under constant current discharge conditions. Beside methanol use, water drag number (WD, the net number of water molecules dragged to the cathode per proton transferred from the anode to the cathode) can also be measured for understanding how well the new membrane or MEA structure can handle the water management. Figure 5.22 summarizes the key measurements of power density (PD) at 0.45 V 70 °C, methanol use and water drag number of MEAs based on several selected HC membranes versus Nafion 115 with different anode catalyst layer designs. Here Nafion 1 and HC 1, 3, 5–13 had a design with anode PtRu loading of 3 mg/cm² and Nafion 2 and HC 2, 4, 6–12 had another design with high PtRu loading of about 8 mg/cm². It is clear that, for similar anode catalyst layer design, some HC membrane MEAs show improved power density and methanol use with reduced water drag number compared with those of Nafion 115.

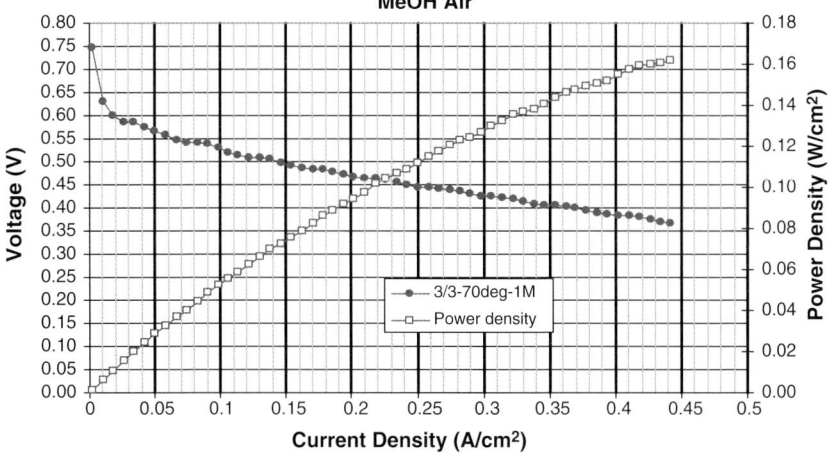

Figure 5.21 Performance of HC membrane based MEA with 60 wt% SCR anode catalyst at 70 °C in DMFC with 1 M methanol and dry air and anode/cathode stoich flows = 3/3.

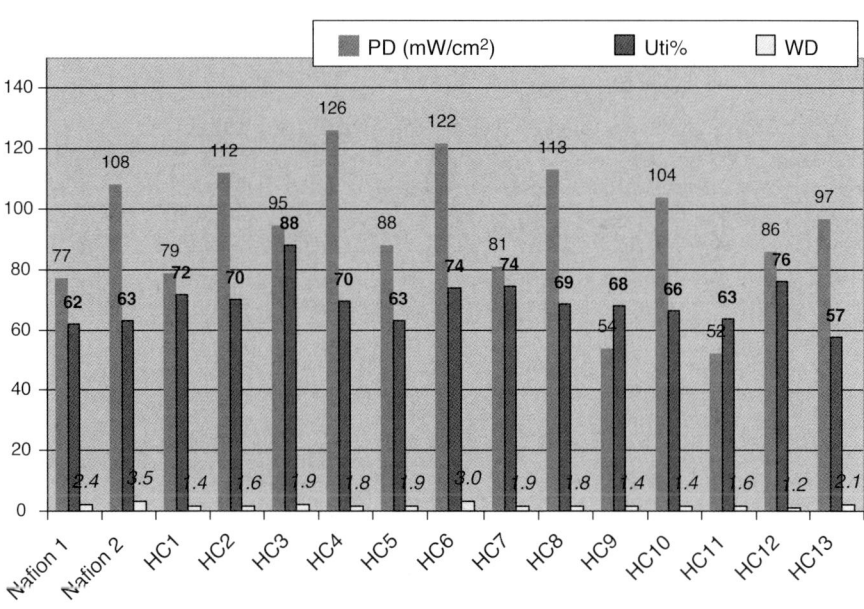

Figure 5.22 MEA performance comparison of some HC membranes vs. Nafion in DMFC. Here PD is power density at 0.45 V measured in a cell polarization, Uti% is methanol use and WD is water drag number, both Uti% and WD are measured by mass balance under constant current discharge.

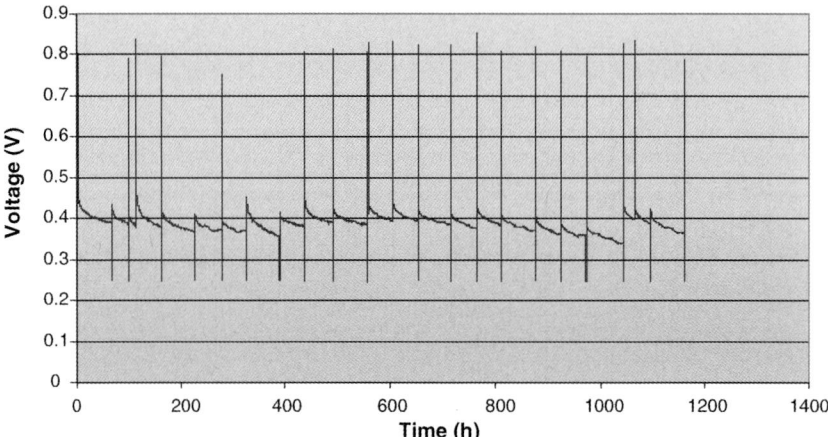

Figure 5.23 Durability testing of HC membrane based MEA over 1000 hours in DMFC. The test was performed under constant current discharge (200 mA/cm^2) with 1 M methanol and dry air at 70 °C. The cell has been interrupted several times to measure the cell performance and mass balance.

Significant progress has been made on HC membrane MEAs for DMFC application as discussed. However, the adoption of HC membranes has been very limited so far, since HC membrane studies are still at the research, development and demonstration stage. One reason is that some HC membrane production is in the validation phase and membranes are not available in large quantities as commercial grade products. The durability of HC membrane is also still under investigation. There is not enough HC MEA durability data published compared to that of Nafion membrane. In fact a recent study performed by Cabot shows that some HC membranes are durable enough for DMFC operations. Figure 5.23 shows an example of one HC membrane MEA over 1000 hours operation in a single cell DMFC with 1 M methanol and 70 °C under constant current discharge. The overall decay rate was just about 20 µV/hr. Cabot's other HC-membrane-based MEAs have been demonstrated by a third party with over 2400 hours durability in a DMFC stack, together with excellent thermal cyclability, indicating great potential to use new HC membranes for DMFC systems in near future.

5.4.3
Other Aspects of DMFC Catalyst Development

Though most of this chapter is focused on anode catalyst development and related MEA performance, the cathode oxygen reduction catalyst is also very important for DMFC. Cabot SCR cathode catalyst such as 60 wt% supported Pt on carbon shows similar spherical structure, and has demonstrated an improved MEA performance against other Pt cathode catalysts commercially available from other companies, as summarized in Figure 5.24. The MEA performance with Cabot SCR cathode shows

5.4 DMFC Performance Advancement via MEA Design

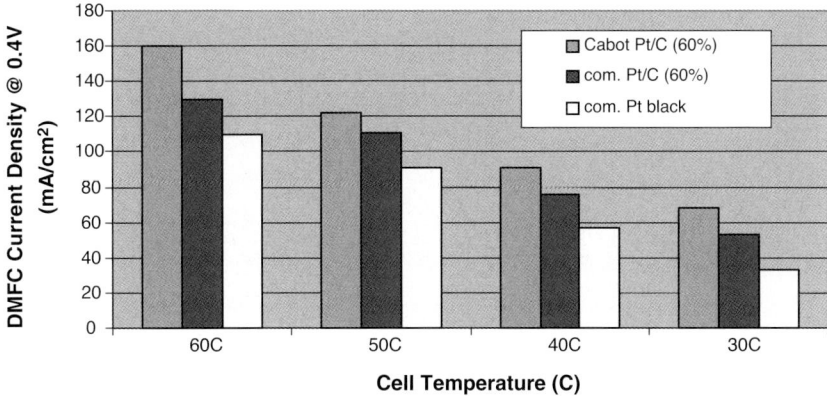

Figure 5.24 Comparison of SCR cathode Pt/C versus other commercially available Pt/C and Pt black catalysts in the market.

20% higher current density at 0.4 V than other commercial 60 wt% Pt/C, almost 40% higher than the unsupported Pt black.

New catalyst development for better kinetics is an eternal topic for fuel cells. Because of the high cost of Pt and PtRu blacks, the supported catalysts represent the future for DMFC catalysts. Though other new compositions should be investigated systematically to advance catalyst activity, the importance of supports, especially the modified chemistry properties of carbon supports, have been less explored. Cabot developed modified carbon black (MCB) technology and started using it for fuel cell catalyst applications. Figure 5.25 shows the concept of using MCB for the next generation of DMFCs. Different functional groups can be attached to carbon to tailor the catalyst hydrophilicity and water retention which are important for both DMFC and PEMFC operations [7, 8].

We should have mentioned that most of the above discussions are based on Active Type DMFC in which methanol and air are 'forced' to supply to the fuel cells for high power output. There is another type called Passive DMFC in which air comes from the environment via 'air-breathing' mode. Passive Type DMFC has much simpler balance-of-plant (BOP) than that of the Active Type DMFC, suitable for small portable devices such as cell phone or laptop chargers. The catalyst development for Passive DMFC could be very different from that of Active DMFC. The use of higher

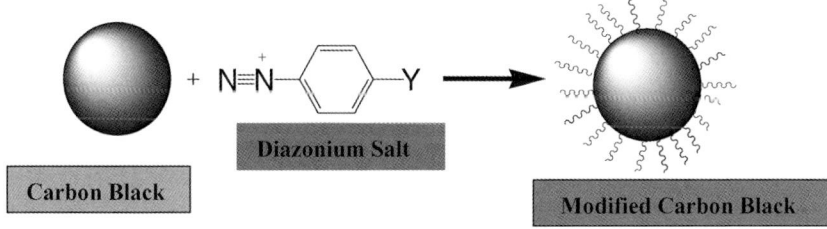

Figure 5.25 Modified carbon black (MCB) for electrocatalyst development.

concentration of methanol, even pure methanol, as the fuel for high energy density, compactness and light weight, mean that the anode catalyst needs more hydrophilic properties to retain water so that the methanol can be diluted to let the methanol oxidation happen at the anode interface. In such a case, a more hydrophobic cathode catalyst layer may be required to help back-diffuse water from cathode to the anode to dilute methanol further. These new operating conditions bring new challenges and opportunities for DMFC catalyst development. It is expected that catalyst surface modifications to tailor the catalyst hydrophilicity properties will play an important role.

5.5
Prospects for DMFC

DMFC technologies for both Active and Passive types have been improved significantly in recent years. Compared to H_2/air fuel cell which has a hydrogen availability/infrastructure problem and current hydrogen storage materials have very low storage capacity, DMFC shows a clear merit due to the readiness of methanol fuel availability. DMFC is a good choice for the next generation supply source especially for portable applications.

One competing technology to DMFC is Li-ion battery which has a high energy density of about 200 Wh/Kg or 500 Wh/L and reasonable sale price in the market. Though methanol has higher energy density (5000 Wh/L) and DMFC can be used for higher energy density devices with almost zero charging time with a fuel cartridge, it is very critical to make the DMFC more compact and cost affordable to meet the market requirements. To reach this goal, it is very important to continue to develop higher activity catalysts so that less amount of Pt is necessary in the MEA to deliver the required power.

Moving forward, since Li-ion batteries have almost reached their capacity upper limit, development of methanol fuel cell will continuously be the focus of the research and development for the new power sources. Next generation DMFC catalysts should possess not only great catalyst activity, but also unique feature in microstructure (such as particle morphology, and pore size and size distribution etc.), surface hydrophilicity properties and improved durability in order to meet different requirements for Active, Passive or semi-active DMFC systems for broad applications.

5.6
Conclusions

Spray conversion reaction (SCR) technology has been demonstrated to create innovative and advanced electrocatalysts for direct methanol fuel cells. The SCR platform not only allows us to manufacture electrocatalysts with high capacity (several Kg per hour) to meet fuel cell commercialization demand, but also brings unique morphology and structure properties for performance and durability advantages. To sum up, new SCR DMFC catalysts have the following features:

- *High catalyst activity*. Because of highly dispersed PtRu alloy crystals on high surface area carbon, Cabot's supported catalysts have a small crystalline size with excellent uniformity which results in high catalyst activity.
- *Excellent durability*. New SCR PtRu/C catalysts are more durable than PtRu blacks and other supported PtRu made by conventional approaches due to much less Ru leaching problem.
- *Low cost*. New SCR PtRu/C reduces precious metal loading by 50% while still delivering excellent MEA performance.
- *Value added property*. Flexibility in control of morphology and hydrophilicity provides new approaches to manage water and methanol problems to help overcome DMFC commercialization barriers.

Acknowledgments

The authors would like to express their appreciation to many Cabot colleagues who have provided assistance in many aspects, in particular Greg Romney, Matthew Ezenyilimba, Ryan Wall, Debbie Schlueter, Gordon Rice, Daniel Salazer, David Wood, Joseph Slanga, Stephen Rice, Angelos Kyrlidis and Jeanne Cambray.

Abbreviations

BET	Brunauer-Emmett-Teller
BOP	balance-of-plant
CCM	catalyst coated membrane
DMFC	Direct methanol fuel cell
EDX	energy-dispersive X-ray spectroscopy
GDE	gas diffusion electrode
GDL	gas diffusion layer
HC	hydrocarbon
ICP	induced coupling plasma
KB	Ketjen black
MCB	modified carbon black
MCO	methanol crossover
MEA	membrane electrode assembly
OCV	open circuit voltage
PEMFC	proton exchange membrane fuel cell
PD	power density
PSD	pore size distribution
RHE	reversible hydrogen electrode
SCR	spray conversion reaction
SEM	scanning electron microscopy
SOA	state-of-the-art
TEM	transmission electron microscopy
THRDE	thin film rotating disk electrode
XRD	X-ray Diffraction

References

1 Olah, G.A., Goeppert, A. and Prakash, G.K.S. (2006) *Beyond Oil and Gas: The Methanol Economy*, Wiley-VCH Verlag GmbH, Weinheim.
2 Lei, H., Suh, S., Gurau, B., Workie, B., Liu, R. and Smotkin, E.S. (2002) *Electrochim. Acta*, **47**, 2913–2919.
3 Lei, H., Atanassova, P., Sun, Y., Wood, D. and Brewster, J. (October 2008) Manufacturing of high performance durable MEAs for portable electronics. Presentation at Fuel Cell Seminar 2008, Phoenix.
4 Lei, H. and Atanassova, P. (June 2008) Innovative inkjetting and spray deposition for low-cost, high-performance fuel cell catalyst coated membrane manufacturing. Poster at 2008 DOE annual review meeting, Washington DC.
5 Lei, H., Atanassova, P., Sun, Y., Blizanac, B. and Wood, D. (May 2008) Advanced DMFC electrocatalyst for portable applications. Presentations at 10th Small Fuel Cells Conference, Atlanta.
6 Lei, H., Atanassova, P., Rice, G. and Sun, Y. (May 2007) High performing durable electrocatalysts for PEM fuel cell applications. Presentation at FCDIC 2007, Tokyo.
7 Lei, H., Atanassova, P., Sun, Y., Rice, G. and Romney, G. (February 2007) Advanced DMFC electrocatalysts and MEAs for portable electronics. Presentation at Fuel Cell Expo 2007, Tokyo.
8 Lei, H., Atanassova, P., Sun, Y. and Rice, G. (October 2006) Advanced electrocatalyst and MEA technologies to boost DMFC performance. Presentation at The 210th Electrochemical Society Meeting, Abstract #646, Cancun, Mexico.
9 Liu, H., Song, C., Zhang, L., Zhang, J., Wang, H. and Wilkinson, D.P. (2006) *J. Power Sources*, **155**, 95–110.
10 Muller, J., Frank, G., Colbow, K. and Wilkinson, D. (2003) *The Handbook of Fuel Cell: Fundamentals, technology and Applications*, vol. 4 (eds W. Vielstich, A. Lamm and H.A. Gasteige), John Wiley & Sons, Ltd, Chichester, pp. 847–855.
11 Neergat, M., Friedrich, K.A. and Stimming, U. (2003) *The Handbook of Fuel Cell: Fundamentals, technology and Applications*, vol. 4 (eds W. Vielstich, A. Lamm and H.A. Gasteige), John Wiley & Sons, Ltd, Chichester, pp. 856–877.
12 Lamm, A. and Muller, J. (2003) *The Handbook of Fuel Cell: Fundamentals, technology and Applications*, vol. 4 (eds W. Vielstich, A. Lamm and H.A. Gasteige), John Wiley & Sons, Ltd, Chichester, pp. 878–893.
13 Narayanan, S.R., Valdaz, T.I. and Rohatgi, N. (2003) *The Handbook of Fuel Cell: Fundamentals, technology and Applications*, vol. 4 (eds W. Vielstich, A. Lamm and H.A. Gasteige), John Wiley & Sons, Ltd, Chichester, pp. 894–904.
14 Narayanan, S.R. and Valdaz, T.I. (2003) *The Handbook of Fuel Cell: Fundamentals, technology and Applications*, vol. 4 (eds W. Vielstich, A. Lamm and H.A. Gasteige), John Wiley & Sons, Ltd, Chichester, pp. 1133–1141.
15 Wasmus, S. and Kuver, A. (1999) *J. Electroanal. Chem.*, **461**, 14–31.
16 Gottesfeld, S. and Zowodzinski, T.A. (1997) *Advances in Electrochemical Sciences and Engineering* (eds R.C. Alkire, H. Gerischer, D.M. Kolb and C.W. Tobias), Wiley-VCH Verlag GmbH, Weinheim, pp. 195–301.
17 Ren, X., Zelenay, P., Thomas, S., Davey, J. and Gottesfeld, S. (2000) *J. Power Sources*, **86**, 111–116.
18 Broussley, M. and Archdale, G. (2004) *J. Power Sources*, **136**, 386–394.
19 Neburchilov, V., Martin, J., Wang, H. and Zhang, J. (2007) *J. Power Sources*, **169**, 221–238.
20 Ren, X., Springer, T.E., Zawodzinski, T.A. and Gottesfeld, S. (2000) *J. Electrochem. Soc*, **147**, 466–474.
21 Ren, X., Springer, T.E. and Gottesfeld, S. (2000) *J. Electrochem. Soc*, **147**, 92–98.

22 Liu, F., Lu, G. and Wang, C.Y. (2006) *J. Electrochem. Soc.*, **153**, A543–A553.
23 Jiang, R. and Chu, D. (2004) *J. Electrochem. Soc.*, **151**, A69–A76.
24 Piela, P., Eickes, C., Brosha, E., Garzon, F. and Zelenay, P. (2004) *J. Electrochem. Soc.*, **151**, A2053–A2059.
25 Eickes, C., Piela, P., Davey, J. and Zelenay, P. (2006) *J. Electrochem. Soc.*, **153**, A171–A178.
26 http://www1.eere.energy.gov/hydrogenandfuelcells/pdfs/merit03/117_lanl_piotr_zelenay.pdf.
27 Knights, S.D., Colbow, K.M., St-Pierre, J. and Wilkinson, D.P. (2004) *J. Power Sources*, **127**, 127–134.
28 Bezerra, C.W.B., Zhang, L., Lee, K., Liu, H., Marques, A.L.B., Marques, E.P., Wang, H. and Zhang, J. (2008) *Electrochim. Acta*, **53**, 4937–4951.
29 Chu, D. and Jiang, R. (2002) *Solid State Ionics*, **148**, 591–599.
30 Antolini, E., Lopes, T. and Gonzalez, E.R. (2008) *J. Alloys Compd.*, **461**, 253–262.
31 Antolini, E., Salgado, J.R.C. and Gonzalez, E.R. (2006) *J. Power Sources*, **160**, 957–968.
32 Papageorgopoulos, D.C., Liu, F. and Conrad, O. (2007) *Electrochim. Acta*, **52**, 4982–4986.
33 Hamnett, A. (1997) *Catal. Today*, **38**, 445–457.
34 Beden, B., Leger, J.-M., Lamy, C. and Bockris, J.O'M. (1992) *Modern Aspects of Electrochemistry*, vol. 22 (eds B.E. Conway and B.E. White), Plenum Press, New York, pp. 97–247.
35 Watanabe, M., Uchida, M. and Motoo, S. (1987) *J. Electroanal. Chem.*, **229**, 395–406.
36 Lee, C.E. and Bergens, S.H. (1998) *J. Phys. Chem. B*, **102**, 193–199.
37 Ticanelli, E., Beery, J.G., Paffett, M.T. and Gottesfeld, S. (1989) *J. Electroanal. Chem.*, **258**, 61–77.
38 Chrzanowski, W. and Wieckowski, A. (1998) *Langmuir*, **14**, 1967–1970.
39 Frelink, T., Visscher, W. and Van Veen, J.A.R. (1995) *Surf. Sci.*, **335**, 353–360.
40 Iwashita, T. and Vielstich, W. (1991) *Advances in Electrochemical Science and Engineering*, vol. 1 (eds R.C. Alkire, H. Gerischer, D.M. Kolb and C.W. Tobias), VCH Publishers, New York, pp. 127–170.
41 Ley, K.L., Liu, R., Pu, C., Fan, Q., Leyarovska, N., Segre, C. and Smotkin, E.S. (1997) *J. Electrochem. Soc.*, **144**, 1543–1548.
42 Reddington, E., Sapienza, A., Gurau, B., Viswanathan, R., Sarangapani, S., Smotkin, E.S. and Mallouk, T.E. (1998) *Science*, **280**, 1735–1739.
43 Gurau, B., Viswanathan, R., Lafrenz, T.J., Liu, R., Ley, L., Smotkin, E.S., Reddington, E., Sapienza, A., Chan, B.C., Mallouk, T.E. and Sarangapani, S. (1998) *J. Phys. Chem. B*, **102**, 9997–10003.
44 Long, J.W., Stroud, R.M., Swider-Lyons, K.E. and Rolison, D.R. (2000) *J. Phys. Chem. B*, **104**, 9772–9776.
45 Dubau, L., Coutanceau, C., Garnier, E., Leger, J. and Lamy, C. (2003) *J. Appl. Electrochem.*, **33**, 419–429.
46 Kinoshita, K. and Stonehart, P. (1977) *Modern Aspects of Electrochemistry*, vol. 12 (eds J.O'.M. Bockris and B.E. Conway), Plenum, New York, pp. 183–266.
47 Kinoshita, K. (1988) *Carbon Electrochemical and Physicochemical Properties*, John Wiley & Sons, Inc., New York, pp 388–396.
48 Hampden-Smith, M., Atanassova, P., Atanassov, P. and Kodas, T.T. (2003) *The Handbook of Fuel Cell: Fundamentals, technology and Applications*, vol. 3 (eds W. Vielstich, A. Lamm and H.A. Gasteige), John Wiley & Sons, Ltd, Chichester, pp. 497–508.
49 Kodas, T. and Hampden-Smith, M. (1999) *Aerosol Processing of Materials*, Wiley-VCH Verlag GmbH, Weinheim.
50 Paolina, P., Atanassov, P., Bhatia, R., Hampdem-Smith, M., Kodas, T. and Napolitano, P. (2002) *Direct-Write Technologies for Rapid Prototyping Applications: Sensors, Electronics, and Integrated Power Sources*, Chapter 1 (eds A. Pique and D.B. Chrisey), Academic Press, San Diego, pp. 55–92.
51 Hyun, M.S., Kim, S.K., Lee, B., Peck, D., Shul, Y. and Jung, D. (2008) *Catal. Today*, **132**, 138–145.

52 Deivaraj, T.C. and Lee, J.Y. (2005) *J. Power Sources*, **142**, 43–49.
53 Yang, B., Lu, Q., Wang, Y., Zhang, L., Lu, J. and Liu, P. (2003) *Chem. Mater.*, **15**, 3552–3557.
54 Kim, T., Takahashi, M., Nagai, M. and Kobayashi, K. (2004) *Electrochem. Acta*, **50**, 813–817.
55 Liu, Z., Ling, X., Lee, J., Su, X. and Gan, L.M. (2003) *J. Mater. Chem.*, **13**, 3049–3052.
56 Watanabe, M. and Motoo, S. (1975) *J. Electroanal. Chem.*, **60**, 267–273.
57 Sasaki, K., Wang, J.X., Balasubramanian, M., McBreen, J., Uribe, F. and Adzic, R.R. (2004) *Electrochim. Acta*, **49**, 3873–3877.
58 Kuk, S.T. and Wieckowski, A. (2005) *J. Power Sources*, **141**, 1–7.
59 Lee, D., Hwang, S. and Lee, I. (2006) *J. Power Sources*, **160**, 155–160.
60 Liu, Z., Ling, X.Y., Guo, B., Hong, L. and Lee, J.Y. (2007) *J. Power Sources*, **167**, 272–280.
61 Liu, Z., Ling, X.Y., Su, X., Lee, J.Y. and Gan, L.M. (2005) *J. Power Sources*, **149**, 1–7.
62 Caillard, A., Coutanceau, C., Brault, P., Mathias, J. and L'eger, J.-M. (2006) *J. Power Sources*, **162**, 66–73.
63 Whitacre, J.F., Valdez, T. and Narayanan, S.R. (2005) *J. Electrochem. Soc.*, **152**, 1780–1789.
64 Denis, M.C., Lef'evre, M., Guay, D. and Dodelet, J.P. (2008) *Electrochim. Acta*, **53**, 5142–5154.
65 Goúerec, P., Denis, M.C., Guay, D., Dodelet, J.P. and Schulz, R. (2000) *J. Electrochem. Soc.*, **147**, 3989–3996.
66 Lu, L. and Lai, M.O. (1998) *Mechanical Alloying*, Kluwer Academic Publishers, Boston.
67 Che, G., Lakshmi, B.B., Martin, C.R. and Fisher, E.R. (1999) *Langmuir*, **15**, 750–758.
68 Li, W., Liang, C., Zhou, W., Qiu, J., Li, H., Sun, G. and Xin, Q. (2004) *Carbon*, **42**, 436–439.
69 Guo, J., Sun, G., Wang, Q., Wang, G., Zhou, Z., Tang, S., Jiang, L., Zhou, B. and Xin, Q. (2006) *Carbon*, **44**, 152–157.
70 Wang, H., Xu, C., Cheng, F., Zhang, M., Wang, S. and Jiang, S.P. (2008) *Electrochem. Comm.*, **10**, 1575–1578.
71 Jeon, M.K., Cooper, J.S. and McGinn, P.J. (2008) *J. Power Sources*, **185**, 913–916.
72 Chu, Y.H., Ahn, S.W., Kim, D.Y., Kim, H.J., Shul, Y.G. and Han, H. (2006) *Catal. Today*, **111**, 176–181.
73 Jiang, R. and Chu, D. (2002) *J. Electroanal. Chem.*, **527**, 137–142.
74 Sun, Y., Brewster, J., Rice, G. and Atanassova, P. (October 2006) Development of durable, low-cost, high-performance Pt alloy electrocatalysts for H_2/air fuel cell application. Presentation at The 210th Electrochemical Society Meeting, Abstract # 573, Cancun, Mexico.
75 Rice, G., Xie, J., Lei, H., Sun, Y., Kyrlidid, A., Napadenski, B. and Atanassova, P. (October 2006) Development of high performance electrocatalysts for hydrogen/air fuel cells and direct methanol fuel cells based on modified carbon black technology. Presentation at The 210th Electrochemical Society Meeting, Abstract # 572, Cancun, Mexico.
76 Atanasova, P. (November 2006) Low precious metal alloy catalysts and durable carbon supports. Presentation at 2006 Fuel Cell Seminar, Hawaii, http://www.fuelcellseminar.com/pdf/2006/Wednesday/3B/Atanassova_Paolina_0835_3B_745(rv3).pdf.
77 Lei, H., (2005) Digital printing for high volume, low cost, high performance fuel cell CCM/MEA manufacturing. Presentation at 1st Symposium on the Manufacturing of MEAs for Hydrogen Applications, Dayton, Ohio, USA, August, 2005.
78 Tripkovic, A.V., Popovic, K.D., Grgur, B.N., Blizanac, B., Ross, P.N. and Markovic, N.M. (2002) *Electrochim. Acta*, **47**, 3707–3714.
79 Schmidt, T.J., Gasteiger, H.A. and Behm, R.J. (1999) *Electrochem. Comm.*, **1**, 1–4.
80 Schmidt, T.J., Noeske, M., Gasteiger, H.A., Behm, R.J., Britz, P., Brijoux, W. and Bönnemann, H. (1997) *Langmuir*, **13**, 2591–2595.

6
Platinum Alloys as Anode Catalysts for Direct Methanol Fuel Cells
Ermete Antolini

6.1
Introduction

The use of methanol as energy carrier and its direct electrochemical oxidation in direct methanol fuel cells (DMFCs) represents an important challenge for polymer electrolyte fuel cell technology, since the complete system would be simpler without a reformer and reactant treatment steps. The use of methanol as the fuel has several advantages in comparison to hydrogen: it is a cheap liquid fuel, easily handled, transported, and stored, and with a high theoretical energy density [1–3]. Although much progress has been made in the development of DMFC, its performance is still limited by the poor kinetics of the anode reaction [3–5]. Methanol oxidation is a slow reaction that requires active multiple sites for the adsorption of methanol and the sites that can donate OH species for the desorption of the adsorbed methanol residues [6]. Methanol oxidation has been extensively investigated since the early 1970s with two main topics: identification of the reaction intermediates, poisoning species and products, and modification of Pt surface in order to achieve higher activity at lower potentials and better resistance to poisoning. The results have been reviewed by several authors [7–9]. The main reaction product is CO_2 [10], although significant amounts of formaldehyde [11, 12], formic acid [10] and methyl formate [12, 13] were also detected. Most studies conclude that the reaction can proceed according to multiple mechanisms. However, it is widely accepted that the most significant reactions are the adsorption of methanol and the oxidation of CO, according to this simplified reaction mechanism:

$$CH_3OH \rightarrow Pt - (CH_3OH)_{ads} \tag{6.1}$$

$$Pt - (CH_3OH)_{ads} \rightarrow Pt - (CO)_{ads} + 4H^+ + 4e^- \tag{6.2}$$

$$Pt - (CO)_{ads} + H_2O \rightarrow Pt + CO_2 + 2H^+ + 2e^- \tag{6.3}$$

Platinum is the most active metal for dissociative adsorption of methanol, but, as is well-known, at room or moderate temperatures it is readily poisoned by

Electrocatalysis of Direct Methanol Fuel Cells. Edited by Hansan Liu and Jiujun Zhang
Copyright © 2009 WILEY-VCH Verlag GmbH & Co. KGaA, Weinheim
ISBN: 978-3-527-32377-7

carbon monoxide, a byproduct of methanol oxidation. To date, the remedy has been to use binary or ternary eletrocatalysts based on platinum, Pt-M_1 and Pt-M_1-M_2 [14–19]. The addition of a second/third element to Pt can improve the reaction of methanol electro-oxidation. The superior performance of these binary and ternary electrocatalysts for the methanol oxidation reaction (MOR) with respect to Pt alone was attributed to the bi-functional effect (promoted mechanism) [20, 21] and/or to the electronic interaction between Pt and alloyed metals (intrinsic mechanism) [9, 20, 22]. According to the promoted mechanism, the oxidation of the strongly adsorbed oxygen-containing species is facilitated in the presence of M oxides by supplying oxygen atoms at an adjacent site at a lower potential than that accomplished by pure Pt. The methanol electro-oxidation mechanism at binary alloys composed of Pt and a second metal (M) and capable of activating H_2O at low potentials is the following [23–27]:

$$CH_3OH \rightarrow Pt - CH_3OH_{ads} \qquad (6.4)$$

$$Pt - CH_3OH_{ads} \rightarrow Pt - CO_{ads} + 4H^+ + 4e^- \qquad (6.5)$$

$$M + H_2O \rightarrow M - H_2O_{ads} \qquad (6.6)$$

$$M - H_2O_{ads} \rightarrow M - OH_{ads} + H^+ + e^- \qquad (6.7)$$

$$Pt - CO_{ads} + M - OH_{ads} \rightarrow Pt + M + CO_2 + H^+ + e^- \qquad (6.8)$$

The intrinsic mechanism postulates that the presence of M modifies the electronic structure of Pt, and, as a consequence, the adsorption of oxygen-containing species or even for dissociative adsorption of methanol. For example, in the case of CO adsorption an electron donation/back donation mechanism takes place. The CO adsorption on Pt is stabilized by two simultaneous effects: electron transfer (donation) from the CO filled 5σ molecular orbital to the empty $d\sigma$ band of Pt; back-donation of electrons from metal $d\pi$ orbital to empty $2\pi^*$ anti-bonding orbital of CO. The generation of an σ type bond strengths the p type bond and vice versa. In the Pt-M alloys a modification of the empty electron states density of Pt occurs, with a shift of the Fermi energy level respect to the energy of CO molecular orbital. In such a situation the synergic mechanism of interaction of Pt–CO bond loses its stabilizing effect. Ipo-electronic metals, like Ru, produce a shift effect and charge redistribution, which strongly influences the CO adsorption phenomena. From the 1960s to 1990s extensive studies have been carried out on Pt-based alloy electrodes for methanol electro-oxidation in acid electrolytes at ambient and elevated temperatures [28–36]. These studies have shown that Pt-Ru alloys are among the best candidate catalysts for methanol electro-oxidation. But also in the case of Pt-Ru anode catalysts the power density of a DMFC is about a factor of 10 lower than that of a proton exchange membrane fuel cell (PEMFC) operated on hydrogen if the same noble metal loading is used. Therefore, the efficiency of the DMFC operating on Pt-Ru alloy catalysts is still insufficient for practical application. Many investigations have been made to improve the performance of DMFC through testing other binary Pt-based catalysts than Pt-Ru or with the incorporation of a third metal, such as W, Mo Sn, Os, in the binary Pt-Ru.

Carbon-supported catalysts are commonly used as anode materials in low temperature fuel cells. Because methanol oxidation occurs on the catalyst surface, a higher surface area of catalysts would result in better catalytic activity. So, the active phase is dispersed on a conductive support such as carbon. The preparation of practical catalysts, that is, supported bimetallic alloys, presents some problems with respect to model catalysts, that is, unsupported bulk alloys. It is important to distinguish between *supported alloy* and *supported bimetallic*: in the latter, two metals are present on the same support and are not alloyed. In many supported catalysts reported in literature the metals were only partially alloyed (*supported bimetallic alloy*). Moreover, in the presence of alloy, we can distinguish between *supported solid solution* (disordered alloy) and *supported intermetallic compound* (ordered alloy).

6.2
Phase Diagram vs. Activity: New Chances for DMFC Anodes

6.2.1
PtRu Catalysts: The Effect of Alloying and Ru Oxide Presence

The crystal structure of pure Pt is face centered cubic (fcc), while that of Ru is hexagonal close packed (hcp). For Ru atomic fractions up to about 0.7, Pt and Ru form a solid solution with Ru atoms replacing Pt atoms on the lattice points of the fcc structure. The lattice constant decreases from 3.923 Å (pure Pt) to 3.83 Å at 0.675 atomic fraction of Ru. Above 0.7 of Ru, another solid solution is formed with Pt atoms replacing Ru in an hcp structure. In carbon-supported catalysts the amount of Ru alloyed with Pt is lower than the nominal Ru content in the material [37], which is the converse to bulk Pt-Ru alloys.

Different researchers have reported that the composition and structure of supported Pt-Ru nanoparticles affects the catalytic activity [37–43]. It has been reported that for supported and unsupported nanoparticle Pt-Ru catalysts, 40–60% Ru gives the optimum catalytic activity for methanol oxidation [13, 42–45]. For example, Chu and Gilman [43] prepared unsupported Pt-Ru alloy nanoparticles with various compositions from Pt : Ru atomic ratio 85 : 15 to 11 : 89. As shown in Figure 6.1 from Ref. [44], they plotted the steady state current densities for the MOR normalized to electrode geometric area vs. Ru content at 25 °C and 60 °C. The highest current densities corresponded to a 50% composition at both temperatures. The same result was found by Watanabe *et al.* [14] for highly dispersed Pt-Ru alloy clusters on carbon-supported electrodes. Similarly, in bulk Pt-Ru alloys the best catalytic activity is found for that one with about 50% Ru [15, 46]. In these articles, the Pt-Ru alloy is in the fcc phase. Indeed, it has been reported that the fcc phase is stabilized in Pt-Ru nanoparticles beyond the bulk stability limit (62% Ru). For example, for 2.5 nm Pt-Ru nanoparticles prepared on carbon supports by the reduction of molecular precursors, mostly fcc nanoparticles are obtained for up to 80 at% Ru [39]. Sputtered films, however, seem to be different from these nanoparticles in that the fcc phase is less stable than in bulk alloy and the hcp phase is formed at a low Ru concentration [47].

230 | *6 Platinum Alloys as Anode Catalysts for Direct Methanol Fuel Cells*

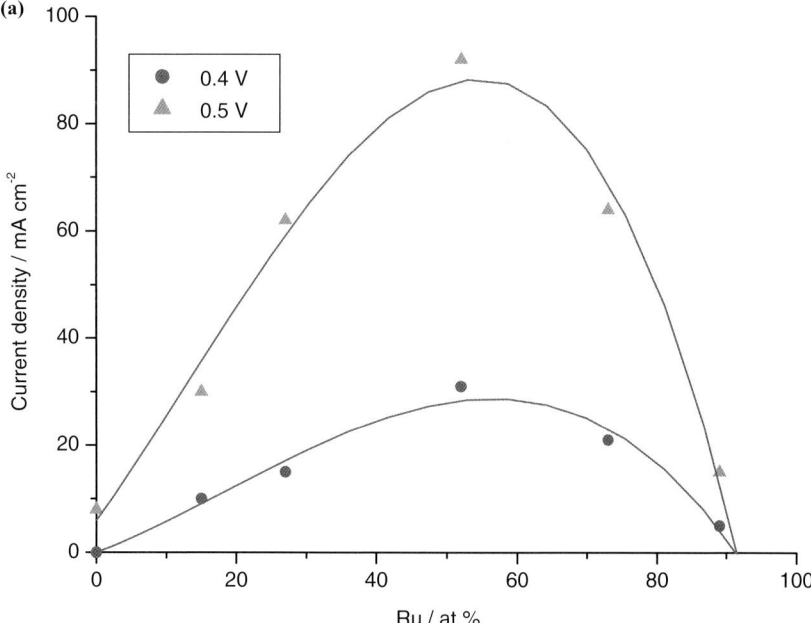

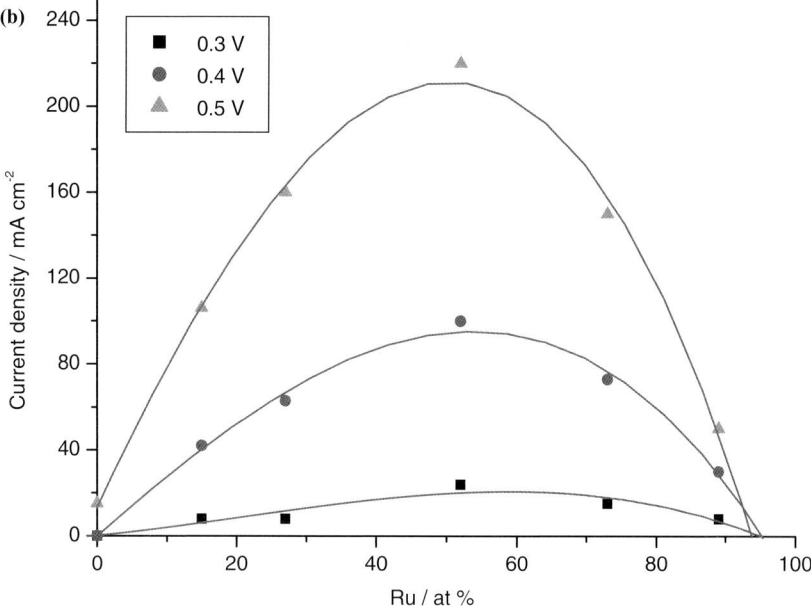

Figure 6.1 Methanol electro-oxidation current densities (electrode geometry surface area) as a function of Ru composition. (a) 25 °C and (b) 60 °C. Reprinted with permission from *J. Electrochem. Soc.* 1996, 143, 1685, Copyright © (1996) The Electrochemical Society.

To understand the structural and electrochemical properties of Pt-Ru for methanol electro-oxidation, Kim et al. [47] prepared sputtered PtRu thin-film electrodes with different compositions and structure. They observed that the Pt-Ru thin-film electrodes evolved from fcc through mixed fcc/hcp to hcp structures as the Ru concentration increased. The phase diagram for sputtered RuPt thin-film electrodes was significantly different from the bulk phase diagram, with the fcc-to-hcp transition in the films occurring at 32–58% Ru, while the bulk transition is at 62–80% Ru. In the electrochemical measurements, the thin-film electrodes showed a broad region of maximum catalytic activity for methanol oxidation at 40–60% Ru. Significantly, the Pt-Ru electrodes with mixed phase showed excellent catalytic activity. They concluded that the presence of hcp Pt-Ru does not adversely affect the catalytic activity of the alloys for methanol oxidation and that the assumption of an fcc structure for Pt-Ru alloys is not necessarily valid for low Ru concentrations. They further showed that the bulk phase diagram is not good to analyze thin-film structure for these materials.

The formation of a Pt-Ru alloy, however, seems not to be an essential requirement for a high MOR activity. According to the bi-functional mechanism, Ru sites are required to locate close enough to the Pt sites to guarantee reaction (6.8). In early studies, Pt-Ru alloy was emphasized because the possible electronic effect between Pt and Ru atoms was also thought to be important, which could reduce the bonding strength between Pt and CO_{ads}, and thus benefit the catalysis. It remains an open question whether the bi-functional effect or the electronic effect is predominant. As observed by Papageorgopolous et al. [48], the CO adsorbed on platinum sites needs to diffuse from platinum to ruthenium. In order for the CO oxidation at Ru to take place at a considerable rate, the platinum and ruthenium sites must be close enough together, but apparently not in the alloyed state. In recent years, Rolison et al. [49–52] strongly advocated avoiding Pt-Ru alloy; they found that the catalytic activity would increase by orders of magnitude if Ru existed in the hydrous oxides form, in comparison with in the alloyed form. The methanol oxidation reaction on Pt-RuOH catalysts occurs in the following way:

$$Pt + CH_3OH \rightarrow Pt - (CO)_{ads} + 4H^+ + 4e^- \quad (6.9)$$

$$Pt - (CO)_{ads} + Ru - (OH)_{ads} \rightarrow Pt - Ru + CO_2 + H^+ + e^- \quad (6.10)$$

The benefits of RuO_xH_y were attributed to the conductibility of its electrons and protons and the innate possession of surface OH groups. A study from Los Alamos National Laboratory (LANL) [53] showed that the higher the RuO_xH_y content in nano Pt–Ru catalyst, the better the DMFC performance; in addition, Nafion was much less required in the anode because of the protonic conductivity of RuO_xH_y. However, they also pointed out that RuO_xH_y was not the prerequisite, high DMFC performance was also achieved in their work using completely alloyed Pt-Ru catalyst, in agreement with the work of Chu and Gilman [43].

The activity of nano Pt-Ru catalysts is a multivariate function of particle size, alloyed degree, oxides composition, and so on. The mono-dependencies of catalytic activity on individual structure parameters (structure–activity relationship, SAR)

Table 6.1 The structure matrix and the activity matrix of different PtRu catalysts.

Sample	Structure matrix			Activity matrix	
	m_{RuOxHy} (%)	$1/\varphi$ (1/nm)	x_{Ru} (%)	Am@0.4 V	Am@0.5 V
1	2.56	0.25	10	32	185
2	2.18	0.23	45	9	63
3	0.90	0.28	30	5	41
4	0.51	0.26	47	9	65
5	3.89	0.36	27	51	295
6	3.55	0.48	38	61	329
7	2.36	0.37	3	22	133

Pt-Ru alloyed degree (Ru fraction in alloy, x_{Ru}), particle size (φ), and contents of hydrous Ru oxides (m_{RuOxHy}). Reproduced with permission from *J. Phys. Chem. B* 2005, 109, 8873–8879. Copyright © (2005) American Chemical Society.

are of great importance but unfortunately unobtainable in practical measurements. A pattern recognition methodology was proposed by Lu et al. [54] to extract SAR information from all relative experimental data, which will hopefully cast a new light on the in-depth understanding of this important catalyst. As a preliminary demonstration, multivariate linear regression and generalized regression neural network were applied to analyze a small dataset for methanol oxidation. As shown in Table 6.1 from Ref. [54], it was found that both increasing the content of hydrous ruthenium oxides and decreasing particle size would benefit the catalytic activity, while the effect of Pt-Ru alloy degree turned out to be unremarkable. Conversely, a negative effect of RuO_x on the methanol oxidation was found by Wu et al. [55]. They prepared electrochemically polarized Pt-Ru/C catalysts. The results indicated that the metallic state PtRu(0)-Ru(0) can be formed during cathodic polarization and contributed to the MOR, while the formation of inactive ruthenium oxide during anodic polarization caused a negative effect on the MOR. These counteracting effects of ruthenium oxide on the MOR can be explained considering that the ruthenium oxides in PtRu/C can be divided into two categories in terms of electrochemical reversibility, one reversible and the other irreversible. An important aspect of the SAR for PtRu/C catalysts is the following: the reversible Ru oxide is beneficial and the irreversible Ru oxide harmful. Lu et al. [56] carried out an anodic treatment of PtRu/C catalysts in 0.5 M sulfuric acid at 1.3 V (vs. RHE) for 0.5 h. This treatment promoted the activity for methanol oxidation by a few tenths to 5 times. As shown in Figure 6.2 from Ref. [56], the activity gained was found to be essentially as stable as the original activity. This anodic activation effect was valid for samples domestically prepared under different conditions and that produced by Johnson-Matthey. Based on the changes of cyclic voltammetry (CV) during the anodic treatment, a model was proposed for the activation effect. According to the model, there are two categories of ruthenium oxides in the catalyst: one is electrochemically reversible and beneficial for catalytic activity while the other is irreversible and harmful. During the anodic treatment, the harmful oxide is decreased while the

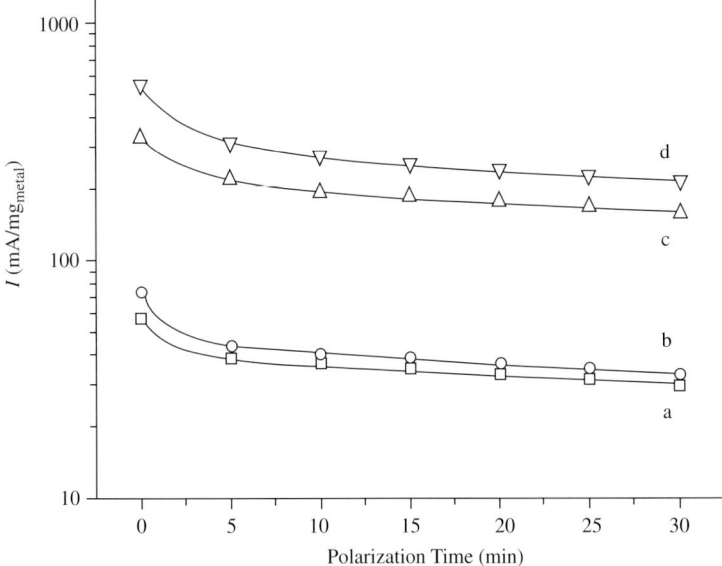

Figure 6.2 Durability of the catalytic activity gained in anodic treatment in 1 M MeOH + 0.5 M H_2SO_4 at 60 °C. Methanol oxidation currents at 0.4 V (a and b) and at 0.5 V (c and d) were recorded before (a and c) and after (b and d) anodic treatments respectively. Reproduced with permission from *J. Phys. Chem. B* 2005, 109, 1715–1722. Copyright © (2005) American Chemical Society.

beneficial oxide either increased or changed only slightly, resulting in a beneficial net change.

6.2.2
PtSn Catalysts: Activity of PtSn Alloys and Non-Alloyed Pt-SnO$_x$

Platinum and tin form five bimetallic intermetallic phases, Pt_3Sn, $PtSn$, Pt_2Sn_3, $PtSn_2$, and $PtSn_4$, of which Pt_3Sn and $PtSn$ are congruently melting compositions. The PtSn phase is commonly known as the mineral niggliite. These intermetallic phases are distinguished by distinct crystalline structures and unique X-ray diffraction (XRD) patterns. Regarding the different values of the lattice parameter of fcc PtSn present in literature [57, 58], Kuznetzov et al. [59] asserted that Pt forms nearly all possible alloys with Sn. Radmilovic et al. [60], instead, attributed the value of the lattice constant of 0.3965 nm, found for a commercial carbon-supported Pt/Sn 1.23/1 catalyst, to a mixture of Pt_9Sn (0.3934 nm) and Pt_3Sn phases.

Pt-Sn/carbon nanocomposites have been studied for decades as anode catalysts for the electro-oxidation of methanol and other small carbohydrate fuels [61]. These nanocomposites have been prepared by a variety of electrochemical or chemical deposition methods, and inconsistencies in catalyst performance have been reported. Indeed, although a superior performance of Pt-Sn catalysts for PEMFC applications is widely proved [62–64], an enhanced catalytic activity for Pt-Sn catalyst in the methanol oxidation reaction [32, 65–69], in contrast to no/negligible enhancement of the MOR rate over Pt-Sn catalysts [63, 70–76], is reported. The results using tin as the secondary element vary in the effects on methanol oxidation from significantly enhancing (co-deposition and bulk alloy [32], underpotential deposition (upd) [65], electrodeposition [66, 67] and mechanical alloying [68], co-deposition [69]) to moderate (immersion and co-precipitation [70, 71], immersion [72]), no (bulk alloy [73]) or even negative (impregnation [64], up and co-deposition [74]).

Wang et al. [63] observed a significant enhancing effect of tin for CO_{ads} oxidation but no effect for methanol oxidation using well-defined alloy surfaces. They also presented a model in which CO_{ads} resulting from the dehydrogenation of methanol is in a different state from that directly adsorbed from gaseous CO.

Recently, Mukerjee and McBreen [77] studied the effect of potential on the electronic and structural characteristics of Pt and Sn when Sn is either alloyed with Pt or is present as upd deposits on the carbon-supported Pt. Alloying of Sn with Pt in a 75 : 25 atomic ratio results in the formation of a $Pt_{75}Sn_{25}$ $L1_2$ phase, a decrease in the Pt d band vacancies, an increase in the lattice parameters, and hence larger Pt–Pt bond distances. Deposition of Sn as upd ad-atoms on Pt/C has minimal effects on the electronic or structural aspects of Pt/C. Both upd Sn on Pt/C and Sn surface atoms in the PtSn/C alloy show similar behavior to exhibit Sn-oxygen interactions that vary with potential. The Sn atoms in both upd Sn on Pt/C and in Pt-Sn/C catalysts are associated with oxygenated species and as a result can initiate the oxidation of adsorbed CO and CHO at lower potentials compared with Pt/C. The main differences in the catalytic activity of Pt-Sn/C and upd Sn on Pt/C for methanol oxidation is due to the differences in the electronic properties of the Pt in these catalysts. The upd Sn does not interfere with the ability of the Pt to adsorbed methanol and dissociate C–H bonds, whereas alloying with Sn inhibits the ability of Pt to carry out these functions.

Colmati et al. [58] investigated the effect of Sn content in Pt-Sn catalysts prepared by the formic acid method (FAM) on the methanol oxidation. The onset potential of methanol oxidation of all the PtSn catalysts was lower than pure Pt, but it increased with Sn content in the alloy. As shown in Figure 6.3 from Ref. [58] (cell potential vs. lattice constant plot), at low current density (0.01 A cm^{-2}) the potential of the cells with Sn-containing catalysts was more than 100 mV higher than Pt. At 0.64 A cm^{-2}, instead the cell potential decreased with the content of Sn in the catalyst. This result can be ascribed to the poor methanol adsorption/dehydrogenation due to alloying of Sn and Pt. Indeed, at low current density, a low amount of methanol is required for the cell operation, then the CH_3OH adsorption/dehydrogenation is less important, and the rate of the MOR is determined by the CO_{ads} oxidation. On increasing the consumption of methanol, the methanol adsorption/dehydrogenation becomes the determining step of the MOR. The maximum power density instead results from an optimum balance of CH_3OH adsorption and CO oxidation. While the lowest onset potential for MOR was obtained by the $Pt_{90}Sn_{10}$/C electrocatalyst, in this case

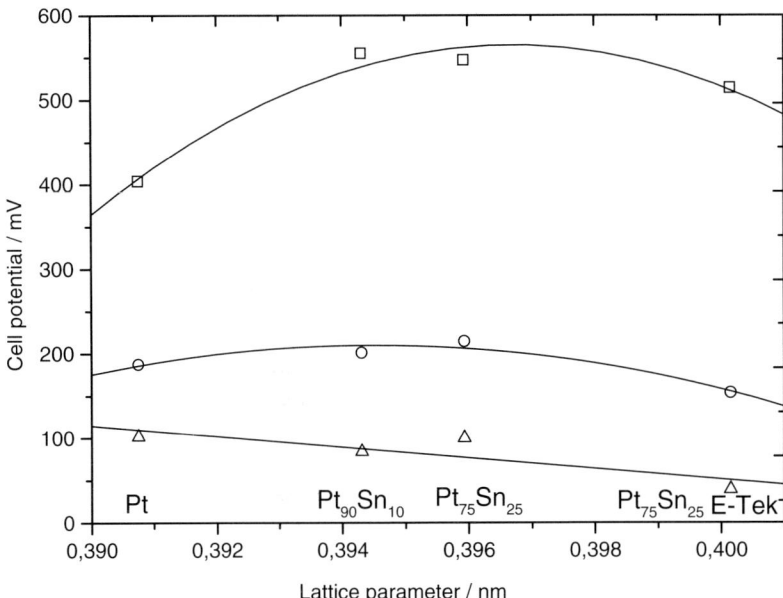

Figure 6.3 Cell potential at various current densities vs. PtSn lattice parameter. Current densities: (□) 0.01 A cm^{-2}; (○) 0.40 A cm^{-2}; (△) 0.64 A cm^{-2}. Reprinted with permission from *Electrochim. Acta* 2005, 50, 5496, Copyright © (2005) Elsevier.

the maximum power density per Pt active area (Figure 6.4 from Ref. [58]) was attained by the cell with the Pt$_{75}$Sn$_{25}$/C catalyst by the FAM. This result was explained by an optimum balance of Sn effect on CO$_{ads}$ (Sn positive effect) and methanol adsorption/dehydrogenation (Sn negative effect).

Summarizing, Pt-Sn electrocatalysts can perform better than pure Pt if the negative effect of the alloying on methanol adsorption/dehydrogenation is minimized under two conditions: (i) moderate degree of alloying and (ii) cell operating at low current density. It has to be remarked, however, that the MOR activity of Pt-Sn catalysts is always lower than that of Pt-Ru catalysts, ascribed to their lower methanol adsorption/dehydrogenation than that of Ru-containing catalysts.

6.2.3
Pt-Co and Pt-Ni Catalysts: Effect of Alloying and CoO$_x$ and NiO$_x$ Presence

In the composition range from 0 to 50 at.% Co (Ni), Pt and Co (Ni) form a substitutional continuous solid solution and two ordered phases. In the region of 75 at.% Pt there are face-centered cubic (fcc) superlattices Pt$_3$Co and Pt$_3$Ni of the Cu$_3$Au (LI$_2$) type. Regular termination of the bulk LI$_2$ structure normal to the three major zone axes produces a variety of surface compositions, from the pure Pt ((200) and (220) planes), 25 at.% Co (Ni) ((111) plane) to 50 at.% Co (Ni) ((100) and (110) planes). To better understand the relationship between the surface composition and

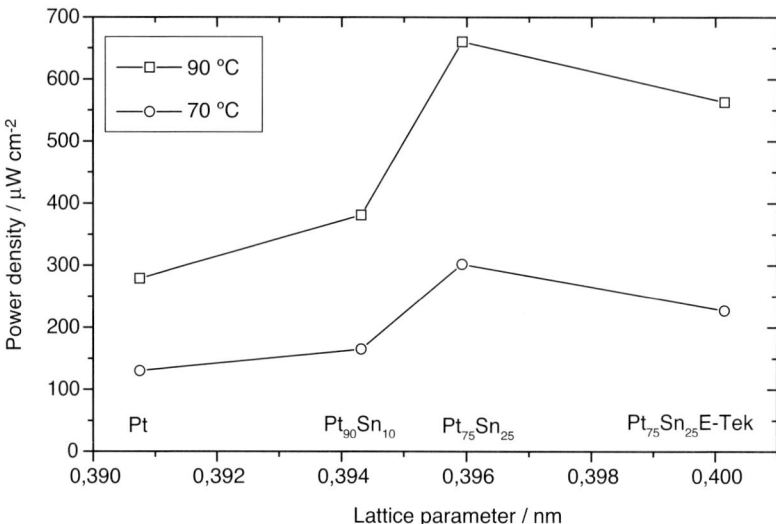

Figure 6.4 Maximum power density normalized by Pt active area vs. PtSn lattice parameter. Reprinted with permission from *Electrochim. Acta* 2005, 50, 5496, Copyright © (2005) Elsevier.

the catalytic activity, it is very important to determine if surface segregation, that is, enrichment of one element at the surface relative to the bulk, takes place during the preparation of these alloy catalysts. Conflicting results regarding the surface segregation on Pt-Co and Pt-Ni alloys are reported in literature. X-ray photoelectron spectroscopy (XPS) data by Shukla *et al.* [78] indicated some surface enrichment of platinum metal on Pt-Co/C (atomic ratio 0.84 vs. 0.72 in the bulk) and Pt-Ni/C (0.86 vs. 0.64) prepared by alloying at high temperature. Conversely, Paulus *et al.* [79] found 70 at.% Pt on the surface of the Pt_3Ni particles and 58 at.% Pt on the surface of the Pt_3Co particles for the $Pt_3Ni(Co)$ catalyst made by E-TEK. The value for Pt_3Ni is very close to the bulk composition of 75 at.% Pt and indicates that no segregation has taken place, whereas in case of the Pt_3Co a slight segregation of Co to the surface is observed. For the Pt-Ni (1 : 1) catalyst made by E-TEK, however, they found only about 20 at.% Pt on the surface Pt and 35 at.% Pt on the surface for the Pt-Co (1 : 1) catalyst made by E-TEK, indicating Ni(Co) segregation to the surface. It is known that, given a similar size, the metal having the lower heat of sublimation tends to surface segregate in binary alloys. The heats of vaporization of Pt and Ni are 509.6 and 370.3 kJ/mol, respectively: as a consequence, an enrichment of Ni at the surface is expected. However, a strong surface enrichment in Pt was found by low-energy ion scattering (LEIS) in Pt-Ni alloys [80, 81]. Further, Mukerjee *et al.* [82] used the electronic theory of d-band density of states of pure components to demonstrate that surface enrichment of Pt in Pt-Ni should occur. This shows that the thermodynamic explanation fails to predict the enrichment behavior.

The actual Pt-Co and Pt-Ni alloy catalysts, particularly the carbon-supported catalysts, are formed by alloyed and non-alloyed Co (Ni) species. The degree of alloying depends on the preparation method of the catalyst. XPS analysis on commercial carbon-supported Pt_3Co and Pt_3Ni electrocatalysts indicated the presence of PtO, CoO and NiO on the surface [83]. In the same way, XPS analysis on unsupported Pt-Ni alloy nanoparticles indicated the presence of metallic Ni, NiO, $Ni(OH)_2$ and NiOOH [84]. Moreover, XPS data suggested that the amount of platinum oxide content in the carbon-supported Pt–Co alloy electrocatalyst is lower than that in Pt and Pt-Ni [78].

Generally Pt-Co and Pt-Ni showed higher activity for the MOR with respect to Pt, but lower activity than Pt-Ru. In some cases, however, Pt-Co and Pt-Ni presented higher MOR activity than Pt-Ru. Pt-Ni, Pt-Co and other transition metal alloys were investigated by Page et al. [85] as low cost alternative catalysts for direct oxidation of methanol and compared with Pt and Pt-Ru using CV. Commercial carbon-supported Pt-Ni and Pt-Co in the atomic ratio 1:1 were used. The alloy catalyst Pt-Co/C was found to be a better catalyst for methanol oxidation in acid solution compared with Pt and other transition metal alloys. Park et al. [84] studied the electro-oxidation of methanol in sulfuric acid solution using unsupported Pt, Pt-Ni (1:1 and 3:1), and Pt-Ru (1:1) alloy nanoparticle catalysts. Methanol oxidation current measured on the Pt-Ni based catalysts in 2.0 M CH_3OH + 0.5 M H_2SO_4 at room temperature exceeded that obtained with pure Pt. They also measured plots of oxidation current vs. time (chronoamperometry, CA) in 2.0 M CH_3OH + 0.5 M H_2SO_4 at 0.42 V, for 3600 s. For each catalyst, the decay in the methanol oxidation is different; for instance, pure platinum nanoparticles require 10 min to reach 70% of the initial current and the oxidation current is reduced steeply. After 1 h, the current decreases below 40% of the initial value. In contrast, Pt-Ni (1:1) and Pt-Ru (1:1) support higher current, and one may conclude that they have higher activity than pure Pt. After 1 h, the order of surface activity for the methanol oxidation is Pt-Ni (1:1) > Pt-Ru (1:1) > Pt. By combining voltammetry and CA, the authors concluded that Pt-Ni (1:1) represent the best alternative candidates for the DMFC anode catalysts with respect to Pt-Ru.

The comparison of the onset potential for the methanol oxidation on Pt-Co and Pt-Ni alloy catalysts with that on Pt presents conflicting results. The disagreement of the results reported in the literature may depend on the Co(Ni) content in the catalyst. To evaluate the effect of the content of the non-precious metal on the onset potential, Antolini et al. [86] plotted the Pt-Co(Ni)/Pt onset potential ratio for the MOR at room temperature vs. the nominal Co (Ni) content in the catalyst for carbon-supported and bulk alloy catalysts. As can be seen in Figure 6.5 from Ref. [86], in the case of carbon-supported catalysts, the onset potential for the methanol oxidation went through a maximum at near 30 at% Co(Ni). For the bulk alloy catalysts, instead, the onset potential for the MOR was always lower than that of Pt, almost independent of the Co (Ni) content for low Co (Ni) content, and decreasing with increasing amounts of the non-precious metal. The different behavior between supported and bulk alloy catalysts may depend on the higher degree of alloying of unsupported alloys than that of carbon-supported alloy catalysts [37]. Low Co (Ni) contents reduce the methanol oxidation by the ensemble effect where the dilution of the active component with the catalytically inert metal

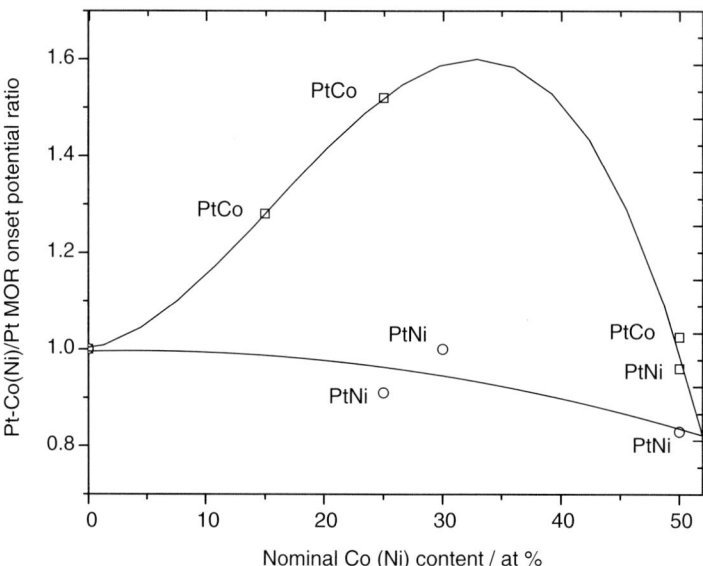

Figure 6.5 Ratio of the onset potential for the MOR for Pt-Co(Ni) and Pt at room temperature vs. the nominal Co (Ni) content in the catalyst. Plots for carbon-supported (□) and bulk (○) alloy catalysts. Reprinted with permission from *Appl. Catal. B* 2006, 63, 137, Copyright © (2006) Elsevier.

reduces the methanol adsorption, while high Co (Ni) contents improve the MOR by electronic effects and by the presence of higher amounts of Co (Ni) oxide species, both enhancing CO oxidation. The linear dependence of the onset potential for the MOR on the amount of alloyed Co (Ni), as shown in Figure 6.6 from Ref. [86], seems to confirm the decrease of the rate of methanol oxidation for low non-precious metal contents.

6.3
Preparation Methods of Pt Alloys

6.3.1
Unsupported Catalysts

Unsupported nanoparticle catalysts are commonly prepared by reduction of Pt and M precursors at low temperature (<100 °C) or at intermediate temperature (250–500 °C) with $NaBH_4$, hydrazine or hydrogen. Chu and Gilman [43] prepared Pt-Ru nanoparticles in the Pt:Ru range 85:15 to 11:89 by flushing H_2PtCl_6 and $RuCl_3$ with H_2/Ar at 250 °C for 20 h. They obtained full alloyed Pt-Ru in the fcc structure up to 52% Ru, in the fcc + hcp structure for 75% Ru, and in the hcp structure only for 89% Ru. The particle size was in the range 13–60 nm. (Pt-Ru (1:1)

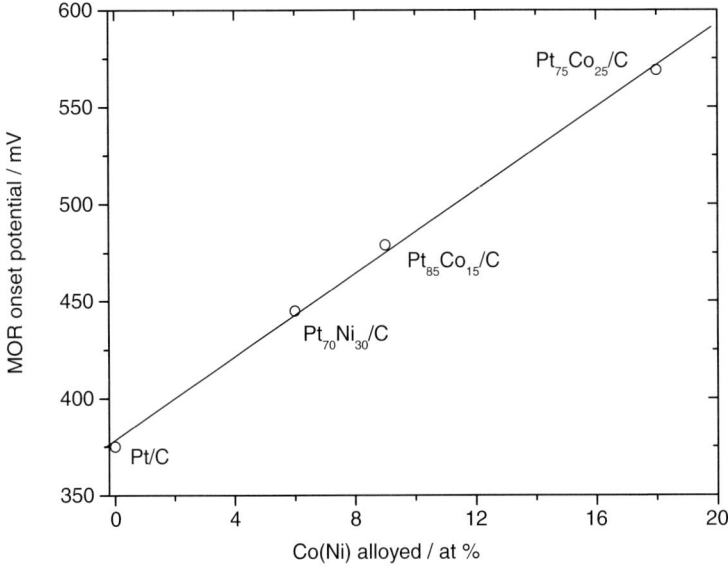

Figure 6.6 Dependence of the onset potential for the MOR on the atomic percentage of alloyed Co (Ni). Reprinted with permission from *Appl. Catal. B* 2006, 63, 137, Copyright © (2006) Elsevier.

and Pt-Ni (3:1 and 1:1) nanoparticles were synthesized at room temperature by Park *et al.* [84] using a conventional reduction method of precursors with $NaBH_4$. The resulting materials were washed with deionized water and dried by freeze-drying. Pt-Ru and Pt-Ni were partially alloyed, and the presence of Ru and Ni oxides was found in the catalysts. The particle size was 3–4 nm. Jiang and Kucernak [87] prepared mesoporous microspheres of Pt-Ru alloy by the electrochemical co-reduction of H_2PtCl_6 and $RuCl_3$ both dissolved in the aqueous domains of the liquid crystalline phase of an oligoethylene oxide surfactant. XRD studies showed that a higher Ru content is alloyed in the material than typically found in chemically prepared Pt-Ru catalysts. Scanning electron micrographs reveal that the microspheres have a narrow size distribution in the range 0.5–1 μm.

The polyol synthesis method seems to be a very efficient way to prepare nanosized clusters of noble metals or noble metal alloys with transition metals [88–90]. Generally, this preparation method consists of H_2PtCl_6 and chloride salts as metal sources, and ethylene glycol as solvent and reducing agent. The reaction occurs at 130–160 °C.

A way to prepare unsupported nanosized Pt-based alloys is the use of organometallic compounds as precursors. By the thermal decomposition or reducing of organometallic precursors, small nanoparticles of metal or alloy with narrow size distribution can be obtained. From the organometallic precursors, Sun *et al.* [91] obtained Pt-Co nanoparticles of about 2 nm. In the same way, nearly monodisperse Pt-Co [92], Pt-Fe [93] and Pt-Mn [94] nanoparticles were prepared by the decomposition of organic precursors.

Finally, bulk binary Pt-Co [95] and ternary Pt-Ru-Mo [96] and Pt-Ru-Co [97] metal nanoparticles were prepared by Zhang. et al. using a water-in-oil reverse microemulsion system, with hydrazine as reducing agent. The main attractive features of the two-step microemulsion reduction technique for preparing mixed metal nanoparticles are the ease and accuracy of composition control. The reduction reaction occurs in a confined reaction zone within the microemulsions. The ultimate nanoparticles should follow the metal composition in the precursor solution, without losing control of particle size.

6.3.2
Supported Catalysts

Carbon-supported catalysts are commonly obtained by adsorption of platinum and metal colloids onto the carbon surface, or by impregnation of carbon support with Pt and M precursor solution.

The colloid methods used to prepare binary and ternary Pt-based catalysts are the Bonnemann method [98] and the polyol synthesis [99, 100] in modified versions. Bonnemann et al. [101, 102] developed a colloidal method to prepare unsupported and supported metals and alloys. Metal salts of Groups 6–12 were reduced using alkali hydrotriorganoborates in hydrocarbons between -20 and $80\,°C$ to give boron free powder metals. The use of tetraalkylammonium hydrotriorganoborates as reducing agents leads to colloidal transition metals in organic phases. These colloids may also be obtained using conventional reducing agents after first reacting the metal salts with the stabilizing tetraalkylammonium halide. According to XRD analysis, the particles were nearly amorphous. The particle size was between 1 and 100 nm, dependent on the metals. By simple co-reduction of suspended metal salts, binary or ternary alloys and intermetallic compounds were obtained. Adsorption of the metal colloids onto supports such as charcoal improves the stability of the resulting catalysts substantially. This method was used to prepare carbon-supported Pt-Ru [103] and Pt-Ru-M (M = Sn, Mo and W) [64, 104, 105] catalysts. Carbon-supported Pt, Pt-Ru, Pt-Pd, Pt-W and Pt-Sn binary electrocatalysts with sharp particle size distribution were prepared via the polyol synthesis method by Zhou et al. [57, 100]. Liu et al. [106, 107] prepared both supported and unsupported Pt-Ru nanocatalysts of around 4.5 nm in size by utilizing the microwave-assisted polyol process. Liang et al. prepared carbon-supported ternary Pt-Ru-Ir (1:1:1) [108] and Pt-Ru-Ni (1:1:1) [109] catalysts having 40 wt.% metal by using a microwave-irradiated polyol plus annealing (MIPA) synthesis method.

Regarding the impregnation method, there are essentially two versions of this method: (i) impregnation of the carbon support with Pt and M precursors and (ii) impregnation of the second metal on preformed Pt/C catalyst, both followed by reduction of the precursors with different reducing agents. The former method was used to prepare carbon-supported Pt-Ru [110], Pt-Co [111] Pt-Ni [112] and Pt-Cr [113] alloy electrocatalysts by impregnating high surface area carbon with Pt and M precursors, followed by the reduction of the precursors with $NaBH_4$. The metal particle size was in the range 3.8–4.8 nm. The latter method of preparation of Pt-M/C

Table 6.2 Pt-Ru data obtained using various baking steps from HRTEM and XRD analyses.

Baking temperature, °C	Lattice constant, nm	x_{Ru}, %	TEM particle size, nm
60	0.38 659	40 ± 2	2.25
100	0.38 687	38 ± 2	2.5
150	0.38 765	31 ± 2	2.5
200	3.8982	14 ± 1	2.75

Reproduced with permission from *J. Phys. Chem. C* 2007, 111, 16 416–16 422.
Copyright © (2007) American Chemical Society.

consists in the formation of carbon-supported platinum followed by the deposition and reduction of the second metal on Pt/C, and alloying at high temperatures. This thermal treatment at high temperatures gives rise to an undesired metal particle growth, by sintering of platinum particles, so different preparation methods at low temperature have been developed [114].

In comparison with the colloidal chemistry method, the impregnation method is relatively simple; in particular, no filtering and washing procedures are required. Yang et al. [115] demonstrated the capability of the impregnation method in the preparation of carbon-supported Pt-Ru. They achieved narrow particle size distribution centered round 1.5 nm at a metal loading as high as 60 wt.% with H_2PtCl_6 and $RuCl_3$ as direct precursors. Wang et al. [116] prepared Pt-Ru/C catalysts applying four different baking temperatures, 60 °C, 100 °C, 150 °C, and 200 °C. The temperature of the baking step is most influential to the alloying degree in the impregnation method for PtRu/C synthesis. As can be seen in Table 6.2 from Ref. [116], raising baking temperature will result in lowering alloying degree of Pt-Ru/C, most possibly owing to partial conversion of $RuCl_3$ to less reducible RuO_x. For samples with equal Pt : Ru ratio in precursor, the alloying degree of Pt-Ru/C can be controlled over a fairly large range mainly via selecting the baking temperature, from an upper limit close to the theoretical value (0.50) to a lower limit slightly below 0.1.

Among the various organometallic precursors used, metal–carbonyl complexes are employed to obtain carbon-supported metal or alloy catalysts. By the reduction of Pt-Ru carbonyl molecular precursors, Nasher et al. [117] and Hills et al. [118] prepared carbon-supported bimetallic Pt-Ru alloys, while by decomposition of Pt and Ru carbonyl in organic solvents Dickinson et al. [119] obtained carbon-supported Pt-Ru alloys. In all the cases, the Pt-Ru catalysts presented small particle size and narrow size distribution. Recently, Manzo-Robledo et al. [120] prepared carbon-supported Pt-Sn catalysts through the carbonyl chemical route. On theses bases, Yang et al. used this method to prepare carbon-supported Pt-Cr [121] and Pt-Ni [122]. The total metal loading of all the carbon-supported catalysts was about 20 wt.%. The energy-dispersive X-ray spectroscopy (EDX) composition of all the catalysts was very close to the nominal value. According to the authors, the nearly linear relationship between the lattice parameter and EDX composition analysis again attests that Cr and Ni are completely alloyed with Pt. The dependence of the lattice parameters of Pt-Cr and Pt-Ni on EDX composition is shown in Figure 6.7 from Ref. [114]. In the same way,

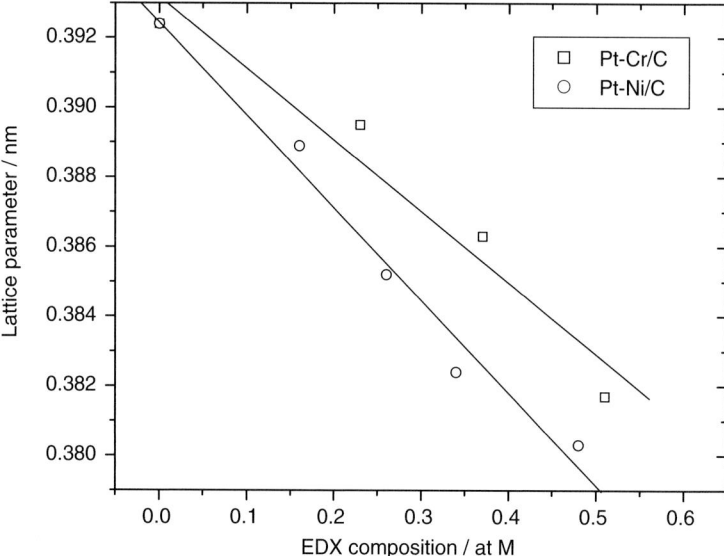

Figure 6.7 Dependence of the lattice parameters of Pt-Cr/C and Pt-Ni/C electrocatalysts with metal content 20 wt.% prepared by the carbonyl route on EDX composition. Reprinted with permission from *Mater. Chem. Phys.* 2007, 101, 395, Copyright © (2007) Elsevier.

Figure 6.8 from Ref. [114] shows the metal particle size from TEM vs. EDX composition plot. For both Pt-Cr and Pt-Ni the metal particle size decreases with the increasing content of non-precious metal in the alloy. Such a lower synthesis temperature could support the formation of carbon-supported alloy nanoparticles with small particle size and narrow size distribution, and thus a good dispersion.

Another way to prepare carbon-supported Pt-M alloy catalysts is based on the reduction of the precursor with formic acid at room temperature. Such a method consists in the treatment of carbon powder with formic acid before Pt and M impregnation. The treated carbon reduces Pt(IV) and M on its surface. carbon-supported Pt-Ru [17], Pt-Mo [123] and Pt-Sn [58] were successful prepared using this method.

Finally, Rojas *et al.* [124] prepared Pt-Ru/C by using the microemulsion technique. They obtained nanosized particle with a narrow size distribution, independent of the metal loading.

6.4
Activity Evaluation of Pt Alloys

The activity for methanol oxidation of Pt-Ru, Pt-Sn, Pt-Ni and Pt-Co alloys has been previously reported in Section 6.2. In this Section we discuss the MOR activity of other Pt-based binary and, particularly, ternary Pt-Ru-based catalysts.

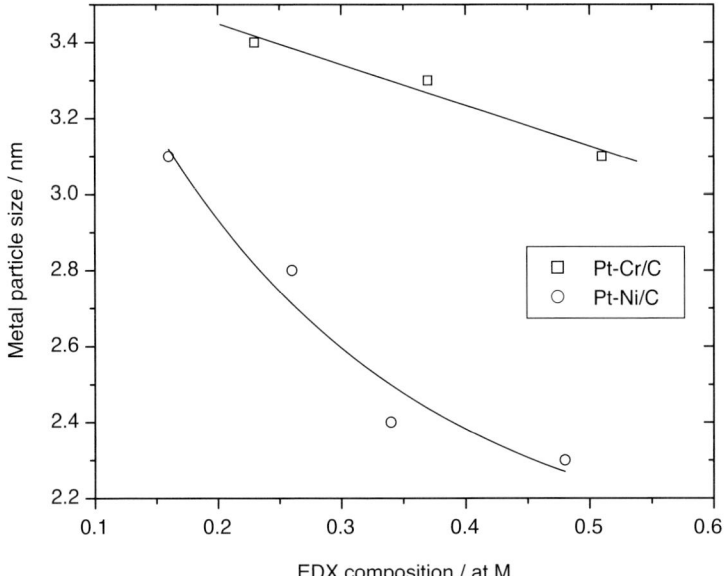

Figure 6.8 Metal particle size from TEM of Pt-Cr/C and Pt-Ni/C electrocatalysts with metal content 20 wt.% prepared by the carbonyl route vs. EDX composition plot. Reprinted with permission from *Mater. Chem. Phys.* 2007, 101, 395, Copyright © (2007) Elsevier.

6.4.1
Pt-Based Binary Catalysts

When Pt-Ru is used as anode electrocatalyst the power density of a DMFC is about a factor of 10 lower than that of a PEMFC operated on hydrogen if the same Pt loading is used. Therefore, alternative elements to Ru, showing a co-catalytic activity for the anodic oxidation of methanol, if used either as platinum alloys or as adsorbate layers on platinum, have been investigated. In the following part of this Section we report the MOR activity of some Pt-based binary catalysts.

6.4.1.1 Pt-W

Improved methanol oxidation activity of WO_x-containing Pt-based catalysts have been reported [64, 125, 126]. Shen and Tseung [125] investigated the MOR on high surface area Pt-WO_x catalysts. They obtained high methanol oxidation currents per geometrical anode area, but the actual turnover number for the MOR per Pt site turned out to be small. Shukla et al. [126] reported superior MOR activity per mass of Pt for Pt-WO_x compared with pure Pt. The question is whether the improved MOR activity for the WO_x-containing catalysts is due to the true catalytic properties and/or whether the addition of WO_x modifies the physical catalyst particle properties, such as particle size and surface area. Indeed, the presence of W in the Pt/C catalyst gives rise to the formation of platinum particles with size larger than pure Pt/C (Pt coalescence during the formation of the catalyst). Then, W presence in Pt/C catalyst affects the morphological characteristics of the material. Moreover, dissolution of WO_x in an acid environment is reported, causing an increase of the Pt surface area. Yang et al. [127] investigated the MOR activity on Pt/C, Pt-WO_x/C, Pt-Ru/C and Pt-Ru-WO_x/C. The activity for methanol oxidation increased in the order Pt/C<Pt-WO_x/C<Pt-Ru/C<Pt-Ru-WO_x/C. McLeod and Birss [128] used WO_x as support for Pt catalysts. They found that, unlike Pt-only catalysts, methanol oxidation currents on WO_x/Pt films did not decay rapidly with potential cycling, indicating the occurrence of co-catalytic behavior. The low resistance exhibited by the WO_x component makes it suitable as an ionically and electronically conducting support for DMFC electrocatalysts.

The positive effect of W presence in ternary Pt-Ru based catalysts was ascribed by some authors [129–132] to its intrinsic co-catalytic effect, supporting the CO oxidation. Strasser et al. [133], instead, adduced that the enhanced electrocatalytic activity in the presence of W is due to a corrosion effect, which increases the surface area of the catalyst. According to Yang et al. [127], the addition of tungsten to Pt-Ru appeared mainly to result in some "physical" modification of the catalytically active Pt and Ru surface components such as differences in electroactive surface area rather than promotion of the CH_3OH oxidation reaction via a true catalytic mechanism.

6.4.1.2 Pt-Mo

The role of surface molybdenum species in methanol oxidation on the platinum electrode was originally investigated by Nakajima and Kita [134]. They found that the redox couple of Mo(III)/Mo(IV) plays a key role in the enhancement of the catalytic activity of the Pt electrode even at a high temperature of 80 °C. An extended analysis of voltammograms to the electrode co-adsorbed with molybdenum and methanol species showed that the highest oxidation current is obtained at an equal coverage of the molybdenum and methanol species. By XPS analysis the presence of the different molybdenum oxidation states in Pt-Mo catalysts supported on highly oriented pyrolytic graphite was observed also by Morante-Catacora et al. [135].

Ordóñez et al. [136] characterized PtMo/C materials by CV in methanol solutions with H_2SO_4 as supporting electrolyte. The highest activity towards anodic methanol oxidation was obtained with materials with low molybdenum loadings.

Mukerjee and Urian [137] investigated the electrocatalysis of CO tolerance and direct methanol oxidation on Pt-Mo/C (3:1) in a PEMFC environment. While a threefold enhancement is observed for CO tolerance when compared with Pt-Ru/C (1:1), no such enhancement occurred for methanol oxidation. In situ XAS at the Pt L and alloying element K edges for Pt/C, Pt-Ru/C and Pt-Mo/C showed that in contrast to Pt-Ru/C, both Mo and Pt surfaces play a distinct role for CO oxidation. While on the Ru surface there is a competition between oxide formation (from activation of water) and CO adsorption, the Mo oxide surface showed no affinity for CO. This provided for efficient CO oxidation at low overpotentials on Pt-Mo/C. However, the corresponding behavior for methanol oxidation showed that Mo oxy-hydroxides were inhibited from efficient removal of CO and CHO species in contrast to Ru oxides. The Mo surface oxides also showed a redox couple involving (V to VI) oxidation states in the presence of both CO and methanol.

Samjeské et al. [138] studied the co-catalytic effect of Mo on the oxidation of adsorbed CO on different types of electrodes using differential electrochemical mass spectrometry (DEMS). They found that the effect of Mo on the oxidation of CO_{ad} formed by adsorbing methanol and on the oxidation of methanol is very small.

Pinheiro et al. [123] investigated the electro-oxidation of methanol on carbon-supported Pt-Ru (80:20), Pt-Mo (60:40), and Pt-Ru-Mo (70:20:10) catalysts. Comparing the steady state responses of the different alloys at the same methanol concentration, it is seen that the best activity is given by Pt-Ru-Mo followed by Pt-Ru and then by Pt-Mo. The results indicated that the Pt-Mo catalysts are not very effective for the initiation of the methanol oxidation reaction when compared with Pt-Ru or Pt-Ru-Mo. However, the increase of the oxidation currents at high potentials denotes a better tolerance of Pt-Mo to the poisoning species (mainly CO) formed as intermediates in the oxidation of methanol. On the other hand, the potentials at which the Pt-Mo alloy is active, are too high for an operational anode in a DMFC.

6.4.1.3 Pt-Au

Platinum–gold nanoparticles provide a synergistic catalytic effect that involves the suppression of adsorbed poisonous species and a change in electronic band structure to modify the strength of the surface adsorption. With such a bimetallic system, one could explore the viability of using Pt as the main dehydrogenation sites, and Au together with Pt to speed up the removal of poisonous species.

The electrocatalytic activity of the bimetallic catalysts for methanol oxidation reaction has been demonstrated for Pt-Au nanoparticles prepared by molecular-capping based chemical synthesis and subsequent assembly on carbon black support and thermal activation treatment [139, 140]. Mott et al. [141] investigated the MOR activity of Pt-Au bimetallic catalysts with different compositions in acid and alkaline environment. One of the significant findings is that the mass activity for the alkaline condition appears to exhibit a maximum around the composition of 65–85% Au, which is remarkably close to the composition range (65% Au) observed for the transition of the band features. While the presence of Au in Pt increases the lattice distance of Pt, the higher electronegativity of Au than Pt could cause an increase of the amount of change being transferred from Pt to Au. The composition was found to significantly modify the electrocatalytic properties of both Au and Pt. The finding that the mass activity in the alkaline electrolyte exhibits a maximum around 65–85% Au is in contrast to the small and gradual increase from no activity of Au to high activity of Pt in the acidic electrolyte. The display of the maximum mass activity for Pt-Au/C catalysts comparable or higher than that for Pt/C catalyst in the alkaline electrolyte is remarkable in view of the fact that only 15–35% Pt was present in the catalyst.

6.4.2
Ternary Pt-Ru-Based Catalysts

In Pt-Ru-M catalysts, the third metal is an oxophile element such as W, Os, Ni, Mo, Ir. The amount of Ru alloyed with Pt depends on the preparation method of the supported catalyst. The degree of alloying of Ru in Pt can decrease in the presence of a third metal compared to the binary Pt-Ru. Regarding the third metal, some metals fully alloy with Pt, some only partially alloy, and some do not form alloy with Pt. Generally, the higher the content of the third metal, the lower is the degree of alloying and the higher is the amount of its oxide.

The question is whether the catalytic activity of a Pt-Ru alloy catalyst, which is the most active system both for methanol oxidation, can be further improved by using ternary systems with elements such as W, Mo, Co, Ni, Os, Sn, having a co-catalytic activity for the anodic oxidation of methanol, if used either as alloys or as adsorbate layers on Pt-Ru. In the following part of this Section the MOR activity of different ternary Pt-Ru-based is compared with that of the binary Pt-Ru catalyst.

Gotz and Wendt [64] prepared various Pt-Ru-M catalysts by the impregnation method and by the colloidal method. In both cases, the carbon-supported Pt-Ru-W was the most active system for operation in DMFCs, followed by Pt-Ru-Mo and Pt-Ru. Conversely, the Pt-Ru-Sn catalyst was less active than Pt-Ru. He et al. [130] investigated the MOR activity of Pt-Ru (1:1), Pt-Ru-W (1:1:1) and Pt-Ru-Pd (1:1:1), prepared by reduction of metal precursors with $Na_2S_2O_3$, in DMFC at 180 °C. The initial performance in methanol at 180 °C was similar to Pt-Ru, Pt-Ru-W and Pt-Ru-Pd, with the Pt-Ru-W only slightly better than the other two. Stability tests indicated that, while the performance of Pt electrodes was found to be poor, this electrode was the most stable for methanol oxidation. The order of stability was: Pt > Pt-Ru-W, Pt-Ru-Pd > Pt-Ru. Pt-Ru/C and Pt-Ru-Mo/C electrocatalysts prepared the colloidal method have been studied by Pinheiro et al. [123] for methanol oxidation as porous thin films on high surface area carbon electrodes. Addition of Mo in the well-established Pt-Ru system is very promising for methanol oxidation. In the potential region around 0.6 V they found that the thermally treated Pt-Ru-Mo/C (1:1:0.5) and the Pt-Ru-Mo/C (1:1:1) electrocatalysts present similar and higher currents than only Pt-Ru by the same method, respectively. The performance of the Pt-Ru-Mo catalysts increased with increasing the Mo content in the catalyst. The catalytic activity for methanol electro-oxidation of a Pt-Ru-Rh (5:4:1) ternary catalyst was investigated by Choi et al. [142] in 0.5 M H_2SO_4/2 M CH_3OH solution and in DMFC, and the results were compared with those related to Pt-Ru (1:1) and Pt-Rh (2:1) catalysts. Pt-Ru-Rh showed the highest catalytic activity, having the lowest onset potential and the largest oxidation current densities with respect to Pt-Ru and Pt-Rh. Lima et al. [143] investigated the activity for the methanol oxidation reaction of various polyaniline (PAni) supported ternary Pt-Ru-M catalysts in $HClO_4$/CH_3OH solutions. They found that the Pt-Ru-Mo ternary catalyst presents enhanced performances compared to Pt-Ru, at low potentials (typically under 500 mV vs. RHE) under stationary conditions. The current densities measured with PAni supported Pt-Ru-Mo during methanol electro-oxidation are up to 10 times greater than those observed with PAni supported Pt-Ru. Moreover, this catalyst exhibits a good stability in a single cell at potentials lower than 550 mV vs. RHE during methanol oxidation. Work performed by the same research group on Pt-Ru-Mo [144] indicated that not only is methanol oxidized at lower potentials on Pt-Ru-Mo, but also the potential at which the adsorbed carbon monoxide (CO_{ads}) is oxidized is lower on Pt-Ru-Mo than on Pt-Ru. Kessler and Castro Luna [145] investigated the methanol oxidation on PAni supported Pt, Pt-Ru (9:15), Pt-Os (9:15), Pt-Ru-Os (9:15:0.1, 9:15:1 and 9:15:10) and Pt-Ru-Mo (9:15:0.1 and 9:15:1). They found that the best PAni-Pt-M_x-M_y catalyst for methanol oxidation is PAni-Pt-Ru-Os (9:15:0.1), meaning that Os as the third component gives a better performance in comparison to Mo.

According to Lasch et al. [131] and Juzyc et al. [132], stationary current-voltage and CV measurements indicated that the introduction of a transition metal-oxide like WO_x, MoO_x and VO_x to Pt-Ru catalysts leads to a decrease in polarization of the methanol electrodes. The ternary compound apparently influences the rate of methanol oxidation and surface oxide formation. They found that the specific surface activity of the catalysts towards methanol oxidation at 60 °C decreases in the order Pt-Ru-VO_x > Pt-Ru-Mo-O_x > Pt-Ru > Pt-Ru-WO_x = Pt-Ru (E-TEK). Liao et al. [146] investigated the methanol oxidation at the carbon-supported and carbon nanotube (CNT) supported Pt-Ru-Ir and Pt-Ru catalysts by CV, all in 0.5 M MeOH + 0.5 M H_2SO_4 solution at room temperature. The peak current densities at 50 mV s^{-1} were 81.7 and 61.2 mA cm^{-2} for the Pt-Ru-Ir/CNT and Pt-Ru/CNT, and 33.4, and 17.4 mA cm^{-2} for Pt-Ru-Ir/C and Pt-Ru/C, respectively. Regarding the role of Ir, the authors supposed that, similar to the role of Ru, OH groups could probably be stabilized at the metallic Ir surface, thus assisting in the oxidation of CO or other adsorbed intermediates. Zhang et al. investigated the electrocatalytic activity for methanol oxidation of Pt-Ru-Mo [96] and Pt-Ru-Co [97] nanoparticles by chronopotentiometry measurements. They found that the Pt-Ru-Mo and Pt-Ru-Co nanoparticles/carbon electrodes give higher activity and longer stability than that of Pt-Ru nanoparticles/carbon electrodes. Kim et al. [147] found that the activity for methanol oxidation of Pt-Ru-Sn/C (3:2:1) is significantly higher than that of Pt/C or Pt-Ru/C catalyst by improving CO tolerance. According to the authors, the higher CO tolerance property of Pt-Ru-Sn/C was caused by the synergic effects of Ru as a water activator and Sn as an electronic modifier of Pt. Park et al. [84] studied the electro-oxidation of methanol in sulfuric acid solution using Pt, Pt-Ni (1:1 and 3:1), Pt-Ru-Ni (5:4:1 and 6:3.5:0.5), and Pt-Ru (1:1) alloy nanoparticle catalysts. The Pt-Ni based catalysts displayed catalytic activity superior to that of pure Pt. The onset potential for methanol oxidation was in the order Pt-Ru-Ni (5:4:1) < Pt-Ru (1:1) < Pt-Ni (1:1) ∼ Pt-Ru-Ni (6:3.5:0.5) < Pt-Ni (3:1) < pure Pt. The authors considered that the electron transfer may contribute to the enhanced CO oxidation (CO generated from methanol decomposition), that is, to the CO tolerance on the Ni-containing composites, in comparison to pure Pt samples. Polarization and power density data in a liquid-feed DMFC unit cell test [148] were in good agreement with the voltammetry and chronoamperometry data, for which Pt-Ru-Ni (5:4:1) showed a higher catalytic activity than Pt-Ru (1:1). Methanol oxidation on Pt-Ru, Pt-Os and Pt-Ru-Os catalysts was studied by Ley et al. [149] on both arc-melted and fuel cell alloys catalysts prepared by the NaBH$_4$ reduction of metal chloride salts. Steady-state voltammetry of the arc-melted alloys at 25 °C confirmed that Pt-Ru-Os (65 : 25 : 10) is more active than Pt-Ru (1 : 1), particularly above 0.6 V. Pt-Ru-Os (65 : 25 : 10) methanol fuel cell performance curves were consistently superior to those of Pt-Ru (1 : 1) (e.g., typically at 90 °C, 0.4 V; 340 mA/cm^2 with Pt-Ru-Os vs. 260 mA/cm^2 with Pt-Ru). According to the authors, the improved performance of these catalysts can be rationalized within a conceptual framework developed for Pt-Ru-Os ternary catalysts, which correlated M−O bond strength and phase stability with performance. The ternary catalysts fit the bi-functional model for methanol oxidation, and the primary role of the alloying elements is to lower the potential for adsorption and activation of water. Siné et al. [150] investigated the methanol electro-oxidation on ternary Pt-Ru-Sn (80 : 10 : 10) nanoparticles synthesized by the microemulsion route and deposited onto boron-doped diamond electrode. The amount of Ru and Sn oxides was not reported. The ternary catalyst exhibited lower onset potential for methanol oxidation than either pure Pt or the corresponding bimetallic Pt-Ru and Pt-Sn catalysts.

6.5
Stability of Pt-Ru Catalysts in DMFC Environment

As previously reported, binary Pt-Ru and ternary Pt-Ru-based catalysts are generally regarded as the most appropriate materials for methanol electro-oxidation. However, Ru is prone to selective dissolution from Pt-Ru materials in normal fuel cell operation conditions. Ru dissolution from Pt-Ru during cycling in acid medium with methanol has also been reported [151]. The Ru dissolution was observed when the catalysts was cycled between reduced and oxidized states of Ru (0–1.3 V RHE) with little or no dissolution observed over much narrower potential ranges. Jeon et al. [152] carried out a stability test on a Pt-Ru black catalyst with an original particle size of 3.3 nm in DMFCs at current densities of 100, 150, and 200 mA cm^{-2}. Each test lasted for 145 h in the three cases. The maximum power densities were 93.9, 79.9, and 55.1% of the initial value after operation at 100, 150, and 200 mA cm^{-2}, respectively. For the membrane electrode assemblies membrane electrode assembly (MEA) operated at 100, 150, and 200 mA cm^{-2}, the Pt-Ru particle sizes increased from the original size to 3.4, 3.9, and 4.2 nm, respectively, while a Pt black catalyst used for the cathode electrode did not change in size. Dissolution of the Ru was observed, and the Pt : Ru ratio changed from 53 : 47 in the case of the fresh MEA, to 54 : 46, 56 : 44 and 73 : 27 for the MEAs after operation at 100, 150, and 200 mA cm^{-2}, respectively.

In the same way, life tests of DMFCs were carried out by Wang et al. [153] with three individual single cells using Pt-Ru as anode catalysts. The tests were operated at a current density of 100 mA cm^{-2} for three different times under ambient pressure and at a cell temperature of 60 °C. XRD results showed that the particle sizes of anodic catalysts increased from an original value of 2.8 to 3.0, 3.2, and 3.3 nm, whereas their lattice parameters first increased and then decreased from an original value of 3.8761 to 3.8879, 3.8777, and 3.8739 Å before and after 117, 210, and 312 working hours, respectively. XPS results indicated that during cell operation the contents of Pt and Ru oxides in anodic catalysts increased, but the metal content gradually decreased with test times. Polarization curves, power density curves, and *in situ* CO-stripping cyclic voltammetric curves were also plotted to evaluate the performances of fuel cells and electrochemically active surface areas (S_{EAS}) of anodic catalysts before and after life tests. After different time tests, the performances of DMFC lowered to different extents and could not recover their initial performances. The S_{EAS} of anodic catalyst decreased slightly by 4.78 and 9.03 $m^2 g^{-1}$ after 117 and 312 working hours, respectively.

In addition to Pt-Ru/C significant changes, the dissolved Ru species are found to be able to travel across the proton exchange membrane and deposit on the Pt cathode electrocatalyst [154]. Park *et al.* [155] investigated the durability behavior of Pt-Ru anode catalysts under DMFC operating conditions. They observed the crossover of ruthenium and platinum from the anode to the cathode due to the decomposition of active Pt-Ru anode catalysts. The Ru crossover measured at the cathode increased linearly with performance drop. The Ru contents at the cathode determined by XPS were less than 0.3 atom%. Sarma *et al.* [156], by Ru K-edge XAS and EDX analysis on the cathode catalyst layer of the faded MEA, found the presence of Ru environment in the cathode catalyst due to the Ru crossover from the anode to the cathode side. Once deposited at the cathode, ruthenium inhibits oxygen reduction kinetics and the catalyst's ability to handle methanol crossover. Depending on the degree of cathode contamination, the overall effect of ruthenium crossover on cell performance may be from as little as ~40 mV up to 200 mV [154]. The relatively rapid and unrecoverable performance decay of many PEMFCs is in part associated with the corrosion of the Pt-Ru anode electrocatalysts. While a significant loss in the CO tolerance of the anode presumes an excessive dissolution of Ru, even a minor level of Ru contamination is demonstrated to have a dramatic impact on the oxygen reduction activity at the Pt cathode. If Ru is present in the electrochemical environment of a thin-film Pt cathode in micromolar concentration, a high Ru coverage builds up on Pt that is found to be largely stable over the electrode potential region relevant to oxygen reduction reaction (ORR). Gancs *et al.* [157] investigated the oxygen reduction activity of a Pt electrocatalyst as a function of Ru contamination by a rotating disk electrode study, which revealed that reaction kinetics can decrease by as much as eight times. Under galvanostatic cathode operation in fuel cells, the effect of Ru contamination can translate to about 160 mV overpotential penalization. A preliminary analysis of the reaction mechanism at low overpotential suggests that Ru ad-atoms on Pt impair ORR by allowing the formation of surface oxides which simply block the surface of the electrode from the electroreduction of molecular oxygen. The facet of Ru dissolution and contamination in PEM fuel cell durability should be addressed by appropriate choice of materials. To overcome this problem, synthesis and application of highly stable Pt-Ru anode materials are currently under research.

The alloying degree plays an important role in the stability of Pt-Ru catalysts. Gancs *et al.* [158] compared the elemental composition as well as CO and MeOH electro-oxidation performance of PtRu samples having the same 50:50 atomic ratio but different degree of alloying. The data suggests that the latter physical property is crucial in maintaining the anodic activity for prolonged time.

Yun et al. [159] observed high Ru dissolution in poorly alloyed PtRu catalysts. On the other hand, a good stability was found in the sample containing the highest amount of Ru in PtRu solid solution. Therefore they concluded that the degree of PtRu solid solution determines the amount of Ru dissolution. Recently, increased stability of Pt-Ru was observed by addition of Au [160].

6.6
Conclusions

Pt-Ru based materials present higher MOR activity than Pt and other Pt-M bimetallic catalysts. Even though the preferable Ru form is still under debate, two conclusions can be drawn from existing studies: first, Pt-Ru alloyed degree and the content of hydrous Ru oxides are two important factors determining the catalytic activity of nano Pt-Ru catalyst; second, the structure of current Pt-Ru catalysts is far from optimization, there is an appreciable space for further improvement in catalytic activity. It has to be remarked that alloyed Ru is more stable against Ru dissolution than non-alloyed Ru. So, even if it seems that Pt/RuO_x is more active for methanol oxidation than Pt-Ru alloy (on the basis of the bi-functional mechanism), due to its higher stability the use of Pt-Ru alloy is recommended.

Experimental and theoretical combinatorial and/or high-throughput screening methods have been used to the development of ternary Pt-Ru-based catalysts with enhanced activity for methanol oxidation [161]. Measurements of the activity for methanol oxidation in half-cell and/or in DMFCs indicated that many ternary Pt-Ru-M catalysts perform better than commercial standard Pt-Ru and/or Pt-Ru prepared by the same method than the ternary catalyst. The improvement of the methanol oxidation of Pt-Ru by the addition of a third metal has been explained by different ways. Liang et al. [108] supposed that in the Pt-Ru-Ir system, RuO_2-IrO_2 interaction promotes the formation of hydroxyl species by dissociating water at a lower potential with respect to the Pt-Ru system. Moreover, they inferred that this interaction could also weaken the bonding between the hydroxyl species and the catalyst surface as compared with the bonding on Pt-Ru nanoparticles. The more weakly adsorbed hydroxyl species further promotes electro-oxidation of CO_{ads} on the active metal sites at a lower potential, thus improving the catalyst performance. Kim et al. [147] ascribed the enhancement of activity for MOR of Pt-Ru-Sn to the synergic effects of Ru as a water activator and Sn as an electronic modifier of Pt. On this basis, in ternary Pt-Ru-M catalysts with almost alloyed Pt-Ru, Ru acts as an electronic modifier of Pt, so M has to be a water activator and, as a consequence, the third metal has to be in non-alloyed form; conversely in poor-alloyed Ru, Ru acts as a water activator, so M has to be an electronic modifier of Pt, and then has to form alloy with Pt.

As in the case of binary Pt-W catalysts, the positive effect of W presence in ternary Pt-Ru based catalysts was ascribed to its intrinsic co-catalytic effect, supporting the CO oxidation, to a corrosion effect, which increases the surface area of the catalyst, and to some "physical" modification of the catalytically active Pt and Ru surface components such as differences in electroactive surface area.

Samieske et al. [138] explained the synergetic effect of Mo and Ru in this way: before OH-adsorption on Ru can set in, some of the CO-molecules adsorbed on Ru have to be oxidized. As long as a weakly adsorbed CO exists on Pt, there is a competition between weakly adsorbed CO (on Pt) and OH from solution for Ru sites. Only when the surface pressure of CO from Pt is reduced due to possible oxidation in the pre-peak can OH adsorb on Ru. (This effect may only be clearly visible at low surface contents of Ru). If this interpretation is true, also the synergetic effect of Ru and Mo is quite understandable: Mo shifts the oxidation of weakly adsorbed CO to lower potentials, and therefore the co-catalytic effect of Ru can set in.

The importance of the amount of the third metal in these catalysts was observed from the point of view theoretical and experimental: in some investigated compositions (Pt-Ru-Mo [105, 145] and Pt-Ru-Ni [84, 148]), at fixed Pt-Ru atomic ratio the MOR activity increased with increasing the content of the third element in the catalyst. On the other hand, the MOR activity of Pt-Ru-Os decreased with increasing Os content [145]. It has to take into account that the presence of MO_x species on the catalyst particle surface decreases the active surface area of Pt-Ru alloy particles. As a consequence, part of the noble metal becomes inactive due to the blocking of the Pt-Ru surface by the oxide species, which counteracts the activity enhancement [132]. Therefore, above a certain amount of the third metal, its presence has a detrimental effect on the performance of the catalyst.

References

1 Lamy, C., Lima, A., Le Rhun, V., Coutanceau, C. and Leger, J.M. (2002) *J. Power Sources*, **105**, 283.

2 Hamnett, A. (1997) *Catal. Today*, **38**, 445.

3 Reddington, E., Sapienza, A., Gurau, B., Viswanathan, R., Sarangapani, S., Smotkin, E.S. and Mallouk, T.E. (1998) *Science*, **280**, 1735.

4 Iwasita, T. and Nart, F.C. (1991) *J. Electroanal. Chem.*, **317**, 291.

5 Jarvi, T.D., Sriramulu, S. and Stuve, E.M. (1997) *J. Phys. Chem. B*, **101**, 3646.

6 Bagotzky, V.S., Vassiliev, Y.B. and Khazova, O.A. (1977) *J. Electroanal. Chem.*, **81**, 229.

7 Parsons, R. and VanderNoot, T. (1988) *J. Electroanal. Chem.*, **257**, 9.

8 Wasmus, S. and Kuver, A. (1999) *J. Electroanal. Chem.*, **461**, 14.

9 Iwasita, T. (2002) *Electrochim. Acta*, **47**, 3663.

10 Iwasita, T. and Vielstich, W. (1986) *J. Electroanal. Chem.*, **201**, 403.

11 Korzeniewski, C. and Childers, C. (1998) *J. Phys. Chem. B*, **102**, 489.

12 Ota, K., Nakagava, Y. and Takahashi, M. (1984) *J. Electroanal. Chem.*, **179**, 179.

13 Jusys, Y., Kaiser, J. and Behm, R.J. (2002) *Electrochim. Acta*, **47**, 3693.

14 Watanabe, M., Uchida, M. and Motoo, S. (1987) *J. Electroanal. Chem.*, **229**, 395.

15 Gasteiger, H.A., Markovic, N.M., Ross, P.N. Jr and Cairns, E.J. (1994) *J. Electrochem. Soc.*, **141**, 1795.

16 Liu, L., Pu, C., Viswanathan, R., Fan, Q., Liu, R. and Smotkin, E.S. (1998) *Electrochim. Acta*, **43**, 3657.

17 Lizcano-Valbuena, W.H., Caldas de Azevedo, D. and Gonzalez, E.R. (2004) *Electrochim. Acta*, **49**, 1289.

18 Chrzanowski, W. and Wieckowski, A. (1997) *Langmuir*, **13**, 5974.

19 Liu, R., Iddir, H., Fan, Q., Hou, G., Bo, A., Ley, K.L., Smotkin, E.S., Sung, Y.E., Kim, H., Thomas, S. and Wieckowski, A. (2000) *J. Phys. Chem. B*, **104**, 3518.

20 Markovic, N.M., Gasteiger, H.A., Ross, P.N., Jiang, X., Villegas, I. and Weaver, M.J. (1995) *Electrochim. Acta*, **40**, 91.

21 Goikovic, S.L., Vidakovic, T.R. and Durovic, D.R. (2003) *Electrochim. Acta*, **48**, 3607.

22 Christensen, P.A., Hamnett, A. and Troughton, G.L. (1993) *J. Electroanal. Chem.*, **362**, 207.

23 Angerstein-Kozlowska, H., Conway, B.E. and Sharp, B.A. (1973) *J. Electroanal. Chem.*, **43**, 9.

24 Deslouis, C., Musiani, M.M., Tribollet, B. and Vorotyntsev, M.A. (1996) *J. Electrochem. Soc.*, **142**, 1902.

25 Xu, Y., Amini, A. and Schell, M. (1994) *J. Phys. Chem.*, **98**, 1258.

26 Verbrugge, M.W. (1989) *J. Electrochem. Soc.*, **136**, 417.

27 Krausa, M. and Vielstich, W. (1994) *J. Electroanal. Chem.*, **397**, 307.

28 Bockris, J.O'M. and Wroblowa, H. (1964) *J. Electroanal. Chem.*, **7**, 428.

29 Petry, G.A., Podlovchenko, B.I., Frumkin, A.N. and Lal, H. (1965) *J. Electroanal. Chem.*, **10**, 253.

30 Watanabe, M. and Motoo, S. (1975) *J. Electroanal. Chem.*, **60**, 259.

31 Watanabe, M. and Motoo, S. (1975) *J. Electroanal. Chem.*, **60**, 275.

32 Janssen, M.M.P. and Molhuysen, J. (1976) *Electrochim. Acta*, **21**, 861.

33 Janssen, M.M.P. and Molhuysen, J. (1976) *Electrochim. Acta*, **21**, 869.

34 Aramata, A., Kodera, T. and Masuda, M. (1988) *J. Appl. Electrochem.*, **18**, 577.

35 Aramata, A. and Masuda, M. (1991) *J. Electrochem. Soc.*, **138**, 1949.

36 Meli, G., Leger, J.M. and Lamy, C. (1993) *J. Appl. Electrochem.*, **23**, 197.

37 Antolini, E. (2003) *Mater. Chem. Phys.*, **78**, 563.

38 Pan, C., Dassenoy, F., Casanove, M.J., Philippot, K., Amiens, C., Lecante, P., Mosset, A. and Chaudret, B. (2003) *J. Phys. Chem. B*, **103**, 10098.

39 Zhang, X. and Chan, K.-Y. (2003) *Chem. Mater.*, **15**, 451.

40 Hills, C.W., Mack, N.H. and Nuzzo, R.G. (2003) *J. Phys. Chem. B*, **107**, 2626.

41 Gurau, B., Viswanathan, R., Liu, R., Lafrenz, T.J., Ley, K.L., Smotkin, E.S., Reddington, E., Sapienza, A., Chan, B.C., Mallouk, T.E. and Sarangapani, S. (1998) *J. Phys. Chem. B.*, **102**, 9997.

42 Goodenough, J.B., Hamnett, A., Kennedy, B.J., Manoharan, R. and Weeks, S.A. (1988) *J. Electroanal. Chem.*, **240**, 133.

43 Chu, D. and Gilman, S. (1996) *J. Electrochem. Soc.*, **143**, 1685.

44 Park, K.-W., Choi, J.-H., Ahn, K.-S. and Sung, Y.-E. (2004) *J. Phys. Chem. B*, **108**, 5989.

45 Arico, A.S., Antonucci, P.L., Modica, E., Baglio, V., Kim, H. and Antonucci, V. (2002) *Electrochim. Acta*, **47**, 3723.

46 Gasteiger, H.A., Ross, P.N. and Cairns, E.J. (1993) *Surf. Sci.*, **293**, 67.

47 Kim, T.W., Park, S.J., Jones, L.E., Toney, M.F., Park, K.W. and Sung, Y.E. (2005) *J. Phys. Chem. B*, **109**, 12845.

48 Papageorgopoulos, D.C., de Heer, M.P., Keijzer, M., Pieterse, J.A.Z. and de Bruijn, F.A. (2004) *J. Electrochem. Soc.*, **151**, A763.

49 Rolison, D.R. (2003) *Science*, **299**, 1698.

50 Stroud, R.A., Long, J.W., Swider-Lyons, K.E. and Rolison, D.R. (2002) *Microsc. Microanal.*, **8**, 50.

51 Long, J.W., Stroud, R.M., Swider-Lyons, K.E. and Rolison, D.R. (2000) *J. Phys. Chem. B*, **104**, 9772.

52 Rolison, D.R., Hagans, P.L., Swider, K.E. and Long, J.W. (1999) *Langmuir*, **15**, 774.

53 Ren, X.M., Wilson, M.S. and Gottesfeld, S. (1996) *J. Electrochem. Soc.*, **143**, L12.

54 Lu, Q., Yang, B., Zhuang, L. and Lu, J. (2005) *J. Phys. Chem. B*, **109**, 8873.

55 Wu, G., Li, L. and Xu, B.Q. (2004) *Electrochim. Acta*, **50**, 1.

56 Lu, Q., Yang, B., Zhuang, L. and Lu, J. (2005) *J. Phys. Chem. B*, **109**, 1715.

57 Zhou, W., Zhou, Z., Song, S., Li, W., Sun, G., Tsiakaras, P. and Xin, Q. (2003) *Appl. Catal. B*, **46**, 273.

58 Colmati, F., Antolini, E. and Gonzalez, E.R. (2005) *Electrochim Acta*, **50**, 5496.

59. Kuznetsov, V.I., Belyi, A.S., Yurchenko, E.N., Smolikov, M.D., Protasova, M.T., Zatolokina, E.V. and Duplayakin, V.K. (1986) *J. Catal.*, **99**, 159.
60. Radmilovic, V., Richardson, T.J., Chen, S.J. and Ross, P.N. (2005) *J. Catal.*, **232**, 199.
61. Gonzalez, M.J., Hable, C.T. and Wrighton, M.S. (1998) *J. Phys. Chem. B*, **102**, 9881.
62. Gasteiger, H.A., Markovic, N.M. and Ross, P.N. (1995) *J. Phys. Chem.*, **99**, 8945.
63. Wang, K., Gasteiger, H.A., Markovic, N.M. and Ross, P.N. (1996) *Electrochim. Acta*, **41**, 2587.
64. Gotz, M. and Wendt, H. (1998) *Electrochim. Acta*, **43**, 3637.
65. Watanabe, M., Furuuchi, Y. and Motoo, S. (1985) *J. Electroanal. Chem.*, **191**, 367.
66. McNicol, B.D., Short, R.T. and Chapman, A.G. (1976) *J. Chem. Soc., Faraday Trans.*, **72**, 2735.
67. Abdel Rahim, M.A., Khalil, M.W. and Hassan, H.B. (2000) *J. Appl. Electrochem.*, **30**, 151.
68. Honma, I. and Toda, T. (2003) *J. Electrochem. Soc.*, **150**, A1689.
69. Cathro, K.J. (1969) *J. Electrochem. Soc.*, **116**, 1608.
70. Frelink, T., Visscher, W. and van Veen, J.A.R. (1994) *Electrochim. Acta*, **39**, 1871.
71. Bittins-Cattaneo, B. and Iwasita, T. (1987) *J. Electroanal. Chem.*, **238**, 151.
72. Campbell, S.A. and Parsons, R. (1992) *J. Chem. Soc., Faraday Trans.*, **88**, 833.
73. Haner, A.N. and Ross, P.N. (1991) *J. Phys. Chem.*, **95**, 3740.
74. Beden, B., Kadirgan, F., Lamy, C. and Leger, J.M. (1981) *J. Electroanal. Chem.*, **127**, 75.
75. Morimoto, Y. and Yeager, E.B. (1998) *J. Electroanal. Chem.*, **444**, 95.
76. Morimoto, Y. and Yeager, E.B. (1998) *J. Electroanal. Chem.*, **444**, 100.
77. Mukerjee, S. and McBreen, J. (1999) *J. Electrochem. Soc.*, **146**, 600.
78. Shukla, A.K., Neergat, M., Bera, P., Jayaram, V. and Hegde, M.S. (2001) *J. Electroanal. Chem.*, **504**, 111.
79. Paulus, U.A., Scherer, G.G., Wokaun, A., Schmidt, T.J., Stamenkovic, V., Radmilovic, V., Markovic, N.M. and Ross, P.N. (2002) *J. Phys. Chem. B*, **106**, 4181.
80. Abraham, F.F., Tsai, N.H. and Pound, G.M. (1979) *Surf. Sci.*, **83**, 406.
81. De Temmerman, L., Creemers, C., Van Hove, M., Neyens, A., Bertolini, J. and Messardier, J. (1986) *Surf. Sci.*, **178**, 888.
82. Mukerjee, S. and Moran-Lopez, J.L. (1987) *Surf. Sci.*, **189–190**, 1135.
83. Mukerjee, S. and Srinivasan, S. (1993) *J. Electroanal. Chem.*, **357**, 201.
84. Park, K., Choi, J., Kwon, B., Lee, S., Sung, Y., Ha, H., Hong, S., Kim, H. and Wieckowski, A. (2002) *J. Phys. Chem. B*, **106**, 1869.
85. Page, T., Johnson, R., Hormes, J., Noding, S. and Rambabu, B. (2000) *J. Electroanal. Chem.*, **485**, 34.
86. Antolini, E., Salgado, J.R.C. and Gonzalez, E.R. (2006) *Appl. Catal. B*, **63**, 137.
87. Jiang, J. and Kucernak, A. (2004) *Chem. Mater.*, **16**, 1362.
88. Wang, Y., Ren, J.W., Deng, K., Gui, L.L. and Tang, Y.Q. (2000) *Chem. Mater.*, **12**, 1622.
89. Toshima, N. and Wang, Y. (1994) *Langmuir*, **10**, 4574.
90. Yan, X.P., Liu, H.F. and Liew, Y. (2001) *J. Mater. Chem.*, **6**, 3387.
91. Sun, S., Murray, C.B., Weller, D., Folks, L. and Moser, A. (2000) *Science*, **287**, 1989.
92. Ould Ely, T., Pan, C., Amiens, C., Chaudret, B., Dassenoy, F., Lecante, P., Casanove, M.J., Mosset, A., Respaund, M. and Broto, J.M. (2000) *J. Phys. Chem. B*, **104**, 695.
93. Park, J.I. and Cheon, J. (2001) *J. Am. Chem. Soc.*, **123**, 5743.
94. Ono, K., Okuda, R., Ishii, Y., Kamimura, S. and Oshima, M. (2003) *J. Phys. Chem. B*, **107**, 1941.
95. Zhang, X., Tsang, K.-Y. and Chan, K.-Y. (2004) *J. Electroanal. Chem.*, **573**, 1.
96. Zhang, X., Zhang, F. and Chang, K. (2004) *J. Mater. Sci.*, **39**, 5845.

97 Zhang, X., Zhang, F. and Chang, K. (2004) *Catal. Commun.*, **5**, 749.
98 Bonnemann, H., Brijoux, W., Brinkmann, R., Dinjus, E., Jouben, T. and Korall, B. (1991) *Angew. Chem. Int. Ed. Engl.*, **30**, 1312.
99 Wang, Y., Ren, J., Deng, K., Gui, L. and Tang, Y. (2000) *Chem. Mater.*, **12**, 1622.
100 Zhou, Z.H., Wang, S.L., Zhou, W.J., Wang, G.X., Jiang, L.H., Li, W.Z., Song, S.Q., Liu, J.Q., Sun, G.Q. and Xin, Q. (2003) *Chem. Commun.*, **1**, 394.
101 Bonnemann, H., Brijoux, W. and Joussen, Th. (1990) *Angew. Chem.*, **102**, 324.
102 Bonnemann, H., Brijoux, W., Brinkmann, R., Fretzen, R., Joussen, Th., Koppler, R., Korall, B., Neiteler, P. and Richter, J. (1994) *J. Mol. Catal.*, **86**, 129.
103 Schmidt, T.J., Noeske, M., Gasteiger, H.A., Behm, R.J., Britz, P., Brijoux, W. and Bonnemann, H. (1997) *Langmuir*, **13**, 2591.
104 Roth, C., Goetz, M. and Fuess, H. (2001) *J. Appl. Electrochem.*, **31**, 793.
105 Oliveira Neto, A., Franco, E.G., Arico, E., Linardi, M. and Gonzalez, E.R. (2003) *J. Eur. Ceram. Soc.*, **23**, 2987.
106 Liu, Z., Ling, X.Y., Lee, J.Y., Su, X. and Gan, L.M. (2003) *J. Mater. Chem.*, **13**, 3049.
107 Liu, Z., Lee, J.Y., Chen, W., Han, M. and Gan, L.M. (2004) *Langmuir*, **20**, 181.
108 Liang, Y., Zhang, H., Zhong, H., Zhou, X., Tian, Z., Xu, D. and Yi, B. (2006) *J. Catal.*, **238**, 468.
109 Liang, Y., Zhang, H., Tian, Z., Zhu, X., Wang, X. and Yi, B. (2006) *J. Phys. Chem. B*, **110**, 7828.
110 Castro Luna, A.M., Camara, G.A., Paganin, V.A., Ticianelli, E.A. and Gonzalez, E.R. (2000) *Electrochem. Commun.*, **2**, 222.
111 Salgado, J.R.C., Antolini, E. and Gonzalez, E.R. (2004) *J. Electrochem. Soc.*, **151**, A2143.
112 Antolini, E., Salgado, J.R.C., dos Santos, A.M. and Gonzalez, E.R. (2005) *Electrochem. Solid-State Lett.*, **8**, A226.
113 Antolini, E., Salgado, J.R.C. and Gonzalez, E.R. (2006) *J. Appl. Electrochem.*, **36**, 355.
114 Antolini, E., Salgado, J.R.C., da Silva, R.M. and Gonzalez, E.R. (2007) *Mater. Chem. Phys.*, **101**, 395.
115 Yang, B., Lu, Q., Wang, Y., Zhuang, L., Lu, J., Liu, P., Wang, J. and Wang, R. (2003) *Chem. Mater.*, **15**, 3552.
116 Wang, D., Zhuang, L. and Lu, J. (2007) *J. Phys. Chem. C*, **111**, 16416.
117 Nasher, M.S., Frenkel, A.I., Somerville, D., Hills, C.W., Shapley, J.R. and Nuzzo, R.G. (1998) *J. Am. Chem. Soc.*, **120**, 8093.
118 Hills, C.W., Nasher, M.S., Frenkel, A.I., Shapley, J.R. and Nuzzo, R.G. (1999) *Langmuir*, **15**, 690.
119 Dickinson, A.J., Carrette, L.P.L., Collins, J.A., Friedrich, K.A. and Stimming, U. (2002) *Electrochim. Acta*, **47**, 3733.
120 Manzo-Robledo, A., Boucher, A.C., Pastor, E. and Alonso-Vante, N. (2002) *Fuel Cells*, **2**, 109.
121 Yang, H., Alonso-Vante, N., Leger, J.M. and Lamy, C. (2004) *J. Phys. Chem. B*, **108**, 1938.
122 Yang, H., Vogel, W., Lamy, C. and Alonso-Vante, N. (2004) *J. Phys Chem. B*, **108**, 11024.
123 Pinheiro, A.L.N., Oliveira-Neto, A., de Souza, E.C., Perez, J., Paganin, V.A., Ticianelli, E.A. and Gonzalez, E.R. (2003) *J. New Mater. Electrochem. Syst.*, **6**, 1.
124 Rojas, S., Garcia-Garcia, F.J., Jaras, S., Martinez-Huerta, M.V., Fierro, J.L.G. and Boutonnet, M. (2005) *Appl. Catal. A*, **285**, 24.
125 Shen, P.K. and Tseung, A.C.C. (1994) *J. Electrochem. Soc.*, **141**, 3082.
126 Shukla, A.K., Ravikumar, M.K., Arico, A.S., Candiano, G., Antonucci, V., Giordano, N. and Hamnett, A. (1995) *J. Appl. Electrochem.*, **25**, 528.
127 Yang, L.X., Bock, C., MacDougall, B. and Park, J. (2004) *J. Appl. Electrochem.*, **34**, 427.
128 McLeod, E.J. and Birss, V.I. (2005) *Electrochim. Acta*, **51**, 684.

129 Venkataraman, R., Kunz, H.R. and Fenton, J.M. (2003) *J. Electrochem. Soc.*, **150**, A278.

130 He, C., Kunz, H.R. and Fenton, J.M. (2003) *J. Electrochem. Soc.*, **150**, A1017.

131 Lasch, K., Jorissen, L. and Garche, J. (1999) *J. Power Sources*, **84**, 225.

132 Jusys, Z., Schmidt, T.J., Dubau, L., Lasch, K., Jorissen, L., Garche, J. and Behm, R.J. (2002) *J. Power Sources*, **105**, 297.

133 Strasser, P., Fan, Q., Devenney, M., Weinberg, W.H., Liu, P. and Norskov, J.K. (2003) *J. Phys. Chem. B*, **107**, 11013.

134 Nakajima, H. and Kita, H. (1990) *Electrochim. Acta*, **35**, 849.

135 Morante-Catacora, T.Y., Ishikawa, Y. and Cabrera, C.R. (2008) *J. Electroanal. Chem.*, **621**, 103–112.

136 Ordóñez, L.C., Roquero, P., Sebastian, P.J. and Ramírez, J. (2005) *Catal. Today*, **107–108**, 46.

137 Mukerjee, S. and Urian, R.C. (2002) *Electrochim. Acta*, **47**, 3219.

138 Samjeské, G., Wang, H., Löffler, T. and Baltruschat, H. (2002) *Electrochim. Acta*, **47**, 3681.

139 Luo, J., Maye, M.M., Kariuki, N.N., Wang, L., Njoki, P., Lin, Y., Schadt, M., Naslund, H.R. and Zhong, C.J. (2005) *Catal. Today*, **99**, 291.

140 Luo, J., Jones, V.W., Maye, M.M., Han, L., Kariuki, N.N. and Zhong, C.J. (2002) *J. Am. Chem. Soc.*, **124**, 13988.

141 Mott, D., Luo, J., Njoki, P.N., Lin, Y., Wang, L. and Zhong, C.J. (2007) *Catal. Today*, **122**, 378.

142 Choi, J., Park, K., Park, I., Nam, W. and Sung, Y. (2004) *Electrochim. Acta*, **50**, 787.

143 Lima, A., Coutanceau, C., Leger, J.-M. and Lamy, C. (2001) *J. Appl. Electrochem.*, **31**, 379.

144 Lima, A., Hahn, F. and Leger, J.-M. (2004) *Russ. J. Electrochem.*, **40**, 326.

145 Kessler, T. and Castro Luna, A.M. (2003) *J. Solid State Electrochem.*, **7**, 593.

146 Liao, S., Holmes, K., Tsaprailis, H. and Birss, V.I. (2006) *J. Am. Chem. Soc.*, **128**, 3504.

147 Kim, T.Y., Kobayashi, K., Takahashi, M. and Nagai, M. (2005) *Chem. Lett.*, **34**, 798.

148 Choi, J., Park, K., Kwon, B. and Sung, Y. (2003) *J. Electrochem. Soc.*, **150**, A973.

149 Ley, K.L., Liu, R., Pu, C., Fan, Q., Leyarovska, N., Segre, C. and Smotkin, E.S. (1997) *J. Electrochem. Soc.*, **144**, 1543.

150 Siné, G., Smida, D., Limat, M., Foti, G. and Comninellis, Ch. (2007) *J. Electrochem. Soc.*, **154**, B170.

151 Holstein, W.L. and Rosenfeld, H.D. (2005) *J. Phys. Chem. B*, **109**, 2176.

152 Jeon, M.K., Lee, K.R., Oh, K.S., Hong, D.S., Won, J.Y., Li, S. and Woo, S.I. (2006) *J. Power Sources*, **158**, 1344.

153 Wang, Z.-B., Wang, X.-P., Zuo, P.-J., Yang, B.-Q., Yin, G.-P. and Feng, X.-P. (2008) *J. Power Sources*, **181**, 93.

154 Piela, P., Eickes, C., Brosha, E., Grazon, F. and Zelenay, P. (2004) *J. Electrochem. Soc.*, **151**, A2053.

155 Park, G.-S., Pak, C., Chung, Y.-S., Kim, J.-R., Jeon, W.S., Lee, Y.-H., Kim, K., Chang, H. and Seung, D. (2008) *J. Power Sources*, **176**, 484.

156 Sarma, L.S., Chen, C.H., Wang, G.R., Hsueh, K.L., Huang, C.P., Sheu, H.S., Liu, D.G., Lee, J.F. and Hwang, B.J. (2007) *J. Power Sources*, **167**, 358.

157 Gancs, L., Murthi, V. and Mukerjee, S. (2007) 211th Electrochemical Society Meeting, 2007 May 6–10, Chicago, Illinois.

158 Gancs, L., Hakim, N., Hult, B. and Mukerjee, S. (2006) *ECS Trans.* **3** (1), 607.

159 Hyun, M., Kim, S., Jung, D., Lee, B., Peck, D. and Shul, Y. (2007) 20th North American Catalysis Society Meeting, 2007, June 17–22, Houston, Texas.

160 Liang, Z.X., Zhao, T.S. and Xu, J.B. (2008) *J. Power Sources*, **185**, 166–170.

161 Cooper, J.S. and McGinn, P.J. (2006) *J. Power Sources*, **163**, 330.

7
Methanol-Tolerant Cathode Catalysts for DMFC

Claude Lamy, Christophe Coutanceau, and Nicolas Alonso-Vante

In this chapter, we will summarize the kinetic behavior of the oxygen reduction reaction (ORR), mainly on platinum electrodes since this metal is the most active electrocatalyst for the ORR in an acidic medium. The discussion will, however, be restricted to the characteristics of this reaction in direct methanol fuel cells (DMFCs) because the possible presence in the cathode compartment of methanol, which can crossover the proton exchange membrane and react with oxygen.

7.1
Introduction

The oxygen electrode has been the subject of many extensive investigations over the past century. The pronounced irreversibility of the cathodic and anodic reactions in aqueous solutions has imposed severe limitations on the information that can be obtained concerning the pathways from electrochemical kinetic studies. In most instances at current densities practical for fuel cell applications, the current–voltage data are not sensitive to the back reaction and hence yield information only up to the rate-determining step (r.d.s.), which usually occurs early in a multiple-step reaction sequence. Further the reduction and oxidation processes are usually studied only at widely separated potentials and thus the surface conditions differ sufficiently, so that the reduction and oxidation pathways are probably not complementary. The situation is still more complicated by the large number of possible pathways and intermediate states for the O_2 electrode reactions.

The surface states of the anodic films play a key role in the reaction mechanisms. They usually depend strongly on potential and also on the time history of the electrode. Thus complex kinetic behavior of the ORR at a Pt electrode has been observed as a function of potential.

Electrocatalysis of Direct Methanol Fuel Cells. Edited by Hansan Liu and Jiujun Zhang
Copyright © 2009 WILEY-VCH Verlag GmbH & Co. KGaA, Weinheim
ISBN: 978-3-527-32377-7

7.2
Thermodynamics and Kinetics of the Oxygen Reduction Reaction (ORR) [1, 2]

7.2.1
The ORR at a Platinum Electrode in a DMFC

At the cathode, the oxygen electroreduction does occur by a four-electron process leading to water, according to the overall reaction (in acid medium):

$$O_2 + 4H^+ + 4e^- \leftrightarrow 2H_2O \quad \text{with} \quad E_1^\circ = 1.23 \text{ V vs. SHE} \quad (7.1)$$

where E_1° is the standard electrode potential versus the standard hydrogen electrode (SHE) under standard conditions. This reaction involves in many cases the formation of an intermediate species, that is hydrogen peroxide, by a two-electron reaction:

$$O_2 + 2H^+ + 2e^- \leftrightarrow H_2O_2 \quad \text{with} \quad E_2^\circ = 0.69 \text{ V vs. SHE} \quad (7.2)$$

followed by its further reduction to water:

$$H_2O_2 + 2H^+ + 2e^- \leftrightarrow 2H_2O \quad \text{with} \quad E_3^\circ = 1.77 \text{ V vs. SHE} \quad (7.3)$$

A possible reaction path involving the formation of hydrogen peroxide, as a two-electron intermediate, is thus the following (Scheme 7.1) [1]:

Several reaction mechanisms, involving different adsorbed species, have been proposed in the literature [1], such as the next one, which is able to interpret most of the experimental results (Scheme 7.2):

The kinetics of the four-electron oxygen reduction is relatively slow (j_o from 10^{-6} to 10^{-10} A cm^{-2}, referred to the geometric surface area, depending on the degree of dispersion of the platinum catalyst). This is the main cause of the high cathodic overpotential η_c ($\eta_c \approx$ about 400 mV for a hydrogen/oxygen PEMFC, working at 500 mA cm^{-2}). This comes from the rather complex reaction mechanism involved in the ORR.

The open-circuit potential usually deviates from the thermodynamic electrode potential and establishes close to 0.9–1.0 V vs. SHE. The deviation from the

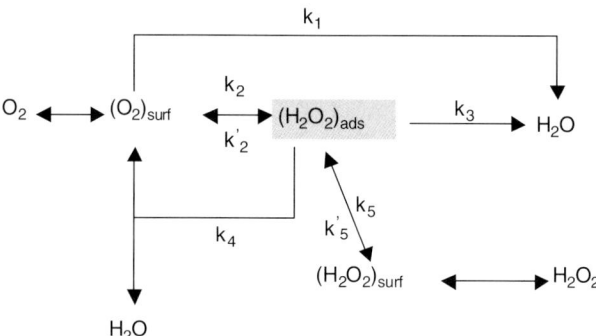

Scheme 7.1 Simple reaction mechanism of the ORR involving H_2O_2 as reaction intermediate or final product.

```
                Pt + O₂ → Pt - O₂ₐds
                Pt - O₂ₐds + H⁺ + e⁻ → Pt - O₂Hₐds
either    ⎧  Pt - O₂Hₐds + H⁺ + e⁻ → Pt + H₂O₂
or        ⎩  Pt + Pt - O₂Hₐds → Pt - Oₐds + Pt - OHₐds
                Pt - Oₐds + H⁺ + e⁻ → Pt - OHₐds
                2 Pt - OHₐds + 2 H⁺ + 2 e⁻ → 2 Pt + 2 H₂O

                Overall reaction: O₂ + 4 H⁺ + 4 e⁻ → 2 H₂O
```

Scheme 7.2 Detailed reaction mechanism of the ORR involving the formation of H_2O_2.

equilibrium potential, even at low current densities, is a consequence of some possible side-reactions:

i) the production of H_2O_2, either as an intermediate in the four-electron reduction of O_2 to water, or as a reaction product. In the latter case the thermodynamic equilibrium potential of reaction (7.2) is 0.695 V/SHE, instead of 1.23 V/SHE for the overall oxygen reduction reaction. Usually in acid medium, such as in a PEMFC, no H_2O_2 is formed at a Pt electrode at an operating potential, E_c, of 0.9 to 0.7 V.

ii) a slow r.d.s., that is,

$$O_2 + H^+ + e^- \rightarrow O_2H_{ads} \tag{7.4}$$

where the adsorbed intermediate (O_2H_{ads}) follows a Temkin isotherm at low overpotential leading to a Tafel slope of $-\frac{RT}{F}$ (low current densities). At high overvoltage (high current densities and low coverage) the adsorption isotherm becomes Langmuirian on the pure platinum surface, and the Tafel slope corresponds to $-\frac{2RT}{F}$, as normally observed.

iii) observance of a mixed potential of about 1.0 V (instead of the equilibrium thermodynamic reversible potential $E_c° = 1.23$ V vs. SHE) due to the formation of surface oxides at the Pt electrode, according to different electrode reactions (see [3] for Equation 7.5a:

$$Pt + H_2O \leftrightarrow PtOH + H^+ + e^- \quad \text{with} \quad E_4° = 0.62 \text{ V} \tag{7.5a}$$

$$Pt + H_2O \leftrightarrow PtO + 2H^+ + 2e^- \quad \text{with} \quad E_5° = 0.88 \text{ V} \tag{7.5b}$$

$$Pt + 2H_2O \leftrightarrow Pt(OH)_2 + 2H^+ + 2e^- \quad \text{with} \quad E_6° = 0.98 \text{ V} \tag{7.5c}$$

$$PtO + H_2O \leftrightarrow PtO_2 + 2H^+ + 2e^- \quad \text{with} \quad E_7° = 1.04 \text{ V} \tag{7.5d}$$

$$Pt(OH)_2 \leftrightarrow PtO_2 + 2H^+ + 2e^- \quad \text{with} \quad E_8° = 1.11 \text{ V} \tag{7.5e}$$

or even to the presence of minute trace of organic impurities undergoing an oxidation reaction, such as methanol:

$$CH_3OH + H_2O \leftrightarrow CO_2 + 6H^+ + 6e^- \tag{7.6}$$

In all these cases, the electrode potential E_m will be determined by a mixed reaction resulting from the reduction of oxygen and the oxidation of the platinum surface or of methanol at the same potential E_m. As the two reactions are quite irreversible, a Tafel behavior is practically always observed. Under these conditions, the current density (j_m) and mixed potential (E_m) are given by the equations:

$$j_m = j_{oa} 10^{\frac{(E_m - E_a^o)}{b_a}} = j_{oc} 10^{-\frac{(E_m - E_c^o)}{b_c}} \tag{7.7a}$$

and:

$$E_m = \frac{b_c E_a^o + b_a E_c^o}{b_a + b_c} + \frac{b_a b_c}{b_a + b_c} \log \frac{j_{oc}}{j_{oa}} \tag{7.7b}$$

where j_{oi} and b_i are the exchange current density and the Tafel slopes, respectively, for both half-cell reactions.

The exchange current density j_{oa} for methanol oxidation depends on the methanol concentration, that is:

$$j_{oa} = nFk_s^o [CH_3OH]^{\alpha_a} \tag{7.8}$$

where the value of the charge transfer coefficient α_a is very probably equal to 0.5. For the sake of simplicity, the standard rate constant k_s^o includes the concentration of the oxidized species (i.e., CO_2). Therefore a small crossover of methanol through the membrane, increasing for example the methanol concentration in the cathodic compartment by a factor of 10^6, will increase the exchange current density j_{oa} by 10^3. This will result in a negative shift, ΔE_m, of the mixed potential at the oxygen cathode, as given by the following equation:

$$\Delta E_m = \frac{b_a b_c}{b_a + b_c} \log \frac{j_{oa}}{j_{o'a}} \approx \frac{120}{2} \log 10^{-3} \approx -180 \text{ mV} \tag{7.9}$$

with $j_{o'a} = 10^3 j_{oa}$ and $b_a \approx b_c \approx 120$ mV/decade for both the oxygen reduction (at high current densities) and the methanol oxidation reactions. If $b_c \approx 60$ mV/decade, then $\Delta E_m \approx -120$ mV. For a higher membrane crossover rate, leading to a methanol concentration in the cathodic compartment of the order of 10^{-2} M for Nafion117, after 5 h of operation, as measured in our laboratory [4], the shift of the oxygen electrode potential will be $\Delta E_m \approx 120/2 \log 10^{-4} \approx -240$ mV. Such cathode potential shifts are effectively observed in a working DMFC-PEMFC (Figure 7.1).

Therefore, one main drawback of the PEMFC configuration with a standard proton exchange membrane (such as Nafion), and a standard Pt gas diffusion cathode is the cathode depolarization due to the occurrence of a mixed potential resulting from the methanol crossover through the membrane. There are two possibilities to overcome these difficulties, the first is to conceive novel electrocatalysts for oxygen reduction

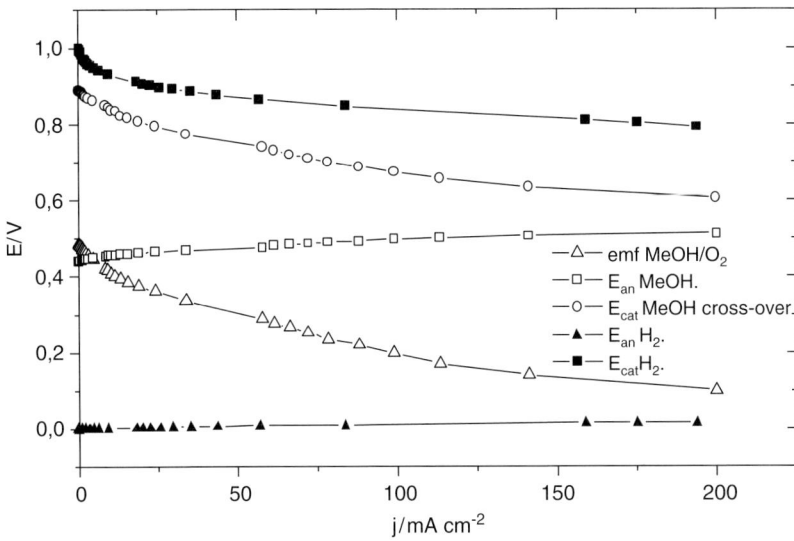

Figure 7.1 Polarization curves of an H_2/O_2 PEMFC and a DMFC in the presence of methanol at the cathodic compartment: (■) potential of the PEMFC cathode; (○) potential of the DMFC cathode; (△) DMFC cell voltage. After Ref. [5].

but which are highly inactive for methanol oxidation, and the second is to develop new membranes that are more methanol tight.

7.2.2
Concepts for Novel Oxygen Reduction Electrocatalysts

One strategy to circumvent the effect of methanol leakage through the ionomeric membrane and to avoid the Pt cathode depolarization, which reduces the cell voltage by about 0.1–0.2 V, is to replace Pt catalysts by transition metal compounds, such as macrocycles or chalcogenides, which are inactive towards the oxidation of methanol.

Among the non-noble metal electrocatalysts suitable for the ORR, organometallic macrocycles have often been considered as an alternative to Pt-based catalysts [6]. In particular, transition metal phthalocyanines and porphyrins, which have a similar square planar structure, were thoroughly examined for their activity towards oxygen electroreduction. Investigation of O_2 reduction electrocatalysts have usually been carried out on thin films of the organometallic complex supported on a carbon, graphite or a metal substrate, such as gold [7–11]. The nature of the central metal ion (Fe, Co, Mn, Ni, and so on) [12] or that of the substitutes of the organic ligands have a strong influence on the catalytic activity [13]. The polymerization of the macrocyclic monomer, which is possible through the highly conjugated organic structures with delocalized π electrons, greatly improved the electrical conductivity of the film, and therefore its electroactivity. Co and Fe porphyrins and/or phthalocyanines were found to be the best non-noble metal electrocatalysts for the oxygen electroreduction. It was found that the electrocatalytic activity is mainly associated with the metallic $M^{II/III}$ redox couple. Many experiments, carried out, in our laboratory and others, showed a very beneficial effect in that these macrocyclic compounds are insensitive to methanol, that is, no methanol oxidation is observed on these compounds, nor is the oxygen cathode depolarized in the presence of methanol.

For practical cathodes of a PEMFC, the macrocyclic compounds are supported on a carbon powder (such as Vulcan XC72) and thermally treated (up to 800 °C), which substantially increases their electrocatalytic activities. After pyrolysis and dispersion of the macrocyclic catalysts on the carbon substrate, the nature and structure of the electrocatalytic sites are difficult to assign and to be correlated with their electroactivity for oxygen reduction. The pyrolysis treatment also destroys the molecular structure. A better approach is to disperse the electrocatalytic entity at the molecular level in a convenient substrate, such as an electron-conducting polymer (polyaniline PAni, or polypyrrole PPy). One of the first ideas was to incorporate a tetrasulfonated cobalt porphyrin (CoTsPP) or a tetrasulfonated iron phthalocyanine (FeTsPc) as a counter-ion during the electropolymerization process of pyrrole [14, 15]. These modified electrodes, particularly those containing FeTsPc, display an excellent electrocatalytic activity for the ORR, somewhat similar to that of Pt electrodes, leading to a four electron process in acid medium [16, 17]. However the stability of these electrodes is poor due to a degradation of the PPy film at the electrode potential of the oxygen cathode. Better stability is obtained when the macrocyclic catalyst is dispersed in a PAni film, and this may be due to a greater resistance to oxidation by small traces of hydrogen peroxide resulting from a two-electron reduction process [16].

Since the macrocycle counter-ion can be expelled from the polymeric matrix at low potentials, an alternative is to directly electropolymerize a modified or substituted metal porphyrin or phthalocyanine, containing an electropolymerizable functional group (e.g., an amino group). This approach was followed by Murray et al., who showed that a poly-tetra(o-aminophenyl)porphyrin (poly-CoTAPP) film, formed by electropolymerization of the corresponding monomer on a glassy carbon electrode, displayed a good electrocatalytic activity for the oxygen reduction reaction [18, 19]. Similarly, the electropolymerization in non-aqueous medium (acetonitrile containing 0.1 M tetraethylammonium perchlorate) of cobalt 4,4′,4″,4‴-tetraaminophthalocyanine (CoTAPc) leads to electrocatalytically active oxygen cathodes [20], which are more efficient for a four-electron reduction process than those prepared from the cobalt porphyrin monomer (CoTAPP).

When the behavior of these modified electrodes towards the oxygen electroreduction are compared together with that of bare Pt (see Figure 7.20 in Section 7.4.2.5), the tetrasulfonated iron phthalocyanine dispersed in a polypyrrole film is the best non-metallic electrocatalyst, displaying an electrocatalytic activity close to that of Pt. The presence of 0.1 M CH_3OH in the electrolytic solution (0.5 M H_2SO_4) does not affect the oxygen reduction wave, except for the Pt electrode, confirming the prospect of these macrocyclic transition metal electrocatalysts being methanol insensitive during oxygen electroreduction in a DMFC.

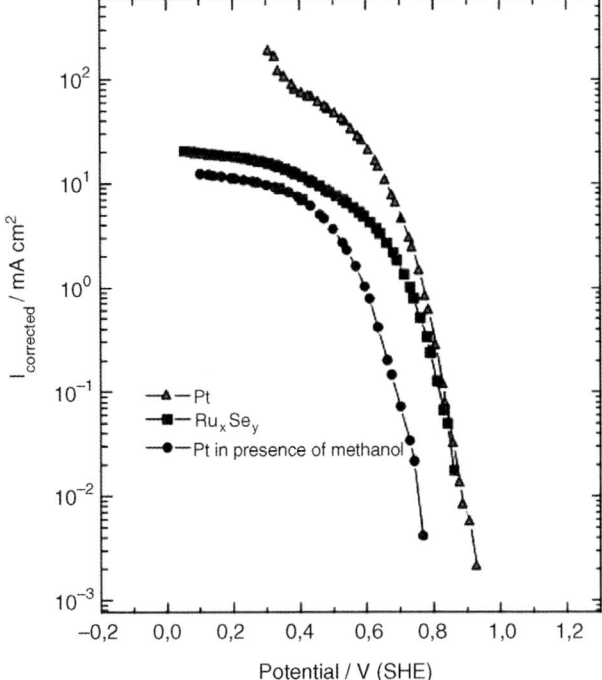

Figure 7.2 Tafel plot of the electroreduction of oxygen in sulfuric acid on a Pt and Ru_xSe_y electrodes in the presence or not of methanol: (▲) Pt; (■) Ru_xSe_y; (●) Pt in presence of methanol.

Other kinds of electrocatalysts could be used for oxygen electroreduction, such as transition metal chalcogenides [21], which seem to be more stable at higher temperature. The comparative behavior of Ru_xSe_y and Pt, using a rotating disk electrode (RDE), for the electroreduction of oxygen, is shown in Figure 7.2. Even if the performances of the chalcogenide compound are slightly weaker than those of Pt electrocatalysts, the most important difference is the behavior of these electrodes in the presence and absence of methanol, which is added to the electrolytic solution to simulate a crossover in the fuel cell. Under these conditions, the electroactivity of Ru_xSe_y is not changed in the presence of methanol, in contrast to the case of Pt, for which a potential shift of 120 to 150 mV in the negative direction is clearly visible. Similar behavior was observed when the electrocatalyst material was embedded in a polymeric matrix, such as polyaniline [22]. These results prove that the chalcogenide of Ru is insensitive to the presence of methanol, in contrast to Pt.

Preliminary experiments using Ru_xSe_y catalysts deposited on a carbon powder (Vulcan XC72) were carried out in a DMFC single cell [23]. They showed clearly that the ruthenium selenide catalyst is active enough for the electroreduction of oxygen and is a possible alternative to Pt as cathode in a DMFC (Figure 7.3). The performances achieved are close to those obtained, with a Pt E-Tek cathode. The overall internal resistance of the cell with a Ru_xSe_y electrode seems to be higher than that

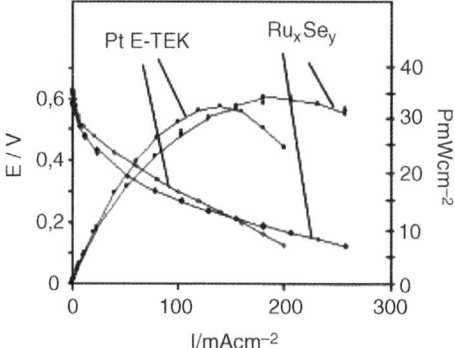

Figure 7.3 Comparison of the behavior of a Ru_xSe_y (2 mg cm^{-2}) cathode and a Pt E-TEK (2 mg cm^{-2}) electrode in a DMFC. $T = 90\,°C$.

with an E-Tek electrode, but it should be noted also that the preparation of fuel cell electrodes with Ru_xSe_y is not yet optimized. The amount of catalyst, the thickness of the active layer, the presence of Nafion, and so on, are parameters which need to be carefully optimized and validated, and it is only after this step that the real behavior of this novel cathode catalyst can be able to replace Pt in a DMFC cathode.

7.3
Experimental Details

7.3.1
Determination of the Methanol Crossover of Proton Exchange Membranes

7.3.1.1 Experimental Procedures

The methanol crossover rate, that is, the membrane permeability to methanol, is one of the key points, which determines DMFC working characteristics: open circuit voltage (OCV) of the cell, fuel utilization, energy and power densities. Therefore the intrinsic permeability coefficient has to be determined in the absence of electrical current, either at the OCV, or in a separate experiment with no electro-osmotic drag.

A two-compartment glass cell, separated by the investigated membrane, was used for methanol permeability measurements. The first compartment (of volume V = 40 mL) was filled with a solution of 1 M methanol prepared with ultrapure water (Millipore MilliQ of resistivity 18 MΩ cm) and the second one (of the same volume V) contained ultrapure water. The membrane (diameter 2 cm, surface area 3.14 cm^2) was clamped between the two compartments, and the liquid tightness was ensured by two Teflon gaskets (Figure 7.4).

Solution samples were withdrawn from the second compartment and analyzed by liquid chromatography (HPLC) using a "Phenomenex" column to determine the methanol concentration as a function of time.

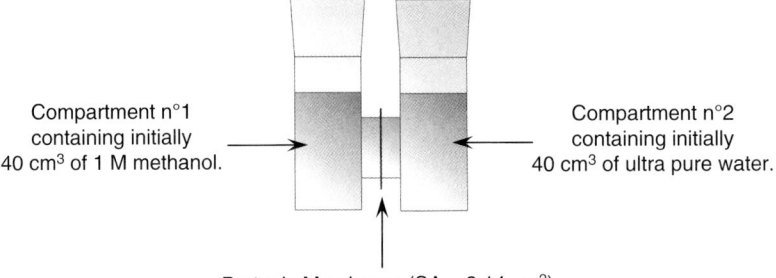

Figure 7.4 Two-compartment glass cell used for measurement of the membrane permeability.

The methanol flow, J, across the membrane, resulting from the concentration difference between the two compartments, is defined as the number of moles of methanol, dN, crossing the membrane (of surface area S) per unit time dt, that is:

$$J = \frac{1}{S}\frac{dN}{dt} \quad (7.10)$$

This flow results from methanol diffusion due to a concentration gradient and can be expressed by the first Fick's law:

$$J = -D\left(\frac{\Delta c}{\Delta x}\right) \quad (7.11)$$

where Δx is the membrane thickness ($\Delta x = e$), D the methanol coefficient diffusion through the membrane and $\Delta c = c_0 - c$, the concentration difference between the two compartments. Since the concentration c in the second compartment is very small compared to that in the first one c_0 ($c < c_0$), one may neglect it and write $\Delta c \approx c_0$, so that:

$$J = D\frac{c_0}{e} = \frac{1}{S}\frac{dN}{dt} = \frac{V}{S}\frac{dc}{dt} \quad (7.12)$$

$$\text{leading to} \quad \frac{dc}{dt} = \frac{SD}{Ve}c_0 \quad (7.13)$$

Assuming D constant inside the membrane, this equation gives, by integration, a linear relationship between the concentration in the second compartment and time, that is:

$$c(t) = \frac{S}{V}\frac{D}{e}c_0(t-t_0) \quad (7.14)$$

where t_0 is a time delay related to diffusion of methanol through the membrane ($t_0 = e^2/6D$) [24, 25].

The methanol concentration in the second compartment is thus a linear function of time, the slope of which allows us to calculate the permeability coefficient of the membrane defined as the number of moles crossing over the membrane of surface S per unit time, that is, the methanol flow.

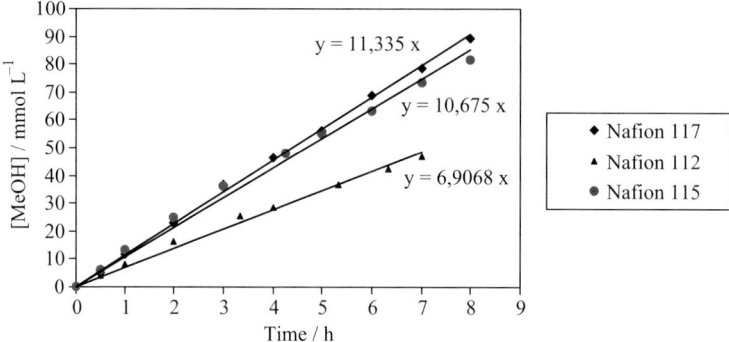

Figure 7.5 Methanol permeability vs. time for different Nafion membranes. MeOH = 1.0 M, normalized thickness of 100 μm, $T = 25\,°C$.

The methanol concentration is measured at different time by liquid chromatography and the curves $c(t)$ are drawn as a function of time up to 8 h.

7.3.1.2 Results

The permeability measurements were carried out with the 3 usual Nafion membranes used in PEMFC (N117, N115 and N112) in order to check the measurement procedure. Because the thickness of these membranes is different the methanol flow is normalized with a thickness of 100 μm, in order to obtain an intrinsic property of the material.

The results give a linear relationship between the concentration c and time t (Figure 7.5), from which the slope is evaluated (linear fitting) and the methanol permeability is calculated (Table 7.1).

In a research project from the European Union [23], the methanol crossover of several high temperature membranes prepared by the Solvay Company was evaluated. These new membranes, called NEM membranes, were based on the radiochemical grafting on a fluorinated film of a barrier polymer (to methanol) and a cross-linking agent, followed by a sulfonation step in order to obtain a good protonic conductivity.

Therefore, we measured the crossover of methanol through about 20 new membranes (NEM series) prepared by Solvay. The amount of methanol crossing

Table 7.1 Evaluation of the methanol flow and permeability for different Nafion membranes.

Membrane	e/μm	$dc/dt/10^{-6}$ $mol\,L^{-1}\,s^{-1}$	$J/10^{-8}$ $mol\,s^{-1}\,cm^{-2}$	$J_{norm}{}^{a}/10^{-7}$ $mol\,s^{-1}\,cm^{-2}$	$D/10^{-7}\,cm^{2}\,s^{-1}$
N 117	175	6.48	7.31	1.28	12.8
N 115	125	10.2	9.52	1.19	11.9
N 112	50	11.1	15.6	0.78	15.6

^{a}The normalized flow is calculated with a 100 μm membrane thickness and 1 M MeOH.

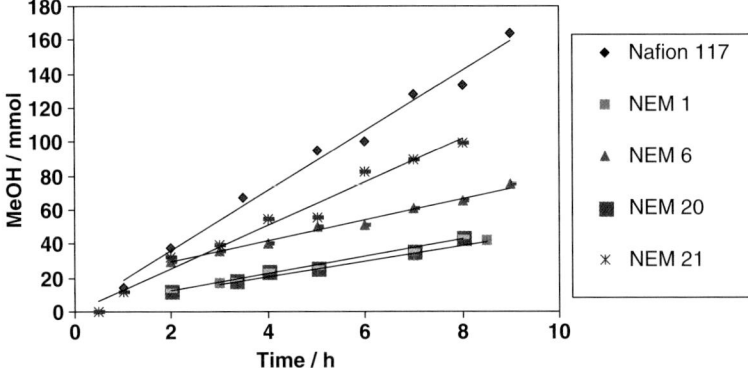

Figure 7.6 Permeability to methanol of different Solvay membranes in comparison with Nafion 117 at room temperature.

the membranes was analyzed quantitatively by liquid chromatography and followed during several hours. The curves obtained are again quasi-linear allowing evaluating the methanol permeability (Figure 7.6).

It appears clearly that all the different NEM membranes prepared by Solvay, have a lower permeability to methanol than Nafion117. Even if there are some differences observed between the NEM membranes, their permeability determined at room temperature are in the same order of magnitude.

The amount of methanol crossing over the membrane varies linearly with time, and the rate of this crossover can be estimated from the slope of the curves (Table 7.2). The lowest permeability measured was that of NEM14.

In another research project from the French network on Fuel Cells [26] we measured the permeability to methanol of several membranes of sulfonated polyetheretherketones (S-PEEK) modified either by SiO_2 (FC series) or by ZrO_2 (S-PEEK-ZrP series). These membranes were prepared by a CNRS Laboratory, the LAMMI, at the University of Montpellier [27].

Table 7.2 Permeability of different Solvay membranes compared with that of Nafion 117 at 25 °C.

Membrane	Permeability to methanol/mol mn^{-1} cm^{-2}	Membrane	Permeability to methanol/mol mn^{-1} cm^{-2}
Nafion 117	9.3×10^{-6}	TGD	2.8×10^{-6}
NEM 1	2.5×10^{-6}	10 205	2.9×10^{-6}
NEM 1A	3.6×10^{-6}	10 220	2.7×10^{-6}
NEM 2	3.0×10^{-6}	NEM 13	3.8×10^{-6}
NEM 3	3.9×10^{-6}	NEM 14	1.9×10^{-6}
NEM 6	4.6×10^{-6}	NEM 20	2.9×10^{-6}
TGA	4.3×10^{-6}	NEM 21	6.8×10^{-6}
TGB	3.2×10^{-6}	NEM 21c	3.9×10^{-6}
TGC	2.6×10^{-6}	NEM 21b2	7.2×10^{-6}

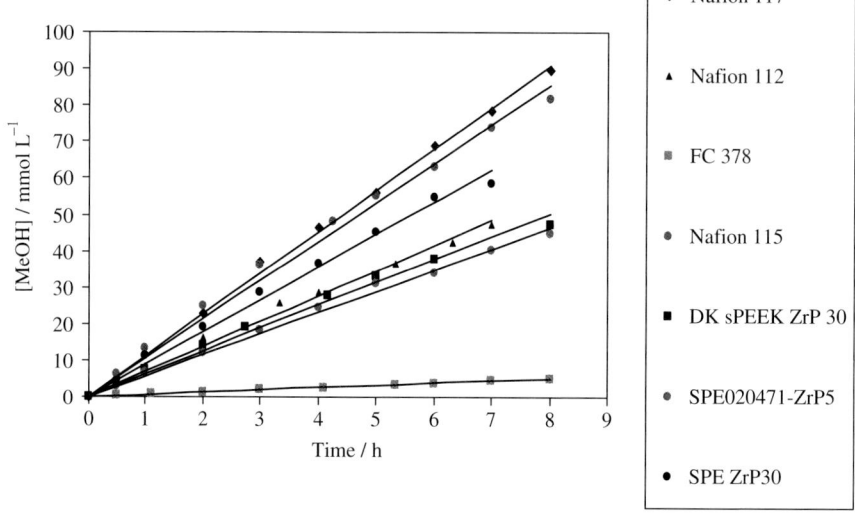

Figure 7.7 Methanol permeability of several membranes prepared by the LAMMI (MeOH = 1.0 M, $T = 25\,°C$, membrane thickness normalized to 100 μm).

The results obtained (Figure 7.7) for all the membranes investigated (FC 378, DK S-PEEK-ZrP 30, SPE ZrP 30, SPE020471-ZrP 5, S-PEEK-ZrP 10 et S-PEEK-ZrP 20) again show a linear behavior with time of the methanol concentration in the second compartment. This allowed us to evaluate their methanol permeability (Table 7.3).

The FC 378 (S-PEEK, 10% SiO_2) membrane is the less permeable to methanol (at least by a factor of 5 – see Table 7.3), but the results obtained in a DMFC are

Table 7.3 Methanol permeability of different LAMMI membranes comparatively to that of Nafion 117 (MeOH = 1,0 M, normalized membrane thickness = 100 μm, $T = 25\,°C$).

Membrane	Measured permeability (1 M MeOH)/mol sec^{-1} cm^{-2}	Normalized permeability (100 μm, 1 M MeOH)/mol sec^{-1} cm^{-2}
Nafion 117	7.2×10^{-8}	1.28×10^{-7}
Nafion 115	9.34×10^{-8}	1.19×10^{-7}
Nafion 112	1.53×10^{-7}	7.80×10^{-8}
FC 378	1.43×10^{-8}	7.15×10^{-9}
DK s-PEEK-ZrP 30	1.40×10^{-7}	7.00×10^{-8}
SPE 020471 ZrP 5	1.29×10^{-7}	6.43×10^{-8}
SPE ZrP 30	1.98×10^{-7}	9.88×10^{-8}
S-PEEK ZrP 10	1.06×10^{-8}	6.37×10^{-9}
S-PEEK ZrP 20	4.25×10^{-8}	2.55×10^{-8}

rather poor (about 30 mW cm^{-2} at 90 °C against 110 mW cm^{-2} at 120 °C for S-PEEK ZrP 20).

7.3.2
Electrochemical Measurements (Voltammetry, RDE, RRDE, etc.)

7.3.2.1 Experimental Set-Up

Cyclic voltammetry and chronoamperometry experiments were carried out at 20 °C with classical equipment consisting of a POS 73 Wenking Potentioscan (including a triangular wave generator) and a SE790 ABB X-Y(t) recorder. A 3-electrode electrochemical cell was used for all experiments with a working electrode disc, a reversible hydrogen electrode (RHE) as reference electrode and a vitreous carbon plate as counter electrode.

For RDE experiments the electrochemical set-up consists of a Voltalab PGZ 402 potentiostat controlled by a computer, a Radiometer Speed Control Unit CTV 101 and a (RDE) Radiometer BM-EDI 101. The working electrode surface area was 0.196 cm^2.

The rotating ring disc electrode (RRDE) experiments were performed using a Pine Instrument Company AFMSRX modulator speed rotator mounted with an AFDT22 electrode including a 0.152 cm^2 Pt ring and a 0.126 cm^2 GC disc with a collection efficiency N = 21%. The counter electrode was a vitreous carbon plate and a RHE was used as the reference electrode. The working electrodes were prepared in the same way as for the RDE experiments. Two Voltalab PGZ–402 potentiostats were connected as a bipotentiostat to control the ring and the disc potentials and to collect the respective currents. The disc potential was swept at 2 mV s^{-1} from 0.9 V to 0.3 V, the ring potential was held at 1200 mV vs. RHE and the rotation rate of the electrode was fixed to 2500 rpm.

The reduction current of oxygen on the different electrocatalysts in an oxygen saturated 0.5 M H_2SO_4 solution was recorded at 293 K under quasi-stationary conditions (voltammetric sweep at 2 mV s^{-1} between 0.9 and 0.0 V vs. RHE) for various rotation speeds Ω (from 0 to 3600 rpm) of the disc electrode, and corrected from the background current (recorded under a pure nitrogen - U quality from L'Air Liquide - atmosphere). The real surface area, A_r, of the dispersed electrode was calculated either from the hydrogen adsorption-desorption region in a cyclic voltammogram recorded on the Pt electrocatalyst under a nitrogen atmosphere, or from the quantity of electricity involved in the oxidation of a monolayer of preadsorbed CO; the latter method was particularly useful for the low Pt loadings when the hydrogen adsorption region is hardly visible in the voltammograms. To perform these calculations the quantity of electricity involved in the voltammetric peaks was calculated by integration of the j vs. E (i.e., vs. time) curves

7.3.2.2 Analysis of the Data

A more general analysis of the data was carried out by separating the contribution of the diffusion of molecular dioxygen from that of the surface processes involved in the oxygen reduction reaction using the $1/j$ vs. $1/\sqrt{\Omega}$ Koutecky–Levich plots. For that purpose a more detailed mechanism of the ORR, involving the formation of H_2O_2, can be written as follows [2]:

$$O_{2sol} \rightarrow O_{2surf} \quad \text{diffusion in the bulk electrolyte (diffusion coefficient D)} \tag{7.15a}$$

$$O_{2surf} \rightarrow O_{2elec} \quad \text{diffusion in the catalyst layer film} \quad \text{(limiting current density } j_l^{film}) \tag{7.15b}$$

$$O_{2elec} \rightarrow O_{2ads} \quad \text{adsorption process (limiting current density } j_l^{ads}) \tag{7.16}$$

$$O_{2ads} + H^+ + e^- \rightarrow HO_{2ads} \quad \text{electron transfer rds} \quad \text{(exchange current } j_o, \text{ Tafel slope } b) \tag{7.17}$$

$$HO_{2ads} + H^+ + e^- \rightarrow [H_2O_2]_{ads} \rightarrow H_2O_2 \quad \text{hydrogen peroxide formation or} \tag{7.18a}$$

$$HO_{2ads} \rightarrow O_{ads} + OH_{ads} \quad \text{oxygen bond splitting} \tag{7.18b}$$

$$O_{ads} + H^+ + e^- \rightarrow OH_{ads} \tag{7.19}$$

$$2(OH_{ads} + H^+ + e^-) \rightarrow 2H_2O \tag{7.20}$$

Assuming reaction (7.17) to be the r.d.s. one can express the equation of the $j(E)$ curves as follows [28, 29]:

$$\frac{1}{|j|} = \frac{1}{j_l^{diff}} + \frac{1}{j_l^{film}} + \frac{1}{j_l^{ads}} + \frac{1}{j_o(\theta/\theta_e)10^{(|\eta|/b)}} \tag{7.21a}$$

with $\eta = E - E_{eq}$, the overvoltage and θ, θ_e the degree of coverage of the Pt surface by oxygen containing adsorbed species at potential E and at the equilibrium potential E_{eq}, respectively. Equation 7.21a assumes that the electron transfer step is first order in an oxygen adsorbed species, and that the backward reaction, that is oxygen evolution, is quite negligible. Furthermore the adsorption process (7.16) being more rapid than the electron transfer step (7.17), one may assume that $\theta \approx \theta_e$ at all electrode potentials.

Because of the porous structure of the dispersed electrode, a resistance can exist due to oxygen diffusion in the catalytic film. The corresponding diffusion limiting current density j_l^{film} is independent of the rotation rate of the electrode and its contribution cannot be separated from the whole process, as is the case for the adsorption limiting current density of molecular oxygen on the catalyst. In fact, j_l is a limiting current density resulting from a mixed control by the diffusion of molecular oxygen inside the catalyst film and by adsorption of dioxygen on the catalytic sites [30].

The diffusion limiting current density, j_l^{diff}, of molecular oxygen in the bulk electrolyte is controlled by the rotation speed Ω, whereas the diffusion limiting current density inside the catalyst layer, j_l^{film}, is independent of Ω. Levich's law gives the diffusion limiting current density in the electrolyte:

$$j_l^{diff} = 0.20 n_t F D^{2/3} \nu^{-1/6} c_o \sqrt{\Omega} = n_t B \sqrt{\Omega} \tag{7.22}$$

where n_t is the total number of exchanged electrons and Ω the rotation speed expressed in revolution per minute (r.p.m.) [31].

The factor B can be calculated at 20 °C assuming $D = 2.1 \times 10^{-5}$ cm^2 s^{-1} for the diffusion coefficient of molecular oxygen, $\nu = 1.075\ 10^{-2}$ cm^2 s^{-1} for the kinematics viscosity of 0.5 M H$_2$SO$_4$, and $c_o = 1.03\ 10^{-3}$ mol dm^{-3} for the bulk concentration of oxygen in a saturated electrolytic 0.5 M H$_2$SO$_4$ solution [30]. This gives $B = 0.0322$ mA cm^{-2} (rpm)$^{-1/2}$, when Ω is expressed in rpm.

Equation (7.21a) then becomes:

$$\frac{1}{|j|} = \frac{1}{n_t B \sqrt{\Omega}} + \frac{1}{j_l} + \frac{1}{j_o 10^{(|\eta|/b)}} \quad \text{with} \quad \frac{1}{j_l} = \frac{1}{j_l^{\text{film}}} + \frac{1}{j_l^{\text{ads}}} \quad (7.21b)$$

j_l is a limiting current density resulting from a mixed control by the diffusion of molecular oxygen inside the catalytic layer, and by adsorption of dioxygen at the catalyst surface. Since these current densities are both independent of the rotation speed Ω and of the electrode potential E, it is not possible to separate their contribution, so that the analysis of data will only give j_l.

Equation 7.21b shows that the plot of the inverse of the current density, corrected from the background current (in the absence of dioxygen), is a linear function of $1/\sqrt{\Omega}$ (Koutecky-Levich plots). The slope of the straight line gives the total number of exchanged electrons, n_t, and the intercept at the origin ($\Omega \to \infty$) gives the inverse of the kinetic current j_k, corrected from the diffusion current, as a function of the overvoltage, that is:

$$\frac{1}{|jk|} = \frac{1}{j_l} + \frac{1}{j_o 10^{(|\eta|/b)}} \quad (7.23)$$

Equation 7.23 thus shows that for high overvoltages ($|\eta| \to \infty$) the quantity j_k^{-1} tends towards j_l^{-1}, so that one can obtain the limiting current density j_l by extrapolating Equation 7.23 at high $|\eta|$. This allowed us to transform Equation 7.23 as follows:

$$\eta = E - E_{eq} = -b\left[\text{Log}\frac{j_l}{j_o} + \text{Log}\left|\frac{j_k}{j_l - j_k}\right|\right] \quad (7.24)$$

showing that a plot of the overpotential η, or the electrode potential E, vs. Log $|(j_k/(j_l - j_k)|$ is a straight line, the slope of which giving b, and the intercept at the origin of which, that is at the equilibrium electrode potential ($E_{eq} = 1.184$ V/RHE, calculated with $c_0 = 1.03 \times 10^{-3}$ M), then giving $[-\text{Log}(j_l/j_o)]$. We may thus evaluate j_o knowing j_l.

Knowing j_k and j_l, one can access to the Tafel slope b and to the exchange current density j_0. The evaluation of the kinetic parameters using this method gives: $n_t = 3.9$–4.0, $b = -124$ mV/decade, $j_0 = -6.5\ 10^{-5}$ mA cm^{-2}, and $j_l = 18.7$ mA cm^{-2}. These results are similar to those obtained by Coutanceau et al. [32] for a Pt loading close to 155 µg cm^{-2}.

The total number of electrons involved during the ORR on FePc can also be evaluated using a RRDE at 2500 rpm rotation rate. Here, it was assumed that the potential scan rate was slow enough and the H$_2$O$_2$ diffusion in the catalytic film was fast enough to obtain a quasi steady state of diffusion. But, a RRDE study realized by E. Claude et al. [33] on oxygen reduction at CoTMPP dispersed on carbon shown that, knowing the collection factor N, this technique was convenient to distinguish and evaluate the two mechanistic pathways for oxygen electroreduction (via a 2- or a 4- electron process). From RRDE experiments, the water formation efficiency $x(H_2O)$ can be calculated as follows [34, 35]:

$$x(H_2O) = \frac{1 - \frac{I_R}{N|I_D|}}{1 + \frac{I_R}{N|I_D|}} \qquad (7.25)$$

where $N = I_R/|I_D|$ is the collection efficiency, I_D the disc current, I_R the ring current.

Values of $x(H_2O)$ lead to the number of exchanged electrons during the oxygen reduction (n_t) using the following equation:

$$n_t = 4x(H_2O) + 2[1 - x(H_2O)] \qquad (7.26)$$

7.4
Synthesis and Characterizations of Nanostructured Catalysts for the ORR

7.4.1
Platinum-Based Catalysts and Electrodes

Platinum is a metal which displays the highest catalytic activity for both methanol oxidation and oxygen reduction reactions. It is then necessary to develop novel Pt-based electrocatalysts that can catalyze the oxygen reduction reaction but are relatively inactive for the oxidation of methanol. For example, studies have shown that carbon-supported Pt–Cr alloy catalyst [36, 37] and bulk $Pt_{70}Ni_{30}$ alloy catalyst [38] exhibited a high methanol tolerance during oxygen reduction in comparison with pure Pt. Many synthesis routes were then developed for the preparation of pure and bimetallic Pt-based catalysts: sputtering methods for the preparation of Pt-Fe, Pt-Ni and Pt-Co alloys [39, 40], modification of commercial Pt/C by transition metal salts of Cr, Mn, Fe, Co, Ni, then reduction of the salt and heat treatment at high temperatures (temperature range from 700 °C to 900 °C) [41–43], polyol process [44], and so on. In a recent review paper Antolini described some of the different methods to synthesize supported Pt-M alloys for low temperature fuel cell application [45]. However, challenges for the preparation of such Pt alloy catalysts with high metal loading include the need for synthesis procedures leading to catalysts with the desirable composition, controlled nanoparticle size and a narrow size distribution even with high metal loadings.

To meet these requirements, the colloidal routes [46–50] and catalyst syntheses through steps involving carbonyl complexes have been shown to provide nanosized Pt-based alloy catalysts with a narrow size distribution [36, 37, 51–54]. We will then focus here on catalyst preparations using the carbonyl complex route and the so-called Bönnemann method.

7.4.1.1 Synthesis of Platinum-Based Catalysts by the Carbonyl Complex Route

Pt–Ni [51] and Pt–Cr [36] alloy nanoparticle catalysts with high metal loadings (20 to 40 wt.%) can be prepared via a carbonyl complex route [55], followed by H_2 reduction at high temperatures. Pt and Ni or Cr carbonyl complexes are synthesized simultaneously using methanol as the solvent through the reaction of Pt and Ni or Cr salts with CO at about 50 °C for 24 h with constant mechanical stirring. The resulting carbonyl complexes are a mixture of Pt and Ni or Cr carbonyl complexes or a Pt–X molecular complex, identified by their infrared spectra [51]. After the synthesis of Pt–X carbonyl complexes, Vulcan XC-72 carbon was added to the mixture under a N_2 gas flow and stirred at about 55 °C for more than 6 h. Subsequently, the solvent was removed and the catalyst powder was subjected to heat treatment at different temperatures under nitrogen and hydrogen, respectively. In the case of Pt-Ni, an alloying temperature of about 300 °C, significantly lower than that given by other authors [45, 56], was found suitable: it is low enough to be beneficial to the formation of small alloy particles with a narrow particle size distribution and high enough to avoid the formation of a mixture of pure Pt phase and an alloy phase, as showed by XRD measurements for $T = 200$ °C. After heat treatment, the sample was washed with ultrapure water until no chlorine ions were detected and then dried under nitrogen at about 130 °C.

7.4.1.2 Synthesis of Platinum-Based Catalysts by the Colloidal Route

The procedure described by Bönnemann and co-workers [57, 58] was slightly modified and adapted for the preparation of mono- and bimetallic catalyst precursors. The synthesis is carried out under controlled atmosphere (argon) free of oxygen and water, with non-hydrated metal salts (99.9% $PtCl_2$, $FeCl_3$, $NiCl_2$, $CrCl_3$ or $CoCl_2$ from Alfa Aesar). The first step consists in the preparation of a tetraalkyltriethylborohydride reducing agent $(Nalk_4)^+ (Bet_3H)^-$, which will also act as a surfactant after the reduction of the metal salt [46, 48], preventing any agglomeration of the metallic particles:

$$MCl_n + n[(Nalk_4)^+(Bet_3H)^-] \rightarrow M[(Nalk_4)^+Cl^-]_n + nBEt_3 + n/2\, H_2 \tag{7.27}$$

The colloidal precursors are then dispersed on a carbon support (Vulcan XC 72) and calcined at 300 °C for 1 h under air atmosphere to remove the organic surfactant. The Bönnemann method allows different possibilities for the synthesis of multimetallic supported catalysts [59]:

i) Synthesis of catalysts with a controlled atomic ratio from a given mixture of the different metal salts before the reduction step and formation of the precursor colloid:

$$xPtCl_2 + yMCl_n + (2x+ny)[(Nalk_4)^+(Bet_3H)^-]$$
$$\rightarrow Pt_xM_y[(Nalk_4)^+Cl^-]_{(2x+ny)} + (2x+ny)BEt_3 + (2x+ny)/2\, H_2 \tag{7.28}$$

ii) Synthesis of catalysts with a controlled atomic ratio by co-deposition of different metal colloids before the calcination step and formation of the catalytic powder:

$$xPt[(Nalk_4)^+Cl^-]_2 + yM[(Nalk_4)^+Cl^-]_n \xrightarrow{THF, 300\,°C, air, C} Pt_x + M_y/C \quad (7.29)$$

7.4.1.3 Physicochemical Characterizations of the Catalysts

Because the reactions involved in a DMFC are structure sensitive, it is very important to characterize the composition and the structure of the Pt-based catalysts. The main characterization techniques used for this purpose are TEM, XRD, EDX, ICP-OES and XPS. The first four methods give only the bulk composition of the catalysts, whereas the last permits estimation of the surface composition. Even if XPS may provide a better knowledge of the surface composition of as-prepared Pt-based catalysts [43], it was clearly shown by Watanabe and coworkers, using XPS measurements, that the surface composition of the PtFe and PtCo catalysts was different before and after the electrochemical experiments [39, 60].

XRD is a powerful method to characterize the structure, bulk composition and crystallite size of a supported Pt-based catalyst. For example, Figure 7.8 shows the XRD patterns of the carbon-supported Pt–Ni alloy catalysts prepared by the carbonyl complex route with a metal loading of 40 wt.% and different Pt/Ni atomic ratios. The first peak located at about 24.8° in all the XRD patterns is associated with the Vulcan XC-72 carbon support. The other four peaks located at 2θ values of about 39.8°, 46.5°, 67.8° and 81.2° are characteristic of face-centered-cubic (fcc) crystalline Pt, and correspond to (111), (200), (220) and (311) crystallographic planes, respectively. The diffraction peaks for the Pt–Ni alloy catalysts are shifted slightly to higher 2θ values. The higher angle shifts of the Pt diffraction peaks reveal the formation of an alloy involving the incorporation of Ni into the fcc structure of Pt.

The broad diffraction peak suggests that the as-prepared Pt–Ni alloys exist in small particle sizes with a relatively narrow particle size distribution and in a disordered form. No peak for pure Ni or its oxides was found. The practical composition of the three Pt–Ni alloy catalysts with 40 wt.% loading was estimated using Vegard's law behavior for solid solution [51] and was found to be very close to the nominal value,

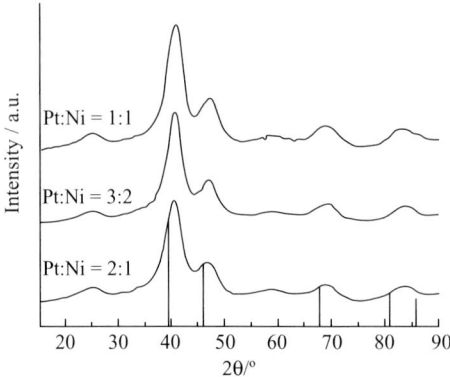

Figure 7.8 X-ray diffraction patterns of carbon-supported 40 wt.% Pt–Ni alloy catalysts.

Table 7.4 Structural parameters of the Pt-Ni/C alloy catalysts (40 wt.%).

Catalyst	Pt true surface area (m^2 g^{-1})	Lattice parameter (nm)	Mean crystallite size (nm)	Estimated Pt/Ni atomic ratio (%)
Pt:Ni (2:1)	44.3	0.3837	3.1	67.1:32.9
Pt:Ni (3:2)	38.1	0.3821	3.2	61.0:39.0
Pt:Ni (1:1)	36.4	0.3804	3.0	54.6:45.4

confirming that Ni is completely alloyed with Pt within all the as-prepared Pt–Ni catalysts. The average size of the Pt–Ni alloy nanoparticles was estimated by using Scherrer's equation [61]:

$$d = 0.94 \lambda_{K\alpha 1} / B_{(2\theta)} \cos \theta_B \tag{7.30}$$

where d is the average crystallite diameter, $\lambda_K \alpha_1$ is the wavelength of the X-ray radiation (0.154056 nm), θ_B is the angle of the (220) crystallographic plane, and $B_{(2\theta)}$ is the width in radians of the diffraction peak at half height. The average crystallite sizes obtained for all the catalysts are given in Table 7.4.

Similar results were obtained by Yang et al. [36] for Pt-Cr alloys prepared by the carbonyl route. In this case XRD results were correlated with EDX measurements in order to evaluate the bulk composition of the catalysts. Correlation between particle size of the catalysts as determined by TEM and results of CO stripping, assuming that CO does not adsorb on Cr, lead also to evaluate the surface composition. Results are given in Table 7.5.

In the case of the Bönnemann method, considering $Pt_{(1-x)}$-Cr_x catalysts as examples, the physical characteristics are given in Table 7.6. It can be seen that the coreduction method leads to the formation of alloyed Pt-Cr catalysts, as evidenced by the variation in the lattice parameter, whereas, the codeposition method leads to the formation of non-alloyed bimetallic Pt-Cr catalysts. The atomic ratio is determined by EDX, whereas volume weighted particle size and crystallite size are determined by TEM and XRD, respectively (see Section 7.4.3.2).

7.4.1.4 Electrochemical Characterization of the Catalysts

Many investigations have shown that some Pt-based alloy catalysts, such as Pt–M (where M = Co, Ni, Fe or Cr) [36–38, 40, 42, 51, 62–72], exhibited an enhanced electrocatalytic activity for the ORR with respect to Pt alone. Comparison of the activity of different $Pt_{0.7}$–$M_{0.3}$ alloys towards the ORR was carried out. The catalysts

Table 7.5 Structural parameters of the Pt-Cr/C alloy catalysts (40 wt.%).

Catalyst	Lattice parameter (nm)	Mean particle size (nm)	EDX Pt/Cr atomic ratio (%)
Pt:Cr (3:1)	0.3895	3.5	77.3:22.7
Pt:Cr (2:1)	0.3863	3.4	62.7:37.3
Pt:Cr (1:1)	0.3817	3.2	49.06:50.94

Table 7.6 EDX and XRD analyses of the different PtCr/C catalysts prepared via the colloidal route (*bimodal particle size distribution).

Catalyst sample	EDX analyses of Pt-Cr catalysts	D_v TEM/Å	Cell parameter /nm	EDX analyses of Pt+Cr catalysts	D_v TEM/Å	D_v XRD/Å	Cell parameter /Å
Pt/C	—	4.4	—	—	4.4	5.9	
$Pt_{0.9}Cr_{0.1}$/XC72	91.9 : 8.1	4.1	0.3916	86.6 : 13.4	5.1	3.3	3.918
$Pt_{0.8}Cr_{0.2}$/XC72	73.5 : 26.5	4.3	0.3898	80.3 : 19.7	5.5	4.5	3.917
$Pt_{0.7}Cr_{0.3}$/XC72	69.4 : 30.6	4.3	0.3890	68.1 : 31.9	5.6	4.5	3.918
$Pt_{0.6}Cr_{0.4}$/XC72	61.3 : 38.7	4.5	0.3878	—	—	—	—

were prepared by coreduction of the metallic salts using the Bönnemann method. Porous electrodes were obtained by mixing ultrasonically ten milligram of catalyst, 0.5 mL of Nafion solution (5 wt.%, Aldrich) and 2.5 mL of water. A measured volume (3 μL) of this ink was transferred via a syringe onto a freshly polished glassy carbon disk (3 mm in diameter). After the solvents were evaporated overnight at room temperature, the prepared electrode served as the working electrode. Each electrode contained about 56 μg cm^{-2} of metal. Figure 7.9 displays the polarization curves for the ORR recorded in a half-cell configuration by linear sweep voltammetry. It appears that in the high cathode potential region (which corresponds to the potential of a working cathode in a Fuel Cell), the catalyst activity, as determined by the higher current densities achieved, decreases as follows: $Pt_{0.7}Cr_{0.3} > Pt_{0.7}Co_{0.3} > Pt_{0.7}Fe_{0.3} >$

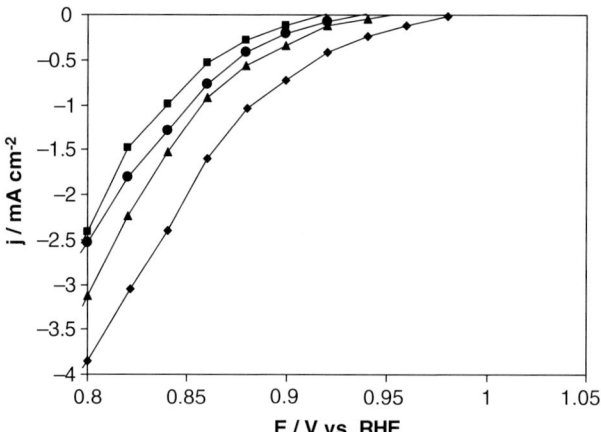

Figure 7.9 Voltammograms of (♦) $Pt_{0.7}Cr_{0.3}$/C, (▲) $Pt_{0.7}Co_{0.3}$/C, (●) $Pt_{0.7}Fe_{0.3}$/C and (■) $Pt_{0.7}Ni_{0.3}$/C catalysts, obtained by coreduction via the Bönnemann method, recorded in an O_2 saturated 0.5 M H_2SO_4 electrolyte ($T = 20$ °C, $v = 5$ mV/s, $\Omega = 2500$ rpm).

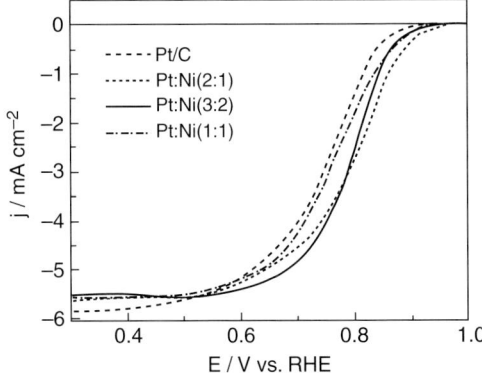

Figure 7.10 Voltammograms of a commercial E-Tek Pt/C and nanosized Pt–Ni alloy catalysts prepared via the carbonyl route, recorded in O_2 saturated 0.5 M H_2SO_4 electrolyte ($T = 20\,°C$, $v = 5\,mV\,s^{-1}$, $\Omega = 2000\,rpm$).

$Pt_{0.7}Ni_{0.3}$. The $Pt_{0.7}Cr_{0.3}$ allows shifting the onset potential of the reduction wave of about 0.1 V towards higher potentials, which means a corresponding decrease of the ORR overpotential.

The polarization curves recorded for different Pt-Ni/C catalysts prepared by the carbonyl route are also compared with that recorded for a Pt/C catalyst in acid medium (Figure 7.10) [73]. The ORR on all the catalysts is diffusion-controlled when the potential is less than 0.6 V versus RHE and is under mixed diffusion-kinetic control in the potential region between 0.6 and 0.85 V. For potentials higher than 0.6 V, the activities of all the Pt–M alloy catalysts are higher than that of pure Pt.

To explain the activity enhancement of Pt–M catalysts towards the oxygen electroreduction, several ideas were proposed. Jalan and Taylor [74] found that PtCr alloy was the most active catalyst for ORR in phosphoric acid medium and explained that the enhancement of the activity was due to a change in the Pt interatomic spacing. This idea was also developed by Toda et al. [63] for PtNi compounds. These authors also proposed, in agreement with Mukerjee et al. [62], that electronic effect could be involved and could explain the activity enhancement. Mukerjee et al. [75] concluded, on the basis of in situ XANES and EXAFS experiments on the electronic state of Pt, that a correlation between the 5d-band vacancy and the activity for the ORR existed, which is influenced by the alloying agent: the Pt–OH bonds, via electronic interactions, are then affected. On the other hand, Tamizhmani and Capuano [71], on the basis of XRD experiments, proposed that dissolution of the non-noble metal occurs and, as a result, the formation of a Raney type Pt catalyst with high surface roughness takes place, which increases the catalytic activity of Pt towards the ORR. Mukerjee and Srinivasan [76] showed that the dissolution of the non-noble metal (Cr) during long time experiments was lower than 5%. Paulus et al. [65] worked on PtNi and PtCo alloys having a Pt atomic composition of 50 and 75%. These authors explained that several factors may contribute to the activity enhancement towards the ORR but can hardly be separated. The alloying component may change the distribution of islands

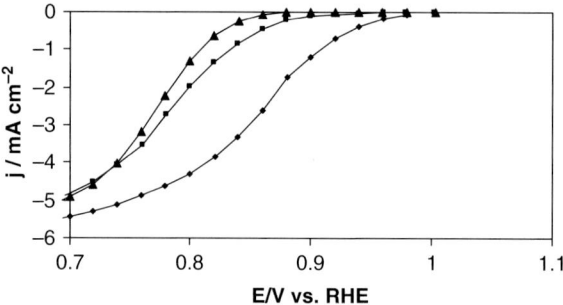

Figure 7.11 Voltammograms of (▲) Pt/C, (■) $Pt_{0.8} + Cr_{0.2}/C$ and (♦) $Pt_{0.8}Cr_{0.2}/C$ catalysts, obtained via the colloidal route, recorded in O_2 saturated 0.5 M H_2SO_4 electrolyte ($T = 20\,°C$, $v = 5$ mV/s, $\Omega = 2500$ rpm).

of a given metal (ensemble effect), it may change the local binding geometry (structure effect) or it may directly modify the activity of platinum surface atoms (electronic effect).

Koffi et al. [48] carried out experiments on non-alloyed $Pt_{(1-x)}$-Cr_x/C catalysts ($x = 0$, 0.1, 0.2 and 0.3) prepared via the Bönnemann method. EDX and XRD confirmed the nominal atomic ratio and the non-alloyed structure of the catalyst, respectively. The dispersed catalysts were deposited on vitreous carbon according to a method developed by Gloaguen et al. [77]. The carbon-supported catalyst powder is added to a mixture of 25 wt.% (based on the powder content) Nafion solution (5 wt.% from Aldrich) and ultrapure water (Millipore MilliQ, 18 MΩ cm). After ultrasonic homogenization of the PtCr/XC72-Nafion ink, a given volume is deposited from a syringe onto a freshly polished glassy carbon substrate yielding a metal loading of 84 µg cm^{-2}. The solvent is then evaporated in a stream of ultrapure nitrogen at room temperature. These catalysts displayed a higher catalytic activity towards ORR in 0.5 M H_2SO_4 medium in the potentials range from 1.0 to 0.75 V versus RHE than the Pt/C catalyst (Figure 7.11). In this case, that is using non-alloyed catalysts, the Pt interatomic spacing is likely not modified by the presence of chromium as demonstrated by the constancy of the XRD peak position that these authors obtained. Thus, electronic effect or interaction between Pt and Cr can be considered. Dissolution of chromium could be involved, but the dissolution of Cr and the formation of a Raney type Pt catalyst could not explain the higher tolerance of the catalyst to the presence of methanol (see Section 7.5.1). Other effects are likely involved, such as changes in the Pt interatomic spacing, since alloyed PtCr catalyst displays higher activity than the non-alloyed one.

7.4.2
Syntheses and Characterization of Transition Metal Macrocycles

The main representative compounds of transition metal macrocycles are phthalocyanines, porphyrins and tetraazaannulenes (Figure 7.12). Transition metal macro-

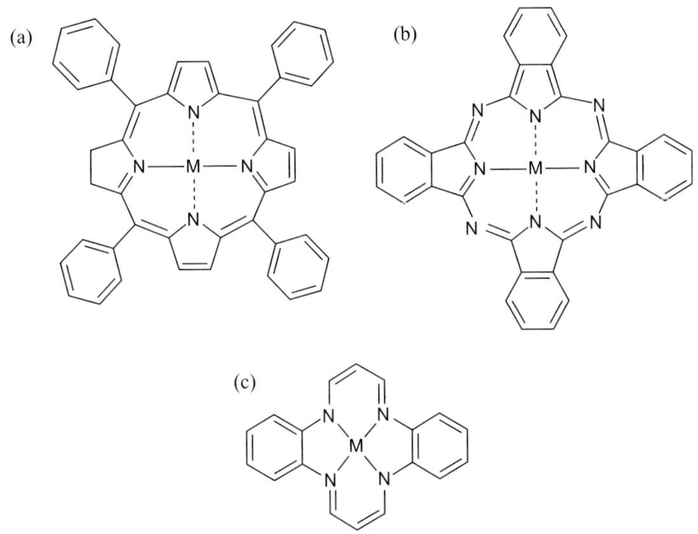

Figure 7.12 Structure of transition metal macrocycles: (a) tetraphenyl-porphyrin (MTPP), (b) phthalocyanine (MPc), (c) tetraazaannulene (MTAA).

cycles (with M = Ru, Pt, Zn, Mn, Co and Fe) are active for the electrocatalytic oxygen reduction reaction [78]. The most active metal centers are cobalt, iron and manganese in acid, neutral and alkaline media [79–81]. However, manganese phthalocyanine, although active, displayed a very high chemical instability [80, 82], so that Co and Fe based macrocycles are the most investigated compounds for the ORR.

7.4.2.1 Syntheses of Transition Metal Phthalocyanines

Metal tetrasulfonated phthalocyanines (Figure 7.13a) are synthesized using the method of Weber and Bush [83] slightly modified. In summary, a solid mixture of metal sulfate or chloride, urea, ammonium chloride, monosodium salt of sulfophthalic acid and ammonium molybdate as catalyst is heated at 180 °C in nitrobenzene for 6 to 8 h. The resulting blue solid reaction product is washed with methanol to remove nitrobenzene, dissolved in water and metal tetrasulfonated phthalocyanine is precipitated with sodium chloride. The last step of purification is performed by dialysis in order to remove sodium chloride. The blue solution of MTsPc in water is then dried using a rotating evaporator. Recrystallization is carried out for 4 h in ethanol at reflux. The final products consist in blue and blue-purple crystals, for FeTsPc and CoTsPc, respectively.

Cobalt 4,4′,4″,4‴-tetraaminophthalocyanine (CoTAPc) (Figure 7.13b) is synthesized using the methods of Achar and coworkers [84, 85]. Cobalt 4,4′,4″,4‴-tetranitrophthalocyanine is first synthesized. Briefly, a solid mixture of cobalt sulfate heptahydrate, nitrophthalic acid, ammonium chloride, excess of urea and ammonium molybdate as catalyst is heated at 185 °C in nitrobenzene for about 5 h. The resulting solid reaction product is washed with methanol to remove nitrobenzene,

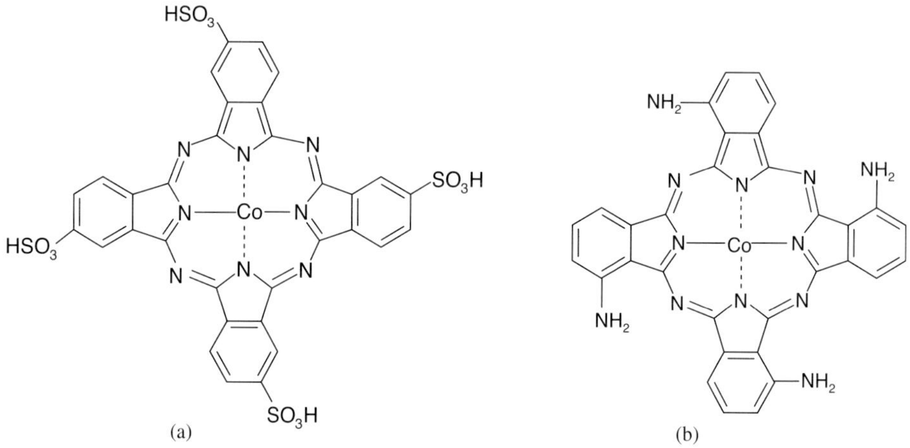

Figure 7.13 Structure of (a) cobalt tetrasulfonated phthalocyanine (CoTsPc), (b) cobalt 4,4′,4″,4‴-tetraaminophthalocyanine (CoTAPc).

dissolved in 1.0 M HCl saturated in NaCl solution, boiled and filtered. The solid is then put in a NaOH + NaCl solution, heated up to 90 °C until ammoniac evolution was finished, and filtered. The $CoT_{NO_2}Pc$ solid product is then treated with 1.0 M HCl solution and separated by centrifugation. At last, it is thoroughly washed by ultrapure water (MilliQ from Millipore, 18 MΩ cm) to remove chlorides, and dried in an oven at 125 °C. The second step consists in the reduction of the nitro groups of the $CoT_{NO_2}Pc$ compounds into amino groups by dissolving it in water in the presence of NaS as reducing agent at 50 °C under stirring for 5 h. The solid compound is then separated by centrifugation, treated by 1.0 M NaOH and separated again by centrifugation. The dark green solid product (CoTAPc in Figure 7.13b) is then washed with ultrapure water to remove NaOH and NaCl, and dried under vacuum.

7.4.2.2 Syntheses of Transition Metal Porphyrins

Metal-free tetra(o-aminophenyl)porphyrin (H_2TAPP) is synthesized using the method of Collman et al. [86] and then metalized with cobalt or iron to give MTAPP adapting the procedure of Adler et al. [87]. First the tetra(o-nitrophenyl)porphyrin ($H_2T_{NO_2}PP$) is synthesized by slowly mixing freshly distilled pyrrole with ortho-nitrobenzaldehyde in a glacial acetic acid solution at reflux. As soon as the solution becomes dark, chloroform is added and the solution is cooled in a chilled bath down to 35 °C. Purple crystals are obtained after filtration and washed with chloroform. $H_2T_{NO_2}PP$ is dried at 100 °C for several hours in an oven. $H_2T_{NO_2}PP$ is dissolved in concentrated HCl solution and an excess of $SnCl_2, 2H_2O$ is added as reducing agent. The resulting green mixture is rapidly heated up to 65–70 °C for 25 min, and then neutralized by adding concentrated NH_4OH solution. Chloroform is added and the mixture is kept under stirring for one hour before separating two liquid phases. The organic phase is recovered and filtered. The filtrate volume is reduced in a rotating evaporator, washed with diluted NH_4OH in water and dried on anhydrous sodium

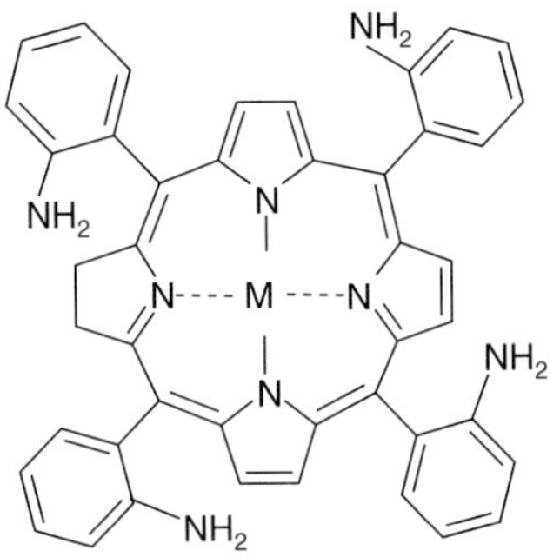

Figure 7.14 Structure of metal tetra(o-aminophenyl)porphyrin (MTAPP).

sulfate. Recrystallization is performed in an ethanol + heptene solution. After evaporation of the solvent dark purple crystals of H_2TAPP are recovered and washed with methanol. At last, H_2TAPP was metalized using either $FeCl_2$ or $CoCl_2$. MCl_2 and H_2TAPP are dissolved in oxygen free DMF at room temperature. The solution is boiled and maintained at reflux for one hour under nitrogen atmosphere and then cooled down to room temperature under inert atmosphere. A solution of aqueous KCl is added and, after decantation, the precipitate is dissolved in dichloromethane and washed with water in a separating funnel. The organic phase is gently concentrated by evaporation overnight and the resulting MTAPP (Figure 7.14) is washed with hexane and dried under vacuum.

7.4.2.3 Synthesis of Transition Metal Tetraazaannulenes

Metal free tetraazaannulene (H_2TAA) is synthesized according to the method developed by Hiller et al. [88]. First, ortho-phenylenediamine is dissolved in DMF under magnetic stirring and nitrogen atmosphere. A solution of propynal (synthesized from propargylic acid according to the method of Sauer [89]) and methanol is added, the mixture is maintained at reflux for one hour under a nitrogen flow and then cooled down to room temperature. The mixture is kept overnight at 4 °C under a nitrogen flow and filtered. The solid is rinsed several times with methanol before being dried for 4 h under reduced pressure in order to obtain a purple powder. The metalation is performed either with $CoCl_2$, $6H_2O$ or $FeCl_2$, $4H_2O$. The metal salt and H_2TAA are dissolved in DMF; the solution is stirred, deaerated and kept at reflux for 30 min. Then, the mixture is cooled down to room temperature and kept at 4 °C overnight to favor the precipitation of the metal complex. After filtration, the solid is rinsed several times with methanol and dried to form a crimson powder of MTAA.

7.4.2.4 Characterization of the Macrocycles

UV-visible absorbance spectroscopy and ESR spectroscopy are often used to characterize transition metal macrocycles.

In the case of CoTsPc, as an example, the UV-visible spectra display a main band located at 663 nm for the dimeric form of the phthalocyanine and at 670 nm for the monomeric form [90], in agreement with the results obtained by Gruen and

Table 7.7 λ_{max} for CoT$_{NO_2}$Pc and CoTAPc from UV-visible spectra recorded in DMF.

Macrocycle compound	CoT$_{NO_2}$Pc	CoTAPc
λ_{max}/nm	325	309
	620	411
	670	644
	770	721

Blagrove [91]. Concerning CoT$_{NO_2}$Pc and CoTAPc the maximum wavelengths (λ_{max}) are reported in Table 7.7. The λ_{max} shift towards lower values for CoTAPc and are in agreement with those proposed in the literature [85].

In the case of H$_2$T$_{NO_2}$PP, the Soret band between 400–450 nm corresponding to the π-π^* transition is characteristic of porphyrin compounds; four other bands are located at 511, 548, 592 and 647 nm, in agreement with the results proposed by other authors [18, 92]. Table 7.8 gives the obtained λ_{max} after reduction of the NO$_2$ groups and metalation of the macrocycles. The Soret band is always present and in the case of CoTAPP the absorption bands are located at λ_{max} in agreement with those reported in the literature [18, 93].

Macrocycle powders are also characterized by ESR spectroscopy. Spectra of CoT$_{NO_2}$Pc, CoTAPP and FeTAPP are given in Figure 7.15. CoT$_{NO_2}$Pc gives a wide signal ($g = 2.31$) and a less intense signal ($g = 1.94$). The first one, comparable to that obtained by Assour [94] with a CoPc and Walker [95] with a Co porphyrin, can be attributed to unpaired electron. The second signal can be attributed to conjugated electrons of the macrocycle. No signal was obtained with CoTAPc and CoTsPc powders at room temperature. For this kind of compound, it seems necessary to carry out ESR measurements in liquid nitrogen to detect the electronic resonance transition [94]. In the case of CoTAPP, the ESR signal characteristic of the porphyrin macrocycles is located at $g = 2.09$. The signal due to cobalt ion is detected, at $g = 2.46$, in agreement with published results [94, 95]. With FeTAPP, three ESR signals are detected at $g = 2.09$ characteristic of unpaired electrons in the porphyrin macrocycle, at $g = 2.14$ and at $g = 2.32$, which can be attributed to the presence of iron ion.

7.4.2.5 Preparation of Macrocycle Electrodes and Characterization of their Activity

Electrodes containing CoPc and FePc (MPc) are prepared from a dispersion of MPc and carbon Vulcan XC 72 in ethanol with a weight ratio MPc/C = 1. A given volume of

Table 7.8 λ_{max} for CoTAPP and FeTAPP from UV-visible spectra recorded in dichloromethane.

Macrocycle compound	CoTAPP	FeTAPP
λ_{max}/nm	400–450	410
	528	513
	613	565

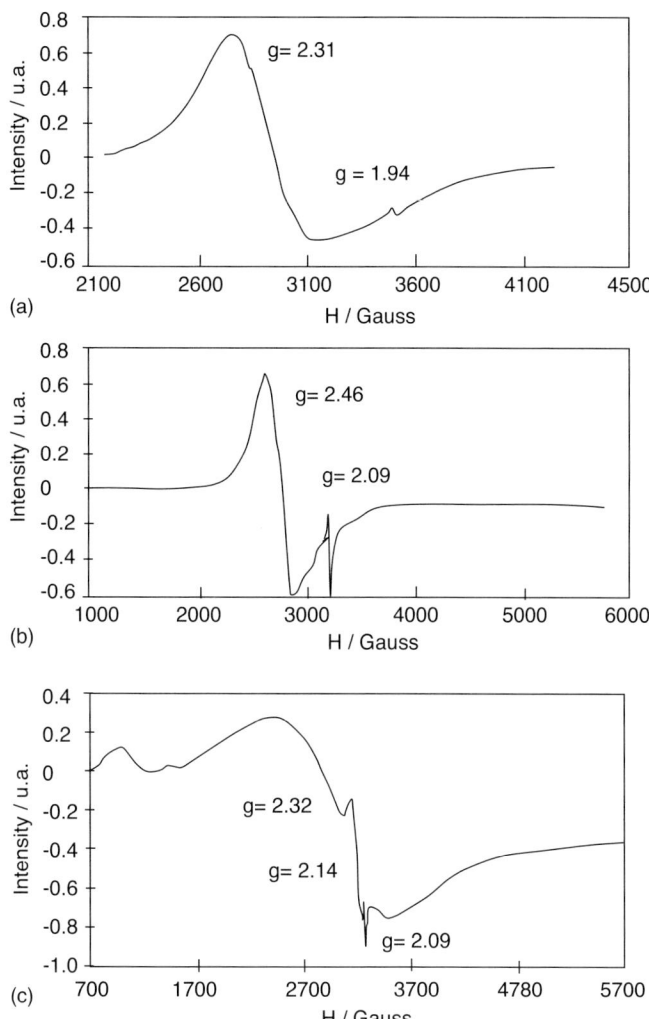

Figure 7.15 ESR spectra recorded at 20 °C from (a) CoT$_{NO_2}$Pc powder, (b) CoTAPP powder and (c) FeTAPP powder.

the dispersion is deposited on a vitreous carbon substrate of 0.071 cm^2 in order to obtain MPc surface concentration of 210 µg cm^{-2}. The electrodes are characterized towards their electroactivity for the electroreduction of oxygen in acid medium (Figure 7.16). FePc/C always displays higher activity than CoPc, since higher current densities are recorded from 900 to 0 mV versus RHE. In acid medium, ORR starts to occur with an overpotential close to 600 mV on a CoPc/C electrode against only 350 mV on a FePc/C electrode. Moreover, it is known that the ORR occurs on a CoPc electrode mainly via a mechanism involving the exchange of 2 electrons per oxygen molecule to form hydrogen peroxide [78], which is harmful for the solid polymer

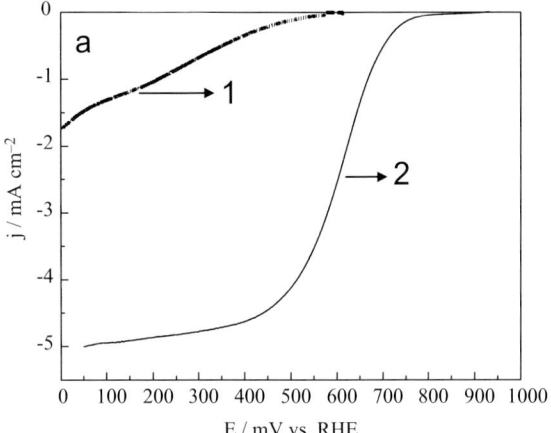

Figure 7.16 ORR curves recorded in H_2SO_4 0.5 M on (1) CoPc/C electrode; (2) FePc/C electrode (O_2-saturated electrolyte, $T = 20\,°C$, $\nu = 5\,mV\,s^{-1}$, $\Omega = 2500\,rpm$).

electrolyte stability in case of use in a DMFC. Indeed, hydrogen peroxide can decompose to give radicals (OH$^\bullet$ or OOH$^\bullet$), which further can attack any H-containing terminal bonds present in the polymer membrane. Peroxide radical attack on H-containing end groups is generally believed to be the principal degradation mechanism of the polymer membrane [96].

In the case of FePc/C, RRDE polarization curve and collected current of H_2O_2 oxidation at the Pt ring electrode (Figure 7.17), which leads to the determination of

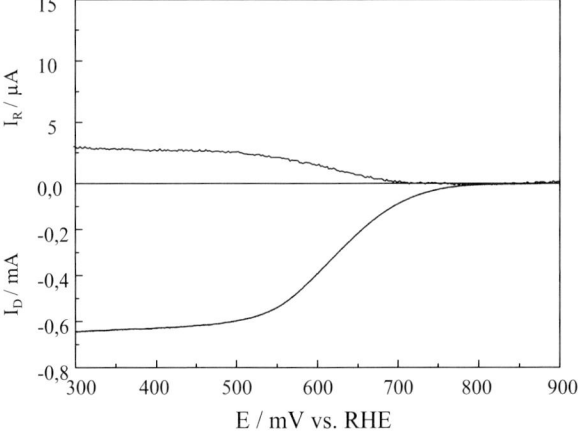

Figure 7.17 Polarization curve of the ORR on a FePc/C disc electrode (I_D) and oxidation current of hydrogen peroxide on the Pt ring (I_R) as a function of potential of the disc electrode recorded in O_2-saturated 0.5 M H_2SO_4 electrolyte ($T = 20\,°C$, $\nu = 5\,mV\,s^{-1}$, $\Omega = 2500\,rpm$).

the water formation efficiency $x(H_2O)$ and then to the calculation of the value of the number of exchanged electrons n [34, 35], indicate that n is greater than 3.92 in the whole studied potential range [97]. This means that the hydrogen peroxide formation does not exceed 4% [98]. Moreover, in a lower overpotential range (from 700 to 850 mV versus RHE) no hydrogen peroxide oxidation current is detected on the Pt ring electrode, which means that the reduction occurs totally via a four-electron process.

SEM observation of a FePc/C catalyst with a loading of 210 μg cm^{-2} has shown that FePc crystals form aggregates of FePc sticks having a size close to 20 nm thickness and 100–200 nm length. These aggregates have an average size close to 1 μm [99]. Because of the size of aggregates, the performance of a fuel cell cathode prepared with this catalyst should not be optimized. The dispersion at a molecular level of the catalyst could be a good alternative to improve electrode efficiency towards ORR. To realize this, metal tetrasulfonated phthalocyanines can be dispersed inside a polymer matrix. These macrocycle compounds are water soluble, which prevents their use alone as cathode catalyst. The molecules have to be fixed in a substrate; the most common strategy consists in insertion of the tetrasulfonated salt phthalocyanine as the counter anion during the electropolymerization process of electron conducting polymers like polypyrrole (PPy) or polyaniline (PAni) [16, 17, 90, 99–101]. Because in the case of MTsPc-PPy systems the polymer seems to play an important role in the ORR at high overvoltage, as shown by *in situ* ESR measurements during the ORR at a CoTsPc-PPy electrode [90], only MTsPc-PAni will be considered here. FeTsPc-PAni electrode was electropolymerized between 50 and 1100 mV versus RHE at a 0.071 cm^2 vitreous carbon disc electrode under the following conditions: 0.5 M H_2SO_4 + 10^{-2} M aniline + 10^{-3} M FeTsPc at $T = 20\,°C$, $v = 50\,mV\,s^{-1}$ and Ω 2500 rpm. The electropolymerization is stopped as soon as the first oxidation peak of PAni reaches 10 mA cm^{-2}. Under these conditions, it was calculated that the maximum surface concentration of FeTsPc is 1.3×10^{-7} mol cm^{-2} [99]. SEM observation of the film performed by Baranton *et al.* [98] showed irregular polymer spheres of 100–200 nm diameter, which led to a rough electrode surface, and therefore to a high surface area. This structure can be an advantage for elaborating an active cathode for fuel cell by increasing the active surface area of the catalytic layer. Figure 7.18 gives a comparison of the electrocatalytic activity of FePc/C (containing 3.7×10^{-7} mol cm^{-2} FePc) with that of FeTsPc-PAni (containing 1.3×10^{-7} mol cm^{-2} FeTsPc) electrodes [99].

Although the surface concentration in active sites is four times lower for the FeTsPc-PAni, the ORR occurs with a lower overvoltage (100 mV lower) than on the FePc/C electrode. The enhancement of activity towards oxygen reduction reaction with FeTsPc species incorporated in the PAni film is likely due to the better dispersion of the catalytic species into a conducting polymer film and further to the higher formation of dimeric FeTsPc species [98].

It is also possible to disperse phthalocyanines at a molecular level by electropolymerization of macrocycles functionalized by amino groups [20, 102], for example. Figure 7.19 shows the electropolymerization of FeTAPP on a vitreous carbon in CH_3CN + 0.1 M tetraethylammonium perchlorate (TEAP) electrolyte, as an

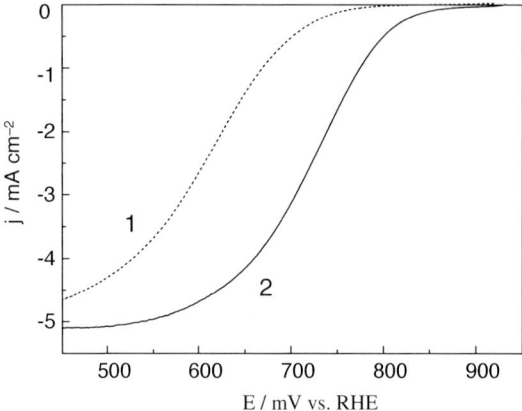

Figure 7.18 Polarization curves of the ORR on (1) FePc/C and (2) FeTsPc-PAni electrodes recorded in O_2-saturated 0.5 M H_2SO_4 electrolyte ($T = 20\,°C$, $v = 5$ mV s^{-1}, $\Omega = 2500$ rpm).

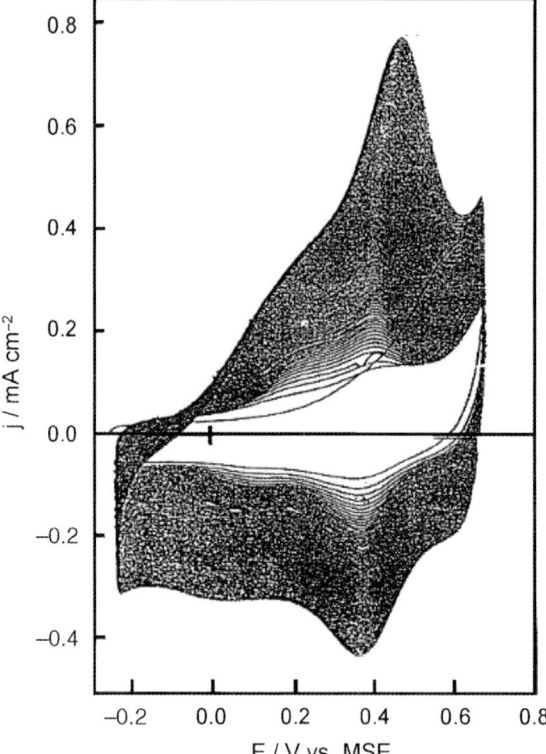

Figure 7.19 Electropolymerization of 1 mM FeTAPP on a vitreous carbon electrode in CH_3CN + 0.1 M tetraethylammonium perchlorate (TEAP) electrolyte ($T = 20\,°C$, $v = 200$ mV s^{-1}, 100 voltammetric cycles).

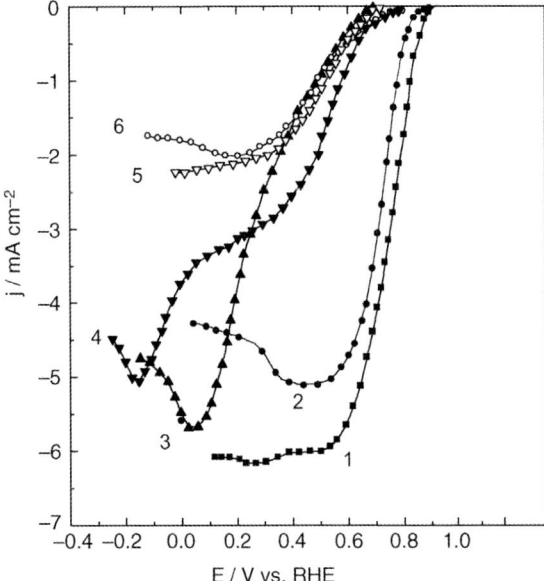

Figure 7.20 Polarization curves of ORR on (1) bare Pt, (2) FeTsPc-PAni, (3) CoTsPc-PPy, (4) poly-CoTAPc, (5) poly-CoTAPP and (6) poly Fe-TAPP electrodes recorded in O_2-saturated 0.5 M H_2SO_4 electrolyte ($T = 20\,^{\circ}C$, $\nu = 5\,mV\,s^{-1}$, $\Omega = 2500\,rpm$).

example. The continuous increase in the amplitude of the voltammetric peaks indicates that a film is formed on the electrode surface, as a consequence of electropolymerization of the attached aniline groups. Besides, other electropolymerized macrocycles were prepared, such as poly-FeTAPP, poly-CoTAPP and poly-CoTAPc [20, 102].

Their electroactivity towards the ORR was compared to that of MTsPc inserted in conducting polymers and to that of bare Pt (Figure 7.20). All electrodes display a lower activity towards the ORR than bare Pt, although FeTsPc-PAni electrode leads to a reduction wave very close to that of Pt. The overpotential is 50 mV higher for FeTsPc-PAni than for bare Pt, whereas it is 200 mV higher for poly-FeTAPP and poly-CoTAPP. Moreover, porphyrin-based electrodes display maximum current densities of oxygen reduction (close to $2\,mA\,cm^{-2}$), which are more than two times lower than the other electrodes. Diffusion limiting current density in the range of 5 to $6\,mA\,cm^{-2}$ for $\Omega = 2500$ rpm are in agreement with a four-electron process of oxygen reduction [48, 103]. With porphyrin-based electrodes, the ORR occurs mainly via a two-electron process as confirmed by El Mouhaid et al. [20] in the case of poly-CoTAPP electrode. For poly Co-TAPc, two reduction waves are visible; in the first reduction wave at lower overpotentials the reduction occurs via a two-electron process, whereas in the reduction wave at higher overpotentials, it occurs mainly via a four-electron process [20].

Porous CoTAA modified carbon electrodes (graphite powder HSAG 300, Lonza) can also be prepared by electrodeposition at 0.8 V versus Ag/AgCl for 1 min from a N_2

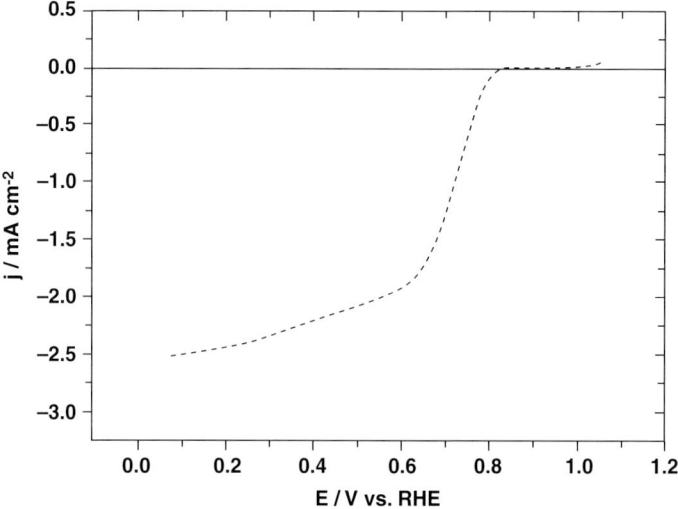

Figure 7.21 Polarization curve of the ORR on a porous CoTAA modified carbon electrode recorded in 0.1 M HClO$_4$ O$_2$-saturated solution ($T = 20\,°C$, $v = 5$ mV s^{-1}, $\Omega = 2500$ rpm).

purged and stirred 0.1 M Bu$_4$NClO$_4$ acetonitrile solution saturated with CoTAA (~2 mM). An electrode with a CoTAA surface concentration close to 2.2×10^{-8} mol cm^{-2} is obtained and characterized by *in-situ* UV-visible reflectance spectroscopy and cyclic voltammetry [104]. Figure 7.21 shows the oxygen reduction curves obtained at a porous CoTAA modified carbon electrode recorded in O$_2$-saturated 0.1 M HClO$_4$. The onset potential of ORR is close to 0.8 V versus RHE and the maximum current density in the diffusion region lies between 2.0 and 2.5 mA cm^{-2} which indicates that a two-electron process is mainly involved in the oxygen reduction, and therefore that hydrogen peroxide is the main reaction product.

But such macrocycle compounds are known to be unstable in acid solution [97, 100–106] and display lower activity towards ORR than Pt. Heat treatments at temperatures in the 400–900 °C range under an inert atmosphere were used to improve their activity and stability [105, 106]. Lalande *et al.* studied different heat-treated and unpyrolyzed FePcs [107] and CoPcs [108] adsorbed on carbon black; they concluded that in the case of FePc, unpyrolyzed catalysts were the most active and less stable, whereas in the case of CoPc, heat treatment between 500 °C and 700 °C led to higher activity. Van den Putten *et al.* [109] proposed that the catalyst stability is improved via the creation by pyrolysis of a chemical bond between the active centers of the macrocycle. But the identification of the active centers was widely discussed in the literature. It has been proposed that after such thermal treatment of CoTAA compounds, the central metal was no longer the active site for the ORR [110], or that in the case of pyrolyzed cobalt tetramethoxyphenyl porphyrin the presence of metallic ions was necessary to reduce oxygen [111]. From XPS measurements before and after duration tests of heat-treated CoTAA dispersed on carbon, Biloul *et al.* [112] proposed that the ORR occurs on nitrogen atom containing fragments, although the possibility that remaining cobalt ions were involved in the whole process could not be eliminated because of the low accuracy of the measurements. According to Scherson et al. [111] partial dissolution of metal species takes place in acid medium; metal species can re-adsorb on carbon surface and form active sites most likely involving nitrogen atoms (in a M−N type bond). This idea was criticized by Van Veen et al. [113]. They proposed, on the basis of EXAFS studies, that MN_4 is the real active site of O_2 reduction at CoTPP/Norit BRX.

However, Baranton et al. [97] showed that with the FePc/C electrode in a gas diffusion configuration, pressed against a solid polymer electrolyte (Nafion 117), the kinetics of the degradation mechanism is very slow: no decrease of the activity of the FePc/C cathode fed with $25\,mL\,mn^{-1}$ O_2 was observed over four days under $25\,mA\,cm^{-2}$ (electrode potential around 150 mV) at room temperature. To explain this result, they proposed on the basis of in-situ infrared reflectance spectroscopy measurements, that the main degradation process of FePc occurred via demetalation by proton in acid solution. Under gas diffusion electrode configuration, protons are preferentially used to assist the oxygen reduction reaction and become less or not available to attack the metal center of the phthalocyanine.

7.4.3
Transition Metal Chalcogenide Catalysts and Electrodes

7.4.3.1 Synthesis of Metal Chalcogenides

One of the strategies to develop chalcogenide materials in nanodivided form is based on the use of carbonyl molecular clusters as precursors [114]. Firstly the decomposition of these clusters by pyrolysis is carried out under mild conditions, that are at temperatures below 200 °C. Secondly the process favors the formation of nanoparticles with highly degree of disorder. Taking as a primary basis the tris-ruthenium dodecacarbonyl complex, the reaction process can be represented by the reaction (7.31):

$$Ru_3(CO)_{12} + yX \rightarrow [Ru_xX_y-CO_z] \rightarrow Ru_xX_y + zCO \quad \text{where} \quad X = S, Se, Te \tag{7.31}$$

The process of synthesis in organic solvents, such as xylene (Xyl), or 1,2 dichlorobenzene (DCB) is schematized in Figure 7.22.

This approach is also valid for other chalcogens, such as sulfur, and tellurium [115]. In fact, for the synthesis of Ru_xSe_y in powder form, the reaction of tris-ruthenium dodecacarbonyl, $Ru_3(CO)_{12}$, and elemental selenium dissolved in an organic solvent leads to a polynuclear chemical precursor in the first stages of reaction (see Figure 7.22, prior to step 4). It is important to notice that the formation of a polynuclear cluster compound $[Ru_xX_y-(CO)_z]$ – see Equation 7.31 – was revealed by ^{13}C NMR analysis [114,116]. The appearance of a chemical shift ($\delta = 198.88$ ppm) in the ^{13}C-NMR spectrum is characteristic of the presence of $Ru_4Se_2(CO)_{11}$. This latter compound represents the real chemical precursor of the Ru_xSe_y electrocatalyst (after step 4), and therefore we can assign the stoichiometry $x = 2$ and $y = 1$. The nature of the solvent is important, since it takes part in the coordination chemistry.

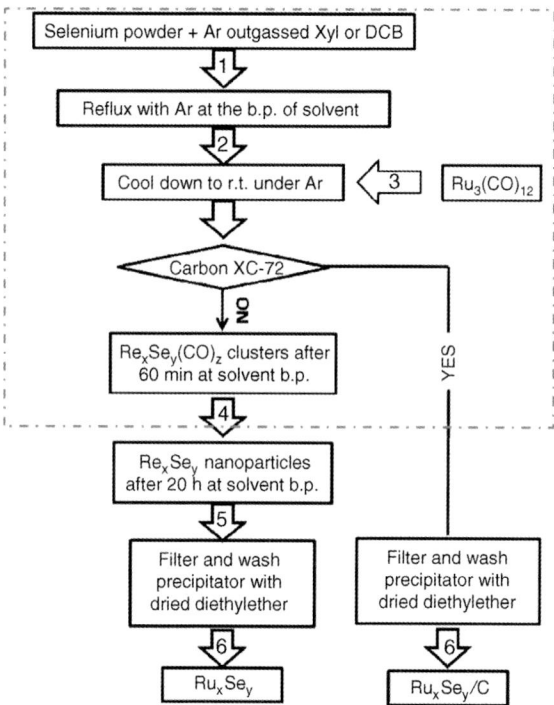

Figure 7.22 Flow chart of ruthenium chalcogenide-based low temperature synthesis by means of reaction between elemental selenium and $Ru_3(CO)_{12}$. In this process, the box prior to step 4 indicates the formation of a high nuclearity cluster compound, which serves as the chemical precursor for Ru_xSe_y synthesis [73].

The formation of $Ru_4Se_2(CO)_{11}$ was found to be very selective in 1,2 dichlorobenzene (DCB). This approach represents the easiest way, chemically speaking, of generating such a cluster compound in comparison with other ways of synthesis reported by Layer et al. [117, 118]. The loss of the carbonyl groups is obtained by keeping the boiling temperature of the solvent in refluxing conditions under argon or nitrogen for 20 h of reaction (step 4 in Figure 7.22). The intermediate cluster compound $Ru_4Se_2(CO)_{11}$ changes continuously with the reaction time to the novel cluster-like material Ru_xSe_y. The addition of Vulcan XC-72 would lead to a Ru_xSe_y supported catalyst. Following alternative reaction processes other workers have succeeded to obtain a fairly good distribution of chalcogenide-ruthenium nanoparticles on carbon support [119, 120].

Using the flow chart schematized in Figure 7.22 as a synthesis basis, other groups report chalcogenide materials in which CO is still involved as a ligand in, for example, $Ru_xFe_ySe_z(CO)_n$, $Mo_xRu_ySe_z(CO)_n$, $W\text{-}Se\text{-}Os(CO)_n$, $Mo_xS_y(CO)_n$, $Mo_xSe_y(CO)_n$ or $W(CO)_n$ [121–136]. These descriptive stoichiometries are most probably the result of an incomplete reaction of pyrolysis of the reactants, and make the understanding of the role of the catalytic sites of cathode materials rather difficult.

The flow chart of Figure 7.22 may also serve as a template for tailoring single- or multi-metal centered catalysts. It has been extended, for example, to tailor Pt-based nanocatalyts, either for CO tolerant anodes [53, 137], or for methanol-tolerant cathodes [36, 37, 51, 55] (see Section 7.4.1.1). However, one should keep in mind that certain conditions must be fulfilled to perform multi-metal centered catalysts. Since reduction of the amount of precious metal is aimed, one can imagine a kind of 'dilution' effect by introducing a proper neighboring coordinating metal center, in such a way as to protect, electronically and chemically, the catalytic reaction site. This phenomenon is, in fact, fulfilled for the Ru_xSe_y, as revealed by measurements using solid-state ^{77}Se-NMR and X-ray photoelectron spectroscopy (XPS) [138]. Therefore, one important goal of the synthesis via carbonyl compounds is the possibility to identify the presence of the intermediate molecular complex if any – see Equation 7.32 – which may produce the metal-mixed chalcogenide, compound, as discussed recently [139].

$$Ru_3(CO)_{12} + X + M_{x'}CO_y \rightarrow [Ru_xM_{x'}X_y \cdot CO_z] \rightarrow Ru_xM_{x'}X_y + zCO \quad (7.32)$$

where X = S, Se, Te; and M = Fe, Co, W, and so forth.

The non-supported ruthenium chalcogenide cluster-like compound is tested for the ORR in acid medium. Figure 7.23 shows an experiment under RDE conditions of an electrode prepared from a suspension of 2 g/L deposited onto glassy carbon disks. The corresponding catalyst charge is 85 µg cm^{-2}. This typical characteristic results after an activation in the potential interval from 0 to 0.83 V/RHE under argon or nitrogen out-gassed electrolyte. Extending the scan at more positive potentials leads to the surface oxidation process of the chalcogenide particles, and therefore to a decrease in the ORR activity, as the result of the increasing amount of

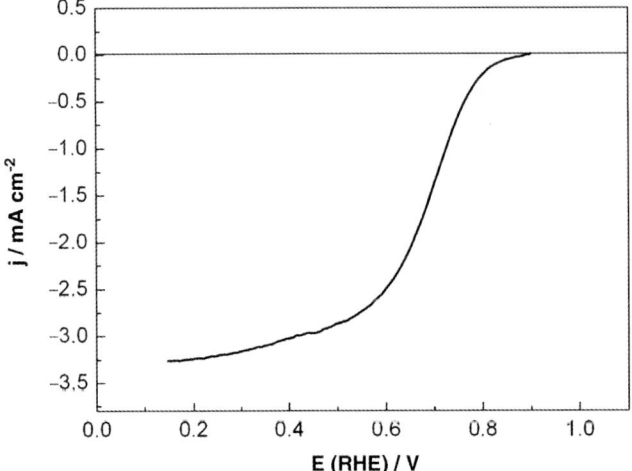

Figure 7.23 Current–potential curves for oxygen reduction on unsupported Ru_xSe_y electrode. The measurements were carried out in an O_2-saturated, 0.5 M H_2SO_4 solution at 5 mV s^{-1}, at the rotation rate of 1600 rpm.

ruthenium-oxide-like surface species. As demonstrated by XPS, selenium is removed in the form of SeO_2 [138]. The surface chemical condition of such Ru_xSe_y clusters are their resistance towards oxidation with oxygen from air, as experimentally revealed via the NMR/XPS [140] and *in situ* WAXS studies [141]. This property cannot be taken as a advantage to decrease the ORR reaction overpotential, since ruthenium and selenium surface atoms in Ru_xSe_y react with water molecules [138].

7.4.3.2 Physicochemical Characterizations

Structural properties of nanocatalysts are of paramount importance. The electronic structure of nanomaterials is another interesting aspect to get a deeper insight in the interfacial electrode activity. Interesting information has been gained with the use of X-ray photoelectron spectrometry (XPS), as well as the use of X-ray absorption spectroscopy in the EXAFS and XANES modes. However, this section will be devoted only to the most popular and accessible techniques, such as the transmission electron microscopy (TEM) and X-ray diffraction (XRD). Some case studies will be presented below.

Transmission electron microscopy (TEM) This technique is very useful for the determination of size distribution and dispersion of supported nanocatalysts. Average particle size can be also estimated from independent methods, see below. Assuming that all particles are spherical, the volume to surface mean diameter can be obtained from particle size distribution: $\bar{d} = \sum_i n_i d_i^3 / \sum_i n_i d_i^2$. Further, the surface average dispersion, D, which is defined as the ratio of the surface atoms to the total atoms within the nanoparticles can be deduced following the model proposed by Borodzinski and Bonarowska [142]. There are a myriad of such examples in the specialized literature [53, 116, 120, 143–149]. Generally TEM images are performed before and after the thermal treatment. One example can be taken from recent results reported by J. Tian *et al.* concerning the thermal stability of $PtRuTiO_x/C$ DMFC anodes prepared by the polyol chemical synthesis route [150]. The TEM pictures given in Figure 7.24 show that agglomeration effect is caused by the thermal treatment. However, it is very distinctly observed that the agglomeration effect is less on catalysts deposited onto an oxide (Figure 7.24(c) to (d)). The oxide substrate confers some thermal stability to the nanocatalysts. This simple observation indicates, qualitatively, that the oxide strengthens the interaction with the nanometallic catalysts.

Another example concerns the synthesis of nanoalloy materials produced by the carbonyl chemical route [51]. Figure 7.25 depicts Pt:Ni (2:1) supported catalyst on carbon, as well as the corresponding size histogram plot, which shows a narrow particle size distribution. This technique proves that the synthetic chemical route is ideal to vary the metal ratio in order to investigate the electrochemical properties of such nanoalloy materials towards the ORR and their tolerance to methanol. In this sense it was shown that the mean particle size of the as-prepared Pt-Ni alloy catalysts decreases from 3.1 nm to 2.3 nm with the decrease of Pt/Ni atomic ratio from 5:1 to 1:1. The different structural parameters of such catalysts are summarized in the Table 7.9.

X-Ray diffraction (XRD) The use of X-ray diffraction, in the wide angle mode (WAXS), for studying metal clusters has been discussed by Vogel [151]. As much

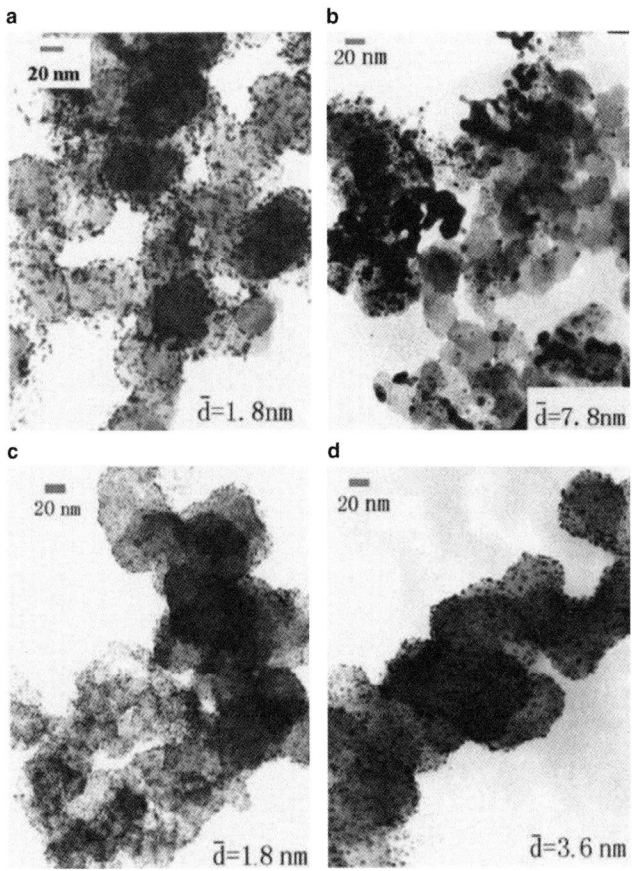

Figure 7.24 TEM images of electrocatalysts: (a) PtRu/C; (b) PtRu/C-500; (c) PtRuTiO$_x$/C; (d) PtRuTiO$_x$/C-500. After ref. [150].

as matter is nanodivided, the diffraction profile of very small coherent domains become broaden and difficult to distinguish from the background. The calculation of experimental diffraction profiles is based using combinations of Debye functions, the so-called Debye function analysis (DFA) simulation. A sequence of model particles (*fcc*, *hcp*, *bcc*, and so on, or non-crystallographic icosahedral or decahedral structure) of different sizes and structures with the experimental function by means of a nonlinear least-square parameters fitting is achieved. It may be recalled that some assumptions about the structure of the particles are necessary to perform the simulation, as well as additional parameters such as (i) the exponent of the Debye–Waller factor, (ii) the average spacing within the individual clusters, and (iii) an additive constant correcting for errors in the background subtraction [151, 152]. The application of the DFA simulation to nanoscale materials in catalysis [141, 152–154], as well as in electrocatalysis [51, 53, 141, 155–161], has been proven to be successful. As an example of application, Figure 7.26 shows the DFA for the Pt-Ni (3 : 1)/C, performed with perfect *fcc*-type model particles.

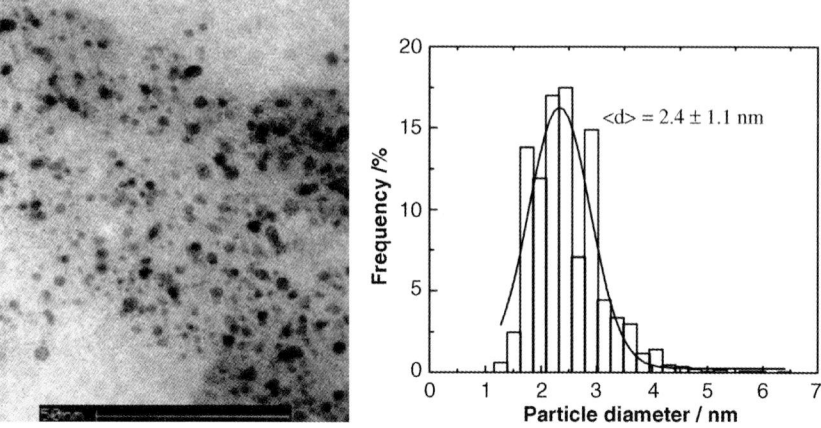

Figure 7.25 TEM image of Pt-Ni(2:1)/C catalyst with 20 wt.% metal loading and its particle size distribution histogram. Adapted from reference [51].

This simulation became even better by adding a fraction of ~20 wt. % of small Pt_3O_4 particles to the model system. One may also notice that the deviation from the simulation is visible at the high order diffraction peaks (311) (222). This is most probably related to some incomplete alloying. The mean lattice parameter is 3.852 Å, and the mass/number weighted mean size is 21/17 Å (see Figure 7.26 insert). This latter is actually in good agreement with the obtained TEM histogram represented for the same material in Figure 7.25. The obtained average crystallite sizes of all the Pt : Ni catalysts are summarized in Table 7.9. This calculation is in good agreement with the TEM observations.

7.4.4
Fuel Cell Tests

In order to calibrate the FC tests, reference catalysts were purchased from E-Tek. The electrodes supplied by E-Tek were composed of 60 wt.% Pt–Ru/C (1/1 atomic ratio) for the anode and 40 wt.% Pt/C for the cathode. The average metal loading of each electrode was 2 mg cm^{-2}. Both electrodes contained 0.8 mg cm^{-2} Nafion in the active

Table 7.9 Composition and mean particle size of as-prepared Pt-Ni/C alloy catalysts with 20 wt.% metal loading.[a]

Catalyst	EDX Pt/Ni composition (%)	Particle size (nm) from TEM	Dispersion D (%)	Crystallite size (nm) from WAXS
Pt : Ni(5 : 1)	84.1 : 15.9	3.1 ± 1.4	33.6	2.8
Pt : Ni(3 : 1)	73.9 : 26.1	2.8 ± 1.2	37.1	2.8
Pt : Ni(2 : 1)	65.8 : 34.2	2.4 ± 1.1	40.4	2.3
Pt : Ni(1 : 1)	52.0 : 48.0	2.3 ± 1.3	38.2	2.8

[a]Adapted from reference [51].

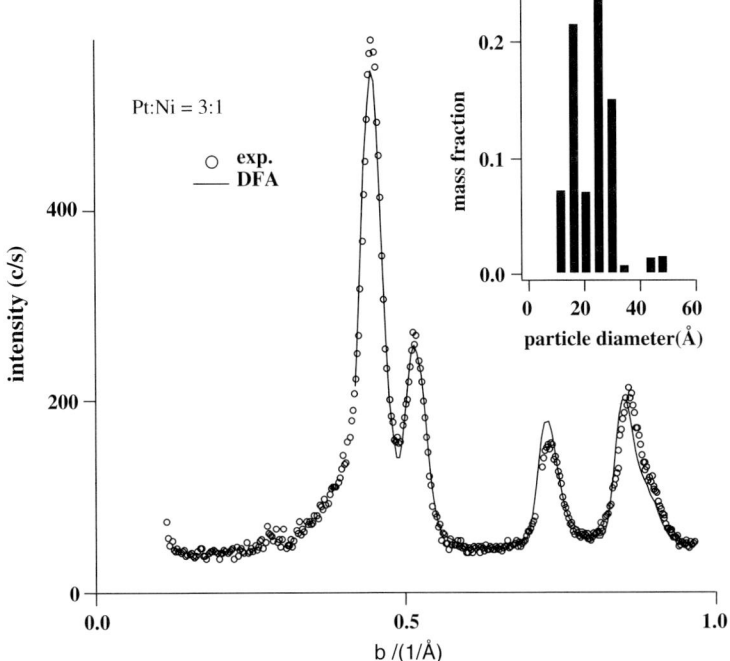

Figure 7.26 DFA of the Pt-Ni(3:1)/C alloy nanoparticles (solid line). The simulation is performed using *fcc* model particles containing no twin defects. The insert shows the mass fractions of the model particles used for the simulation versus their diameter. The diffraction experiments were performed with a Guinier powder diffractometer (HUBER) set at the 45° transmission position. A Johansson type Ge monochromator produces a focused monochromatic Cu-K$_{\alpha 1}$ primary beam ($\lambda = 0.15406$ nm).

layer and 30 wt.% Teflon in the diffusion layer. The homemade PtM/C electrodes for DMFC were prepared from an ink composed of a Nafion solution (5 wt.% from Aldrich), isopropanol and the catalytic powder, brushed on a carbon gas diffusion electrode. Carbon gas diffusion electrodes were homemade using a carbon cloth from Electrochem. Inc., on which was brushed an ink made of Vulcan XC72 carbon powder and PTFE dissolved in isopropanol. The gas diffusion electrodes were loaded with 3.5 mg cm^{-2} of a mixture of carbon powder and 20 or 30 wt.% PTFE for the anode and the cathode, respectively. The metal loading was close to 2 mg cm^{-2} and the Nafion loading was 0.8 mg cm^{-2}. Nafion based MEAs were prepared either by hot pressing of the electrodes onto a Nafion membrane at 130 °C for 90 s under a pressure of 3.43 MPa (35 kg cm^{-2}), or without mechanical pressing for the other membranes. Indeed because of the high glass transition temperature of sulfonated polyimide membranes, hot pressing was not possible, as it would require too high pressing temperatures. As a consequence, for polyimide membrane based MEAs, electrodes were only mechanically pressed at room temperature in the DMFC. On the other hand recast Nafion membrane were used as solid polymer electrolyte for mini-DMFC working at room temperature, whereas Nafion 117 was used in DMFCs working at temperatures higher than 90 °C.

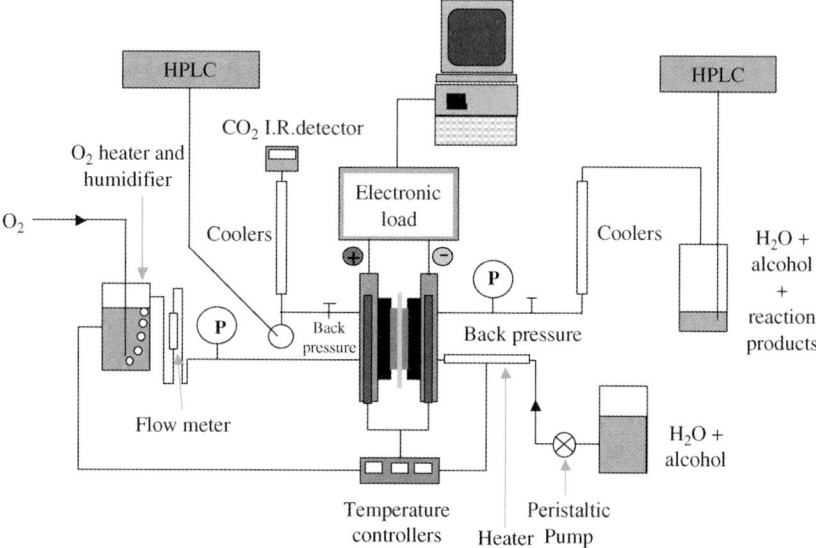

Figure 7.27 Schematic diagram of the complete fuel cell test set-up, including the analysis of reaction intermediates and products.

DMFC tests were carried out in a 5 cm² single cell using a Globe Tech Inc. fuel cell test station, equipped with humidification bottles having pressure and flow rate controls of both reactants: methanol and pure oxygen. In the case of mini-DMFC experiments, the fuel cell tests were carried out at room temperature under 1 bar of oxygen (30 mL min^{-1}) and 1 bar of 5 M methanol solution (2 mL min^{-1}). In the case of DMFC working at temperatures higher than 90 °C, methanol concentration and flow rate were 2 M and 2 mL mn^{-1}, respectively, whereas pure oxygen flow rate was set at 120 mL mn^{-1}.

The E versus j and P versus j curves were recorded using a high power potentiostat (Wenking model HP 88) interfaced with a microcomputer to apply the current sequences and to store the data, and a variable resistance in order to fix the applied current to the cell.

The analysis of reaction products was carried out by high-pressure liquid chromatography (HPLC) for the liquid compounds, whereas the CO_2 amount was evaluated by using an infrared detector (Figure 7.27).

7.5
Catalyst Tolerance in the Presence of Methanol

7.5.1
Behavior of PtM/C Catalysts for the ORR in the Presence of Methanol

It is known that the crossover of methanol from the anode to the Pt-based cathode can lead to a decrease of the cell voltage by about 200–300 mV, particularly when practical

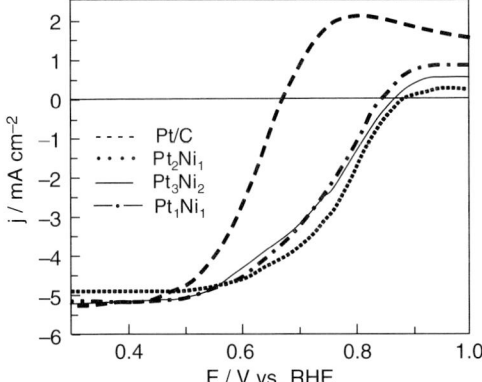

Figure 7.28 Voltammograms of the Pt/C and carbon-supported nanosized Pt–Ni alloy catalysts recorded in 0.5 M H_2SO_4 + 0.5 M CH_3OH solution saturated with pure oxygen ($T = 20\,°C$, $\nu = 5\,mV\,s^{-1}$, $\Omega = 2000\,rpm$).

air flows are used in a DMFC. The ORR activity of Pt/C and nanosized Pt–Ni alloy catalysts in the presence of 0.5 M CH_3OH is shown in Figure 7.28. All Ni containing catalysts displayed a lower overpotential in the presence of methanol than pure Pt catalysts.

In the case of pure Pt catalyst, the significant increase in the ORR overpotential in the presence of MeOH is definitely due to simultaneous oxygen reduction and methanol oxidation, which leads to the formation of a mixed potential (see Section 7.2.1). With Pt–Ni alloy catalysts, a decrease in activity towards the ORR in methanol-containing electrolyte is also observed. But, the potential loss is only about 30–60 mV in comparison with that in pure acid solution. The ORR activity of the Pt–Ni alloy catalysts in methanol-containing solution is much higher than that of pure Pt, indicating that the Pt–Ni alloy catalysts exhibit a higher methanol tolerance during the ORR compared to pure Pt catalyst. In the diffusion-controlled region, the current densities of the ORR in the presence of methanol are slightly lower than those recorded in the absence of methanol (see Figure 7.10, Section 7.4.1.4), which could be ascribed to the fact that some of the catalytically active sites for the ORR have been blocked or covered by adsorbed intermediates from methanol or by methanol molecules, because methanol oxidation and oxygen reduction occurs in parallel at different sites of Pt-based catalysts. Other authors also observed such behavior. Li et al. [162] found that a Pt–Fe alloy catalyst is very active for the ORR in pure acid electrolyte but no more active in methanol-containing electrolyte as compared to a Pt/C catalyst. However, Shukla et al. [163] found that an ordered Pt–Fe alloy catalyst exhibited significantly high oxygen-reduction activity in the presence of methanol. The difference in the methanol-tolerant ORR catalyst could be ascribed to their different structures.

Koffi et al. have shown that in the case of non-alloyed Pt-Cr/C catalysts prepared by the Bönnemann method, the more tolerant Pt-Cr catalyst displayed a Pt : Cr atomic ratio close to 0.7 : 0.3 [48]. It appeared that with the coreduced Pt-Cr/C catalyst

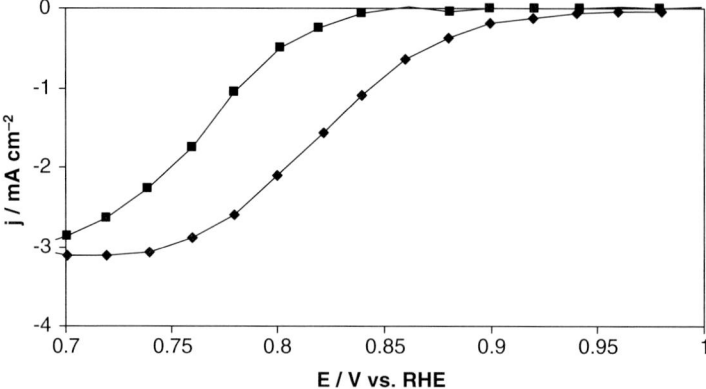

Figure 7.29 Voltammograms of (■) non alloyed $Pt_{0.7} + Cr_{0.3}$/C and (♦) alloyed $Pt_{0.7}Cr_{0.3}$/C catalysts, obtained via the colloidal route, recorded in O_2 saturated 0.5 M H_2SO_4 electrolyte containing 0.5 M MeOH ($T = 20\,°C$, $\nu = 5$ mV/s, $\Omega = 2500$ rpm).

obtained via the colloidal route, the same ratio leads to the best performance for the ORR in the presence of 0.5 M methanol [49]. From 0.95 to 0.8 V versus RHE, the addition of chromium to Pt leads to higher oxygen reduction current densities. The less active catalyst is Pt/C. Although by using the colloidal method, the best ratio Pt:Cr ratio for the ORR in the presence of methanol does not change between the alloyed and the non-alloyed Pt-Cr catalysts, the alloyed one leads to the best electrocatalytic activity towards the ORR in the presence of methanol (Figure 7.29).

After these previous results a 50 μm sulfonated polyimide membrane was chosen to realize a MEA based on a PtCr(7:3)/C catalyst at the cathode and PtRu(4:1)/C catalyst at the anode, both prepared from the Bönnemann method, by coreduction and codeposition, respectively. This MEA was tested at 20 °C under mini-DMFC conditions [49]. As shown in Figure 7.30, comparing the values obtained with an E-Tek cathode, a significant enhancement of the electrical performances could be recorded. The OCV was higher with the PtCr catalyst containing MEA, which may be due to the higher tolerance of PtCr catalyst towards the presence of methanol. Moreover, the maximum power density with commercial catalysts was close to 6 mW cm^{-2}, whereas it reached 18 mW cm^{-2} with a PtCr catalyst at the cathode.

Experiments in half-cell tests have shown that Pt–Ni alloy catalysts exhibited an enhanced catalytic activity for the ORR both in the absence and presence of methanol as compared to a Pt/C catalyst. It is then interesting to check the behavior of such catalysts under DMFC working conditions: oxygen pressure and high temperature. Figure 7.31 shows a comparison of the DMFC polarization curves recorded at 100 °C with different cathode catalysts and the same anode catalyst (Pt-Ru (1:1)/C from E-Tek, 40 wt.% metal loading). The open circuit voltage (OCV) for Pt–Ni alloy catalysts is higher than that for pure Pt. The highest OCV was found with the Pt–Ni (2:1)/C catalyst. All Ni containing catalysts display higher activity than pure Pt catalyst for current density lower than 200 mA cm^{-2}. However, conversely to that was observed

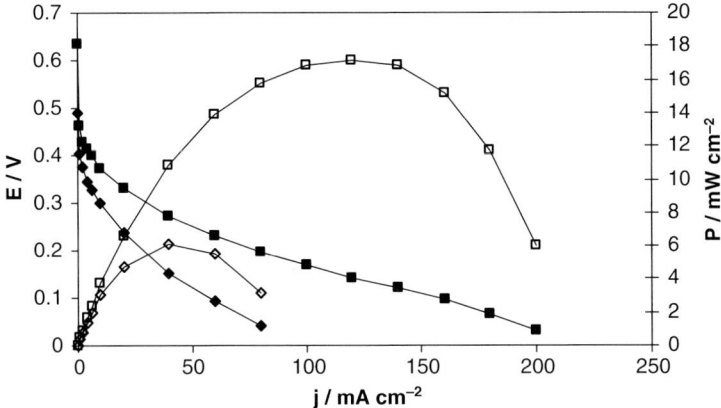

Figure 7.30 Polarization curves E(j) and power density curves P(j) recorded with a 50 μm thickness sulfonated polyimide membrane in a single 5 cm² DMFC at $T = 25\,°C$ (5 M MeOH, $P_{MeOH} = 1$ bar, 2 mL min⁻¹; $P_{O_2} = 1$ bar, 20 mL min⁻¹). The MEA was made either with (♦) E-Tek electrodes (cathode: Pt/C, 2.0 mg cm⁻² Pt loading; anode: PtRu(1:1)/C, 2.0 mg cm⁻² metal loading) or (■) home made electrode (cathode: PtCr(7:3)/C, 2.0 mg cm⁻² metal loading; anode: PtRu(8:2)/C, 2.0 mg cm⁻² metal loading) without hot pressing.

in a half-cell, Pt–Ni (1:1)/C lead to lower performance for higher current densities. This may be due to some possible changes in the surface composition and/or particle sizes of such a catalyst occurring at high current density, or to lower accessible Pt sites due to higher Ni surface concentration, which could lead to mass transport limitation. The maximum activity is obtained with the Pt–Ni (2:1)/C catalyst. The decrease in the overvoltage with such a cathode at the same current density is about 50 mV as

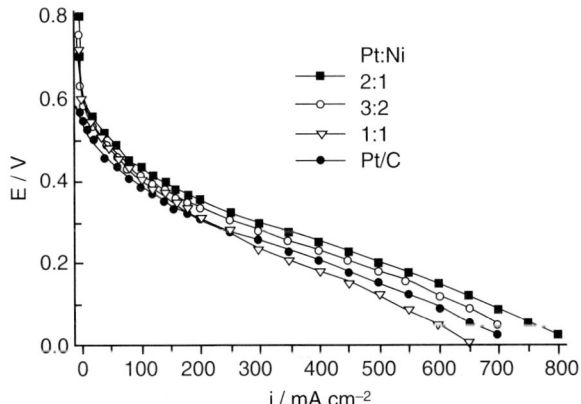

Figure 7.31 Cell voltage against current density curves recorded at 100 °C in a single DMFC using different Pt-Ni catalysts ([MeOH] = 2 M; $P_{CH_3OH} = 1.9$ bar; $P_{O_2} = 2.5$ bar, Nafion 117 membrane).

compared with that of a Pt/C catalyst, which is consistent with the previous half-cell measurements.

More recently, Lee et al. developed novel methanol-tolerant catalysts for the ORR based on Ir-Co alloys [164]. Among them Ir_xCo_{1-x}/C catalysts, with x in the range 0.7–0.8 exhibited the highest activity for the ORR with high tolerance to the presence of methanol. Moreover these alloys can catalyze the ORR by a four-electron transfer reaction leading to water.

7.5.2
Behavior of Transition Metal Macrocycles for the ORR in the Presence of Methanol

Transition metal macrocycles display lower activity than Pt catalyst towards the ORR. However, Pt is a good electrocatalyst not only for the ORR, but also for the dissociative adsorption of methanol and its oxidation [165]. Now, the solid polymer electrolyte used in a DMFC (particularly Nafion membranes) is not totally tight to methanol [49]. Methanol crossover results in a significant loss in coulombic efficiency and voltage efficiency of a DMFC because methanol would be oxidized at the commonly used Pt/C cathode [166, 167]. To avoid this problem, one strategy is the development of novel membranes or Nafion modified membranes, which are less permeable to methanol [168–171]. However, all these methods are able to reduce the methanol crossover to a certain extent, but none can prevent methanol crossover and all of these methods also increase the impedance for proton transport through the membrane [172]. Therefore, developing cathode catalysts totally tolerant to methanol is of great interest. The tolerance to the presence of methanol was established by Chu et al. [173, 174] on pyrolyzed Co and Fe-tetraphenyl porphyrins. But few results are available on non heat-treated macrocycles. Baranton et al. evaluated the tolerance of a FePc catalyst [97]. Figure 7.32 displays as an example the oxygen reduction curves recorded at a FePc/C electrode and at a Pt/C electrode in the absence and in the presence of methanol. In the presence of 1.0 M MeOH, the current densities of oxygen reduction on a Pt/C electrode are observed at potentials lower than 700 mV versus RHE that is with overpotentials close to 300 mV higher than in the absence of methanol. In the potential range from 850 to 700 mV, competition between methanol oxidation and oxygen reduction reactions is even in favor of the former, as oxidation current densities are observed. In the case of a FePc/C electrode, the addition of methanol to the electrolyte seems not to affect the ORR, as the two curves are almost superimposed. In the presence of methanol, the FePc electrode becomes more active than the Pt/C electrode towards oxygen reduction between 800 and 400 mV versus RHE.

However, it is important to check that the presence of methanol has no effect on the kinetics and on the mechanism of ORR at FePc/C. Koutecky-Levich plots (1/j versus $1/\Omega^{1/2}$) were used to determine j_0 the exchange current density, j_l the limiting current density resulting from a mixed control by the diffusion of molecular oxygen inside the FePc/C film and by adsorption of dioxygen on the FePc catalyst, n the number of electrons exchanged per reduced oxygen molecule [32, 97], and b the Tafel slopes [98]. Table 7.10 reports the kinetics data obtained at a FePc/C electrode in the presence and

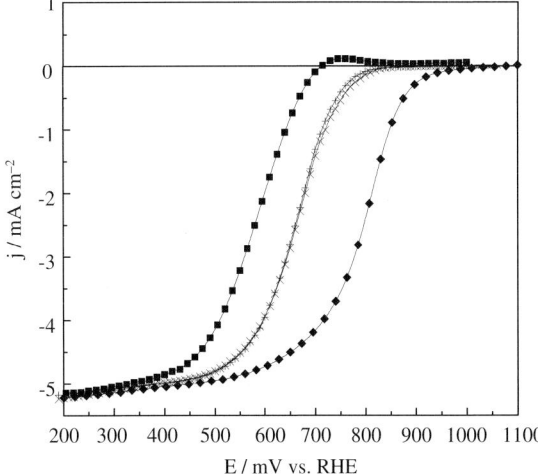

Figure 7.32 Polarization curve of the ORR recorded in 0.5 M H$_2$SO$_4$ O$_2$-saturated electrolyte on (♦) Pt/C electrode in the absence and (■) Pt/C electrode in the presence of 1.0 M MeOH, respectively, (×, +) FePc/C electrode in the absence and the presence of 1.0 M MeOH, respectively ($T = 20\,°C$, $\nu = 5\,\text{mV s}^{-1}$, $\Omega = 2500\,\text{rpm}$).

in the absence of 1.0 M MeOH. Comparison of the kinetics data obtained in the two media seems to indicate that the presence of methanol has no effect on the kinetics and mechanistic pathway for oxygen reduction on FePc/C electrode. But, the number of exchanged electrons could not be confirmed and determined more precisely by RRDE method because the presence of methanol leads to distort the oxidation current measurements on the Pt ring. However, FePc is clearly tolerant to the presence of methanol, and it can be recognized as a methanol-insensitive catalyst.

The behavior of macrocycle-based electrodes has hardly been investigated under DMFC configuration. Lalande et al. [107, 108] studied the activity and stability of

Table 7.10 Kinetic parameters of ORR in O$_2$-saturated 0.5 M H$_2$SO$_4$ electrolyte determined from Koutecky-Levich curves [97] in the presence and in the absence of 1.0 M MeOH ($T = 20\,°C$, $\nu = 5\,\text{mV s}^{-1}$, $\Omega = 400, 900, 1600, 2500\,\text{rpm}$).

H$_2$SO$_4$ electrolyte	j_0 (mA cm^{-2})	j_l (mA cm^{-2})	n	Tafel slope b (low overpotential region)	Tafel slope b (high overpotential region)
MeOH-free	-6.5×10^{-5}	-19	4–3.9	$-65\,\text{mV dec}^{-1}$	$-121\,\text{mV dec}^{-1}$
1.0 M MeOH	-5.4×10^{-5}	-20	~4	$-60\,\text{mV dec}^{-1}$	$-115\,\text{mV dec}^{-1}$

Tafel slopes were determined from ORR polarization curves at $\Omega = 2500\,\text{rpm}$, after correction of the diffusion contribution [98].

CoPc/C, FePc/C and FePc/C under PEMFC conditions (with pure hydrogen and oxygen reactants) at 50 °C and 80 °C. Biloul et al. [112] also carried out PEMFC (H_2/O_2 and H_2/air) experiments with heat-treated Co-porphyrins and CoTAA. In each case, the activity of the macrocycles was much lower than that of Pt/C catalysts. The main interest of using macrocycle-based catalyst is their tolerance to the presence of alcohol, namely methanol. Baranton et al. [99] studied the behavior of a PAni-FeTsPc electrode under DMFC conditions at room temperature. FeTsPc-PAni electrodes were prepared by electropolymerization of aniline between 50 and 1100 mV versus RHE at a 7.8 cm^2 carbon gas diffusion layer under the same conditions as those used with a vitreous carbon electrode, except that magnetic stirring of the solution was used instead of electrode rotation. Then a Nafion solution was pulverized on the electrode to obtain a loading in ionomer of 1 mg cm^{-2}. The cathode was then put in a DMFC together with a Nafion 117 membrane and an E-TEK anode (2.0 mg cm^{-2} PtRu(1 : 1)/C, 60 wt.% metal loading, 0.8 mg cm^{-2} Nafion). The DMFC was fed with pure oxygen at the cathode side and 2, 5 and 10 M MeOH at the anode side. In Figure 7.33a, the cathode potential is plotted as a function of the current density for a DMFC working either with a Pt/C or a PAni-FeTsPc cathode, and fueled with different concentrations of methanol. The Pt/C cathode potential at $j = 0$, achieves 820 mV versus RHE when feeding the cell with 2 M MeOH, and decreases down to 740 mV and only 575 mV versus RHE for 5 and 10 M MeOH, respectively. At the same time, the cathode potential at $j = 0$ of the PAni-FeTsPc electrode remains invariant with the methanol concentration of the anode feed. For the same MeOH concentration in the anode feed (5 M), the PAni-FeTsPc cathode displays a better activity towards ORR from $j = 0$ to 20 mA cm^{-2} and becomes the best cathode over the whole current density range for higher methanol concentrations. This fact makes this kind of cathode catalysts good alternatives to Pt for applications needing high methanol concentration, like mini- and micro-DMFC [49, 99].

Polarization and power density curves of DMFCs working with Pt/C and PAni-FeTsPc cathodes and fueled with 5 M MeOH are compared (Figure 7.33b). A PAni-FeTsPc cathode delivers 6 mW cm^{-2} as maximum power density, whereas a Pt/C cathode leads to achieve 14 mW cm^{-2}. Moreover, with a PAni-FeTsPc cathode, the stability is rather poor. According to numerous studies on the stability of FePc in acid medium [98], the FeN$_4$ active center is not affected by the ORR in solid acid electrolyte medium, that is no demetalation of the macrocycle and no destruction of the organic structure occurs, at least in the time scale of several hours. And, effectively, during prolonged experiments under fuel cell conditions, the water at the cathode outlet acquires a deep blue color, which corresponds to the color of FeTsPc in water solution. The loss of stability of the cathode seems more due to desorption of the iron macrocycles from the polymeric film than to the destruction or demetalation of the phthalocyanine.

7.5.3
Transition Metal Chalcogenides

The use of the cluster compound concept, in preparing chalcogenide materials as cathode catalysts for the ORR, was first reported in 1986 [175]. It was soon recognized

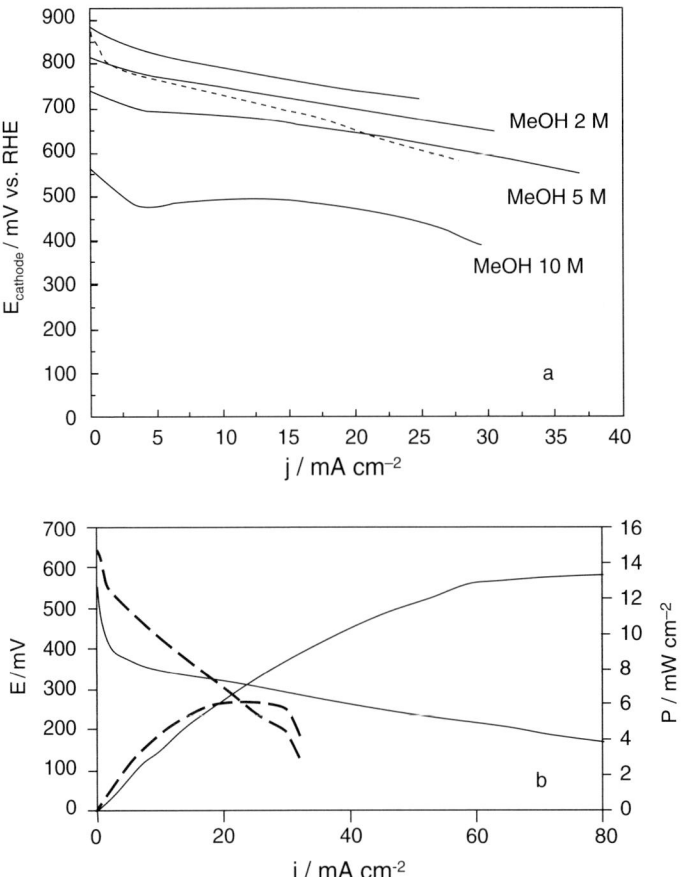

Figure 7.33 (a) Potentials versus RHE of a cathode Pt/C, 2 mg$_{Pt}$ cm^{-2} (--), and of a cathode PAni-FeTsPc (---), recorded under working DMFC conditions vs. current density. [MeOH] = 2 to 10 M, 10 mL min^{-1}; pure O$_2$, 25 mL min^{-1}, T = 20 °C. (b) mini-DMFC electrical characteristics recorded at 20 °C with different cathodes: (—) Pt/C, 2.0 mg cm^{-2}, (---) PAni-FeTsPc; anode: PtRu (1 : 1)/C, 2.0 mg cm^{-2} metal loading (P$_{O_2}$ = 1 bar; P$_{MeOH}$ = 1 bar; [MeOH] = 5 M; Nafion 117 membrane).

that transition metal centered chalcogenides in the massive forms, belonging to the Chevrel phase structure, for example, Mo$_4$Ru$_2$Se$_8$, sustained an ORR electrocatalytic current without degradation in the presence of methanol (1–3.5 M) [176]. This behavior was further verified with nanodivided ruthenium chalcogenide materials (Ru$_x$Se$_y$) [21], and a review in this subject matter concerning the carbonyl generated catalysts was given in 2003 [177].

The performance in a mixed-reactant feed simple strip fuel cell was tested for ruthenium–selenium chalcogenide electrocatalyst [178]. The authors showed that better fuel efficiency of the DMFC is obtained at low current density, which is achieved with the selective electrodes. In the same line of research other groups

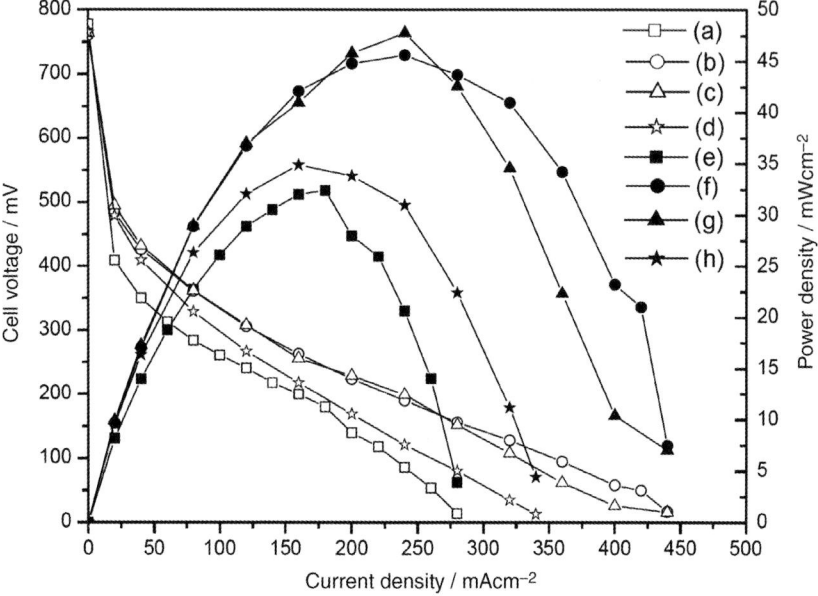

Figure 7.34 Polarization curves obtained for mixed-reactant DMFC at 90 °C for (a) RuSe/C (1 mg cm^{-2}), (b) RuSe/C (2 mg cm^{-2}), (c) RuSe/C (2.5 mg cm^{-2}), and (d) RuSe/C (3 mg cm^{-2}). The corresponding power density data are shown in (e), (f), (g), and (h), respectively. Adapted from reference [181].

reported mixed-reactant feed fuel-cell performances [179–181]. Indeed, Scott et al. [181] reported maximum power density of 48 mW cm^{-2}. The system was based on anode reactants (methanol + oxygen) and cathode reactants (methanol + oxygen) using as anode PtRu/C (loading of 1.46 mg cm^{-2}), and as cathode RuSe/C with a loading of 2.5 mg cm^{-2}. Under the same conditions, the authors tested cathode catalysts such as FeCoTMPP/C, CoTMPP/C, CoTMPP/C, and FeTMPP/C. These systems delivered lower power densities as shown in Figure 7.34.

In the case of ruthenium chalcogenides, it seems that the composition and the distribution onto the carbon support is not yet fully optimized.

7.5.4
Other Non Pt-Based Catalysts

The electrocatalytic activity of chalcogenides of different transition metals, such as Co, Ni, W, Mo, V, Ti and Cr, as well as that of bimetallic chalcogenides towards ORR has been reported by Baresel et al. [182]. Co-S and Co-Ni-S compounds displayed the highest activity for ORR in 2 M H_2SO_4 at 70 °C. The results of fuel cell tests, however, indicated that the durability of all examined materials is insufficient as a cathodic catalyst for fuel cells with acid electrolytes. Later on, Behret et al. [183] reported that Co_3S_4 had the highest activity among Me(a)Me$_2$(b)X$_4$ spinels in 1 M H_2SO_4.

However, in both cases the electrocatalytic activity for the ORR decreases in the order S > Se > Te, possibly as the results of the geometric and electrostatic conditions in the spinel structure. Co or Fe-based chalcogenides have still attracted a lot of interest due to their low cost and their abundance, although these materials have low chemical and electrochemical stabilities at high overpotentials in acid medium, as compared with Pt- or Ru-based cathode catalysts. Further, calculations predicting a reversible potential of about 0.74 V and 0.5 V vs. RHE for Co_9S_8 and Co_9Se_8 have been reported by the Anderson's group [184,185]. These authors considered that the (202) surface of Co_9S_8 and Co_9Se_8 is active towards the ORR. Other non-precious metals catalysts for ORR in acid medium, in the form of thin films, such as $Co_{1-x}Se_x$ [186], FeS_2 and (Co, Fe)S_2 [187], have been reported. More recently, the electrocatalytic activity of 20 wt.% $CoSe_2$/C nanoparticles towards ORR in acid medium was reported [145,188]. An OCV of about 0.7 V vs. RHE was observed for 20 wt.% $CoSe_2$/C and a Tafel slope of -125 mV dec^{-1} was obtained.

Figure 7.35 compares the ORR current-potential curves recorded on $CoSe_2$/C, Ru/C, Ru_xSe_y/C and Pt/C. The half-wave potentials on each system correspond to 0.52 V versus RHE for $CoSe_2$ and Ru; 0.73 V and 0.88 V versus RHE for Ru_xSe_y and Pt, respectively. Under these conditions, one can observe that the activities of $CoSe_2$ and Ru are close together and that the $j(E)$ curves are 0.36 V more negative than those for Pt. Despite the low activity towards ORR, $CoSe_2$ material is cheaper than Ru-based material, and its development is worthwhile, since this material alike the Ru_xSe_y is also a methanol-tolerant catalyst.

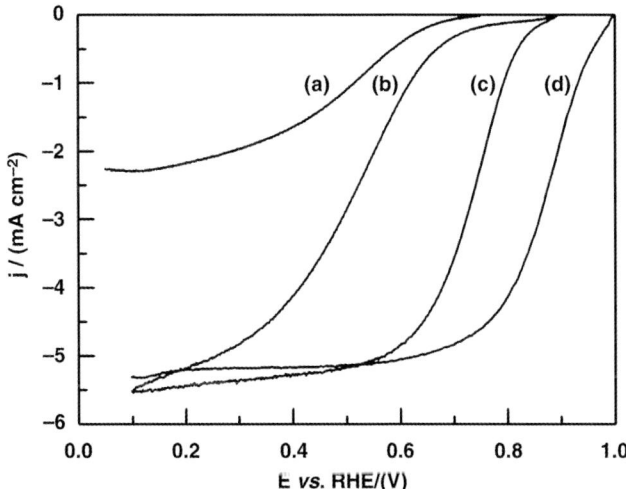

Figure 7.35 Molecular oxygen electroreduction current–potential curves obtained on a rotating disk electrode (RDE) in H_2SO_4 and O_2 saturated solution at 1600 rpm. (a) $CoSe_2$/C (50 wt %; 28 µg cm^{-2}; 0.5 M H_2SO_4) (b) Ru/C (20 wt %; 56 µg cm^{-2}; 0.1 M H_2SO_4) (c) Ru_xSe_y/C (20 wt %; 56 µg cm^{-2}; 0.1 M H_2SO_4) (d) Pt/C (10 wt %; 56 µg cm^{-2}; 0.1 M H_2SO_4). Scan rate of 5 mV s^{-1}.

Lee et al. obtained similar results with Ir_xSe_y chalcogenides [189]. The synthesized compound, Ir_4Se, has high electrocatalytic activity towards the ORR and a strong methanol-tolerance. However the ORR was not complete, since 10% of H_2O_2 were detected by RRDE experiments.

Recently, novel non-noble MeN type electrocatalysts were prepared and displayed interesting activity towards the ORR and good methanol tolerance. For example, Xia et al. [190] synthesized methanol-tolerant MoN carbon-supported catalysts by heat-treating carbon-supported molybdenum tetraphenylporphyrin in a flow NH_3 atmosphere. These novel catalysts were characterized by XRD, TEM, EDX and XPS. These materials, with an average particle size close to 4 nm, displayed a high activity towards the oxygen reduction reaction together with a strong methanol tolerance. The ORR was found to proceed approximately via a four-electron reduction process, the main reaction product being water.

7.6
Summary and Outlook

The oxygen reduction reaction (ORR) is one of the main electrocatalytic reaction involved in fuel cell technologies. It is also the most difficult reaction with low kinetics responsible for the loss of electric performances at the air cathode due to a high overvoltage (at least 0.3 to 0.4 V) leading to a third of the losses in energy and power densities.

When associated with a methanol anode in a DMFC the loss of energy and power are even greater (about two thirds of the theoretical values) due to both the low kinetics of methanol oxidation and the deleterious effect, at the cathode side, of methanol crossover through the protonic membrane. The latter effect results from a mixed potential at the Pt cathode. Therefore a great challenge for developing DMFCs is to find novel cathode electrocatalysts that are more active for the ORR and more tolerant to the presence of methanol.

Several kinds of electrocatalysts are discussed in this chapter, for example Pt-based alloys, PtM, with M a noble metal or a transition metal, transition metal macrocycles, such as porphyrins, phthalocyanines, tetraazaannulenes, and transition metal chalcogenides, mostly selenides. Different synthesis methods are described, together with the different physicochemical and electrochemical methods to characterize these electrocatalysts. If transition metal macrocycles and chalcogenides are quite insensible to the presence of methanol, their electrocatalytic activity is lower than that of Pt-based catalysts, by at least two or three orders of magnitude, since the ORR wave is negatively shifted by about 100 to 200 mV. On the other hand, PtM electrocatalysts (with M = Fe, Co, Ni, Cr, etc.) have a similar or even better activity than Pt for the ORR, but their methanol tolerance is lower than that of transition metal macrocycles and chalcogenides.

To conclude we may emphasize that the ORR electrocatalysis needs to be carried on with great efforts, in order to discover and develop new electrocatalysts both active for the ORR and methanol-tolerant, since it seems difficult to find methanol-tight protonic membranes.

Acknowledgments

The authors would like to acknowledge the contribution of many PhD students and co-workers. They also acknowledge the financial support of CNRS, and of the French Ministry of Research through OPTIMET (00 S 0060 contract) and MICROMET (01 S 0886 contract) research projects. They are also indebted to the European Commission through the NEMECEL project (JOE3-CT97-003 contract).

Abbreviations

CoPc	cobalt phthalocyanine
CoTsPc	tetrasulfonated cobalt phthalocyanine
CoTAPP	cobalt tetra(o-aminophenyl)porphyrin
CoTAPc	cobalt tetra(o-aminophenyl)phthalocyanine
DFA	Debye function analysis
DMFC	direct methanol fuel cell
EDX	energy dispersive X-ray fluorescence
ESR	electron spin resonance
EXAFS	extended X-ray absorption fine structure
fcc	face-centered cubic
FePc	iron phthalocyanine
FeTsPc	tetrasulfonated iron phthalocyanine
FeTAPP	iron tetra(o-aminophenyl)porphyrin
HPLC	high-performance liquid chromatography
H_2TAPP	metal-free tetra(o-aminophenyl)porphyrin
ICP-OES	inductively coupled plasma optical emission spectroscopy
MEA	membrane electrode assembly
MTsPc	metal tetrasulfonated phthalocyanines
MPc	metal phthalocyanine
MTAA	metal tetraazaannulene
MTPP	metal tetraphenyl-porphyrin
$(Nalk_4)^+(Bet_3H)^-$	tetraalkyltriethylborohydride
NEM	new membranes
NMR	nuclear magnetic resonance
OCV	open circuit voltage
ORR	oxygen reduction reaction
PAni	polyaniline
PEMFC	proton exchange membrane fuel cell
PPy	polypyrrole
PTFE	polytetrafluoroethylene
RDE	rotating disk electrode
r.d.s.	rate-determining step
r.p.m.	revolution per minute
RRDE	rotating ring disk electrode

RHE	reversible hydrogen electrode
SHE	standard hydrogen electrode
S-PEEK	sulfonated polyetheretherketones
TEAP	tetraethylammonium perchlorate
TEM	transmission electron microscopy
WAXS	wide angle X-ray scattering
XANES	X-ray absorption near edge structure
XAS	X-ray absorption spectroscopy
XRD	X-ray diffraction
XPS	X-ray photoelectron spectroscopy

References

1 Kinoshita, K. (1992) *Electrochemical Oxygen Technology*, John Wiley and Sons, Inc., New York, pp. 19–32.
2 Tarasevich, M.R., Sadkowski, A. and Yeager, E. (1983) *Comprehensive Treatise of Electrochemistry*, vol. 7 (eds B.E. Conway, J.O'M. Bockris, E. Yeager, S.U.M. Khan and R.E. White), Plenum Press, New York, pp. 301–398.
3 Anderson, A.B. (2002) *Electrochim. Acta*, **47**, 3759–3763.
4 Lamy, C. and Léger, J.-M. (1995) Proceedings of the 1st International Symposium on New Materials for Fuel Cell Systems (eds O. Savadogo, P.R. Roberge and T.N. Veziroglu), Montreal, pp.296–309.
5 Lamy, C., Léger, J.-M. and Srinivasan, S. (2001) Direct methanol fuel cells: FROM fundamental aspects to technological development, in *Modern Aspects of Electrochemistry*, vol. 34 (eds J.O'M. Bockris, and B.E. Conway), Plenum Press, New York, pp. 53–117.
6 Zagal, H. (1992) *Coordin. Chem. Rev.*, **119**, 89.
7 Jasinski, R. (1965) *J. Electrochem. Soc.*, **122**, 526.
8 Alt, H., Binder, H. and Sandstede, G. (1973) *J. Catal.*, **28**, 8.
9 Zagal, J., Sen, R.K. and Yeager, E. (1977) *J. Electroanal. Chem.*, **83**, 207.
10 Magner, G., Savy, M. and Scarbeck, G. (1980) *J. Electrochem. Soc.*, **127**, 1076.
11 Jahnke, H.G. (1968) *Ber. Bunsen Ges. Phys. Chem.*, **72**, 1053.
12 Alt, H., Binder, H., Lindner, W. and Sandstede, G. (1971) *J. Electroanal. Chem.*, **31**, 19.
13 Beck, F., Zammert, W., Heiss, J., Miller, H. and Polster, R. (1973) *Z. Naturforsch.*, **28**, 1009.
14 Bull, R.A., Fan, F.R. and Bard, A.J. (1984) *J. Electrochem. Soc.*, **131**, 687.
15 Okabayashi, K., Ikeda, O. and Tamura, H. (1983) *J. Chem. Soc., Chem. Commun.*, 684.
16 El Hourch, A., Belcadi, S., Moisy, P., Crouigneau, P., Léger, J.-M. and Lamy, C. (1992) *J. Electroanal. Chem.*, **339**, 1.
17 Elzing, A., Van Der Putten, A., Visscher, W. and Barendrecht, E. (1987) *J. Electroanal. Chem.*, **233**, 113.
18 Bettelheim, A., White, B.A., Raybuck, S.A. and Murray, R.W. (1987) *Inorg. Chem.*, **26**, 1009.
19 Bettelheim, A., White, B.A. and Murray, R.W. (1987) *J. Electroanal. Chem.*, **217**, 271.
20 El Mouahid, O., Coutanceau, C., Belgsir, E.M., Crouigneau, P., Léger, J.-M. and Lamy, C. (1997) *J. Electroanal. Chem.*, **426**, 117.
21 Alonso-Vante, N. (1996) *J. Chim. Phys.*, **93**, 702.
22 Alonso-Vante, N., Cattarin, S. and Musiani, M. (2000) *J. Electroanal. Chem.*, **481**, 200–207.

23 Lamy, C., Léger, J.-M., Alonso-Vante, N., Coutanceau, C., Le Rhun, V., Lima, A. and Rakotondrainibe, A. (July 2002) Conception and Realization of a New Low Cost Direct Methanol Fuel Cell (NEMECEL), Research Project no. JOE3-CT97-003, European Commission, Final Report.

24 Tricoli, V., Caretta, N. and Bartolozzi, M. (2000) *J. Electrochem. Soc.*, **147**, 1286–1290.

25 Ramya, K. and Dhathathreyan, K.S. (2008) *J. Membr. Sci.*, **311** (1–2), 121–127.

26 Lamy, C., Léger, J.-M., Coutanceau, C., Yang, H. and Dubau, L. (June 2004) Optimisation de la Conception d'une Pile à Méthanol Direct (OPTIMET). Research Project no. 00 S 0060, French Ministry of Research, Final Report.

27 Jones, D.J., Rozière, J., Marrony, M., Lamy, C., Coutanceau, C., Léger, J.-M., Hutchinson, H. and Dupont, M. (October 2005) *Fuel Cells Bull.*, **2005** (10), 12–15.

28 Croissant, M.J., Napporn, W.T., Léger, J.-M. and Lamy, C. (1998) *Electrochim. Acta*, **43**, 2447.

29 Schmidt, T.J., Gasteiger, H.A. and Behm, R.J. (1999) *J. Electrochem. Soc.*, **146**, 1296.

30 Jakobs, R.C.M., Janssen, L.J.J. and Barendrecht, E. (1985) *Electrochim. Acta*, **30**, 1085.

31 Zagal, J., Bindra, P. and Yeager, E. (1980) *J. Electrochem. Soc.*, **127**, 1506.

32 Coutanceau, C., Croissant, M.J., Napporn, T. and Lamy, C. (2000) *Electrochim. Acta*, **46**, 579.

33 Claude, E., Addou, T., Latour, J.-M. and Aldebert, P. (1998) *J. Appl. Electrochem.*, **28**, 57.

34 Elzing, A., van der Putten, A., Visscher, W. and Barendrecht, E. (1987) *J. Electroanal. Chem.*, **233**, 99.

35 Vork, F.T.A. and Barendrecht, E. (1990) *Electrochim. Acta*, **35**, 135.

36 Yang, H., Alonso-Vante, N., Léger, J.-M. and Lamy, C. (2004) *J. Phys. Chem. B*, **108**, 1938.

37 Yang, H., Alonso-Vante, N., Lamy, C. and Akins, D.L. (2005) *J. Electrochem. Soc.*, **152**, A704.

38 Drillet, J.-F., Ee, A., Friedemann, J., Kötz, R., Schnyder, B. and Schmidt, V.M. (2002) *Electrochim. Acta*, **47**, 1983.

39 Toda, T., Igarashi, H. and Watanabe, M. (1999) *J. Electroanal. Chem.*, **460**, 258.

40 Toda, T., Igarashi, H., Uchida, H. and Watanabe, M. (1999) *J. Electrochem. Soc.*, **146**, 3750.

41 Shim, J., Yoo, D.-Y. and Lee, J.-S. (2000) *Electrochim. Acta*, **45**, 1943.

42 Min, M.-K., Cho, J., Cho, K. and Kim, H. (2000) *Electrochim. Acta*, **45**, 4211.

43 Shukla, A.K., Neergat, M., Bera, P., Jayaram, V. and Hegde, M.S. (2001) *J. Electroanal. Chem.*, **504**, 111.

44 Santiago, E.I., Varanda, L.C. and Villullas, H.M. (2007) *J. Phys. Chem. C*, **111**, 3146.

45 Antolini, E. (2003) *Mater. Chem. Phys.*, **78**, 563.

46 Dubau, L., Hahn, F., Coutanceau, C., Léger, J.-M. and Lamy, C. (2003) *J. Electroanal. Chem.*, **554–555**, 407.

47 Xiong, L. and Manthiram, A. (2005) *Electrochim. Acta*, **50**, 2323.

48 Koffi, R.K., Coutanceau, C., Garnier, E., Léger, J.-M. and Lamy, C. (2005) *Electrochim. Acta*, **50**, 4117.

49 Coutanceau, C., Koffi, R.K., Léger, J.-M., Marestin, K., Mercier, R., Nayoze, C. and Capron, P. (2006) *J. Power Sources*, **160**, 334.

50 Yano, H., Kataoka, M., Yamashita, H., Uchida, H. and Watanabe, M. (2007) *Langmuir*, **23**, 6438.

51 Yang, H., Vogel, W., Lamy, C. and Alonso-Vante, N. (2004) *J. Phys. Chem. B*, **108**, 11024.

52 Dickinson, A.J., Carrette, L.P.L., Collins, J.A., Friedrich, K.A. and Stimming, U. (2002) *Electrochim. Acta*, **47**, 3733.

53 Boucher, A.C., Alonso-Vante, N., Dassenoy, F. and Vogel, W. (2003) *Langmuir*, **19**, 10885.

54 Huang, J.J., Yang, H., Huang, Q.H., Tang, Y.W., Lu, T.H. and Akins, D.L. (2004) *J. Electrochem. Soc.*, **151**, A1810.

55 Yang, H., Coutanceau, C., Léger, J.-M., Alonso-Vante, N. and Lamy, C. (2005) *J. Electroanal. Chem.*, **576**, 305.

56 Murthi, V.S., Urian, R.C. and Mukerjee, S. (2004) *J. Phys. Chem. B*, **108**, 11011.
57 Bönnemann, H., Brijoux, W., Brinkmann, R., Dinjus, E., Joussen, T. and Korall, B. (1991) *Angew. Chem. Int. Engl.*, **30**, 1312.
58 Bönnemann, H., Brijoux, W., Brinkmann, R., Fretzen, R., Joussen, T., Köppler, R., Korall, B., Neiteler, P. and Richter, J. (1994) *J. Mol. Catal.*, **86**, 129.
59 Coutanceau, C., Brimaud, S., Dubau, L., Lamy, C., Léger, J.-M., Rousseau, S. and Vigier, F. (2008) *Electrochim. Acta.*, **53**, 6865.
60 Wakisaka, M., Mitsui, S., Hirose, Y., Kawashima, K., Uchida, H. and Watanabe, M. (2006) *J. Phys. Chem. B*, **110**, 23489.
61 Matyi, R.J., Schwartz, L.H. and Butt, J.B. (1987) *Catal. Rev.-Sci. Eng.*, **29**, 41.
62 Mukerjee, S., Srinivasan, S., Soriaga, M.P. and Mc Breen, J. (1995) *J. Phys. Chem. B*, **99**, 4577.
63 Toda, T., Igarashi, H. and Watanabe, M. (1998) *J. Electrochem. Soc.*, **145**, 4185.
64 Paulus, U.A., Wokaun, A., Scherer, G.G., Schmidt, T.J., Stamenkovic, V., Radmilovic, V., Markovic, N.M. and Ross, P.N. (2002) *J. Phys. Chem. B*, **106**, 4181.
65 Paulus, U.A., Wokaun, A., Scherer, G.G., Schmidt, T.J., Stamenkovic, V., Markovic, N.M. and Ross, P.N. (2002) *Electrochim. Acta*, **47**, 3787.
66 Stamenkovic, V., Schmidt, T.J., Ross, P.N. and Markovic, N.M. (2002) *J. Phys. Chem. B*, **106**, 11970.
67 Glass, G.T., Cahen, G.L., Stoner, G.E. and Taylor, E.J. (1987) *J. Electrochem. Soc*, **134**, 158.
68 Paffet, M.T., Beery, G.J. and Gottesfeld, S. (1988) *J. Electrochem. Soc.*, **135**, 1431.
69 Beard, B.C. and Ross, P.N. (1990) *J. Electrochem. Soc.*, **137**, 3368.
70 Antolini, E., Passos, R.R. and Ticianelli, E.A. (2002) *Electrochim. Acta*, **48**, 263.
71 Tamizhmari, G. and Capuano, G.A. (1994) *J. Electrochem. Soc.*, **141**, 968.
72 Li, W., Zhou, W., Li, H., Zhou, Z., Zhou, B., Sun, G. and Xin, Q. (2004) *Electrochim. Acta*, **49**, 1045.
73 Alonso-Vante, N. (2006) *Fuel Cells*, **6**, 182–189.
74 Jalan, V. and Taylor, E.J. (1983) *J. Electrochem. Soc.*, **130**, 2299.
75 Mukerjee, S., Srinivasan, S., Soriaga, M.P. and McBreen, J. (1995) *J. Electrochem. Soc.*, **142**, 1409.
76 Mukerjee, S. and Srinivasan, S. (1993) *J. Electroanal. Chem.*, **357**, 201.
77 Gloaguen, F., Andolfatto, N., Durand, R. and Ozil, P. (1994) *J. Appl. Electrochem.*, **24**, 863.
78 Zagal, J. (2003) Macrocycles, *Handbook of Fuel Cell – Fundamentals, Technology and Application*, John Wiley and Sons, Ltd., Chichester, p 544.
79 Savy, M., Andro, P., Bernard, C. and Magner, G. (1973) *Electrochim. Acta*, **18**, 191.
80 Randin, J.-P. (1974) *Electrochim. Acta*, **19**, 83.
81 Zagal, J., Paez, M., Tanaka, A.A., Dos santos, J.R. Jr and Linkous, C.A. (1992) *J. Electroanal. Chem.*, **339**, 13.
82 Moser, F.M. and Thomas, A.L. (1963) *Phthalocyanines compounds*, Reinhold, New York, p. 134.
83 Weber, J.H. and Bush, D.H. (1965) *Inorg. Chem.*, **25**, 469.
84 Achar, B.N., Fohlen, G.M. and Parker, J.A. (1982) *J. Polym. Sci.*, **20**, 2773.
85 Achar, B.N., Fohlen, G.M., Parker, J.A. and Keshavayya, J. (1987) *Polyhedron*, **6**, 1463.
86 Collman, J.P., Gagne, R.R., Reed, C.A., Halbert, T.R., Lang, G. and Robinson, W.T. (1975) *J. Am. Chem. Soc.*, **97**, 1427.
87 Adler, A.D., Longo, F.R., Kampas, F. and Kim, J. (1970) *J. Inorg. Nucl. Chem.*, **32**, 2443.
88 Hiller, H., Dimroth, P. and pfitzner, H. (1968) *Liebigs Ann. Chem.*, **717**, 137.
89 Sauer, J.C. (1956) *Org. Synth.*, **36**, 66.
90 Coutanceau, C., Rakotondrainibe, A., Crouigneau, P., Léger, J.M. and Lamy, C. (1995) *J. Electroanal. Chem.*, **386**, 173.
91 Gruen, L.C. and Blagrove, R.J. (1973) *Aust. J. Chem.*, **26**, 319.

92 Little, R.G. (1978) *J. Heterocyclic Chem.*, **15**, 203.
93 Bied-Charreton, C., Mérienne, C. and Gaudemer, A. (1987) *New J. Chem.*, **11**, 633.
94 Assour, J.M. (1965) *J. Am. Chem. Soc.*, **87**, 4701.
95 Walker, F.A. (1970) *J. Am. Chem. Soc.*, **92**, 7235.
96 Curtin, D.E., Lousenberg, R.D., Henry, T.J., Tangeman, P.C. and Tisack, M.E. (2004) *J. Power Sources*, **131**, 41.
97 Baranton, S., Coutanceau, C., Roux, C., Hahn, F. and Léger, J.-M. (2005) *J. Electroanal. Chem.*, **577**, 223.
98 Baranton, S., Coutanceau, C., Garnier, E. and Léger, J.-M. (2006) *J. Electroanal. Chem.*, **590**, 100.
99 Baranton, S., Coutanceau, C., Roux, C., Capron, P. and Léger, J.-M. (2005) *Electrochim. Acta*, **51**, 517.
100 Coutanceau, C., El Hourch, A., Crouigneau, P., Léger, J.M. and Lamy, C. (1995) *Electrochim. Acta*, **40**, 2739.
101 Coutanceau, C., Crouigneau, P., Léger, J.M. and Lamy, C. (1994) *J. Electroanal. Chem.*, **379**, 389.
102 El Mouahid, O., Rakotondrainibe, A., Crouigneau, P., Léger, J.-M. and Lamy, C. (1998) *J. Electroanal. Chem.*, **455**, 209.
103 Maillard, F., Martin, M., Gloaguen, F. and Léger, J.-M. (2002) *Electrochim. Acta*, **47**, 3431.
104 Convert, P., Coutanceau, C., Crouigneau, P., Gloaguen, F. and Lamy, C. (2001) *J. Appl. Electrochem.*, **31**, 945.
105 Yeager, E. (1986) *J. Mol. Catal.*, **38**, 5.
106 Biloul, A., Coowar, F., Contamin, O., Scarbeck, G., Savy, M., Van den Ham, D., Riga, J. and Verbist, J.J. (1990) *J. Electroanal. Chem.*, **289**, 189.
107 Lalande, G., Faubert, G., Côté, R., Guay, D., Dodelet, J.P., Weng, L.T. and Bertrand, P. (1996) *J. Power Sources*, **61**, 227.
108 Lalande, G., Côté, R., Tamizhmani, G., Guay, D., Dodelet, J.P., Dignard-Bailey, L., Weng, L.T. and Bertrand, P. (1995) *Electrochim. Acta*, **40**, 2635.
109 Van den Putten, A., Elzing, B., Visscher, W. and Barendrecht, E. (1986) *J. Electroanal. Chem.*, **205**, 233.
110 Gruening, G., Wiesener, K., Gamburtsev, S., Iliev, I. and Kaisheva, A. (1983) *J. Electroanal. Chem.*, **159**, 155.
111 Scherson, D.A., Gupta, S.L., Fierro, C., Yeager, E.B., Kordesch, M.E., Eldridge, J., Hoffman, R.W. and Blue, J. (1983) *Electrochim. Acta*, **28**, 1205.
112 Biloul, A., Gouérec, P., Savy, M., Scarbeck, G., Besse, S. and Riga, J. (1996) *J. Appl. Electrochem.*, **26**, 1139.
113 Van Veen, J.A.R., Colijn, H.A.Z. and Van Baar, J.F. (1988) *Electrochim. Acta*, **33**, 801.
114 Alonso-Vante, N. (2003) *Catalysis and Electrocatalysis at Nanoparticle Surfaces* (eds A. Wieckowski, E.R. Savinova and C.G. Vayenas), Marcel Dekker, Inc, New York, Basel, pp. 931–958.
115 Alonso-Vante, N., Malakhov, I.V., Nikitenko, S.G., Savinova, E.R. and Kochubey, D.I. (2002) *Electrochim. Acta*, **47**, 3807–3814.
116 Le Rhun, V. and Alonso-Vante, N. (2000) *J. New Mater. Electrochem. Syst.*, **3**, 331–336.
117 Johnson, B.F.G., Layer, T.M., Lewis, J., Raithby, P.R. and Wong, W. (1993) *J. Chem. Soc., Dalton Trans.*, 973–980.
118 Layer, T.M., Lewis, J., Martin, A., Raithby, P.R. and Wong, W.T. (1992) *J. Chem. Soc., Dalton Trans.*, 3411–3417.
119 Zaikovskii, V.I., Nagabhushana, K.S., Kriventsov, V.V., Loponov, K.N., Cherepanova, S.V., Kvon, R.I., Bonnemann, H., Kochubey, D.I. and Savinova, E.R. (2006) *J. Phys. Chem. B*, **110**, 6881–6890.
120 Zehl, G., Bogdanoff, P., Dorbandt, I., Fiechter, S., Wippermann, K. and Hartnig, C. (2007) *J. Appl. Electrochem.*, **37**, 1475–1484.
121 Castellanos, R.H., Campero, A. and Solorza-Feria, O. (1998) *Int. J. Hydrogen Energy*, **23**, 1037–1040.

122 Castellanos, R.H., Ocampo, A.L., Moreira-Acosta, J. and Sebastian, P.J. (2001) *Int. J. Hydrogen Energy*, **26**, 1301–1306.

123 Pattabi, M., Castellanos, R.H., Castillo, R., Ocampo, A.L., Moreira, J., Sebastian, P.J., McClure, J.C. and Mathew, X. (2001) *Int. J. Hydrogen Energy*, **26**, 171–174.

124 Pattabi, M., Castellanos, R.H., Sebastian, P.J. and Mathew, X. (2000) *Electrochem. Solid-State Lett.*, **3**, 431–432.

125 Castellanos, R.H., Ocampo, A.L. and Sebastian, P.J. (2002) *J. New Mater. Electrochem. Syst.*, **5**, 83–90.

126 Ocampo, A.L., Castellanos, R.H. and Sebastian, P.J. (2002) *J. New Mater. Electrochem. Syst.*, **5**, 163–168.

127 Gonzalez-Cruz, R. and Solorza-Feria, O. (2003) *J. Solid State Electrochem.*, **7**, 289–295.

128 Sebastian, P.J. (2000) *Int. J. Hydrogen Energy*, **25**, 255–259.

129 Rodriguez, F.J. and Sebastian, P.J. (1999) *J. New Mater. Electrochem. Syst.*, **2**, 107–110.

130 Romero, T., Solorza, O., Rivera, R. and Sebastian, P.J. (1999) *J. New Mater. Electrochem. Syst.*, **2**, 111–114.

131 Sebastian, P.J. and Rodriguez, F.J. (2000) *Surf. Eng.*, **16**, 43–46.

132 Rodriguez, F.J. and Sebastian, P.J. (2000) *Int. J. Hydrogen Energy*, **25**, 243–247.

133 Tributsch, H., Bron, M., Hilgendorff, M., Schulenburg, H., Dorbandt, I., Eyert, V., Bogdanoff, P. and Fiechter, S. (2001) *J. Appl. Electrochem.*, **31**, 739–748.

134 Sebastian, P.J., Ocampo, A.L. and Moreira, J. (2001) *J. New Mater. Electrochem. Syst.*, **4**, 3–6.

135 Sebastian, P.J., Ocampo, A.L. and Moreira, J. (2001) *Int. J. Hydrogen Energy*, **26**, 139–143.

136 Sebastian, P.J., Rodriguez, F.J. and Ocampo, A.L. (1999) *J. New Mater. Electrochem. Syst.*, **2**, 103–106.

137 Manzo-Robledo, A., Boucher, A.C., Pastor, E. and Alonso-Vante, N. (2002) *Fuel Cells*, **2**, 109–116.

138 Lewera, A., Inukai, J., Zhou, W.P., Cao, D., Duong, H.T., Alonso-Vante, N. and Wieckowski, A. (2007) *Electrochim. Acta*, **52**, 5759–5765.

139 Alonso-Vante, N. (2008) *Pure Appl. Chem.*, **80**, 2103–2114.

140 Babu, P.K., Lewera, A., Jong, H.C., Hunger, R., Jaegermann, W., Alonso-Vante, N., Wieckowski, A. and Oldfield, E. (2007) *J. Am. Chem. Soc.*, **129**, 15140–15141.

141 Vogel, W., Le Rhun, V., Garnier, E. and Alonso-Vante, N. (2001) *J. Phys. Chem. B*, **105**, 5238–5243.

142 Borodzinski, A. and Bonarowska, M. (1997) *Langmuir*, **13**, 5613–5620.

143 Song, S., Wang, Y., Tsiakaras, P. and Shen, P.K. (2008) *Appl. Catal. B-Environ.*, **78**, 381–387.

144 Mokrane, S., Makhloufi, L. and Alonso-Vante, N. (2008) *ECS Trans.*, **6**, 93–103.

145 Feng, Y. and Alonso-Vante, N. (2008) *Phys. Status Solidi B*, **245**, 1792–1806.

146 Delacote, C., Johnston, C., Zelenay, M.P. and Alonso-Vante, N. (2008) *ECS Trans.*, **6**, 289–296.

147 Wippermann, K., Richter, B., Klafki, K., Mergel, J., Zehl, G., Dorbandt, I., Bogdanoff, P., Fiechter, S. and Kaytakoglu, S. (2007) *J. Appl. Electrochem.*, **37**, 1399–1411.

148 Lee, J.W. and Popov, B.N. (2007) *J. Solid State Electrochem.*, **11**, 1355–1364.

149 Vracar, L.M., Krstajic, N.V., Radmilovic, V.R. and Jaksic, M.M. (2006) *J. Electroanal. Chem.*, **587**, 99–107.

150 Tian, J., Sun, G., Jiang, L., Yan, S., Mao, Q. and Xin, Q. (2007) *Electrochem. Commun.*, **9**, 563–568.

151 Vogel, W. (1997) *Cryst. Res. Technol.*, **33**, 1140–1154.

152 Gnutzmann, V. and Vogel, W. (1990) *J. Phys. Chem.*, **94**, 4991–4997.

153 Vogel, W., Rosner, B. and Tesche, B. (1993) *J. Phys. Chem.*, **97**, 11611–11616.

154 Vogel, W., Knözinger, H., Carvill, B.T., Sachtler, W.M.H. and Zhang, Z.C.

(1998) *J. Phys. Chem. B*, **102**, 1750–1758.

155 Bönnemann, H., Braun, G., Brijoux, W., Brinkmann, R., Tilling, A.S., Seevogel, K. and Siepen, K. (1996) *J. Organomet. Chem.*, **520**, 143–162.

156 Vogel, W., Britz, P., Bönnemann, H., Rothe, J. and Hormes, J. (1997) *J. Phys. Chem. B*, **101**, 11029–11036.

157 Reetz, M.T., Winter, M., Breinbauer, R., Thurn-Albrecht, T. and Vogel, W. (2001) *Chem. Eur. J.*, **7**, 1084–1094.

158 Dassenoy, F., Vogel, W. and Alonso-Vante, N. (2002) *J. Phys. Chem. B*, **106**, 12152–12157.

159 Alonso-Vante, N. and Vogel, W. (2005) ECS Meeting Abstracts. 207th Meeting of The Electrochemical Society, Quebec City (Canada), May 15–20, p. 1586.

160 Vogel, W. and Alonso-Vante, N. (2005) *J. Catal.*, **232**, 395–401.

161 Habrioux, A., Sibert, E., Servat, K., Vogel, W., Kokoh, K.B. and Alonso-Vante, N. (2007) *J. Phys. Chem. B*, **111**, 10329–10333.

162 Li, W., Zhou, W., Li, H., Zhou, Z., Zhou, B., Sun, G. and Xin, Q. (2004) *Electrochim. Acta*, **49**, 1045.

163 Shukla, A.K., Raman, R.K., Choudhury, N.A., Priolkar, K.R., Sarode, P.R., Emura, S. and Kumashiro, R. (2004) *J. Electroanal. Chem.*, **563**, 181.

164 Lee, K., Zhang, L. and Zhang, J. (2007) *J. Power Sources*, **170**, 291–296.

165 Cruickshank, J. and Scott, K. (1998) *J. Power Sources*, **70**, 40–47.

166 Arico, A., Creti, P., Antonucci, P.L. and Antonucci, V. (1998) *Electrochem. Solid-State Lett.*, **1**, 66.

167 Gurau, B. and Smotkin, E.S. (2002) *J. Power Sources*, **11**, 339.

168 Ren, X., Springer, T.E. and Gottesfeld, S. (2000) *J. Electrochem. Soc*, **147**, 92.

169 Le Ninivin, C., Balland-Longeau, A., Demattei, D., Coutanceau, C., Lamy, C. and Léger, J.-M. (2004) *J. Appl. Electrochem.*, **34**, 1159.

170 Silva, V.S., Schirmer, J., Reissner, R., Ruffmann, B., Silva, H., Mendes, A.,
Madeira, L.M. and Nunes, S.P. (2005) *J. Power Sources*, **140**, 41.

171 Lin, H.-L., Leon Yu, T., Huang, L.-N., Chen, L.-C., Shen, K.-S. and Jung, G.-B. (2005) *J. Power Sources*, **150**, 11.

172 Han, J. and Liu, H. (2007) *J. Power Sources*, **164**, 166.

173 Chu, D., Jiang, R. and Walker, C. (2000) Proc. Power Sources Conf, 39, 140.

174 Jiang, R. and Chu, D. (2000) *J. Electrochem. Soc.*, **147**, 4605.

175 Alonso-Vante, N. and Tributsch, H. (1986) *Nature*, **323**, 431–432.

176 Alonso-Vante, N., Schubert, B. and Tributsch, H. (1989) *Mater. Chem. Phys.*, **22**, 281–307.

177 Alonso-Vante, N. (2003) *Handbook of fuel cells*, vol. 2 (eds W. Vielstich, A. Lamm and H. Gasteiger), John Wiley & Sons, Ltd., Chichester, pp. 534–543.

178 Barton, S.C., Patterson, T., Wang, E., Fuller, T.F. and West, A.C. (2001) *J. Power Sources*, **96**, 329–336.

179 Papageorgopoulos, D.C., Liu, F. and Conrad, O. (2007) *Electrochim. Acta*, **53**, 1037–1041.

180 Papageorgopoulos, D.C., Liu, F. and Conrad, O. (2007) *Electrochim. Acta*, **52**, 4982–4986.

181 Scott, K., Shukla, A.K., Jackson, C.L. and Meuleman, W.R.A. (2004) *J. Power Sources*, **126**, 67–75.

182 Baresel, D., Sarholz, W., Scharner, P. and Schmitz, J. (1974) *Ber. Bunsen Ges.*, **78**, 608–611.

183 Behret, H., Binder, H. and Sandstede, G. (1975) *Electrochim. Acta*, **20**, 111–117.

184 Vayner, E., Sidik, R.A., Anderson, A.B. and Popov, B.N. (2007) *J. Phys. Chem. C*, **111**, 10508–10513.

185 Sidik, R.A. and Anderson, A.B. (2006) *J. Phys. Chem. B*, **110**, 936–941.

186 Susac, D., Sode, A., Zhu, L., Wong, P.C., Teo, M., Bizzotto, D., Mitchell, K.A.R., Parsons, R.R. and Campbell, S.A. (2006) *J. Phys. Chem. B*, **110**, 10762–10770.

187 Susac, D., Zhu, L., Teo, M., Sode, A., Wong, K.C., Wong, P.C., Parsons, R.R.,

Bizzotto, D., Mitchell, K.A.R. and Campbell, S.A. (2007) *J. Phys. Chem. C*, **111**, 18715–18723.

188 Feng, Y., He, T. and Alonso-Vante, N. (2008) *Chem. Mater.*, **20**, 26–28.

189 Lee, K., Zhang, L. and Zhang, J. (2007) *J. Power Sources*, **165**, 108–113.

190 Xia, D., Liu, S., Wang, Z., Chen, G., Zhang, L., Zhang, L., Hui, S. and Zhang, J. (2008) *J. Power Sources*, **177**, 296–302.

8
Carbon Nanotube-Supported Catalysts for the Direct Methanol Fuel Cell
Chen-Hao Wang, Li-Chyong Chen, and Kuei-Hsien Chen

8.1
Introduction

Fuel cells generate power by converting chemical energy to electricity via electrochemical reactions, which involve eco-friendly reactants and products, in comparison to conventional power generators based on the internal combustion engine. Accordingly, the fuel cell is an attractive and promising power generator with a wide range of applications in sensors, portable devices, automotives, and stationary power systems. The direct methanol fuel cell (DMFC) resembles the proton exchange membrane fuel cell (PEMFC) in its power-generating unit, and the membrane electrode assembly (MEA) with a proton exchange-membrane that is sandwiched between two electrodes. However, instead of hydrogen gas, which is used as fuel in a PEMFC, the DMFC uses a methanol solution as the fuel feeding the anode. The methanol is oxidized to CO_2, via an electrochemical process, at the anode and the air is reduced at the cathode to generate the electricity. With methanol solution as fuel, the DMFC system can be operated without a humidification unit, which is essential for the pure hydrogen PEM systems. The fact that methanol solution can be easily stored and refilled and hence can provide a relatively high energy density of 6.08 Whg^{-1} (neat methanol) makes the DMFC a potential solution as a mobile energy source. Therefore, the DMFC system, rather than a secondary battery (such as Li-ion battery), is regarded as a next-generation power source in a portable system.

Despite all the advantages of DMFC, the sluggish methanol oxidation rate and the methanol crossover detrimentally affect its performance. A key technology for the development of DMFC is the development of a highly efficient MEA, in which the methanol can be easily and rapidly oxidized by the catalysts before it arrives at the membrane. As evident from literature, the uniform deposition of carbon-supported nanoscale catalysts, Pt-Ru with an atomic ratio of 1 : 1, coated on the gas diffusion layer (i.e., carbon cloth and carbon paper) has been widely adopted to provide the catalyst layer of the anode in the DMFC [1–3]. Besides the catalysts, the carbon support also markedly interferes with the MEA performance due to its role in

Electrocatalysis of Direct Methanol Fuel Cells. Edited by Hansan Liu and Jiujun Zhang
Copyright © 2009 WILEY-VCH Verlag GmbH & Co. KGaA, Weinheim
ISBN: 978-3-527-32377-7

stabilizing the catalysts, providing diffusion channels and conduction of the output electrons. Traditionally, carbon blacks (CBs), possessing moderate conductivity and large surface area, are widely adopted as the carbon supports. For example, Vulcan XC-72 with an electrical conductivity of 0.25 S cm^{-1} and a BET surface area of 237 m^2 g^{-1}, is extensively used as carbon supports [4]. Although the CBs is normally selected as the support for the fuel cell, a new carbon support with higher conductivity and larger surface area is desired. Recently, carbon nanotubes (CNTs) with a large surface area, high electrical conductivity, and unique shape [5, 6] have emerged as potential carbon supports in fuel cell applications. Numerous works have attempted to disperse platinum (Pt/CNT) or platinum-ruthenium (Pt-Ru/CNT) clusters on CNTs, and the extensive studies exhibited that the CNT-supported catalysts certainly can improve fuel cell performance [7–9].

This chapter will review the recent studies of the CNT-supported catalysts, covering their syntheses, geometry, electrochemical behavior and MEA performance. The MEA that employs the CNT-supported catalysts outperforms that using the CB-supported catalysts, because of the unique surface structures of the former and its advantage of possessing fast electron-transfer characteristics. Additionally, a new approach to grow CNTs directly on carbon cloth as a catalyst support has been demonstrated, along with its application for favorable activity with very low catalyst loading. In overall view, the CNT-supported catalysts have excellent characteristics for fabricating highly efficient MEAs, and hence can potentially promote the commercialization of the DMFC.

8.2
Preparation of Carbon Nanotube-Supported Catalysts

The preparing route of CNT-supported catalysts briefly comprises a functionalization step, followed by a reduction step and a deposition step. High-performance CNT-supported catalysts must have (1) a narrow distribution of size in nanometric scale with an average particle diameter of less than 4 nm; (2) a high-degree binary alloy; (3) uniform dispersion on CNT surface; (4) uniform composition throughout the binary catalysts. Numerous important approaches have been implemented to develop some innovative and cost-effective preparation methods that meet these criteria and optimize by controlling the synthetic procedure and conditions. The methods most usually adopted are the impregnation method, the colloidal method and the electrodeposition method. These three methods include the reduction step in the formation of the catalysts and the deposition step to disperse the catalysts on the CNTs.

8.2.1
Functionalization of Carbon Nanotubes

The distribution of functional groups on the carbon support that serve as nucleation sites is one of the key technologies for the uniform dispersion of the catalysts on the carbon support. From an earlier study of CB-supported catalysts, it has been established that the chemical treatment, which strongly affects the catalyst size as well as the amount of catalyst loading on the CBs, also influences the surface characteristics and the degree of aggregation of CBs [10]. According to its atomic configuration, a CNT is a graphene sheet rolled into a cylinder of nanometer-size diameter; hence planar sp^2 bonding, a characteristic of graphite, can be expected to play a significant role in CNTs characteristics. In order to improve the adhesion of catalyst particles to CNT surfaces, the aromatic rings of CNT sidewalls must be disrupted. An aggressive treatment is required to purify CNTs to enhance the efficiency of surface functionalization of CNTs, i.e., to increase the number of functional groups on CNT surfaces [9]. Briefly, the CNTs are dispersed in concentrated nitric acid for a period to remove the impurities, such as the metallic catalysts, used during the CNTs synthesis, and the residual amorphous carbons. After filtration by abundant water, the CNTs are immersed into a concentrated H_2SO_4–HNO_3 (1:1 v/v) solution with refluxing at 140 °C, creating surface-functional groups, such as carboxyl (–COOH), carbonyl (–CO), and hydroxyl (–OH) [11, 12]. As a result, these oxygen-containing groups stabilize the CNTs in water or any other polar solvents by developing the negative surface charges, and promoting the dispersion of CNTs via electrostatic repulsion among the CNTs.

As stated before, the functionalization step can critically influence the distribution of catalysts on the CNT surface. Beside this, the functional groups also affect the hydrophilic property of the carbon support. Guo et al. noted that the Pt-Ru catalysts supported by the treated carbon nanofibers (CNFs) that had undergone aggressive acid treatment (Pt-Ru/OCNF) outperformed the Pt-Ru catalysts supported on pristine CNFs (Pt-Ru/RCNF) [13]. The polarization curve of the DMFC test indicated that the Pt-Ru/OCNF exhibited a better performance at low and moderate current densities, revealing distinguishable activation and ohmic regions, respectively, because of its substantial hydrophilicity, compared with Pt-Ru/RCNF. However, Pt-Ru/OCNF performed slightly worse at high current densities, corresponding to the mass-transfer region, due to the excessive adherence of CO_2 bubbles to the OCNFs. Hence, the carboxylic functional groups have a large impact on the wetting property of carbon supports, and hence potentially affect the performance of the DMFC.

8.2.2
Polymer-Modified CNTs

In principle, the functional polymers, deposited on the CNTs, would reduce the interfacial resistance between the catalysts and the support, and increase the adsorptive force of catalyst precursors on the supports. To meet these criteria, functional polymers should possess adequate electrical conductivity with the chemical stability during the fuel cell operation. Hsin et al. reported the use of poly(vinylpyrrolidone) (PVP) prior to the deposition of the catalysts on the surface-modified carbon supports (PVP-carbon supports) [14]. They also adopted acid treatment to modify the surface of the carbon supports (AO-carbon support). They found methanol oxidation performance followed the order: Pt/PVP-GNF$_H$ > Pt/PVP-CNT$_{arc}$ > Pt/AO-CNT$_{arc}$ > Pt/PVP-CNT$_{CVD}$ > Pt/AO-CNT$_{CVD}$ > Pt/CB(XC-72) > Pt/AO-GNF$_H$, such that the Pt/PVP-GNF$_H$ performed 270% better than the Pt/CB(XC-72). Here, GNF$_H$, CNT$_{arc}$, CNT$_{CVD}$ and CB(XC-72) stands for the herringbone graphite carbon nanofibers, the multiwalled carbon nanotubes (MWCNTs) from arc discharge, the MWCNTs from chemical vapor deposition (CVD), and the commercial carbon blacks (Vulcan XC-72R), respectively. These results showed the importance of surface modification, which is crucial to improve the DMFC performance of PVP-modified carbon supports because it affects the electrical conductivity of the carbon supports. Selvaraj et al. employed the conducting polymer films of polypyrrole (PPy), for the surface functionalization of the carbon supports, prior to the deposition of Pt and/or Pt-Ru (alloy) catalysts [15]. The methanol oxidation activities follow the order Pt-Ru/PPy-MWCNT > Pt-Ru/MWCNT > Pt-Ru/CB(commercial) > Pt-Ru/CB(home-made) > Pt-Ru/CNF, indicating that Pt-Ru/PPy-MWCNT exhibits an excellent catalytic activity and stability in methanol oxidation reaction (MOR). Accordingly, the composite films are more promising support materials for fuel cell applications.

To enhance adsorption between the catalysts and the CNTs, Wang et al. adopted the poly(diallyldimethylammonium chloride) (PDDA) that can create cationic charges on modified CNT surfaces, and hence induce electronic adsorption of metal catalysts on the negatively charged surface, causing the electrostatic assembly of nanoparticles thereon [16]. Pan et al. demonstrated polyoxmetalate-modified MWCNTs (PMo12-MWCNT) as the carbon supports [17]. Polyoxmetalate (PMo12) is highly proton conductive and its stable redox activity in acid solution is as follows (Equation 8.1).

$$PMo_{12}O_{42}^{7-} + ne^- + nH^+ \leftrightarrow H_nPMo_{12}O^{7-} \quad (n = 2, 4, 6) \tag{8.1}$$

The catalysts, coated on the PMo12-MWCNT, provided a 1.4 times higher exchange current density, a 1.5 times higher specific activity, and better cycle stability than those on the untreated MWCNTs. Polyaniline (PAni) polymer, which is not only an electrically conducting polymer but also a proton conductor, after dispersion on the MWCNTs as a composite to form the catalyst support, provided a considerable enhancement in MOR activity [18–21].

8.2.3
Impregnation Method

The impregnation method is the most popular procedure for preparing CNT-supported catalysts; it is a simple and straightforward route, in which the reducing agents convert the catalytic precursors to the metals catalysts on the carbon supports. Figure 8.1 presents the brief procedure of the impregnation method, whose main steps are the impregnation step and the chemical reduction step. During the

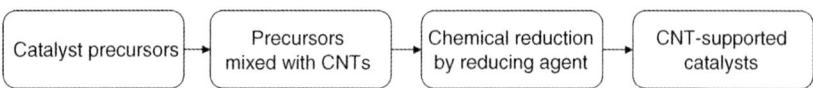

Figure 8.1 Preparing route of impregnation method.

impregnation step, the catalytic precursors are mixed with the functionalized CNTs in an aqueous solution, and a homogenous mixture is formed by powerful stirring and ultrasound. In the chemical reduction step, the reduction is conducted by either the liquid-phase agents (such as $NaBH_4$, $Na_2S_2O_3$, $Na_4S_2O_5$, N_2H_4, and formic acid) or the gas-phase agents (like CO, H_2 etc.), at elevated temperature. Studies have established that the synthetic conditions, including the process temperature, the choice of the species of metal precursor and the reducing agent are essential to the impregnation efficiency. Table 8.1 lists some important reports on CNT-supported catalysts that were synthesized by the impregnation method [22–35].

A typical example of the impregnation method, using the liquid-phase reducing agent, is as follows: the functionalized CNTs are mixed with the metallic precursors (such as H_2PtCl_6 and $RuCl_3$) in an aqueous solution with strong stirring; the weighed percentage of metal loading is ranged of 20%–60% according to the desired amount. The pH value of mixed solution is adjusted in the range 10–11 by adding 1 M NaOH. The metallic precursors are reduced by slowly adding a reducing solution at elevated temperature. Once the reaction is complete, the mixed solution is washed and filtered with hot water to remove the residual by-products (e.g., NaCl etc.). Finally, the filtered ink is dried at 80 °C in a vacuum oven. In the liquid-phase impregnation, the solution environment strongly affects the as-prepared Pt and Ru nanoparticles (NPs) formed on the CNT surface. Clearly, the pH of the solution environment markedly affects the potential of zero charge (PZC) of a double layer structure at the interface between the CNTs and the solution, indicating that the isoelectric point (IEC) of the interface affects the catalytic deposition rate and uniformity. Therefore, NaOH is added to maintain an alkaline environment to alter the IEC of the interface. To prevent the agglomeration of NPs during the deposition, which can lead to the worsening of the uniformity in NP distribution and the formation of large catalysts, an additional stabilizer is introduced into the reaction [23, 26, 30]. For instance, Liu *et al.* introduced EDTA-2Na as the stabilizer, which has four carboxyl anions and two nitrogen-containing groups [26]. The EDTA-2Na molecules can adsorb on the surface of metal particles and exert either steric hindrance or coulombic effects on the metal particles, and stabilize them. Additionally, EDTA-2Na can be easily removed from the particle surfaces by washing in water or by extraction. Figure 8.2 presents Pt/MWCNT prepared by adding EDTA-2Na: the Pt NPs are uniformly dispersed on the MWCNTs with a diameter of 1.5 nm and the particle size distribution is very narrow. Therefore, it suggests that the application of the additional stabilizer is very effective and offers sufficient protection against the aggregation of NPs. Chien *et al.* stated that the use of additives like $NaHSO_3$ and $Ca(OH)_2$ increased the conversion efficiency of Ru^{3+} to Ru metal to 2–3 times that without the additives [36]. SO_3^{2-} is a stronger ligand than Cl^-, and so can form more stable complex ions with Pt^{4+} and Ru^{3+}, improving the dispersion and narrowing the distribution of CNT-supported catalysts.

Gas-phase impregnation is a faster process than liquid-phase impregnation and less labor intense. In the conventional method, CNTs that are mixed with the precursors are reduced by hydrogen gas at high temperature. However, the resulting particles are found to be larger than 5 nm, because of the agglomeration at high-reducing

Table 8.1 Summary of the reported CNT-supported catalysts synthesized by different preparation route of the impregnation method.

wt.% catalyst/carbon support, metal loading	Precursor	Reducing agent, stabilizer	Particle size	Activity	Measurement protocol	Ref.
A: 20% $Pt_{50}Ru_{50}$/MWCNT, 2.5 mg cm^{-2}	H_2PtCl_6, $RuCl_3$	$NaBH_4$	3–4 nm	60 mA cm^{-2} @ 0.4 V	A single DMFC, 1 M MeOH and O_2, 70 °C	[22]
C: 20% Pt/MWCNT + 20% Pt/C, 5.0 mg cm^{-2}				P_{max}: 35 mW cm^{-2}		
A: 20% $Pt_{50}Ru_{50}$/MWCNT	H_2PtCl_6, $RuCl_3$	$NaBH_4$	3–4 nm	45 mA mg^{-1} @ 0.4 V	A single DMFC, 2 M MeOH and O_2, 70 °C	[23]
C: 40% Pt/C, 2.1 mg cm^{-2}				P_{max}: 26 mW mg^{-1}		
30% $Pt_{48}Ru_{52}$/MWCNT	$Pt(acac)_2$, $Ru(acac)_3$	H_2 with supercritical fluid CO_2	2.8 nm	$i_{p,a}$: 680 mA mg^{-1} Pt @ 0.913 V (vs. NHE)	2 M MeOH + 1 M H_2SO_4, v: 60 mV s^{-1}	[112]
15% Pt/MWCNT-Chitosan	H_2PtCl_6	Ethylene glycol	5.0 nm	$i_{p,a}$: 33 mA mg^{-1} Pt @ 0.760 V (vs. SCE)	0.25 M MeOH + 0.1 M H_2SO_4, v: 5 mV s^{-1}	[25]
20% Pt/MWCNT	H_2PtCl_6	Fomaldehyde, EDTA-2Na	1.5 nm	$i_{p,a}$: 1250 mA mg^{-1} Pt @ 0.800 V (vs. Ag/AgCl)	0.5 M MeOH + 0.5 M H_2SO_4, v: 50 mV s^{-1}	[26]
20% $Pt_{53}Ru_{47}$/MWCNT	H_2PtCl_6	Ethylene glycol	2.3 nm	i_a: 135 mA mg^{-1} Pt @ 0.700 V (vs. RHE)	2 M MeOH + 1 M H_2SO_4, v: 5 mV s^{-1}	[113]
10% Pt/MWCNT	H_2PtCl_6	H_2	10 nm	i: 6.1 mA mg^{-1} Pt	1 M MeOH + 1 M H_2SO_4, polarization test @ 0.6 V (vs. SCE), 60 °C	[28]
		HCOONa	5.3 nm	i: 20 mA mg^{-1} Pt		
		$NaBH_4$	4.6 nm	i: 25 mA mg^{-1} Pt		
20% $Pt_{50}Fe_{50}$/MWCNT	H_2PtCl_6, $FeCl_3$	$NaBH_4$	1.9 nm	$i_{p,a}$: 240 mA mg^{-1} Pt @ 0.677 V (vs. Ag/AgCl)	1.0 M MeOH + 0.5 M H_2SO_4, v: 50 mV s^{-1}	[29]

Catalyst	Precursor	Reducing agent	Size	Performance	Conditions	Ref.
34% $Pt_{45}Ru_{45}Ir_{10}$/MWCNT	H_2PtCl_6, $RuCl_3$, $IrCl_3$	Ethylene glycol, citric acid	1.1 nm	$i_{p,a}$: 2549 mA mg^{-1} Pt @ 1.192 V and i_a: 852 mA mg^{-1} Pt @ 0.700 V (vs. RHE)	0.5 M MeOH + 0.5 M H_2SO_4, v: 50 mV s^{-1}	[30]
20% $Pt_{50}Ru_{50}$/MWCNT	H_2PtCl_6, $RuCl_3$	$NaBH_4$, tetraoctylammonium bromide (TOAB)	3.2 nm	$i_{p,a}$: 300 mA mg^{-1} Pt @ 0.750 V (vs. SCE)	1.0 M MeOH + 0.5 M H_2SO_4, v: 50 mV s^{-1}	[31]
A: 20% $Pt_{50}Ru_{50}$/MWCNT, 2.1 rg cm^{-2} C: 40% Pt/C, 2.0 mg cm^{-2}	H_2PtCl_6, $RuCl_3$	Ethylene glycol	3.5–4.0 nm	40 mA cm^{-2} @ 0.4 V P_{max}: 32 mW cm^{-2}	A single DMFC, 2 M MeOH and O_2, 70 °C	[32]
A: 20% $Pt_{50}Ru_{50}$/MWCNT, 4.0 rg cm^{-2} C: Pt black, 4.0 mg cm^{-2}	H_2PtCl_6, $RuCl_3$	Ethylene glycol	3.5–4.0 nm	80 mA cm^{-2} @ 0.4 V P_{max}: 62 mW cm^{-2}	A single DMFC, 1 M MeOH and O_2, 60 °C	[32]
6.4% $Pt_{45}Ru_{55}$/SWCNT	$Pt(acac)_2$, $Ru(acac)_2$	H_2 with supercritical fluid CO_2	5–10 nm	$i_{p,a}$: 833 mA mg^{-1} Pt @ 0.700 V (vs. SCE)	1.0 M MeOH + 2.0 M H_2SO_4, v: 50 mV s^{-1}	[33]
A: 30% $Pt_{50}Ru_{50}$/C, 2.0 rg cm^{-2} C: 10% Pt/MWCNT, 1.0 rg cm^{-2}	H_2PtCl_6	Ethylene glycol	2.5 nm	190 mA cm^{-2} @ 0.4 V P_{max}: 105 mW cm^{-2}	A single DMFC, 1 M MeOH and O_2, 90 °C	[27]
40% Pt/MWCNT	H_2PtCl_6	HCOOH	2.9 nm	$i_{p,a}$: 430 mA mg^{-1} Pt @ 0.680 V (vs. SCE)	1.0 M MeOH + 0.5 H_2SO_4, v: 50 mV s^{-1}	[34]
A: 30% $Pt_{50}Ru_{50}$/MWCNT, 4.0 rg cm^{-2} C: 40% Pt/C, 3.0 mg cm^{-2}	H_2PtCl_6, $RuCl_3$	Ethylene glycol	3.5 nm	50 mA cm^{-2} @ 0.4 V P_{max}: 117 mW cm^{-2}	A single DMFC, 2 M MeOH and O_2, 70 °C	[35]

MWCNT: Multi-walled CNT; DWNT: Double-walled CNT; SWCNT: Single-walled CNT.
A: anode; C: cathode.
P_{max}: the maximum power of a single DMFC test.
i_a: forward current of CV scan.
$i_{p,a}$: forward peak current of CV scan.
v: scan rate of CV scan.

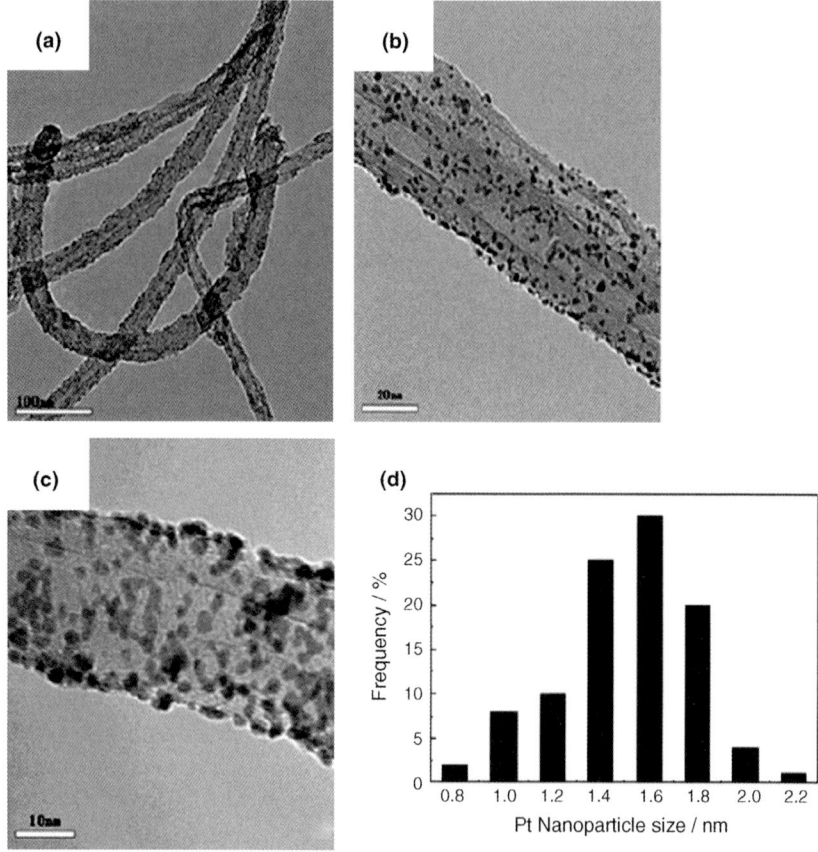

Figure 8.2 (a) TEM images of Pt/MWCNT prepared at EDTA-2Na/Pt molar ratio of 1:1. (b) and (c) HRTEM image of the Pt/MWCNT shown in panel 6a; (d) Size histogram of Pt nanoparticles in the Pt/MWCNT prepared at EDTA-2Na/Pt molar ratio of 1:1 [26].

temperature (>200 °C) [28]. To prevent the catalyst agglomeration, Yen *et al.* adopted supercritical CO_2 as the medium and H_2 as the reducing agent [24]. When the final pressure of the reduction step has been finely controlled at a high pressure (about 250 atm), the catalyst size was effectively reduced to 2.8 nm.

8.2.4
Colloidal Method

The colloidal method is another extensively used route for preparation of the CNT-supported catalysts, in which the catalyst precursors are reduced into the colloids in the solution, followed by the adsorption of colloids onto the CNT surfaces, as depicted in Figure 8.3. Table 8.2 represents some relevant works [37–39].

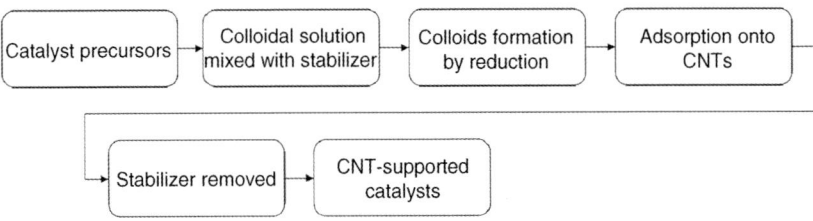

Figure 8.3 Preparing route of colloidal method.

To control the particle growth, and simultaneously to prevent the agglomeration during the preparation, an additional stabilizer must be used. As shown in Table 8.2, a surfactant is generally selected as a stabilizer in the process, because it can reduce the surface tension or interfacial tension between as-prepared colloids and fluid. As a result, the smaller colloids are formed in the suspension. Cao et al. prepared Pt NPs on single-walled carbon nanotubes (SWCNTs) with a small particle size of 2.2 nm and a high metal loading of 60 wt.% by introducing sodium dodecylbenzene sulfonate (SDBS) as the stabilizer and CO as the reducing agent [38]. The brief procedure is as follows: SWCNTs are mixed with SDBS ultrasonically to form surfactant-assisted water-soluble SWCNTs. To prepare the Pt–carbonyl solution, K_2PtCl_6 and CH_3COONa, in methanol solution, were subjected to CO environment in a sealed flask, at a temperature of 50 °C for 24 h, and then the green colored Pt–carbonyl complex is yielded. Mixing the surfactant-assisted solubilized SWCNTs and the Pt–carbonyl complex yields a homogeneous solution with a complete disappearance of green color, suggesting the bonding of Pt clusters onto the surface of SWCNTs via a stronger interaction with SWCNTs. Pt NPs are observed to be almost monodispersed on the SWCNT surface, as presented in Figure 8.4. Mu et al. employed an alternative approach for preparing well-dispersed Pt NPs on MWCNTs [39]. The Pt precursors are reduced by the ethylene glycol in NaOH solution at elevated temperature, forming Pt colloids. The ethylene glycol is normally oxidized to the glycolic acid during the reduction, and the glycolic acid is subsequently deprotonated to form a glycolic anion in alkaline solution. Since the glycolic anion is a good stabilizer, it stabilizes the Pt colloids by potentially forming chelate-type complexes via its carboxyl groups [40]. Then, the final colloidal solution is mixed with a solution of triphenylphosphine (PPh_3) in toluene, forming a PPh_3-modified Pt/toluene solution. To prepare Pt/MWCNT, the mixtures of MWCNTs and PPh_3-modified Pt are immersed into the toluene solution with ultrasonic stirring for several days, and the PPh_3 acts as a crosslinker to connect Pt and MWCNT.

8.2.5
Electrodeposition Method

The electrodeposition method is straightforward; its precursors are directly reduced on the CNTs via electrochemical deposition process. Figure 8.5 shows the main steps, which include the dissolution of the precursors and the subsequent electrochemical reduction. Table 8.3 lists some interesting and relevant investigations [41–44].

Table 8.2 Summary of the reported CNT-supported catalysts synthesized by different preparation route of the colloidal method.

wt.% catalyst/carbon support, metal loading	Precursor	Reducing agent, stabilizer	Particle size	Activity	Measurement protocol	Ref.
20% Pt-CeO$_2$/MWCNT	H$_2$PtCl$_6$	Ethylene glycol, itself	2.5 nm	$i_{p,a}$: 490 mA mg^{-1} Pt @ 0.750 V (vs. SCE)	1 M MeOH + 1 M HClO$_4$, v: 50 mV s^{-1}	[37]
60% Pt/SWCNT	Na$_2$PtCl$_6$	CO, sodium dodecylbenzene sulfonate (SDBS)	2.2 nm	$i_{p,a}$: 893 mA mg^{-1} Pt @ 0.780 V (vs. SCE)	0.5 M MeOH + 0.5 H$_2$SO$_4$, v: 50 mV s^{-1}	[38]
24% Pt/SWCNT	H$_2$PtCl$_6$	Ethylene glycol, triphenylphosphine (PPh$_3$)	2.7 nm	$i_{p,a}$: 150 mA mg^{-1} Pt @ 0.740 V (vs. SCE)	2.0 M MeOH + 1.0 H$_2$SO$_4$, v: 20 mV s^{-1}	[39]

MWCNT: Multi-walled CNT; DWNT: Double-walled CNT; SWCNT: Single-walled CNT.
A: anode; C: cathode.
P_{max}: the maximum power of a single DMFC test.
i_a: forward current of CV scan.
$i_{p,a}$: forward peak current of CV scan.
v: scan rate of CV scan.

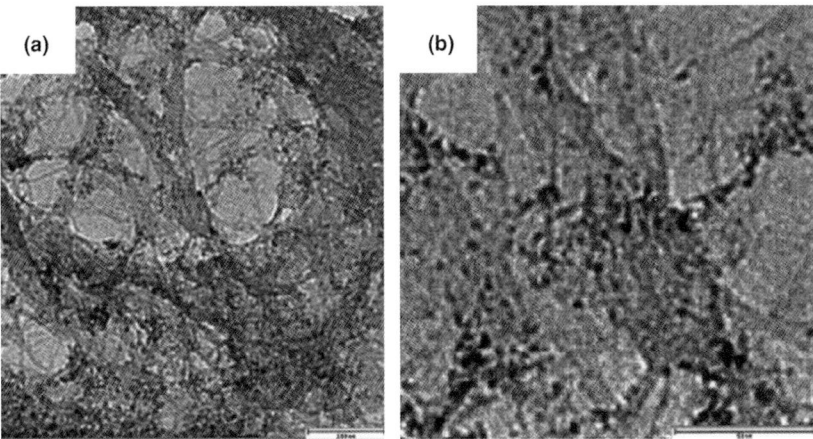

Figure 8.4 TEM images of Pt/SWCNT with 60 wt.% Pt loading. (a) length scale bar is 100 nm; (b) length scale is 50 nm [38].

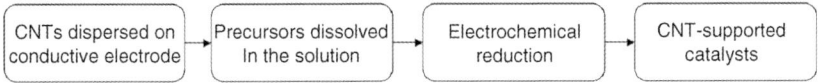

Figure 8.5 Preparing route of electrodeposition method.

Some reports indicate that the resultant catalyst sizes produced by either the constant potential method or the cyclic voltammetry (CV) method are larger than 10 nm [41, 43]. It seems that the constant potential and CV methods easily induces the agglomeration of catalysts. Recently, Zhan et al. adopted the potential step method with a pulse width of 0.001 s to avoid the agglomeration during the preparation of CNT-supported catalysts [42]. They followed three steps: (1) the electrochemical treatment of CNTs, (2) the electrochemical oxidation of Pt^{2+} to Pt^{4+}, and finally (3) the electrochemical reduction of Pt^{4+} to Pt NPs onto the MWCNT surfaces. For comparison, they also used CV method to prepare CNT-supported catalysts, revealing that the methanol oxidation peak current of Pt/MWCNT prepared by the potential step method was ten times stronger than that of Pt/MWCNT prepared by the CV method.

8.3
Characteristics of the Carbon Nanotube Electrode

8.3.1
Electrochemical Behavior of the CNT Electrode

In 1991, Iijima first observed the CNTs [45], and explored their unique mechanical, optical and electronic properties. Henceforth, the other interesting characteristics/phenomena of CNTs have been explored and exploited in numerous applications [6, 46–49]. One attractive property is their very low resistivity in the range of 10^{-2}

Table 8.3 Summary of the reported CNT-supported catalysts synthesized by different preparation route of the electrodeposition method.

wt.% catalyst/carbon support, metal loading	Precursor	Condition	Particle size	Activity	Measurement protocol	Ref.
Pt/MWCNT	H_2PtCl_6, H_2SO_4	−0.05 V (vs. RHE)	10.2 nm	$i_{p,a}$: 225 mA mg^{-1} Pt @ 1.080 V (vs. RHE)	0.5 M MeOH + 0.5 H_2SO_4, v: 10 mV s^{-1}	[41]
Pt/SWCNT			9.1 nm	$i_{p,a}$: 170 mA mg^{-1} Pt @ 1.080 V (vs. RHE)		
Pt/MWCNT	K_2PtCl_4, H_2SO_4	Potential-step method, 0.001 s pulse, 1.0 V and −0.26 V	2–4 nm	$i_{p,a}$: 18 mA cm^2 @ 0.680 V (vs. Ag/AgCl)	2.0 M MeOH + 0.1 H_2SO_4, v: 100 mV s^{-1}	[42]
Pt$_{50}$Ru$_{50}$/MWCNT	H_2PtCl_6, $RuCl_3$, H_2SO_4	−0.25 V (vs. SCE)	60–80 nm	$i_{p,a}$: 12.5 mA cm^2 @ 0.65 V (vs. SCE)	1.0 M MeOH + 0.6 H_2SO_4, v: 50 mV s^{-1}	[43]
Pt/SWCNT	H_2PtCl_6, H_2SO_4	A duty factor of 0.5 at −0.35 V (vs. SCE) with pulses of 0.012 s duration	20 nm	$i_{p,a}$: 371 mA mg^{-1} Pt @ 0.800 V (vs. SCE)	1.8 M MeOH + 1.0 H_2SO_4, v: 100 mV s^{-1}	[44]

MWCNT: Multi-walled CNT; DWNT: Double-walled CNT; SWCNT: Single-walled CNT.
A: anode; C: cathode.
P_{max}: the maximum power of a single DMFC test.
i_a: forward current of CV scan.
$i_{p,a}$: forward peak current of CV scan.
v: scan rate of CV scan.

to $10^{-4}\,\Omega$-cm [50], which value is two to four orders of magnitude lower than that of conventional CBs (Vulcan XC-72), which is approximately $5\,\Omega$-cm [4]. Therefore, CNTs are regarded as potential conductors, suitable for electronic applications [51–54].

In electrochemical applications, the electron transfer rate of an electrode depends on several factors, such as the structure and morphology of the carbon material used in the electrodes. Nugent et al. [55] showed that the micron-size electrodes made of MWCNTs exhibited the peak potential separation ($\geq E_p$) of the MWCNT-electrode is 59 mV, at all sweep rates examined (20–50 mV/s), for a redox reaction in 5 mM $Fe(CN)_6^{3-/4-}$ and sulfuric acid solution (Figure 8.6). Such observation indicates

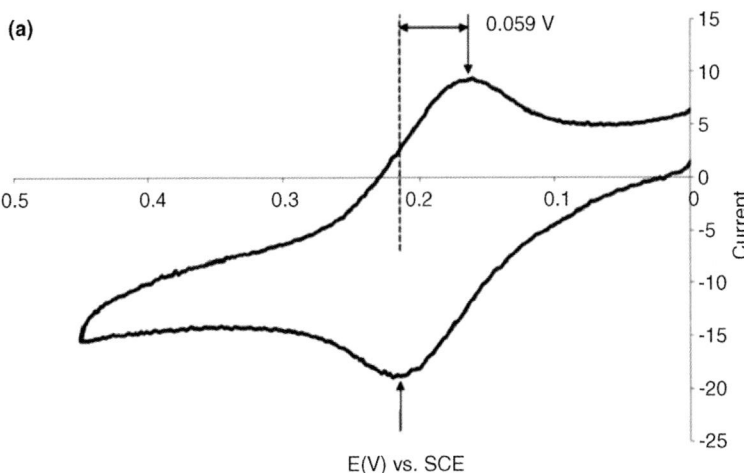

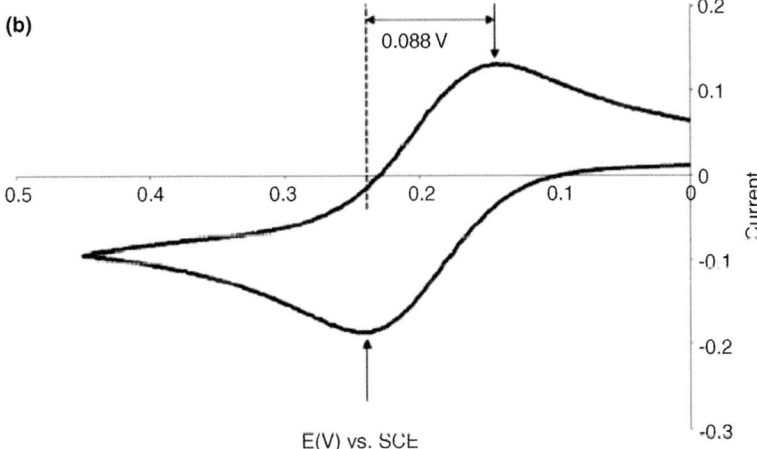

Figure 8.6 CV curves of 5 mM $K_4Fe(CN)_6$ (a) on a MWCNT electrode and (b) on a Pt electrode [55].

that nearly reversible reactions occur on the MWCNT-electrode. Restated, the MWCNT-electrode exhibits Nernstian behavior and fast electron transfer kinetics associated with electrochemical reactions, because of either the ballistic electron transport phenomena [56] or the possession of high density of states in the CNTs [55]. Accordingly, the CNT electrode has promising electrochemical activity that can be exploited in electrochemical-based applications.

Interestingly, the modified CNT electrodes exhibited significantly enhanced performance relative to the unmodified CNT electrode in redox reactions [57–63]. For example, Hu et al. adopted carboxyl-modified MWCNT films as an electrode, by involving the reduction of the carboxylic acid groups on the CNTs [61], which exhibited a stable and enhanced CV behavior in 1 mM Fe^{2+}/0.2 M $HClO_4$ solution. They concluded that, because of their high aspect ratio, the modified MWCNT films have a much larger surface area than the conventional electrodes, such as glassy carbon and highly oriented pyrolytic graphite (HOPG) electrodes, and therefore could exhibit comparatively more efficient charge transfer and hence higher current. Barisci et al. performed a series of studies on the CV responses and capacitive behavior of SWCNT paper [57–60]. When the SWCNT paper was pretreated by thermal annealing, it showed significant changes including increase of hydrophobicity and elimination of electroactive surface-functional groups and other impurities; because treated SWCNT paper has a lower double-layer capacitance, no Faradaic response or associated pseudo-capacitance and a better frequency response. Yang et al. adopted two electrodes: one of opened MWCNT and the other of closed MWCNT [61]. They were employed in organic solution with Li ions, and the associated kinetic mechanisms were elucidated by the electrochemical impedance spectra (EIS). Li ions perform faster diffusion for the opened MWCNT-electrode than the closed MWCNT-electrode: the diffusion of Li ion into the opened MWCNT occurs around the inner core and the outer surface, and so is rapid. The decrease of the double-layer capacitance of the opened MWCNT also promotes the ion diffusion herein. Papakonstantinou et al. stated that the acid-treated MWCNTs substantially improved the electron transfer kinetics [62]. Acid treatment modifies the charge carrier density on the surface by introducing surface states and hence enhances the electron transfer.

Two methods are preferred for the synthesis of MWCNTs: one is the arc discharge method (Arc-MWCNT), and the other is chemical vapor deposition technique (CVD-MWCNT). Which procedure is suitable for preparing a CNT electrode? The basis of the electrochemical behavior of the CNT electrode is well established and suggested to be attributed entirely to the presence of edge-plane defects at the open end of the CNT, rather than the basal plane graphite of the CNT sidewall, as presented in Figure 8.7 [64]. Evidence has demonstrated that the CVD-MWCNT has a greater catalytic activity than the Arc-MWCNT, as the latter has closed ends and thus resembles basal plane graphite [47, 65]. However, after the Arc-MWCNT is subjected to an aggressive treatment (by plasma etching, anodic activation, and acidic treatment), the basal-plane end caps of Arc-MWCNT suffer rupture exposing the edge-plane defects, and hence the electrochemical activity markedly improves [65].

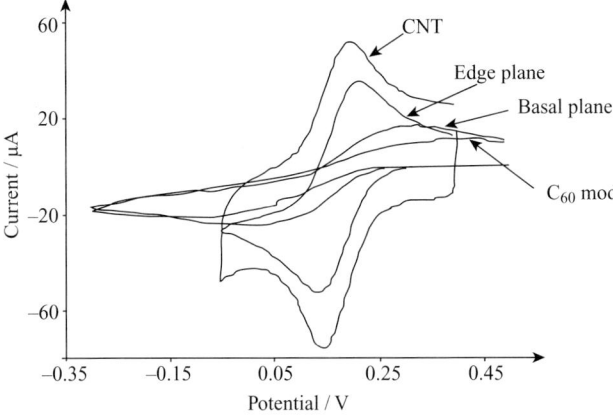

Figure 8.7 CV scans for the reduction of 1 mM ferricyanide for CNT and C_{60} film-modified bppg electrodes. Also shown is the response of a bare basal plane electrode and an edge plane pyrolytic graphite electrode. All scan rates of 100 mV s^{-1} [64].

8.3.2
Durability of the CNT Electrode

The 'corrosion' of carbon materials proceeds by the following steps (Equations 8.2 and 8.3) [66]:

$$C_x \rightarrow C_x^+ + e^- \tag{8.2}$$

$$C_x^+ + yH_2O \rightarrow C_xO_y + 2yH^+ \tag{8.3}$$

These steps are the oxidation of carbon atoms, followed by hydrolysis, and then disproportionation, which produces stable surface oxides and CO_2. Some studies have noted that the reaction of water molecules yields the oxygen that forms surface oxides or CO_2 in electrochemical oxidations at elevated temperature [67–69]. Since PEMFC and DMFC systems normally operate under 100 °C, the corrosion problem associated with the oxidation of carbon materials is mild [70]. However, the cathode is usually operated at a high potential, which may accelerate the corrosion of carbon. As a result, CO_2 may be produced by the further oxidation, causing a loss of mass and surface area of the carbon supports:

$$C + 2H_2O \rightarrow CO_2 + 4H^+ + 4e^- \quad E = 0.207\ V_{RHE} \tag{8.4}$$

Roen et al. characterized the effect of Pt on the corrosion of the carbon supports by on-line mass spectroscopy during CV scans, with varying Pt mass fraction, species of catalyst and temperature. They found that the CO_2 generation rate increased with Pt mass fraction ~0, 10 and 39%, with the balance CBs (Vulcan XC-72) [71]. Obviously, the existence of Pt accelerates the corrosion rate of the carbon support. On the other hand, the degradation of fuel cell performance due to unprotected start up and shut

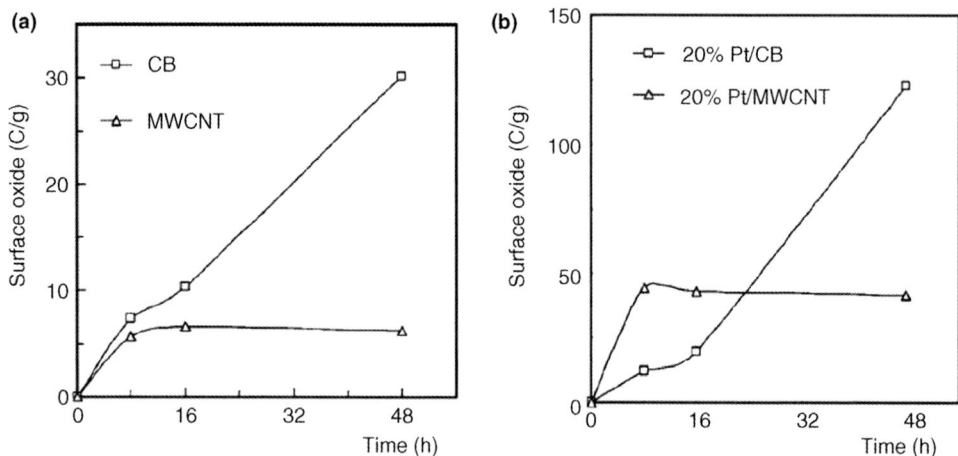

Figure 8.8 (a) The amount of surface oxides in terms of anodic change transfer for electrochemically oxidized MWCNT and CB under (a) noncatalyzed conditions and (b) under catalyzed conditions, showing that electrochemical oxidation for the MWCNTs was stabilized after about 8 h under both conditions [73].

down can be accelerated by the corrosion/oxidation of the carbon supports at high potential [72].

Li et al. studied noncatalyzed MWCNTs and catalyzed MWCNTs (Pt/MWCNT) in 1.0 M sulfuric acid at a constant potential of 1.2 V [73]. For comparison, CB (Vulcan XC-72) was also treated under the same conditions. Figure 8.8a and b present the corrosive trends of the noncatalyzed supports (CBs and MWCNTs) and the catalyzed supports (Pt/CB and Pt/MWCNT). The noncatalyzed MWCNTs and catalyzed MWCNTs are stabilized after an initial period, although the oxidation of the former occurs much more slowly than that of the latter. In contrast, the oxidation of noncatalyzed and catalyzed CBs is more aggressive. The difference in the corrosion behavior between MWCNTs and CBs can be explained by the difference between their structures: CB comprises mostly amorphous carbon but MWCNT comprises rolled-up graphene sheets. Amorphous carbon is well known to be more easily oxidized than graphitic carbon [4], and so the turbostratic structure of CB can be easily oxidized. The MWCNTs consist of a basal graphitic plane since it is of tubular structure, except at its end edge. Therefore, operation at high potential can damage the one or two of outermost graphene layers, generating surface defects with edges on the MWCNT surface. Figure 8.9 presents the corrosion mechanisms, revealing that the swelling of CB leads to an in-depth electrochemical oxidation, while the swelling of MWCNT can occur only at the ends of the nanotubes, making only a very limited contribution to the overall oxidation. Wang et al. [74] and Shao et al. [75] also demonstrated that MWCNTs were electrochemically more stable than CBs (Vulcan XC-72) with the formation of fewer surface oxides and a lower corrosion current under the investigated conditions. Studies also reported that, even

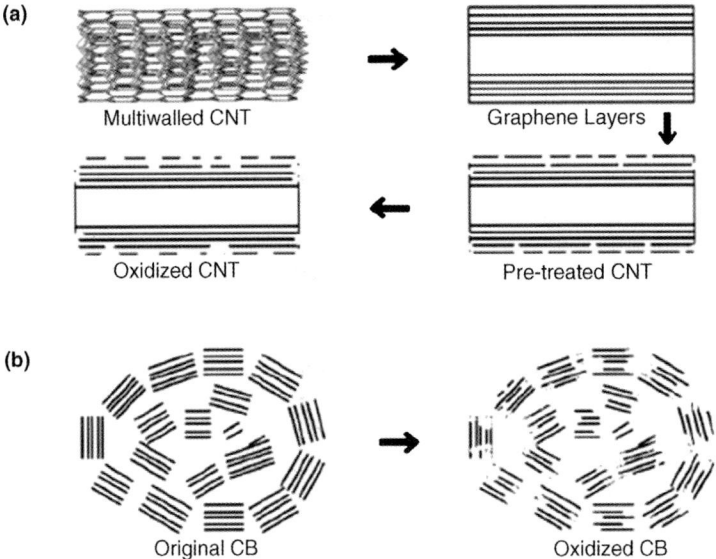

Figure 8.9 Schematic illustration of the structural changes of (a) a MWCNT and (b) a CB particle, before and after electrochemical oxidation. For MWCNTs, corrosion is limited to the top defect layers. The inside three basal graphene layers remain intact except at the two ends; for CB, corrosion progresses into the inside of the CB particle [73].

at an elevated temperature (80 °C), MWCNTs still possessed more stability than CBs [76].

In order to compare the corrosion effects on SWCNTs and MWCNTs supports, Wang et al. estimated the oxygen/carbon atomic ratios (O/C ratios) of carbon supports after aggressive oxidation treatment [77]. The O/C ratio of SWCNT was increased from 2.6 to 9.1% whereas that of MWCNT was increased from 2.6 to 6.7%, indicating that the MWCNTs was more electrochemically stable than the SWCNTs. The SWCNTs has a high BET surface area, which is effectively related to the corrosion rate. Since the local strain energy increases along the entire length of the tube because the radius of curvature declines [78], the smaller tube diameter (~2 nm) of SWCNT should correspond to a higher oxidation rate.

8.4
Electrochemical Behavior of Carbon Nanotube-Supported Catalysts

8.4.1
Methanol Oxidation Reaction

Methanol oxidation is a six-electron transfer reaction, specified as in Equation 8.5:

$$CH_3OH + H_2O \rightarrow CO_2 + 6H^+ + 6e^- \tag{8.5}$$

At room temperature, the reversible potential is 0.046 V. However, the MOR does not proceed by a simple route, but along a variety of routes [79]. The 'preferred' route can be divided into three steps (Equations 8.6, 8.7, 8.8).

$$1^{st} \text{ step}: CH_3OH \rightarrow CH_2O + 2H^+ + 2e^- \tag{8.6}$$

$$2^{nd} \text{ step}: CH_2O + H_2O \rightarrow HCOOH + 2H^+ + 2e^- \tag{8.7}$$

$$3^{rd} \text{ step}: HCOOH \rightarrow CO_2 + 2H^+ + 2e^- \tag{8.8}$$

Accordingly, MOR has a high overpotential and is a more sluggish reaction, inducing a high onset potential and a low oxidation current for the DMFC. A survey of suitable catalysts indicates that only Pt and Pt-based alloy can provide with reliable activity and stability of MOR [79–81]. The $-CO$, one of the intermediates in MOR, can adsorb onto the Pt surface, consequently blocking the active reaction sites, and hence reducing the rate of reaction. The presence of a second promoting metal, such as ruthenium (Pt-Ru), provides $-OH$ by water hydrolysis at low potential, which reacts with $-CO$, and leads to the formations of CO_2 and water, via a bifunctional mechanism [1] and a ligand (electronic) effect [82].

According to literature [70, 83–85], the current consensus is that the optimal ratio of the Pt-Ru catalyst is 1:1, with a high degree of alloying, and a nanoscale size of particles (2–5 nm preferred) in order to optimize catalyst use. The anodic catalyst loading for the practical DMFC operation is 2–8 mg cm^{-2}, which is much higher than the anodic catalyst loading of the practical PEMFC (0.2–0.4 mg cm^{-2}). The high catalyst loading of DMFC not only accelerates the sluggish response of MOR but also reduces the methanol crossover. Moreover, the use of large amounts of noble catalyst is costly, obstructing the commercialization of DMFC.

The performance of a DMFC depends substantially on the electrical conductivity of the carbon support and the electron-transfer capacity of the catalyst along the catalyst layer through the gas diffusion layer to the current collector of MEA. According to the above discussion, the CNT has higher electrical conductivity in individual graphene sheets and faster electron transfer on the sidewall, than the conventional CB. Therefore, catalysts coated on the CNTs are believed to outperform the CBs. Liu et al. have employed the Pt-Ru/MWCNT with particle size of 3.5 nm and the Pt-Ru/CB with particle size of 3.8 nm for MOR, revealing that the former had a lower onset potential than the latter in the CV response [35]. Moreover, the Pt-Ru/MWCNT also exhibited a higher ratio of the forward peak current to the backward peak current than the Pt-Ru/CB, due to the high surface area of MWCNTs and the nanostructure of Pt-Ru catalysts, as observed by Lin et al. [33].

In order to gain a better understanding of the influence of the atomic structure of different carbon supports on the catalytic activity of Pt-Ru alloy, toward the MOR, Prabhuram et al. [31] have employed the most widely used carbon supports, such as CBs (Vulcan XC-72) and MWCNTs, for Pt-Ru catalysts. The Pt:Ru atomic ratios are specified as 1:1 and 7:3, and the MOR activities were systematically investigated for different DMFC anodes, namely Pt-Ru(1:1)/MWCNT, Pt-Ru(1:1)/CB, Pt-Ru(7:3)/MWCNT, and Pt-Ru(7:3)/CB. As observed form XRD patterns (Figure 8.10), the 2θ

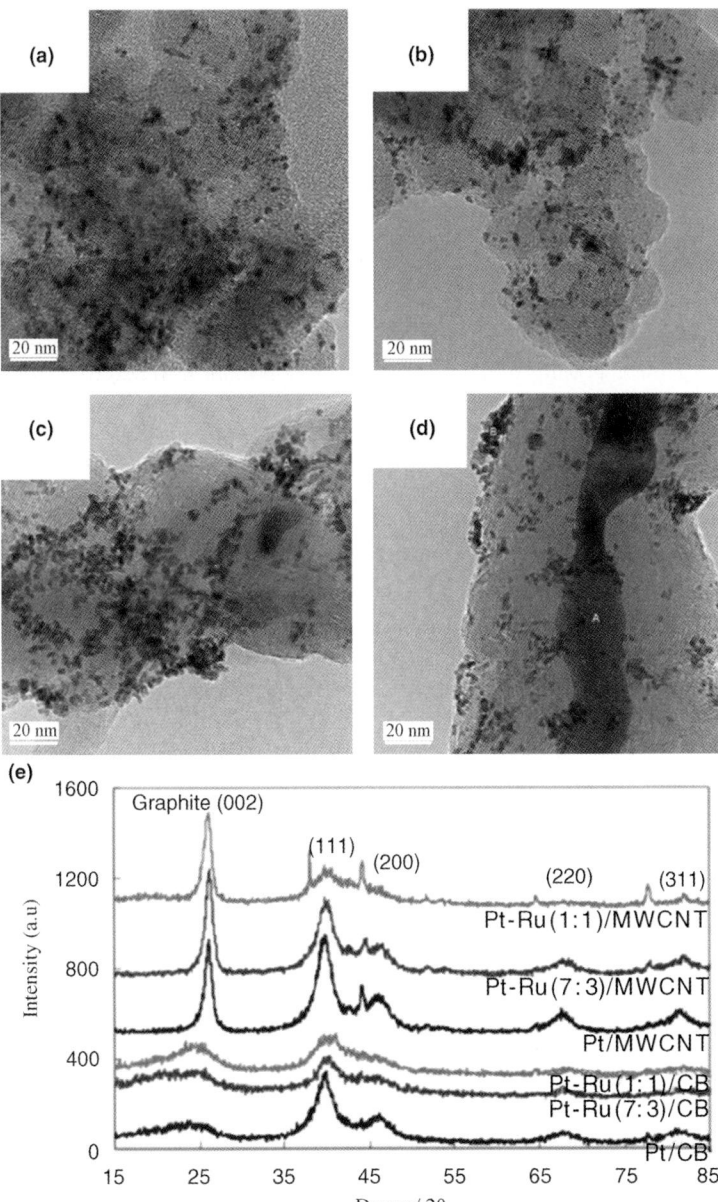

Figure 8.10 TEM images of (a) Pt-Ru(1:1)/CB; (b) Pt-Ru(7:3)/CB; (c) Pt-Ru(1:1)/MWCNT;, and (d) Pt-Ru(7:3)/MWCNT; and (e) corresponding XRD patterns [31]. The mean particle sizes and atomic ratio of catalysts obtained from TEM and XRD information are shown in the inset table.

values of the (111) peaks for Pt-Ru(7 : 3)/CB and Pt-Ru(7 : 3)/MWCNT are 40.6° and 40.3°, respectively, whereas those for Pt-Ru(1 : 1)/CB and Pt-Ru(1 : 1)/MWCNT are slightly shifted to the higher values, 41.0° and 40.75°, respectively. The observation indicates the Pt-Ru with the atomic ratio of 1 : 1 forming a higher degree of alloying than that with the atomic ratio of 7 : 3. The inset table of Figure 8.10 compares the particle sizes of Pt-Ru estimated from TEM observation and those calculated by Scherrer's equation from the (220) diffraction peaks, revealing that all the catalysts are ranged about 3.0 nm. Figure 8.11a and b shows the CV scans for the abovementioned carbon supports with Pt-Ru having the atomic ratio of 1 : 1 and 7 : 3, respectively, in which the MOR current densities are normalized to the amount of Pt-Ru loading. Comparison of the MOR activity, represented by the peak of forward scan, showed the following order: Pt-Ru(1 : 1)/MWCNT > Pt-Ru(7 : 3)/MWCNT > Pt-Ru(1 : 1)/CB > Pt-Ru(7 : 3)/CB. Beside the forward peak current, the onset potential is also another important index corresponding to the MOR activity; the Pt-Ru supported on MWCNTs exhibited an earlier onset than that on CBs. Figure 8.11c shows the polarization curves of the various preparations for the anodes. The Pt-Ru (1 : 1)/MWCNT yields a maximum power density of $32\,mW\,cm^{-2}$, which is about 35% higher than that of the Pt-Ru(1 : 1)/C. Similarly, the maximum power density of the Pt-Ru(7 : 3)/MWCNT is around 39% higher than that of the Pt-Ru(7 : 3)/CB. The results from polarization curves are consistent with the mass activity from CV scans. As mentioned above, the CNT electrode improves the activity of the catalyst in MOR, in which the large electrochemical surface area and the high electrical conductivity of CNTs play the decisive roles.

Since the Pt crystallite phase strongly affects the MOR activity [86], it can be expected that the graphene layer of CNT may affect or control the distinctive crystallite phase of Pt particles. Bessel et al. demonstrated that the Pt NPs supported on the 'platelet type' (Pt/pGNF) and 'ribbon type' (Pt/rGNF) graphite nanofibers are observed to be significantly less susceptible to CO poisoning than Pt/CB [87]. Such improvement in MOR activities, for Pt/pGNF and Pt/rGNF, is believed to be associated with the fact that the metal NPs have specific crystallographic orientations when dispersed on the highly tailored graphite nanofiber structures.

Carmo et al. have adopted three different catalysts supports, CBs, SWCNTs and MWCNTs, by applying the Pt-Ru catalysts by impregnation [88], with the particle sizes of 6.39, 4.06 and 4.62 nm, respectively, and with a metal loading of $0.4\,mg\,cm^{-2}$, employing as an anode of MEA. The polarization curves indicate that the MEA using Pt-Ru/MWCNT yields higher power density than those using Pt-Ru/CB or Pt-Ru/SWCNT. Li et al. have selected MWCNTs, double-walled CNTs (DWCNTs) and SWCNTs, with an uniform deposition of Pt-Ru NPs with diameters in the range 2–3 nm and a narrow particle size distribution [89]. CV scans reveal that these three CNT-supported catalysts have the same onset of about 0.1 V, as presented in Figure 8.12. However, the forward peak current of Pt-Ru/DWCNT is approximately $0.037\,mA\,cm^{-2}$ Pt, which significantly exceeds the Pt-Ru/MWCNT ($0.019\,mA\,cm^{-2}$ Pt), Pt-Ru/SWCNT ($0.012\,mA\,cm^{-2}$ Pt) and Pt-Ru/CB ($0.017\,mA\,cm^{-2}$ Pt). Restated, Pt-Ru/DWCNT generates a higher current, and thus possesses higher specific activity than others possess. Further clarification by the cross-sectional

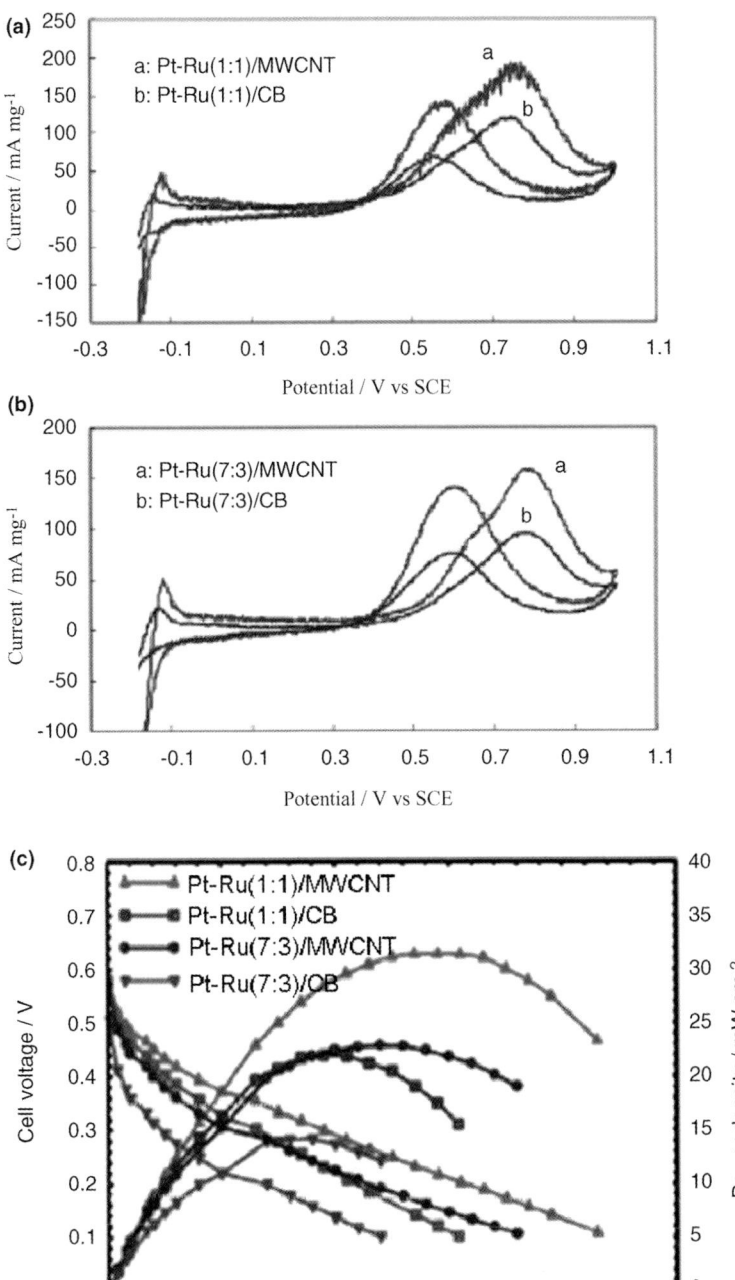

Figure 8.11 CV curves of (a) Pt-Ru(1:1) supported on MWCNT and CB; (b) PtRu(7:3) supported on MWCNT and CB in 0.5 M H_2SO_4 + 1 M CH_3OH with scan rate of 50 mV s^{-1} under N_2; (c) comparison of DMFC performance at the anode with 2 M methanol at 70 °C [31].

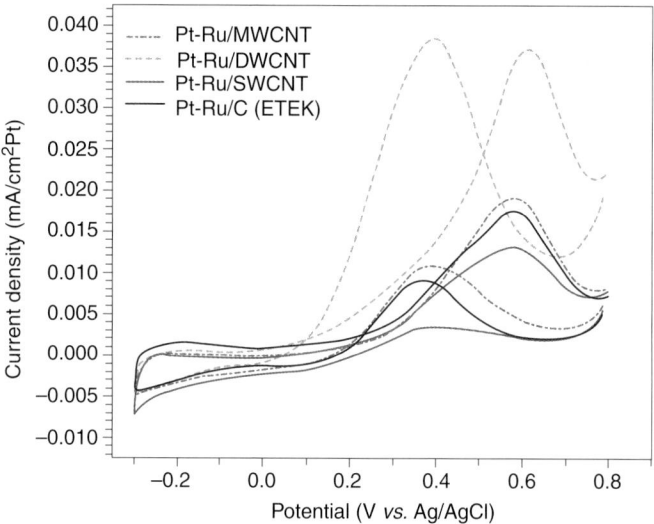

Figure 8.12 Methanol oxidation activity of Pt-Ru/SWCNT, Pt-Ru/DWCNT, Pt-Ru/MWCNT, and Pt/C in 0.5 M CH_3OH + 0.5 M H_2SO_4 at room temperature. Scan rate: 50 mV/s [89].

SEM images and the corresponding EDX of Pt-Ru/MWCNT and Pt-Ru/DWCNT, coated on Nafion membranes, reveal that the MWCNTs in the Pt-Ru/MWCNT layer do not adopt any preferred orientation (Figure 8.13). The orientation is directly associated with the mutual interaction between the hydrophobicity of MWCNTs and the hydrophilicity of filter paper; however the Nafion ionomers make the MWCNT film more hydrophilic during the conventional filtration process. Thus, the MWCNTs seemed to 'stand up,' disfavoring the interaction. At a metal loading of 0.5 mg cm^{-2}, the thicknesses of Pt-Ru/MWCNT and Pt-Ru/DWCNT are about 7–8 and 15–20 μm, respectively. Figure 8.14 plots the polarization curves of single DMFCs with various anodes. The open circuit voltage (OCV) exhibits significantly higher value, 0.627 V, for the Pt-Ru/DWCNT (30 wt.%, 0.5 mg cm^{-2}), compared with those for Pt-Ru/MWCNT (0.610 V) and Pt-Ru/SWCNT are (0.583 V). For Pt-Ru/CB, even when the metal loading increased to 2.0 mg cm^{-2}, the OCV remains approximately 0.620 V. Since the value of OCV can reflect the catalytic activation at the anode (because the same cathode was used), the highest OCV shown by the Pt-Ru/DWNTs catalyst suggests that it has the highest MOR activity, which is consistent with the previous results. The better performance of Pt-Ru/DWCNT can be attributed to its high electrical conductivity and high surface area. The small diameter of the DWCNT is responsible for the unique interaction between the catalysts and the DWCNT. Accordingly, the charges can be easily transferred from Pt to the DWCNT, increasing the specific activity of Pt.

Wu et al. provided a detailed comparison between MWCNTs and SWCNTs, and claimed that the Pt/SWCNT is preferable as the carbon support in DMFC [41]. The CO stripping indicates that the Pt/SWCNT has a lower onset potential and a lower

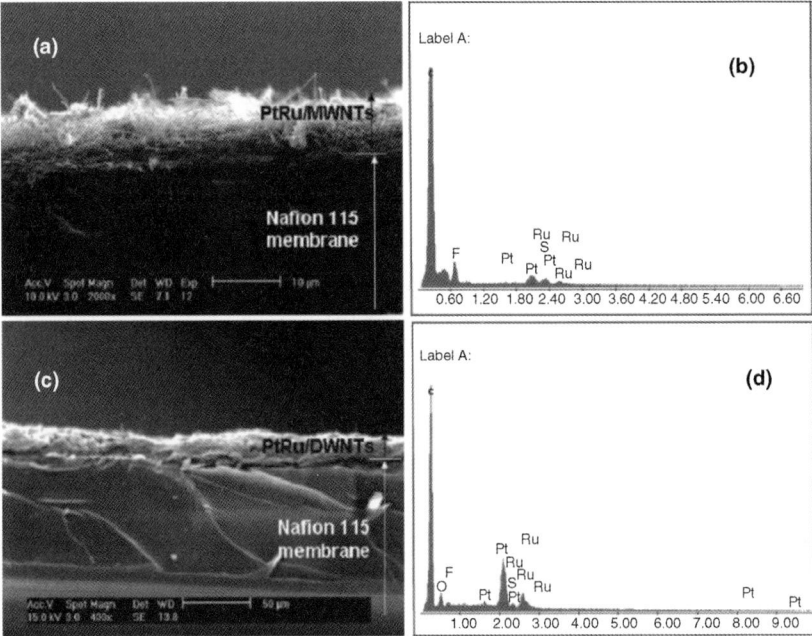

Figure 8.13 SEM micrographs and EDX analysis of CNT-supported catalysts (0.5 mg Pt-Ru/cm^2) coated on Nafion 115 membrane: (a and b) for Pt-Ru/MWCNT (30 wt.% Pt-Ru), (c and d) for Pt-Ru/DWCNT (50 wt.%Pt-Ru) [89].

peak potential than the Pt/MWCNT. Also, compared with Pt/MWCNT, Pt/SWCNT exhibits significantly higher forward peak current, as evident from the CV scans. Both the observations established the evidences in favor of the claim. The EIS reveal that Pt/SWCNT has lower charge-transfer resistance and lower interfacial resistance than Pt/MWCNT. The excellent performance of the SWCNT-based supports is associated with its sound graphitic crystallinity, the abundant oxygen-containing surface-functional groups and the highly mesoporous 3-D structure.

Related investigations demonstrate that various factors can affect the performance of the CNT-supported catalysts, and so no clear conclusion regarding which form of CNT is mostly suitable for carbon supports can be drawn. However, it has been established that any CNT-supported catalyst is clearly superior to a CB-supported catalyst.

8.4.2
Oxygen Reduction Reaction

The oxygen reduction reaction (ORR) and its kinetics are of fundamental interest in relation to fuel cell applications. ORR is well known to have two reaction pathways in acid solution: one is the direct four-electron pathway and the other is the two electron pathway (the peroxide pathway, Equations 8.9 and 8.10):

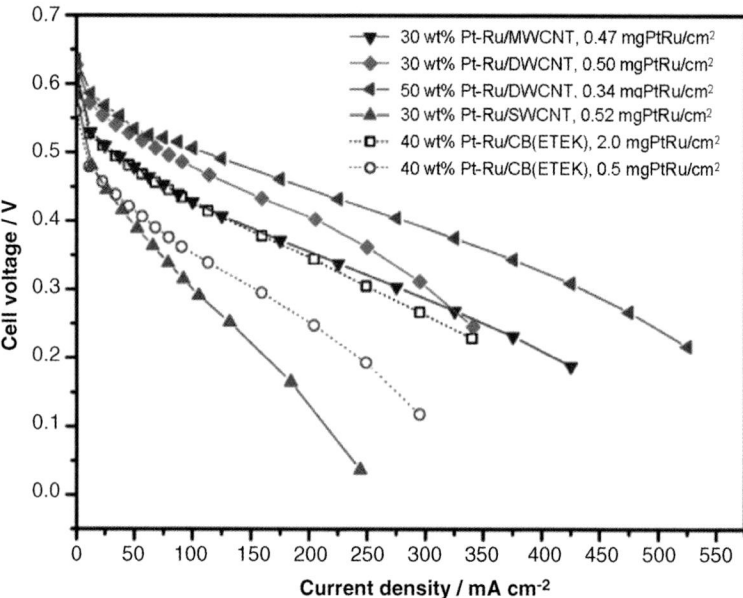

Figure 8.14 Polarization curves of DMFC with low noble metal loading (about 0.5 mg Pt-Ru/cm²) anode catalyst. Cathode, 2.0 Pt mg/cm² (80 wt.% Pt/C, E-Tek); membrane, Nafion 115; cell, 90 °C; O2, 75 °C, 0.2 MPa, 200 mL/min; methanol, 1.0 M, 2.0 mL/min [89].

$$O_2 + 4H^+ + 4e^- \rightarrow 2H_2O (E^0 = 1.23 \text{ V vs. SHE}) \qquad (8.9)$$

$$O_2 + 2H^+ + 2e^- \rightarrow H_2O_2 (E^0 = 0.67 \text{ vs. SHE}) \qquad (8.10)$$

The two-electron pathway not only reduces the ORR efficiency but also causes damage to the Nafion membrane due to the presence of peroxide. Methanol crossover, at which the methanol diffuses through the Nafion membrane from the anode to the cathode, is responsible for the mixed potential and the increase in the cathodic overpotential. To eliminate these unfavorable routes, the cathodic metal loading, in DMFC, is typically >2 mg cm^{-2}, which is one order of magnitude higher than that of PEMFC (>0.2 mg cm^{-2}).

The high porosity of SWCNTs facilitates the diffusion of the reactant and interaction with the Pt surface [90]. The accelerated durability tests demonstrate that SWCNTs increase the stability of the catalyst [90]. The lower energy of CO adsorption associated with the Pt/SWCNT also demonstrates the CO-tolerance of the catalyst and thus supports the use of SWCNTs in the DMFC [90]. Girishkumar et al. coated Pt/SWCNT on an optically transparent electrode (OTC) by electrophoretic deposition, and revealed that its methanol oxidation and the ORR exceed those of unsupported Pt [44]. Li et al. prepared MWCNT-supported catalysts by impregnation using ethylene glycol (Pt/MWCNT(A)) and HCHO (Pt/MWCNT(B)) as the reducing

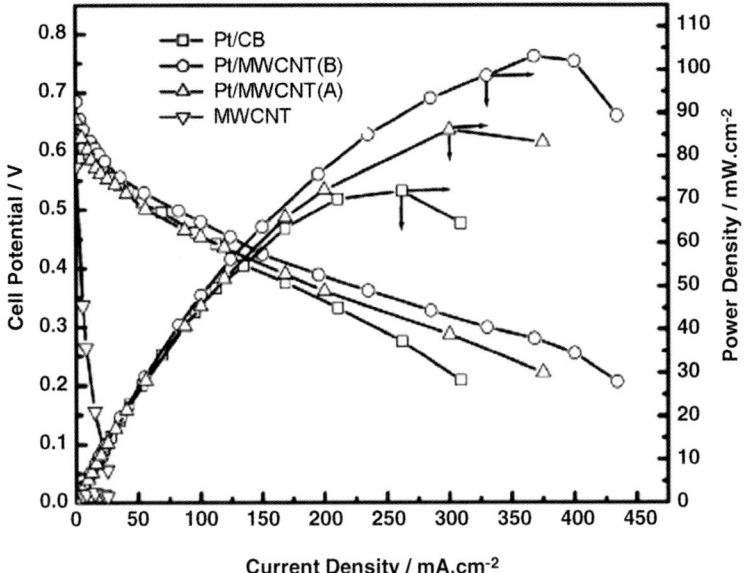

Figure 8.15 Comparison of the polarization curves for the ORR in the DMFC at Pt/CB, Pt/MWCNT(A), Pt/MWCNT(B), and blank test of MWCNT supports. Anode: Pt-Ru/CB (20 wt.% Pt, 10 wt.% Ru, Johnson Matthey Corp.; catalyst loading: 2.0 mg Pt-Ru/cm^2); electrolyte membrane: Nafion 115; operating temperature: 90 °C; methanol concentration: 1.0 M; flow rate: 1.0 mL/min; oxygen pressure: 0.2 MPa [27].

agents [27]. The sizes of Pt particles on Pt/MWCNT(A) and Pt/MWCNT(B) are 3.1 nm and 2.5 nm, respectively. The polarization curves show that the cathode using the Pt/MWCNT(B) outperforms (103 mW cm^{-2}) the Pt/MWCNT(A) (87 mW cm^{-2}) and the Pt/CB (70 mW cm^{-2}) under the same test conditions, as plotted in Figure 8.15. The mass activities at 0.7 V on Pt/C, Pt/MWCNT(A), and Pt/MWCNT(B) are 2.2 A g^{-1}, 5.2 A g^{-1} and 14.7 A g^{-1}, respectively. The fact that the low degree of agglomeration of particles in the Pt/MWCNT(B) gives the reactants an easy access to the catalytic active sites, further improving the mass transport in the cell system, is considered. This result indicates that the CNT-supported catalysts increase the ORR activity and reduce Pt usage.

Reshetenko et al. used MWCNTs as an additive to the cathode catalyst layer, improving the performance at the intermediate MWCNT/catalyst ratio of 0.05, as evident from the increase in the BET surface area and the porosity of the catalyst layer associated with the filamentous morphology of MWCNTs [91]. The structure of the conventional catalyst layer has two distinctive pore distributions with a boundary at 100 nm. The smaller pores (primary pores) are the spaces among the primary catalyst particles inside the clusters, and the larger pores (secondary pores) are the spaces among the clusters [92]. Studies revealed that the pores of 20 to 40 nm are the primary pores and those of 40 nm to 1 μm are the secondary ones [93]. Reshetenko et al. found that the MWCNT-containing cathode, in which mesopores of sizes up to 100 nm and the macropores of sizes from 100 nm to 2 μm were the primary and secondary pores, respectively, yielded favorable channels for reactant transfer and low ohmic loss. However, the excess cathode porosity exhibit disfavors the ionic conductivity.

8.4.3
Electrochemical Impedance Analysis

Electrochemical impedance analysis is a powerful method for measuring the charge-transfer resistance, the mass transfer and the interfacial resistance involved in an electrochemical process. Muller *et al.* characterized MOR on a CB-supported catalyst by using EIS measurements and its analysis [94, 95]. Figure 8.16a shows a typical Nyquist plot of a DMFC anode, in which an inductive loop is presented in the complex plane by the most extensively accepted mechanism of MOR. Figure 8.16b presents the equivalent circuit model, in which the circuit elements C_d, R_c, R_{ct}, and L_{co} are corresponding to the double-layer capacitor, the part of the current response occurring without a change in CO coverage for the electron-transfer resistance, the

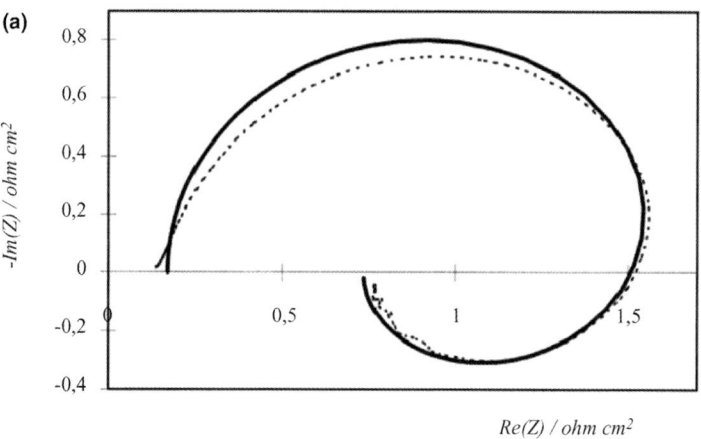

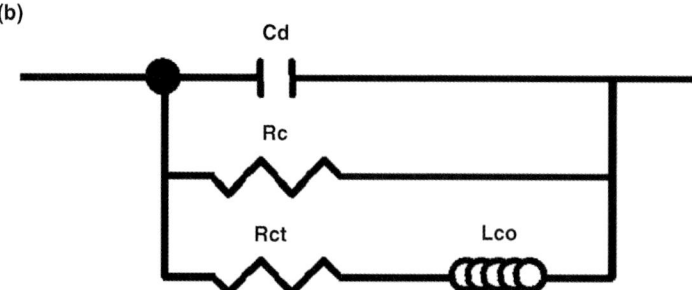

Figure 8.16 (a) A typical Nyquist plot of a DMFC anode; (b) the equivalent circuit model for simulating (a); dashed line: measured spectrum, and solid line: simulated spectrum [95].

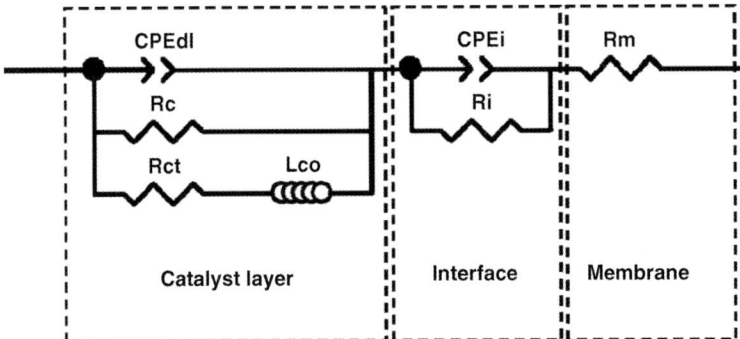

Figure 8.17 The equivalent circuit model for the impedances of a DMFC anode with CPE incorporated model of membrane + interface + catalyst layer [97].

modified resistance for the phase-delay significance, and postulating slowness of CO coverage relaxation for the phase-delay significance of inductance. A fitted line is also plotted in the Figure 8.16a, which agrees well with the experimental observations.

The equivalent circuit model, in Figure 8.16b, follows a simple model, without considering the interfacial impedances between catalysts and the membrane. Hence, a new model of the CNT electrode of DMFC has been proposed [96, 97], as presented in Figure 8.17, which comprises two parallel RC sections to simulate two interfaces – one between the membrane and the catalyst and the other between the catalyst and the carbon support. A constant phase element (CPE) rather than the conventional capacitance element is employed, increasing the accuracy and usefulness of the practical results. The CPE is defined in Equation 8.11.

$$z = \frac{1}{Q(j\omega)^p} \tag{8.11}$$

where Q is the admittance constant that corresponds to the capacitance and p is the adjustment parameter. EIS measurements and simulations show that the catalyst support markedly affects the performance of the catalyst in CO oxidation: the dehydrogenation of methanol dominates at low methanol concentration and low potentials, while CO oxidation dominates at high potential and high concentrations. CNT-supported catalysts have greater catalytic activity than CB-supported catalysts for CO and methanol oxidation. Based on an impedance analysis, Jeng et al. indicated that Pt-Ru/MWCNT with a high metal loading (Pt: 40 wt.% and Ru 20 wt.%) had lower charge-transfer and mass-transfer resistances than one with low metal loading (Pt: 20 wt.% and Ru 10 wt.%) [98].

8.5
Direct Growth of Carbon Nanotubes as Catalyst Supports

The conventional catalyst layer is placed on the gas diffusion layer to serve as a current collector and the gas diffusion layer allows a ready access of the fuel to the catalysts layer. After methanol is converted to carbon dioxide, electrons are transferred via the carbon supports to the gas diffusion layer and then to the outlet circuit. During this electron-transfer process, energy is lost due to the internal resistance of the carbon supports, the interfacial barriers between the carbon supports and the contact resistance between the carbon supports and the gas diffusion layer. CNTs as the support are expected to reduce such energy loss because of their low resistance. However, catalyst ink is generally adopted to prepare a catalyst layer on the diffusion layer, by a process called catalyst ink-brushing: in which the addition of Nafion ionomers in the catalyst layer, for the proton transport, tends to isolate the carbon-supported catalysts, resulting in low use thereof.

An electrode with a new structure was recently introduced to solve the above-mentioned problem. CNTs are directly grown the carbon cloth (CNT-CC), as supports for catalyst deposition to form the electrode in the DMFC. Accordingly, the catalysts that are dispersed on the CNT can be more efficiently involved in electrochemical reactions, and the electrons can transfer directly via CNTs to the carbon cloth (CC), reducing the energy loss and improving the performance. The growth and performance of CNT-CC supports will be discussed later in next section.

8.5.1
Direct Growth of CNTs on Carbon Cloth (CNT-CC)

Wang et al. grew MWCNTs directly on CC via an iron-assisted catalytic growth process, using microwave-plasma-enhanced chemical vapor deposition (MPECVD); Pt-Ru catalysts were then deposited on the CNT-CC using magnetron sputtering [99, 100]. Prior to the MWCNT growth, the iron NPs were sputtered on the CC by ion beam sputtering deposition to form a catalyst layer. The working pressure in an atmosphere of argon was maintained at 5×10^{-4} Torr during the deposition and a Kaufman ion source was operated at a beam voltage of 1250 V and a current of 20 mA. The optimal deposition time of the iron catalyst layer was 10 min. Then the CC coated with an iron catalyst layer was introduced to MPECVD system. The hydrogen plasma treatment was conducted at 1 kW at a chamber pressure of 28 Torr for 10 min, in order to induce the transformation of the iron catalyst layer into nanoparticles. Hydrogen plasma treatment not only reduced the iron oxides on the surface of the iron catalyst layer to the iron metal state, but also generated more active catalyst sites. Finally, the MWCNTs were grown on the conductive carbon fibers, in a gas mixture of $H_2/CH_4/N_2$ (80 : 20 : 80) at a microwave power of 2 kW, a chamber pressure of 45 Torr and a substrate temperature of 900 °C for 10 min. Figure 8.18 exhibits the nondirectional growth of MWCNTs, with a diameter and length of approximately 20 nm and 5–10 μm, respectively.

Figure 8.19a presents a cross-sectional TEM image of iron-coated CC in the early stage of growth, after the hydrogen plasma treatment. Notably, the amorphous carbon film is observed to encapsulate the iron NPs. While no external carbon source is introduced at this early stage, high-energy ions and free radicals in the hydrogen plasma bombard the carbon fibers, producing some hydrocarbon ions or radicals. A cross-sectional TEM image (Figure 8.19b) of the MWCNT, grown directly on CC,

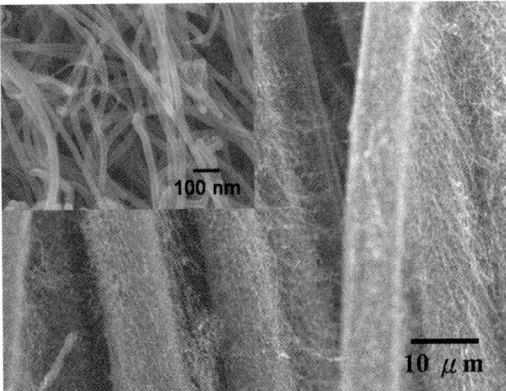

Figure 8.18 SEM images showing the MWCNTs directly grown on the carbon fibers of carbon cloth [99].

indicates that the grapheme sheets of CNTs not only adhere to CC but also directly conjoin with CC. Figure 8.19c demonstrates that the sample is a MWCNT with a bamboo-like structure. The proportion of nitrogen in CNTs is approximately 8%, as revealed by EDX. Addition of N_2 to the process gas is believed to promote the

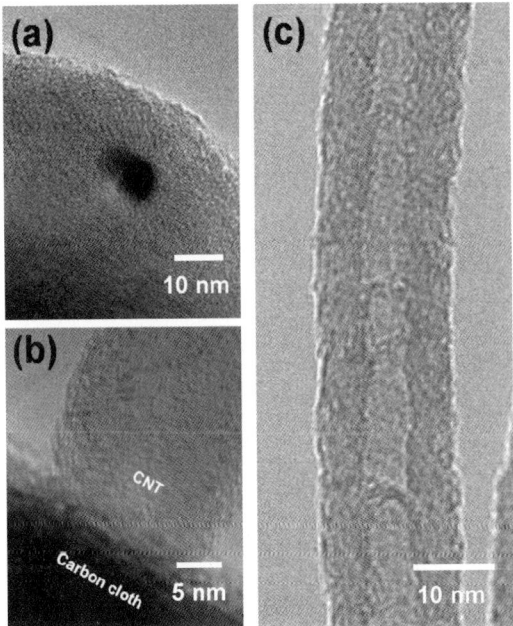

Figure 8.19 TEM image showing (a) the iron catalyst encapsulated in amorphous carbon film at early stage of growth; (b) the cross-sectional view of CNT directly conjoin with carbon cloth; and (c) the microstructural CNI with bamboo-like structure [99].

formation of a bamboo-like structure having a surface with a small radius of curvature [101–103]. Moreover, the incorporation of nitrogen atoms into graphene sheets with a bamboo-like structure favors the formation of pentagons and heptagons, and increases the reactivity of the neighboring carbon atoms, yielding the decorated nanotubes. Detail discussions on the advantages of such nitrogen substitution phenomena and its relevant activities in electrochemical reactions will come later in Section 5.3. Tsai et al. have observed the similar result when they prepared MWCNTs on CC by thermal CVD using $C_2H_4/NH_3/Ar$ as the precursors [104–106]. The as-grown MWCNT also shows a bamboo-like structure due to the incorporation of nitrogen from the decomposition of ammonia.

A redox reaction of $Fe(CN)_6^{3-/4-}$ is used as an electrochemical benchmark to characterize the CNT-CC and CC electrodes. The CV results demonstrate that the separation between anodic and cathodic peak potentials of CNT-CC is 60 mV, indicating that an almost reversible redox reaction occurs on the electrode. The peak current of CNT-CC exceeds that of CC. Such characteristics suggest the possession of a large reaction surface area of CNT-CC and rapid electron transfer on its surface [55]. Furthermore, EIS analyses indicate that the charge transfer resistance of CNT-CC ($0.46\,\Omega\,cm^2$) is much lower than that of CC ($3121\,\Omega\,cm^2$).

8.5.2
Appearance of CNT-CC-Supported Catalysts

The sputtering method [99, 100], the impregnation method [107], and the electrodeposition method [104–106] were employed to deposit catalysts on the CNT-CC. The resulting catalytic distribution on the CNTs was uniform. Figure 8.20a and b present the images of the Pt-Ru catalysts on the CNT-CC (Pt-Ru/CNT-CC), which present a homogenous dispersion of NPs on the CNT surface. The rings in the corresponding selected area diffraction (SAD) pattern, in the inset in Figure 8.20b, show that the catalysts are polycrystalline and randomly oriented. The degree of alloying of the Pt-Ru, which forms homogeneous alloy nanoclusters on CNT-CC, is 47.74%, according to XRD analysis.

8.5.3
Electrochemical Behavior of CNT-CC-Supported Catalysts

Figure 8.21 displays CV scans of Pt-Ru/CNT-CC and Pt-Ru/CC, in 1 M methanol solution and 1 M sulfuric acid, recorded at a scan rate of $10\,mV\,s^{-1}$. The CV scans present the typical MOR features: an anodic peak during the forward sweep at approximately 0.55 V corresponds to MOR activity and an anodic peak during the reverse sweep at approximately 0.45 V is associated with the removal of incompletely oxidized carbonaceous species that are formed during the forward sweep [108]. At a given metal loading of $0.4\,mg\,cm^{-2}$, the onset potential of the MOR at the Pt-Ru/CNT-CC is less than that at the Pt-Ru/CC, in the anodic sweep, suggesting that the Pt-Ru/CNT-CC reduces the activation overpotential. Oxidation peaks associated with the Pt-Ru/CNT-CC and the Pt-Ru/CC, in the anodic sweep, are

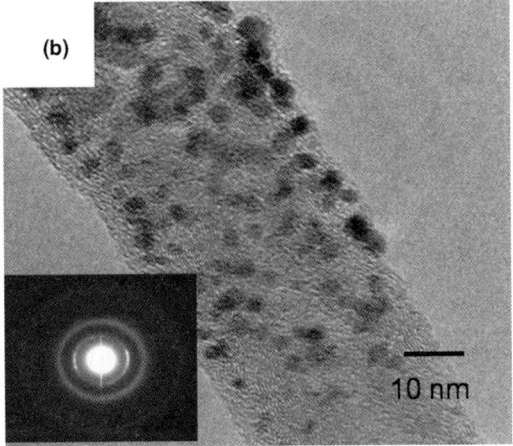

Figure 8.20 Pt-Ru catalysts on the CNT-carbon cloth: (a) SEM image and (b) TEM image with a SAD pattern in the inset [99].

observed at 0.56 and 0.68 V, respectively. The oxidation peak of the Pt-Ru/CNT-CC is more negative, indicating that MOR is energetically more favorable at Pt-Ru/CNT-CC than at Pt-Ru/CC. Additionally, Pt-Ru/CNT-CC has a much higher current density, seven times that of the Pt-Ru/CC at 0.4 V, after the application of the onset potential. The strong performance of the Pt-Ru/CNT-CC is attributed not only to the fast electron transport on CNTs but also to the lower interfacial resistance between CNTs and CC.

As discussed in Section 8.5.1, the substituted nitrogen sites on the decorated CNTs may be the initial nucleation sites of the deposition of the catalysts, due to the enhanced reactivity of the neighboring carbon atoms. The incorporation of 3–8% nitrogen in CNT has been demonstrated to increase MOR activity by improving the catalytic dispersion [109]. Sun et al. found that Pt NPs formed predominantly on the graphite layers of the outer surface of CNT [110], and they demonstrated that the homogeneous N-doping caused atomic scale structural deformation in CNT, increasing the tube diameter [111].

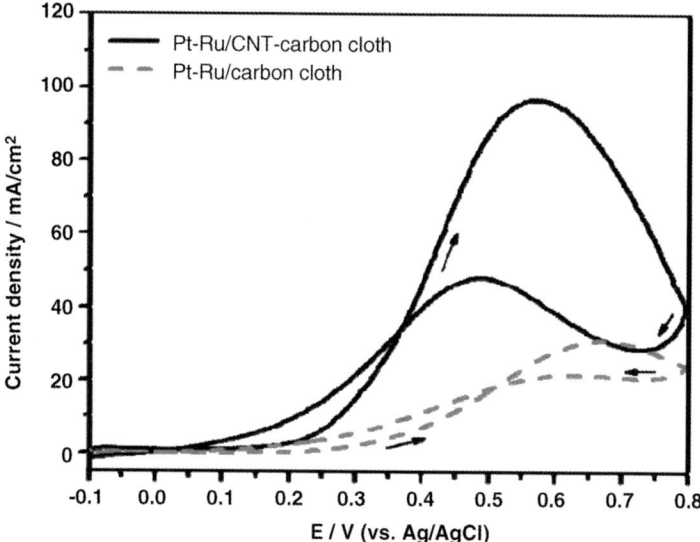

Figure 8.21 CV scans for Pt-Ru/CNT-carbon cloth and Pt-Ru/carbon cloth in 1M CH_3OH + 1M H_2SO_4 at a scan rate of 10 mV s^{-1} [99].

Pt-Ru/CNT-CC, Pt-Ru/CC and Pt-Ru/CB with metal loadings of 0.4, 0.4, and 3.0 mg cm^{-2}, respectively, were used as the anodes in the DMFC tests, while the MEA used 3.0 mg cm^{-2} Pt black and Nafion 117 as the cathode and membrane, respectively. Figure 8.22 plots polarization curves at a cell temperature of 60 °C.

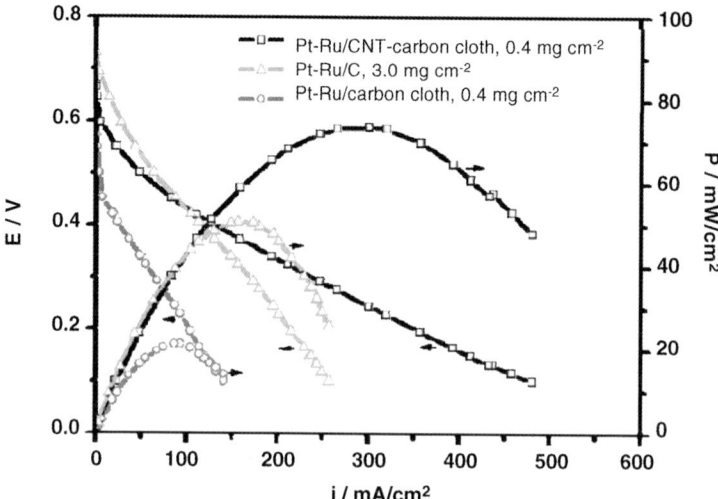

Figure 8.22 Polarization curves of DMFCs using various anodes operated at 60 °C with 1M CH_3OH and oxygen feeding to the anode and the cathode, respectively [99].

A higher OCV was obtained for Pt-Ru/CB, because the higher metal loading is associated with the lower activation overpotential. In the middle cell voltage range, the Ohmic overpotential controls the electrochemical reaction. The significant enhancement in the maximum power density of Pt-Ru/CNT-CC (73.6 mW cm^{-2}), compared with Pt-Ru/CC (21.3 mW cm^{-2}) and Pt-Ru/CB (50.6 mW cm^{-2}), suggests that the improvement in the performance is associated with the involvement of CNT-CC. At low cell voltage, where the mass transport overpotential dominates the reaction, this behavior was not evident.

Figure 8.23 represents the cross-sectional SEM images of the microstructures of MEAs, fabricated by hot-pressing technique, using Pt-Ru/CC or Pt-Ru/CNT-CC as the anodes, in which the thickness of catalyst layer of Pt-Ru/CC is typically 50–100 μm and that of Pt-Ru/CNT-CC is thinner than 10 μm. The MEA using Pt-Ru/CNT-CC has a thinner catalyst layer and the Pt-Ru NPs are homogeneously

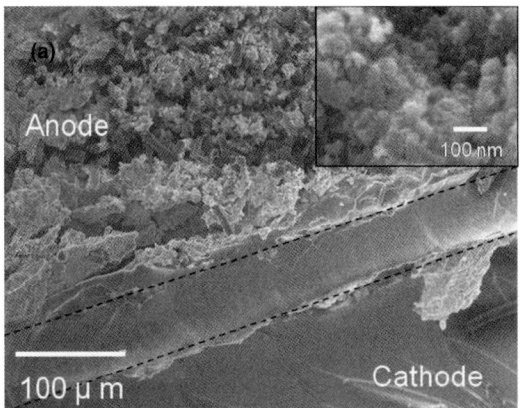

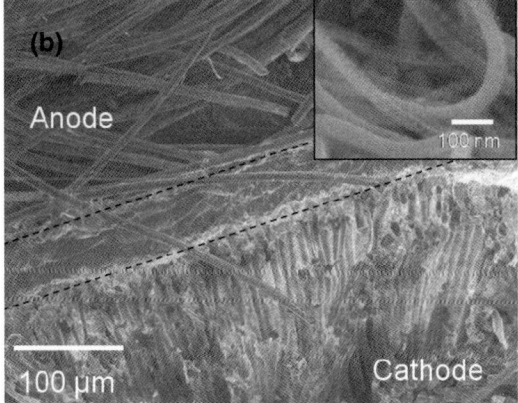

Figure 8.23 Cross-sectional SEM images of microstructural MEAs and using (a) Pt-Ru/C with metal loading of 3.0 mg cm^{-2} and (b) Pt-Ru/CNT-carbon cloth with metal loading of 0.4 mg cm^{-2} as anodes [99].

coated on CNT, as displayed in the inset of Figure 8.23b. For CNT-CC composite-based electrode, the electrons from the catalysts are readily transferred to the highly conductive CNT and then directly to the CC, hence the Pt-Ru/CNT-CC, even with a thinner catalyst layer, exhibits significantly more efficient electron and proton transport. Besides, a thinner catalyst layer can also provide less mass transport loss, which is not observed in Figure 8.22.

However, a high metal loading (>1 mg cm^{-2}) is difficult to achieve because of the shadow effect, in which the catalyst precursors are easily deposited on the top of the CNT layer. Therefore, the optimization of the design and performance of CNT-CC based MEAs still need close attention in the future.

8.6
Conclusion

DMFC is regarded as a next generation power source for portable devices, but the sluggish electrochemical activities undermine its implementation. This chapter introduces the highly conducting CNTs, as a preferable choice for catalysts supports, which markedly improve the electrochemical activity of the supported catalysts. Recent progresses in CNT-CC electrodes have led one step ahead exhibiting an excellent performance, even at a low catalyst loading, which can be attributed to the high specific area and the high conductivity of the CNT supports together with the direct transport of electrons and the short transport of protons.

References

1 Watanabe, M. and Motoo, S. (1975) Electrocatalysis by ad-atoms: Part II. Enhancement of the oxidation of methanol on platinum by ruthenium ad-atoms. *J. Electroanal. Chem.*, **60**, 267–273.

2 Nakagawa, N., Suzuki, Y., Watanabe, T., Takei, T. and Kanamura, K. (2007) Preparation of Pt-Ru nanoparticles with a uniform size distribution on a mesoporous carbon and their activity towards methanol electro-oxidation. *Electrochemistry*, **75**, 172–174.

3 Behm, R.J., Jusys, Z. and Kaiser, J. (2002) Composition and activity of high surface area PtRu catalysts towards adsorbed CO and methanol electrooxidation - A DEMS study. *Electrochim. Acta*, **47**, 3693–3706.

4 Kinoshita, K. (1988) *Carbon: Electrochemical and Physicochemical Properties*, John Wiley & Sons, Inc., New York, **xiii**, p. 533.

5 Endo, M., Iijima, S. and Dresselhaus, M.S. (1996) *Carbon Nanotubes*, 1st edn, Pergamon, Oxford; Tarrytown, N.Y., **x**, p. 183.

6 Saito, R., Dresselhaus, G. and Dresselhaus, M.S. (1998) *Physical Properties of Carbon Nanotubes*, Imperial College Press, London, **xii**, p. 259.

7 Che, G., Lakshmi, B.B., Fisher, E.R. and Martin, C.R. (1998) Carbon nanotubule membranes for electrochemical energy storage and production. *Nature*, **393**, 346–349.

8 Che, G., Lakshmi, B.B., Martin, C.R. and Fisher, E.R. (1999) Metal-nanocluster-filled carbon nanotubes: Catalytic properties and possible applications in

electrochemical energy storage and production. *Langmuir*, **15**, 750–758.

9 Yu, R., Chen, L., Liu, Q., Lin, J., Tan, K.L., Ng, S.C., Chan, H.S.O., Xu, G.Q. and Hor, T.S.A. (1998) Platinum deposition on carbon nanotubes via chemical modification. *Chem. Mater.*, **10**, 718–722.

10 Kim, S. and Park, S.-J. (2006) Effects of chemical treatment of carbon supports on electrochemical behaviors for platinum catalysts of fuel cells. *J. Power Sources*, **159**, 42–45.

11 Esumi, K., Ishigami, M., Nakajima, A., Sawada, K. and Honda, H. (1996) Chemical treatment of carbon nanotubes. *Carbon*, **34**, 279–281.

12 Shaffer, M.S.P., Fan, X. and Windle, A.H. (1998) Dispersion and packing of carbon nanotubes. *Carbon*, **36**, 1603–1612.

13 Guo, J., Sun, G., Wang, Q., Wang, G., Zhou, Z., Tang, S., Jiang, L., Zhou, B. and Xin, Q. (2006) Carbon nanofibers supported Pt-Ru electrocatalysts for direct methanol fuel cells. *Carbon*, **44**, 152–157.

14 Hsin, Y.L., Hwang, K.C. and Yeh, C.T. (2007) Poly(vinylpyrrolidone)-modified graphite carbon nanofibers as promising supports for PtRu catalysts in direct methanol fuel cells. *J. Am. Chem. Soc.*, **129**, 9999–10010.

15 Selvaraj, V. and Alagar, M. (2007) Pt and Pt-Ru nanoparticles decorated polypyrrole/multiwalled carbon nanotubes and their catalytic activity towards methanol oxidation. *Electrochem. Commun.*, **9**, 1145–1153.

16 Wang, J., Xi, J., Bai, Y., Shen, Y., Sun, J., Chen, L., Zhu, W. and Qiu, X. (2007) Structural designing of Pt-CeO2/CNTs for methanol electro-oxidation. *J. Power Sources*, **164**, 555–560.

17 Pan, D., Chen, J., Tao, W., Nie, L. and Yao, S. (2006) Polyoxometalate-modified carbon nanotubes: New catalyst support for methanol electro-oxidation. *Langmuir*, **22**, 5872–5876.

18 Qiao, Y., Li, C.M., Bao, S.-J. and Bao, Q.-L. (2007) Carbon nanotube/polyaniline composite as anode material for microbial fuel cells. *J. Power Sources*, **170**, 79–84.

19 Santhosh, P., Gopalan, A. and Lee, K.-P. (2006) Gold nanoparticles dispersed polyaniline grafted multiwall carbon nanotubes as newer electrocatalysts: preparation and performances for methanol oxidation. *J. Catal.*, **238**, 177–185.

20 Niu, L., Li, Q., Wei, F., Chen, X. and Wang, H. (2003) Electrochemical impedance and morphological characterization of platinum-modified polyaniline film electrodes and their electrocatalytic activity for methanol oxidation. *J. Electroanal. Chem.*, **544**, 121–128.

21 Hu, Z.A., Ren, L.J., Feng, X.J., Wang, Y.P., Yang, Y.Y., Shi, J., Mo, L.P. and Lei, Z.Q. (2007) Platinum-modified polyaniline/polysulfone composite film electrodes and their electrocatalytic activity for methanol oxidation. *Electrochem. Commun.*, **9**, 97–102.

22 Jha, N., Leela Mohana Reddy, A., Shaijumon, M.M., Rajalakshmi, N. and Ramaprabhu, S. (2008) Pt-Ru/multi-walled carbon nanotubes as electrocatalysts for direct methanol fuel cell. *Int. J. Hydrogen Energy*, **33**, 427–433.

23 Prabhuram, J., Zhao, T.S., Liang, Z.X. and Chen, R. (2007) A simple method for the synthesis of PtRu nanoparticles on the multi-walled carbon nanotube for the anode of a DMFC. *Electrochim. Acta*, **52**, 2649–2656.

24 Yen, C.H., Shimizu, K., Lin, Y.-Y., Bailey, F., Cheng, I.F. and Wai, C.M. (2007) Chemical fluid deposition of Pt-based bimetallic nanoparticles on multiwalled carbon nanotubes for direct methanol fuel cell application. *Energy and Fuels*, **21**, 2268–2271.

25 Gutierrez, M.C., Hortiguela, M.J., Manuel Amarilla, J., Jimenez, R., Ferrer, M.L. and Del Monte, F. (2007) Macroporous 3D architectures of self-assembled MWCNT surface decorated with pt nanoparticles as anodes for

a direct methanol fuel cell. *J. Phys. Chem. C*, **111**, 5557–5560.

26 Liu, J.M., Meng, H., Li, J.L., Liao, S.J. and Bu, J.H. (2007) Preparation of high performance Pt/CNT catalysts stabilized by ethylenediaminetetraacetic acid disodium salt. *Fuel Cells*, **7**, 402–407.

27 Li, W., Liang, C., Zhou, W., Qiu, J., Zhou, Z., Sun, G. and Xin, Q. (2003) Preparation and characterization of multiwalled carbon nanotube-supported platinum for cathode catalysts of direct methanol fuel cells. *J. Phys. Chem. B*, **107**, 6292–6299.

28 Figueiredo, J.L., Pereira, M.F.R., Serp, P., Kalck, P., Samant, P.V. and Fernandes, J.B. (2006) Development of carbon nanotube and carbon xerogel supported catalysts for the electro-oxidation of methanol in fuel cells. *Carbon*, **44**, 2516–2522.

29 Xu, J., Hua, K., Sun, G., Wang, C., Lv, X. and Wang, Y. (2006) Electrooxidation of methanol on carbon nanotubes supported Pt-Fe alloy electrode. *Electrochem. Commun.*, **8**, 982–986.

30 Liao, S., Holmes, K.-A., Tsaprailis, H. and Birss, V.I. (2006) High performance PtRuIr catalysts supported on carbon nanotubes for the anodic oxidation of methanol. *J. Am. Chem. Soc.*, **128**, 3504–3505.

31 Prabhuram, J., Zhao, T.S., Tang, Z.K., Chen, R. and Liang, Z.X. (2006) Multiwalled carbon nanotube supported PtRu for the anode of direct methanol fuel cells. *J. Phys. Chem. B*, **110**, 5245–5252.

32 Jeng, K.-T., Chien, C.-C., Hsu, N.-Y., Yen, S.-C., Chiou, S.-D., Lin, S.-H. and Huang, W.-M. (2006) Performance of direct methanol fuel cell using carbon nanotube-supported Pt-Ru anode catalyst with controlled composition. *J. Power Sources*, **160**, 97–104.

33 Lin, Y., Cui, X., Yen, C.H. and Wai, C.M. (2005) PtRu/carbon nanotube nanocomposite synthesized in supercritical fluid: a novel electrocatalyst for direct methanol fuel cells. *Langmuir*, **21**, 11474–11479.

34 Tian, Z.Q., Jiang, S.P., Liang, Y.M. and Shen, P.K. (2006) Synthesis and characterization of platinum catalysts on multiwalled carbon nanotubes by intermittent microwave irradiation for fuel cell applications. *J. Phys. Chem. B*, **110**, 5343–5350.

35 Liu, Z., Lee, J.Y., Chen, W., Han, M. and Gan, L.M. (2004) Physical and electrochemical characterizations of microwave-assisted polyol preparation of carbon-supported PtRu nanoparticles. *Langmuir*, **20**, 181–187.

36 Chien, C.-C. and Jeng, K.-T. (2006) Effective preparation of carbon nanotube-supported Pt-Ru electrocatalysts. *Mater. Chem. Phys.*, **99**, 80–87.

37 Wang, J., Deng, X., Xi, J., Chen, L., Zhu, W. and Qiu, X. (2007) Promoting the current for methanol electro-oxidation by mixing Pt-based catalysts with CeO2 nanoparticles. *J. Power Sources*, **170**, 297–302.

38 Cao, J., Du, C., Wang, S.C., Mercier, P., Zhang, X., Yang, H. and Akins, D.L. (2007) The production of a high loading of almost monodispersed Pt nanoparticles on single-walled carbon nanotubes for methanol oxidation. *Electrochem. Commun.*, **9**, 735–740.

39 Mu, Y., Liang, H., Hu, J., Jiang, L. and Wan, L. (2005) Controllable Pt nanoparticle deposition on carbon nanotubes as an anode catalyst for direct methanol fuel cells. *J. Phys. Chem. B*, **109**, 22212–22216.

40 Bock, C., Paquet, C., Couillard, M., Botton, G.A. and MacDougall, B.R. (2004) Size-selected synthesis of PtRu nano-catalysts: reaction and size control mechanism. *J. Am. Chem. Soc.*, **126**, 8028–8037.

41 Wu, G. and Xu, B.-Q. (2007) Carbon nanotube supported Pt electrodes for methanol oxidation: a comparison between multi- and single-walled carbon nanotubes. *J. Power Sources*, **174**, 148–158.

42 Zhao, Y., Fan, L., Zhong, H., Li, Y. and Yang, S. (2007) Platinum nanoparticle

clusters immobilized on multiwalled carbon nanotubes: Electrodeposition and enhanced electrocatalytic activity for methanol oxidation. *Adv. Func. Mater.*, **17**, 1537–1541.
43 He, Z., Chen, J., Liu, D., Zhou, H. and Kuang, Y. (2004) Electrodeposition of Pt-Ru nanoparticles on carbon nanotubes and their electrocatalytic properties for methanol electrooxidation. *Diam. Relat. Mater.*, **13**, 1764–1770.
44 Girishkumar, G., Vinodgopal, K. and Kamat, P.V. (2004) Carbon nanostructures in portable fuel cells: Single-walled carbon nanotube electrodes for methanol oxidation and oxygen reduction. *J. Phys. Chem. B*, **108**, 19960–19966.
45 Iijima, S. (1991) Helical microtubules of graphitic carbon. *Nature*, **354**, 56.
46 Dresselhaus, M.S., Dresselhaus, G. and Eklund, P.C. (1996) *Science of Fullerenes and Carbon Nanotubes*, Academic Press, San Diego, xviii, p. 965.
47 Satishkumar, B.C., Thomas, P.J., Govindaraj, A. and Rao, C.N.R. (2000) Y-junction carbon nanotubes. *Appl. Phys. Lett.*, **77**, 2530–2532.
48 Wang, T., Huang, J.-L., Kim, P. and Lieber, C.M. (2000) Structure and electronic properties of carbon nanotubes. *J. Phys. Chem. B*, **104**, 2794–2809.
49 Baughman, R.H., Zakhidov, A.A. and de Heer, W.A. (2002) Carbon nanotubes–the route toward applications. *Science*, **297**, 787–792.
50 Nalwa, H.S. (2004) *Encyclopedia of Nanoscience and Nanotechnology*, American Scientific Publishers, Stevenson Ranch, Calif., **10**.
51 Avouris, P., Appenzeller, J., Martel, R. and Wind, S.J. (2003) Carbon nanotube electronics. *Proc. IEEE*, **91**, 1772–1783.
52 Blanchet, G.B., Fincher, C.R. and Gao, F. (2003) Polyaniline nanotube composites: a high-resolution printable conductor. *Appl. Phys. Lett.*, **82**, 1290–1292.
53 Chen, K.H., Wu, J.J., Chen, L.C., Wen, C.Y., Kichambare, P.D., Tarntair, F.G., Kuo, P.F., Chang, S.W. and Chen, Y.F. (2000) Comparative studies on field emission properties of carbon-based materials. *Diamond Relat. Mater.*, **9**, 1249–1256.
54 Schonenberger, C. (2006) Charge and spin transport in carbon nanotubes. *Semicond. Sci. Technol.*, **21**, 1–9.
55 Nugent, J.M., Santhanam, K.S.V., Rubio, A. and Ajayan, P.M. (2001) Fast electron transfer kinetics on multiwalled carbon nanotube microbundle electrodes. *Nano Lett.*, **1**, 87–91.
56 Chico, L., Benedict, L.X., Louie, S.G. and Cohen, M.L. (1996) Quantum conductance of carbon nanotubes with defects. *Phys. Rev. B*, **54**, 2600.
57 Barisci, J.N., Wallace, G.G. and Baughman, R.H. (2000) Electrochemical studies of single-wall carbon nanotubes in aqueous solutions. *J. Electroanal. Chem.*, **488**, 92–98.
58 Barisci, J.N., Wallace, G.G. and Baughman, R.H. (2000) Electrochemical quartz crystal microbalance studies of single-wall carbon nanotubes in aqueous and non-aqueous solutions. *Electrochim. Acta.*, **46**, 509–517.
59 Barisci, J.N., Wallace, G.G. and Baughman, R.H. (2000) Electrochemical characterization of single-walled carbon nanotube electrodes. *J. Electrochem. Soc.*, **147**, 4580–4583.
60 Barisci, J.N., Wallace, G.G., Chattopadhyay, D., Papadimitrakopoulos, F. and Baughman, R.H. (2003) Electrochemical properties of single-wall carbon nanotube electrodes. *J. Electrochem. Soc.*, **150**, 409–415.
61 Hu, C.G., Wang, W.L., Wang, S.X., Zhu, W. and Li, Y. (2003) Investigation on electrochemical properties of carbon nanotubes. *Diamond Relat. Mater.*, **12**, 1295–1299.
62 Papakonstantinou, P., Kern, R., Robinson, L., Murphy, H., Irvine, J., McAdams, E., McLaughlin, J. and McNally, T. (2005) Fundamental electrochemical properties of carbon nanotube electrodes. *Fullerenes*,

Nanotubes, *Carbon Nanostruct.*, **13**, 91–108.
63 Yang, Z.-h. and Wu, H.-q. (2001) The electrochemical impedance measurements of carbon nanotubes. *Chem. Phys. Lett.*, **343**, 235–240.
64 Banks, C.E., Moore, R.R., Davies, T.J. and Compton, R.G. (2004) Investigation of modified basal plane pyrolytic graphite electrodes: definitive evidence for the electrocatalytic properties of the ends of carbon nanotubes. *Chem. Commun.*, 1804–1805.
65 Musameh, M., Lawrence, N.S. and Wang, J. (2005) Electrochemical activation of carbon nanotubes. *Electrochem. Commun.*, **7**, 14–18.
66 Binder, H., Köhling, A., Richter, K. and Sandstede, G. (1964) Über die anodische oxydation von aktivkohlen in wässrigen elektrolyten. *Electrochim. Acta*, **9**, 255–274.
67 Kinoshita, K. and Bett, J.A.S. (1974) Determination of carbon surface oxides on platinum-catalyzed carbon. *Carbon*, **12**, 525–533.
68 Scholta, J. and Wendt, H. (1992) *Phosphoric Acid Fuel Cells. Material Problems, Process Techniques and Limits of the Technology*, Inst of Chemical Engineers, Loughborough, UK.
69 Pyun, S.-I. Lee, E.-J., Kim, T.-Y., Lee, S.-J., Ryu, Y.-G. and Kim, C.-S. (1994) Role of surface oxides in corrosion of carbon black in phosphoric acid solution at elevated temperature. *Carbon*, **32**, 155–159.
70 Hoogers, G. (2003) *Fuel Cell Technology Handbook [electronic resource]*, CRC Press, Boca Raton, USA.
71 Roen, L.M., Paik, C.H. and Jarvi, T.D. (2004) Electrocatalytic corrosion of carbon support in PEMFC cathodes. *Electrochem. Solid-State Lett.*, **7**, 19–22.
72 Tang, H., Qi, Z., Ramani, M. and Elter, J.F. (2006) PEM fuel cell cathode carbon corrosion due to the formation of air/fuel boundary at the anode. *J. Power Sources*, **158**, 1306–1312.
73 Li, L. and Xing, Y. (2006) Electrochemical durability of carbon nanotubes in noncatalyzed and catalyzed oxidations. *J. Electrochem. Soc.*, **153**, 1823–1828.
74 Wang, X., Li, W., Chen, Z., Waje, M. and Yan, Y. (2006) Durability investigation of carbon nanotube as catalyst support for proton exchange membrane fuel cell. *J. Power Sources*, **158**, 154–159.
75 Shao, Y., Yin, G., Zhang, J. and Gao, Y. (2006) Comparative investigation of the resistance to electrochemical oxidation of carbon black and carbon nanotubes in aqueous sulfuric acid solution. *Electrochim. Acta*, **51**, 5853–5857.
76 Li, L. and Xing, Y. (2008) Electrochemical durability of carbon nanotubes at 80 °C. *J. Power Sources*, **178**, 75–79.
77 Wang, J., Yin, G., Shao, Y., Wang, Z. and Gao, Y. (2008) Electrochemical durability investigation of single-walled and multi-walled carbon nanotubes under potentiostatic conditions. *J. Power Sources*, **176**, 128–131.
78 Lu, X., Ausman, K.D., Piner, R.D. and Ruoff, R.S. (1999) Scanning electron microscopy study of carbon nanotubes heated at high temperatures in air. *J. Appl. Phys.*, **86**, 186–189.
79 Hamnett, A. (1997) Mechanism and electrocatalysis in the direct methanol fuel cell. *Catal. Today*, **38**, 445–457.
80 Liu, H., Song, C., Zhang, L., Zhang, J., Wang, H. and Wilkinson, D.P. (2006) A review of anode catalysis in the direct methanol fuel cell. *J. Power Sources*, **155**, 95–110.
81 Parsons, R. and VanderNoot, T. (1988) Oxidation of small organic molecules. A survey of recent fuel cell related research. *J. Electroanal. Chem. Interfacial Electrochem.*, **257**, 9–45.
82 Frelink, T., Visscher, W. and van Veen, J.A.R. (1995) On the role of Ru and Sn as promotors of methanol electro-oxidation over Pt. *Surf. Sci.*, **335**, 353–360.
83 Dillon, R., Srinivasan, S., Arico, A.S. and Antonucci, V. (2004) International Activities in DMFC R and D: Status of

84 Hogarth, M.P. and Ralph, T.R. (2002) Catalysis for low temperature fuel cells - Part III: challenges for the direct methanol fuel cell. *Platinum Met. Rev.*, **46**, 146–164.

85 Larminie, J. and Dicks, A. (2003) *Fuel Cell Systems Explained*, 2nd edn, John Wiley & Sons, Ltd, Chichester, **xxii**, p. 406.

86 Herrero, E., Franaszczuk, K. and Wieckowski, A. (1994) Electrochemistry of methanol at low index crystal planes of platinum: an integrated voltammetric and chronoamperometric study. *J. Phys. Chem.*, **98**, 5074–5083.

87 Bessel, C.A., Laubernds, K., Rodriguez, N.M. and Baker, R.T.K. (2001) Graphite nanofibers as an electrode for fuel cell applications. *J. Phys. Chem. B*, **105**, 1115–1118.

88 Carmo, M., Paganin, V.A., Rosolen, J.M. and Gonzalez, E.R. (2005) Alternative supports for the preparation of catalysts for low-temperature fuel cells: the use of carbon nanotubes. *J. Power Sources*, **142**, 169–176.

89 Li, W., Wang, X., Chen, Z., Waje, M. and Yan, Y. (2006) Pt-Ru supported on double-walled carbon nanotubes as high-performance anode catalysts for direct methanol fuel cells. *J. Phys. Chem. B*, **110**, 15353–15358.

90 Kongkanand, A., Kuwabata, S., Girishkumar, G. and Kamat, P. (2006) Single-wall carbon nanotubes supported platinum nanoparticles with improved electrocatalytic activity for oxygen reduction reaction. *Langmuir*, **22**, 2392–2396.

91 Reshetenko, T.V., Kim, H. T. and Kweon, H.-J. (2008) Modification of cathode structure by introduction of CNT for air-breathing DMFC. *Electrochim. Acta*, **53**, 3043–3049.

92 Watanabe, M., Tomikawa, M. and Motoo, S. (1985) Experimental analysis of the reaction layer structure in a gas diffusion electrode. *J. Electroanal. Chem.*, **195**, 81–93.

93 Uchida, M., Aoyama, Y., Eda, N. and Ohta, A. (1995) Investigation of the Microstructure in the catalyst layer and effects of both perfluorosulfonate ionomer and PTFE-loaded carbon on the catalyst layer of polymer electrolyte fuel cells. *J. Electrochem. Soc.*, **142**, 4143–4149.

94 Mueller, J.T. and Urban, P.M. (1998) Characterization of direct methanol fuel cells by ac impedance spectroscopy. *J. Power Sources*, **75**, 139–143.

95 Mueller, J.T., Urban, P.M. and Hoelderich, W.F. (1999) Impedance studies on direct methanol fuel cell anodes. *J. Power Sources*, **84**, 157–160.

96 Ocampo, A.L., Miranda-Hernandez, M., Morgado, J., Montoya, J.A. and Sebastian, P.J. (2006) Characterization and evaluation of Pt-Ru catalyst supported on multi-walled carbon nanotubes by electrochemical impedance. *J. Power Sources*, **160**, 915–924.

97 Hsu, N.-Y., Yen, S.-C., Jeng, K.-T. and Chien, C.-C. (2006) Impedance studies and modeling of direct methanol fuel cell anode with interface and porous structure perspectives. *J. Power Sources*, **161**, 232–239.

98 Jeng, K.-T., Chien, C.-C., Hsu, N.-Y., Huang, W.-M., Chiou, S.-D. and Lin, S.-H. (2007) Fabrication and impedance studies of DMFC anode incorporated with CNT-supported high-metal-content electrocatalyst. *J. Power Sources*, **164**, 33–41.

99 Wang, C.H., Du, H.Y., Tsai, Y.T., Chen, C.P., Huang, C.J., Chen, L.C., Chen, K.H. and Shih, H.C. (2007) High performance of low electrocatalysts loading on CNT directly grown on carbon cloth for DMFC. *J. Power Sources*, **171**, 55–62.

100 Wang, C.-H., Shih, H.-C., Tsai, Y.-T., Du, H.-Y., Chen, L.-C. and Chen, K.-H. (2006) High methanol oxidation activity of electrocatalysts supported by directly grown nitrogen-containing carbon nanotubes on carbon cloth. *Electrochim. Acta*, **52**, 1612–1617.

101 Chan, L.H., Hong, K.H., Xiao, D.Q., Hsieh, W.J., Lai, S.H., Shih, H.C., Lin, T.C., Shieu, F.S., Chen, K.J. and Cheng, H.C. (2003) Role of extrinsic atoms on the morphology and field emission properties of carbon nanotubes. *Appl. Phys. Lett.*, **82**, 4334–4336.

102 Chan, L.H., Hong, K.H., Xiao, D.Q., Lin, T.C., Lai, S.H., Hsieh, W.J. and Shih, H.C. (2004) Resolution of the binding configuration in nitrogen-doped carbon nanotubes. *Phys. Rev. B*, **70**, 125407–125408.

103 Chen, L.C., Wen, C.Y., Liang, C.H., Hong, W.K., Chen, K.J., Cheng, H.C., Shen, C.S., Wu, C.T. and Chen, K.H. (2002) Controlling steps during early stages of the aligned growth of carbon nanotubes using microwave plasma enhanced chemical vapor deposition. *Adv. Func. Mater.*, **12**, 687–692.

104 Tsai, M.-C., Yeh, T.-K. and Tsai, C.-H. (2006) An improved electrodeposition technique for preparing platinum and platinum-ruthenium nanoparticles on carbon nanotubes directly grown on carbon cloth for methanol oxidation. *Electrochem. Commun.*, **8**, 1445–1452.

105 Tsai, M.-C., Yeh, T.-K., Chen, C.-Y. and Tsai, C.-H. (2007) A catalytic gas diffusion layer for improving the efficiency of a direct methanol fuel cell. *Electrochem. Commun.*, **9**, 2299–2303.

106 Tsai, M.-C., Yeh, T.-K., Juang, Z.-Y. and Tsai, C.-H. (2007) Physical and electrochemical characterization of platinum and platinum-ruthenium treated carbon nanotubes directly grown on carbon cloth. *Carbon*, **45**, 383–389.

107 Du, H.Y., Wang, C.H., Hsu, H.C., Chang, S.T., Chen, U.S., Yen, S.C., Chen, L.C., Shih, H.C. and Chen, K.H. (2008) Controlled platinum nanoparticles uniformly dispersed on nitrogen-doped carbon nanotubes for methanol oxidation. *Diamond Relat. Mater.*, **17**, 535–541.

108 Mancharan, R. and Goodenough, J.B. (1992) Methanol oxidation in acid on ordered NiTi. *J. Mater. Chem.*, **2**, 875–887.

109 Rajesh, B., Karthik, V., Karthikeyan, S., Ravindranathan Thampi, K., Bonard, J.M. and Viswanathan, B. (2002) Pt-WO3 supported on carbon nanotubes as possible anodes for direct methanol fuel cells. *Fuel*, **81**, 2177–2190.

110 Sun, C.-L., Chen, L.-C., Su, M.-C., Hong, L.-S., Chyan, O., Hsu, C.-Y., Chen, K.-H., Chang, T.-F. and Chang, L. (2005) Ultrafine platinum nanoparticles uniformly dispersed on arrayed CN x nanotubes with high electrochemical activity. *Chem. Mater.*, **17**, 3749–3753.

111 Sun, C.L., Wang, H.W., Hayashi, M., Chen, L.C. and Chen, K.H. (2006) Atomic-scale deformation in N-doped carbon nanotubes. *J. Am. Chem. Soc.*, **128**, 8368–8369.

112 Yen, C.H., Shimizu, K., Lin, Y.Y., Bailey, F., Cheng, I.F. and Wai, C.M. (2007) Chemical fluid deposition of Pt-based bimetallic nanoparticles on multiwalled carbon nanotubes for direct methanol fuel cell application. *Energy Fuels*, **21**, 2268–2271.

113 Li, L. and Xing, Y. (2007) Pt-Ru nanoparticles supported on carbon nanotubes as methanol fuel cell catalysts. *J. Phys. Chem. C*, **111**, 2803–2808.

9
Mesoporous Carbon-Supported Catalysts for Direct Methanol Fuel Cells

Chanho Pak, Ji Man Kim, and Hyuk Chang

9.1
Introduction

Recent developments in information technology and convergence of multifunctions in personal mobile devices have caused an increase of more than 350 times in consumption of per-person portable power in the last two decades and this is expected to exceed 10 000 Wh/year/person at 2010 as listed in Table 9.1. In addition, recently developed 4G mobile devices will require advanced functions such as fast exchanging data, mobile internet, digital mobile broadcasting, which demand more power and energy capacity (Table 9.2). To overcome these issues, new concepts of portable power source are required, and the direct methanol fuel cell (DMFC) is one of the most promising candidates owing to its high energy capacity as well as quick charging time of methanol liquid fuel.

To increase the competitiveness of the DMFC system to the current technology of Li-ion battery, there are many requirements such as low cost, small volume and high energy efficiency, energy density and power density. Especially to mitigate the cost issue, the amount of Pt catalyst in the electrode should be reduced by increasing the activity of catalyst itself and/or the utilization ratio of Pt in the electrode. Finding more active element or alloys than Pt, which has been the most active component for several decades, is very challenging. Relatively, improving Pt use is easier due to versatility of methods of increasing the use of catalytic nanoparticles. As a simple approach, using a carbon support, the dispersion of catalyst can be increased or its size can be reduced to increase exposed surface areas of the active components. Among the various carbon supports, mesoporous carbon (MC), defined by the size of pore inside the carbon particles, is dealt with in view of novel support for electrocatalyst in this chapter. We introduce MC in Section 9.2, in which the concept, preparation and characteristics of MC are presented. As a preparation method, both the hard-templating and soft-templating methods are discussed. In Section 9.3, the preparation and characterization methods for supported catalysts using MC are covered. Examples of the application and approaches of MC for DMFC catalysts

Table 9.1 Average power consumption per year based on the personal equipment for one person over 10 years.

Year	1980	1990	2000	2010
Wh/year/person	10	500	3500	10 500[a]

[a]Expected value.

Table 9.2 Load and power demand of 4G mobile phone.

Functions	Call	DMB	MP3	Game	VOD	DC	Bell	Etc.	Total
Load (mA)	400	500	225	490	390	250	350	120	—
Power (W)	1.45	1.8	0.8	1.8	1.4	0.9	1.3	—	—
Operation (hr)	2	1.5	1	1	0.5	0.5	0.05	17.45	24
Capacity (mAh)	800	750	225	490	180	125	20	1200	3790

are summarized in Section 9.4. The last Section 9.5 provides prospects for other applications of this novel material.

9.2
Mesoporous Carbon

9.2.1
General Aspects of Mesoporous Carbon

Porous carbon materials have attracted much attention in the fields of fundamental studies as well as industrial applications such as the electrode materials for batteries [1] and fuel cells [2, 3], as the sorbents for separation processes [4–6], gas storage [7], and removal of pollutants [2, 8], and as the catalysts themselves and/or catalyst supports for many important catalytic processes [9–11] due to their unique physical and chemical properties. According to the IUPAC (International Union of Pure and Applied Chemistry) definition, the porous materials are classified depending on their pore sizes as microporous (<2 nm), mesoporous (2–50 nm) and macroporous (>50 nm) materials [12]. Although conventional carbon materials such as activated carbon and carbon black have been widely utilized in the fields of adsorption, separation and catalysis, they have some limitations for the applications due to their relatively broad pore-size distributions in the ranges of micropore, mesopore, and macropore. In order to produce the porous carbon materials with desired porosity, there are several synthesis pathways, such as the activation of pre-formed carbon materials by physical and/or chemical methods [13], the carbonization of polymer composites containing thermosetting and thermally unstable components [14], and the carbonization of aerogels [15]. However, these methods typically give porous carbon materials with broad pore-size distributions similar to the conventional carbon materials.

Recently, a new type of ordered mesoporous carbon (OMC) material with highly ordered mesostructures has been synthesized via a nano-casting technique using ordered mesoporous silica (OMS) materials as the hard templates [16–21]. Insertion of carbon precursors within the ordered mesopores of the OMS templates, their carbonization by thermal treatment at high temperature under inert conditions and subsequent removal of the silica templates result in the manufacture of OMC materials. The physical and chemical properties of the OMC materials, such as pore connectivity, pore-size, morphology, overall particles size, surface functionality, electric conductivity and thermal stability, can be controlled by adjusting OMS template, carbon precursor, carbonization temperature, heating environment, post-treatment, and so on [19, 20].

The other approach, in order to obtain the OMC materials directly, has been developed by using soft-templates such as surfactants and amphiphilic block copolymers [21–23], which are very similar with the sol-gel synthesis of mesoporous inorganic materials by synergistic self-assembly in the presence of organic templates [24–28]. However, direct synthesis of OMC materials by the self-assembly approach has been hindered due to the nature of the organic framework and template, even though there are some attempts to achieve this goal and they obtain the ordered mesostructures between polymer frameworks and organic templates [29–31]. The mesostructures thus obtained, are deconstructed during the carbonization of polymer frameworks and the removal of organic templates. Very recently, these issues to get the OMC materials have been solved by choosing the proper polymeric frameworks, which are carbon precursors, and by adjusting the chemical interactions between templates and carbon precursors [21–23].

Compared with typical porous carbon materials, these OMC materials promise to be suitable as adsorbents, catalyst supports, and materials for advanced electronics applications, due to not only their high porosity and surface area but also their controllable uniform mesopore system [19, 20]. In this regards, considerable effort has been made to prepare the OMC materials with various pore structures and framework properties. There have been vast advances in the preparation of OMC materials with modified surfaces, graphitic frameworks, and pore structures, which enables their use for various kinds of applications.

The focus of this section concerns the synthesis of OMC materials via the nano-casting method as well as via the direct self-assembly approach, and the characteristics of OMC materials for the applications, especially for DMFCs.

9.2.2
Synthesis of Mesoporous Carbon

9.2.2.1 Synthesis of OMC Materials via Nano-Casting Method

The concept for the synthesis of mesoporous carbon materials via the nano-casting method, including the impregnation of carbon precursor in the presence of pre-synthesized inorganic hard template, the carbonization, and the subsequent

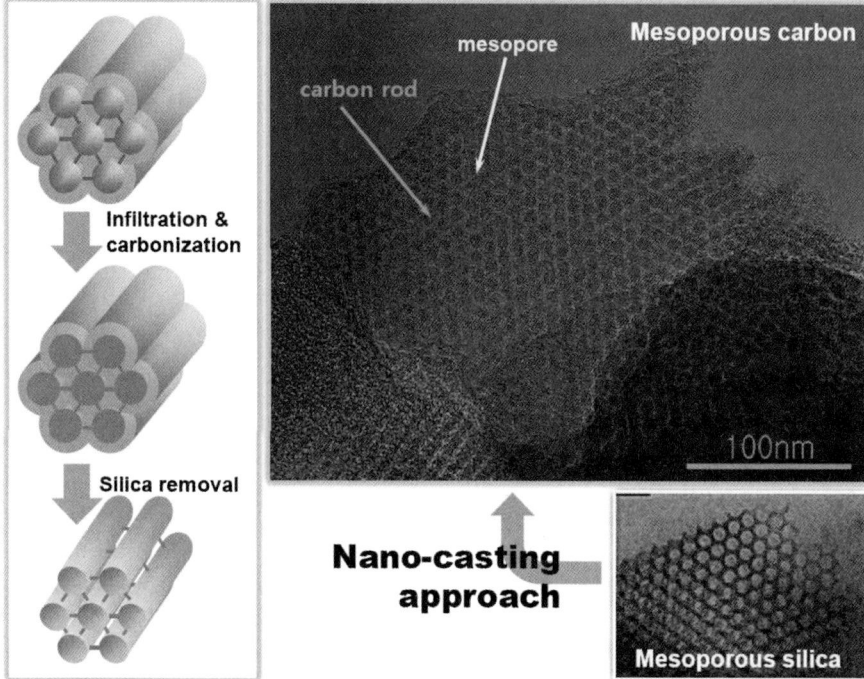

Figure 9.1 A schematic diagram of the nano-casting approach for the synthesis of OMC material, and TEM images of mesoporous silica template and OMC material.

removal of template by selective chemical etching, has been developed with spherical solid gel as the template by Knox and coworkers early 1980s [32]. In 1999, Ryoo and coworkers expanded this synthesis strategy to the preparation of OMC materials by using a highly ordered mesoporous silica template, as shown in Figure 9.1 [16]. About few months apart, Hyeon and coworkers also reported the synthesis of OMC materials with the same mesoporous silica template [18]. After these pioneering works, there are numerous efforts on the preparation and application of OMC materials with various mesostructures, pore sizes, particle morphologies, and framework structures.

According to the synthesis strategy for the OMC materials as shown in Figure 9.1, the first step is the selection of mesoporous silica templates, determining the mesostructures of OMC materials thus obtained because their structures are the inverse replicas of the silica templates. One should choose a proper mesoporous silica template with three dimensionally interconnected pore structure, in order to maintain the mesostructured carbon frameworks and exhibit ordered mesoporous topologies upon the removal of silica template. Therefore, the OMC materials with different mesostructures can be obtained by using various kinds of mesoporous silica templates such as MCM-48 (bicontinuous cubic *Ia3d*), SBA-15 (2-D hexagonal *P6mm*), SBA-1 (discontinuous cubic *Pm3n*), SBA-16 (large pore discontinuous cubic $Im3m$), and KIT-6 (large pore bicontinuous cubic $Ia3d$) [16, 33–45]. The first synthesis of OMC material (CMK-1) is carried out by using MCM-48 with bicontinuous cubic $Ia3d$ symmetry as the template. It is noteworthy that the CMK-1 is not a true inverse replica of the MCM-48, but exhibits tetragonal $I41/a$ (or lower) mesostructure [16]. This structural transformation upon the silica removal can be explained in terms of a displacement model. When the silica frameworks of MCM-48 are partially etched, giving interconnecting pores within the wall, the $Ia3d$ symmetry is retained in the carbon replica (CMK-4) [34]. The CMK-6 material, from KIT-6 silica template with interconnecting micropores within the silica frameworks, also exhibits the same mesostructure as that of silica template [43–45]. All the OMC materials, except of CMK-1, exhibit true inversely replicated mesostructures from those of mesoporous silica templates. Interestingly, the stable OMC materials can be obtained by using the SBA-15 silica template with 2-D arrays of meso-channels because the material possesses lots of interconnecting micropores within the silica frameworks [41]. The OMC materials can also be synthesized from the MCM-41 silica template if the interconnecting channels within the silica frameworks are generated by using a microwave digestion or by using a surfactant mixture as the template [46, 47]. The SBA-15 results in two types of OMC materials, CMK-3 and CMK-5, depending on the preparation method. The CMK-3 material, which is composed of carbon nanorods arranged in a hexagonal pattern and connected with bridges between the rods, can be obtained when sucrose and sulfuric acid are utilized as the carbon source and catalyst, respectively. When the Al-SBA-15 template and furfuryl alcohol are utilized for the synthesis of OMC material, the acidic aluminum sites can act as the catalyst for the selective polymerization of furfuryl alcohol only on the pore wall, so that the resulting OMC materials exhibits a hexagonally ordered array of carbon nanopipes (CMK-5) [17, 48].

9.2.2.2 Synthesis of OMC Materials via Direct Self-Assembly Approach

The nano-casting approach discussed above seems to be fairly complex, time consuming and cost-ineffective because it is necessary to synthesize the mesoporous silica templates prior to the process for the OMC synthesis. Recently, much effort has been made to develop a direct self-assembly approach method (soft-templating approach) for OMC materials via an organic-organic self-assembly between carbon precursors and organic templates. The chemical interactions between templates and carbon precursors play a key role in the success of the soft-template synthesis. Dai and coworkers [21] first reported the successful preparation of highly ordered mesoporous carbon materials by the carbonization of ordered mesostructures between resorcinol-formaldehyde (RF) resin and polystyrene-block-poly(4-vinylpyridine) (PS-P4VP). The RF resin can be converted to the rigid carbon frameworks while the PS-P4VP part inside the composite is decomposed during the carbonization under an inert atmosphere. Nishiyama and coworkers [22] also reported the successful synthesis of ordered porous carbon films (COU-1) by using RF and triethyl orthoacetate as the carbon sources and a Pluronic amphiphilic block copolymer, F127, as the template. The synthesis of highly ordered mesoporous carbon films

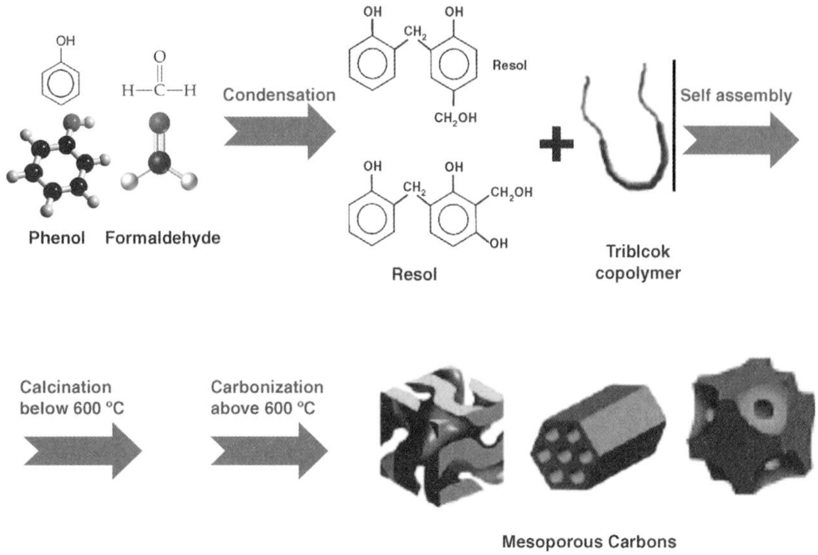

Figure 9.2 A schematic diagram of the direct self-assembly approach for the synthesis of OMC materials.

with hexagonal and cubic $Im3m$ mesostructures using Pluronic triblock copolymers as templates and a soluble low molecular weight polymer of phenol and formaldehyde as precursors has been also reported by Zhao and coworkers [49]. More importantly, they have developed the synthesis process for the OMC materials in powder form by performing the direct assembly of polymer templates (F127, F108, and P123) and resol in an aqueous media. The synthesis pathway, involving preparation of resol, formation of resol/template mesostructure, curing of resol by thermopolymerization, removal of template and carbonization as shown in Figure 9.2, is very similar to those in the synthesis of mesoporous inorganic materials. However, there might be lots of pending questions such as the control of morphologies and reproducibility in order to apply the OMC materials, synthesized by the direct self-assembly approach, to practical applications.

9.2.3
Characteristics of Mesoporous Carbon

Since the first report for the synthesis of ordered mesoporous carbon (OMC) using a mesoporous silica as the template [16, 18], it has been of great interest due to the regular mesopore size ($\sim$ 3 nm), high surface area (700–2000 $m^2\,g^{-1}$) and high pore volume (1–2 $cm^3\,g^{-1}$) [2, 19, 20, 50]. There have been vast advances in the preparation of OMC materials, which enables their use for various applications such as adsorbents, catalyst supports, nanotemplates, materials for advanced electronics, and so on. For the applications of OMC materials, the rational control of their structural parameters such as pore size, particle morphology, and primary particle size is of great importance. Especially, in order to utilize the OMC materials as the catalyst support for fuel cell applications, the materials should satisfy several requirements [3]. Even though the OMC materials already exhibit very high surface areas, the materials should have highly developed and designable mesoporosity to allow a facile diffusion of reactant and by-product gases or liquids. High electrical conductivity for providing electrical pathways, stabilizing ability for noble catalyst nanoparticles, and electrochemical stability for long-term operation are also necessary.

It is believed that a rational pathway for the synthesis of OMC materials with controlled pore sizes would greatly increase their utility for diverse applications. The OMC materials with large pore sizes are highly desirable for applications involving large molecules and fast mass transfer. However, the pore size control of OMC materials is hitherto limited, while that of mesoporous silica is relatively successful. Although there has been much effort to control the pore size of OMC materials, few successes have been reported in the literature thus far. Ryoo and co-workers have systematically synthesized a series of SBA-3 (2-D hexagonal) silica materials with tailored silica wall thicknesses by using mixtures of cationic and neutral surfactants. They have reported the synthesis of an OMC material with controllable pore sizes ranging from 2.2 to 3.3 nm using these mesoporous silica templates [47]. Syntheses of OMC materials with tunable pore sizes (3.0–5.1 nm) using mesoporous silica templates with larger unit cell parameters while maintaining the regularity of mesopores have been also reported [51–55]. All of these trials are based on the design of the starting mesoporous silica template, since the silica framework creates the mesopores of the OMC materials. Thus, the silica template needs to be designed properly to obtain OMC materials with controlled pore sizes. However, in general, pore size control of OMC materials, based on the design of mesoporous silica template, has distinct limitations due to the difficulty of controlling the properties of silica frameworks [56, 57].

As an alternative solution, different approaches have been reported to obtain OMC materials with hierarchical porosity in order to satisfy the needs for OMC materials with controlled mesopore sizes. For example, bimodal mesoporous carbon materials, such as OMC material constructed with nanopipes [17, 48], dual porous OMC material via a dual templating route [23, 51, 58–62], and precise control of mesoporosity by spontaneous phase separation [63] have been reported. The CMK-5 material that has a hexagonally ordered array of amorphous carbon nanopipes shows two maxima in pore size distributions, one at 5.9 nm from the pores inside the nanopipes and the other at 4.2 nm from the pores between the adjacent nanopipes. The CMK 5 exhibits very high surface area of about $2000\,m^2\,g^{-1}$ and total pore volume of about $1.5\,cm^3\,g^{-1}$. The OMC material, named as NCC-1, has also been synthesized by Schüth and coworkers via the nano-casting process using SBA-15 as the template and diluted furfuryl alcohol solution as the carbon source [64]. The NCC-1 shows a mesostructure similar to CMK-5, a bimodal pore size distribution in the mesopore range and an extremely high pore volume up to $3.2\,cm^3\,g^{-1}$. Since the pore systems in NCC-1 are independently generated at different stages during the synthesis, they can also be modified independently.

Gierszal and Jaroniec have reported a mesoporous carbon with bimodal distributions formed by co-imprinting of spherical silica colloids and hexagonally ordered mesoporous particles of SBA-15 into mesophase pitch particles and subsequent silica dissolution [58]. A facile dual-templating approach is also demonstrated to prepare hierarchically ordered macro-/mesoporous carbons by Zhao and coworkers [59]. Mono-dispersed silica colloidal crystals are used as a hard template, amphiphilic block copolymer as a soft template, and soluble resols as a carbon source. The porous carbons obtained have a highly ordered face-centered cubic macrostructure with tunable pore sizes of 230–430 nm and interconnected windows with a size of 30–65 nm. The rigid silica hard templates can prevent the shrinkage of the mesostructure during the thermosetting and carbonization process, resulting in large cell parameters (similar to 18 nm) and pore sizes (similar to 11 nm). Hyeon and coworkers have also reported the synthesis of bimodal mesoporous carbon having 4 nm sized framework mesopores and similar to 30 nm sized textural pores using bimodal mesoporous silica as a template [60]. Hollow macroporous core/mesoporous shell-type carbon materials have also been synthesized using a silica template with a solid core/mesoporous shell structures [23, 51, 62]. A dual templating approach using both mesoporous silica template and silica nanoparticle, of which average sizes are about 1 nm, has been also developed by Kim and coworkers, as shown in Figure 9.3 [61]. Interestingly, they have maximized the microporosity within the carbon framework by combining both nanoreplication and nano-imprinting techniques using dual silica templates, which provides a simple way to synthesize ordered mesoporous carbons with bimodal pore size distributions (similar to 1.5 nm and similar to 3.5 nm).

Very recently, a rational synthesis pathway for OMC materials with controllable 3 to 10 nm pores has been reported by Kim and coworkers [63]. This new synthesis strategy to precisely control the pore sizes of OMC materials uses an inorganic pore expanding agent such as boric acid during the infiltration of carbon precursors within the mesopores of silica template. The proposed synthesis mechanism is demonstrated in Figure 9.4, which involves the formation of borosilicate and boron oxide nanolayers between carbon framework and silica surface within the mesopores

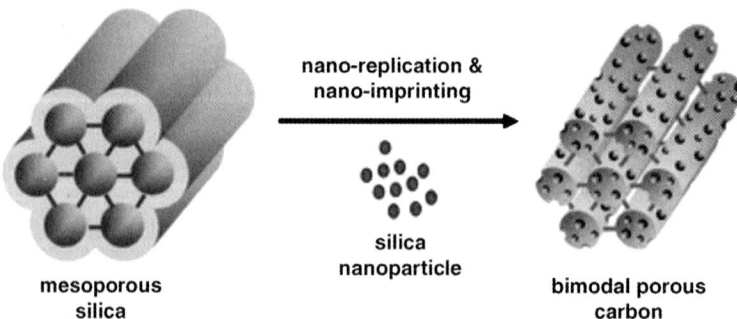

Figure 9.3 A schematic diagram of the dual templating approach using both mesoporous silica template and silica nanoparticles.

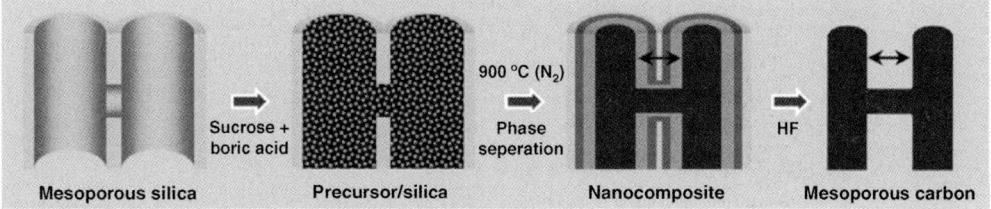

Figure 9.4 A schematic diagram for the synthesis of large pore OMC materials via simple and spontaneous phase separation during carbonization.

of the silica template caused by spontaneous phase separation and subsequent solid-state reaction during carbonization at high temperature.

The OMCs were fabricated in various forms including rods [65], spheres [62], thin films [21, 22, 49], and monoliths [66, 67] by nano-casting OMS templates with the same morphologies. Along with the pore size and particle morphology, the primary particle size of porous materials also represents an important limiting factor in several applications controlled by intramolecular diffusions such as catalysis, adsorption, and separation processes. For example, in the cathode reaction of fuel cells, which is the main application target of the present work, for the electrode reaction to occur, three components (catalyst particles, proton conducting materials, and oxygen) should simultaneously be in contact with each other. Hence, the diffusion and transport of each reaction component are of significant importance, which might be limited or controlled by adjusting the primary particle size of the carbon supports. During the OMC replication process from an OMS, in most cases, the external morphology of the primary particles is preserved, while the internal porous structure of the OMS templates is inversely replicated. Thus, the OMCs with various particle sizes can simply be prepared by nano-casting OMSs with different particle sizes. However, thus far, the synthesis of OMCs with controlled particle sizes has rarely been reported [68], due to the limited controllability of OMS particle sizes. On the other hand, Lin *et al.* [35] and Zhao *et al.* [69] prepared OMCs with film-like and fiber-like morphologies, respectively, which have short channel lengths compared with conventional OMCs. The supported catalysts using these OMCs showed higher electrocatalytic activities in the methanol oxidation reaction than commercial catalysts or conventional OMCs supported catalysts.

9.3
Mesoporous Carbon-Supported Catalyst

9.3.1
Concepts of Mesoporous Carbon-Supported Catalyst

Nowadays, there are two types of commercial catalyst for DMFC. One is a black catalyst, which is a very fine unsupported powder catalyst containing only a catalytically active component such as Pt or PtRu. The other is a supported catalyst, in which the active metal components are uniformly dispersed on the support surface which usually has a large surface area.

Up to date, the black catalysts are widely used for the DMFCs due to their high activity at low temperature. To increase the catalytic activity of the black catalyst, addition of a new metal element has been investigated and introduced to the pristine component such as Pt and PtRu [70–72]. In addition, especially for the cathode Pt catalyst, efforts have been made to decrease the particle size to enlarge the electrochemically active surface area and to increase the tolerance for methanol penetrated from the anode side through the membrane has been attempted [3]. Although the activities of the black catalyst increase with time, a large amount of catalyst is needed to compensate for the low use of catalyst particles and to ensure the life time of the membrane electrode assembly (MEA).

To minimize the amount of catalyst in the electrode by increasing the use of the catalyst, the supported catalyst has been developed and adopted for the DMFC [73, 74]. It is expected that the particle size of metals, especially Pt, decreases by dispersing the metals on the large surface area of carbon supports. As a carbon support for the DMFC, there are several requirements: (i) a high electrical conductivity, (ii) a large surface area, (iii) adequate water-handling property, (iv) good corrosion resistance and (v) easy mass-transport [75, 76].

Up to now, highly conductive carbon blacks such as Vulcan XC, Ketjen black, and Denka black have been adopted to the electrocatalyst. To improve the performance of supported catalyst for DMFCs, nanostructured carbon materials have been exploited. Such nanocarbons include carbon nanotubes [77], carbon nanofibers [69], graphitic carbon nanocoils [78–80] and mesoporous carbons including macroporous and OMCs. Electrocatalysts adopting these carbon supports showed promising activities for the oxygen reduction and methanol oxidation reactions, which will be discussed in detail later. This was attributed to the unique structural features of the carbon materials such as large surface area, highly conductive framework structures, and periodic pore structures in the mesoporous and macroporous ranges.

Among the new nanocarbon supports, mesoporous carbon is the focus of this Chapter as a promising support, due to the large surface area from well-defined pore structures and the unique array of mesopores inside the carbon particles as mentioned in Section 9.2. The former can afford the preparation of highly dispersed catalytic nanoparticles, which is the subject of Section 9.3.2. The latter provide facile transport of reactants and by-products of the fuel cell electrode reactions.

9.3.2
Preparation Methods for Mesoporous Carbon-Supported Catalyst

Selection of the preparation method for a supported catalyst is very important especially for the electrocatalyst for DMFCs because the size, distribution and morphology of catalytic nanoparticles depend on the preparation method [74]. In the design of a DMFC catalyst layer by using a supported catalyst, a thinner catalyst layer in the MEA is favored for efficient diffusion and transport of fuels, and low interfacial resistance [81]. To this end, very high metal loading usually above 50 wt.% is required. However, the size of metal nanoparticles is easy to increase with the metal loading. Thus, high dispersion of metal nanoparticles in the supported catalyst is the aim of many researchers. Relatively, the PtRu nanoparticles are easily distributed uniformly with high loading on the carbon support [51, 82, 83], whereas it is hard to obtain a high dispersion of the Pt nanoparticles.

Since the carbon-supported Pt catalyst has been used as an electrocatalyst for the fuel cell in the 1960s [84], there have been continuing efforts to optimize the preparation method for the electrocatalysts having high metal dispersion and large active surface area [74, 85]. The colloidal and impregnation routes are the most well known methods for the preparation of supported catalysts. In the colloid method, metal nanoparticles are formed in the presence of protective agent, such as surfactant and polymer, and the nanoparticles obtained exhibit uniform particle size distribution. However, if the nanoparticles are to function as real catalyst, the organic protecting shell should be removed by washing in an appropriate solvent or by heating at elevated temperature, during which the nanoparticles are susceptible to agglomeration.

On the other hand, the impregnation method has been the easiest method for the preparation of electrocatalysts over the decades. The reduction of impregnated metal ion is usually performed by chemically or electrochemically using borohydride, hydrazine, formic acid, or hydrogen gas as reducing agents. However, this method usually produces metal nanoparticles with wider particle size distribution, compared with a colloid method, and the size increases with high metal loading. Other than colloid and impregnation routes, alternative preparation methods based on radiolytic reduction [86], polyol process [39, 87, 88], microwave irradiation [89, 90] and pre-precipitation method [72, 91] have also been suggested to improve the dispersion and catalytic activity of supported catalysts.

For example, Shen and Tian [89] reported a novel method based on intermittent microwave heating to prepare highly dispersed Pt/C catalysts with 60 wt.% Pt. Recently, Joo et al. [92] reported that the preparation of 60 wt.% Pt catalyst supported on mesocellular carbon forms, which have a very large pore size (>10 nm), by a solution reduction method using ethanol as solvent and sodium ethoxide as a reducing agent. Similarly, the 50 wt.% Pt-supported catalyst on CMK-3, which is one of the famous OMCs, was prepared by the solution reduction method using paraformaldehyde as a new reducing agent [40]. On the other hand, a decade ago, 2–3 nm Pt particles, supported on large surface area carbon support, was prepared by radiolytic reduction of carbonyl cluster of Pt with above the 50 wt.% loading [86].

We developed a novel preparation method for highly dispersed Pt catalyst supported on the OMC by incipient wetness method using small amount of acetone and H_2PtCl_6 precursor [93]. We pursued the high loading of Pt in the supported catalyst using OMC as support, because the amount of Pt required in the electrode of DMFC caused a thick catalyst layer if we use low loading supported catalyst. Figure 9.5 shows the SEM image of the 60 wt.% Pt supported on OMC, which is prepared by using OMS having primary particle size of 300 nm. The uniformly distributed white dot is the Pt nanoparticles on the surface of OMC support. The large black holes are the mesopores at the surface of the OMC support.

9.3.3
Characterization of Mesoporous Carbon-Supported Catalyst

For the characterization of physical properties such as size, morphology and distribution of metal nanoparticles in the mesoporous carbon-supported catalyst, the general X-ray diffraction (XRD), scanning electron microscopy (SEM) and transmission electron microscopy (TEM) methods are used, in a similar way to the heterogeneous supported catalyst. From the XRD, the average size of nanoparticles can be estimated by Scherrer equation:

$$L = 0.9\lambda/B_{2\theta} \cos \theta_{max} \quad (9.1)$$

where L is the average particle, λ is the X-ray wavelength, $B_{2\theta}$ is the full width at the half maximum in radians and θ_{max} is the maximum angle of given peak in radians. The position of this peak can be used to estimate the lattice constant and degree of alloy, especially for the alloy (PtRu) nanoparticles. The particle size can be estimated by counting directly from the SEM and TEM images with high magnification. In addition, SEM and TEM can provide the distribution state on the surface of mesoporous carbon support like Figure 9.5. From the images, the uniformity of particle distribution can be decided. High resolution TEM can determine the crystallinity of nanoparticles and lattice parameter for each facet.

For the characterization of electrochemical properties, two configurations can be used to measure the performance of supported catalyst. One is the half-cell configuration, which measures the mass activity and stability for the methanol oxidation reaction (MOR) and oxygen reduction reaction (ORR) of supported catalyst itself in the electrolyte solution using rotating disk electrode or carbon paper electrode.

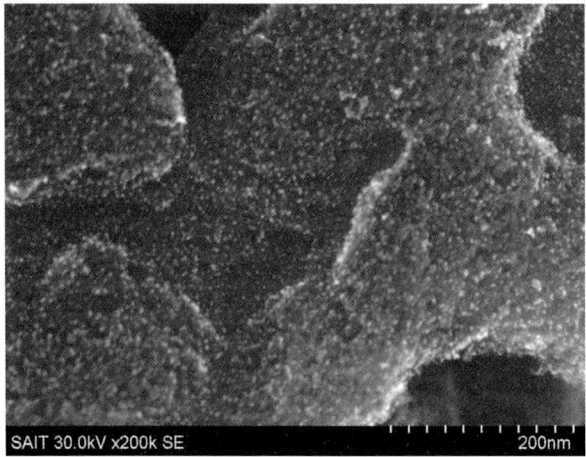

Figure 9.5 SEM images of Pt nanoparticles in supported catalyst using OMC support.

Table 9.3 Physical properties of Pt supported on POMC and SOMC catalysts.

Samples	Pt particle size[a] (nm)	ECSA[b] (m^2/g_{Pt})	Mass activity for ORR[c] (A/g_{Pt})	Sheet resistance[d] ($m\Omega/cm^2$)
Pt/POMC	2.7	55.1	1.01	54
Pt/SOMC	2.5	66.2	0.95	202

[a] Estimated from XRD.
[b] Electrochemically active surface area estimated from CV.
[c] Estimated by linier sweep voltammetry at 0.75 V vs. NHE.
[d] Obtained by four-point probe method at the pressure of 75.4 kg/cm^2.

This configuration can provide the electrochemically active surface area of metal nanoparticles in the supported catalyst by cyclic voltammetry (CV), mass or specific activity of the catalyst by linear sweep voltammetry and degradation rate of catalyst performance with time by chronoamperometry. Many researchers for developing electrocatalysts prefer this method because it is easy to screen many candidates in a short time compared with the single cell configuration. To test the developed catalyst in a single cell, the fabrication of MEA should be optimized by preparing a good catalyst slurry and choosing a process to obtain a suitable catalyst layer. The final MEA performance provided by single cell evaluation is very important and vital information for deciding whether the developed catalyst is really improved or not.

Recently, we showed the importance of measurement of performance in a single cell during a study about the higher conductive OMC support [94]. As listed in the Table 9.3, the physical properties for Pt supported on POMC and SOMC (Pt/POMC and Pt/SOMC) which is prepared using phenanthrene and sucrose as carbon precursors, respectively, is very similar. The big difference is only the resistance of OMC supports. The mass activity for the ORR is very similar for both Pt/POMC and Pt/SOMC catalysts. However, the single cell performance showed very significant difference as shown in Figure 9.6. The power density at 0.4 V and 323 K is increased from 35 mW cm^{-2} for Pt/SOMC catalyst to 48 mW cm^{-2} for Pt/POMC catalyst. Despite the fact that two Pt catalysts supported on SOMC and POMC showed very similar ORR activity and particle size, the MEA of Pt/POMC catalyst showed superior performance by 50% to the Pt/SOMC catalyst. Thus, it is suggested that the effect of lower resistance (higher electrical conductivity) of POMC on the performance is developed significantly at the electrode in the single cell MEA.

9.4
Fuel Cell Performance of Mesoporous Carbon-Supported Catalyst

The possibility of application using mesoporous carbon as a new support for fuel cell catalysts is suggested from the report of Pt supported on CMK-5 carbon [17], which is the ordered mesoporous carbon having hexagonal structure and dual mesopore, and prepared by nano-casting method as described in previous

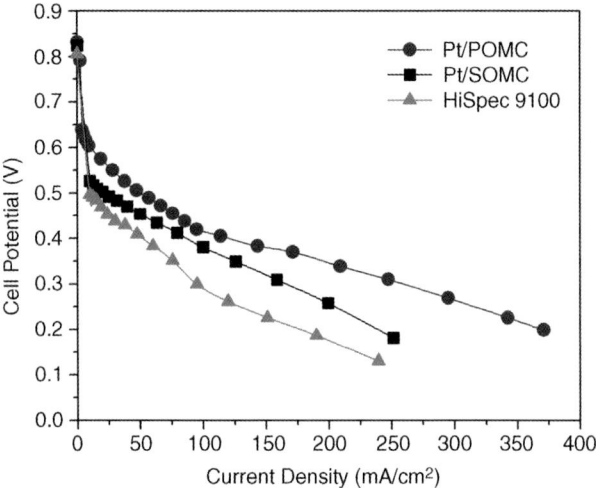

Figure 9.6 Potentiodynamic polarization plots of DMFC single cells at 323 K using Pt/POMC, Pt/SOMC and HiSpec 9100 as cathode catalysts, respectively.

Section 9.2. Since then, several groups have reported that mesoporous carbon-supported catalyst is a promising support for the fuel cell catalyst.

Nam et al. [95] prepared the Pt catalyst supported on CMK-3 carbon via an incipient wetness method and measured the performance of a single cell in which the Pt catalyst was used in the cathode for DMFC. Ding et al. [39] prepared the Pt-supported catalyst using pre-formed Pt particles inside the pore of silica template and the PtRu-supported catalyst by the polyol process. The Pt supported on CMK-3 catalyst with 10 wt.% Pt showed higher activity for the ORR than the commercial Pt-supported catalyst, whereas PtRu supported on the CMK-3 catalyst showed inferior MOR activity compared with the PtRu commercial catalyst from E-Tek.

Choi et al. [96] reported the methanol tolerance of their Pt-supported catalyst, which was prepared by simultaneous carbonization and introduction of Pt and carbon precursor into the template. At a final supported catalyst, most Pt nanoparticles were studded inside the nanorods of mesoporous carbon. To confirm the methanol tolerance of this catalyst, the open circuit voltage (OCV) was measured in the single cell with various methanol concentrations. At high concentration of methanol, the OCV of the MEA using Pt-supported catalyst was 40–60 mV higher than those of commercial Pt catalyst. Recently, Liu et al. [97] prepared Pt- and PtRu-supported catalyst using tube-type OMC structures as support, by using a similar method. They showed that this catalyst has a higher tolerance towards CO poisoning. However, in this approach, the loading of Pt in the supported catalyst is not enough to make a practical MEA. This approach is adopted to make the methanol-tolerant cathode catalyst for DMFC by Wen et al. [98]. The authors, at first distributed the fine Pt nanoparticle inside the mesopore of SBA-15 and then followed the nanocasting strategy using glucose to generate Pt nanoparticles embedded in mesoporous carbon. They evaluated the durability of their catalyst through repeated CV cycles in an O_2-saturated electrolyte with 0.5 M methanol. The variation of the current density was only about 4% after 40 cycles, which indicate that their catalyst has a considerably stable activity for the ORR in the presence of methanol.

Su et al. [99] synthesized the OMC with increased graphitic character by chemical vapor deposition (CVD) of benzene into SBA-15 pure silica template. The Pt-supported catalyst (20 wt.% Pt) was then prepared by sodium borohydride reduction method using benzene-derived OMC as support, which showed higher specific activities for methanol oxidation than that of the commercial Pt catalyst. Recently, the same group reported the synthesis of mesoporous carbon microfibers (MCMF) using porous alumina microfiber as template via CVD method [100]. This MCMF has a core-shell structure in which the core structure was composed of mesopores with 8.8 nm pore and the shell had 15–20 nm dense walls with parallel discrete graphene layers. The authors suggested that the Pt supported on MCMF displayed a better CO tolerance and activity for methanol oxidation at room temperature. Similar mesoporous carbon nanofiber (MCNF) was synthesized by the dual templating approach using anodic aluminum oxide membrane as hard template and block copolymer surfactant as soft template [69]. This MCNF showed larger pore size (10.2 nm) and higher conductivity (71 S cm^{-1}) compared with conventional OMC. The Pt-supported catalyst using MCNF support displayed superior performance toward methanol oxidation in sulfuric acid solution. The CVD method utilized to prepare the CMK-5 type mesoporous carbon using the SBA-15 as template and ferrocene as both precursor and catalyst for carbon [48]. They optimized the CVD time and temperature for highly ordered mesoporous carbon support and used prepared support for the deposition of 20 wt.% Pt nanoparticles. It was suggested that the Pt supported on mesoporous carbon showed generally stronger hydrogen adsorption peaks relative to the commercial catalyst and enhanced methanol oxidation activity.

Yi and coworkers prepared mesoporous carbon with a disordered pore structure by using colloidal silica particles as the template and resorcinol-formaldehyde as the carbon source [101, 102]. As the pH of the synthesis mixture increased, the pore sizes of the resulting porous carbon decreased, whereas the portion of mesopores increased. The particle sizes of the supported PtRu nanoparticles gradually decreased with the increase of mesopore volume in the porous carbon, and accordingly single cell performances of supported catalysts showed enhanced activities.

As mentioned in the previous Section 9.2, direct synthesis of mesoporous carbon without using a template, occurs through self-organization of surfactants and carbon precursor. Recently, there are some attempts to use this directly-prepared mesoporous carbon as a catalyst support [46, 103, 104]. Zhou et al. [46] dispersed Pt nanoparticles onto the mesoporous carbon, which is similarly prepared in previous report, by a facile cetyltrimethylammonium bromide (CTAB) assisted microwave synthesis process. The dispersion and electrochemical active surface area were changed by the amount of CTAB in the initial mixture. Hayashi et al. [103] prepared Pt catalysts (30 wt.%) deposited on the mesoporous carbon support, which were synthesized using Pluronic F127 as pore former and resorcinol as carbon precursor. They investigated the effect of solvent dispersing Nafion ionomer on the ORR activity with half cell configuration. The more hydrophobic solvent, 2-propanol showed higher performance than other solvents like water and ethanol mixture and ethanol because of the hydrophobicity of pore in the mesoporous carbon. In addition, the authors showed the precursor of Pt effects on the performance toward ORR. When they used the platinum(II) acetylacetonate, the Pt catalyst showed a clear hydrogen adsorption and desorption peak in the CV and a higher activity for ORR compared with the Pt catalyst prepared using tetrammineplatinum (II) nitrate. Similar mesoporous carbon was modified with rare earth oxide to improve the Pt use [104]. After modification of mesoporous carbon by PrO_x, Pt-supported catalyst was synthesized in a microwave oven. The results showed that PrO_x effectively separated Pt nanoparticles and improved hydrogen adsorptive property of mesoporous carbon.

Yu and coworkers prepared ordered mesoporous or macroporous carbon structures by using periodic arrays of uniform silica spheres as template. They used this carbon as catalyst support for DMFC application [51, 82, 83, 105]. The PtRu-supported catalyst with 80 wt.% metal loading was prepared on macroporous carbon with about 200 nm pores, which showed superior performance of single cell MEA for the DMFC compared with the commercial PtRu-supported catalyst. Also, they changed the pore size in the range of 20–500 nm of porous carbon by changing the size of template particle size and reported that the single cell performance of the PtRu-supported catalysts increased with the decrease of pore size. The authors claimed that higher surface areas and larger pore volumes with decreasing pore sizes can provide the higher degree of catalyst dispersion, which consequently resulted in the observed tendency of the single cell performance with pore sizes.

Lee et al. [106] used mesophase pitch as carbon precursor to produce porous carbon using silica particles, similarly to the method of Yu and coworkers. Also, they attempted to modify the surface of porous carbon by introducing $-SO_3H$ groups. PtRu nanoparticles were supported on the pitch-based porous carbons, and used as anode catalyst in the DMFC. MEA using this catalyst showed improved power density compared with a commercial 60 wt.% PtRu supported on Vulcan XC-72R. Nam et al. also prepared the more conductive porous carbon using mesophase pitch as carbon precursor by changing the size of the silica sphere [107]. The authors deposited 60 wt.% PtRu nanoparticles on conductive porous carbons using the sodium borohydride reduction method. Based on half-cell and single cell performance measurement, the authors suggested that the porous carbon from 20 nm silica sphere is the most suitable support for the PtRu-supported catalyst in their investigation.

Recently, PtRu nanoparticles on mesoporous carbon were produced directly during the carbonization process using SBA-15 as template, furfuryl alcohol as carbon source and platinum and ruthenium acetylacetonates as the metal precursors [97], which resembles the method in a previous report for producing methanol-tolerant Pt catalyst [96]. The $Pt_{50}Ru_{50}$-supported catalyst was found to have the best catalytic activity and long-term stability. Lin et al. [35] compared the activities of PtRu-supported catalyst on various carbon supports such as thin film OMC (TFC), CMK-3, and XC-72. TFC was prepared using a vertically aligned OMS template and the CMK-3 carbon support was synthesized by following the reported method [16]. The enhanced methanol oxidation from PtRu/TFC catalyst was ascribed to increased use by short channels of TFC.

During several years in our laboratory, Pt-supported catalysts on OMC support have been developed to apply to the DMFC system by changing the physical properties of OMC such as primary particle size, conductivity and pore sizes [3, 93, 94].

The mass transport of reactants and by-products in MEAs has a significant effect on the single-cell performance of fuel cells, which is significantly affected by the primary particle size of the carbon support. Commercial carbon supports such as carbon black and activated carbon consist of spherical particles of about 30–60 nm as shown in Figure 9.7(a). The particle sizes of OMC are dependent on the size of the OMS templates in most cases. For fuel cell applications, similar to other areas, smaller particle sizes are favorable for the efficient diffusion and transport of reactants and by-products. Furthermore, each primary particle should be well separated without significant agglomeration. With this consideration, the synthesis of OMS templates was performed by varying the Na to Si ratio in the silica source solution, sodium silicate [93]. The result showed that silica templates with primary particle size of 300 nm were obtained. Thus, OMS of 300 nm particle size was adopted as the template for the synthesis of OMC, and the resulting OMC as observed in Figure 9.7(b) was further used as a support for the cathode catalyst.

A catalyst of 60 wt.% Pt loaded OMC (Pt/OMC-300) was prepared through the modified incipient wetness method, which consists of two successive incipient wetness impregnations and gas reductions [3, 93]. The TEM images of Pt/OMC-300 sample are shown in Figure 9.8. The images by conventional TEM in Figure 9.8(a) show that small Pt nanoparticles of 3 nm are scattered on the OMC particles. TEM images for the Pt/OMC-300 particles were taken in three-dimensions

Figure 9.7 SEM images of (a) conventional carbon black and (b) OMC support prepared from OMS having 300 nm of primary particle size.

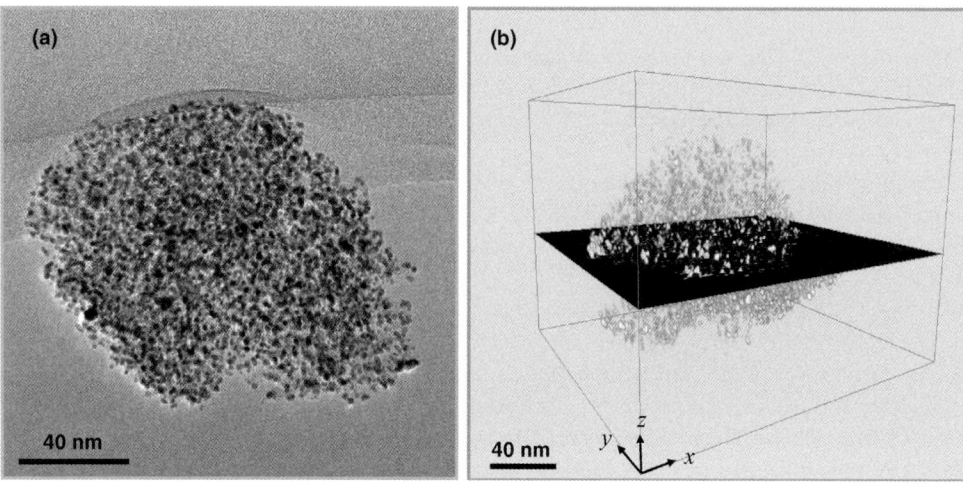

Figure 9.8 (a) TEM images Pt/OMC and (b) Volume rendering of Pt nanoparticles through 3-D TEM reconstruction.

(3-D) to observe the distribution of the Pt particles in the OMC as displayed in Figure 9.8(b). The images show that Pt particles of about 3 nm are 3-D spatially distributed in a uniform and homogeneous manner within the OMC particle, which indicates that the Pt nanoparticles are supported inside the mesopores of the OMC. The polarization curve of a DMFC single cell adopting Pt/OMC-300 as a cathode catalyst and PtRu black as an anode catalyst displayed superior performance to that of a single cell with Pt black catalyst (HiSpec 6000) from Johnson-Matthey [93]. The higher activity can be attributed to the highly dispersed state of the Pt nanoparticles and the uniform mesopore arrays of the OMC.

The effects of the graphitic character of the OMC supports on the DMFC performance have already been discussed in Section 9.3.3 [94]. The electrical conductivity of a carbon support has a significant effect on the performance, which is induced from the materials property and interfacial resistance between the support particles. A new synthetic scheme has been developed as a route to decrease the interparticle resistance [108], where the conductive polymer, polypyrrole is located selectively at the outside of the OMC particle as an electrical bridge. Also, we tried to modify the OMC support to attach the sulfonic group by direct sulfonation using ammonium sulfate. PtRu nanoparticles with 2.9 nm size were supported on the sulfonated OMC support with 70 wt.% loading and showed superior mass activity for the MOR compared with the commercial PtRu catalysts [109].

Recently, the optimization of the cathode catalyst layer based on a supported 60 wt. % Pt/OMC catalyst to maximize the performance of MEA for DMFC was conducted by controlling the ionomer amount in the catalyst slurry and compression pressure on the catalyst layer [110]. The power density at 0.45 V and 323 K of MEA using optimized cathode layer was increased to 104 mW cm^{-2} from that (91 mW cm^{-2}) of MEA using Pt black catalyst as shown in the Figure 9.9. It is worth noting that the Pt loading in the catalyst layer was reduced from 6 mg cm^{-2} for Pt black catalyst

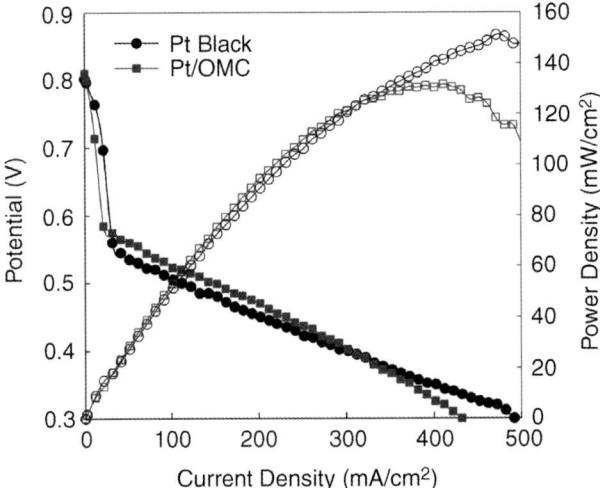

Figure 9.9 Potentiodynamic polarization and power density plots for DMFC MEA at 343 K prepared using Pt/OMC (2 mg cm^{-2}) or Pt Black (6 mg cm^{-2}) as cathode catalysts.

to 2 mg cm^{-2} for Pt/OMC supported catalyst, which lead to a significant reduction in the cost of MEA.

For a real application of OMC as a new support, the 60 wt.% Pt and 70 wt.% PtRu nanoparticles were supported on the OMC prepared using phenanthrene to be applied as anode and cathode catalysts, respectively in the DMFC systems. The MEA was fabricated using OMC-supported catalysts assembled with diffusion layers and a proton-conducting membrane, and subsequently the stack was constructed by assembling 20 MEAs, each of which has a 25 cm^2 active area. The performance of the MEAs in this stack showed 80 mW cm^{-2} power density at 8 V using a 0.75 M methanol solution and air under ambient conditions. This stack was adopted to make the power system for Note PC (SENS Q1, Samsung Electronics) with several other BOP (balance of plant) components as shown in Figure 9.10 [111]. This DMFC system can operate Note PC for 10 h using pure 100 mL methanol, without connecting to the grid line.

9.5
Summary and Prospect

In this chapter, recent advances in the synthesis and applications of mesoporous carbon materials have been discussed, from the viewpoints of the prospective materials as the supports for electrocatalyst in fuel cell applications. The development of the nano-casting approach, to obtain the mesoporous carbon materials from well-organized mesoporous inorganic templates, has opened a breakthrough for the new type of carbon materials. Later, an alternative synthesis route for the mesoporous

Figure 9.10 20W DMFC system connected to Note PC.

carbon materials has also been reported by synergistic self-assembly between organic template and polymers, followed by controlled carbonization. There are some discrepancies between the reported results because the properties of mesoporous carbon-supported catalysts depend on several factors such as primary particle size, pore structure and crystallinity of the materials as well as the preparation method for catalyst particles on the supports. However, in most cases, the mesoporous carbon materials supporting electrocatalysts are known to exhibit enhanced activities in half-cell reactions (methanol oxidation or oxygen reduction) or in single cell configurations, compared with the commercial catalysts; this is probably due to their unique pore structures, controllable surface properties and high surface areas of mesoporous carbon materials.

A fuel cell stack using MEAs adopting mesoporous carbon materials has been developed and applied to a prototype power system for Note PC. The application of these materials can be extended to other mobile devices such as cellular phones and PDAs in the form of either a mobile charger or direct power source as suggested in the previous report [3]. Furthermore, the use of mesoporous carbon materials in fuel cells for residential and vehicle applications may be also viable in the future. And also, mesoporous carbon materials can be used in other electrochemical energy devices such as Li-ion batteries [112,113] and electric double layer capacitors (EDLCs) [114,115].

Many issues remain for development in the design and synthesis. Directions on the evolution of structural characteristics of mesoporous carbon materials for enhancing the performance of energy devices are suggested: (i) more graphitic framework structure, (ii) wider pore size (~ 10 nm) with 3-D interconnected pore structures, which would be more favorable for mass transport than 1-D or 2-D pore structures and (iii) smaller (~ 100 nm) primary particle size. Catalyst preparation methods using mesoporous carbon supports should be also developed enabling support metal (Pt or PtRu) nanocatalysts with a loading level above 90 wt.% while maintaining the particle sizes below than 3 nm. With more intensive and interdisciplinary efforts enhanced performance will be demonstrated in electrochemical energy devices by the use of mesoporous carbon materials. And it will suggest the promise as innovative materials for such applications, which would eventually lead to commercialization in the near future.

References

1 Cheng, F., Tao, Z., Liang, J. and Chen, J. (2008) *Chem. Mat.*, **20**, 667.
2 Lee, J., Kim, J. and Hyeon, T. (2006) *Adv. Mater.*, **18**, 2073.
3 Chang, H., Joo, S.H. and Pak, C. (2007) *J. Mater. Chem.*, **17**, 3078.
4 Pyrzynska, K. (2007) *Anal. Sci.*, **23**, 631.
5 Guillarme, D. and Heinisch, S. (2005) *Sep. Purif. Rev.*, **34**, 181.
6 Saufi, S.M. and Ismail, A.F. (2004) *Carbon*, **42**, 241.
7 Lozano-Castello, D., Alcaniz-Monge, J., de la Casa-Lillo, M.A., Cazorla-Amoros, D. and Linares-Solano, A. (2002) *Fuel*, **81**, 1777.
8 Dias, J.M., Alvim-Ferraz, M.C.M., Almeida, M.F., Rivera-Utrilla, J. and Sanchez-Polo, M. (2007) *J. Environ. Manage.*, **85**, 833.
9 Song, H.H., Li, L.X. and Chen, X.H. (2006) *New Carbon Mater.*, **21**, 374.
10 Wan, Y., Yang, H.F. and Zhao, D.Y. (2006) *Accounts Chem. Res.*, **39**, 423.
11 Rolison, D.R. and Dunn, B. (2001) *J. Mater. Chem.*, **11**, 963.
12 McCusker, L.B., Liebau, F. and Engelhardt, G. (2001) *Pure Appl. Chem.*, **73**, 381.
13 Hu, Z.H., Srinivasan, M.P. and Ni, Y.M. (2000) *Adv. Mater.*, **12**, 62.
14 Kowalewski, T., Tsarevsky, N.V. and Matyjaszewski, K. (2002) *J. Am. Chem. Soc.*, **124**, 10632.
15 Tamon, H., Ishizaka, H., Yamamoto, T. and Suzuki, T. (1999) *Carbon*, **37**, 2049.
16 Ryoo, R., Joo, S.H. and Jun, S. (1999) *J. Phys. Chem. B*, **103**, 7743.
17 Joo, S.H., Choi, S.J., Oh, I., Kwak, J., Liu, Z., Terasaki, O. and Ryoo, R. (2001) *Nature*, **412**, 169.
18 Lee, J., Yoon, S., Hyeon, T., Oh, S.M. and Kim, K.B. (1999) *Chem. Commun.*, 2177.
19 Lu, A.H. and Schuth, F. (2006) *Adv. Mater.*, **18**, 1793.
20 Liang, C., Li, Z. and Dai, S. (2008) *Angew. Chem. Int. Ed.*, **47**, 3696.
21 Liang, C.D., Hong, K.L., Guiochon, G.A., Mays, J.W. and Dai, S. (2004) *Angew. Chem. Int. Ed.*, **43**, 5785.
22 Tanaka, S., Nishiyama, N., Egashira, Y. and Ueyama, K. (2005) *Chem. Commun.*, 2125.
23 Zhang, F.Q., Meng, Y., Gu, D., Yan, Y., Yu, C.Z., Tu, B. and Zhao, D.Y. (2005) *J. Am. Chem. Soc.*, **127**, 13508.
24 Beck, J.S., Vartuli, J.C., Roth, W.J., Leonowicz, M.E., Kresge, C.T., Schmitt, K.D., Chu, C.T.W., Olson, D.H., Sheppard, E.W., McCullen, S.B., Higgins, J.B. and Schlenker, J.L. (1992) *J. Am. Chem. Soc.*, **114**, 10834.
25 Kresge, C.T., Leonowicz, M.E., Roth, W.J., Vartuli, J.C. and Beck, J.S. (1992) *Nature*, **359**, 710.
26 Monnier, A., Schuth, F., Huo, Q., Kumar, D., Margolese, D., Maxwell, R.S., Stucky, G.D., Krishnamurty, M., Petroff, P., Firouzi, A., Janicke, M. and Chmelka, B.F. (1993) *Science*, **261**, 1299.
27 Huo, Q.S., Margolese, D.I., Ciesla, U., Feng, P.Y., Gier, T.E., Sieger, P., Leon, R., Petroff, P.M., Schuth, F. and Stucky, G.D. (1994) *Nature*, **368**, 317.
28 Yang, P.D., Zhao, D.Y., Margolese, D.I., Chmelka, B.F. and Stucky, G.D. (1998) *Nature*, **396**, 152.
29 Moriguchi, I., Ozono, A., Mikuriya, K., Teraoka, Y., Kagawa, S. and Kodama, M. (1999) *Chem. Lett.*, 1171

30 Li, Z.J., Yan, W.F. and Dai, S. (2004) *Carbon*, **42**, 767.
31 Meng, Y., Gu, D., Zhang, F.Q., Shi, Y.F., Cheng, L., Feng, D., Wu, Z.X., Chen, Z.X., Wan, Y., Stein, A. and Zhao, D.Y. (2006) *Chem. Mat.*, **18**, 4447.
32 Knox, J.H., Kaur, B. and Millward, G.R. (1986) *J. Chromatog.*, **352**, 3.
33 Ryoo, R., Joo, S.H., Kruk, M. and Jaroniec, M. (2001) *Adv. Mater.*, **13**, 677.
34 Kaneda, M., Tsubakiyama, T., Carlsson, A., Sakamoto, Y., Ohsuna, T., Terasaki, O., Joo, S.H. and Ryoo, R. (2002) *J. Phys. Chem. B*, **106**, 1256.
35 Lin, M.L., Huang, C.C., Lo, M.Y. and Mou, C.Y. (2008) *J. Phys. Chem. C*, **112**, 867.
36 Vinu, A., Hossian, K.Z., Srinivasu, P., Miyahara, M., Anandan, S., Gokulakrishnan, N., Mori, T., Ariga, K. and Balasubramanian, V.V. (2007) *J. Mater. Chem.*, **17**, 1819.
37 Vinu, A., Srinivasu, P., Takahashi, M., Mori, T., Balasubramanian, V.V. and Ariga, K. (2007) *Microporous. Mesopoorous Mater.*, **100**, 20.
38 Wu, W., Cao, J.M., Chen, Y. and Lu, T.H. (2007) *Chin. J. Catal.*, **28**, 17.
39 Ding, J., Chan, K.Y., Ren, J.W. and Xiao, F.S. (2005) *Electrochim. Acta*, **50**, 3131.
40 Wu, W., Cao, J., Chen, Y. and Lu, T. (2007) *Chin. J. Catal.*, **28**, 17.
41 Solovyov, L.A., Shmakov, A.N., Zaikovskii, V.I., Joo, S.H. and Ryoo, R. (2002) *Carbon*, **40**, 2477.
42 Guo, W.P. and Zhao, X.S. (2005) *Nanoporous Materials IV*, Stud. Surf. Sci. Catal., vol. 156 (eds M. Jaroniec and A. Sayari), Elsevier, Amsterdam, p. 551.
43 Vinu, A., Anandan, S., Anand, C., Srinivasu, P., Ariga, K. and Mori, T. (2008) *Microporous Mesoporous Mater.*, **109**, 398.
44 Kim, T.W. and Solovyov, L.A. (2006) *J. Mater. Chem.*, **16**, 1445.
45 Gierszal, K.P., Jaroniec, M., Kim, T.W., Kim, J. and Ryoo, R. (2008) *New J. Chem.*, **32**, 981.
46 Zhou, J.-H., He, J.-P., Ji, Y.-J., Dang, W.-J., Liu, X.-L., Zhao, G.-W., Zhang, C.-X., Zhao, J.-S., Fu, Q.-B. and Hu, H.-P. (2007) *Electrochim. Acta*, **52**, 4691.
47 Lee, J.S., Joo, S.H. and Ryoo, R. (2002) *J. Am. Chem. Soc.*, **124**, 1156.
48 Lei, Z., Bai, S., Xiao, Y., Dang, L., An, L., Zhang, G. and Xu, Q. (2008) *J. Phys. Chem. C*, **112**, 722.
49 Meng, Y., Gu, D., Zhang, F., Shi, Y., Yang, H., Li, Z., Yu, C., Tu, B. and Zhao, D. (2005) *Angew. Chem. Int. Ed.*, **44**, 7053.
50 Lee, J., Kim, J. and Hyeon, T. (2006) *Int. J. Nanotechnol.*, **3**, 253.
51 Yu, J.-S., Kang, S., Yoon, S.B. and Chai, G. (2002) *J. Am. Chem. Soc.*, **124**, 9382.
52 Fuertes, A.B. (2004) *Microporous Mesoporous Mater.*, **67**, 273.
53 Alvarez, S. and Fuertes, A.B. (2004) *Carbon*, **42**, 433.
54 Vinu, A., Streb, C., Murugesan, V. and Hartmann, M. (2003) *J. Phys. Chem. B*, **107**, 8297.
55 Sun, C.W., Xie, Z., Xia, C.R., Li, H. and Chen, L.Q. (2006) *Electrochem. Commun.*, **8**, 833.
56 Sayari, A., Liu, P., Kruk, M. and Jaroniec, M. (1997) *Chem. Mat.*, **9**, 2499.
57 Mokaya, R. (1999) *J. Phys. Chem. B*, **103**, 10204.
58 Gierszal, K.P. and Jaroniec, M. (2004) *Chem. Commun.*, 2576.
59 Deng, Y.H., Liu, C., Yu, T., Liu, F., Zhang, F.Q., Wan, Y., Zhang, L.J., Wang, C.C., Tu, B., Webley, P.A., Wang, H.T. and Zhao, D.Y. (2007) *Chem. Mat.*, **19**, 3271.
60 Lee, J., Kim, J. and Hyeon, T. (2003) *Chem. Commun.*, 1138.
61 Lee, H.I., Pak, C., Shin, C.H., Chang, H., Seung, D., Yie, J.E. and Kim, J.M. (2005) *Chem. Commun.*, 6035.
62 Xia, Y.D. and Mokaya, R. (2004) *Adv. Mater.*, **16**, 886.
63 Lee, H.I., Kim, J.H., You, D.J., Lee, J.E., Kim, J.M., Ahn, W.S., Pak, C., Joo, S.H., Chang, H. and Seung, D. (2008) *Adv. Mater.*, **20**, 757.
64 Lu, A.-H., Schmidt, W., Spliethoff, B. and Schüth, F. (2003) *Adv. Mater.*, **15**, 1602.
65 Yu, C., Fan, J., Tian, B., Zhao, D. and Stucky, G.D. (2002) *Adv. Mater.*, **14**, 1742.

66 Taguchi, A., Smått, J.-H. and Lindén, M. (2003) *Adv. Mater.*, **15**, 1209.
67 Wang, X., Bozhilov, K.N. and Feng, P. (2006) *Chem. Mater.*, **18**, 6373.
68 Fuertes, A.B. (2003) *J. Mater. Chem.*, **13**, 3085.
69 Zhao, G., He, J., Zhang, C., Zhou, J., Chen, X. and Wang, T. (2008) *J. Phys. Chem. C*, **112**, 1028.
70 Reddington, E., Sapienza, A., Gurau, B., Viswanathan, R., Sarangapani, S., Smotkin, E.S. and Mallouk, T.E. (1998) *Science*, **280**, 1735.
71 Park, K.-W., Choi, J.-H., Lee, S.-A., Pak, C., Chang, H. and Sung, Y.-E. (2004) *J. Catal.*, **224**, 236.
72 Liu, H., Song, C., Zhang, L., Zhang, J., Wang, H. and Wilkinson, D.P. (2006) *J. Power Sources*, **155**, 95.
73 Hogarth, M.P. and Ralph, T.R. (2002) *Platinum Metals Rev.*, **46**, 146.
74 Chan, K.Y., Ding, J., Ren, J.W., Cheng, S.A. and Tsang, K.Y. (2004) *J. Mater. Chem.*, **14**, 505.
75 Antolini, E. (2008) *Appl. Catal. B: Environ.*, doi: 10.1016/j.apcatb.2008.09.030.
76 Dicks, A.L. (2006) *J. Power Sources*, **156**, 128.
77 Hsu, N.-Y., Chien, C.-C. and Jeng, K.-T. (2008) *Appl. Catal. B: Environ.*, **84**, 196.
78 Xia, B.Y., Wang, J.N., Wang, X.X., Niu, J.J., Sheng, Z.M., Hu, M.R. and Yu, Q.C. (2008) *Adv. Funct. Mater.*, **18**, 1790.
79 Park, K.-W., Sung, Y.-E., Han, S., Yun, Y. and Hyeon, T. (2004) *J. Phys. Chem. B*, **108**, 939.
80 Wang, J.N., Zhang, L., Niu, J.J., Yu, F., Sheng, Z.M., Zhao, Y.Z., Chang, H. and Pak, C. (2007) *Chem. Mat.*, **19**, 453.
81 You, D.J., Lee, S.-A., Joo, S.H., Pak, C., Chang, H. and Seung, D. (2006) *Scientific Bases for the Preparation of Heterogeneous Catalysts*, Stud. Surf. Sci. Catal., vol. 162 (eds E.M. Gaigneaux, M. Devillers, D.E. De Vos, S. Hermans, P.A. Jacobs, J.A. Martens and P. Ruiz), Elsevier, Amsterdam, p. 537.
82 Chai, G.S., Yoon, S.B., Yu, J.S., Choi, J.H. and Sung, Y.E. (2004) *J. Phys. Chem. B*, **108**, 7074.
83 Chai, G.S., Shin, I.S. and Yu, J.S. (2004) *Adv. Mater.*, **16**, 2057.
84 Steele, B.C.H. (2001) *J. Mater. Sci.*, **36**, 1053.
85 Yu, X. and Ye, S. (2007) *J. Power Sources*, **172**, 133.
86 Gratiet, B.L., Remita, H., Picq, G. and Delcourt, M.O. (1996) *J. Catal.*, **164**, 36.
87 Zhou, Z., Wang, S., Zhou, W., Wang, G., Jiang, L., Li, W., Song, S., Liu, J., Sun, G. and Xin, Q. (2003) *Chem. Commun.*, 394.
88 Zhou, Z., Zhou, W., Wang, S., Wang, G., Jiang, L., Li, H., Sun, G. and Xin, Q. (2004) *Catal. Today*, **93–95**, 523.
89 Shen, P.K. and Tian, Z. (2004) *Electrochim. Acta*, **49**, 3107.
90 Liu, Z., Gan, L.M., Hong, L., Chen, W. and Lee, J.Y. (2005) *J. Power Sources*, **139**, 73.
91 Liu, C., Xue, X., Lu, T. and Xing, W. (2006) *J. Power Sources*, **161**, 68.
92 Joo, J.B., Kim, P., Kim, W. and Yi, J. (2008) *Catal. Today*, **131**, 219.
93 Pak, C., You, D.J., Lee, S.-A., Kim, J.M. and Chang, H. (2005) *Samsung J. Innovative Technol.*, **1**, 239.
94 Joo, S.H., Pak, C., You, D.J., Lee, S.A., Lee, H.I., Kim, J.M., Chang, H. and Seung, D. (2006) *Electrochim. Acta*, **52**, 1618.
95 Nam, J.H., Jang, Y.Y., Kwon, Y.U. and Nam, J.D. (2004) *Electrochem. Commun.*, **6**, 737.
96 Choi, W.C., Woo, S.I., Jeon, M.K., Sohn, J.M., Kim, M.R. and Jeon, H.J. (2005) *Adv. Mater.*, **17**, 446.
97 Liu, S.H., Yu, W.Y., Chen, C.H., Lo, A.Y., Hwang, B.J., Chien, S.H. and Liu, S.B. (2008) *Chem. Mat.*, **20**, 1622.
98 Wen, Z.H., Liu, J. and Li, J.H. (2008) *Adv. Mater.*, **20**, 743.
99 Su, F.B., Zeng, J.H., Bao, X.Y., Yu, Y.S., Lee, J.Y. and Zhao, X.S. (2005) *Chem. Mat.*, **17**, 3960.
100 Su, F., Zeng, J., Bai, P., Lv, G., Guo, P.-Z., Sun, H., Li, H.L., Yu, J., Lee, J.Y. and Zhao, X.S. (2007) *Ind. Eng. Chem. Res.*, **46**, 9097.

101 Kim, P., Kim, H., Joo, J.B., Kim, W., Song, I.K. and Yi, J. (2005) *J. Power Sources*, **145**, 139.
102 Kim, H., Kim, P., Joo, J.B., Kim, W., Song, I.K. and Yi, J. (2006) *J. Power Sources*, **157**, 196.
103 Hayashi, A., Notsu, H., Kimijima, K., Miyamoto, J. and Yagi, I. (2008) *Electrochim. Acta*, **53**, 6117.
104 Zhou, J., He, J., Zhao, G., Zhang, C., Wang, T. and Chen, X. (2008) *Electrochem. Commun.*, **10**, 76.
105 Chai, G.S., Yoon, S.B., Kim, J.H. and Yu, J.S. (2004) *Chem. Commun.*, 2766.
106 Lee, J.-B., Park, Y.-K., Yang, O.-B., Kang, Y., Jun, K.-W., Lee, Y.-J., Kim, H.Y., Lee, K.-H. and Choi, W.C. (2006) *J. Power Sources*, **158**, 1251.
107 Nam, K., Jung, D., Kim, S.K., Peck, D. and Ryu, S. (2007) *J. Power Sources*, **173**, 149.
108 Choi, Y.S., Joo, S.H., Lee, S.A., You, D.J., Kim, H., Pak, C., Chang, H. and Seung, D. (2006) *Macromolecules*, **39**, 3275.
109 Pak, C., Joo, S.H., You, D.J., Kim, J.M., Chang, H. and Seung, D. (2007) *Recent Progress in Mesostructured Materials*, Stud. Surf. Sci. Catal., vol. 165 (eds D. Zhao, S. Qiu, Y. Tang and C. Yu), Elsevier, Amsterdam, p. 401.
110 Kim, H.T., You, D.J., Yoon, H.K., Joo, S.H., Pak, C., Chang, H. and Song, I.S. (2008) *J. Power Sources*, **180**, 724.
111 Chang, H., Song, I. and Seung, D. (2006) DMFC for Note PC and Mobile Phone: From Materials to System, in *The 8th Annual International Symposium; Small Fuel Cell*, Knowledge Press, Washington D.C., USA.
112 Zhou, H., Zhu, S., Hibino, M., Honma, I. and Ichihara, M. (2003) *Adv. Mater.*, **14**, 2107.
113 Fan, J., Wang, T., Yu, C., Tu, B., Jiang, Z. and Zhao, D. (2004) *Adv. Mater.*, **16**, 1432–1436.
114 Li, L., Song, H. and Chen, X. (2006) *Electrochim. Acta*, **51**, 5715.
115 Numao, S., Judai, K., Nishijo, J., Mitzuuchi, K. and Nishi, N. (2008) *Carbon*, doi: 10.1016/J.carbon.2008.10.012.

10
Proton Exchange Membranes for Direct Methanol Fuel Cells[1)]
Dae Sik Kim, Michael D. Guiver, and Yu Seung Kim

10.1
Introduction

Fuel cells are now widely recognized as one solution to increasing power demands, mitigation of air pollutants, and efficient use of fossil fuels. It is increasingly recognized that portable direct methanol fuel cells (DMFC)s will likely be the first fuel cells commercially available to the general public; there has been a recent increase in the number of pre-commercial portable devices powered by DMFCs that have been demonstrated in the press and at conferences.

Extensive efforts have been made to develop alternative polymer electrolyte membranes (PEM)s to overcome the drawbacks of the current widely used perfluorinated sulfonic acid (PFSA) copolymer membrane (e.g., Nafion) such as performance deterioration at high temperature and high methanol permeability, which leads to lower cell voltage and decreased fuel efficiency in DMFCs. Nafion is a statistical copolymer comprising a highly hydrophobic perfluorinated backbone that contains a number of short, flexible pendant side chains with single strongly hydrophilic sulfonic acid groups. This structure produces nanophase separated membrane morphology and thus they show excellent thermal, mechanical and electrochemical properties. The flexibility of the side chain of Nafion allows for the aggregation of the superacid fluoroalkyl sulfonic sites into channels, which conduct protons well. However, the proton conductivity of Nafion is reduced above 100 °C due to morphological relaxations.

Significant research has been conducted on developing lower methanol permeable electrolytes. Various polymer types such as polyarylenes, polyimides, polyphosphazenes, radiation grafted polystyrenes, polystyrene block copolymers and polyvinyl alcohols (PVA)s have been developed over the last 10 years. Much of the research on polyarylenes has been performed on poly(arylene ether sulfone)s (PES) [1–8], poly (arylene ether ketone)s (PAEKs) [9–11], and other polyphenylene copolymers [12–14]. These polymers are traditional engineering polymers and are known for their good thermal/mechanical properties, oxidative stability and processibility [15]. In their

---

1) NRCC publication number 50886.

Electrocatalysis of Direct Methanol Fuel Cells. Edited by Hansan Liu and Jiujun Zhang
Copyright © 2009 WILEY-VCH Verlag GmbH & Co. KGaA, Weinheim
ISBN: 978-3-527-32377-7

sulfonated form, they have shown good mechanical properties, proton conductivities and relatively low methanol permeabilities compared with Nafion. The goals of this chapter are (i) to give an overview of non-PFSA alternative hydrocarbon-based PEMs investigated specifically for DMFCs. In order to achieve this goal, we consider that there are two categories of alternative PEMs. One contains aliphatic polymer such as polystyrene sulfonic acid (PS), and PVA. The other one comprises PEMs having aromatic polymer backbones such as PES, PAEK, and poly(arylene ether nitriles); (ii) to present currently reported properties and performance of DMFCs, particularly those using alternative membranes. In order to achieve this goal, we first present the *ex situ* membrane properties required for a successful fuel cell electrolyte, such as high proton conductivity, low methanol permeability, chemical stability, and good mechanical properties. Finally, DMFC performance, the ultimate goal of any novel electrolyte, is presented using reported performance of a number of polymers.

10.2
Synthesis of Polymer Electrolyte Membranes for DMFC

A wide variety of PEMs have been developed for DMFC over the last 10 years and there is continued effort in developing new materials. The discussion presented here is divided into two categories of PEMs: those containing aliphatic polymer backbones and those containing aromatic polymer backbones, such as poly(arylene ether) copolymers.

10.2.1
Synthesis and Properties of PEMs Containing Aliphatic Polymers

The most commonly known and studied PEMs are based on non-aromatic perfluorinated hydrocarbons such as Nafion, Aciplex, Flemion. The current state-of-the-art PEM is Nafion, a DuPont product that was developed in the late 1960s, primarily as a permselective separator in chlor-alkali electrolyzers [16]. Nafion is a free radical initiated copolymer of a crystallizable hydrophobic tetrafluoroethylene (TFE) backbone sequence (~87 mol% at 1100 equivalent weight) with a comonomer, which ultimately has pendant side chains of perfluorinated vinyl ethers terminated by perfluorosulfonic acid groups. The reported chemical structure of Nafion for PEM membranes is shown in Figure 10.1.

Polymers based on styrene, and particularly its fluorinated derivatives, have been used to form PEMs. Styrenic monomers are widely available and easy to modify, and

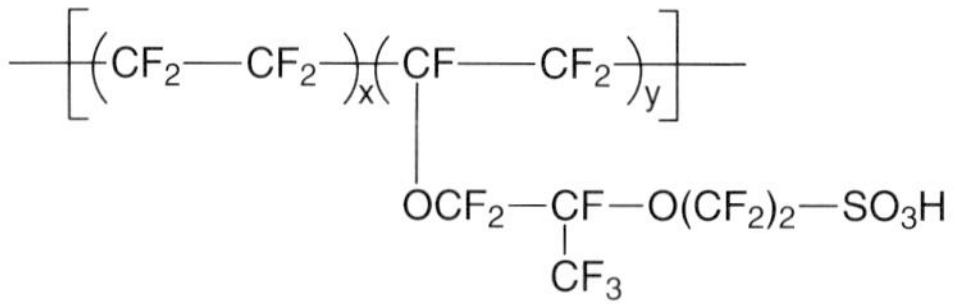

$$-[(CF_2-CF_2)_x(CF-CF_2)_y]-$$
$$\quad\quad\quad\quad\quad\quad |$$
$$\quad\quad\quad\quad OCF_2-CF-O(CF_2)_2-SO_3H$$
$$\quad\quad\quad\quad\quad\quad\quad |$$
$$\quad\quad\quad\quad\quad\quad\quad CF_3$$

Figure 10.1 Chemical structure of Nafion.

their polymers are easily synthesized via conventional free radical and other polymerization techniques. One of the primary concerns with this type of polymer is loss of ion exchange capacity (IEC) due to backbone degradation of polystyrene under fuel cell conditions [17, 18], caused by the presence of the labile α-proton. Some literature reports of these materials have shown reasonable lifetimes, up to 2000 h, with little performance degradation in DMFC testing [19, 20]. Several researchers have suggested that sulfonated styrene copolymers could be used for DMFC membranes. Copolymers, such as poly(vinylidene fluoride) grafted styrene sulfonic acid (PVDF-g-PSSA) [21, 22], sulfonated poly(styrene-isobutylene-styrene) [23–25], sulfonated poly([vinylidene difluoride-co-chlorotrifluoro ethylene]-g-styrene [P(VDF-co-CTFE)-g-SPS]) [26] have been synthesized and characterized. The Holdcroft group [27] compared graft and diblock copolymers that were synthesized so as to contain a similar ratio of fluorous to styrene components, and similar chemical compositions, but distinctly different macromolecular structures: P(VDF-co-CTFE)-g-SPS consisted of a hydrophobic fluorous backbone with ionic sulfonated styrenic side chains, P(VDF-co-HFP)-b-SPS possessed a hydrophobic fluorous segment linearly connected to an ionic sulfonated styrenic segment as shown in Figure 10.2. A comparison of the two macromolecular systems showed that the graft copolymers yield membranes which tolerate much higher ionic contents without excessive swelling and dissolution, and which leads to membranes that possess highly concentrated, isotropically connected ionic domains.

Early commercialization of DMFC requires new proton-conducting membranes that can significantly reduce methanol permeability while having suitable proton conductivity. In the past, several attempts have been made to minimize methanol crossover by developing new membranes from non-fluorinated and partially fluorinated hydrocarbon-based polymers.

Specific attempts to reduce the excessive swelling of the membranes have been made to modify membranes, such as using chemical crosslinking structure and higher molecular weight (MW) polymers [28–31].

Poly(vinyl alcohol) (PVA) membranes are used in pervaporation-based dehydration of alcohols because they preferentially permeate water and retain alcohol [32–34]. Taking advantage of its high selectivity, PVA based membranes have been investigated for DMFCs. To introduce the proton exchange site and reduce water swelling, several techniques have been used such as crosslinking [35–37], or blending with

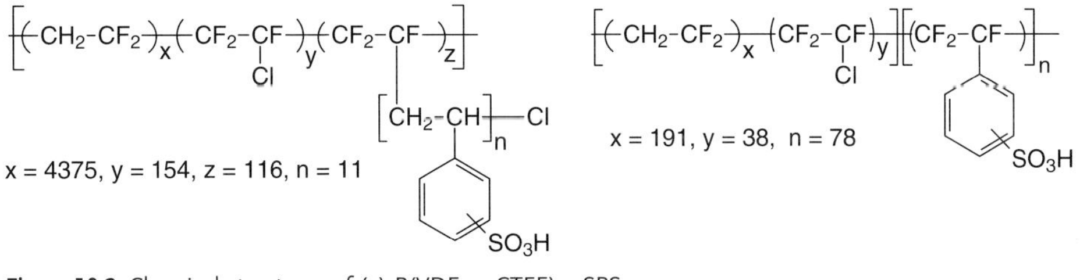

Figure 10.2 Chemical structures of (a) P(VDF-co-CTFE)-g-SPS graft and (b) P(VDF-co-HFP)-b-SPS diblock copolymers [27].

inorganic filler [38–42]. PVA itself does not have any negatively charged ions such as carboxylic and sulfonic acid groups. Rhim's group reported that sulfonic acid groups were introduced into the PVA matrix by modifications of the chemical structure of the PVA through esterification with sulfosuccinic acid (SSA) [35], and poly(styrene sulfonic acid-co-maleic acid) (PSSA-co-MA) [36], and poly(acrylic acid-co-maleic acid) (PAA-co-MA) [37], which have sulfonic acid and/or carboxylic groups. These materials containing ionic groups were used as a crosslinking agent and as a donor of hydrophilic $-SO_3H$ and/or $-COOH$ groups. Figure 10.3 shows the structure of PVA-based membranes. Kannan reported that a series of semi-interpenetrating network (SIPN) membranes was synthesized by using PVA with crosslinking agent SSA and PSSA-co-MA as proton sources, and their physico-chemical and electrochemical characterizations were reported [43]. Although a power density value of over 100 mW/cm^2 was obtained for a SIPN membrane-based membrane electrode assembly (MEA) at 80 °C, the PVA-based membranes have some issues with chemical and/or electrochemical stability.

Following this brief review of various types of PEMs containing aliphatic polymers developed for DMFC applications, we now focus attention on the comparative evaluation of these membranes. Numerous studies have investigated the methanol

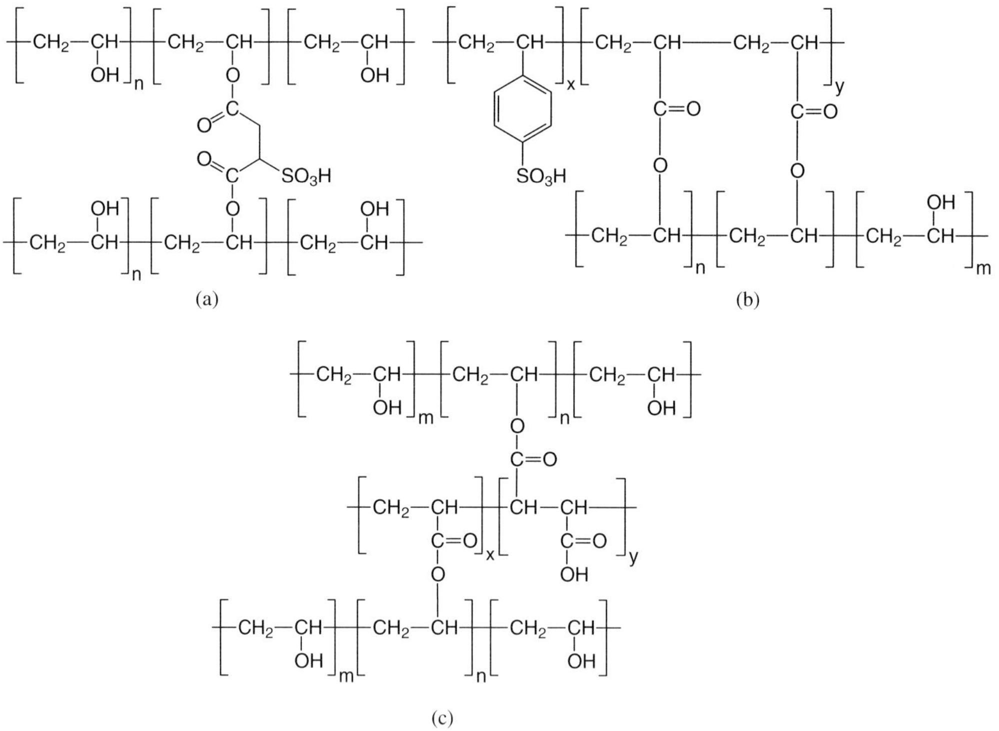

Figure 10.3 Chemical structures of crosslinked PVA with (a) sulfosuccinic acid [35] (b) PSSA-MA [36] and (c) PAA-co-MA [37].

10.2 Synthesis of Polymer Electrolyte Membranes for DMFC | 383

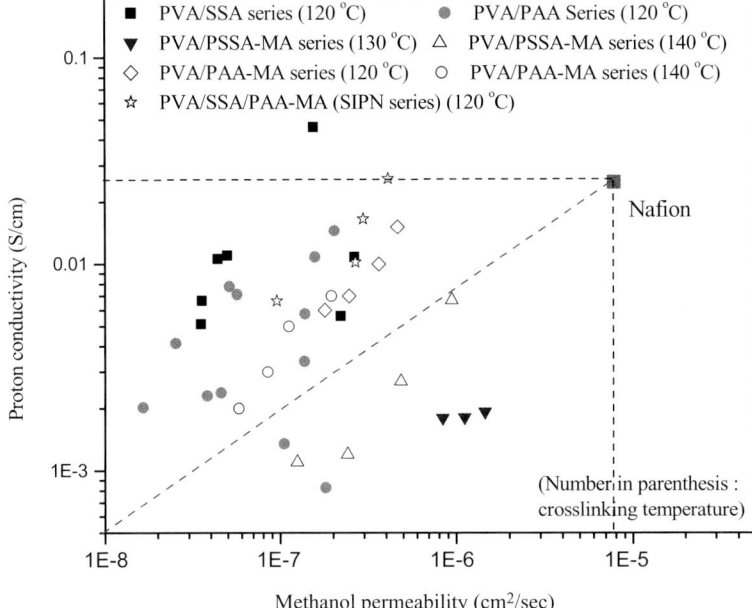

Figure 10.4 Methanol permeability versus proton conductivity for PEMs based PVA membranes (Refs. [35–38, 43]).

permeability and proton conductivity of alternative PEMs and compared them with Nafion in an attempt to show improved properties for DMFCs. Figure 10.4 shows the relationship of proton and methanol permeability for selected PVA membranes at 30 °C. Proton conductivity has a strong trade-off in its relationship with methanol permeability. The methanol permeabilities of the PVA based membranes are lower than that of Nafion. However, the proton conductivities are also lower than that of Nafion, even though the sulfonic acid group was introduced into PVA matrix.

Water uptake of PEMs is an important evaluation parameter for both membrane-electrode compatibility and mechanical properties of the membrane. Water uptake versus conductivity has rarely been plotted in the multitude of literature references although water uptake values are often tabulated along with conductivity data. Generally, plots of IEC versus conductivity are presented; however, these do not allow direct comparison of membranes having different compositions. The relationship between water uptake and proton conductivity can be described by two different methods. One uses a relative value; the other uses an upper bound relationship. Figure 10.5a shows the relative conductivity of various PEMs based on PVA as a function of relative water uptake. In this study, the conductivity and water uptake of selected membranes have been normalized to the values reported of Nafion. Hence, Nafion is defined as having a relative conductivity and water uptake of 1. Figure 10.5b shows the relationship between proton conductivity and water uptake reported in the temperature range of 20–30 °C [35–39, 43]. The data shown in Figure 10.5b demonstrate a strong empirical upper-bound relationship. The upper

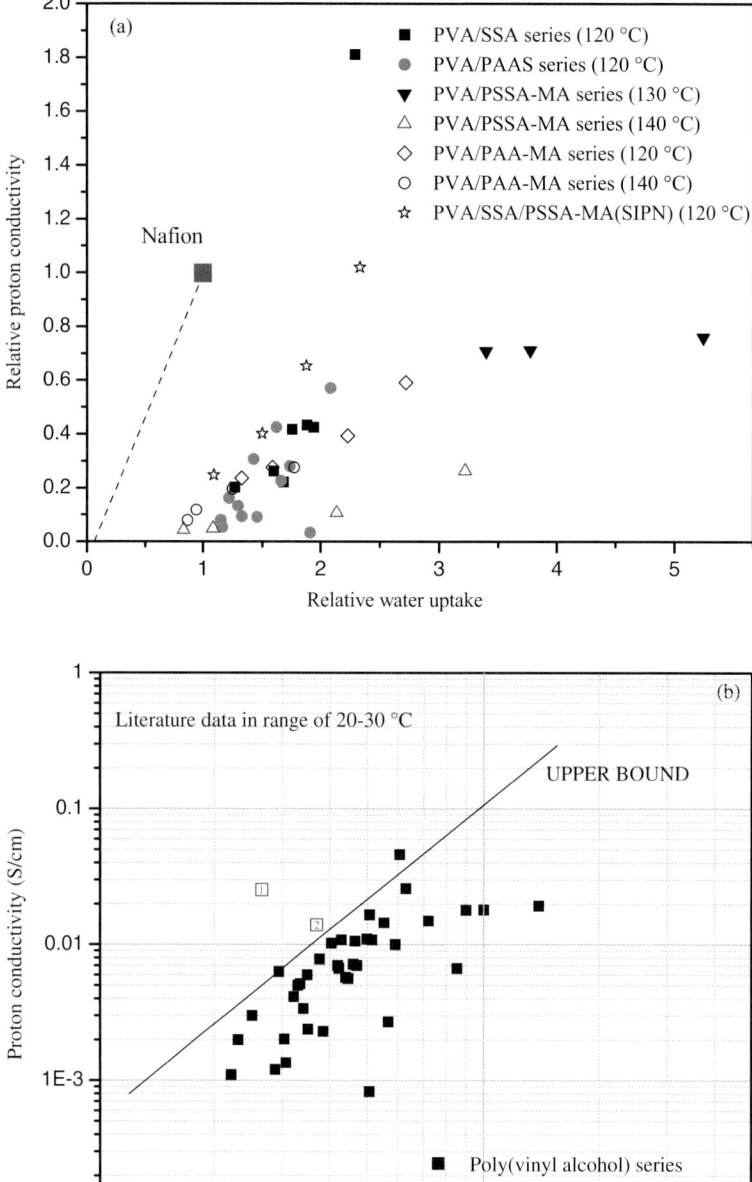

Figure 10.5 (a) Relative proton conductivity versus relative water uptake (b) upper bound relationship for PEMs based PVA membranes (Refs. [35–38, 43]).

bound in this case is an experimental 'state-of-the-art' limit for this type of polymer system of conductivity expressed as a function of water uptake. McGrath's group also showed a similar upper bound relationship for PEM materials [44]. As shown in Figure 10.5a and b, most of the PEMs membranes based on PVA have up to two times higher water uptake and lower proton conductivity, in spite of the introduction of ionic groups in the crosslinking agent. Although there is no clear guideline for the maximum allowable water uptake, PEMs with higher water uptake and lower proton conductivity values typically lead to difficulties in the preparation, durability and performance of MEAs.

10.2.2
Synthesis of Sulfonated Poly(aryl ether) Copolymers

Poly(arylene ether)s such as poly(arylene ether sulfone) (PES), poly(arylene ether ether ketone) (PEEK), and their derivatives are considered as more promising materials on which to base high performance PEMs because of their better oxidative and hydrolytic stability in the fuel cell environment and because many polymer structural variants are possible. At least two methodologies have been used to introduce proton exchange sites, usually sulfonic acid or carboxylic acid, into poly (arylene ether)s. Both post-sulfonation of existing polymers and direct copolymerization of sulfonated monomers are widely used [45], and are discussed below.

10.2.2.1 Post-Sulfonation of Polymers
The sulfonation reaction is an electrophilic aromatic substitution reaction that normally occurs on the site of benzene rings that have a high electron density; therefore, its efficacy depends upon the substituents present on the aromatic ring. Electron-donating substituents favor sulfonation, whereas electron-withdrawing substituents retard or prevent sulfonation. Concentrated sulfuric acid, fuming sulfuric acid, chlorosulfonic acid and trimethylsilylchlorosulfonic acid have been commonly employed as sulfonating agents.

Post-sulfonation is attractive because of the availability of inexpensive commercial high MW thermoplastic starting materials as well as relatively simple reaction procedures, enabling the process to be readily scaled up. However, experimental difficulties are sometimes encountered in achieving precise control over the site of sulfonation and the degree of sulfonation (DS), resulting in a random or less-defined distribution of sulfonic acid groups along the polymer chain. The latter may lead to excessive dimensional swelling in the presence of water, resulting in lower durability of MEAs. Furthermore, consideration must be given to reaction conditions that do not result in polymer chain degradation, which compromise mechanical properties.

The sulfonation reaction of commercial Victrex has been thoroughly studied by several groups [10, 46–48]. Poly(ether ether ketone) (PEEK) shows little solubility in organic solvents due to crystallinity. By introducing sulfonic acid groups to the main chain, the crystallinity decreases and solubility increases. The sulfonation of PEEK has been reported to be a second-order reaction, which takes place at the aromatic ring flanked by two ether links, due to the higher electron density of the ring [10].

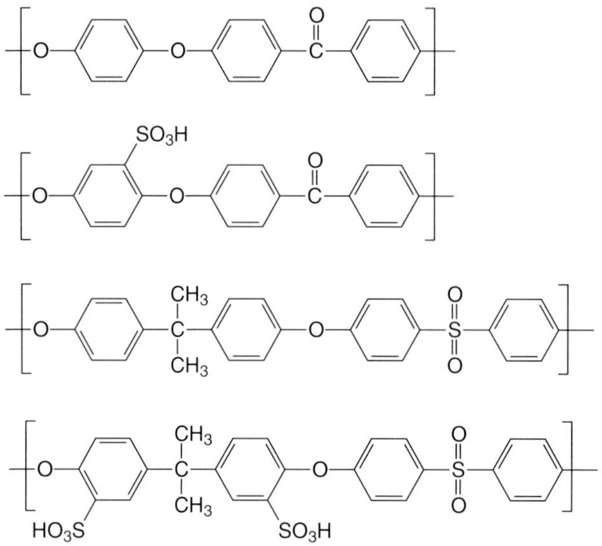

Figure 10.6 General chemical structures for unsulfonated and sulfonated poly(arylene ether sulfone or ketone) copolymers.

Sulfonated PESs as well as PEEK have been investigated intensively using different sulfonating agents such as chlorosulfonic acid, and a sulfur trioxide-triethyl phosphate complex. General chemical structures for the unsulfonated and sulfonated analogs are shown in Figure 10.6.

Guiver's group reported a novel class of PEM materials based on poly(phthalazinone)s (PPs). Poly(phthalazinones) including poly(phthalazinone ether sulfone) (PPES), poly(phthalazinone ether ketone) (PPEK) and poly(phthalazinone ether sulfone ketone) (PPESK) are new high performance polymers in the early stages of commercialization. Among other advantages, this class of polymers is distinguished by excellent chemical and oxidative resistance, mechanical strength, high thermal stability and very high glass transition temperatures (295, 263 and 278 °C, respectively) [49]. The structures of PPESK and PPEK are shown in Figure 10.7. Membrane obtained from highly sulfonated PPs showed proton conductivity above 10^{-2} S/cm at room temperature and elevated temperature [49–51].

A series of PAEKs with structurally different phenyl pendant groups was sulfonated with regular reagent concentrated sulfuric acid (95–98%) at room temperature [52]. In comparison to the sulfonation of PEEK, the phenylated and 4-methyphenylated PEEKs have a considerable advantage in having a much faster sulfonation reaction rate (Figure 10.8). For these two PAEKs, the sulfonation reaction proceeded rapidly in concentrated sulfuric acid at room temperature, with polymers having DS of 1.0 being obtained within several hours. It is also relevant to note that sulfonation was site-specific; only one substitution site on the pendant benzene ring per repeat unit occurred for short reaction times, essentially resulting in homopolymer-like structures. Extended reaction times resulted in additional site-specific ring sulfonation,

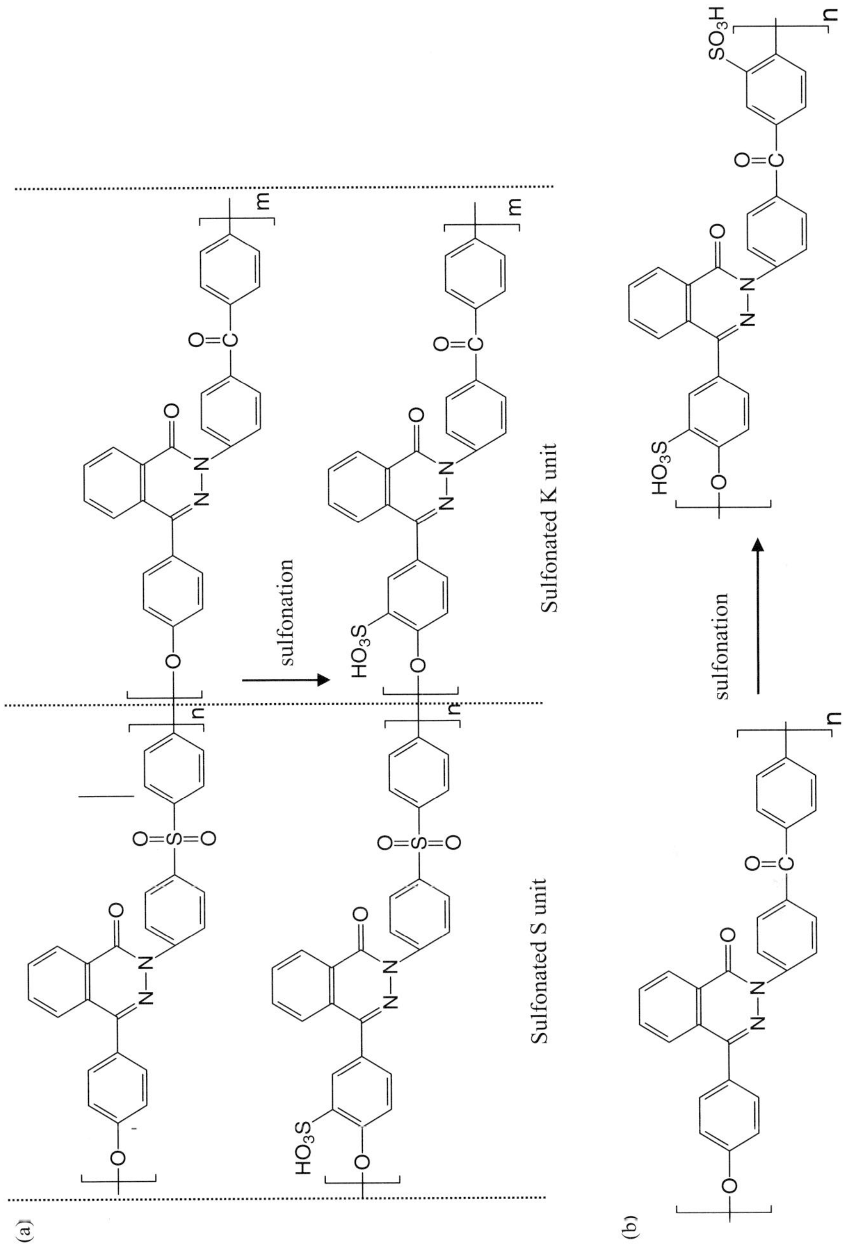

Figure 10.7 Sulfonation reaction of (a) SPPESK, (b) SPPEK.

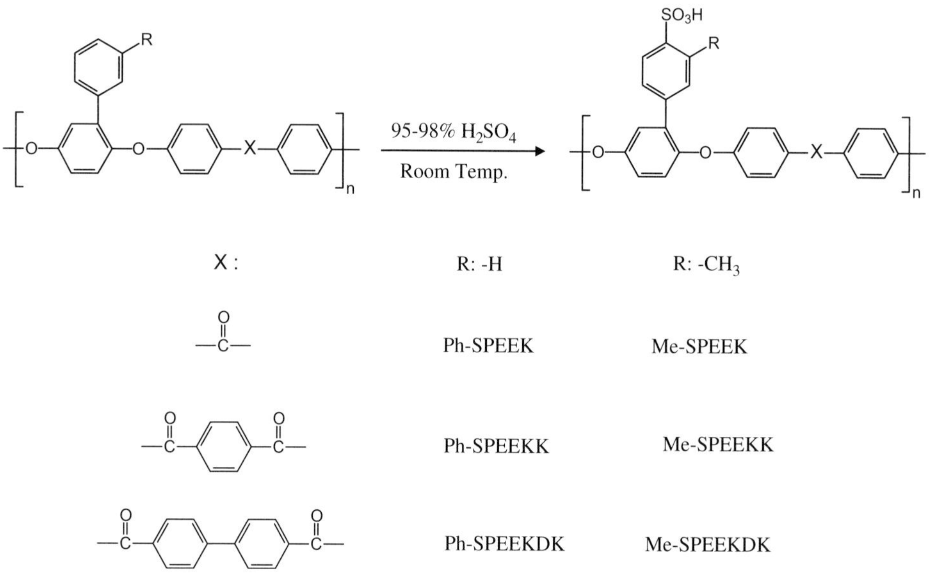

Figure 10.8 Sulfonation reactions of SPAEK polymers.

leading to homopolymer-like structures with a DS of 2.0 [53]. Recently, sulfonated poly(arylene ether)s containing phenyl pendant groups were prepared using chlorosulfonic acid [53b, c] (Figure 10.9a–c). The copolymers fluorinated sulfonated poly (arylene ether) (Figure 10.9a) and sulfonated poly(arylene ether nitrile)s (Figure 10.9b), each having a DS of 1.0 had high proton conductivities of 135.4 and 140.1 mS/cm at 80 °C, and acceptable volume-based water uptake of 44.5–51.9 vol% at 80 °C, respectively, compared with Nafion [53b]. The group of Watanabe reported PES-based ionomers containing sulfofluorenyl groups having proton conductivity of 0.3 S/cm at 80 °C and 93% relative humidity(RH) (Figure 10.9d) [53d].

10.2.2.2 Direct Copolymerization of Sulfonated Monomers

In comparison with the post-sulfonation of a pre-formed polymer, the direct copolymerization of a sulfonated monomer is an alternative approach with some distinct advantages with respect to the controllability of sulfonic acid content (SC) and site of sulfonation. To some degree, the incorporation of sulfonated or functional monomers allows a closer control of molecular design of the resulting copolymer [54].

Typically cited drawbacks of post-modification include the lack of control over the degree and site of functionalization, which is a common problem in macromolecular chemistry. It is of interest to investigate the effect of introducing sulfonic acids, for example, onto the deactivated sites of the repeat units, since one might expect enhanced stability and higher acidity from sulfonic acid groups that are attached to electron-deficient aromatic rings rather than those bonded to electron-rich aromatic rings as in the case of post-modification [55]. The possibilities of controlling and/or increasing MW to enhance durability are not feasible in the case of post-reaction on an existing commercial product.

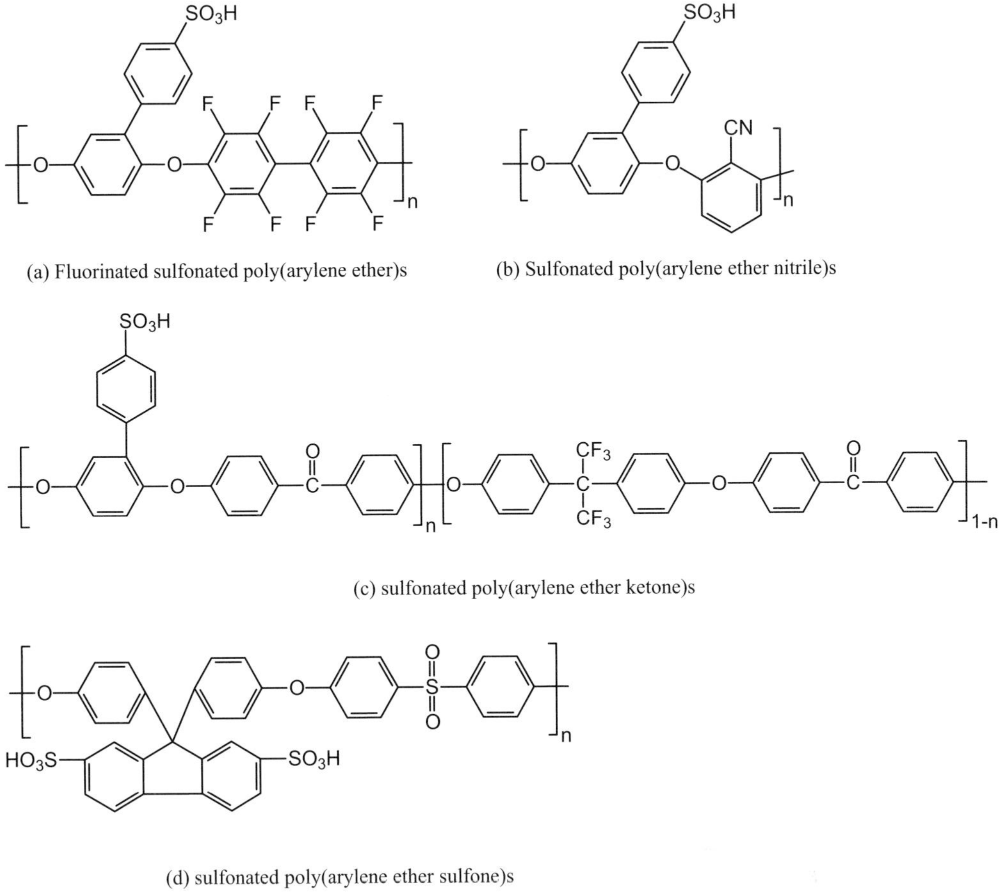

(a) Fluorinated sulfonated poly(arylene ether)s

(b) Sulfonated poly(arylene ether nitrile)s

(c) sulfonated poly(arylene ether ketone)s

(d) sulfonated poly(arylene ether sulfone)s

Figure 10.9 Structure of various sulfonated poly(arylene ether) copolymers.

Since the first report of a sulfonated dihalo monomer by Robeson and Matzner [56], Ueda et al. [57], and Wang et al. [58] reported the sulfonation of 4,4-dichlorodiphenyl sulfone, and 4,4-difluorodiphenyl ketone, prepared via an electrophilic reaction using fuming sulfuric acid. Sulfonated 1,4-bis(4-fluorobenzoyl)benzene (1,4-BFBB) was reported by several groups [59, 60]. Powerful sulfonation conditions of fuming sulfuric acid at a relatively high temperature (100 °C) were necessary to sulfonate these monomers.

Guiver's group also synthesized directly copolymerized sulfonated PEKs and PESs using several sulfonated bisphenol monomers available commercially [61–65]. Since it is not easy to control the substitution sites of the small-MW compounds in the sulfonation process and to purify the sulfonated monomers, only a few sulfonated monomers have been developed so far (Figure 10.10). A variety of sulfonated copolymers based on these sulfonated monomers have been prepared via typical aromatic nucleophilic substitution polycondensation, which is the same type of

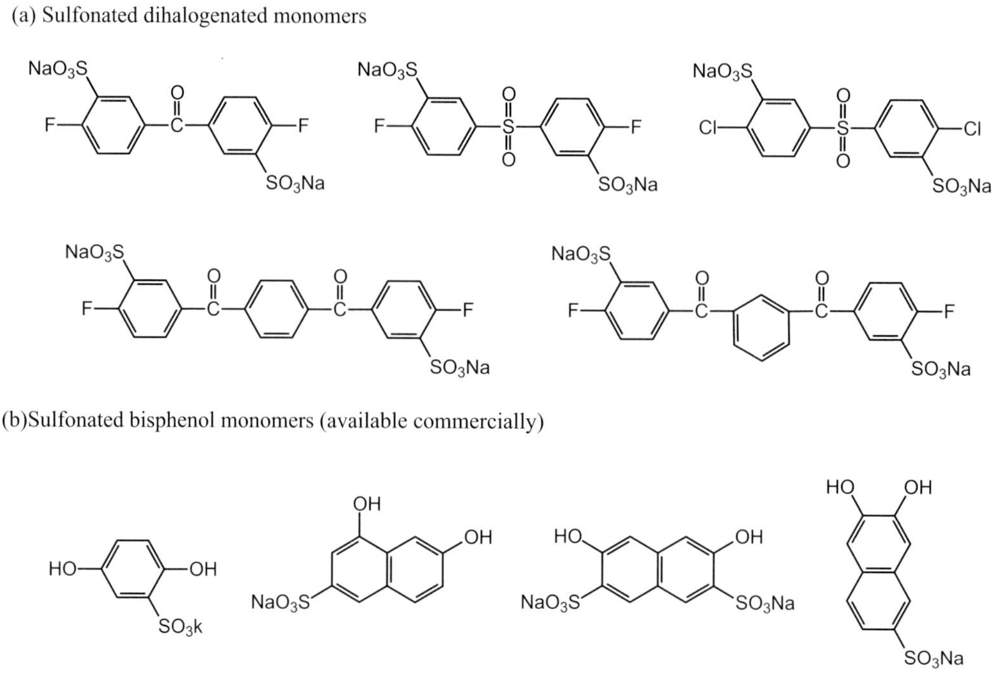

(a) Sulfonated dihalogenated monomers

(b) Sulfonated bisphenol monomers (available commercially)

Figure 10.10 Reported sulfonated monomers.

reaction used for the polymerization of non-sulfonated poly(aryl ether)-type polymers. It is important to note that the polymerization of sulfonated monomers requires the salt formation of both the phenoxides and the sulfonates, which in some cases may reduce the solubility of the monomer and growing polymer chain, thereby limiting MW.

Poly(arylene ether sulfone)s Directly copolymerized sulfonated polymers were produced under similar reaction conditions employed for many years for the synthesis of unsulfonated poly(arylene ether)s, using a weak base such as potassium carbonate. Only moderately higher reaction temperatures and longer times were needed to obtain high MW copolymers, due to the sterically decreased activity of the sulfonated dihalide monomer [66]. McGrath's group reported a sulfonated polyethersulfone copolymer series having different biphenol and/or polar functional groups, and some of their structures are shown in Figure 10.11. It was shown that the bisphenol structure and SC influenced the properties of the sulfonated copolymers, including the solubility, IEC, and water uptake [5, 67–69]. Four bisphenols including (a) hydroquinone, (b) 4,4′-biphenol, (c) hexafluoroisopropylidene bisphenol (bisphenol 6F), (d) bisphenol A in Figure 10.11, were used for the synthesis of poly(arylene ether)s containing ion conducting units. The copolymers based on bisphenol 6F were especially promising in initial fuel cell tests. It is believed that their fluorine content promotes adhesion and electrochemical compatibility with Nafion-based electrodes and reduces swelling.

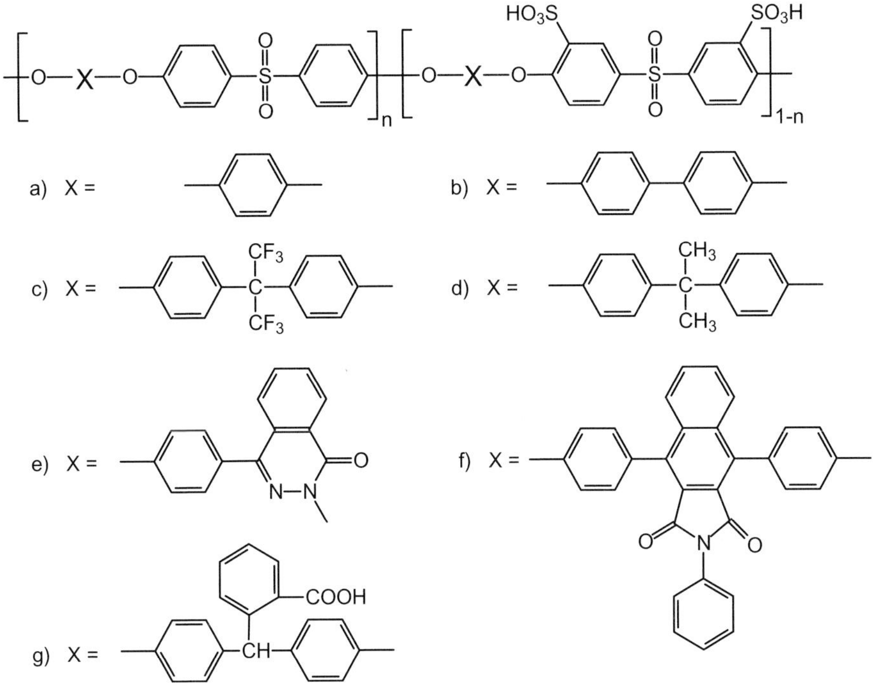

Figure 10.11 Typical copolymers derived from sulfonated 4,4′-dichlorodiphenylsulfone.

Several research groups reported sulfonated PES copolymers containing different biphenols such as (e) phthalazinone, (f) imidoaryl biphenol, and (g) phenolphthalein, as shown in Figure 10.11 [70–72]. Phenolphthalein is a biphenol with a lactone group between two phenolic groups, and it has been used to synthesize aromatic polyesters, PAEKs or sulfones [72]. Phthalazinone has a N—H group that behaves as a phenolic OH group so that the monomer reacts like a bisphenol in nucleophilic aromatic substitution reactions.

McGrath's group and Dang's group also reported poly(arylene sulfide sulfone) disulfonated copolymers using 4,4-thiobisbenzenethiol monomer (Figure 10.12a) and series of poly(aryl ether)-type block copolymers (Figure 10.12b and c) [73–77]. Block copolymer architecture containing hydrophobic/hydrophilic segments is believed to enhance phase-separated morphology, which is expected to improve the PEM properties for FC applications. A typical method to prepare block copolymers is by synthesis of hydrophobic and hydrophilic oligomers with reactive end-groups, followed by their copolymerization. Sulfonated-fluorinated multiblock copolymers were synthesized by copolymerization of activated fluorine-terminated hydrophobic oligomer and hydroxyl-terminated hydrophilic oligomer as shown in Figure 10.12.

Poly(arylene ether ketone)s Poly(arylene ether ketone)PAEK copolymers are produced by nucleophilic aromatic substitution, in a similar manner to the sulfone analogs. The polymerization is typically conducted in high-boiling polar aprotic

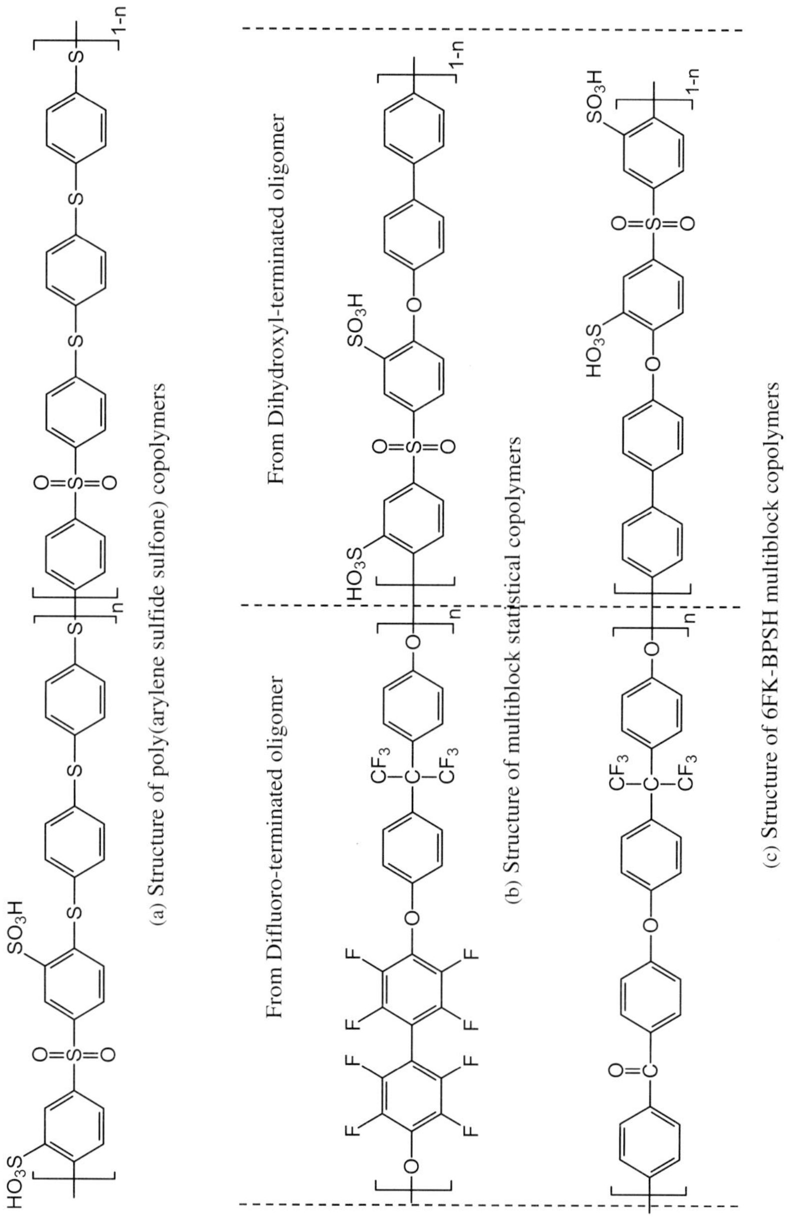

(a) Structure of poly(arylene sulfide sulfone) copolymers

(b) Structure of multiblock statistical copolymers

(c) Structure of 6FK-BPSH multiblock copolymers

Figure 10.12 Structure of sulfonated polymer.

solvents, in the presence of excess K$_2$CO$_3$ at elevated temperature. SC, and ultimately IEC, is adjusted by the feed molar ratios of sulfonated to non-sulfonated monomers. To obtain the membranes in the proton form, the copolymers (usually after casting films) in salt form (sodium or potassium cations) are immersed in acid (e.g., dilute sulfuric acid or hydrochloric acid) for ion exchange, often under boiling conditions. To do that, the membranes are immersed in dilute sulfuric acid at 100 °C for 2 h followed by similar water treatment at 100 °C for 2 h [55].

The preparation of many directly copolymerized sulfonated PAEKs is possible by employing a sulfonated dihalide ketone monomer, as first reported by Wang et al. [58]. Using 3,3'-disulfonated 4,4'-difluorodiphenyl ketone (DFBP) (Figure 10.13), Wang et al. produced high MW copolymers from phenolphthalein and non-sulfonated 4,4'-difluorodiphenyl ketone comonomers [78] (Figure 10.13a). The choice of bisphenol-type monomers for the polymerization of PAEKs is large. Guiver's group also reported sulfonated poly(arylene ether ketones) derived from DFBP and

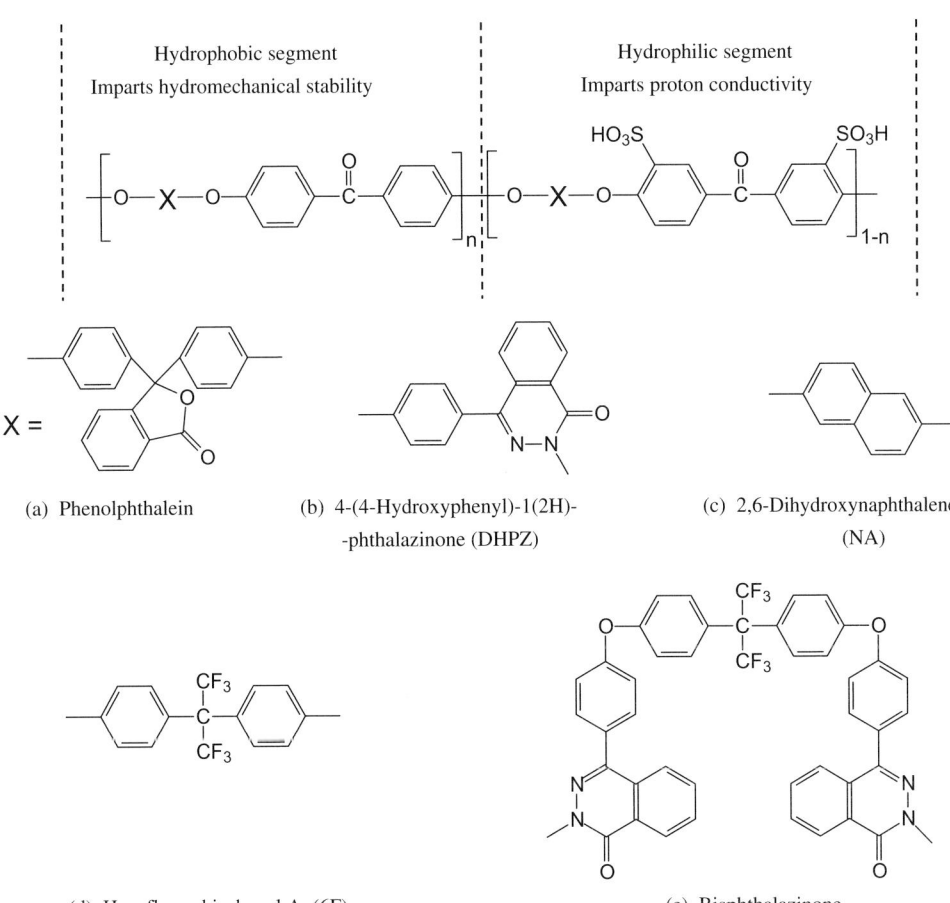

Figure 10.13 Typical copolymers derived from sulfonated 4,4'-difluorodiphenyl ketone.

sulfonated DFBP using various non-sulfonated comonomers such as (b) 4-(4-hydroxyphenyl)-1(2H)-phthalazinone (DHPZ), (c) 2,6-dihydroxynaphthalene (NA), and (d) bisphenol 6F as shown in Figure 10.13 [70, 79, 80]. The rigid planar aromatic NA group was incorporated into the polymers' backbone in order to improve the hot water stability of sulfonated PAEKs having a high SC. Introducing dissymmetric monomers such as NA into the backbone disrupts the polymer chain packing, having implications for free volume, decreasing the melting temperature (T_m) and crystallinity of PAEK, while improving organic solvent solubility [80].

Meng's group reported a new approach to the preparation of ionomers from poly (phthalazinone ether ketone)s that were synthesized by a N–C coupling reaction [11] (Figure 10.13e). These polymers are claimed to exhibit improved oxidative resistance by the Fenton's test when compared with other sulfonated polymers.

From the molecular design viewpoint, the incorporation of long and regular ether-ether-ketone-ketone (EEKK) moieties should increase the relative length of non-sulfonated hydrophobic segments, thereby giving greater separation to the hydrophilic segments, which could lead to possible improvements in the mechanical properties, and methanol and hot water dimensional stability. A series of sulfonated aromatic polymers comprising rigid PEEKK backbones (associated with hot-water stability and low methanol permeability) and bulky pendant fluorenyl group (associated with free volume, and thereby water uptake and proton conductivity) were prepared by polymerization of sulfonated monomer. As shown in Figure 10.14, SPEEKK copolymers derived from sulfonated and non-sulfonated 1,4-BFBB both containing bulky fluorenyl groups and hydrophobic (a) 4,4'-(9-fluorenylidene) diphenol (FDP), (b) bisphenol 6F, (c) ditrifluoromethylphenyl (6F-PH) and (d) 3,3',5,5'-tetramethyl diphenyl-4,4'-diol (TMDPD) groups have been reported [60, 81, 82].

SPEEKK copolymers derived from 1,3-bis(4-fluorobenzoyl) benzene (1,3-BFBB) were also reported to increase the statistical length of non-sulfonated segments in order to improve the mechanical strength of the membranes, as shown in Figure 10.15 [63, 83]. Tensile strength measurement indicated that these SPAEEKKs membranes are tough and strong at ambient conditions [83].

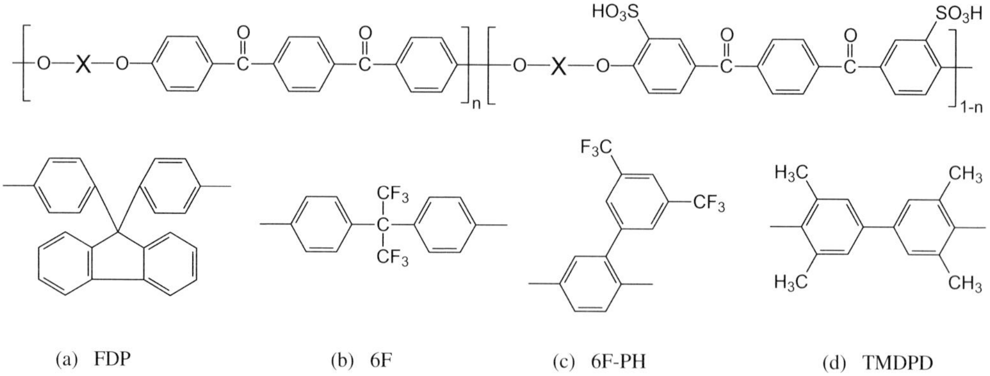

(a) FDP (b) 6F (c) 6F-PH (d) TMDPD

Figure 10.14 Typical copolymers derived from sulfonated 1,4-bis(4-fluorobenzoyl)benzene.

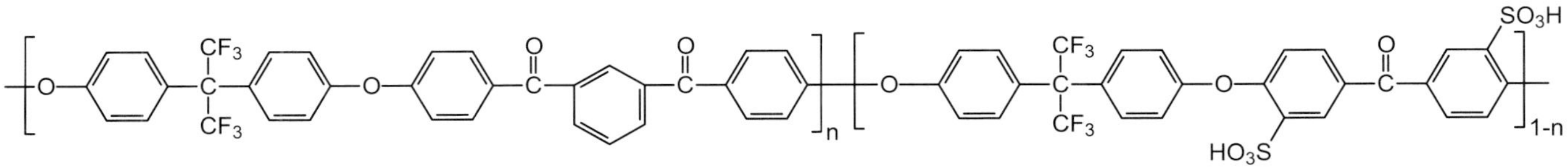

Figure 10.15 Structures of SPEEKK derived from 1,3-bis(4-fluorobenzoyl) benzene.

(a) Structure of sulfonated poly(aryl ether ketone) copolymers

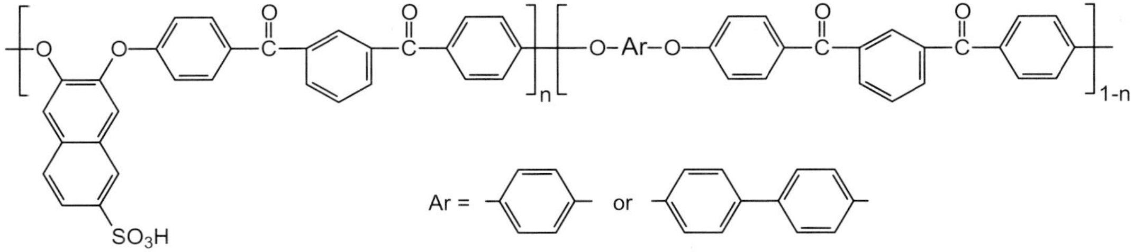

(b) Structure of sulfonated poly(arylene ether ether ketone ketone) copolymers

Poly(arylene ether nitrile)s Poly(arylene ether nitrile)s are a class of high performance thermoplastic polymers prepared by polycondensation of bisphenols and dihalobenzonitrile or dinitrobenzonitriles in dipolar aprotic solvent [61]. A combination of their good mechanical properties, high chemical and thermal resistance, and strongly polar pendant nitrile groups, make poly(arylene ether nitrile)s good candidates for matrices in advanced composites for the aerospace industry. It is also found that the introduction of nitrile groups into proton conductive sulfonated thermoplastics decreased their moisture absorption [84]. In addition, it is also believed that the presence of nitriles have the potential to promote adhesion of the polymers to heteropolyacids incorporated into composite membranes or to electrodes in the preparation of MEAs.

Poly(arylene ether nitrile)s containing sulfonic acid groups from 2,6-difluorobenzonitrile (2,6-DFBN), potassium 2,5-dihydroxybenzenesulfonate (SHQ), and additional non-sulfonated comonomers 4,4′-biphenol or hydroquinone were reported [62] (Figure 10.16a). The resulting copolymers exhibit lower water uptake than sulfonated PAEKs and PESs of similar equivalent weight and IEC values, and with proton conductivities close to, or higher than that of Nafion 117, reaching 10^{-1} S/cm. However, for this series of poly(arylene ether nitrile)s having comparable conductivities to Nafion, the water swelling still exceeded that of Nafion, particularly in hot water. Guiver's group also reported that a more rigid and hydrophobic naphthalene-based bisphenol containing sulfonic acid groups bonded meta- to the ether linkage was used instead of the more flexible and hydrophilic hydroquinone to prepare an additional series of poly(arylene ether nitrile)s. This is the first example of using this inexpensive commercially available monomer 2,8-dihydroxynaphthalene-6-sulfonate sodium salt (2,8-DHNS-6) as a monomer in a polycondensation reaction [61] (Figure 10.16b). It was incorporated in anticipation of improving in-water dimensional stability and the mechanical properties of the films. The sulfonic acid groups, which are not in a deactivated position, could potentially improve the proton conductivity and hydrolytic stability of the polymer chain. The DMFC performance of this copoly(arylene ether nitrile) derived from 2,8-DHNS-6 was reported [85]. Comparative PEMFC and DMFC performance was reported for poly(arylene ether ether nitrile) copolymers containing sulfonic acid group bonded in structurally different ways (HQ-SPAEEN: Figure 10.16a, m-SPAEEN: Figure 10.16. (b), p-SPAEEN: Figure 10.16c), which are non-fluorinated and have nitrile groups in both the hydrophilic and hydrophobic repeat units [86].

Sulfonated poly(phthalazinone ether ketone nitrile) copolymers (SPPEKN) were prepared by copolymerization of disodium 3,3′-disulfonate-4,4′-difluorobenzophenone (SDFB-Na), 2,6-difluorobenzonitrile (2,6-DFBN), and DHPZ and potassium carbonate at 160 °C in N-methyl-2-pyrrolidinone (NMP) [87] (Figure 10.16d). McGrath's group also reported a sulfonated poly(arylene ether) copolymer containing aromatic nitriles [84] (Figure 10.16e).

The presence of highly polar nitrile groups in sulfonated poly(arylene ether)-type copolymer has the apparent effect of reducing water uptake and dimensional swelling through increases in inter-chain molecular forces.

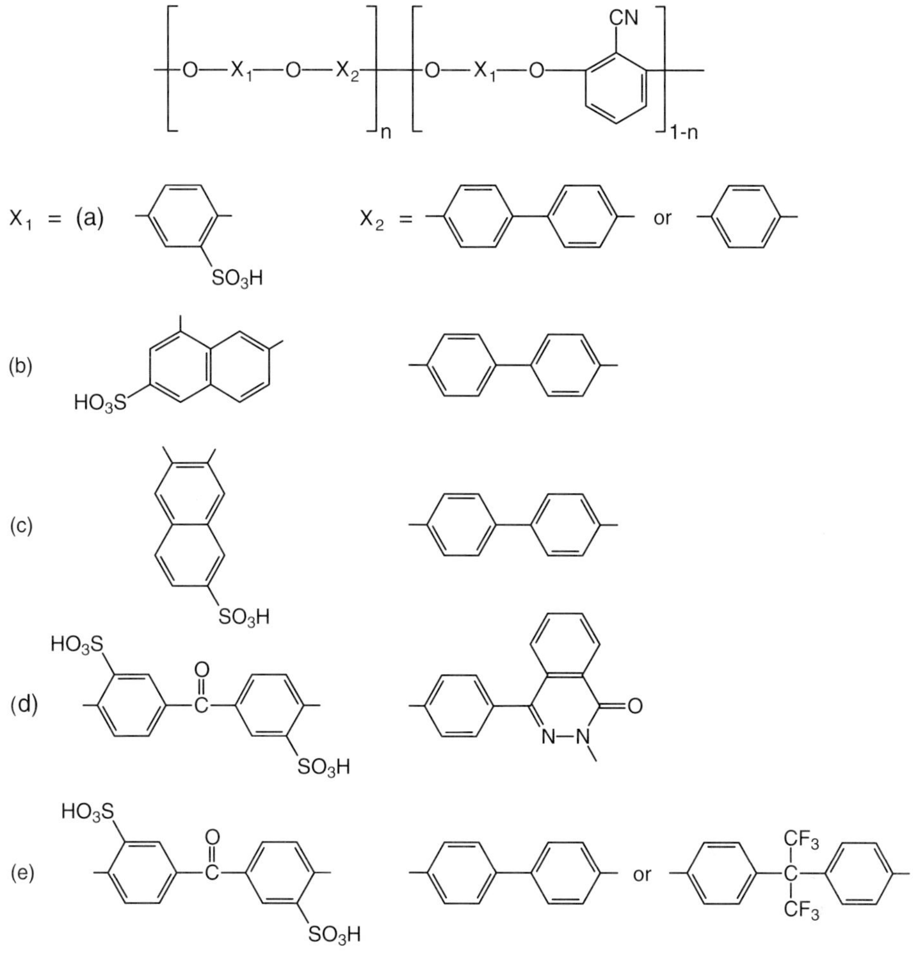

Figure 10.16 Typical copolymers containing nitrile groups.

10.2.2.3 Other Synthetic Strategies: Introducing Sulfonic Acid Groups

In the previous section, we briefly reviewed sulfonated polymers prepared by direct polymerization of sulfonated monomer or by post-sulfonation. The majority of this work is based on poly(arylene ether)s that contain the sulfonic acid groups located along the polymer backbone. The sulfonated aromatic polymers can be divided into two types according to the position of the sulfonic acid groups attached, main-chain-type and side-chain-type, in which the sulfonic acid groups are attached to the polymer backbone and side chains, respectively. In general, main-chain-type polymers show suitable conductivities only at high IECs, resulting in excessive water uptake above a critical temperature (percolation threshold), or a dramatic loss of mechanical properties that makes them unsuitable for practical PEM applications. Kreuer et al. [88] reported that these sulfonated polymers are unable to form defined hydrophilic domains, as the rigid polyaromatic backbone prevents continuous ionic clustering from occurring. This discussion covers other synthetic strategies for alternative PEM materials. One promising way to lower the water uptake and enhance properties in terms of PEM performance is to design materials having distinctly separate hydrophilic sulfonic acid group regions from the hydrophobic polymer main chain, by placement of the sulfonic acid groups on side chains grafted onto the polymer main chain [89]. For side-chain-type sulfonated polymers, short pendant side chains provide spacing between the polymer main chain and the sulfonic acid groups, which may facilitate the formation of nanophase separation of hydrophilic and hydrophobic domains, thereby leading to possible improvements in hydrolytic stability and water uptake. A variety of side-chain-type sulfonated polymers have been prepared by the chemical grafting method or by post-sulfonation on activated pendants of the corresponding parent polymers. Ding *et al.* reported the properties of graft polymers comprising graft chains of a macromonomer poly (sodium styrenesulfonate) and a polystyrene backbone [90]. If flexible pendant side chains linking the polymer main chain and the sulfonic acid groups exist in the polymer structure, nanophase separation between hydrophilic and hydrophobic domains may be improved [91].

The group of Jannasch reported [92] convenient methods for preparing polysulfones containing sulfonated side-chains, as shown in Figure 10.17. A side-group-sulfonic acid polysulfone was prepared via lithiation, followed by reaction of the resulting lithiated polysulfone intermediate with sulfobenzoic acid cyclic anhydride (Figure 10.17a). A polysulfone with pendant sulfonated aromatic side chains was also reported (Figure 10.17b), having proton conductivities of 11 to 32 mS/cm at 120 °C [91]. Lithiated polysulfone was first reacted with 4-fluorobenzoyl chloride to introduce 4-fluorobenzoyl side chains to the polymer main chain. The resulting activated fluoro groups were then reacted with 4-sulfophenolate or 7-sulfo-2-naphtholate via a nucleophilic substitution reaction.

Einsla *et al.* reported that sulfonated PES copolymers with pendant sulfonic acid groups were prepared using barium pentafluorobenzenesulfonate and 4-nitrobenzenesulfonyl chloride; the PEMs showed lower proton conductivity of 1 to 8 mS/cm [89]. Guiver's group reported a new class of comb-shaped polymers comprising a rigid, partially fluorinated hydrophobic backbone and monodisperse α-methyl polystyrene hydrophilic side chains prepared by anionic polymerization (Figure 10.18a) [93].

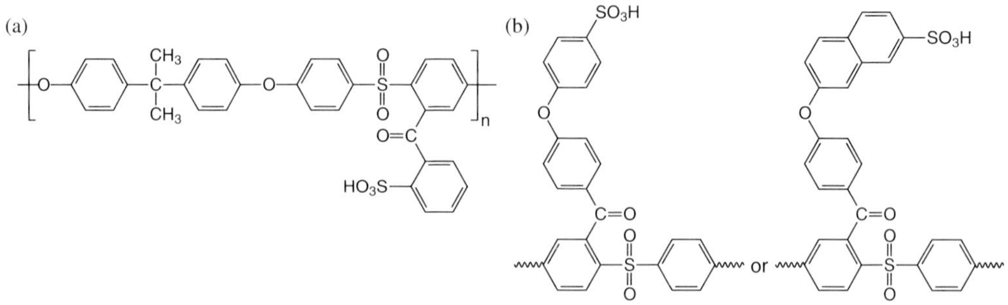

Figure 10.17 Polysulfones containing sulfonated side-groups.

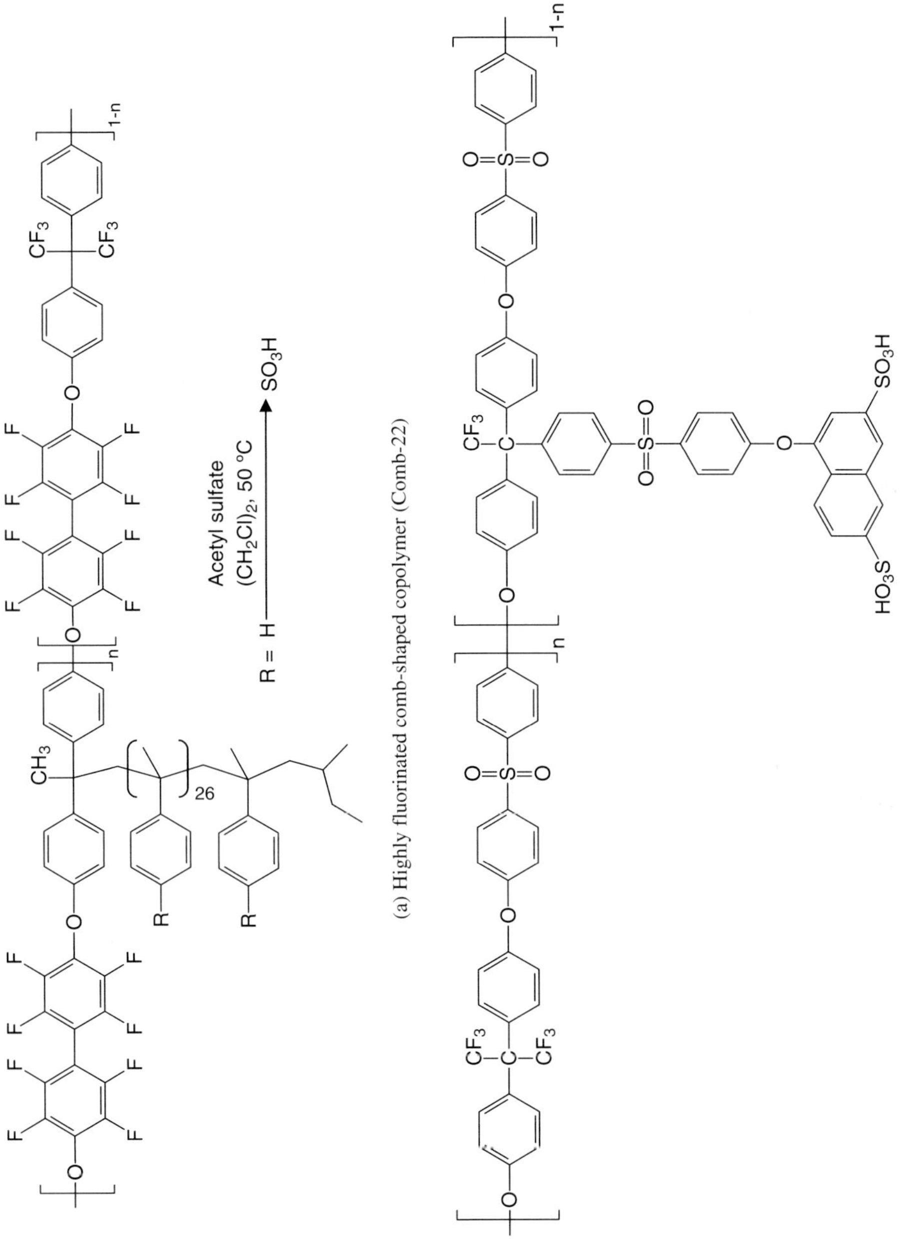

10.2 Synthesis of Polymer Electrolyte Membranes for DMFC | 399

(a) Highly fluorinated comb-shaped copolymer (Comb-22)

(b) Sulfonated poly(ether sulfone)s with grafting sulfonated groups

Figure 10.18 Comb- or pendant-type copolymer.

A new bisphenol monomer containing a masked grafting site, 1,1-bis(4-hydroxyphenyl)-1-(4-(4-fluorophenyl)thio)phenyl-2,2,2-trifluoroethane (3FBPT), (Figure 10.18b) was also reported [94]. Copolymers containing 4-fluorophenyl sulfide pendant groups were obtained via conventional aromatic nucleophilic substitution polycondensation using this monomer. Simple oxidation from the sulfide to sulfone activates the *para* fluorine on the polymer pendant group for nucleophilic attack by phenolates and other species in order to introduce arylsulfonic acid.

The group of Jiang [95, 96] also reported a new sulfonated copolymer derived from sulfonated monomers such as sodium-3-(4-(2,6-difluorobenzoyl) phenyl) propane-1-sulfonate (DFPPS) and sodium 4-(4-(2,6- difluorobenzoyl) phenoxy) benzenesulfonate (SDFBS) (Figure 10.19). It was reported that all PSA-SPAE membranes exhibited reasonable flexibility and tensile strength in the range of 41–78 MPa. PSA-SPAE copolymers bearing sulfonic acid groups on flexible side chains showed considerably reduced dimensional swelling and improved proton conductivities.

More recently, Kim and Guiver reported [97] comb-shaped PESs as PEMs (Figure 10.20). A sulfonated side-chain grafting unit containing two or four sulfonic

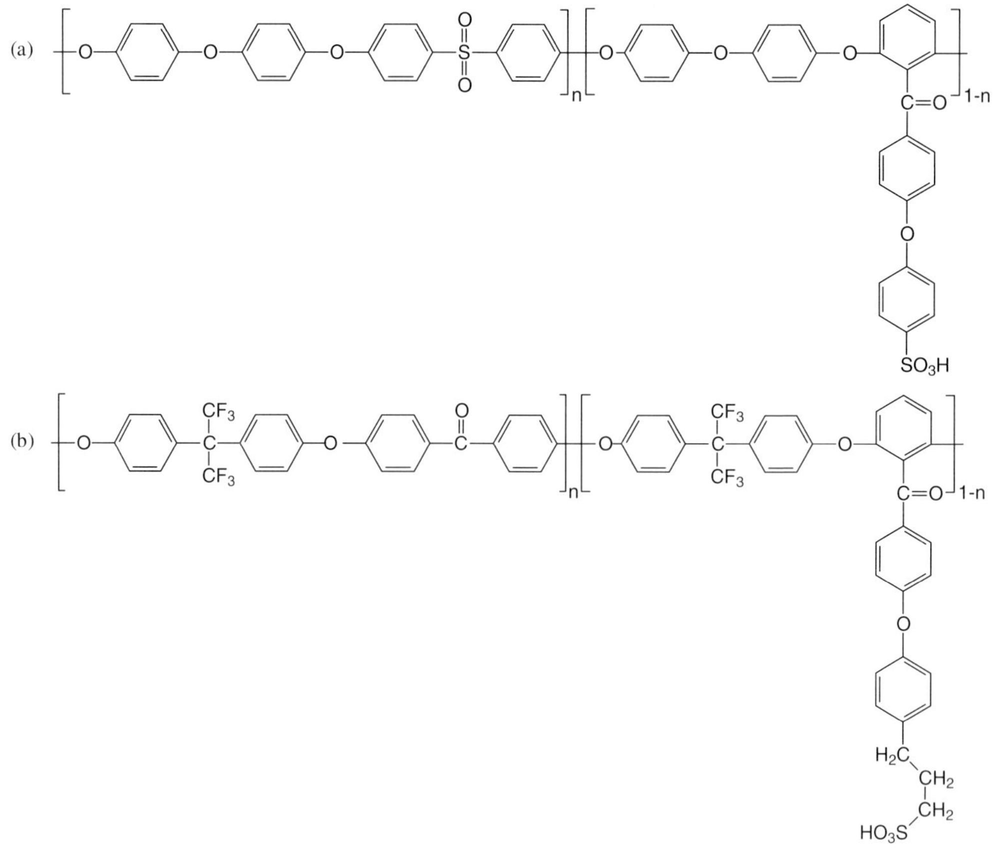

Figure 10.19 Sulfonated poly(arylene ether)s with pendant sulfonated groups.

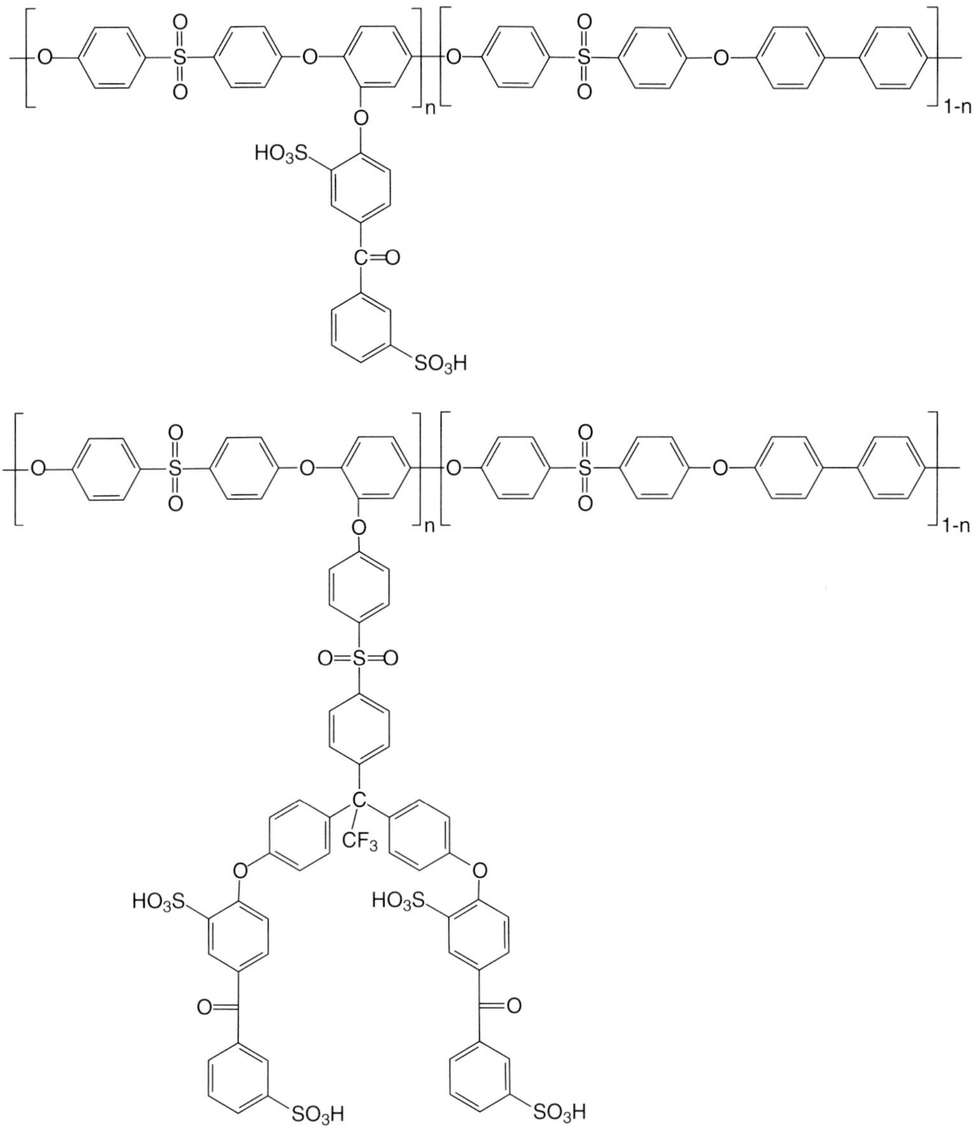

Figure 10.20 Sulfonated poly(arylene ether sulfone)s with pendant sulfonated groups.

acid groups was synthesized using sulfonated 4-fluorobenzophenone (FBP) and 3FBPT. Poly(arylene ether sulfone) containing methoxy group was synthesized. After deprotecting the methoxy group to hydroxyl group, the sulfonated side chains were grafted onto the −OH functionalized copolymer to make the comb-shaped sulfonated PES copolymers. The comb-shaped copolymers with two or four sulfonic acid groups show high proton conductivities in the range of 34 to 147 mS/cm and

63 to 125 mS/cm, respectively. The methanol permeabilities of these copolymers were in the range of 8.2×10^{-7} to 5.6×10^{-8} cm^2/s.

10.2.2.4 Properties of Sulfonated Poly(arylene ether) Copolymers

Common themes critical to all high performance PEMs include (i) high protonic conductivity; (ii) low permeability to fuel and oxidant; (iii) low water transport through diffusion and electro-osmosis; (iv) oxidative and hydrolytic stability; (v) good mechanical properties in both the dry and hydrated states; (vi) cost; and (vii) capability for fabrication into MEAs [66]. To date, much research has been limited to the polymer synthesis and characterization of stand-alone membranes, while far fewer MEA studies of hydrocarbon-based sulfonated copolymers have been conducted because of issues with dimensional swelling, high methanol permeability and oxidative and hydrolytic stability under fuel cell operating conditions [66, 67, 98–100]. Nearly all existing membrane materials for PEM fuel cells rely on absorbed water and its interaction with acid groups to enable it to conduct protons. Due to the large fraction of absorbed water in the membrane, both mechanical properties and water transport become key issues.

The difficulties in preparing high performance MEAs as a result of membrane-electrode incompatibility is one of the primary reasons that improved membrane properties have not always led to improved DMFC performance. The group of Pivovar [101] attributed membrane-electrode interfacial resistance to differential swelling between the membrane and electrodes, leading to electrode delamination. Water uptake of PEMs is important for both membrane-electrode compatibility and mechanical properties of the membranes.

Generally, polymer composition or IEC is plotted versus conductivity; however, these plots do not allow a direct comparison between different membrane materials. Membrane water uptake is known to have a profound effect on proton conductivity and methanol permeability, and provides a better comparative basis for materials. Although literature data have been compiled for proton conductivity data versus methanol permeability, proton conductivity versus water uptake has rarely been plotted in the multitude of literature references. The group of McGrath suggest that an upper bound relationship may exist, expressed by a linear log-log plot of water content versus proton conductivity of the membrane [44]. Figure 10.21 shows the proton conductivities of various PEMs as a function of water uptake (wt%) at 80 °C. The water uptake values in Figure 10.21 have been reported on a mass basis, because density data for the polymers were not available in many cases. The changes in length scale (reflected in volume measurements) are expected to be the most appropriate comparison basis because electrochemical properties such as proton conductivity and permeability occur over length scales under operating conditions independent of mass [69]. Figure 10.22a shows the proton conductivity as function of volume-based water uptake (vol%) using the previous reported data having density data. Figure 10.22b shows the relative water uptake as function of relative conductivity. In this study the conductivity and water uptake (vol%) of alternative membranes have been normalized to the values reported for Nafion, which is defined as having a relative conductivity and water uptake of 1. The data points of the copolymer

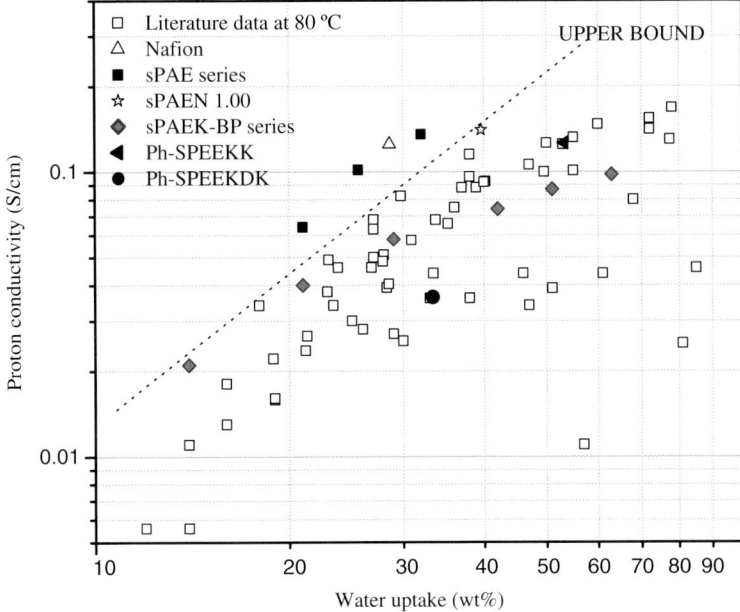

Figure 10.21 Proton conductivity and water uptake (wt.%) at 80 °C (Refs. [52, 53a, 60, 64, 65, 79, 80, 85, 93, 96, 97, 102–106].

membranes (sPAE series and sPAEN 1.0 as illustrated in Figure 10.9a and b) are located above the upper bound line in the area of potential high performance (i.e., high proton conductivity coupled with low water uptake) in the trade-off plot for the relationship between PEM proton conductivity versus water uptake (weight based or volume based) shown in Figures 10.21 and 10.22a. It is often the case that hydrocarbon membranes having the same water uptake as Nafion have lower conductivity, or membranes with similar conductivity have higher water uptake based on weight or volume. Figure 10.23 shows the proton conductivity of sulfonated poly(vinyl alcohol) membrane series and sulfonated poly(arylene ether) copolymer series as a function of water uptake (vol%). The sulfonated poly(arylene ether) copolymer series shows superior properties of proton conductivity and water uptake. Higher water uptakes often lead to difficulties preparing robust and high performance MEAs or decreased durability during cycling between different levels of hydration.

10.2.3
Single Cell Performances

The performance of PEMs is typically evaluated by comparison with Nafion membrane under identical test conditions. This is largely because there is no standard testing protocol for PEMs. Although these studies allow membrane performance comparisons under the same test conditions, one must keep in mind that the test conditions employed may not be equivalent for each membrane. The performance of

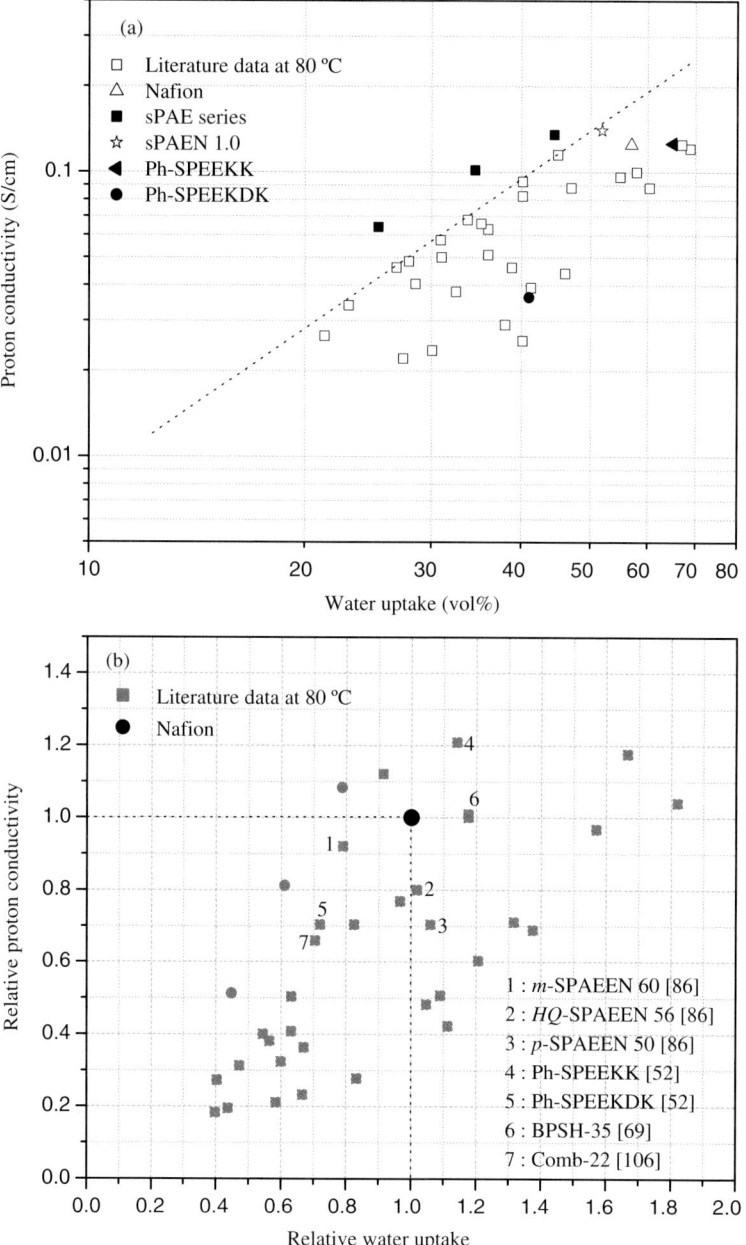

Figure 10.22 (a) Proton conductivity and water uptake (vol.%) at 80 °C (b) relative proton conductivity and relative volume water uptake (vol.%) (Refs. [52, 53a, 65, 85, 86, 97, 102, 105, 106]).

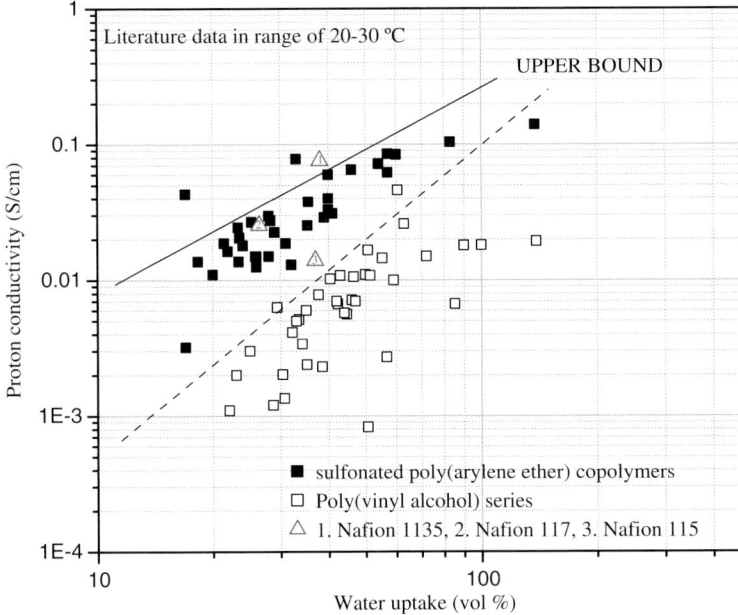

Figure 10.23 Proton conductivity and water uptake (vol.%) at 20–30 °C.

PEM materials is highly variable between various research groups due to the MEA fabrication techniques and fuel cell test conditions. Therefore, in this section, we discuss the performance of MEAs that are measured under the same conditions. To do that, we selected PEMs such as sulfonated PAEKs, PESs, poly(arylene ether nitrile)s, and comb-type copolymers.

Most hydrocarbon-based PEM research has been limited to reporting the polymer synthesis and the characterization of stand-alone membranes. There are relatively fewer MEA studies conducted for fuel cell applications because of the issues with dimensional swelling, high methanol permeability and oxidative and hydrolytic stability under fuel cell operating conditions [86]. Membranes with excessive water uptake tend to be (i) less effective in proton conduction and (ii) mechanically fragile and subject to dimensional changes under dehydration/hydration cycling. Therefore, several hydrocarbon PEMs were selected for further MEA study based on superior *ex situ* membrane properties presented in Figure 10.22b. The seven PEMs were selected based on a favorable combination of proton conductivity and water uptake (vol.%). The structures of sulfonated poly(arylene ether nitrile) copolymers (*m*-SPAEEN-60, *HQ*-SPAEEN-56, and *p*-SPAEEN-50) are shown in Figure 10.16. The structures of Ph-SPEEKK, Ph-SPEEKDK are shown in Figure 10.8. The BPSH-35 and Comb 22 are shown in Figures 10.11b and 10.18, respectively. The properties of these seven membranes are listed in Table 10.1. The changes in length scale (reflected in volume measurements) are expected to directly impact observed properties, because electrochemical properties such as proton conductivity and permeability occur over length scales under operating conditions independent of mass [69]. The

Table 10.1 Properties of the membranes.

Copolymer	Density (g/cm^3)^a	IEC$_w$ (meq./g)^b	IEC$_v$ (meq./cm^3)^c dry	IEC$_v$ (meq./cm^3)^c wet	Water uptake wt%^{d}	Water uptake vol%^{e}	Proton Conductivity (mS/cm) 30°C	Proton Conductivity (mS/cm) 80°C
1. m-SPAEEN-60	1.18	1.91	2.26	1.70	28	33	78	115
2. HQ-SPAEEN-56	1.29	1.90	2.45	1.80	28	36	60	100
3. p-SPAEEN-50	1.55	1.60	1.75	1.20	30	46	65	88
4. Ph-SPEEKK	1.22	1.80	2.20	1.57	33	40	38	151
5. Ph-SPEEKDK	1.23	1.60	1.97	1.49	25	32	23	88
6. BPSH-35	1.33	1.87	2.49	1.63	40	53	78	126
7. Comb 22	1.35	1.20	1.62	1.38	13	18	43	82
Nafion 1135	1.98	0.91	1.78	1.29	19	38	76	125

^{a}Based on dry state.
^{b}Based on weigh of dry membrane.
^{c}Based on volume of dry and/or wet membranes (IEC$_v$ (wet) = IEC$_v$(dry)/(1 + 0.01 WU)).
^{d}WU (mass %) = (W$_{wet}$ − W$_{dry}$)/W$_{dry}$ × 100.
^{e}WU (vol%) = (W$_{wet}$ − W$_{dry}$)/δ_w)/(W$_{dry}$/δ_m) × 100, (W$_{wet}$ and W$_{dry}$ are the weights of the wet and dry membranes, respectively; δ_w is the density of water (1 g/cm^3), and δ_m is the membrane density in the dry state; water uptake measured at room temp.

water uptake directly affects the sulfonic acid concentrations within the polymer matrix under hydrated conditions, which is gauged by comparing wet volume-based IEC (IEC$_V$ (wet)) values with IEC$_W$ values. As listed in Table 10.1, it was found that the introduction of nitrile groups into sulfonated copolymers reduced water uptake and dimensional swelling when compared with polymers that did not contain nitrile, at similar IEC values [84, 87]. We suggest that a plausible factor for the low water uptake of copolymers containing nitrile groups is the presence of strong nitrile dipole interchain interactions occurring in certain polymer structural configurations that combine to limit swelling in water [85, 86]. In addition, nitrile-sulfonic acid group interactions also appear to be important, as nitrile groups have been found to associate with sulfonic acid groups through bridging water molecules in specific spectroscopic studies [107]. The introduction of highly polar nitrile has been suggested to improve mechanical strength of the polymers and promote their adhesion to various substrates.

We compare the voltage-current characteristics (i.e., H$_2$/air and DMFC polarization curves) of MEAs using the seven copolymers, and Nafion 212 and 1135. Figure 10.24 shows the H$_2$/air performance of the selected membranes, since it is useful to compare the performance of similar materials in both the DMFC and the hydrogen system. For comparison, we tested relatively thin membranes in order to reduce cell ohmic resistance. MEAs using nitrile copolymers showed lower performance in comparison with the Nafion 212 MEA, but higher performance compared with the MEA using sulfonated polysulfone BPSH-35. However, it should be noted that with the exception of the comb 22 PEM, the thickness of all hydrocarbon

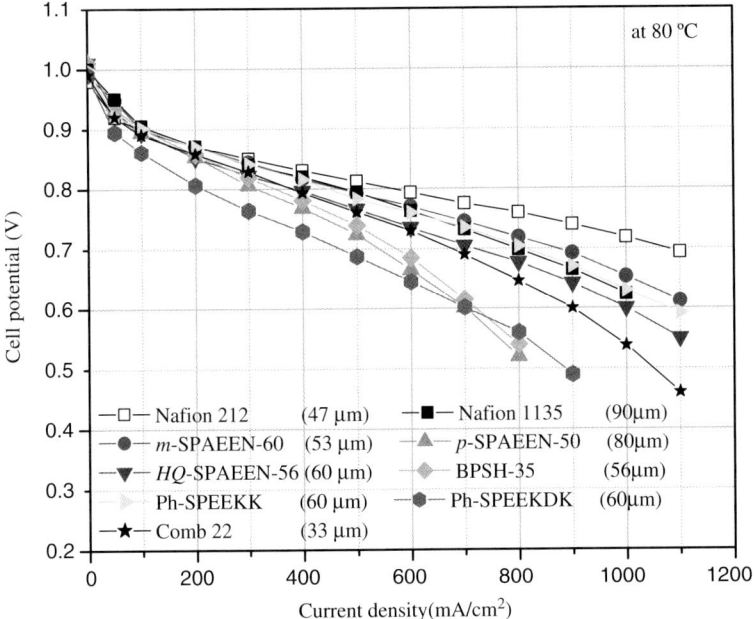

Figure 10.24 H$_2$/air fuel cell performance of selected PEM (value in parenthesis denote membrane thickness).

PEMs including m-SPAEEN-60 membrane (53 μm) were greater than that of Nafion 212 (47 μm). The slightly lower performance of the cell using p-SPAEEN compared with the cell using BPSH-35 may be due to the increased resistance of the thicker membrane. Among the cells incorporating nitrile copolymers, the MEA using m-SPAEEN-60 showed the best performance. Although the thickness of Nafion 1135 (90 μm) was higher than that of m-SPAEEN-60, the performance of m-SPAEEN-60 is higher than that of Nafion 1135. Qualitatively good correlation between cell resistance and polarization characteristics indicates that interfacial incompatibility between membrane and electrode and resulting performance loss is minor, which we expect from relatively low water uptake of the tested membranes.

Methanol crossover and cell resistance play a major role in determining DMFC performance. Methanol crossover in the MEA was estimated by measuring the limiting methanol crossover current [108–110]. The total cell resistance is composed of the sum of membrane resistance, electronic resistances of the fuel cell components (flow field, current collectors, and gas diffusion layers), the resistance of the electrodes and interfacial resistances associated with the interfaces between electronic components and between the electrode and the membranes. Therefore, cell resistance from these cumulative factors is always higher than the associated free-standing membrane resistance. Figure 10.25 shows the high frequency resistance (HFR) and methanol crossover limiting current of single cells using the hydrocarbon copolymers and Nafion at 80 °C under DMFC operating conditions (0.5 M MeOH). A membrane with ideal properties should have very low HFR (ohmic losses) and low methanol crossover

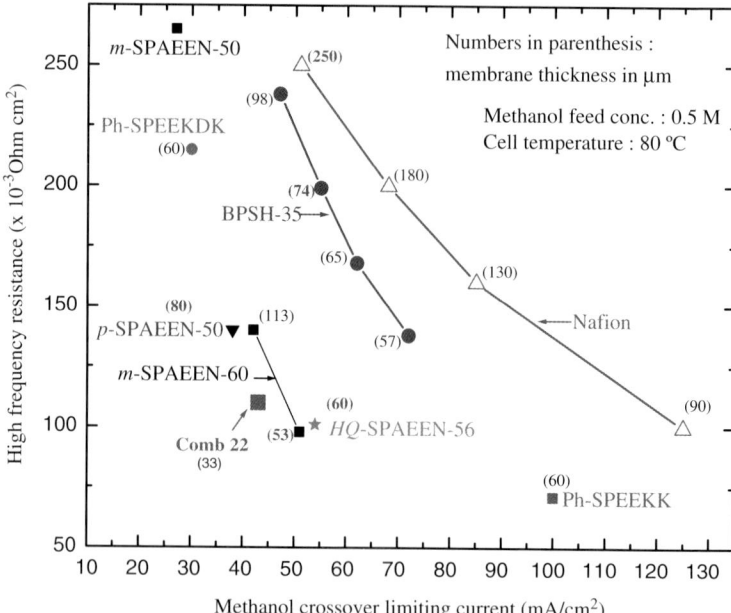

Figure 10.25 HFR vs. methanol crossover limiting current of selected PEMs and Nafion as a function of membrane thickness measured in DMFC mode at 80 °C.

(low crossover losses). Of the selected polymers shown, Nafion shows the poorest DMFC performance, having the lowest selectivity resulting from a combination of the higher membrane thickness (resulting in higher HFR) that is required to compensate for high methanol crossover. The selectivity for the PEMs of various thicknesses is listed in Table 10.2. Compared with the selectivities of Nafion, the selected copolymers show significantly improved selectivities, having much lower methanol crossover limiting current at comparable high frequency resistance. From these results, one would expect improved DMFC performance using selected copolymers compared with Nafion.

Although polarization curves are the most popular method for evaluating DMFC performance, making reasonable and relevant performance comparisons across different types of membranes is difficult because cell properties depend on membrane thickness. While DMFC cells using hydrocarbon membranes show the same thickness effect, the optimum thickness of each specific membrane polymer family should be different, due to differences in the conductivity and methanol permeability of the membrane. Apart from the effect of membrane thickness, there are several other factors that make accurate comparative performance evaluation difficult using single polarization curves: (i) methanol crossover (fuel use) is not fully interpreted by polarization curves, and (ii) optimum operating conditions may be very different for different systems [85]. In order to lessen the uncertainty caused by methanol crossover and to provide a meaningful comparison, we selected membranes for the fuel cell tests as having an appropriate thickness for which methanol crossover

10.2 Synthesis of Polymer Electrolyte Membranes for DMFC

Table 10.2 Electrochemical properties of membrane and Nafion at 80 °C (0.5 M MeOH).

Copolymer	Thickness (μm)	HFR (mΩ cm^2)	MeOH limiting current (mA/cm^2)	Selectivity (α)
1. m-SPAEEN-60	53	97	52	198
	113	140	42	170
2. HQ-SPAEEN-56	60	101	54	183
3. p-SPAEEN-50	80	140	38	188
4. Ph-SPEEKK	60	71	100	141
5. Ph-SPEEKDK	60	215	30	155
6. BPSH-35	65	168	62	96
	74	199	55	91
	98	238	47	89
7. Comb 22	33	110	43	211
Nafion 112	50	70	152	94
Nafion 1135	90	100	125	80
Nafion 1110	250	250	51	78

MEA selectivity (α): $\alpha(\text{HFR}^{-1}\,\text{MeOH current}^{-1}) = \frac{1}{\text{HFR}\times\xi_{\text{lim}}}$, ($\xi_{\text{lim}}$ limiting methanol crossover current of membrane).

limiting currents were similar (35~55 mA/cm^2) across different polymer systems, as shown in Figure 10.25.

Figures 10.26 and 10.27 show the cell performance of the MEAs using selected copolymers having similar methanol crossover limiting current at methanol feed

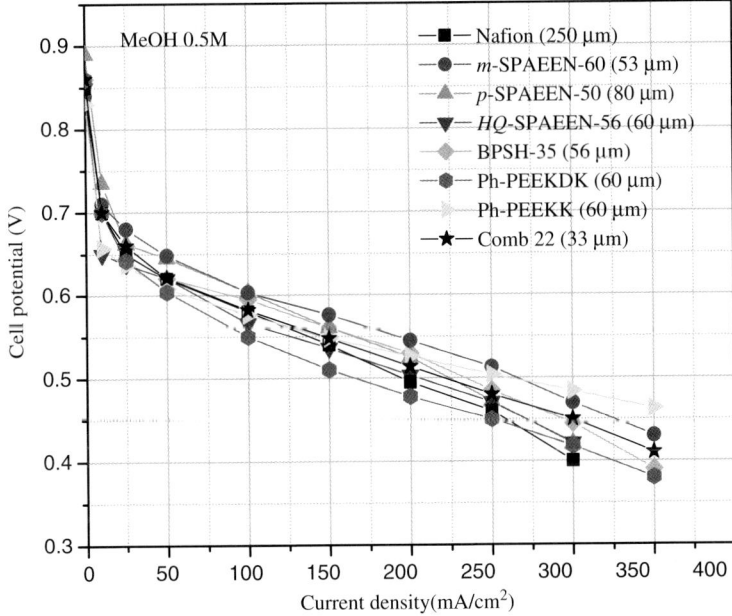

Figure 10.26 DMFC performance of selected PEM materials at 0.5 M (Cell temperature: 80 °C).

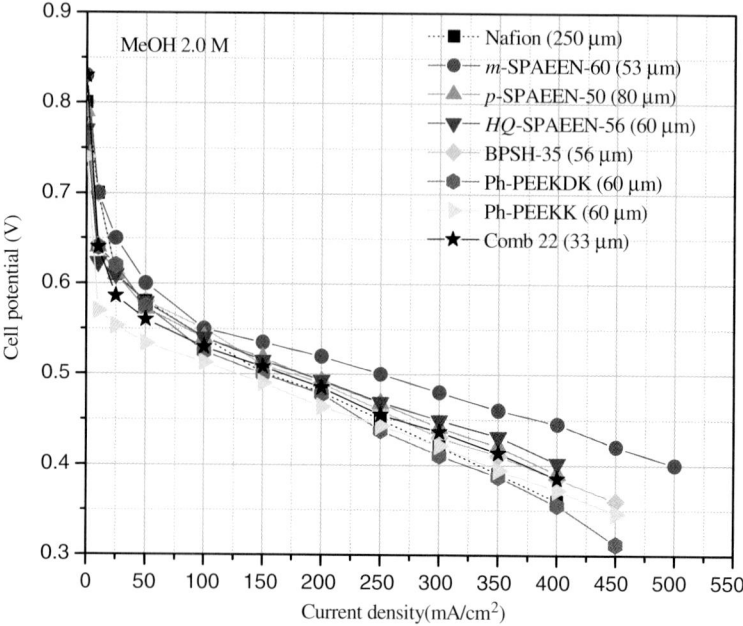

Figure 10.27 DMFC performance of selected PEM materials at 2.0 M (Cell temperature: 80 °C).

concentrations of 0.5 and 2.0 M. The performance of the MEAs using SPAEEN copolymers and BPSH-35 were superior to that of the Nafion MEA (0.5 M MeOH) (Figure 10.26). The performance of BPSH-35 was superior to the p- and HQ-SPAEEN at 0.5 M methanol concentration, while the performance of p-, and HQ-SPAEEN was superior to the BPSH-35 at the higher 2.0 M methanol concentration.

Among the selected membranes and Nafion, the m-SPAEEN-60 shows the best performance. For example, the current density of the MEA using m-SPAEEN-60 at 0.5 V and 0.5 M methanol was 265 mA/cm^2, whereas the current densities of the MEAs using BPSH-35 and Nafion were 230 and 195 mA/cm^2 (Figure 10.26). At 2.0 M methanol, open circuit potential and mass transport limitations for all MEAs decreased, but the performance trend remained remarkably similar to that at 0.5 M methanol. The current density of the MEA using m-SPAEEN-60 at 0.5 V and 2.0 M methanol was 245 mA/cm^2, whereas the power densities of the MEAs using BPSH-35 and Nafion were 195 and 170 mA/cm^2, respectively. The maximum power density of the MEA using m-SPAEEN-60 was shown in 1.0 M methanol. The maximum power density at 550 mA/cm^2 and 1.0 M methanol was 220 mW/cm^2.

It was reported that the sulfonated hydrocarbon polymer membranes showed better performance than Nafion [67, 100]. However, interfacial incompatibility between hydrocarbon membrane and Nafion PFSA-based electrodes can limit long-term performance. In order to investigate the interfacial stability of the MEA, the DMFC performance for before and after life test for various hydrocarbon-based cells was measured, as shown in Figure 10.28. These results indicate that the interfacial compatibilities of m-SPAEEN-60 and Ph-SPEEKK are likely good using

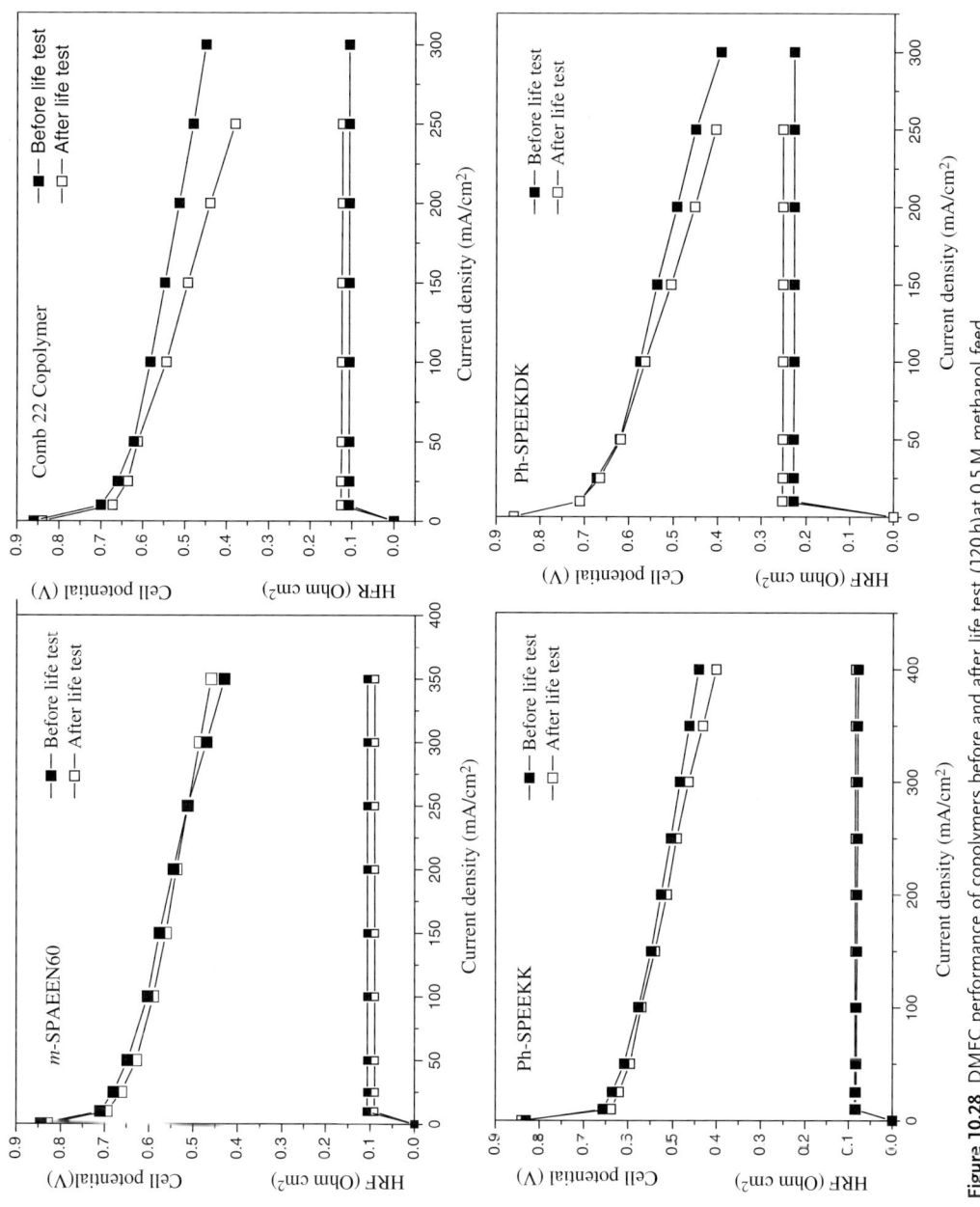

Figure 10.28 DMFC performance of copolymers before and after life test (120 h) at 0.5 M methanol feed concentration (cell temperature: 80 °C).

Nafion-bonded electrodes, since there was no increase in HFR observed over the experimental time period. In fact, a slight decrease in HFR was observed, which can be attributed to morphological reorganization resulting in an increase in proton conductivity of the membrane [111]. It is known that nitrile groups in polymers promote adhesion of polymer to substrates, possibly through polar interactions with other functional groups. This may be a reason for good adhesion to the catalyst-electrode layer [112].

Although the current density of Comb 22 copolymer showed some loss after the life test, it is not considered to be symptomatic of a hydrocarbon membrane-electrode interfacial problem. The reason is that the performance degradation is too fast and there is linear increase behavior for HFR (Figure 10.28). The most plausible explanation is that the poly(α-methyl styrene) side chain is susceptible to chemical degradation under fuel cell conditions, which caused a loss of the proton-conducting sulfonic acid sites, thereby resulting in decreasing membrane performance. However, the structural architecture of graft and comb-shaped polymers serves to illustrate their potential to improve fuel cell performance.

10.3
Conclusions

In this chapter, we focused on a review of the synthesis and performance of predominately hydrocarbon-based copolymers containing sulfonic acid groups for DMFC application, largely based on the authors' research work. A number of mechanically stable alternative hydrocarbon PEMs with high selectivity has been developed. Many research groups have observed that certain structural design strategies for PEM materials can lead to improved microstructure and performance, by achieving good proton conductivity while maintaining good mechanical properties through controlling water uptake. Our perspective in the design of PEM materials has led us to conclude that two effective strategies for making PEMs with high proton conductivity and lower methanol permeability are (i) introduction of nitrile groups into sulfonated polymers that limit membrane swelling due to nitrile-nitrile dipole interaction; (ii) morphological structures in comb- and/or pendant-type copolymers that are capable of providing materials with high proton conductivity, high mechanical strength and good dimensional stability upon swelling. These materials showed good performance under DMFC conditions.

References

1 Haubold, H.G., Vad, T., Jungbluth, H. and Hiller, P. (2001) *Electrochim. Acta.*, **46**, 1559.
2 Nolte, R., Ledjeff, K., Bauer, M. and Mulhaupt, R. (1993) *J. Membr. Sci.*, **83**, 211.
3 Kerres, J., Cui, W. and Reichle, S. (1996) *J. Polym. Sci. Part A: Polym. Sci.*, **34**, 2421.
4 Wang, F., Hickner, M., Kim, Y.S., Zawodzinski, T.A. and McGrath, J.E. (2002) *J. Membr. Sci.*, **197**, 231.

5 Harrison, W., Wang, F., Mecham, J.B., Bhanu, V., Hill, M., Kim, Y.S. and McGrath, J.E. (2003) *J. Polym. Sci., Part A: Polym. Chem.*, **41**, 2264.
6 Poppe, D., Frey, H., Kreuer, K.D., Heinzel, A. and Mulhaupt, R. (2002) *Macromolecules*, **35**, 7936.
7 Li, L. and Wang, Y.X. (2005) *J. Membr. Sci.*, **246**, 167.
8 Karlsson, L.E. and Jannasch, P. (2004) *J. Membr. Sci.*, **230**, 61.
9 Li, L., Zhang, J. and Wang, Y.X. (2003) *J. Membr. Sci.*, **226**, 159.
10 Xing, P.X., Robertson, G.P., Guiver, M.D., Mikahailenko, S.D., Wang, K.P. and Kaliaguine, S. (2004) *J. Membr. Sci.*, **229**, 95.
11 Chen, Y.L., Meng, Y.Z. and Hay, A.S. (2005) *Macromolecules*, **38**, 3564.
12 Ninivin, C.L., Balland-Longeau, A., Demattei, D., Coutanceau, C., Lamy, C. and Leger, J.M. (2004) *J. Appl. Electrochem.*, **34**, 1159.
13 Ramya, K., Vishnupriya, B. and Dhathathreyan, K.S. (2001) *J. New Mater. Electrochem. Sys.*, **4**, 115.
14 Yang, T. and Shi, P. (2008) *J. Power Sources*, **175**, 390.
15 Wang, S. and McGrath, J.E. (2003) *Synthesis of Poly(arylene ether)s, Synthetic Methods in Step-Growth Polymers*, John Wiley & Sons, Inc., New York, p. 327.
16 Resnick, P.R. and Grot, W.G., du Pont de Nemours and Company, Wilmington, DE, Sept 12, 1978; U.S. Patent 4,113,585.
17 Assink, R.A., Arnold, C. and Hollandsworth, R.P. (1991) *J. Membr. Sci.*, **56**, 143.
18 Becker, W. and Schmidt-Naake, G. (2002) *Chem. Eng. Technol.*, **25**, 373.
19 Saarinen, V., Kallio, T., Paronen, M., Tikkanen, P., Rauhala, E. and Kontturi, K. (2005) *Electrochim. Acta*, **50**, 3453.
20 Shen, M., Roy, S., Kuhlmann, J.W., Scott, K., Lovell, K. and Horsfall, J.A. (2005) *J. Membr. Sci.*, **251**, 121.
21 Lehtinen, T., Sundholm, G., Holmberg, S., Sundholm, F., Björnbom, P. and Burdell, M. (1998) *Electrochim. Acta*, **43**, 1881.
22 Qiu, X.P., Li, W.Q., Zhang, S.C., Liang, H.Y. and Zhu, W.T. (2003) *J. Electrochem. Soc.*, **150**, A917.
23 Elabd, Y.A., Napadensky, E., Sloan, J.M., Crawford, D.M. and Walker, C.W. (2003) *J. Membr. Sci.*, **217**, 227.
24 Elabd, Y.A. and Napadensky, E. (2004) *Polymer*, **45**, 3037.
25 Storey, R.F., Chisholm, B.J. and Lee, Y. (1997) *Polym. Eng. Sci.*, **37**, 73.
26 Zhang, M. and Russell, T.P. (2006) *Macromolecules*, **39**, 3531.
27 Tsang, M.W.E., Zhang, Z., Shi, Z., Soboleva, T. and Holdcroft, S. (2007) *J. Am. Chem. Soc.*, **129**, 15106.
28 Won, J., Park, H.H., Kim, Y.J., Choi, S.W., Ha, H.Y., Oh, I.-H., Kim, H.S., Kang, Y.S. and Ihn, K.J. (2003) *Macromolecules*, **36**, 3228.
29 Zhong, S., Cui, X., Cai, H., Fu, T., Zhao, C. and Na, H. (2007) *J. Power Sources*, **164**, 65.
30 Li, Y., Wang, F., Yang, J., Liu, D., Roy, A., Case, S., Lesko, J. and McGrath, J.E. (2006) *Polymer*, **47**, 4210.
31 Qiao, J., Ono, H., Oishi, T. and Okada, T. (2006) *ECS Trans.*, **3**, 97.
32 Chiang, W.Y. and Chen, C.L. (1998) *Polymer*, **39**, 2227.
33 Rhim, J.W. and Kim, Y.K. (2000) *J. Appl. Polym. Sci.*, **75**, 1699.
34 Rhim, J.W., Sohn, M.Y., Joo, H.J. and Lee, K.H. (1993) *J. Appl. Polym. Sci.*, **50**, 679.
35 Rhim, J.W., Park, H.B., Lee, C.S., Jun, J.H., Kim, D.S. and Lee, Y.M. (2004) *J. Membr. Sci.*, **238**, 143.
36 Kim, D.S., Guiver, M.D., Nam, S.Y., Yun, T.I., Seo, M.Y., Kim, S.J., Hwang, H.S. and Rhim, J.W. (2006) *J. Membr. Sci.*, **281**, 156.
37 Kim, D.S., Park, H.B., Lee, C.H., Lee, Y.M., Moon, G.Y., Nam, S.Y., Hwang, H.S., Yun, T.I. and Rhim, J.W. (2005) *Macromol. Res.*, **13** (4), 314.
38 Kim, D.S., Park, H.B., Rhim, J.W. and Lee, Y.M. (2004) *J. Membr. Sci.*, **240**, 37.

39 Kim, D.S., Park, H.B., Rhim, J.W. and Lee, Y.M. (2005) *Solid State Ionics*, **176**, 117.
40 Sawa, H. and Shimada, Y. (2004) *Electrochem.*, **72**, 111.
41 Lin, C.W., Thangamuthu, R. and Yang, C.J. (2005) *J. Membr. Sci.*, **253**, 23.
42 Li, L., Xy, L. and Wang, Y.X. (2003) *Mater. Lett.*, **57**, 1406.
43 Lin, C.W., Huang, Y.F. and Kannan, A.M. (2007) *J. Power Sources*, **171**, 340.
44 Robeson, L.M., Hwu, H.H. and McGrath, J.E. (2007) *J. Membr. Sci.*, **302**, 70.
45 Kim, D.S. and Guiver, M.D. (2008) Development of sulfonated poly(ether ether ketone)s for PEMFC and DMFC in *Polymer Membranes for Fuel Cells* (eds S.M.J. Zaidi and T. Matsuura), Springer Science Inc., USA, **4**, 51.
46 Bishop, M.T., Karasz, F.E., Russo, P.S. and Langley, K.H. (1985) *Macromolecules*, **18**, 86.
47 Shibuya, N. and Porter, R.S. (1992) *Macromolecules*, **25**, 6495.
48 Bailly, C., Williams, D.J., Karasz, F.E. and MacKnight, W.J. (1987) *Polymer*, **28**, 1009.
49 Gao, Y., Robertson, G.P., Guiver, M.D. and Jian, X. (2003) *J. Polym. Sci., Part A: Polym. Chem.*, **41**, 497.
50 Gao, Y., Robertson, G.P., Guiver, M.D., Jian, X., Mikhailenko, S.D., Wang, K. and Kaliaguine, S. (2003) *J. Membr. Sci.*, **227**, 39.
51 Dai, Y., Jian, X., Liu, X. and Guiver, M.D. (2001) *J. Appl. Polym. Sci.*, **79**, 1685.
52 Liu, B., Robertson, G.P., Kim, D.S., Guiver, M.D., Hu, W. and Jiang, Z. (2007) *Macromolecules*, **40**, 1934.
53 (a) Liu, B. Kim, Y.S., Hu, W., Robertson, G.P., Pivovar, B.S. and Guiver, M.D. (2008) *J. Power Sources*, **185**, (2), 899–903; (b) Kim, D.S., Robertson, G.P., Kim, Y.S. and Guiver, M.D. (2009) *Macromolecules*, **42**, 957–963; (c) Jeong, M.H., Lee, K.S., Hong, Y.T. and Lee, J.S. (2008) *J. Membr. Sci.*, **314**, 212; (d) Miyatake, K., Chikashige, Y., Higuchi, E. and Watanabe, M. (2007) *J. Am. Chem. Soc.*, **129**, 3879.
54 Harrison, W.L.Ph.D. dissertation, Virginia Polytechnic Institute and State University, USA.
55 Kim, Y.S., Wang, F., Hickner, M., McCartney, S., Hong, Y.T., Zawodzinski, T.A. and McGrath, J.E. (2003) *J. Polym. Sci., Part B: Polym. Phys.*, **41**, 2816.
56 Robeson, L.M. Matzner, M. (1983) Flame retardant polyarylate compositions U.S. Patent 4,380,598.
57 Ueda, M., Toyota, H., Ochi, T., Sugiyama, J., Yonetake, K., Masuko, T. and Teramoto, T. (1993) *J. Polym. Sci., Part A: Polym. Chem.*, **31**, 853.
58 Wang, F., Chen, T. and Xu, J. (1998) *Macromol. Chem. Phys.*, **199**, 1421.
59 Li, X.F., Na, H. and Lu, H. (2004) *J. Appl. Polym. Sci.*, **94**, 1569.
60 Liu, B., Kim, D.S., Murphy, J., Robertson, G.P., Guiver, M.D., Mikhailenko, S., Kaliaguine, S., Sun, Y.M., Liu, Y.L. and Lai, J.Y. (2006) *J. Membr. Sci.*, **280**, 54.
61 Gao, Y., Robertson, G.P., Guiver, M.D., Mikhailenko, S.D., Li, X. and Kaliaguine, S. (2006) *Polymer*, **47**, 808.
62 Gao, Y., Robertson, G.P., Guiver, M.D., Mikhailenko, S.D., Li, X. and Kaliaguine, S. (2005) *Macromolecules*, **38**, 3237.
63 Gao, Y., Robertson, G.P., Guiver, M.D., Mikhailenko, S.D., Li, X. and Kaliaguine, S. (2004) *Macromolecules*, **37**, 6748.
64 Gao, Y., Robertson, G.P., Kim, D.S., Guiver, M.D., Mikhailenko, S.D., Li, X. and Kaliaguine, S. (2007) *Macromolecules*, **40**, 1512.
65 Kim, D.S. and Guiver, M.D. (2008) *J. Polym. Sci., Part A: Polym. Chem.*, **46**, 989.
66 Hickner, M.A., Ghassemi, H., Kim, Y.S., Einsla, B.R. and McGath, J.E. (2004) *Chem. Rev.*, **104**, 4587.
67 Harrison, W.L., Hickner, M.A., Kim, Y.S. and McGrath, J.E. (2005) *Fuel Cells*, **5**, 201.
68 Hickner, M.A. and Pivovar, B.S. (2005) *Fuel Cells*, **5**, 213.
69 Kim, Y.S., Einsla, B., Sankir, M., Harrison, W. and Pivovar, B.S. (2006) *Polymer*, **47**, 4026.

70 Gao, Y., Robertson, G.P., Guiver, M.D., Jian, X., Mikhailenko, S.D., Wang, K. and Kaliaguine, S. (2003) *J. Polym. Sci., Part A: Polym. Chem.*, **41**, 2731.

71 Kim, D.S., Park, H.B., Jang, J.Y. and Lee, Y.M. (2005) *J. Polym. Sci., Part A: Polym. Chem.*, **43**, 5620.

72 Kim, D.S., Shin, K.H., Park, H.B., Chung, Y.S., Nam, S.Y. and Lee, Y.M. (2006) *J. Membr. Sci.*, **278**, 428.

73 (a) Wiles, K.B., Wang, F. and McGrath, J.E. (2005) *J. Polym. Sci., Part A: Polym. Chem.*, **43**, 2964. (b) Khalfan, A.N., Sanchez, L.M., Kodiweera, C., Greenbaum, S.G., Bai, Z. and Dang, T.D. (2007) *J. Power Sources*, **173**, 853.

74 Ghassemi, H., McGrath, J.E. and Zawodzinski, T.A. (2006) *Polymer*, **47**, 4132.

75 Wang, H. and McGrath, J.E. (2005) *Prepr. Symp. – Am. Chem. Soc., Div. Fuel, Chem.*, **50**, 581.

76 Yu, X., Roy, A. and McGrath, J.E. (2005) *Prepr. Symp. – Am. Chem. Soc., Div. Fuel, Chem.*, **50**, 577.

77 Li, Y., Roy, A., Badami, A.S., Hill, M., Yang, J., Dunn, S. and McGrath, J.E. (2007) *J. Power Sources*, **172**, 30.

78 Wang, F., Li, J., Chen, T. and Xu, J. (1999) *Polymer*, **40**, 795.

79 Xing, P., Robertson, G.P., Guiver, M.D., Mikhailenko, S.D. and Kaliaguine, S. (2004) *Macromolecules*, **37**, 7960.

80 Xing, P., Robertson, G.P., Guiver, M.D., Mikhailenko, S.D. and Kaliaguine, S. (2004) *J. Polym. Sci., Part A: Polym. Chem.*, **42**, 2866.

81 Liu, B., Robertson, G.P., Guiver, M.D., Sun, Y.M., Liu, Y.L., Lai, J.Y., Mikhailenko, S. and Kaliaguine, S. (2006) *J. Polym. Sci Part B: Polym. Phys.*, **44**, 2299.

82 Li, X., Liu, C., Lu, H., Zhao, C., Wang, Z., Xing, W. and Na, H. (2005) *J. Membr. Sci.*, **255**, 149.

83 Xing, P., Robertson, G.P., Guiver, M.D., Mikhailenko, S.D. and Kaliaguine, S. (2005) *Polymer*, **46**, 3257.

84 Sumner, M.J., Harrison, W.L., Weyers, R.M., Kim, Y.S., McGrath, J.E., Riffle, J.S., Brink, A. and Brink, M.H. (2004) *J. Membr. Sci.*, **239**, 199.

85 Kim, Y.S., Kim, D.S., Liu, B., Guiver, M.D. and Pivovar, B.S. (2008) *J. Electrochem. Soc.*, **155**, B21.

86 Kim, D.S., Kim, Y.S., Guiver, M.D. and Pivovar, B.S. (2008) *J. Membr. Sci.*, **321**, 199.

87 Gao, Y., Robertson, G.P., Guiver, M.D., Wang, G., Jian, X., Mikhailenko, S.D., Li, X. and Kaliaguine, S. (2005) *J. Memb. Sci.*, **278**, 26.

88 Kreuer, K.D. (2001) *J. Membr. Sci.*, **185**, 29.

89 Einsla, B.R. and McGrath, J.E. (2004) *Prepr. Symp. – Am. Chem. Soc., Div, Fuel, Chem.*, **49**, 616.

90 Ding, J., Cuy, C. and Holdcroft, S. (2002) *Macromolecules*, **35**, 1348.

91 Lafitte, B., Puchner, M. and Jannasch, P. (2005) *Macromol. Rapid Comm.*, **26**, 1464.

92 Lafitte, B., Karlsson, L.E. and Jannasch, P. (2002) *Macromol. Rapid Comm.*, **23**, 896.

93 Norsten, T.B., Guiver, M.D., Murphy, J., Astill, T., Navessin, T., Holdcroft, S., Frankamp, B.L., Rotello, V.M. and Ding, J. (2006) *Adv. Funct. Mater.*, **16**, 1814.

94 Li, Z., Ding, J., Robertson, G.P. and Guiver, M.D. (2006) *Macromolecules*, **39**, 6990.

95 Pang, J., Zhang, H., Li, X., Wang, L., Liu, B. and Jiang, Z. (2008) *J. Membr. Sci.*, **318**, 271.

96 Pang, J., Zhang, H., Li, X. and Jiang, Z. (2007) *Macromolecules*, **40**, 9435.

97 Kim, D.S., Robertson, G.P. and Guiver, M.D. (2008) *Macromolecules*, **41**, 2126.

98 Yang, B. and Manthiram, A. (2003) *Solid-State Lett.*, **6**, A229.

99 Miyatake, K., Zhou, H., Matsuo, T., Uchida, H. and Watanabe, M. (2004) *Macromolecules*, **37**, 4961.

100 Fu, Y.Z. and Manthiram, A. (2006) *J. Power Sources*, **157**, 222.

101 Pivovar, B.S. and Kim, Y.S. (2007) *J. Electrochem. Soc.*, **154** (8), B739.

102 Kim, D.S., Robertson, G.P., Guiver, M.D. and Lee, Y.M. (2006) *J. Membr. Soc.*, **281**, 111.

103 Wang, Z., Li, X., Zhao, C., Ni, H. and Na, H. (2006) *J. Power Sources*, **160**, 969.

104 Wang, A., Ni, H., Zhao, C., Li, X., Zhang, G., Shao, K. and Na, H. (2006) *J. Membr. Sci.*, **285**, 239.

105 Kim, D.S., Liu, B. and Guiver, M.D. (2006) *Polymer*, **47**, 7871.

106 Kim, D.S., Kim, Y.S., Guiver, M.D., Ding, J. and Pivovar, B.S. (2008) *J. Power Sources*, **182**, 100.

107 Saha, S. and Hamaguchi, H. (2006) *J. Phy. Chem. B.*, **110**, 2777.

108 Kim, Y.S., Sumner, M.J., Harrison, W.L., Riffle, J.S., McGrath, J.E. and Pivovar, B.S. (2004) *J. Electrochem. Soc.*, **151**, A2150.

109 Ren, X., Springer, T.E. and Gottesfeld, S. (2000) *J. Electrochem. Soc.*, **147**, 92.

110 Ren, X., Springer, T.E., Zawodzinski, T.A. and Gottesfeld, S. (2000) *J. Electrochem. Soc.*, **147**, 466.

111 Kim, Y.S., Dong, L., Hickner, M.A., Pivovar, B.S. and McGrath, J.E. (2003) *Polymer*, **44**, 5729.

112 Rao, V.L., Saxena, A. and Ninan, K.N. (2002) *J. Macromolecular Sci. Part C: Polym. Rev.*, **42**, 513.

11
Fabrication and Optimization of DMFC Catalyst Layers and Membrane Electrode Assemblies

Liang Ma, Yunjie Huang, Ligang Feng, Wei Xing, and Jiujun Zhang

11.1
Introduction

Proton exchange membrane fuel cells (PEMFCs) (or polymer electrolyte membrane fuel cells) have attracted a great deal of attention recently as promising energy converting devices for clean and renewable energy applications, due to their high power density, high efficiency, and low/zero emissions. There are two major types of PEMFC: the hydrogen fuel cell is fed by hydrogen gas, while the direct fuel cell is fed by a liquid fuel such as methanol, ethanol, or formic acid. Among the latter, the direct methanol fuel cell (DMFC) is the most promising device close to being commercialized. Unfortunately, several challenges still hinder DMFC commercialization, including low performance, high cost, and insufficient reliability/durability. Currently, the majority of the efforts in DMFC research and development are focused on overcoming these challenges in order to accelerate the commercialization process.

With respect to DMFC performance improvement, two major efforts can be identified. One is to improve both the kinetics of the anodic methanol oxidation reaction (MOR) and the cathodic oxygen reduction reaction (ORR) by developing breakthrough catalysts, catalyst layers, as well as the corresponding membrane electrode assemblies (MEA), and the other is to reduce methanol crossover by developing innovative membranes. The slow kinetics of the MOR and ORR appear to be largely responsible for the low performance of present-day DMFCs.

Due to the complexity of both the MOR and the ORR in a DMFC, the interactions among the catalyst activities, the catalyst layer structures, and the corresponding MEAs play a large role in performance improvement and optimization. For example, a catalyst that shows excellent activity towards the MOR or ORR in a half-cell test will not necessarily result in superior fuel cell performance when it is integrated into the catalyst layer. Therefore, when using this catalyst some optimization of the

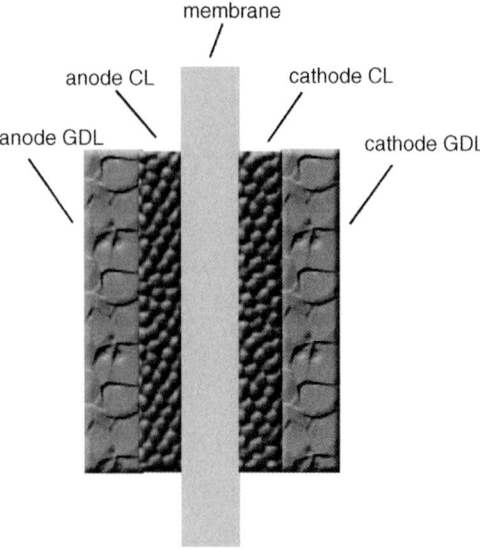

Figure 11.1 Schematic diagram of an MEA.

catalyst layer and MEAs is definitely needed in order to obtain reasonable fuel cell performance.

A DMFC membrane electrode assembly, as shown in Figure 11.1, consists of an anode gas diffusion layer (GDL), anode catalyst layer, membrane, cathode catalyst layer, and cathode GDL. Although GDLs and membranes have some effect on performance, the catalyst layer where the electrochemical reaction (MOR or ORR) takes place seems to play a more important or even dominant role in determining DMFC performance. An ideal catalyst layer should have maximum catalyst/ionomer/reactant interface or three-phase interface, and excellent mass transport properties. In order to achieve this, the catalyst layers and their corresponding MEAs should be optimized with respect to the complicated interactions and trades-off taking place between them. In practice, the catalyst layers and MEAs should be tailored carefully by employing the proper components, optimal structures, and innovative fabrication methods.

It is worth noting that sometimes a membrane's effect is significant. For example, methanol is prone to cross the membrane from the anode to the cathode side, resulting in lower fuel use and at the same time reducing the cathode potential. Therefore, methanol crossover should be minimized by developing novel membranes. In addition, as the DMFC cathode suffers serious flooding problems compared with the hydrogen-fed PEMFC, an optimized MEA should have good water management capabilities, either draining the water or facilitating water backflux to the anode.

In this chapter, the key aspects of DMFC catalyst layer and MEA design, fabrication, and performance optimization are reviewed and presented, based on recent research and developments reported in the literature.

11.2
Components for DMFC Catalyst Layer Optimization

11.2.1
Catalysts for the Methanol Oxidation Reaction (MOR)

DMFC performance is strongly dependent on the electrochemical MOR and ORR activities of the catalysts. A highly active catalyst is definitely necessary if one wants to fabricate a high-performance catalyst layer. Currently, platinum-ruthenium (PtRu) alloy is the most effective and practical anode electrocatalyst for the MOR in a DMFC, although its performance may need further improvement [1–6]. The catalytic activity of PtRu catalyst can be affected by its bulk composition [7, 8], dispersion [9–13], surface states [14–20], and morphology [21, 22]. Although numerous studies have been conducted to investigate the effects of these factors, a full understanding of the catalytic mechanisms has not yet been achieved. Here it is infeasible to present a complete picture of these effects due to the complexity of this topic. Here, we only focus on the effects of catalyst microscopic composition and structure, which may be the dominant factors interacting with other components and affecting the three-phase interface in the catalyst layer.

It is generally recognized that the generation of active oxygen-containing species on the PtRu catalyst surface is vital for improving the MOR. These species are necessary for the removal of CO-like poisoning intermediates on the Pt surface sites. To achieve this, we need an optimal amount of oxygen-containing species on the catalyst surface for the MOR process. It was reported that a practical PtRu catalyst is not a uniform single-phase alloy but is actually composed of a mixture of metal particles, different forms of ruthenium oxides, and platinum oxides [14, 23–25]. Reported conclusions about the identification of Ru active sites on the PtRu catalyst for the MOR seem to be inconsistent. Rolison et al. [23] recommended that hydrous ruthenium oxide, RuO_xH_y (or $RuO_2 \cdot (H_2O)_x$), which is a mixed proton and electron conductor, should be the preferred species for active sites rather than Ru metal or anhydrous RuO_2. They further showed that the activity of the as-received PtRu black (50:50 atomic ratio) for methanol oxidation could be reduced by two orders of magnitude if the catalyst was reduced in a stream of 10% of hydrogen at 100 °C for 2 hours. However, if this reduced catalyst was re-oxidized in water-saturated oxygen at 100 °C for 20 hours, the activity could be partially recovered, as shown in Figure 11.2 [26]. Therefore, they concluded that bulk quantities of electron-proton conducting RuO_xH_y are required to achieve high activity for the MOR. However, Dinh et al. [25] concluded from their in-situ CO stripping experiments that Ru oxide species seem to have an inhibiting effect on the surface catalytic activity of PtRu catalyst for the MOR, through blocking the active metal alloy sites. They suggested that the surface metal alloy domains with an atomic ratio close to 1 : 1 were key for higher DMFC anode activity. This was also supported by Briss et al. [27], who reported that the MOR activity of as-received PtRu black catalyst (50:50 atomic ratio) could be improved by reducing this catalyst in hydrogen at 100 °C for 2–3 hours. This inconsistency may be interpretable according to the work conducted by Zhuang

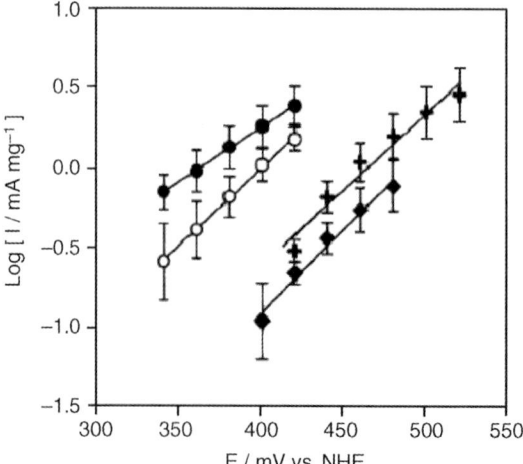

Pt-Ru black treatment		Mass-normalized exchange current (mA/g)
●	As-received	5.5
◆	Reduced in H_2	0.020
○	Reduced in H_2; re-oxidized in O_2/H_2O	0.41
✚	300 °C anneal	0.064

Figure 11.2 Tafel plots for methanol oxidation activity measured by potential-step chronoamperometry in 1 M methanol/1 M H_2SO_4 electrolyte for the as-received and temperature/atmosphere-treated Pt-Ru blacks. [26].

et al. [28]. They reported that the activity of carbon-supported PtRu catalyst (PtRu/C) could be remarkably enhanced by anodic treatment, and this enhancement effect could be attributed to the favorable changes in Ru oxides. Based on the changes in the cyclic voltammogram of the catalyst during the anodic treatment, they identified two types of ruthenium oxides on the catalyst surface: one was electrochemically reversible and could be beneficial to catalytic activity enhancement, while the other was irreversible and harmful. During the anodic treatment, it was observed that the harmful oxide gradually decreased, while the beneficial oxide either increased or changed only slightly. This observation may be useful in understanding the inconsistent conclusions regarding catalyst active site identification. Different reduction conditions used in the work of Briss et al. and Rolison et al. might produce different surface sites, resulting in different catalytic activities. Zhuang et al. [29] also found increasing the hydrous ruthenium oxides to be beneficial, and further proposed a pattern-recognition methodology to elucidate the structure-activity relationship of the PtRu catalyst. Therefore, it is believed that mixed proton-electron conductive active oxygen-containing surface species can indeed improve this catalyst's activity towards the MOR.

It has also been recognized that the existence of hydrous ruthenium oxides on the PtRu catalyst surface can be necessary for fabricating a better anode catalyst layer. Thomas et al. [24] showed that hydrous ruthenium oxides in the anode catalyst layer could provide sufficiently high proton conductivity, which if allowed to reduce or even eliminate the addition of an ionomer into the catalyst layer resulted in a thinner catalyst layer with better catalyst utilization and mass transport properties. Scheiba et al. [30] investigated the hydrous ruthenium oxides as a novel support material for electrocatalysts in DMFCs. The results indicated that this catalyst showed inferior activity towards the MOR in half-cell measurements compared with the MOR catalyzed by a commercially available Johnson-Matthey 20% Vulcan XC72R-supported PtRu catalyst. However, if this catalyst was used to fabricate a catalyst layer in a full cell arrangement, the proton conductivity and catalyst utilization were both improved, resulting in enhanced fuel cell performance. Similar to the work of Zhuang et al. [28], Jeon et al. [31] showed that the performance of DMFC single cells could be improved by anodic treatment in a potential range of 0.4 to 0.7 V. Using electrochemical impedance spectroscopy (EIS) analysis and X-ray photoelectron spectroscopy (XPS), they concluded that the improved performance arose from the formation of RuO_xH_y, resulting in increased MOR activity, as shown in Figure 11.3.

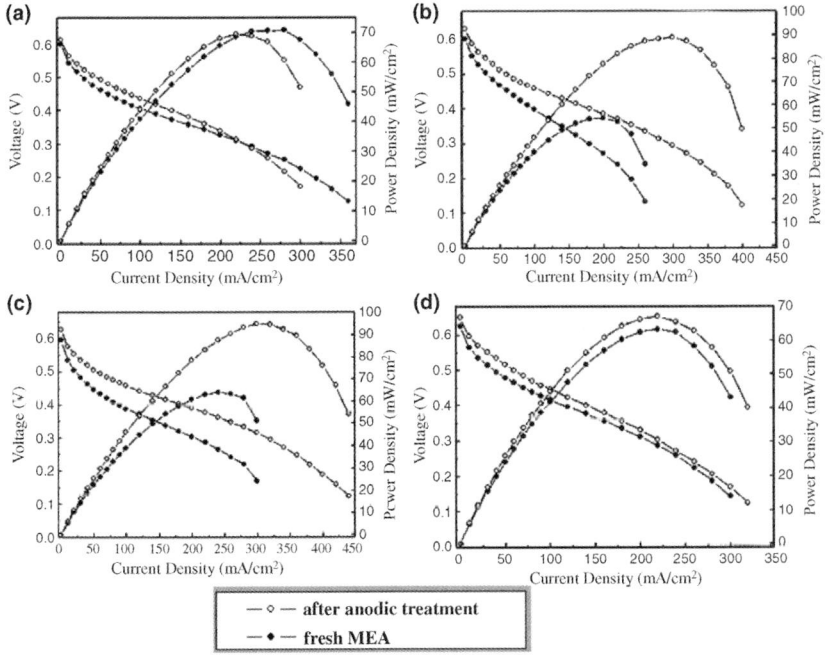

Figure 11.3 Current-voltage curves of single cells before and after the anodic treatment at (a) 0.1, (b) 0.4, (c) 0.7, and (d) 1.0 V (vs. RHE). Filled circles are for fresh MEAs and hollow circles for MEAs after anodic treatment [31].

Besides the roles of surface active species in catalyst activity, catalyst morphology is also important in catalyst layer optimization. Liu et al. [32] showed that a supported PtRu catalyst exhibited a much higher specific activity for the MOR than an unsupported PtRu catalyst. However, the fuel cell performance of the supported catalyst showed no improvement when catalyst loading exceeded 0.5 mg/cm^2, while that of the unsupported catalyst showed no such limitation even when the loading exceeded 8 mg/cm^2. Moreover, the MEA fabricated with supported PtRu presented a thicker catalyst layer and increased methanol crossover when compared with unsupported PtRu. Choi et al. [21] synthesized a bimetallic PtRu nanowire network (Pt-Ru-NW) in an SBA-15 template and compared it with commercially available Johnson-Matthey PtRu black. As presented in Figure 11.4a, obtained using half-cell measurements, the Pt-Ru-NW shows a slightly lower activity at low potentials (E < 0.55 V vs. RHE), and dramatically higher activity at higher potentials (E > 0.55 V

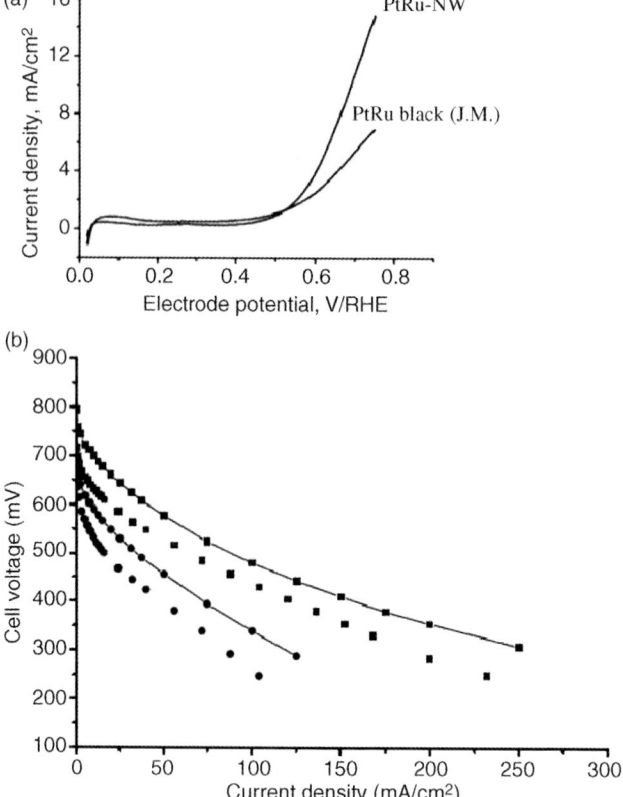

Figure 11.4 (a) Cyclic voltammograms for methanol electro-oxidation and (b) DMFC current density–voltage curves at 40 °C (circles) and 80 °C (squares). The performance of Pt–Ru-NW (connected data point) and that of Johnson Matthey commercially available Pt–Ru black (unconnected data point) in a single-cell test are also shown for comparison (catalyst loading: 5 mg cm^{-2}) [21].

vs. RHE) than commercially available PtRu black. However, the fuel cell using this catalyst shows a higher performance when compared to that employing a commercially available PtRu catalyst, probably indicating that the three-dimensional PtRu nanowire network could lead to effective mass transport in the catalyst layer, although the MOR kinetics catalyzed by this catalyst presents no visible improvement.

11.2.2
Catalysts for the Oxygen Reduction Reaction (ORR)

In DMFC cathode catalyst development, while there has been some exploration of platinum alloy and non-platinum materials for electrocatalysis, Pt black and/or carbon-supported platinum (Pt/C) catalysts are still the most widely used and practical choices. This is mainly due to their intrinsic high activity and stability in acidic solutions. Two major properties should be considered for a DMFC cathode catalyst: catalytic activity towards the ORR and methanol-tolerance capability. With respect to the former, catalyst particle size seems to play a considerable role. Some researchers have reported that in a phosphoric acid medium, the maximum mass activity of Pt/C catalyst for the ORR occurred when the particle size was 3 nm. At this scale, the Pt particles have maximum fractions of (111) and (100) surface atoms with a cubo-octahedral geometry [33]. Takasu et al. [34] found that decreasing the platinum particle size also decreased the specific activity for the MOR. Therefore, there should be an optimum platinum particle size for both ORR activity and methanol-tolerance capability.

Platinum catalysts can achieve high activity and methanol-tolerance capability simultaneously, through modification of the platinum catalyst morphology or chemical characteristics. Wen et al. [35] reported a facile template route to entrap in-situ highly distributed Pt/C nanoparticles into the nanochannels of mesoporous carbon, as shown in Figure 11.5. Here, each Pt nanoparticle is covered by a carbon

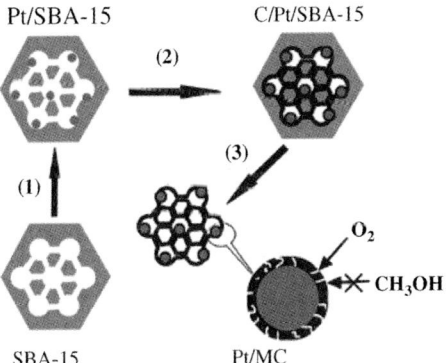

Figure 11.5 Schematic illustration of the synthetic procedure of Pt@C/MC: (1) deposition of Pt in the nanochannels of SBA-15; (2) polymerization of glucose; and (3) carbonization and removal of SBA-15 [35].

film that contains micropores, allowing only oxygen to access the Pt nanoparticle activity sites and hindering methanol access. In this way, the formed catalyst (Pt@C/MC) showed not only a high activity and considerable stability for the ORR but also a strong methanol-tolerance capability. In addition, the adsorption of anions such as chloride on the Pt catalyst was also found to be beneficial for enhancing methanol-tolerance capability. Normally, strongly adsorbing chloride ions could reduce the activity of a Pt cathode and cause substantial amounts of H_2O_2 to be formed [36]. However, in the presence of methanol (as in the case of a DMFC cathode), modifying the Pt cathode by adding a certain amount of chloride ions during operation of the DMFC could improve both performance and stability [37]. This improvement may be ascribable to the surface blocking effect of adsorbed Cl^-, which could suppress the adsorption of oxygen species on the Pt surface and subsequently reduce the methanol oxidation activity on the cathode Pt surface. This reduced methanol oxidation activity can effectively reduce the depressing effect arising from the direct reaction between the Pt catalyst and methanol crossover from the anode side.

11.2.3
Ionomer in the Catalyst Layer

In addition to the catalyst itself, other components are also critical for the performance of a catalyst layer. As the electrochemical reaction (MOR or ORR) takes place within a unique three-phase interface, it is important to ensure that the active catalyst sites are sufficiently accessed by electrons, protons, and reactants. The inclusion of an ionomer in the catalyst layer can facilitate proton conductivity, resulting in significant improvement in catalyst layer performance. However, this ionomer can also have negative effects such as reduced hydrophobicity, resulting in a mass transport problem. Therefore, the ionomer content of the catalyst layer should be carefully controlled. The optimum ionomer content (normally Nafion) in a catalyst layer varies with the catalysts used [24, 38–40]. Two main aspects should be considered when dealing with ionomer content: first, whether or not the ionomer would interact with the catalyst and affect its MOR catalytic activity; second, whether or not the ionomer's distribution inside the catalyst layer is uniform.

With respect to the effect of the ionomer on catalyst activity, Gojkovic et al. [41] compared the MOR activities of two electrode surfaces, one a bare Pt electrode and the other a Pt electrode covered by Nafion ionomer film. The results showed that the MOR rate was reduced in the presence of Nafion ionomer, while the kinetic parameters of the reaction were not affected. This observation is understandable in that the Nafion could block the active sites. In the work of Chu et al. [42], the effect of Nafion ionomer on the MOR activity catalyzed by PtRu was investigated using a combinatorial electrochemical method. They prepared a PtRu-Nafion catalyst by reducing the metal precursor-Nafion mixture, and observed that the addition of Nafion ionomer could improve the electrode catalytic activity. They claimed a synergistic effect between Nafion-active metal catalysts for methanol electro-oxidation. Similarly, Sarma et al. [43] also found that the addition of Nafion during catalyst preparation could enhance the MOR activity. Structural characterization was carried out using X-ray adsorption spectroscopy. The results suggested that the enhanced catalytic activity could be attributed to both better dispersion of and increased d-band vacancy in the Pt catalyst.

With respect to ionomer distribution inside the catalyst layer, the uniformity of such a distribution is strongly dependent on catalyst usage of catalysts. Ideally, all Pt catalyst particles are impregnated (or contacted) by the ionomer. For example, using an unsupported PtRu catalyst, a more homogeneous Nafion distribution could be obtained than when using a supported PtRu catalyst [44]. This is due to the effect of the carbon support. In general, Pt particles can be finely dispersed on the surface of 10–40 nm carbon particles to form a supported catalyst. Unfortunately, these supported catalyst particles may agglomerate via intermolecular interaction between their surfaces to form aggregations, as shown in Figure 11.7. The ionomer can also form agglomerates with a size distribution ranging from 20 nm to 2 µm, depending on the different dielectric constants (ε) of the solvent [45–49]. Therefore, in a catalyst layer there should be two distinct pore size distributions, namely, the primary and the secondary pores [46]. The primary pores with a size of 20–40 nm can be defined as the nano spaces/pores inside the catalyst aggregations, while the secondary pores between 40–1000 nm can be defined as the spaces/pores between the catalyst and Nafion ionomer agglomerates. It is very difficult for an ionomer to penetrate into the smaller spaces/pores (<about 20 nm) inside the catalyst agglomerates, and thus the Pt catalyst particles inside the agglomerate cannot be used, resulting in poor ionomer distribution and low catalyst utilization.

Several solutions have been proposed to achieve a better ionomer distribution. One involves intentionally modifying the carbon support before loading the catalyst. Park et al. [50] modified the carbon supports by applying a coat of Nafion ionomer before dispersing the catalyst. In this way, the Pt particles would not be deposited into undesirable nanopores on the carbon support surface, but instead into larger pores or in the vicinity of the ionomer, as shown in Figure 11.6. In fuel cell validation tests,

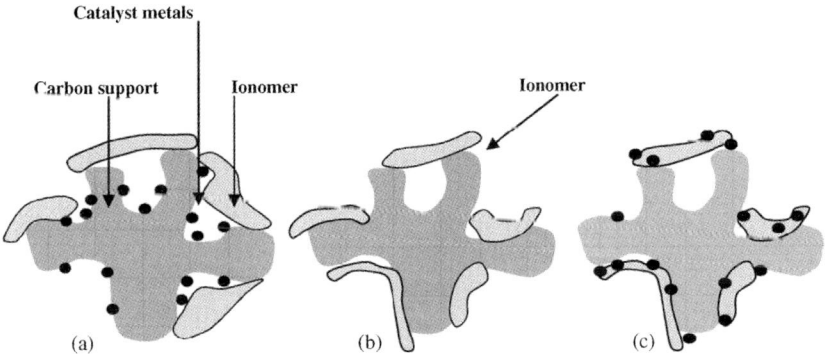

Figure 11.6 Diagrammatic representation of:
(a) catalyst–ionomer interaction in electrode using plain carbon as a support; (b) ionomer-coated carbon support; and
(c) catalyst–ionomer interaction on ionomer-coated carbon support [50].

they found that the produced carbon-supported PtRu catalyst showed better performance than the unmodified catalyst. In addition, the optimum amount of ionomer required in the catalyst layer was also reduced by half as compared with the plain carbon-supported catalyst.

Another solution for obtaining a better ionomer distribution has been to tailor the carbon surface by grafting proton-conductive functional groups. Lee *et al.* [51] synthesized a pitch-based porous carbon. This carbon support surface had $-SO_3H$ groups created by carbonization of mesophase pitch with sulfonation. They found that their pitch-based porous carbon-supported catalyst showed a fuel cell performance comparable with that of commercially available 60% PtRu/Vulcan XC72 catalyst, while the amount of Nafion ionomer needed in the catalyst layer was greatly reduced. In a similar way, Kuroki *et al.* [52] grafted methylsulfonic acid groups ($-CH_2SO_3H$) to the carbon support (Figure 11.7). They found that the chemical connections between the grafted groups and the carbon surface were stable up to around 380 °C. Furthermore, the groups could be homogeneously introduced into both the primary and the secondary pores without significant structural change in the latter. Using grafting catalysts, the group fabricated MEAs, and in a DMFC test these showed superior performance over a conventional catalyst (Figure 11.8).

Another solution has been to use carbon supports with preferred pore distribution. The main intention was to ensure good contact between the Nafion ionomer and the Pt catalyst particles inside the agglomerates. Rao *et al.* [53] used a supported PtRu catalyst to systematically investigate the effects of both carbon support porosity and

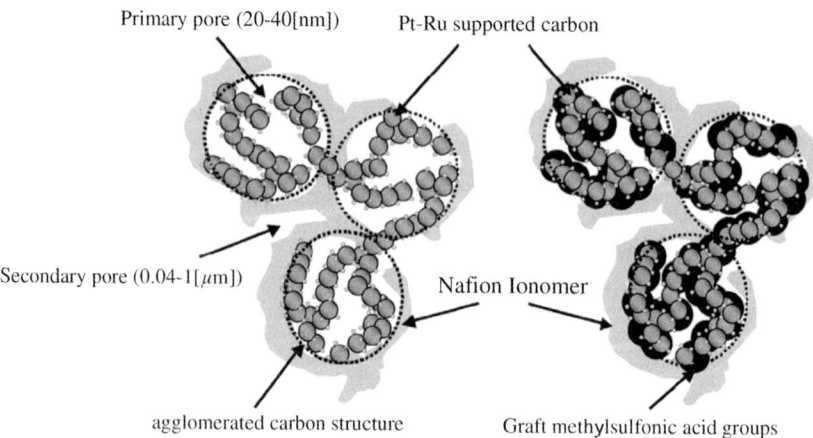

Figure 11.7 Schematic illustration of the internal structure of the catalyst layer: (a) catalyst layer made by nongrafted Pt–Ru/Ketjen black; (b) catalyst layer made by grafted Pt–Ru/Ketjen black. Nano-sized Pt–Ru catalysts were dispersed on each 30–40 nm sized carbon black surface, and the carbon black particles were agglomerated. Nafion ionomers were partly covered on the agglomerated structure [52].

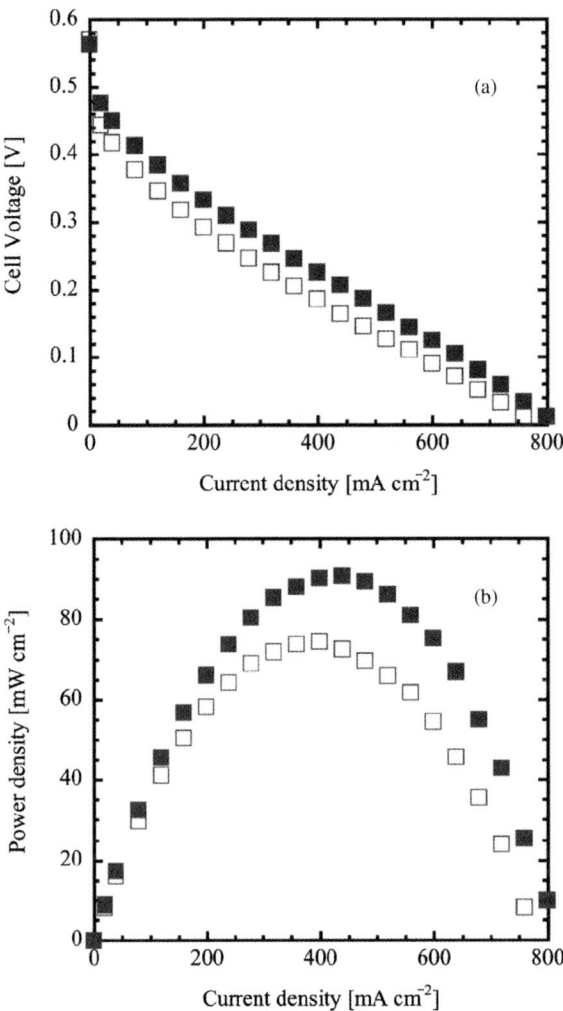

Figure 11.8 (a) Current-voltage and (b) current-power curves of the MEA made by nongrafted Pt–Ru/Ketjen black (□) and the MEA made by grafted Pt–Ru/Ketjen black (■). Operating conditions: cell temperature 50 °C; atmospheric pressure; fuel flow rate 2.5 mol L^{-1} of methanol solution; anode Pt–Ru loading about 1.0 mg cm^{-2} [52].

specific surface area on anode performance in a DMFC. As shown in Figure 11.9, the low surface area Sibunit support, with a minimal amount of pores below 20 nm, yielded amazingly high catalyst utilization. If this material was used to fabricate a carbon-supported catalyst, a mass-specific activity could be obtained that exceeded by nearly a factor of 3 that of commercially available E-Tek 20% PtRu/Vulcan XC72. Using cryo- and xerogel carbons with high mesoporosity in the pore size range of

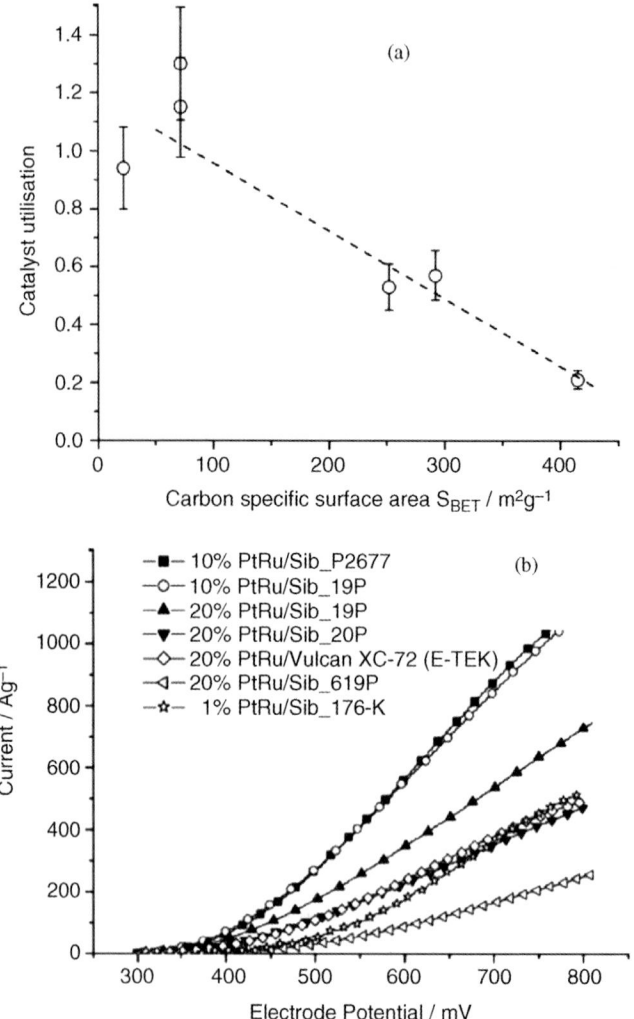

Figure 11.9 (a) Catalyst use factors plotted vs. SBET of carbon supports. (b) Current–potential characteristics for PtRu anode catalysts at 50 °C and 0.5 mVs^{-1} scan rate, measured in a half-cell DMFC. Y-axis shows current normalized to the metal loading [53].

20–40 nm, Arbinzzani et al. [54] showed that DMFC performance could be greatly enhanced as compared with using Vulcan carbon as a support. They attributed this improvement to better contact between the catalyst particles and the Nafion ionomer agglomerates.

Besides the approaches described above, an additional solution has been recommended by Wu et al. [55], who added different amounts of multi-wall carbon nanotubes (MWCNTs) to the anode catalyst layer of a DMFC and found that the catalyst layer showed enhanced proton conductivity and performance. They suggested

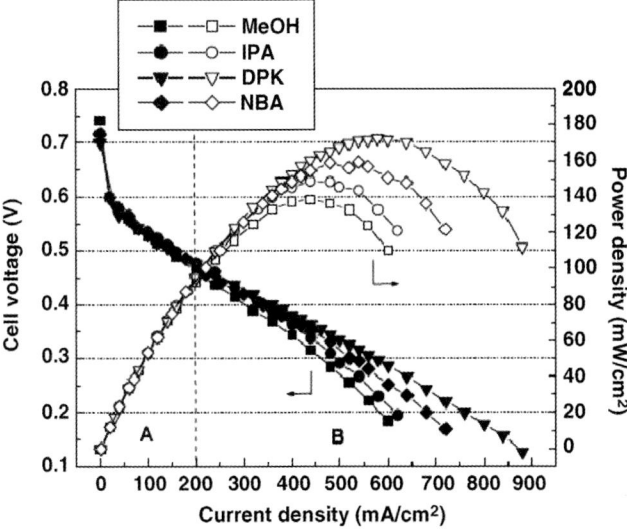

Figure 11.10 Polarization curves for the MEAs with anodes using different solvents for the catalyst ink preparation. DMFC was operated using 2 M methanol and O_2 gas at 80 °C and atmospheric pressure [48].

that the well-dispersed one-dimensional MWCNTs with Nafion ionomer on their surface played a conductive wiring role between the multiple catalyst aggregates.

Nafion ionomer agglomerate sizes can also affect catalyst layer performance, depending upon the ionomer solvent; using a proper solvent can improve catalyst utilization. Kim et al. [48] showed that DMFC performance could be improved when a less polar solvent was used in the catalyst ink (Figure 11.10). Using electrochemical impedance spectroscopy analysis, they found this performance improvement could be attributed to better proton conductivity and thus better catalyst utilization. Tominaka et al. [56] showed that the MOR current of the catalyst layer prepared with ethylene glycol dimethyl ether was twice that of a catalyst layer prepared with a conventional isopropyl alcohol aqueous solution. This could be attributed to the large improvement in proton conductivity (85 times greater than for a conventional solvent). Wang et al. [49] reported that the large aggregation particles in a Nafion aqueous solution could be significantly suppressed by adding NaOH to the solution. In this way, the large ionic aggregations of the ionomer could be dissociated into small aggregations. The resulting solution contained about 75% Nafion ionomer particles with a size distribution of 30–40 nm. Using this solvent, the resulting cathode catalyst layer had a uniform dispersion of catalyst and Nafion ionomer, and thus a high DMFC performance. However, it should be noted that the solvent used for the catalyst ink preparation affected not only the Nafion ionomer dispersion but also the structure formation process of the catalyst layer [48, 57, 58]. Solvent selection should involve a compromise between different properties such as dielectric constant, viscosity, boiling point, and so on.

11.2.4
Components Related to Mass Transport

It is known that DMFCs can encounter severe mass transport problems in the high current density region, particularly when a carbon-supported catalyst is used and/or when the fuel cell is operated using a low methanol concentration. In general, the mass transport property of the catalyst layer is affected mainly by its pore distribution and hydrophilic/hydrophobic properties. These microstructural features are closely related to the various components used in the catalyst layer.

From a mass transport point of view, PtRu and Pt black catalysts are preferable for DMFCs. However, from the point of view of catalyst cost reduction, carbon-supported catalysts are better. Practically speaking, a carbon-supported catalyst requires a high catalyst loading, which has a significant effect on DMFC performance. For example, Lee et al. [59] showed that at the same catalyst load, the performance of 80 wt% PtRu/C was superior to that of 40 and 60 wt% PtRu/C in the high current density region, while in the low current density region, all the catalysts showed similar activities (Figure 11.11). Han et al. [60] prepared high metal loading catalysts with fine metal particle sizes of less than 3 nm. They found that 90 wt% Pt–Ru/Ketjen Black showed the best anode performance in a DMFC, although 80 wt% Pt–Ru/Ketjen Black showed the highest activity for MOR in electrochemical half-cell tests. In the effort to reduce the catalyst loading, Liu et al. [32] showed that the DMFC performance of their 0.46 mg/cm^2 PtRu/C anode was comparable to an anode using 2 mg/cm^2 unsupported PtRu black.

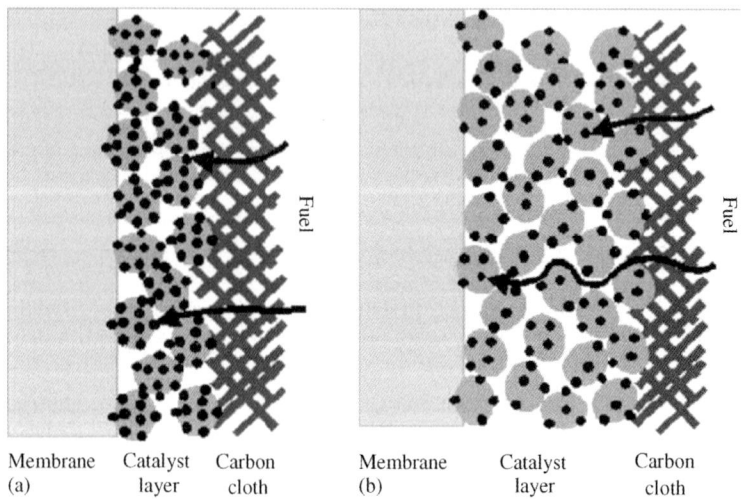

Figure 11.11 Schematic planner representation of catalyst layers with the same metal loading: (a) at high metal ratio catalyst, fuel could penetrate to the catalyst layer easily while (b) at the low metal ratio catalyst, the fuel paths are more complex and longer than those in (a) [59].

11.2 Components for DMFC Catalyst Layer Optimization

The mass transport in a DMFC is somewhat different from that in a hydrogen PEMFC. In a DMFC, effective convective transport of the methanol supply and the carbon dioxide gas produced at the anode is desirable. Therefore, some researchers have tried to give the catalyst layer some degree of hydrophobicity. Nordlund et al. [61] showed that by adding a certain amount of PTFE to the anode catalyst layer, pores smaller than 40 nm were unaffected while the volume fraction of pores 40–1000 nm was increased. Therefore, the PTFE in the catalyst layer could only affect the secondary pores, forming continuous hydrophobic channels for gaseous carbon dioxide to leave and neighboring hydrophilic channels for carbon dioxide to induce convection. These morphology changes could create some room for carbon dioxide to evolve as a gas and induce convection in the catalyst layer. In this way, carbon dioxide evolution was facilitated and mass transport was improved. Wei et al. [62] also added PTFE to the catalyst layer and found that methanol crossover decreased.

Mass transport can also be improved by using proper carbon supports. Mesoporous carbon and three-dimensional ordered carbon have recently come into use as fuel cell catalyst supports because of their unique mass transport properties and high potential for designing ordered MEAs. For example, Liu et al. [63] dispersed PtRu catalyst on a mesocarbon microbead (MCMB) support with a particle diameter of 1–40 µm and a low specific surface area. This MCMB-supported PtRu catalyst showed lower polarization characteristics than a Vulcan XC72-supported catalyst. Using cross-section SEM analysis, they showed large pores and channels favoring mass transport.

A series of uniform ordered mesoporous carbon (OMC) frameworks with pore sizes in the range of 15–1000 nm were synthesized by Chai et al. [64] via carbonization of a carbon precursor in removable colloidal silica crystalline templates. Using this porous carbon as support, they prepared a series of Pt(50)-Ru(50) catalysts with different pore size characteristics. In DMFC tests, these uniform ordered porous carbons yielded improved catalytic activity for the MOR; the porous carbon ~25 nm in diameter (PtRu-C-25) showed the highest performance. They ascribed this overall high activity to the large surface areas and pore volumes as well as the three-dimensionally interconnected uniform pore structures of the carbon support, which allowed a higher degree of catalyst dispersion and efficient reagent diffusion. Bang et al. [65] synthesized two types of micrometer-sized porous spherical carbons with ultrasonic spray pyrolysis (USP), and used these materials as catalyst supports in a catalyst layer. BET analysis showed individual porous carbon spheres I (PC-I) with mesopores ranging from 5 to 50 nm, while porous carbon spheres II (PC-II) showed only some macropores. The catalyst layer with PtRu/Vulcan showed slightly higher performance in the low current density region, while in the high current density region the catalyst layer with PtRu/PC-I showed better performance. Further, the catalyst layer with a mixture of PtRu/Vulcan and PtRu/PC-I (weight ratio 2 : 1, 33 wt% PC-I) yielded even better performance. This result could be attributed to the synergic effect of two carbons (PC-I and Vulcan XC-72): the mixture could improve the effective mass transport of reactant methanol and products while maintaining a reasonably high conductivity.

With respect to the DMFC cathode, water management should be considered. As this cathode can experience more water than a hydrogen PEMFC cathode, water flooding may become a problem. In order to overcome this issue, some researchers have added pore-forming agents into the cathode catalyst layer to tailor the pore distributions and therefore improve the mass transport properties. For example, Reshetenko et al. [66] fabricated MEAs by adding a pore-forming agent, ammonium carbonate or ammonium hydrocarbonate, into the cathode catalyst layers. It was found that ammonium carbonate predominantly produced macropores while ammonium hydrocarbonate was effective in forming mesopores. The obtained extra-porosity could promote oxygen transport to the catalyst surface and thereby enhance cell performance. However, it was also observed that cell performance decreased when excessive pore-forming agent was added. This was probably because the increased amount of pore-forming agents could lead to a decrease in the Nafion ionomer fraction and subsequently reduce the effective ionic conductivity. Therefore, the amount of pore-forming agent needs to be optimized for MEA water management. In another approach, Reshetenko et al. [67] also introduced carbon nanotubes (CNTs) into the cathode catalyst layer as pore forming agents. The results demonstrated that BET surface area and electrode porosity were enlarged due to the filamentous morphology of the CNTs. This morphology could produce both a large electro-active area and enhanced oxygen transport in the catalyst layer. As shown in Figure 11.12, when the weight ratio of CNTs to Pt was 0.05, the cell performance was distinctly increased in all the current density ranges as compared with the reference. In addition, Bang et al. [65] prepared some carbon powders composed of porous micrometer-sized spheres (the PC-II described above) as a pore-forming agent in the cathode catalyst layer. Adding a relatively small amount of PC-II into the

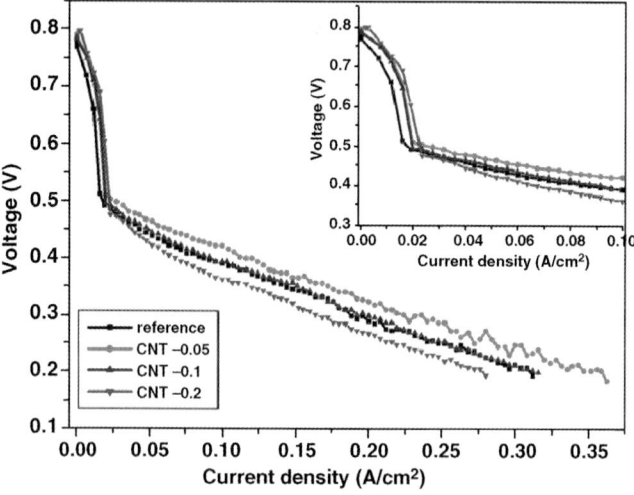

Figure 11.12 Polarization curves for the samples modified by CNTs (T = 50 °C, 3 M methanol, fourth day of evaluation) [67].

commercially available Pt/C catalyst significantly improved the cell performance, due to enhanced oxygen diffusion despite the low airflow rate.

11.3
Catalyzed DMFC Electrode Structure and Fabrication Process

11.3.1
Fabrication Process of DMFC Electrode

In general, there are two methods of fuel cell electrode fabrication: the first is to apply a thin layer of catalyst ink (catalyst layer) onto the gas diffusion layer substrate to form a catalyzed diffusion medium (CDM), and the other is to apply a thin layer of catalyst ink (catalyst layer) onto each side of a proton exchange membrane (PEM) to form a catalyst coated membrane (CCM). For CDM-based MEA preparation, a PEM is placed between two pieces of CDM (one for the anode and the other for the cathode), with catalyst layers in contact with the PEM surfaces. Then this assembly is hot-pressed to form a MEA. For CCM-based MEA preparation, the CCM is placed between two pieces of gas diffusion layer to form an assembly. This may be hot-pressed to form a MEA; in practice, this hot-press process may be unnecessary when using a CCM-based MEA, and the formed assembly can be directly incorporated into the fuel cell or stack hardware without hot-pressing. In terms of MEA performance, the catalyst ink preparation, electrode fabrication procedure and configuration, MEA hot pressing, and subsequent treatment conditions all significantly influence the catalyst layer microstructure and thus its performance and durability.

The catalyst ink preparation method and procedure play a critical role in the formation and corresponding MEA performance of the catalyst layer. For example, Lim et al. [68] used the ultrasonicating or ball-milling method to make PtRu black catalyst inks in two different types of solvent (deionized water (DI) and isopropyl alcohol (IPA)) and also did some comparative studies in order to optimize the catalyst ink preparation. The ink prepared by ball-milling in IPA yielded a catalyst layer with the thickest and most porous structure. This anode catalyst layer also showed enhanced electrocatalytic activity towards the MOR and a lower methanol crossover. In order to prepare a catalyst layer with high catalyst utilization, Song et al. [69] used an autoclave treatment on the catalyst ink. As shown in Figure 11.13, a mixture of supported Pt catalyst and Nafion ionomer solution were heated in an autoclave at 200 °C, followed by a quenching step to form the catalyst ink. The microstructure analysis indicated that this autoclave treatment could promote effective introduction of Nafion ionomer into primary pores of the Pt/carbon black agglomerates. It was shown that the catalyst layer prepared using this catalyst ink exhibited very high performance. This could be attributed to high Pt utilization and improved gas diffusion compared with a conventional catalyst layer.

With respect to the effect of electrode preparation procedures on fuel cell performance, Song et al. [70] used a decal transfer method (DTM) and found that during the electrode fabrication process, the decal transfer process caused Pt black

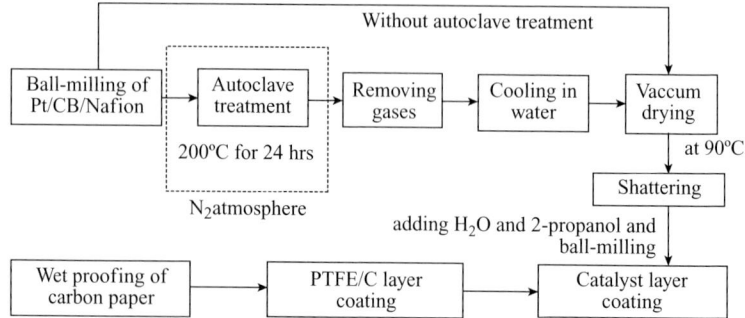

Figure 11.13 Flowchart for preparation of the gas diffusion electrode [69].

particles to grow larger, while the PtRu black particles were unaffected. However, the surface of PtRu black contained more metal oxides, leading to a smaller CO stripping charge and thus having a detrimental effect on DMFC performance. Nevertheless, the DTM process could still provide MEAs with higher DMFC performance when compared to the conventional method. This could be because the DTM produced better contact between the catalyst layer and the membrane and thus reduced the protonic and electronic resistance. Moreover, the DTM could also make thinner catalyst layers and thus improve mass transportation.

Catalyst layer configuration is also an important factor affecting fuel cell performance. Reshetenko et al. [71] studied the effects of cathode configuration on DMFC performance, and found that the CCM cathode had a main microporous/mesoporous structure, whereas the CDM cathode showed macroporosity together with micropores/mesopores. The CDM cathode also showed superior performance due to enhanced oxygen transport through the macropores. Mao et al. [72] found that the CCM configuration could improve both anode and cathode performance, and showed that the CCM anode favored anode polarization more than the CDM. In terms of methanol crossover, the combination of brush painting the CDM anode and spraying the CCM cathode configuration could yield the highest performance.

In order to optimize the composition and configuration of the catalyst layer, some researchers have also developed the concept of a multiple catalyst layer. For example, Wei et al. [62] reported that the performance of their multi-layer electrode could be enhanced and the stability improved by adding two hydrophilic catalyst thin films to the traditional electrode. Liu et al. [73] investigated the effect of different cathode catalyst layer configurations on fuel cell performance. The results indicated that the CCM catalyst layer could give higher cell open circuit voltages and higher cell voltages at lower current densities, while the CDM catalyst layer could give a better performance at high current densities. The cathode catalyst layer with a composite structure, consisting of both CCM and CDM, showed better performance in both kinetic and mass transport limitation regions, due to its better porosity distribution and catalyst utilization. Kim et al. [74] proposed the concept of a microstructured MEA, featuring a double catalyst layer and a micropatterned interface between the membrane and the catalyst layer. As shown in Figure 11.14, the double catalyst layer consists of a dense CCM first layer and a porous CDM second layer. The

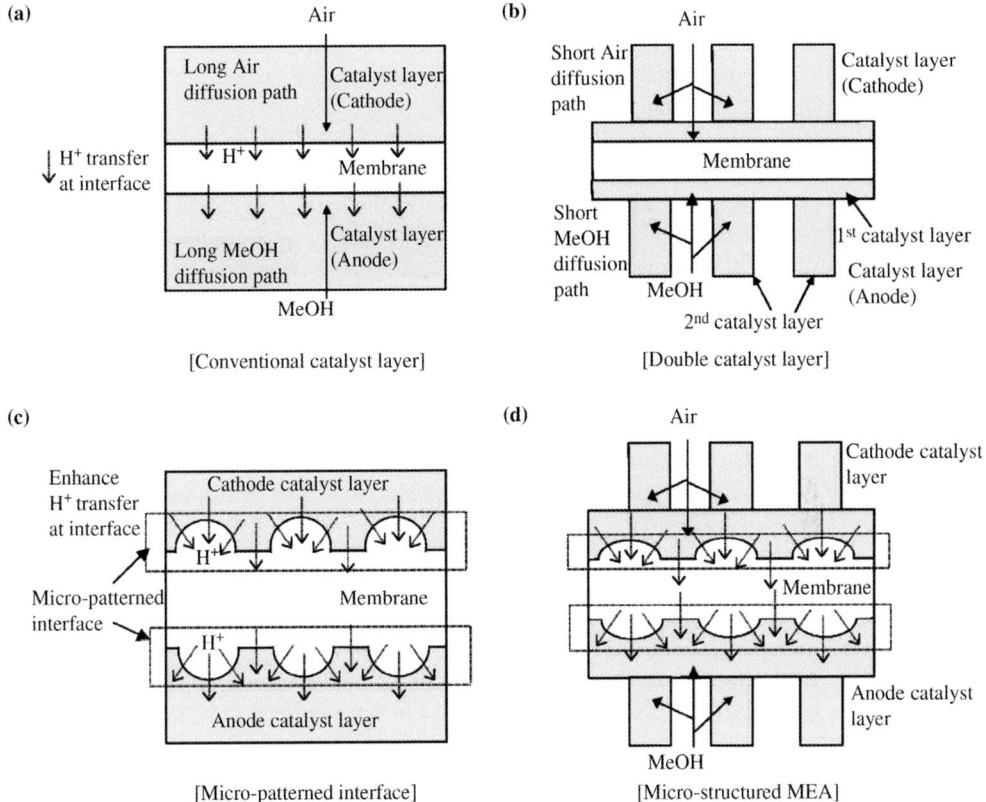

Figure 11.14 Design of microstructured MEA: (a) conventional single catalyst layer without micropattern; (b) double catalyst layer; (c) micropatterned interface; and (d) double catalyst layer with micropatterned interface [74].

micropatterned interface was generated by forming a micropattern on the membrane surface prior to catalyst coating. Using current-voltage polarization and electrochemical impedance spectroscopy, they reported that the mass transfer in the catalyst layer was significantly improved owing to the large pores in the second catalyst layer, and that the cathode reaction rate was not influenced by the micropatterned interface while the anode reaction rate was profoundly increased. The introduction of a double-layer structure resulted in a power density increase of 18%, and the generation of the micropatterned interface promoted an additional power density increase up to 23% at 50 °C and 0.4 V.

During the electrode fabrication process, hot-pressing and subsequent treatment are also vital for MEA performance and durability. Liang et al. [75] investigated the effect of hot-pressing duration on MEA performance and durability. They found that a longer duration (about 60 min) yielded significantly improved MEA durability over a typical shorter one (about 3 min). This was attributed to the fact that the longer duration could result in better interfacial binding between the electrodes and the membrane, which could reduce the risk of catalyst layer delamination. Also, they found that the cathode catalyst particle size could be increased in the case of longer hot-pressing, thus reducing the cathode performance. Jung et al. [76] also reported that the annealing of MEA at 130 °C, which was around the glass transition temperature of the recast Nafion binder, gave the electrode's highest proton conductivity and thus the highest performance. Liu et al. [77] investigated the effects of heat treatment and cell conditioning on the catalyst layer microstructure and performance, and found that during cell conditioning, the Nafion ionomer in the catalyst layer expanded and hence the pores shrank, leading to enhanced MOR kinetics and severe current limitation. Also, the internal resistance initially dropped dramatically but eventually leveled off, indicating that a steady-state of ionomer swelling had been reached. From CO stripping it was found that the Ru oxides at the catalyst surface were continuously reduced during cell conditioning, resulting in enhanced MOR kinetics. They also reported that heat treatment could alter the ionomer crystallinity and hence affect its swelling behavior, while the interfacial properties between the catalyst and ionomer, that is, at the triple-phase boundary, remained unaltered.

11.3.2
Novel Structures with Extended Reaction Zone

At the DMFC anode, a large volume of carbon dioxide is produced during cell operation. The conventional electrode, which consists of a catalyst layer, a microporous layer and a carbon diffusion substrate, can encounter severe mass transport problems [78]. In a conventional electrode structure, the catalyst ink may sink into the diffusion layer, resulting in lower catalyst utilization. In order to remedy this problem, some researchers have developed integrated electrode structures to extend the reaction zone to the diffusion layer. For example, Tsai et al. [79] developed PtRu decorated carbon nanotubes (CNTs) that were directly grown on the carbon cloth to form a catalyzed electrode. Using a thermal chemical vapor deposition method (CVD), dense CNTs 20–50 nm in diameter and 2–3 μm in length were successfully grown on the carbon cloth. PtRu nanoparticles 4–6 nm in diameter were then electrodeposited onto the surfaces of these CNTs. The electrode fabricated in such a way was used to make DMFC MEAs, and the performance results indicated a 27% improvement in specific peak power density at ambient temperature. They further reported that their electrode structure also performed better if the PtRu catalyst was sputtered directly onto the CNTs [80]. This enhanced performance was attributed to better electrocatalyst dispersion, a thinner electrocatalyst layer, as well as lower internal and interfacial resistances.

In order to facilitate the removal of CO_2 at the DMFC anode, Ti mesh with an open structure has been explored as an electrode. For example, Lim et al. [81] fabricated a Ti mesh-based anode by thermal decomposition, and the results showed that the thermally decomposed catalyst layer had large catalyst particle sizes with reduced Ru content (Figure 11.15). However, this catalyzed Ti mesh anode could yield a much higher power density in 0.5 M methanol solution as compared with a carbon-based

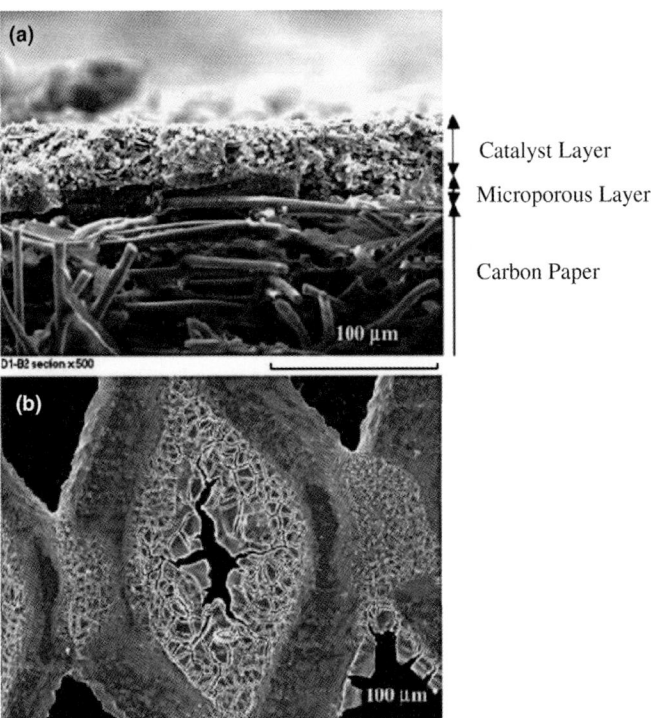

Figure 11.15 SEM micrographs of macroscopic view of fuel cell anodes: (a) PtRu black catalyst-coated GDL (microporous layer/Toray 090) and (b) thermal-decomposed Pt–Ru on Ti mesh [81].

anode. The same group also reported that the electrodeposited Ti mesh anode had a similar performance to that of a thermally decomposed one [82]. Shao et al. [83] also electrodeposited a thin PtRu catalyst layer onto Ti mesh to form PtRu/Ti anodes with varying Pt:Ru ratios. These electrodes were found to be very active for the MOR. The Ru surface coverages were optimized at about 9 at% for a DMFC operating at 20 °C and 11 at% at 60 °C. Their PtRu/Ti anode showed a performance comparable to that of the conventional carbon-based anode in a DMFC operating with 0.25 M or 0.5 M methanol solution and atmosphere oxygen gas at 90 °C. However, this kind of electrode structure was not suitable for high concentration methanol operation [81, 84].

Different three-dimensional substrates, such as reticulated vitreous carbon (RVC), uncompressed graphite felt (UGF), and Ti mesh, were also used as anode catalyst supports by Cheng et al. [85]. They electrodeposited the PtRu alloy onto three-dimensional substrates using Triton X-100/isopropanol aqueous micellar media and perforated counter electrodes. The resulting catalyst layers were found to be very mesoporous. Among the electrodes produced, PtRu/RVC yielded the highest power density. They also claimed that their electrode design could open up the possibility of reducing precious metal catalyst loading to 1 mg cm^{-2} in DMFC anodes. The synergy between the three-dimensional support and the PtRu catalyst preparation method could reduce the catalyst loading fourfold and at the same time generate improved power density compared with the CCM and CDM electrode structures [86].

11.4
Other Electrode Fabrication Methods for DMFCs

Catalyst layers prepared by the conventional ink-based hot-pressing method have some drawbacks, such as low catalyst utilization, a non-uniform catalyst layer, high resistance due to poor contact between the catalyst and the membrane, severe delamination after long operation, and so on. Many researchers have proposed new fabrication methods to enhance MEA performance, reproducibility, and durability. The following section briefly describes these new fabrication methods.

11.4.1
Electrodeposition

As mentioned above, the catalytic reaction takes place only in the three-phase interface within the catalyst layer. Using an electrodeposition method, the catalysts can be selectively deposited onto the electrode regions where ions and electrons are accessible. This is because only in those regions are there electron transport channels and metal ion transport channels resulting from the hydrophilicity of Nafion. Liang *et al.* [87] electrodeposited Pt nanowire particles into a Nafion membrane to form a platinum nanowire network in their Pt-Nafion integrated electrode, which not only could provide electron conduction pathways but also could function as the catalyst for the MOR, while the remaining part of the membrane without deposited Pt could continue to serve as the electrolyte. The results showed that this Pt-Nafion integrated electrode exhibited a higher electrochemical active surface area and a lower rate of methanol crossover, and therefore improved performance compared to a commercially available E-Tek electrode.

11.4.2
Sputtering

In order to reduce the cost of DMFC MEAs, low or ultra-low levels of catalyst loading are highly desirable. Many researchers have employed the sputtering technique to deposit the catalyst onto the surface of a membrane or carbon layer. A great advantage of the direct sputtering technique is that the catalyst remains at the membrane/electrode interface. Makino *et al.* [88] evaluated sputter-deposited Pt electrodes as cathodes in MEAs of DMFCs, and found that the mass activities of a sputter-deposited Pt electrode are 10 times higher than those of a paste-spread electrode. Xiu *et al.* [89] reported that adding a nanoscale Pt layer to the anode catalyst layer using the sputtering technique could significantly enhance cell performance.

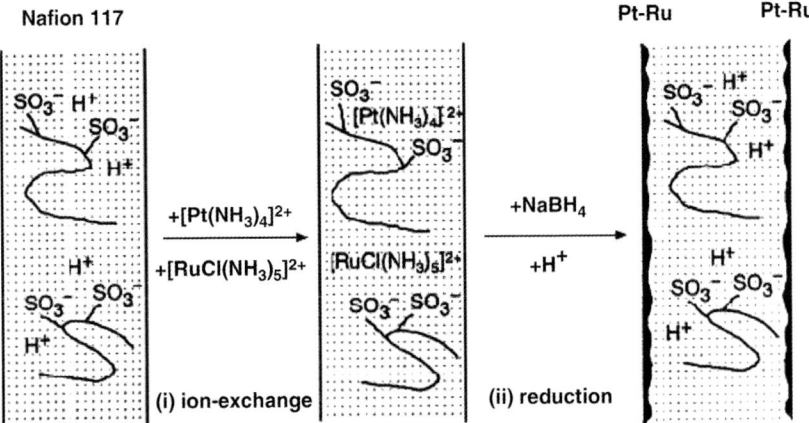

Figure 11.16 Schematic representation of the PtRu-PEM composite prepared by the IR method: (i) the ion-exchange step; (ii) the reduction step [90].

11.4.3
Chemical Reduction Method

In the chemical reduction method (Figure 11.16), a cation-exchange membrane with a pre-exchanged cationic catalyst metal species was immersed in a reduction solution. After the chemical reduction, the catalyst metal particles were formed at the outer surface of the membrane [90]. The DMFC MEA formed by this method could facilitate the release of evolving CO_2 and exhibit excellent adhesion and durability. This was mainly because the metal particles were embedded in the membrane surface. Although the active surface area of such a PtRu layer was small (a few microns in thickness), the DMFC anode with 3.2 mg cm^{-2} of catalyst loading exhibited a power density of 0.14 W cm^{-2} when the fuel cell was operated at 120 °C and 0.3 MPa.

11.4.4
Dry Production Techniques

Gülzow et al. [91] developed a low-cost MEA production technique (Figure 11.17) based on dry production techniques. Catalyst material prepared in a knife mill was atomized and sprayed in a nitrogen stream through a slit nozzle directly onto a membrane to form the electrode. Then this membrane electrode was hot rolled or pressed with the gas diffusion layers to form the MEA. This new fabrication technique allows simple production of MEAs and electrodes, and has a high degree of automation units which can be scaled up for industrial requirements. This process does not employ solvents and is both environmental friendly and time-saving. An additional advantage is that the process is very flexible and can meet other requirements of DMFC electrode fabrication, such as gradated layer structures or other materials.

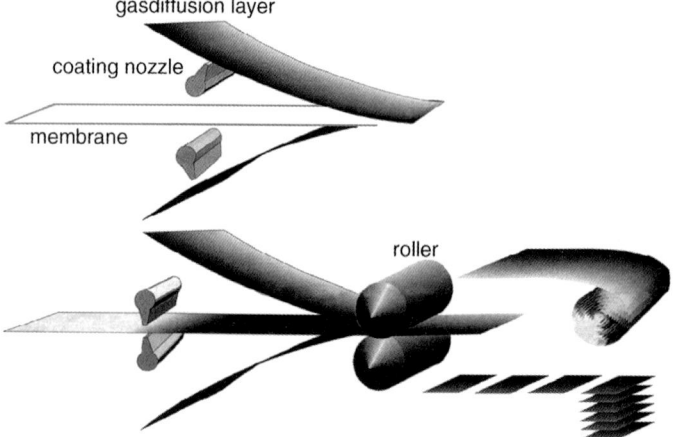

Figure 11.17 Schematic of the dry production technique for PEMFC and DMFC MEAs [91].

11.4.5
Glue Method

It has been reported [92] that after a long period of DMFC operation, MEA delamination was severe at both the anode and cathode interfaces if the MEA was fabricated by the conventional hot-pressing method. In order to achieve and retain better interfacial contact between the electrode and membrane, Liang [93] presented a simple glue method for fabricating DMFC MEAs, in which a binding agent consisting of Nafion solution was introduced between the polymer electrolyte membrane and the anode/cathode. The introduced binding agent could provide better adhesion and a stronger binding force between the membrane and electrode, thereby facilitating better interfacial contact between the electrode and the Nafion membrane without conventional dry hot-pressing at high pressure and temperature. It was shown that the MEA fabricated by this glue method was more stable in cell performance than that fabricated by the conventional hot-pressing method.

11.4.6
Sedimentation Method

In order to produce a homogeneous catalyst layer, DMFC electrodes have also been prepared by a sedimentation method [94]. In this process, a suspended catalyst ink was filtered through a polycarbonate film, yielding a thin catalyst layer on this film. This thin catalyst film was transferred onto a gas diffusion layer by applying pressure to the assembly, and the filter film was then peeled off. The reproducibility of this method is very high. Compared to the conventional brushing technique, the output power density of the MEA fabricated by the sedimentation method was 70% higher.

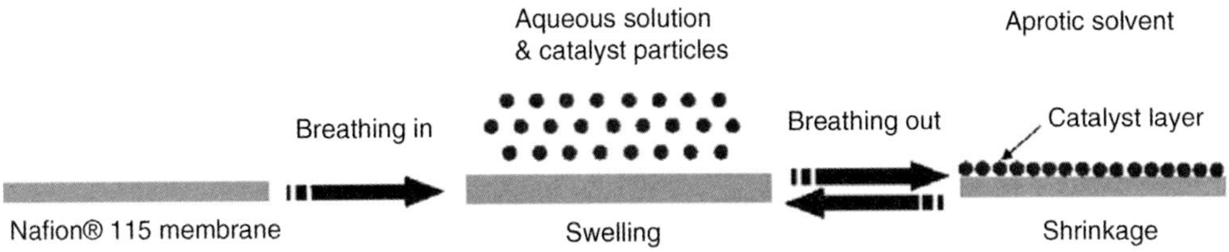

                    Aqueous solution   Aprotic solvent
                    & catalyst particles

        Breathing in   • • • • • • • •   Breathing out   Catalyst layer
                       • • • • • • • •   ●●●●●●●●●●●●●
                       • • • • • • •
         ────▶   ────────────   ◀────   ─────────────
Nafion® 115 membrane   Swelling   Shrinkage

Figure 11.18 Schematic illustration for loading the catalysts onto the proton exchange membrane by the breathing process [95].

11.4.7
Breathing Process

Yu et al. [95] introduced a "breathing process" for producing highly active and reproducible catalyst layers. As shown in Figure 11.18, a Nafion 115 membrane was swollen by immersion in an aqueous solution containing the PtRu catalyst (breathing in), and then the swollen Nafion 115 membrane was shrunk by placing it into an aprotic solvent to drive the aqueous solution out (breathing out). The amount of catalyst loaded in the membrane was controlled by repeating the breathing processes. This method could produce a membrane electrode with good connection between the electrolyte membrane and the catalyst particles. The MEA prepared with this membrane electrode showed a much better performance when compared to one produced by the conventional loading method.

11.5
Summary

DMFC performance is mainly determined by the structures and properties of the catalyst layers. The proper selection of catalyst layer components, structures, and fabrication methods are the most important factors in obtaining a MEA with high performance and stability. The pore structure and hydrophilic/hydrophobic property of the catalyst layer should be carefully tailored in order to improve DMFC performance. In addition, structural optimization of the catalyst layer and MEA with respect to DMFC performance must take into consideration the complex interactions between different components.

References

1 Hogarth, M.P. and Hards, G.A. (1996) Direct methanol fuel cells: technological advances and further requirements. *Platinum Met. Rev.*, **40**, 150–159.

2 Hamnett, A. (1997) Mechanism and electrocatalysis in the direct methanol fuel cell. *Cat. Today*, **38**, 445–457.

3 Arico, A.S., Srinivasan, S. and Antonucci, V. (2001) DMFCs: from fundamental aspects to technology development. *Fuel Cells*, **1**, 133–161.

4 Lamy, C., Leger, J.M. and Srinivasan, S. (2001) Direct methanol fuel cells: from a twentieth century electrochemist's dream

to a twenty-first century emerging technology, in *Modern Aspects of Electrochemistry*, **34** (eds J.O'.M. Bockris, B.E. Conway and R.E. White), Kluwer Academic Pub., New York, pp. 53–115.

5 Hogarth, M.P. and Ralph, T.R. (2002) Catalysis for low temperature fuel cells Part III: challenges for the direct methanol fuel cell. *Platinum Met. Rev.*, **46**, 146–164.

6 Dillon, R., Srinivasan, S., Arico, A.S. and Antonucci, V. (2004) International activities in DMFC R&D: status of technologies and potential applications. *J. Power Sources*, **127**, 112–126.

7 Gasteiger, H.A., Markovic, N., Ross, J.P.N. and Cairns, E.J. (1994) Temperature-dependent methanol electro-oxidation on well-characterized Pt-Ru Alloys. *J. Electrochem. Soc.*, **141** (7), 1795–1803.

8 Arico, A.S., Antonucci, P.L., Modica, E., Baglio, V., Kim, H. and Antonucci, V. (2002) Effect of Pt—Ru alloy composition on high-temperature methanol electro-oxidation. *Electrochim. Acta*, **47** (22–23), 3723–3732.

9 Takasu, Y., Itaya, H., Iwazaki, T., Miyoshi, R., Ohnuma, T., Sugimoto, W. and Murakami, Y. (2001) Size effects of ultrafine Pt-Ru particles on the electrocatalytic oxidation of methanol. *Chem. Commun.*, **4**, 341–342.

10 Takasu, Y., Kawaguchi, T., Sugimoto, W. and Murakami, Y. (2003) Effects of the surface area of carbon support on the characteristics of highly-dispersed Pt—Ru particles as catalysts for methanol oxidation. *Electrochim. Acta*, **48** (25–26), 3861–3868.

11 Bock, C., Paquet, C., Couillard, M., Botton, G.A. and MacDougall, B.R. (2004) Size-selected synthesis of PtRu nano-catalysts: reaction and size control mechanism. *J. Am Chem. Soc.*, **126** (25), 8028–8037.

12 Su, Y., Xue, X.Z., Xu, W.L., Liu, C.P., Xing, W., Zhou, X.C., Tian, T. and Lu, T.H. (2006) Structure-activity relationship of surfactant for preparing DMFC anodic catalyst. *Electrochim. Acta*, **51** (20), 4316–4323.

13 Park, G.G., Yang, T.H., Yoon, Y.G., Lee, W.Y. and Kim, C.S. (2003) Pore size effect of the DMFC catalyst supported on porous materials. *Int. J. Hydrogen Energy*, **28** (6), 645–650.

14 Arico, A.S., Creti, P., Kim, H., Mantegna, R., Giordano, N. and Antonucci, V. (1996) Analysis of the electrochemical characteristics of a direct methanol fuel cell based on a Pt-Ru/C anode catalyst. *J. Electrochem. Soc.*, **143** (12), 3950–3959.

15 Frelink, T., Visscher, W. and van Veen, J.A.R. (1996) Measurement of the Ru surface content of, electrocodeposited PtRu electrodes with the electrochemical quartz crystal microbalance: implications for methanol and CO electrooxidation. *Langmuir*, **12**, 3702–3708.

16 Chrzanowski, W. and Wieckowski, A. (1997) Ultrathin films of ruthenium on low index platinum single crystal surfaces: an electrochemical study. *Langmuir*, **13** (22), 5974–5978.

17 Chrzanowski, W. and Wieckowski, A. (1998) Surface structure effects in platinum/ruthenium methanol oxidation electrocatalysis. *Langmuir*, **14** (8), 1967–1970.

18 Arico, A.S., Creti, P., Modica, E., Monforte, G., Baglio, V. and Antonucci, V. (2000) Investigation of direct methanol fuel cells based on unsupported Pt-Ru anode catalysts with different chemical properties. *Electrochim. Acta*, **45** (25–26), 4319–4328.

19 Arico, A.S., Baglio, V., Di Blasi, A., Modica, E., Antonucci, P.L. and Antonucci, V. (2003) Analysis of the high-temperature methanol oxidation behaviour at carbon-supported Pt-Ru catalysts. *J. Electroanal. Chem.*, **557**, 167–176.

20 Bock, C., Blakely, M.A. and MacDougall, B. (2005) Characteristics of adsorbed CO and CH_3OH oxidation reactions for complex Pt/Ru catalyst systems. *Electrochimica Acta*, **50** (12), 2401–2414.

21 Choi, W.C. and Woo, S.I. (2003) Bimetallic Pt-Ru nanowire network for anode

material in a direct-methanol fuel cell. *J. Power Sources*, **124** (2), 420–425.

22 Peng, X., Koczkur, K. and Chen, A. (2007) Synthesis and characterization of ruthenium-decorated nanoporous platinum materials. *Nanotechnology*, **18** (30), 305605.

23 Rolison, D.R., Hagans, P.L., Swider, K.E. and Long, J.W. (1999) Role of hydrous ruthenium oxide in Pt-Ru direct methanol fuel cell anode electrocatalysts: the importance of mixed electron/proton conductivity. *Langmuir*, **15** (3), 774–779.

24 Thomas, S.C., Ren, X. and Gottesfeld, S. (1999) Influence of ionomer content in catalyst layers on direct methanol fuel cell performance. *J. Electrochem. Soc.*, **146** (12), 4354–4359.

25 Dinh, H.N., Ren, X., Garzon, F.H., Piotr, Z. and Gottesfeld, S. (2000) Electrocatalysis in direct methanol fuel cells: in-situ probing of PtRu anode catalyst surfaces. *J. Electroanal. Chem.*, **491** (1–2), 222–233.

26 Long, J.W., Stroud, R.M., Swider-Lyons, K.E. and Rolison, D.R. (2000) How to make electrocatalysts more active for direct methanol oxidation avoid PtRu bimetallic alloys!. *J. Phys. Chem. B*, **104** (42), 9772–9776.

27 Sirk, A.H.C., Hill, J.M., Kung, S.K.Y. and Birss, V.I. (2004) Effect of redox state of PtRu electrocatalysts on methanol oxidation activity. *J. Phys. Chem. B*, **108** (2), 689–695.

28 Lu, Q., Yang, B., Zhuang, L. and Lu, J. (2005) Anodic activation of PtRu/C catalysts for methanol oxidation. *J. Phys. Chem. B*, **109** (5), 1715–1722.

29 Lu, Q., Yang, B., Zhuang, L. and Lu, J. (2005) Pattern recognition on the structure-activity relationship of nano Pt-Ru catalysts: methodology and preliminary demonstration. *J. Phys. Chem. B*, **109** (18), 8873–8879.

30 Scheiba, F., Scholz, M., Cao, L., Schafranek, R., Roth, C., Cremers, C., Qiu, X., Stimming, U. and Fuess, H. (2006) On the suitability of hydrous ruthenium oxide supports to enhance intrinsic proton conductivity in DMFC anodes. *Fuel cells*, **6** (6), 439–446.

31 Jeon, M.K., Won, J.Y. and Woo, S.I. (2007) Improved performance of direct methanol fuel cells by anodic treatment. *Electrochem. Solid-State Lett.*, **10** (1), B23–B25.

32 Liu, L., Pu, C., Viswanathan, R., Fan, Q., Liu, R. and Smotkin, E.S. (1998) Carbon supported and unsupported Pt-Ru anodes for liquid feed direct methanol fuel cells. *Electrochim. Acta*, **43** (24), 3657–3663.

33 Giordano, N., Passalacqua, E., Pino, L., Arico, A.S., Antonucci, V., Vivaldi, M. and Kinoshita, K. (1991) Analysis of platinum particle size and oxygen reduction in phosphoric acid. *Electrochim. Acta*, **36** (13), 1979–1984.

34 Takasu, Y., Iwazaki, T., Sugimoto, W. and Murakami, Y. (2000) Size effects of platinum particles on the electro-oxidation of methanol in an aqueous solution of $HClO_4$. *Electrochem. Commun.*, **2** (9), 671–674.

35 Wen, Z., Liu, J. and Li, J. (2008) Core/shell Pt/C nanoparticles embedded in mesoporous carbon as a methanol-tolerant cathode catalyst in direct methanol fuel cells. *Adv. Mater.*, **20** (4), 743–747.

36 Schmidt, T.J., Paulus, U.A., Gasteiger, H.A. and Behm, R.J. (2001) The oxygen reduction reaction on a Pt/carbon fuel cell catalyst in the presence of chloride anions. *J. Electroanal. Chem.*, **508** (1–2), 41–47.

37 Uhm, S., Noh, T., Kim, Y.D. and Lee, J. (2008) Enhancement of methanol tolerance in DMFC cathode: addition of chloride ions. *ChemPhysChem*, **9** (10), 1425–1429.

38 Kim, J.H., Ha, H.Y., Oh, I.H., Hong, S.A., Kim, H.N. and Lee, H.I. (2004) Electrochemical studies of DMFC anodes with different ionomer content. *Electrochim. Acta*, **50** (2–3), 801–806.

39 Zhao, X., Fan, X., Wang, S., Yang, S., Yi, B., Xin, Q. and Sun, G. (2005) Determination of ionic resistance and optimal composition in the anodic catalyst layers of

DMFC using AC impedance. *Int. J. Hydrogen Energy*, **30** (9), 1003–1010.
40 Lee, J.B., Park, Y.K., Yang, O.B., Kang, Y., Jun, K.W., Lee, Y.J., Kim, H.Y., Lee, K.H. and Choi, W.C. (2006) Synthesis of porous carbons having surface functional groups and their application to direct-methanol fuel cells. *J. Power Sources*, **158** (2), 1251–1255.
41 Gojkovic, S.L. and Vidakovic, T.R. (2001) Methanol oxidation on an ink type electrode using Pt supported on high area carbons. *Electrochim. Acta*, **47** (4), 633–642.
42 Chu, Y.H., Shul, Y.G., Choi, W.C., Woo, S.I. and Han, H.S. (2003) Evaluation of the Nafion effect on the activity of Pt-Ru electrocatalysts for the electro-oxidation of methanol. *J. Power Sources*, **118** (1–2), 334–341.
43 Sarma, L.S., Lin, T.D., Tsai, Y.W., Chen, J.M. and Hwang, B.J. (2005) Carbon-supported Pt-Ru catalysts prepared by the Nafion stabilized alcohol-reduction method for application in direct methanol fuel cells. *J. Power Sources*, **139** (1–2), 44–54.
44 Aricò, A.S., Shukla, A.K., El-Khatib, K.M., Cretì, P. and Antonucci, V. (1999) Effect of carbon-supported and unsupported Pt–Ru anodes on the performance of solid-polymer-electrolyte direct methanol fuel cells. *J. Appl. Electrochem.*, **29** (6), 673–678.
45 Uchida, M., Aoyama, Y., Eda, N. and Ohta, A. (1995) New preparation method for polymer-electrolyte fuel cells. *J. Electrochem. Soc.*, **142** (2), 463–468.
46 Uchida, M., Fukuoka, Y., Sugawara, Y., Eda, N. and Ohta, A. (1996) Effects of microstructure of carbon support in the catalyst layer on the performance of polymer-electrolyte fuel cells. *J. Electrochem. Soc.*, **143** (7), 2245–2252.
47 Arico, A.S., Creti, P., Antonucci, P.L., Cho, J., Kim, H. and Antonucci, V. (1998) Optimization of operating parameters of a direct methanol fuel cell and physico-chemical investigation of catalyst-electrolyte interface. *Electrochim. Acta*, **43** (24), 3719–3729.

48 Kim, J.H., Ha, H.Y., Oh, I.H., Hong, S.A. and Lee, H.I. (2004) Influence of the solvent in anode catalyst ink on the performance of a direct methanol fuel cell. *J. Power Sources*, **135** (1–2), 29–35.
49 Wang, S., Sun, G., Wu, Z. and Xin, Q. (2007) Effect of Nafion(R) ionomer aggregation on the structure of the cathode catalyst layer of a DMFC. *J. Power Sources*, **165** (1), 128–133.
50 Park, C.H., Scibioh, M.A., Kim, H.J., Oh, I.H., Hong, S.A. and Ha, H.Y. (2006) Modification of carbon support to enhance performance of direct methanol fuel cell. *J. Power Sources*, **162** (2), 1023–1028.
51 Lee, J.B., Park, Y.K., Yang, O.B., Kang, Y., Jun, K.W., Lee, Y.J., Kim, H.Y., Lee, K.H. and Choi, W.C. (2006) Synthesis of porous carbons having surface functional groups and their application to direct-methanol fuel cells. *J. Power Sources*, **158** (2), 1251–1255.
52 Kuroki, H. and Yamaguchi, T. (2006) Nanoscale morphological control of anode electrodes by grafting of methylsulfonic acid groups onto platinum--ruthenium-supported carbon blacks. *J. Electrochem. Soc.*, **153** (7), A1417–A1423.
53 Rao, V., Simonov, P.A., Savinova, E.R., Plaksin, G.V., Cherepanova, S.V., Kryukova, G.N. and Stimming, U. (2005) The influence of carbon support porosity on the activity of PtRu/Sibunit anode catalysts for methanol oxidation. *J. Power Sources*, **145** (2), 178–187.
54 Arbizzani, C., Beninati, S., Manferrari, E., Soavi, F. and Mastragostino, M. (2007) Cryo- and xerogel carbon supported PtRu for DMFC anodes. *J. Power Sources*, **172** (2), 578–586.
55 Wu, P., Li, B., Du, H., Gan, L., Kang, F. and Zeng, Y. (2008) The influences of multi-walled carbon nanotube addition to the anode on the performance of direct methanol fuel cells. *J. Power Sources*, **184** (2), 381–384.
56 Tominaka, S., Akiyama, N., Momma, T. and Osaka, T. (2007) An impedance analysis on properties of DMFC catalyst

layers based on primary and secondary pores. *J. Electrochem. Soc.*, **154** (9), B902–B909.

57 Fernandez, R., Ferreira-Aparicio, P. and Daza, L. (2005) PEMFC electrode preparation: influence of the solvent composition and evaporation rate on the catalytic layer microstructure. *J. Power Sources*, **151**, 18–24.

58 Malek, K., Eikerling, M., Wang, Q., Navessin, T. and Liu, Z. (2007) Self-organization in catalyst layers of polymer electrolyte fuel cells. *J. Phys. Chem. C*, **111** (36), 13627–13634.

59 Lee, J.S., Han, K.I., Park, S.O., Kim, H.N. and Kim, H. (2004) Performance and impedance under various catalyst layer thicknesses in DMFC. *Electrochim. Acta*, **50** (2–3), 807–810.

60 Han, K., Lee, J. and Kim, H. (2006) Preparation and characterization of high metal content Pt-Ru alloy catalysts on various carbon blacks for DMFCs. *Electrochim. Acta*, **52** (4), 1697–1702.

61 Nordlund, J., Roessler, A. and Lindbergh, G. (2002) The influence of electrode morphology on the performance of a DMFC anode. *J. Appl. Electrochem.*, **32** (3), 259–265.

62 Wei, Z., Wang, S., Yi, B., Liu, J., Chen, L., Zhou, W., Li, W. and Xin, Q. (2002) Influence of electrode structure on the performance of a direct methanol fuel cell. *J. Power Sources*, **106** (1–2), 364–369.

63 Liu, Y.C., Qiu, X.P., Huang, Y.Q. and Zhu, W.T. (2002) Mesocarbon microbeads supported Pt-Ru catalysts for electrochemical oxidation of methanol. *J. Power Sources*, **111** (1), 160–164.

64 Chai, G.S., Yoon, S.B., Yu, J.S., Choi, J.H. and Sung, Y.E. (2004) Ordered porous carbons with tunable pore sizes as catalyst supports in direct methanol fuel cell. *J. Phys. Chem. B*, **108** (22), 7074–7079.

65 Bang, J.H., Han, K., Skrabalak, S.E., Kim, H. and Suslick, K.S. (2007) Porous carbon supports prepared by ultrasonic spray pyrolysis for direct methanol fuel cell

electrodes. *J. Phys. Chem. C*, **111** (29), 10959–10964.

66 Reshetenko, T.V., Kim, H.T. and Kweon, H.J. (2007) Cathode structure optimization for air-breathing DMFC by application of pore-forming agents. *J. Power Sources*, **171** (2), 433–440.

67 Reshetenko, T.V., Kim, H.T. and Kweon, H.J. (2008) Modification of cathode structure by introduction of CNT for air-breathing DMFC. *Electrochim. Acta*, **53** (7), 3043–3049.

68 Lim, C., Allen, R.G. and Scott, K. (2006) Effect of dispersion methods of an unsupported Pt-Ru black anode catalyst on the power performance of a direct methanol fuel cell. *J. Power Sources*, **161** (1), 11–18.

69 Song, J.M., Suzuki, S., Uchida, H. and Watanabe, M. (2006) Preparation of high catalyst utilization electrodes for polymer electrolyte fuel cells. *Langmuir*, **22** (14), 6422–6428.

70 Song, S., Zhou, W., Liang, Z., Cai, R., Sun, G. and Xin, Q. (2005) The effect of methanol and ethanol cross-over on the performance of PtRu/C-based anode DAFCs. *Appl. Catal. B: Environ.*, **55** (1), 65–72.

71 Reshetenko, T.V., Kim, H.T., Lee, H., Jang, M. and Kweon, H.J. (2006) Performance of a direct methanol fuel cell (DMFC) at low temperature: cathode optimization. *J. Power Sources*, **160** (2), 925–932.

72 Mao, Q., Sun, G., Wang, S., Sun, H., Wang, G., Gao, Y., Ye, A., Tian, Y. and Xin, Q. (2007) Comparative studies of configurations and preparation methods for direct methanol fuel cell electrodes. *Electrochim. Acta*, **52** (24), 6763–6770.

73 Liu, F. and Wang, C.Y. (2006) Optimization of cathode catalyst layer for direct methanol fuel cells: Part I. experimental investigation. *Electrochim. Acta*, **52** (3), 1417–1425.

74 Kim, H.T., Reshentenko, T.V. and Kweon, H.J. (2007) Microstructured membrane electrode assembly for direct methanol

fuel cell. *J. Electrochem. Soc.*, **154** (10), B1034–B1040.
75 Liang, Z.X., Zhao, T.S., Xu, C. and Xu, J.B. (2007) Microscopic characterizations of membrane electrode assemblies prepared under different hot-pressing conditions. *Electrochim. Acta*, **53** (2), 894–902.
76 Jung, H.Y., Cho, K.Y., Lee, Y.M., Park, J.K., Choi, J.H. and Sung, Y.E. (2007) Influence of annealing of membrane electrode assembly (MEA) on performance of direct methanol fuel cell (DMFC). *J. Power Sources*, **163** (2), 952–956.
77 Liu, F. and Wang, C.Y. (2005) Variations in interfacial properties during cell conditioning and influence of heat-treatment of ionomer on the characteristics of direct methanol fuel cells. *Electrochim. Acta*, **50** (6), 1413–1422.
78 Argyropoulos, P., Scott, K. and Taama, W.M. (1999) Carbon dioxide evolution patterns in direct methanol fuel cells. *Electrochim. Acta*, **44** (20), 3575–3584.
79 Tsai, M.C., Yeh, T.K., Chen, C.Y. and Tsai, C.H. (2007) A catalytic gas diffusion layer for improving the efficiency of a direct methanol fuel cell. *Electrochem. Commun.*, **9** (9), 2299–2303.
80 Wang, C.H., Du, H.Y., Tsai, Y.T., Chen, C.P., Huang, C.J., Chen, L.C., Chen, K.H. and Shih, H.C. (2007) High performance of low electrocatalysts loading on CNT directly grown on carbon cloth for DMFC. *J. Power Sources*, **171** (1), 55–62.
81 Lim, C., Scott, K., Allen, R.G. and Roy, S. (2004) Direct methanol fuel cells using thermally catalysed Ti mesh. *J. Appl. Electrochem.*, **34** (9), 929–933.
82 Allen, R.G., Lim, C., Yang, L.X., Scott, K. and Roy, S. (2005) Novel anode structure for the direct methanol fuel cell. *J. Power Sources*, **143** (1–2), 142–149.
83 Shao, Z.G., Zhu, F., Lin, W.F., Christensen, P.A. and Zhang, H. (2006) PtRu/Ti anodes with varying Pt:Ru ratio prepared by electrodeposition for the direct methanol fuel cell. *Phys. Chem. Chem. Phys.*, **8** (23), 2720–2726.
84 Shao, Z.G., Zhu, F., Lin, W.F., Christensen, P.A., Zhang, H. and Yi, B. (2006) Preparation and characterization of new anodes based on Ti mesh for direct methanol fuel cells. *J. Electrochem. Soc.*, **153** (8), A1575–A1583.
85 Cheng, T. and Gyenge, E. (2008) Direct methanol fuel cells with reticulated vitreous carbon, uncompressed graphite felt and Ti mesh anodes. *J. Appl. Electrochem.*, **38** (1), 51–62.
86 Cheng, T.T. and Gyenge, E.L. (2008) Efficient anodes for direct methanol and formic acid fuel cells: the synergy between catalyst and three-dimensional support. *J. Electrochem. Soc.*, **155** (8), B819–B828.
87 Liang, Z.X. and Zhao, T.S. (2007) New DMFC anode structure consisting of platinum nanowires, deposited into a nafion membrane. *J. Phys. Chem. C*, **111** (22), 8128–8134.
88 Makino, K., Furukawa, K., Okajima, K. and Sudoh, M. (2005) Optimization of the sputter-deposited platinum cathode for a direct methanol fuel cell. *Electrochim. Acta*, **51** (5), 961–965.
89 Xiu, Y.K. and Nakagawa, N. (2004) Performance of a DMFC with a sputtered Pt layer on the electrode/electrolyte interface of the anode. *J. Electrochem. Soc.*, **151** (9), A1483–A1488.
90 Fujiwara, N., Yasuda, K., Ioroi, T., Siroma, Z. and Miyazaki, Y. (2002) Preparation of platinum-ruthenium onto solid polymer electrolyte membrane and the application to a DMFC anode. *Electrochim. Acta*, **47** (25), 4079–4084.
91 Gulzow, E., Schulze, M., Wagner, N., Kaz, T., Reissner, R., Steinhilber, G. and Schneider, A. (2000) Dry layer preparation and characterisation of polymer electrolyte fuel cell components. *J. Power Sources*, **86** (1–2), 352–362.
92 Liu, J., Zhou, Z., Zhao, X., Xin, Q., Sun, G. and Yi, B. (2004) Studies on performance degradation of a direct methanol fuel cell (DMFC) in life test. *Phys. Chem. Chem. Phys.*, **6** (1), 134–137.

93 Liang, Z.X., Zhao, T.S. and Prabhuram, J. (2006) A glue method for fabricating membrane electrode assemblies for direct methanol fuel cells. *Electrochim. Acta*, **51** (28), 6412–6418.

94 Liu, J.H., Jeon, M.K., Choi, W.C. and Woo, S.I. (2004) Highly-optimized membrane electrode assembly for direct methanol fuel cell prepared by sedimentation method. *J. Power Sources*, **137** (2), 222–227.

95 Yu, K.C., Kim, W.J. and Chung, C.H. (2006) Utilization of Pt/Ru catalysts in MEA for fuel cell application by breathing process of proton exchange membrane. *J. Power Sources*, **163** (1), 34–40.

12
Local Current Distribution in Direct Methanol Fuel Cells
Andrei A. Kulikovsky and Klaus Wippermann

12.1
Introduction

The density of a liquid is roughly three orders of magnitude larger than the density of a gas under normal atmospheric conditions. This simple physics stands behind the growing worldwide interest in liquid-fed direct methanol fuel cells (DMFCs). The word 'direct' in this context emphasizes the fact that these cells convert methanol to electric current directly, without an intermediate methanol–hydrogen conversion (although sometimes this happens *inside* a DMFC, as we will see below).

Since 1992 the number of publications on DMFC science and technology has shown steady exponential growth, approximately doubling every two years [1] (Figure 12.1). This Moore's law is a signature of emerging technology.

DMFCs have a number of generic problems, though. Having the advantage of being liquid, methanol yields to hydrogen in terms of fuel-to-electricity efficiency. The kinetics of hydrogen oxidation is extremely fast; breaking an H–H bond and ionization of an H atom does not take much energy. In contrast, ionization of a CH_3OH molecule is much more costly. The second problem is crossover: methanol easily penetrates through the polymer electrolyte membrane and burns down on the cathode side without production of useful current. Last but not least, a big problem hindering the wide-spread use of DMFCs is their relatively fast rate of aging.

This chapter is devoted to local current distribution (LCD) in DMFCs. This distribution is of great interest since the rates of virtually all aging processes in the cell increase with the local current (see below). Ideally, the cell should operate at uniform LCD; in reality, however, LCD is usually strongly nonhomogeneous.

Several factors contribute to this nonhomogeneity. Methanol oxidation reaction produces one CO_2 molecule per each molecule of methanol. At typical working temperatures in the range 60–90 °C the solubility of CO_2 in water is small and gaseous CO_2 bubbles occupy a substantial fraction of the anode volume. Literally speaking, a DMFC anode looks like a boiling kettle: CO_2 bubbles produced in the anode catalyst layer enter the anode flow field and leave the cell through the

---

Electrocatalysis of Direct Methanol Fuel Cells. Edited by Hansan Liu and Jiujun Zhang
Copyright © 2009 WILEY-VCH Verlag GmbH & Co. KGaA, Weinheim
ISBN: 978-3-527-32377-7

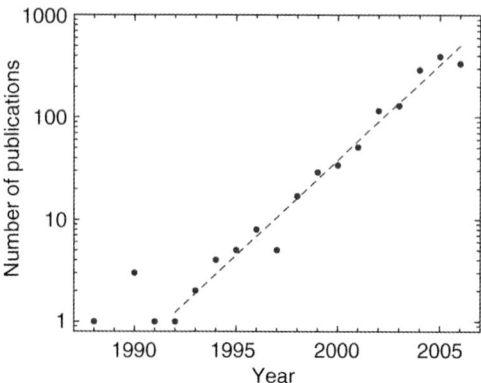

Figure 12.1 Number of publications on DMFC science and technology (data from Scopus; the search string is 'DMFC').

anode exhaust. The details of this picture strongly depend on the flow field configuration, mean current density and operating conditions.

Electrochemical and chemical reactions in DMFC produce heat; local cell temperature thus depends on local current. The growth in local temperature, in turn, increases the rate of reactions thereby increasing local current. This positive feedback loop can provoke thermal instability of cell operation [2]. CO_2 bubbles may stop instability development, blocking methanol transport to the unstable domains. One, therefore, may expect chaotic formation of high-current and 'dead' spots over the cell active area.

This brief discussion shows that in general one cannot expect uniform distribution of local current in a DMFC. Homogenization of this distribution is a difficult task, which requires coordinated efforts of experimentalists and theorists. The situation, however, is not so dramatic if the cell operates at lower temperatures and smaller currents. At a temperature about 30–40 °C the solubility of CO_2 in water is high and the effect of gaseous bubbles on cell performance is much smaller. This is also true for the regime with small currents, since the rate of CO_2 production is proportional to cell current.

In recent years a number of models of DMFCs based on computational fluid dynamics (CFD) have been developed [3–7]. These models include a rough description of two-phase transport of reactants in the anode backing layer, Tafel [4, 7] or non-Tafel [3, 5, 6] kinetics of electrochemical reactions on both sides of the cell, and thermal effects [3]. However, due to a lack of experimental data no attempts have been made to validate simulated distributions of local parameters in DMFCs against measured values.

Due to its enormous flexibility the CFD approach creates a temptation to take into account all imaginable processes in a cell. This leads to a model with a large number of poorly known parameters (e.g., a very detailed model [3] contains nearly 60 transport and kinetic parameters). In this situation expensive and time-consuming CFD modeling does not fully justify hopes, giving a qualitative rather than a quantitative picture of the processes in the cell.

The analytical models we will use below do not pretend to describe real cells with complex flow fields in practically relevant operating conditions. Instead, these models consider the simplest 'test' cell with straight channels on both sides. These models employ a set of simplified assumptions most adequate to the situation under consideration. This approach leaves aside many details important for practical cell engineering. On the other hand, models of that type highlight the most essential processes, which determine cell function. Analysis of the resulting equations gives useful insights into the basic mechanisms of cell operation; sometimes this analysis reveals novel effects.

In this chapter we will focus on the low-current regime of cell operation. Study of this regime is a first step towards a general understanding of nonuniformities in DMFC. Below we will see that even in the 'modest' low-current regime, DMFCs exhibit a number of unexpected effects, which may impact the technology. Here we report experimental and modeling studies of these effects.

12.2
Model

12.2.1
General Description

The DMFC model constructed below belongs to a class of quasi-2D models. Models of that type give analytical expressions or numerical curves for voltage losses due to the activation of electrochemical reactions and due to the transport of reactants across the cell and along the feed channels. Evidently, in the limit of infinite stoichiometries any quasi-2D model must transform into a 1D through-plane model.

Our goal is to understand the coupling of species transport across the cell to their transport along the feed channels. The simplest system which allows us to do that is a cell with two straight channels on both sides (Figure 12.2). The general approach to model construction is as follows.

First we formulate and solve a through-plane model, assuming that the local current j and local oxygen c_{ox} and methanol c_M concentrations in the respective channels are known. The next step is to introduce a second dimension to the problem. This is done taking into account the variation of j, c_{ox} and c_M with the coordinate along the channels z (Figure 12.2). For $c_{ox}(z)$ and $c_M(z)$ we write simple mass balance equations; the equation for $j(z)$ follows from the condition of equipotentiality of electrodes. These equations relate $c_{ox}(z)$ and $c_M(z)$ to their inlet values, and $j(z)$ to J, the mean current density in the cell.

The through-plane part of the model can usually be solved analytically. In certain cases an analytical solution can then be obtained for the channel model. These cases are of special interest, since fully analytical solutions to a 2D problem give a transparent physical picture of the interplay of small-scale (x-directed) and large-scale (z-directed) transport processes (Figure 12.2).

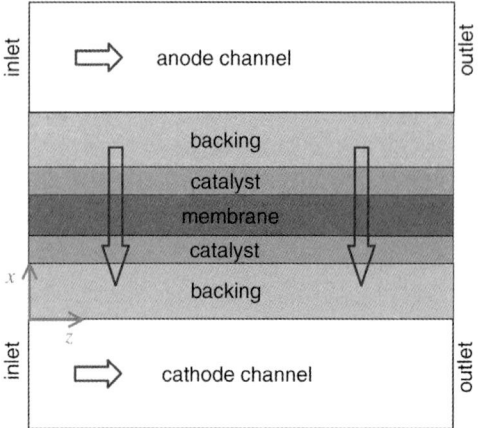

Figure 12.2 Cell cross-section with the straight channels on either side.

12.2.2
Basic Assumptions

The assumptions to the through-plane part of the model are [8]

1. The mechanism of methanol and oxygen transport in the respective backing layer is diffusion due to the concentration gradient.
2. Transport of methanol through the membrane is due to the concentration gradient. For typical DMFC currents below 200 mA cm^{-2} the electro-osmotic flux of methanol in the membrane is small compared with the diffusion flux [8].
3. The variation of oxygen and methanol concentration across the respective catalyst layer is neglected.
4. Kinetics of methanol oxidation and oxygen reduction obey Tafel's law.
5. Permeated methanol is completely consumed in the cathode catalyst layer in a direct catalytic combustion with oxygen [9].

The discussion of these assumptions is given in [8].

The assumptions made to describe the transport of reactants in the feed channels are as follows.

1. We assume plug flow conditions in channels on both sides of the cell, that is reactant concentration is uniform across the respective channel and flow velocity is constant. Note that the latter assumption is especially restrictive: on the anode side it is valid only at small currents. At large temperatures it fails even at moderate cell currents due to gaseous CO_2 bubbles, which enter the anode channel and dramatically accelerate the flow [10].

2. Pressure in the cathode channel is constant. This assumption is valid in all cases for air-fed DMFCs. If the cell is fed with oxygen it is valid for sufficiently large oxygen stoichiometries.

3. Cell resistivity R_c (the sum of contact and membrane resistivities) is constant along z and the respective voltage loss is estimated as $R_c J$.
4. Cell electrodes are highly conductive and thus cell voltage does not depend on z.
5. The cell is isothermal.

All the relationships below are written in terms of dimensionless variables:

$$\tilde{z} = \frac{z}{L}, \quad \tilde{j} = \frac{j}{j_{\lim}^{a0}}, \quad \tilde{\eta} = \frac{\eta}{b^a}, \quad \tilde{\psi} = \frac{\psi}{\psi^0}, \quad \tilde{\xi} = \frac{\xi}{\xi^0} \tag{12.1}$$

where L is the channel length,

$$j_{\lim}^{a0} = \frac{6FD_b^a c^a \psi^0}{l_b^a} \tag{12.2}$$

is the limiting current density due to methanol transport through the anode backing layer, η is the half-cell polarization voltage, b is the Tafel slope, ψ is the methanol molar fraction in the anode channel, ξ is the oxygen molar fraction in the cathode channel, c is the total molar concentration in the flow, the superscripts 'a' and 'c' refer to the anode and the cathode sides, respectively, and the superscript '0' indicates the values at the channel inlet.

A DMFC is usually run at constant oxygen and methanol stoichiometries λ^a and λ^c, respectively. By definition

$$\lambda^a = \frac{6Fh^a v^a c^a \psi^0}{LJ}, \quad \lambda^c = \frac{4Fh^c v^c c^c \xi^0}{LJ} \tag{12.3}$$

where J is the mean current density in a cell, h is the channel height and v is the flow velocity.

12.2.3
Through-Plane Relations

The through-plane model leads to the following expressions for polarization voltages η^a and η^c of the anode and the cathode sides of DMFC [8]:

$$\tilde{\eta}^a = \ln\left(\frac{\tilde{j}}{q\tilde{\psi}}\right) - \ln\left(1 - \frac{\tilde{j}}{\tilde{\psi}}\right) + \ln(1+\beta) \tag{12.4}$$

$$\tilde{\eta}^c = p\left[\ln\left(\frac{\tilde{j}}{\alpha q \tilde{\xi}}\right) - \ln\left(1 - \frac{\tilde{j} + \tilde{j}_{cross}}{\gamma \tilde{\xi}}\right)\right] \tag{12.5}$$

where

$$q = \frac{l_t^a i_*^a l_b^a}{6FD_b^a c_{ref}^a} \tag{12.6}$$

$$\alpha = \frac{l_t^c i_*^c c_{ref}^a}{l_t^a i_*^a c_{ref}^c}\left(\frac{c^c \xi^0}{c^a \psi^0}\right) \tag{12.7}$$

$$\gamma = \frac{2D_b^c l_b^a}{3D_b^a l_b^c}\left(\frac{c^c \xi^0}{c^a \psi^0}\right) \tag{12.8}$$

are constant dimensionless parameters,

$$p = \frac{b^c}{b^a} \tag{12.9}$$

is the ratio of Tafel slopes and

$$\tilde{j}_{\text{cross}} = \beta_*(\tilde{\psi} - \tilde{j}) \tag{12.10}$$

is the equivalent current density of methanol crossover.
Here

$$\beta = \frac{D_m l_b^a}{D_b^a l_m} \tag{12.11}$$

is the ratio of mass transfer coefficients of methanol in the membrane and in the anode backing layer (BL), D_b^c is the diffusion coefficient of oxygen in the cathode backing layer of the thickness l_b^c, D^m is the diffusion coefficient of methanol in the membrane of the thickness l^m, D_b^a is the diffusion coefficient of methanol in the anode BL of the thickness l_b^a, and

$$\beta_* = \frac{\beta}{1+\beta} \tag{12.12}$$

varies between 0 and 1.

The first term on the right side of Equation 12.4 describes voltage loss due to activation of electrochemical reaction of methanol oxidation. The second term is voltage loss due to methanol transport through the anode backing layer. The last term describes the loss due to methanol crossover: the flux of methanol through the membrane reduces the amount of methanol available for useful current production, which leads to the constant shift of anodic overpotential.

The first term in the square brackets in Equation 12.5 describes voltage loss for activation of oxygen reduction reaction (ORR). The second term accounts for a voltage loss due to oxygen transport through the cathode backing layer [8]. Physically, transport loss on the cathode side is determined by the sum of useful and crossover currents, both of which consume oxygen.

Cell voltage is given by:

$$\tilde{V}_{\text{cell}} = \tilde{V}_{oc} - \tilde{\eta}^a - \tilde{\eta}^c - \tilde{R}_c \tilde{J} \tag{12.13}$$

where J is the mean current density.

12.2.4
Equations Along the Channel

Mass conservation equations along the anode and the cathode channels allow us to relate local molar fractions to their inlet values. These equations are [11]:

$$\lambda^a \tilde{J} \frac{\partial \tilde{\psi}}{\partial \tilde{z}} = -(\tilde{j} + \beta_*(\tilde{\psi} - \tilde{j})), \quad \tilde{\psi}(0) = 1 \tag{12.14}$$

$$\lambda^c \tilde{J} \frac{\partial \tilde{\xi}}{\partial \tilde{z}} = -(\tilde{j} + \beta_*(\tilde{\psi} - \tilde{j})), \quad \tilde{\xi}(0) = 1 \tag{12.15}$$

Physically, Equation 12.14 describes methanol consumption along the anode channel. The rate of consumption (the right side) is a sum of the useful and crossover currents. The product $\lambda^a \tilde{J}$ is proportional to the flow velocity, which is assumed to be independent of $\tilde{z}$.

Quite analogously, Equation 12.15 describes decay of oxygen molar fraction in the cathode flow. The rate of oxygen consumption is also proportional to the sum of useful and crossover current densities. Importantly, the dimensionless variables in Equations 12.14 and 12.15 have the same form; different stoichiometry factors for methanol and oxygen appear to be incorporated into the respective value of λ.

The problem is completed with the equation expressing equipotentiality of cell electrodes. Assuming that OCV does not vary along the channel,[1] from Equation 12.13 we obtain:

$$\tilde{\eta}^a + \tilde{\eta}^c = \tilde{E}^0 \tag{12.16}$$

where

$$\tilde{E}^0 = \tilde{V}_{oc} - \tilde{V}_{cell} - \tilde{R}_c \tilde{J} \tag{12.17}$$

is independent of $\tilde{z}$.

Using expressions (12.4) and (12.5) in (12.16), we get:

$$\ln\left(\frac{\tilde{j}}{q\tilde{\psi}}\right) - \ln\left(1 - \frac{\tilde{j}}{\tilde{\psi}}\right) + \ln(1+\beta) + p\left[\ln\left(\frac{\tilde{j}}{\alpha q \tilde{\xi}}\right) - \ln\left(1 - \frac{(\tilde{j} + \tilde{j}_{cross})}{\gamma \tilde{\xi}}\right)\right] = \tilde{E}^0 \tag{12.18}$$

To simplify numerical procedure and further analysis, it is necessary to convert (12.18) into a differential equation for $\tilde{j}$. Differentiating (12.18) with respect to $\tilde{z}$ and using (12.14) and (12.15) to exclude derivatives $\partial \tilde{\psi}/\partial \tilde{z}$ and $\partial \tilde{\xi}/\partial \tilde{z}$, after simple algebraic manipulations we come to:

$$\frac{\partial \tilde{j}}{\partial \tilde{z}} = \frac{A_j}{B_j}, \quad \tilde{j}(0) = \tilde{j}^0 \tag{12.19}$$

where

$$A_j = \tilde{j}(\tilde{j} + \beta \tilde{\psi})[\gamma(1+\beta)\lambda^c \tilde{\xi} + (p\gamma(1+\beta)\lambda^a - \beta(1+p)\lambda^c)\tilde{\psi} \\ + ((p\beta - 1)\lambda^c - p\gamma(1+\beta)\lambda^a)\tilde{j}] \tag{12.20}$$

$$B_j = \lambda^a \lambda^c \tilde{J}(1+\beta)[(p\gamma(1+\beta)\tilde{\xi} + (1-p\beta)\tilde{\psi})\tilde{j} + (1+p)(\beta \tilde{\psi} - \gamma(1+\beta)\tilde{\xi})\tilde{\psi}] \tag{12.21}$$

---

[1] Below we will see, that this assumption is valid only in the galvanic domain.

An equation for local current at the channel inlet $\tilde{j}^0$ is obtained if we substitute inlet values $\tilde{j} = \tilde{j}^0$ and $\tilde{\psi} = \tilde{\xi} = 1$ into (12.18):

$$\ln\left(\frac{\tilde{j}^0}{q}\right) - \ln(1-\tilde{j}^0) + \ln(1+\beta) + p\left[\ln\left(\frac{\tilde{j}^0}{\alpha q}\right) - \ln\left(1 - \frac{\tilde{j}^0 + \beta_*(1-\tilde{j}^0)}{\gamma}\right)\right] = \tilde{E}^0 \quad (12.22)$$

Note, that (12.19) does not contain α and q. These parameters appear in the solution only through the value of $\tilde{j}^0$ (12.22). In other words, α and q simply shift the curve $\tilde{j}(\tilde{z})$ as a whole, not affecting its shape.

The relationships (12.14), (12.15), (12.19) and (12.22) form a quasi-2D model of DMFC with straight channels. Local current $\tilde{j}$ is subject to the following constraint:

$$\int_0^1 \tilde{j} d\tilde{z} = \tilde{J} \quad (12.23)$$

This constraint prescribes the following sequence of numerical calculations. A value of voltage loss $\tilde{E}^0$ is set first. Equation 12.22 is then solved to find initial condition $\tilde{j}^0$ for Equation 12.19. Solution of Equations 12.14, 12.15 and 12.19 gives local values $\tilde{\psi}$, $\tilde{\xi}$ and $\tilde{j}$. Mean current density is then obtained by calculating the integral (12.23). The pair $(\tilde{J}, \tilde{E}^0)$ gives a point on cell polarization curve.

Sometimes it is convenient to set mean current density in the cell $\tilde{J}$ rather than voltage loss $\tilde{E}^0$. The solution of the problem is then obtained iteratively.

12.2.5
Large Methanol Stoichiometry, Small Current

Analysis of the system (12.14), (12.15), (12.19) and (12.22) for this case leads to the following approximate solution [12]:

$$\tilde{\psi} = 1 \quad (12.24)$$

$$\tilde{\xi} = 1 - \frac{\tilde{z}}{\tilde{z}_{ox}} \quad (12.25)$$

$$\tilde{j} = \tilde{j}^0 \left(1 - \frac{\tilde{z}}{\tilde{z}_0}\right)^{\frac{p}{1+p}} \quad (12.26)$$

Here

$$\tilde{z}_{ox} = \frac{\lambda^c \tilde{J}}{\beta_*} \quad (12.27)$$

is a point where oxygen fraction $\tilde{\xi}$ vanishes and

$$\tilde{z}_0 = \lambda^c \tilde{J} \left(\frac{1}{\beta_*} - \frac{1}{\gamma}\right) \quad (12.28)$$

is a point of zero local current density.

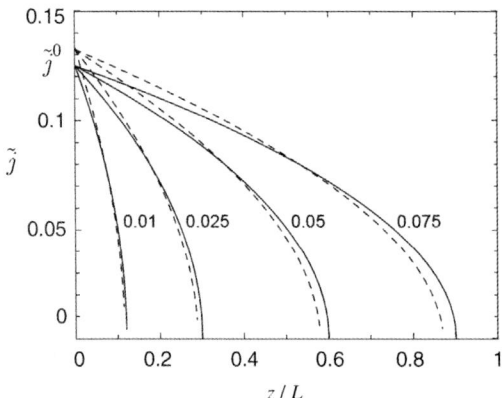

Figure 12.3 The shapes of the local current density along the channel for indicated values of mean current density in a cell $\tilde{J}$. Solid lines: analytical curves (12.26), dashed lines: numerical solutions of the full system of equations. Parameters for the calculations are: $\beta=1$, $p=1$, $\gamma=2$, $\lambda^c=4$. Note that current density at the inlet $\tilde{j}^{\tilde{z}_0}$ does not change; variation of total current in a cell occurs due to bridge 'shrinking'.

Local current density (12.26) for several values of average cell current density $\tilde{J}$ is shown in Figure 12.3 along with the respective numerical solutions to the system of Equations 12.14, 12.15 and 12.19. As seen, the approximate analytical result (12.26) well describes the 'exact' numerical curves.

Solution (12.26) makes sense for $\tilde{z} \leq \tilde{z}_0$. Figure 12.3 shows that when $\tilde{z}_0 < 1$, current is localized in the galvanic domain (bridge) $0 \leq \tilde{z} \leq \tilde{z}_0$ (Figure 12.3). At $\tilde{z} > \tilde{z}_0$ solution to a system discussed has no physical meaning: below we will see that in this region $\tilde{V}_{oc}$ dramatically drops and the model above needs to be corrected.

Oxygen concentration vanishes at $\tilde{z}_{ox} > \tilde{z}_0$. In the region $\tilde{z}_0 < \tilde{z} \leq \tilde{z}_{ox}$ no useful current is produced and all available oxygen there is consumed in the reaction with permeated methanol. The length of this region

$$\delta_{tr} \equiv \tilde{z}_{ox} - \tilde{z}_0 = \frac{\lambda^c \tilde{J}}{\gamma} \tag{12.29}$$

increases with $\tilde{J}$.

In the case of $\tilde{z}_0 < 1$, current density at the inlet $\tilde{j}^0$ is obtained from condition $\int_0^{\tilde{z}_0} \tilde{j} d\tilde{z} = \tilde{J}$. Integrating (12.26) we find [2]

$$\tilde{j}^0 = \frac{\gamma \beta_*(1+2p)}{\lambda^c(\gamma-\beta_*)(1+p)} \tag{12.30}$$

Thus, $\tilde{j}^0$ does not depend on $\tilde{J}$, which is directly seen in Figure 12.3.

---

2) Note that our model neglects formation of a domain with negative (electrolytic) current in the region $\tilde{z} > \tilde{z}_0$ (see below). In the latter case $\tilde{J} = \int_0^1 \tilde{j} d\tilde{z}$, since part of the positive current is consumed by the electrolytic domain.

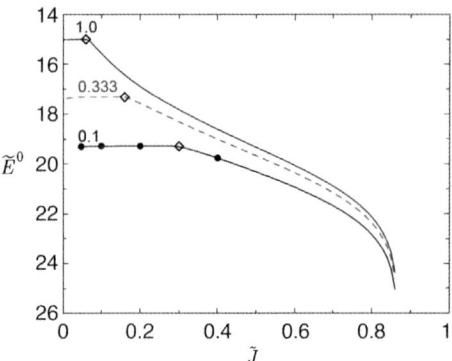

Figure 12.4 Total voltage loss $\tilde{E}^0$ vs mean current density $\tilde{J}$ for indicated values of crossover parameter β. Stoichiometries are $\lambda^a = 4$, $\lambda^c = 2$. The other parameters are $\alpha = 20.04$, $\gamma = 2.803$, $q = 1.727 \times 10^{-5}$. The profiles of local current density in the black points are shown in Figure 12.5. Diamonds: the points where the bridge forms ($\tilde{z}_0 = 1$, see the text).

These results help to clarify the physics of bridge formation and to rationalize the behavior of the cell polarization curve in this regime. The flux of methanol crossover increases linearly with the decrease in $\tilde{j}$, Equation 12.10. Thus, at $\tilde{z} \geq \tilde{z}_0$ this flux is maximal and oxygen in this region is consumed in the reaction with permeated methanol. In other words, part of the cell to the right of $\tilde{z}_0$ experiences oxygen 'starvation' and this leads to localization of current at the channel inlet (at $\tilde{z} < \tilde{z}_0$). Below we will see that the domain $\tilde{z} > \tilde{z}_0$ may turn to electrolysis mode.

The bridge manifests itself as a plateau on the cell polarization curve. Suppose that we are moving along the polarization curve for $\beta = 0.1$ from large to small $\tilde{J}$ (Figure 12.4). As soon as $\tilde{z}_0$ becomes equal to 1, local current at the channel inlet $\tilde{j}^0$ ceases to change (Figure 12.5).

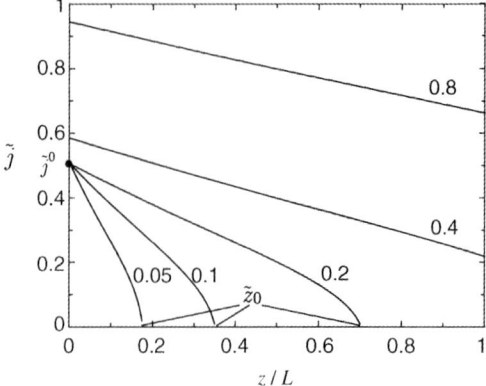

Figure 12.5 The profiles of local current density for indicated values of dimensionless mean current density in a cell and $\beta = 1$. The other parameters are the same as in Figure 12.4. The polarization curve for $\beta = 1$ is shown in Figure 12.4.

12.3 The Bifunctional Regime of DMFC Operation

Further decrease in $\tilde{J}$ is provided by the bridge shrinking, that is, by the shift of $\tilde{z}_0$ to zero at a constant $\tilde{j}^0$ (Figures 12.3 and 12.5). Since total voltage loss $\tilde{E}^0$ depends only on $\tilde{j}^0$, this leads to formation of a plateau on the cell polarization curve (Figure 12.4). The mean current density $\tilde{J}_{crit}$, at which $\tilde{z}_0 = 1$ is clearly seen: it is a point where the polarization curve goes into a plateau (diamonds in Figure 12.4).

Note that the results discussed are obtained assuming that the cell operates at constant oxygen stoichiometry. It is easy to show that at constant λ^c critical current density $\tilde{J}_{crit}$ is equivalent to the critical air flow rate. Indeed, the product $\lambda^c \tilde{J}$ is

$$\lambda^c \tilde{J} = \frac{4Fh^c v^c c^c \xi^0}{LJ}\left(\frac{J}{\tilde{j}_{lim}^{a0}}\right) = \frac{4Fw^c h^c v^c c^c \xi^0}{Lw^c \tilde{j}_{lim}^{a0}} = \frac{4Ff^c c_{ox}^0}{Lw^c \tilde{j}_{lim}^{a0}} \qquad (12.31)$$

where c_{ox}^0 is the oxygen concentration at the inlet and

$$f^c = w^c h^c v^c \qquad (12.32)$$

is the total volumetric airflow (cm³ s⁻¹), w^c is the channel width. Note that we ignore the channel/land structure of the flow field and assume that the channel covers the whole cell active area.

Experiments confirmed the effect of the critical flow rate f_{crit}^c [13]. Measured polarization curves reproduce the sharp bend at a critical air flow remarkably well (see also below). However, experiments [13] were performed with the standard nonsegmented cell and thus the distribution of local current over the cell surface has not been measured.

What happens in the cell to the right of the point $\tilde{z}_0$? To answer this question we have designed a segmented cell with a single straight channel and performed measurements of local current distribution at various air flow rates. The results are discussed in the next section.

12.3
The Bifunctional Regime of DMFC Operation

12.3.1
Experimental

A sketch of the single channel flow field is shown in Figure 12.6. 20 graphite segments with the dimension 0.5 cm × 1.5 cm were embedded into a polysulfone plate. Then the single channel with the dimension 1 × 1 × 104 mm was engraved in the flow field plate (Figure 12.6). The contact area between each segment and electrode is about 5 mm × 3 mm.

The functional layers of the membrane electrode assembly (MEA) were prepared by knife-coating onto carbon cloth ('A' Cloth, E-Tek, thickness of 350 μm). The backing layer with a thickness of about 100 μm consisted of 60 wt.% of carbon (Vulcan XC 72) and 40 wt.% of PTFE, and it was sintered at 350 °C. The inks used for the preparation of the catalyst layers contained isopropanol, catalyst particles

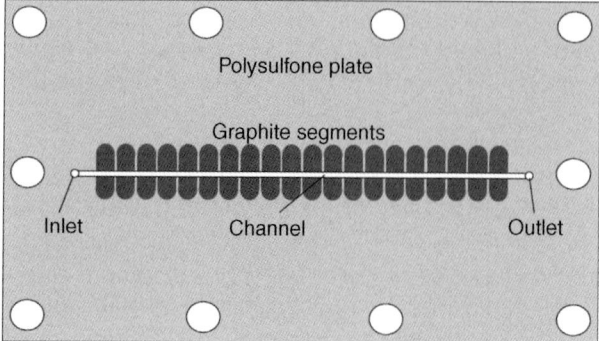

Figure 12.6 Sketch of the segmented flow field. 20 graphite segments are positioned along the single channel 10 cm in length; channel cross section is 1 mm × 1 mm. The flow field on the other side is a copy of the one shown. MEA stripe of width 3 mm is clamped between the two flow fields.

and Nafion. The anode catalyst consisted of 60 wt.% Pt/Ru and 40 wt.% carbon (Johnson Matthey 'HiSpec10000'). The Pt/Ru loading of the anodes was about 2 mg cm^{-2}. The cathode catalyst also had a composition of 60 wt.% Pt and 40 wt. % carbon (Johnson Matthey 'HiSpec9000' or 'HiSpec9100'), with a Pt loading of about 2 mg cm^{-2}. Both anode and cathode had an area of about 3 cm^2 (3 mm × 104 mm). Finally, the electrode stripes were hot-pressed on both sides of either a Nafion 117 or Nafion 115 membrane (6.3 cm × 5.1 cm). The hot-pressing temperature was 130 °C and the pressure 0.5 kN cm^{-2}. More details of the preparation procedure can be found in [14].

The complete electrochemical set-up including the measuring electronics was developed by the Institute of Power Electronics and Electrical Drives (ISEA) of Aachen University. The adapter printed circuit board is shown in Figure 12.20. Twenty gold-plated copper contacts provide a direct electrical connection of each segment to the measuring device and allow the current and impedance distribution to be measured. The adapter circuit board was inserted on the anode side of the cell between the flow field and the endplate. Independently for each segment, this device provides a shunt resistor and an amplifier that translates the current into a voltage signal. The latter can be recorded by multimeter or by multichannel data logger. Additionally, the device contains a circuit for active compensation of the voltage drop across the leads of the printed circuit board and the shunt resistor, thus providing an equal potential at all segments and eliminating the undesired feedback effect of shunt resistors on the current distribution. A detailed description of the measuring system can be found in [15].

All measurements were carried out at temperatures of 70 °C or 80 °C under ambient pressure. The anode was continuously fed with 1 M methanol solution and the cathode was supplied with air. The methanol flow was kept constant, corresponding to a varying stoichiometry λ^a in the range 6–300 depending on the current. The bifunctional (BF) regime (see below) is observed under excess of methanol on the anode side and depletion of air on the cathode side. Depending on operation mode, either the current was kept constant and the air flow varied or vice versa. Further information concerning the experimental conditions is indicated in the figures and figure captions.

12.3.2
Experimental Results

All curves shown below were obtained at a fixed current of 8.5 mA in the external load. Due to the large ratio of MEA length to width the exact active surface of the cell is not known and below we plot the directly-measurable local current of individual segments, rather than a current density.

Figure 12.7 shows the distribution of local current for the values of air flow rate indicated (ml min^{-1}). This figure depicts four repeated measurements with essentially the same operating conditions: an attempt to obtain reproducible results.

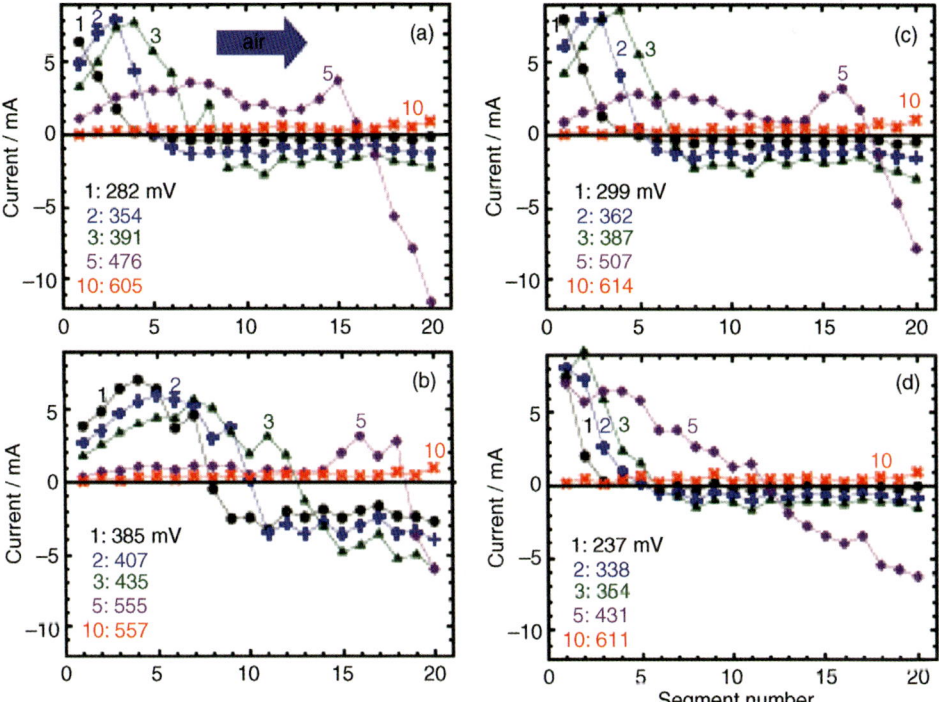

Figure 12.7 The distribution of local current along the channels for indicated air flow rate (ml min^{-1}). Inlet of both channels is at segment 1, co-flow conditions. Flow rate of 1 M methanol solution is fixed at 0.1 ml min^{-1}, current in the external load in all cases is 8.5 mA. The cell potentials for each air flow rate are shown in the lower left corner of the graph. The four graphs show repeated measurements under the same conditions.

At low air flow rates f^c (below $\simeq 7$ ml min^{-1}) the cell turns to the BF regime. Consider the curve 2 ($f^c = 2$ ml min^{-1}) in Figure 12.7a: segments 1–4 work in a galvanic regime; current in segment 5 is close to zero; while current in segments 6–20 is negative, which indicates that these segments operate in the electrolysis mode.

Current produced in the galvanic domain (GD) is, therefore, divided between two consumers: the external load and the electrolytic domain (ED). Therefore, fixing the current in the external load does not guarantee the fixing of total current produced in the GD. We, thus, may expect that the state of the cell in this regime is not unique.

The curves and data in Figure 12.7 show that at least three states of the cell correspond to the same current in the load. Consider the curves 1, 2 and 3 in Figure 12.7a–d. The respective curves in Figure 12.7a and c are practically the same and the respective cell potentials differ by less than 17 mV; for the curves 2 and 3 the difference does not exceed 4 mV. This state of the cell will be referred to as state A.

Curves 1, 2 and 3 in Figure 12.7b differ from their counterparts in Figure 12.7a and c. In this state (B) the GD is twice as long as in state A (cf. e.g., Figure 12.7b and c). Current produced in the GD and current consumed in the ED are also larger than the respective currents in state A. Surprisingly, cell voltage in this state is also significantly larger than in state A (cf. Figure 12.7a and b).

The third state is represented by curves 1, 2 and 3 in Figure 12.7d. In this state (C) the length of the GD is the smallest, the respective cell voltages are the lowest and current consumed in the ED is also minimal. When $f^c = 1$ ml min^{-1}, the total current consumed in the ED is much smaller than the current in the external load; thus, practically all current produced in the GD goes to the load resistor.

We may carefully assume that in the BF regime the cell has many states corresponding to the same current in the load. The curves in Figure 12.7 show that the length of GD increases with the growth of f^c, in accordance with Equation 12.28 (note that $f^c \sim \lambda^c J$). At a critical air flow rate f^c_{crit} the ED disappears and GD covers the whole cell active area. It occurs when f^c is between 5 and 10 ml min^{-1} (Figure 12.7); this critical air flow rate is of particular interest.

12.3.3
Polarization Curves at Constant Oxygen Stoichiometry

Figure 12.8 depicts cell polarization curves measured at the three oxygen stoichiometries $\lambda^c = 2$, 4 and 8. These measurements were performed stepwise in a galvanostatic mode. The cell voltage for each current step was taken after a period of two minutes. The feature of all curves in Figure 12.8 is an abrupt change in the slope of the curve at some current (cf. Figure 12.4).

Figure 12.9 shows the shapes of the local current, which correspond to the several points on polarization curves in Figure 12.8. Consider the curve for $\lambda^c = 8$ (Figure 12.8) and suppose that we are moving along this curve from large to small currents. Figure 12.9c shows the respective variation of local current (note the

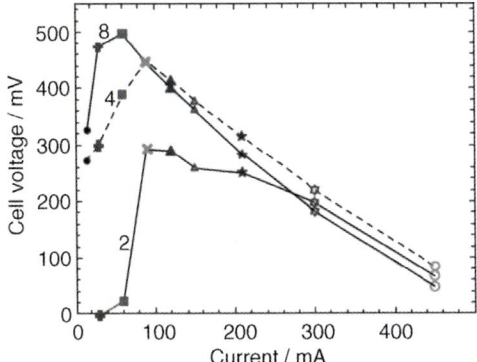

Figure 12.8 Cell polarization curves for indicated oxygen stoichiometries λ^c. For conditions please see the caption to Figure 12.9.

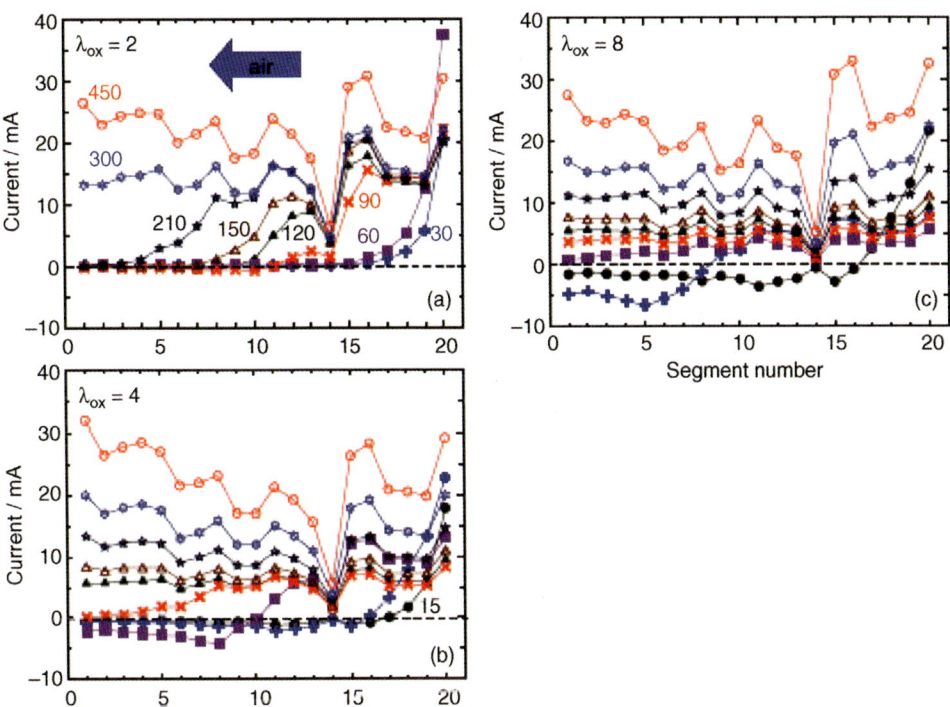

Figure 12.9 The distribution of local current along the channel for different total currents in the load. Oxygen stoichiometry: (a) $\lambda^c = 2$, (b) $\lambda^c = 4$ and (c) $\lambda^c = 8$. The respective polarization curves are shown in Figure 12.8. Inlet of both channels is at segment 20, co-flow conditions (note the change in the direction of air and methanol flow as compared to Figure 12.7). Flow rate of 1 M methanol solution is 0.31 ml min^{-1}. Numbers on the curves indicate total current in the load (see the respective symbols in Figure 12.8). The drop of current in segment 14 is presumably caused by high contact resistance.

direction of airflow in Figure 12.9)[3]. At large currents ($I \geq 60$ mA) all segments work in the galvanic regime. At $I = 60$ mA current at the outlet (first segment) vanishes and the cell enters the BF regime. At $I < 60$ mA part of the segments operate in the electrolysis mode (Figure 12.9c).

Inspection of the curves in Figure 12.9a and b and of the corresponding polarization curves in Figure 12.8 shows that in all cases transition to the BF regime is accompanied with a drastic change in the slope of the polarization curve. Variation of λ^c changes only the current when this transition occurs (Figure 12.8). Note that the same behavior of polarization curves was detected in experiments with the standard square-shaped DMFC with nonsegmented electrodes [13]. The model [12] allows us to explain this effect.

12.3.4
Critical Air Flow Rate

In all experiments reported above, the inequality $\lambda^a \gg \lambda^c$ is fulfilled. Under the excess of methanol the length of the GD $\tilde{z}_0$ is given by Equation 12.28. Oxygen stoichiometry λ^c can be expressed in terms of f^c, Equation 12.32 as

$$\lambda^c = \frac{4 F f^c c_{ox}^0}{LwJ}, \tag{12.33}$$

where Lw is the cell active area.

Using (12.2), (12.8) and (12.3) it is easy to verify that

$$\lambda^c \tilde{J} = \gamma \tilde{f}^c, \tag{12.34}$$

where the dimensionless air flow rate $\tilde{f}^c = f^c/f_*$ and

$$f_* = \frac{Lw D_b^c}{l_b^c} \tag{12.35}$$

is the characteristic flow rate.

Using (12.34) in (12.28) we get

$$\tilde{z}_0 = \gamma \tilde{f}^c \left(\frac{1}{\beta_*} - \frac{1}{\gamma} \right). \tag{12.36}$$

Oxygen concentration in the cathode channel vanishes at $\tilde{z}_{ox}$, Equation 12.27, which can be transformed to

$$\tilde{z}_{ox} = \frac{\gamma \tilde{f}^c}{\beta_*} \tag{12.37}$$

Comparing (12.36) and (12.37) we find

$$\tilde{z}_0 = \tilde{z}_{ox} - \frac{\gamma \tilde{f}^c}{\beta_*} \tag{12.38}$$

---

3) Low current in segment.14 is presumably due to the high local contact resistance in this segment.

Equations (12.36)–(12.38) show that the coordinates of zero current $\tilde{z}_0$ and zero oxygen concentration $\tilde{z}_{ox}$ increase with the growth of air flow rate $\tilde{f}^c$. The growth of $\tilde{z}_0$ with $\tilde{f}^c$ is directly seen in Figure 12.7a–d.

Suppose that λ^c is fixed and we increase $\tilde{J}$. At certain value $\tilde{J}^{crit}$ the ED disappears and we thus have $\tilde{z}_0 = 1$. Equation 12.36 then gives

$$\gamma \tilde{f}^c_{crit} \left(\frac{1}{\beta_*} - \frac{1}{\gamma} \right) = 1. \tag{12.39}$$

Note, that in that state all current produced in the GD goes to the load. Solving Equation 12.39 for $\tilde{f}^c_{crit}$ we find

$$\tilde{f}^c_{crit} = \left(\frac{\gamma}{\beta_*} - 1 \right)^{-1}. \tag{12.40}$$

DMFCs typically operate at $\gamma \geq 3$ and $\beta_* \simeq 0.5$. In that case $\gamma/\beta_* \gg 1$ and Equation 12.40 simplifies to

$$\tilde{f}^c_{crit} \simeq \frac{\beta_*}{\gamma}. \tag{12.41}$$

In dimension variables this relationship takes the form

$$f^c_{crit} = \frac{3\beta_* D^a_b c^0_M Lw}{2 l^a_b c^0_{ox}} = \frac{\beta_* j^{a0}_{lim} Lw}{4 F c^0_{ox}} \tag{12.42}$$

where $c^0_M = c^a \xi^0$ is inlet methanol concentration. Equation 12.42 says that for given oxygen and methanol inlet concentrations there exists the critical (minimal) air flow rate f^c_{crit}, which provides galvanic regime over the whole length of the cell. Note that f^c_{crit} does not depend on cell current [4].

The curves in Figure 12.9a–c directly confirm this result. Inspection of these curves shows that at $\lambda^c = 2$ the ED disappears when $I \simeq 210$ mA (the curve with the stars in Figure 12.9a), at $\lambda^c = 4$ this happens at $I \simeq 100$ mA (the curve with the diagonal crosses in Figure 12.9b) and at $\lambda^c = 8$ the ED vanishes at $I \simeq 60$ mA (the curve with the filled squares in Figure 12.9c). We, thus have $f^c_{crit} \simeq (\lambda^c I)_{crit} \simeq 2 \times 210 = 420$, $4 \times 90 = 360$ and $8 \times 60 = 480$, respectively. All three values are close to each other. Taking into account a number of simplifications made to derive Equation 12.28, we can conclude that the model well describes this critical behavior of DMFC.

---

4) In Ref. [16] an incorrect statement was made that f^c_{crit} is inversely proportional to the oxygen diffusion coefficient in the cathode backing layer D^c_b. In fact, D^c_b does not appear in Equation 12.42, and $f^c_{crit} \sim D^a_b$ instead. This means that the onset of the BF regime does not depend on the rate of oxygen transport to the CCL, though it depends on the rate of methanol transport to the ACL. Physically, D^a_b determines methanol flux through the cell, including the flux due to methanol crossover: the larger D^a_b, the more methanol penetrates through the membrane. According to model assumption, this methanol is consumed in the CCL regardless of the rate of oxygen transport to the CCL and this makes j^c_{crit} independent of D^c_b.

12.4
Direct Methanol–Hydrogen Fuel Cells (DMHFCs)

12.4.1
Experiment

In general, the distribution of local current in the BF regime *is not* reproducible. Figure 12.7 shows that only at an air flow rate of 10 ml min^{-1}, when all segments operate in the galvanic regime, are the shapes of local current reproducible. At lower f^c the ED forms and the state of the cell (the distribution of local current) appears to be undetermined.

Total currents in the GD and ED are related as

$$I_{GD} + I_{ED} = I_{load} \tag{12.43}$$

where I_{load} is the current in the external resistor. Fixing I_{load} we fix only the right side in this equation. Therefore, I_{GD} and I_{ED} are allowed to have arbitrary values provided that their sum is constant.

Clearly, to characterize the cell we should use I_{GD} instead of I_{load}. Physically, the electrolytic domain can be treated as an additional load for the galvanic domain and thus the state of the cell can be characterized by I_{GD}.

In this section we analyze polarization curves corresponding to the local current distributions in Figure 12.7. Figure 12.10 shows the cell polarization curves corresponding to each run in Figure 12.7. For every curve in Figure 12.7 the total galvanic current I_{GD} was calculated summing the currents of the galvanic segments. Together with the respective cell voltage V_{cell}, the pair (I_{GD}, V_{cell}) gives a point on the polarization curve. Five curves in each graph in Figure 12.7 thus give five points in Figure 12.10.

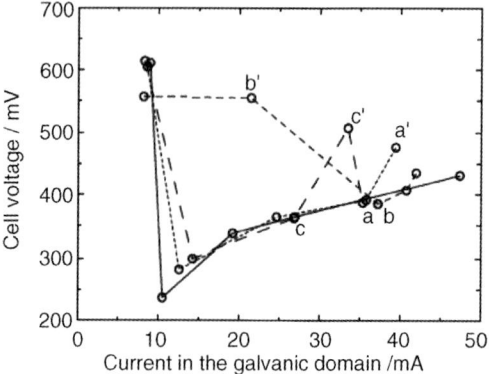

Figure 12.10 Cell polarization curves (cell voltage versus current produced in the galvanic domain). Each curve corresponds to the data in one of the graphs in Figure 12.7. Short-dashed curve: data from Figure 12.7a, medium-dashed curve: data from Figure 12.7b, long-dashed curve: data from Figure 12.7c, and solid curve: data from Figure 12.7d.

The curves in Figure 12.10 exhibit rather irregular behavior. However, comparison of Figures 12.7 and 12.10 helps to understand the situation. The four points in the top left corner in Figure 12.10 correspond to the four curves with $f^c = 10\,\text{ml min}^{-1}$, when all segments operate in a galvanic mode (Figure 12.7). Thus, these points correspond to 'pure' DMFC regime of cell operation.

The other points in Figure 12.10 fall into two groups. The points b', c' and a' represent the state of the cell with high cell voltage; all other points group along the 'low-voltage' curve. This suggests, that the 'high'- and 'low-voltage' points in Figure 12.10 exhibit *two* physically different polarization curves, which describe two different regimes of cell operation.

To understand the difference between these two regimes consider the shapes of local current corresponding to the transition from the low-voltage state a to the high-voltage state a' (Figure 12.10). The respective currents are shown in Figure 12.11a.

In the state a (air flow rate $3\,\text{ml min}^{-1}$, Figure 12.11a) the ED is long (segments 9–20) and the current of each segment in the ED is not large [5]. The GD in this state is relatively short (segments 1–8, Figure 12.11a) and the average GD current is quite large.

In contrast, in state a' (air flow rate $5\,\text{ml min}^{-1}$, Figure 12.11a) the ED is short (segments 17–20) and the average current there is large. The GD in this state is large (segments 1–16) and the average GD current is smaller than in the state a. Of special interest is the region near the interface between the GD and ED (segments 13–16, Figure 12.11a). A distinct peak of positive local current in these segments is clearly seen.

Quite a similar picture is shown by the transitions from the state b to b' (Figures 12.10 and 12.11b) and from c to c' (Figures 12.10 and 12.11c). Again, in the states b and c the GD is short, the ED is long and the average ED current is not large (Figure 12.11 b and c). In the states b' and c' the GD is long, the ED is short and the average ED current is large. Note the peak of local positive current at the GD/ED interface in these states (Figure 12.11b and c).

This peak is a signature of hydrogen oxidation on the anode side. Large negative current in the ED in the states a', b' and c' (Figure 12.11) means that the local rate of hydrogen production in the ED is large. H_2 permeates towards the GD where it is oxidized according to

$$H_2 \to 2H^+ + 2e^-$$

In the next section we will see that between the GD and ED a transition region (TR) forms, which operates as a hydrogen cell. This region manifests itself as a peak of positive current on the curves a', b' and c' (Figure 12.11).

The kinetics of H_2 oxidation are fast; we thus may assume that all hydrogen permeated to the TR is oxidized. This additional source of current dramatically

---

5) Here we are discussing absolute values of all currents.

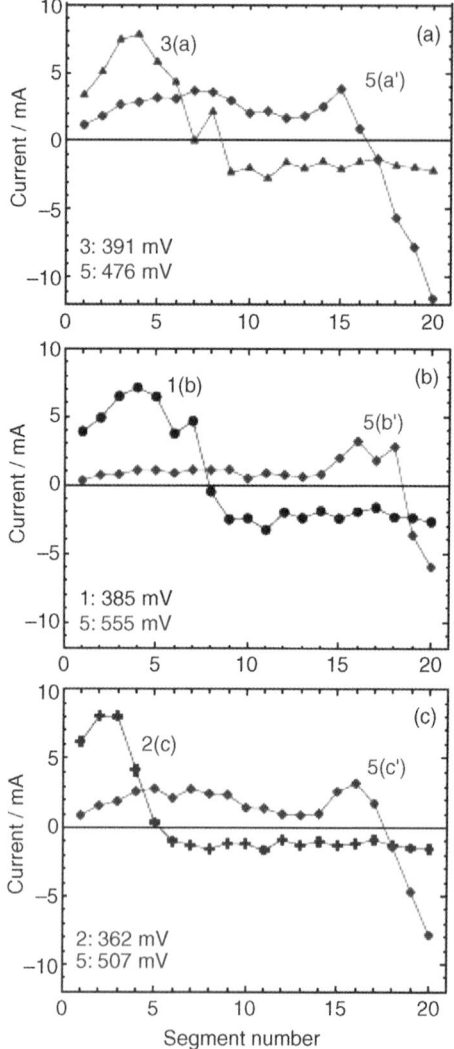

Figure 12.11 The distributions of local current corresponding to 'jumps' on polarization curves in Figure 12.10. Indicated are air flow rates (ml min^{-1}) and cell potentials (mV).

improves the overall cell performance. Furthermore, formation of a TR improves oxygen utilization, so that less (or no) oxygen is available in the ED. This increases the rate of methanol electrolysis. A schematic of the processes in the cell is shown in Figure 12.12.

This effect leads us to the following interpretation of the polarization curves (Figure 12.13). The four points in the left top corner represent the cell in the 'pure' DMFC state.

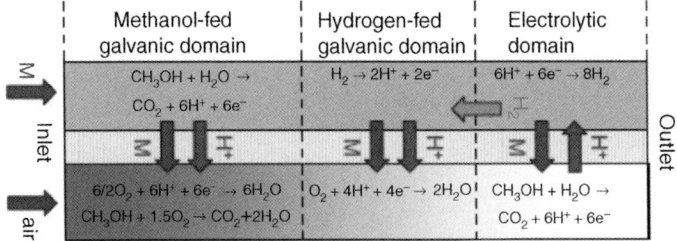

Figure 12.12 Sketch of the processes in a DMHFC. Hydrogen produced in the electrolytic domain permeates along the anode side to the galvanic domain and is used as a fuel there.

The circles with crosses correspond to the state of the cell with the ED and small (or no) effect of hydrogen. In this state, the local rate of hydrogen production at the GD/ED interface is small and hydrogen is simply removed through the anode exhaust.

Filled circles correspond to the direct methanol–hydrogen cell. In this state, the local rate of hydrogen production in the ED is high, hydrogen penetrates along the anode side to the TR and serves as a fuel there.

The mechanism and pathway of hydrogen transport to the TR is not yet clear. Presumably, gaseous H_2 diffuses through the system of small pores in the catalyst and/or backing layers. Alternatively, hydrogen can be transported due to the pressure gradient caused by a high rate of hydrogen production in the ED.

Current in the external load I_{load} is fixed; therefore, if we plot potentials of all points in Figure 12.13 versus I_{load}, we would get a vertical line. Clearly, maximal useful power density is provided in the 'pure' DMFC regime, since the respective

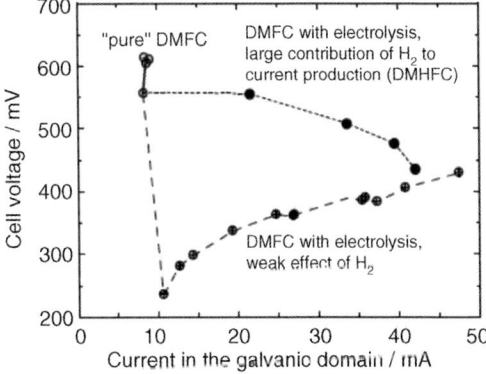

Figure 12.13 Interpretation of the data in Figure 12.10 based on the effect of formation of a hydrogen-fed domain. The points corresponding to different states of the cell are grouped into three curves. Solid bold line: 'pure' DMFC state of the cell, long-dashed line: DMFC with electrolytic domain and weak effect of hydrogen on cell performance, short-dashed line: DMFC with electrolytic and hydrogen-fed domains (DMHFC).

points have the highest cell potential. However, running the cell in a 'pure' DMFC state requires the highest air flow rate (10 ml min^{-1}). The regime with hydrogen oxidation in the TR may be preferential, since the respective points on the polarization curve correspond to at least twice lower f^c. Thus, the overall efficiency of DMHFC system may be higher.

12.4.2
DMHFC: The Mechanism of Functioning

Consider the example of local current distribution for air flow rate 5 ml min^{-1}, corresponding to the cell operation in a DMHFC state (Figure 12.14). Segments 1–17 produce current, while the segments 18–20 consume it.

As discussed above, in segments 1–17 the mechanism of current production is different in different segments. Segments 1–14 operate in normal DMFC mode, while segments 15–17 use hydrogen produced in segments 18–20. Following the discussion of the previous section, hydrogen permeates along the anode side to segments 15–17 and serves as a fuel there. Peak of local current in segments 15–17 and relatively high cell voltage are the signatures of this process.

What distribution of polarization voltages corresponds to the schematic in Figure 12.12? The analysis below aims to answer this question [17].

12.4.2.1 Potentials Across the Cell
Consider first the potential distribution across the cell in the GD and ED. Cell voltage V_{cell} is open-circuit voltage V_{oc} minus losses V_{loss}:

$$V_{cell} = V_{oc} - V_{loss} \qquad (12.44)$$

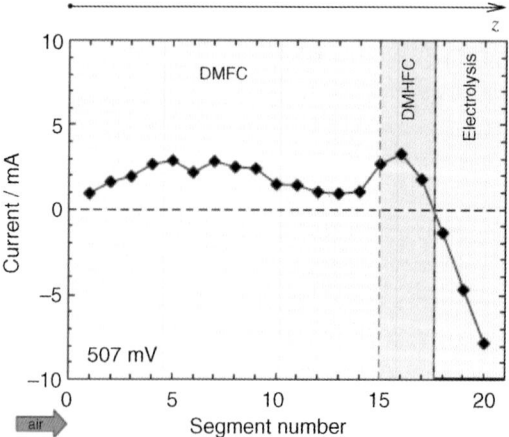

Figure 12.14 The example of a distribution of local current along the channel (the curve from Figure 12.7). Air flow rate is 5 ml min^{-1}; inlet of both channels is at segment 1, co-flow conditions. Flow rate of 1 M methanol solution is 0.1 ml min^{-1}; current in the external load is 8.5 mA; cell potential is 507 mV.

In the GD we have

$$V_{loss}^{GD} = \eta^a + \eta^c + R_m j \tag{12.45}$$

(Figure 12.15). Here $\eta^a = \varphi'^a - \varphi_m$ and $\eta^c = \varphi_m - \varphi^c$ are the polarization voltages of the anode and the cathode side of the cell, respectively, $\varphi'^a = V_{oc} - \varphi_{real}^a$ is the shifted potential of the carbon phase on the anode side,[6] $\varphi^c = 0$ is the potential of this phase on the cathode side, φ_m is the potential of the electrolyte phase and R_m is the membrane resistivity. Note that $V_{loss}^{GD} > 0$ (Figure 12.15).

In the ED the cell consumes current produced in the GD. Two reactions take place in the ED: electrochemical oxidation of methanol on the cathode side and hydrogen evolution on the anode. The thermodynamic open-circuit potential of methanol electro-oxidation is very small [18]; that potential of hydrogen evolution is zero. Thus, V_{oc} in the ED is small. Furthermore, polarization voltage of a hydrogen evolving electrode (anode) is also small and can be neglected.

In the ED we, therefore, have

$$V_{loss}^{ED} = \eta^c + R_m j \tag{12.46}$$

Both terms on the right side of this equation are negative (Figure 12.15). Indeed, as discussed above, in the ED $V_{oc} \simeq 0$ and Equation 12.44 reduces to

$$V_{cell} = -V_{loss}^{ED} \tag{12.47}$$

All segments are equipotential, that is, $V_{cell} > 0$ is constant along the channel. Therefore, V_{loss}^{ED} is negative. Physically, to split methanol the ED needs an 'external' voltage, which is provided by the GD; formally this is equivalent to negative V_{loss} (Figure 12.15).

The picture in Figure 12.15 is consistent with the direction of proton current in the GD and ED. In the GD protons move from left to right, opposite to the gradient $\partial \varphi_m / \partial x$ (Figure 12.15). In the ED this gradient changes sign and protons move from right to left.

Near the GD/ED interface the distribution of potentials in the GD transforms into the distribution in the ED (inset in Figure 12.15). This transformation occurs in the transition region (TR), which plays an important role in cell operation.

12.4.2.2 Potentials Along the Channel

As before, axis z is directed along the channel (Figure 12.14). In our experiments special care has been taken to provide equipotentiality of segments. Thus, the left side of Equation 12.44 is constant along z, while the terms on the right side vary with z.

---

6) $\varphi'^a \equiv V_{loss}$ (Figure 12.15) does not include cell open-circuit voltage, which is explicitly accounted for in Equation 12.44. Measurable potential of the carbon phase on the anode side is $\varphi_{real}^a = V_{oc} - \varphi'^a$. Note that the cathode is assumed to be grounded ($\varphi^c = 0$).

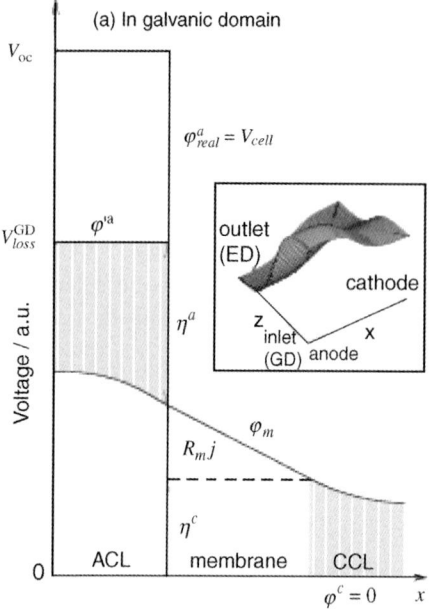

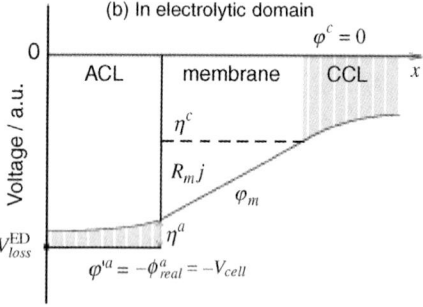

Figure 12.15 Sketch of voltage losses in (a) galvanic and (b) electrolytic domains. ACL and CCL abbreviate the anode and the cathode catalyst layer, respectively. Axes x and z are directed across the membrane electrode assembly and along the oxygen channel, respectively. The inset in (a) shows qualitatively the 2D surface of membrane phase potential. The cathode is assumed to be grounded ($\varphi^c = 0$); the real measurable potential of the anode carbon phase (cell potential) is φ^a_{real}.

We will consider continuous distribution of $V_{oc}(z)$ assuming that the number of segments is large. Following Nernst's law, the dependence of OCV on oxygen concentration can be represented as

$$V_{oc}(z) = V^0_{oc} - V_N(z) \tag{12.48}$$

where V^0_{oc} is OCV at $z=0$ and

$$V_N = \frac{RT}{4F} \ln\left(\frac{\xi^0}{\xi(z)}\right) \tag{12.49}$$

is the Nernst correction due to oxygen exhaustion along the channel. Writing (12.49) we assume that pressure variation along the oxygen channel is small. In Equation 12.48 molar concentrations of the other components and activity coefficients are included in V_{oc}^0.

Due to oxygen consumption V_N increases with $\tilde{z}$ and at certain point we get $V_{oc} = 0$. Below we will show that this point is very close to the point of zero oxygen concentration z_{ox}. Therefore, at all $z > z_{ox}$, due to the absence of oxygen OCV is nearly zero:

$$V_{oc} = 0, \quad z \geq z_{ox} \tag{12.50}$$

In a DMFC the membrane is fully humidified and the term $R_m j$ in Equations 12.45 and 12.46 can be neglected. Writing (12.44) for the GD and ED and taking into account Equations 12.45–12.50 we get

$$V_{cell} = V_{oc}^0 - \frac{RT}{4F} \ln\left(\frac{\xi^0}{\xi(z)}\right) - \eta^a(z) - \eta^c(z), \quad z < z_{ox} \tag{12.51}$$

$$V_{cell} = -\eta^c, \quad z \geq z_{ox} \tag{12.52}$$

We see that in the ED $-\eta^c$ is constant and equals cell voltage.

With the dimensionless variables (12.1), Equation 12.51 can be written as

$$-\tilde{V}_* \ln\tilde{\xi} + \tilde{\eta}^a + \tilde{\eta}^c = \tilde{E}^0 \tag{12.53}$$

where

$$\tilde{V}_* = \frac{RT}{4Fb^a} \tag{12.54}$$

and

$$\tilde{E}^0 = \tilde{V}_{oc}^0 - \tilde{V}_{cell} = \text{const} \tag{12.55}$$

is the voltage loss.

At $\tilde{z} = 0$ we have $\tilde{\xi} = 1$; the Nernst correction vanishes and (12.53) yields

$$\tilde{\eta}^a(0) + \tilde{\eta}^c(0) = \tilde{E}^0 \tag{12.56}$$

However, the Nernst correction alters the shapes of $\tilde{\eta}^a$ and $\tilde{\eta}^c$ along the channel, as discussed in the next section.

12.4.2.3 Potentials in the Galvanic Domain

Substituting Equations 12.24–12.26 into expressions for $\tilde{\eta}^a$ (12.4) and $\tilde{\eta}^c$ (12.5) we get the explicit dependencies of $\tilde{\eta}^a$ and $\tilde{\eta}^c$ on the distance in the region $0 \leq \tilde{z} < \tilde{z}_0$ (Figure 12.16b). In the GD $\tilde{\eta}^c$ increases with $\tilde{z}$ and $\tilde{\eta}^a$ decreases (Figure 12.16b). Oxygen concentration decays with $\tilde{z}$ faster than the local current (Figure 12.16a); thus, to support the required rate of local current production $\tilde{\eta}^c$ increases.

Nernst correction in the GD is small (see below) and (12.53) reduces there to $\tilde{\eta}^a + \tilde{\eta}^c = \tilde{E}^0$. Thus, the growth of $\tilde{\eta}^c$ induces the decrease in $\tilde{\eta}^a$ (Figure 12.16b). Physically, anode polarization decreases with $\tilde{z}$ following the decay of local current.

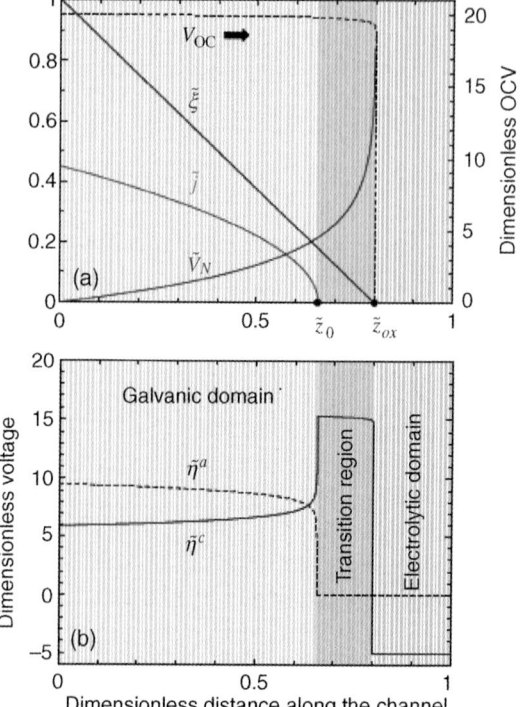

Figure 12.16 (a) Solid curves: oxygen molar fraction $\tilde{\xi}$, local current density in the galvanic domain $\tilde{j}$ and Nernst correction $\tilde{V}_N$ to cell open-circuit voltage along the channel. Dashed curve: cell open-circuit voltage $\tilde{V}_{oc}$. (b) solid line: polarization voltage of the cathode side $\tilde{\eta}^c$, dashed line: polarization voltage of the anode side $\tilde{\eta}^a$. All variables are dimensionless.

12.4.2.4 The Transition Region: Hydrogen Cell

The region $[\tilde{z}_0, \tilde{z}_{ox}]$ between the points of zero current and zero oxygen concentration, (Figure 12.16a) is the transition region (TR) between the GD and ED. On the right side of this region the Nernst correction dramatically grows.

Substituting (12.25) into (12.49) we see that the Nernst correction increases with the distance as

$$\tilde{V}_N = -\tilde{V}_* \ln\left(1 - \frac{\tilde{z}}{\tilde{z}_{ox}}\right) \tag{12.57}$$

On the left side of the TR (at $\tilde{z} = \tilde{z}_0$), $\tilde{V}_N$ is small (Figure 12.16a). Indeed, substituting $\tilde{z} = \tilde{z}_0$ (12.28) into (12.57) and taking into account (12.37) we get

$$\tilde{V}_N(\tilde{z}_0) = -\tilde{V}_* \ln\left(\frac{\beta_*}{\gamma}\right) \tag{12.58}$$

With the parameters in Table 12.1 we get $V_N(\tilde{z}_0) \simeq 12\,\text{mV}$, which is small. This confirms the validity of model above for the GD.

Table 12.1 Parameters.

	Anode side	Cathode side
Temperature (C)	70	70
Pressure (atm)	2	2
D_b (cm² s⁻¹)	2×10^{-5}	3×10^{-3}
i_* (A cm⁻³)	10^{-2}	1
c_{ref} (mol cm⁻³)	10^{-3}	7×10^{-5}
Oxygen molar fraction ξ^0	—	0.2
Methanol molar fraction ψ^0	0.018 (1 M)	—
Tafel slope b (V)	0.05	0.05
Catalyst layer thickness l_t (cm)	10^{-3}	10^{-3}
Backing layer thickness l_b (cm)	2×10^{-2}	2×10^{-2}
Membrane thickness l_m (cm)	10^{-2}	
D_m (cm² s⁻¹) for $\beta = 0.1, 0.333, 1$	$1 \times 10^{-6}, 3.33 \times 10^{-6}, 10^{-5}$	

Dramatic growth of $\tilde{V}_N$ (and the respective decay of $\tilde{V}_{oc}$) occurs in close vicinity to $\tilde{z}_{ox}$ (Figure 12.16a). Estimate shows that the width of the region where OCV drops from about 90% of inlet value to zero is negligibly small. In other words, due to the growth of V_N the shape of OCV is very close to a step function, which at $\tilde{z} = \tilde{z}_{ox}$ changes from $\tilde{V}_{oc}^0$ to zero (Figure 12.16a).

In the TR $\tilde{\eta}^a \simeq 0$ and Equation 12.53 reduces to

$$\tilde{V}_N + \tilde{\eta}^c = \tilde{E}^0 \tag{12.59}$$

Since $\tilde{E}^0$ is constant, the increase in $\tilde{V}_N$ induces the decrease in $\tilde{\eta}^c$ (Figure 12.16b). Near $\tilde{z}_{ox}$ the value of $\tilde{\eta}_c$ becomes negative and for all $\tilde{z} \geq \tilde{z}_{ox}$ we have Equation 12.52 (Figure 12.16b).

The length of the TR $\delta_{tr} \equiv \tilde{z}_{ox} - \tilde{z}_0$ (Figure 12.16a) is given by Equation 12.29. With the parameters from Table 12.1 and $J = 0.034$ A cm⁻² (the curve in Figure 12.14) we get $\delta_{tr} \simeq 0.1$. This is approximately 2 segments in the flow field used in our experiments.

Following our model, in the TR galvanic current is zero (Figure 12.16a). However, the experimental shape of local current exhibits a peak of positive current there (Figure 12.14, segments 15–17). This suggests that in the TR a mechanism of current production exists, which is not accounted for by the analysis above. Since anode polarization voltage in the TR is small, the only possible mechanism of proton generation there is hydrogen oxidation. The source of hydrogen is located nearby: the nearest segments of the ED produce hydrogen on the anode side. Thus, to explain the peak of positive current in the TR we can only assume that hydrogen produced in the ED diffuses towards the TR and serves as a fuel there.

Note that protons produced in the ED cannot reach the TR moving along the z axis: the electric field in the membrane phase pushes them back (inset in Figure 12.15a). Thus, the loop of proton transport in the ED and TR is closed by the diffusion of neutral hydrogen along the anode side, as shown in Figure 12.12.

Hydrogen oxidation reaction (HOR) requires very small anodic overpotential η^a; thus, the shapes of overpotentials in Figure 12.16 support TR operation as a hydrogen cell. Due to small η^a the electrochemical oxidation of methanol in the TR is not activated. In other words, TR operates as a 'pure' hydrogen cell. The whole cell thus works as a system with internal reforming of methanol.

12.5
Bifunctional Activation of DMFC

12.5.1
Single Channel Cell

BF regime of cell operation is not only of academic interest. This regime reveals an interesting option for the recovery of cell performance. In this section we discuss the method and the physics of recovery.

Figure 12.17a shows the distribution of local current density along the channel when air inlet is at the segment 1. Numbers indicate air flow rate f^c (ml min^{-1}); all three curves correspond to subcritical flow rates, when the cell operates in a BF regime.

In this experiment f^c was increased in steps from 1 to 2 and then to 3 ml min^{-1}. Each curve corresponds to a steady-state regime of cell operation that is, the curves were registered when the transient effects vanished. Typically, steady-state distribution of local current establishes in 2–3 minutes of cell operation. Thus, in the course of measurements every segment in the range 9–20 for several minutes has been operated in electrolytic mode.

Then the cell was stopped and the air supply was connected to segment 20, that is, air inlet and outlet were interchanged (Figure 12.17b). The curves in Figure 12.17b correspond to the same three values of air flow rate of 1, 2 and 3 ml min^{-1}. Now peak current in all cases exceeds 10 mA; for $f^c = 1$ ml min^{-1} it reaches 20 mA (segment 20, Figure 12.17b). This suggests that when the segments operate in electrolytic mode, they exhibit better galvanic performance.

To verify this result we have run the cell in normal DMFC mode providing super-critical air flow at a current in the external load of 150 mA. In this (base-case) regime all segments operate in galvanic mode. Every 30–40 minutes we performed an 'activation': air flow was reduced to a subcritical value of 3 ml min^{-1} and the current in the load was decreased to 10 mA. In this regime nearly half of the segments turn into electrolysis mode. The cell was kept in this activation regime for 30 seconds and then returned to the base-case state. Cell voltage versus time is shown in Figure 12.18. In this experiment periodical activation improves the cell galvanic performance by 3–7%.

Note that the actual value of performance improvement varies from cycle to cycle. This may be caused by the variation in a number of segments which turn into electrolysis mode. As discussed above, in the BF regime multiple states of the cell correspond to the same current in the load; thus fixing I_{load} does not guarantee reproducibility of the ED length.

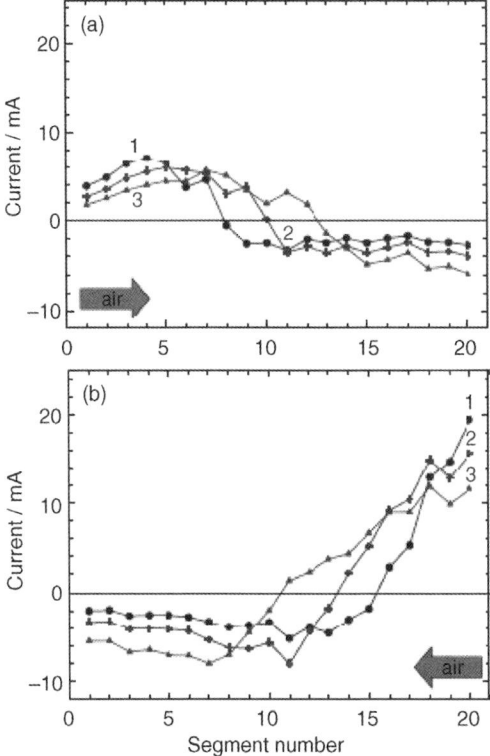

Figure 12.17 Local current density along the channel. (a) Air inlet is connected to segment 1, 'remote' segments operate in electrolysis mode (negative local current). (b) air inlet is then switched to segment 20; better galvanic performance of the segments operated in electrolysis mode is clearly seen. Number on the curve indicates air flow rate (ml min^{-1}).

In this experiment only half of the segments were periodically activated; we thus may expect that activation of all segments would give much better results in terms of performance improvement. This, however, would require a more sophisticated procedure, including reverse of air flow direction ('two-way' activation), as in Figure 12.17.

A similar effect of performance improvement at periodical *full break* of air supply to DMFC was reported in [19]. Following [19], full break of oxygen supply decreases the cathode potential to the value required for cleaning the Pt surface of oxides, which poison the CCL during normal DMFC operation.

Above we have shown that *lowering* of air flow rate below a critical value, turns the cell into the BF regime; operation in this regime, thus, partially recovers cell performance. It is, therefore, important to understand the difference between our method of cell recovery and the method proposed in [19].

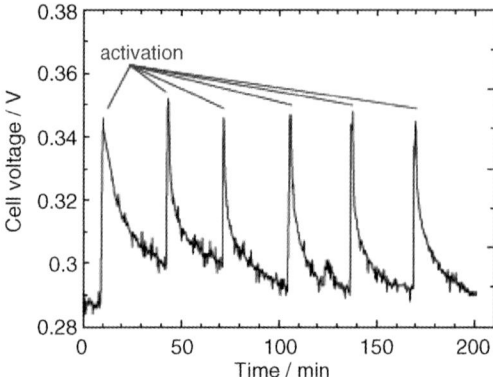

Figure 12.18 The effect of periodical bifunctional activation on cell voltage. The cell is run in normal DMFC mode at 150 mA in the load and super-critical air flow rate of 11 ml min^{-1} ($\lambda^c \simeq 5$). Each 30–40 minutes the cell is activated: for 30 sec air flow is decreased down to a subcritical value of 3 ml min^{-1} and current in the load is decreased to 10 mA. Note that no external power supply is used.

The experiments [19] were performed with a 100 cm^2 DMFC.[7] The cell was not segmented and thus no local measurements of current distribution were performed. During the air-breaking interval, the current in the cell was kept constant with the aid of an external power supply and special electronic equipment. No hydrogen evolution on the anode side was reported. This suggests that at the time of air break an electrolytic domain in the cell was not formed.

In our experiments, however, lowering of air flow creates a steady-state electrolytic domain in the cell. In the ED this reverses the sign of the cathode overpotential and induces hydrogen evolution on the anode side. In the galvanic domain the signs of the cathode and the anode half-cell potentials remain unchanged.

However, thanks to the detailed study [19] we may assume that negative polarization induces Pt cleaning on the cathode side. The argument in favor of this conjecture is as follows.

Oxide film growth on the Pt surface is exponential with time and reaches saturation in about 50 min (Figure 12.7 of Ref. [19]). This correlates with the exponential-like shape of voltage relaxation after activation in Figure 12.18. Moreover, the characteristic time of this relaxation (30–50 min) correlates with the time of saturation of oxide film growth [19].

Another mechanism also contributes to cell recovery: hydrogen evolution on the anode side [20]. Following [20], running a DMFC anode in hydrogen evolution mode increases the number of Pt/Ru sites connected to the feeding transport system that

---

7) Unfortunately, no details of the cell and flow field geometry were reported.

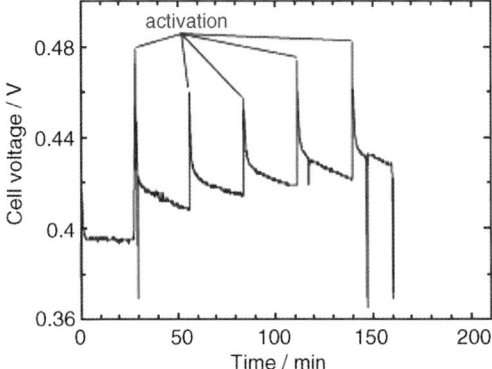

Figure 12.19 The effect of periodical activation on cell voltage. The cell is run in normal DMFC mode at 468 mA in the load; air flow rate is 100 ml min^{-1} ($\lambda^c \simeq 15$). Each $\simeq$ 30 minutes the cell is activated: for 30 sec air is replaced with nitrogen at the same flow rate and simultaneously current in the load is reversed to $I_{load} = -100$ mA using the external power supply.

is, the number of electrochemically active sites. Note that in [20] the cell anode was turned into hydrogen evolution mode using the external power supply.

One more possible mechanism of anode recovery is the improvement of transport properties of the anode backing layer during evolution of gaseous hydrogen. In our laboratory we got evidence that intense formation of gaseous (CO_2) bubbles at high current densities improves the transport properties of the anode backing layer, thereby improving anode performance.

We, therefore, may assume that our method of DMFC activation simultaneously uses recovery mechanisms of the cathode [19] and the anode [20]. The advantage of this method is that it does not require an external power source: the galvanic domain serves as a power supply for the electrolytic domain [16].

To prove that the effect of activation is due to the formation of an electrolytic domain we have performed the following measurements. The cell was run in normal DMFC mode and periodically switched to electrolysis mode using the external power supply. More specifically, for 30 seconds the air flow was replaced with the nitrogen flow and simultaneously current in the load was reversed to $I_{load} = -100$ mA. This procedure turns the whole cell active area to electrolysis mode.

The result is shown in Figure 12.19; the effect of activation is clearly seen. Moreover, the characteristic time of cell voltage relaxation in each cycle is roughly the same as in Figure 12.18. This suggests that the mechanisms of activation in both these methods are the same.[8]

---

8) Note that the gain in cell power in Figure 12.19 is higher, than in Figure 12.18, since cell voltage now relaxes to the higher level. This effect is probably due to MEA conditioning during the measurements.

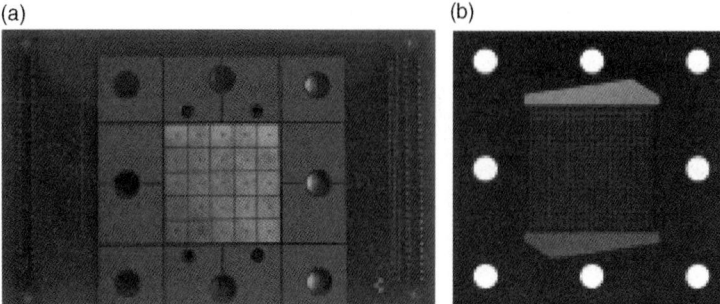

Figure 12.20 Pictures of the adapter circuit board and the endplate used for square-shaped cells. (a) PCB with 25 gold segments. (b) graphite flow field with grid structure.

So far we did not see any signs of electrode or membrane degradation due to the activation. However, more experiments are needed to clarify the long-term effect of this procedure.

12.5.2
Activation of Square-Shaped Cells

Bifunctional activation can be organized in square-shaped cells with the standard flow fields. To show this we performed experiments with the cell having active area of approximately $4.2 \times 4.2 \simeq 18\,cm^2$. The cell was equipped with the flow fields of a grid structure with the dimension of each rib $0.1 \times 0.1 \times 0.1\,cm$ (Figure 12.20). The current was collected using a printed circuit board having 5×5 gold contacts, with a contact area of $0.7\,cm^2$ for each segment (Figure 12.20).

The current densities of five selected segments (4 in the corners and 1 in the center of cell) before, during and after the 'two-way' activation procedure are shown in Figure 12.21. In all cases mean current density in the cell was kept at $10\,mA\,cm^{-2}$.

Before activation the cell is run under normal DMFC operation conditions, that is, under excess of both methanol and oxygen (Figure 12.21a). Air is supplied from the upper part of the cell at a rate of $443\,ml\,min^{-1}$. The segment currents scatter around $10\,mA\,cm^{-2}$ (Figure 12.21a) and the cell voltage is about 600 mV.

When air flow is reduced down to $20\,ml\,min^{-1}$, oxygen depletion in the bottom part of the cell turns the cell into the BF mode (Figure 12.21b). The current densities of segments 1 and 5 are shifted to high positive values (galvanic domain) and the current densities of segments No. 20 and 25 drop down to negative values (electrolytic domain, Figure 12.21b). The current in segment 13, which is located in the transition region between galvanic and electrolytic domains amounts to approximately zero. In this regime the cell voltage drops to a value of about 350 mV.

The distribution of local current in this regime is also shown in Figure 12.22a. The upper part of this figure shows the color map of the full range of currents, while the lower part depicts only signs of currents that is, the galvanic (red) and electrolytic (green) domains.

12.5 Bifunctional Activation of DMFC

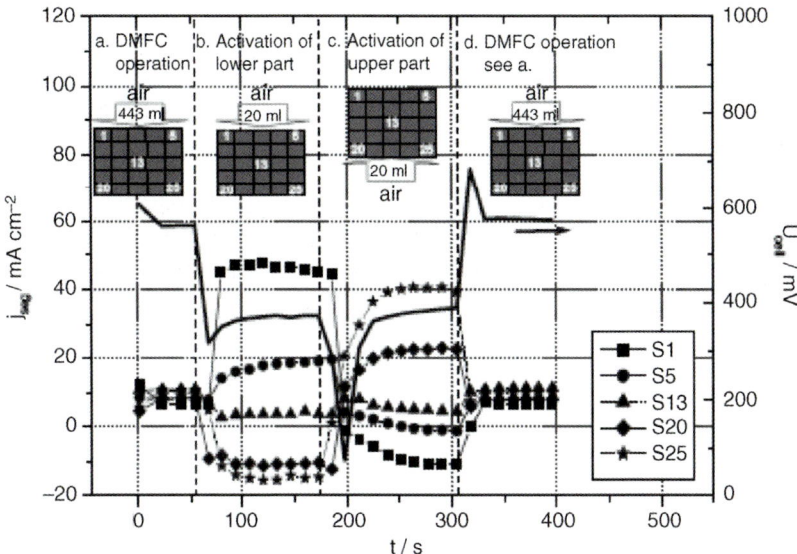

Figure 12.21 Activation of the square-shaped cell with grid flow field. Shown are current densities from five selected segments during the activation process; $T = 70\,°C$, $J = 10\,mA\,cm^{-2}$, 1 M MeOH at 2.56 ml min^{-1}. a and d: DMFC operation, airflow rate 443 ml min^{-1}, b and c: Activation, airflow rate 20 ml min^{-1}. Bold solid line without symbols shows cell voltage.

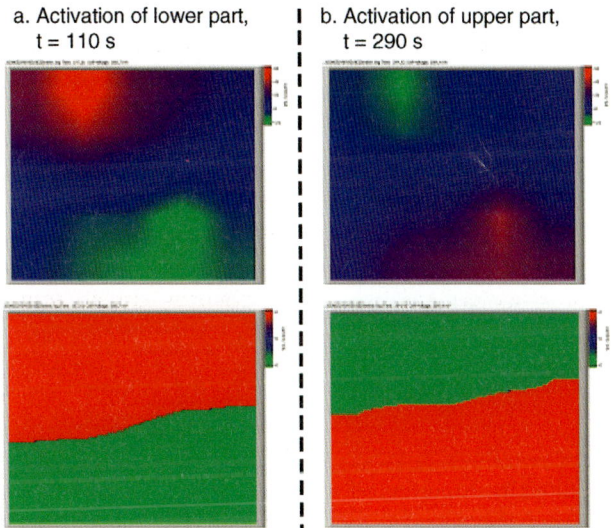

Figure 12.22 Map of local current during bifunctional activation of the square-shaped cell with grid flow field. Left side: activation of lower part of the cell, air inlet at the top of the cell, data taken after 110 s of activation process initiation (see Figure 12.21). Right side: activation of upper part of the cell, air inlet at the bottom of the cell, data taken after 290 s of activation process (see Figure 12.21). Upper part of Figure: a full range of local current densities in the cell. Lower part of Figure: the shapes of galvanic domain (positive currents, red) and electrolytic domain (negative currents, green).

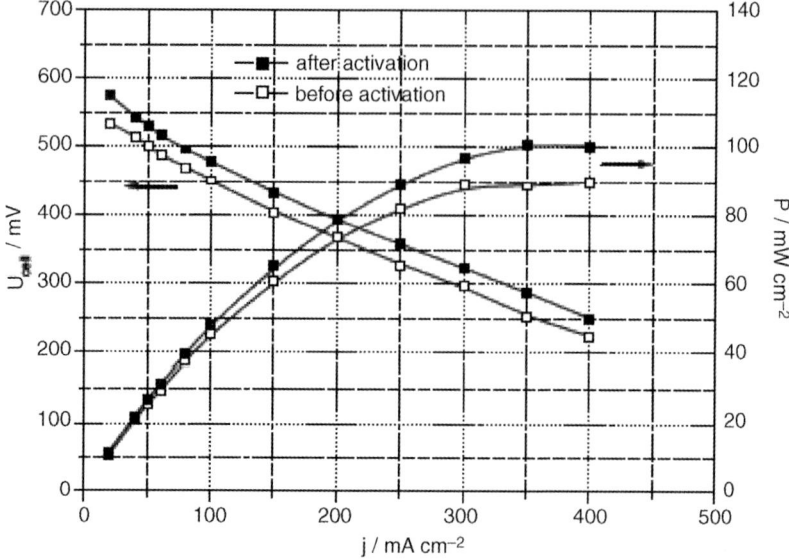

Figure 12.23 Voltage–current and power–current characteristics of a square-shaped cell with grid flow field before and after activation. Operation conditions: $T = 70\,°C$, $\lambda = 4$ for both methanol and air, ambient pressure.

After three minutes of operation in regime b, air inlet and outlet were interchanged. Figure 12.21c shows the relaxation of local currents to new values, corresponding to formation of galvanic domain at the bottom of the cell (close to the oxygen inlet) and of electrolytic domain at the top. This 'inversion' of domains is also seen in Figure 12.22b.

Upon returning to normal DMFC operation conditions (Figure 12.21d) the segment currents return to their values before activation. However, the cell voltage appears to be slightly higher indicating a small activation effect. Voltage–current and power–current characteristics recorded before and after the activation procedure display a performance gain up to 10% (Figure 12.23). This clearly shows the applicability of BF activation to practically relevant DMFC designs and conditions.

12.6
Conclusions

At low oxygen flow rate a direct methanol fuel cell turns into the BF regime, when part of the cell close to the oxygen channel inlet produces current, while the remaining part of the cell consumes it to produce hydrogen. Our 1D + 1D model gives simple relations for the critical air flow rate for onset of this regime and for the position of the interface between galvanic and electrolytic domains. A more detailed analysis, which takes into account variation of open-circuit voltage along the oxygen channel reveals the existence of a transition region between galvanic and electrolytic domains. Under certain (not well defined) conditions this region works as a hydrogen cell, using hydrogen produced in the electrolytic domain.

This analysis was supported by the detailed measurements of local current density distribution in a single-channel DMFC with segmented electrodes. Special electronic equipment provided equipotentiality of all electrodes, thereby simulating a fuel cell with continuous highly conductive endplates.

The measurements confirmed the main predictions of our model. Moreover, they showed the possibility of recovering cell performance by periodically switching the cell into BF mode. In other words, periodical lowering of air flow rate below the critical value partially recovers cell performance. The effect is presumably due to the cleaning of the cathode catalyst from oxides and due to the improvement of anode transport properties under hydrogen evolution conditions.

Many features of the BF regime yet are not fully understood. What are the conditions for the formation of the transition region between GD and ED? What is the mechanism of hydrogen transport from the ED to the TR? How transport properties of the anode side depend on the rate of hydrogen evolution? These and other questions need to be addressed to fully exploit the benefits of DMFC operation in the bifunctional regime.

12.7
List of symbols

$\sim$	Marks dimensionless variables
b	Tafel slope (V)
c	Total molar concentration in the channel (mol cm^{-3})
c_{ref}	Reference molar concentration (mol cm^{-3})
D_b	Diffusion coefficient of feed molecules in the respective backing layer (cm^2 s^{-1})
D_m	Diffusion coefficient of methanol in membrane (cm^2 s^{-1})
E	Total voltage loss (V)
F	Faraday constant (9.6495 × 10^4 Coulomb mol^{-1})
f^c	Flow rate (cm^3 s^{-1})
I	Total cell current (A)
i_*	Exchange current density per unit volume (A cm^{-3})
J	Mean current density in a cell (A cm^{-2})
j	Local current density in a cell (A cm^{-2})
j_{cross}	Equivalent crossover current density in a cell (A cm^{-2})
j_*	Characteristic current density (A cm^{-2})
j_{lim}	limiting current density (A cm^{-2})
h	Channel height (cm)
L	Channel length (cm)
l_t	Thickness of the catalyst layer (cm)
l_b	Thickness of the backing layer (cm)
l_m	Thickness of the membrane (cm)

L	Channel length (cm)
N_{cross}	Molar flux of methanol through the membrane (mol cm^{-2} s^{-1})
p	Ratio of Tafel slopes ($p = b^c/b^a$)
q	Dimensionless parameter (12.6)
R_c	The sum of contact and membrane resistances (Ω cm^2)
R_m	Membrane resistance (Ω cm^2)
T	Cell temperature (K)
V_{cell}	Cell voltage (V)
V_{loss}	Voltage loss (V)
V_N	Nernst correction (V)
V_{oc}	Thermodynamic open circuit voltage (V)
v	Flow velocity in the channel (cm s^{-1})
w	Channel width (cm)
x	Coordinate across the cell (cm)
z	Coordinate along the channel (cm)
z_0	Point of zero local current (cm)
z_{ox}	Point of zero oxygen concentration (cm)

12.7.1
Superscripts

0	at the channel inlet (at $z = 0$)
a	anode side
c	cathode side
ED	electrolytic domain
GD	galvanic domain

12.7.2
Subscripts

*	characteristic value
0	at zero current
b	backing layer
crit	critical
cross	crossover
ED	electrolytic domain
GD	galvanic domain
lim	limiting
load	in external load
m	in membrane
N	Nernst correction
ox	oxygen
t	catalyst layer
tr	transition region

12.7.3
Greek Symbols

α	Dimensionless parameter (12.7)
β	Dimensionless parameter (12.11)
β_*	$\beta_* = \beta/(1 + \beta)$
δ_{tr}	The length of transition region (cm)
γ	Dimensionless parameter (12.8)
η	Polarization voltage (V)
λ	Stoichiometry of feed flow
ξ	Oxygen molar fraction in the cathode channel
ψ	Methanol molar fraction in the anode channel

Acknowledgments

The authors are grateful to B. Fricke and T. Sanders for design of excellent electronic equipment and invaluable help with measurements. Useful discussions with J. Mergel and D.–U. Sauer are gratefully acknowledged.

References

1 Kulikovsky, A.A. (2008) Optimal temperature for DMFC stack operation. *Electrochimica Acta*, **53**, 6391–6396.
2 Kulikovsky, A.A. (2008) Analysis of thermal stability of DMFC stack operation. *J. Electrochem. Soc.*, **155**, B509–B516.
3 Divisek, J., Fuhrmann, J., Gärtner, K. and Jung, R. (2003) Performance modeling of a direct methanol fuel cell. *J. Electrochem. Soc.*, **150**, A811–A825.
4 Wang, Z.H. and Wang, C.Y. (2003) Mathematical modeling of liquid–feed direct methanol fuel cells. *J. Electrochem. Soc.*, **150** (4), A508–A519.
5 Birgersson, E., Nordlund, J., Ekström, H., Vynnycky, M. and Lindbergh, G. (2003) Reduced two–dimensional one–phase model for analysis of the anode of a DMFC. *J. Electrochem. Soc.*, **150**, A1368–A1376.
6 Birgersson, E., Nordlund, J., Vynnycky, M., Picard, C. and Lindbergh, G. (2004) Reduced two–phase model for analysis of the anode of a DMFC. *J. Electrochem. Soc.*, **151**, A2157–A2172.
7 Yang, W.W. and Zhao, T.S. (2007) A two–dimensional, two–phase mass transport model for liquid–feed DMFCs. *Electrochimica Acta*, **52**, 6125–6140.
8 Kulikovsky, A.A. (2002) The voltage current curve of a direct methanol fuel cell: 'exact' and fitting equations. *Electrochem. Comm.*, **4**, 939–946.
9 Vielstich, W., Paganin, V.A., Lima, F.H.B. and Ticianelli, E.A. (2001) Non–electrochemical pathway of methanol oxidation at a platinum–catalyzed oxygen gas diffusion electrode. *J. Electrochem. Soc.*, **148**, A502–A505.
10 Kulikovsky, A.A. (2005) Model of the flow with bubbles in the anode channel and performance of a direct methanol fuel cell. *Electrochem. Comm.*, **7**, 237–243.
11 Kulikovsky, A.A. (2005) On the nature of mixedpotential in a DMFC. *J. Electrochem. Soc.*, **152** (6), A1121–A1127.

12. Kulikovsky, A.A. (2004) 1D + 1D model of a DMFC: Localized solutions and mixed potential. *Electrochem. Comm.*, **6**, 1259–1265.
13. Kulikovsky, A.A., Klafki, K. and Wippermann, K. (2005) Experimental verification of the effect of bridge formation in a direct methanol fuel cell. *Electrochem. Comm.*, **7**, 394–397.
14. Wippermann, K., Richter, B., Klafki, K., Mergel, J., Zehl, G., Dorbandt, I., Bogdanoff, P., Fiechter, S. and Kaytakoglu, S. (2007) Carbon supported Ru–Se as methanol tolerant catalysts for DMFC cathodes. Part ii: preparation and characterization of MEAs. *J. Appl. Electrochem.*, **37**, 1399–1411.
15. Sauer, D.-U., Sanders, T., Fricke, B., Baumhöfer, Th., Wippermann, K., Kulikovsky, A.A., Schmitz, H. and Mergel, J. (2008) Measurement of the current distribution in a direct methanol fuel cell – confirmation of parallel galvanic and electrolytic operation within one cell. *J. Power Sources*, **176**, 477–483.
16. Kulikovsky, A.A., Schmitz, H., Wippermann, K., Mergel, J., Fricke, B., Sanders, T. and Sauer, D.U. (2006) DMFC: Galvanic or electrolytic cell? *Electrochem. Comm.*, **8**, 754–760.
17. Kulikovsky, A.A. (2008) Direct methanol–hydrogen fuel cell: The mechanism of functioning. *Electrochem. Comm.*, **10**, 1415–1418.
18. Jiang, R. and Chu, D. (2002) A combinatorial approach toward electrochemical analysis. *J. Electroanal. Chem.*, **527**, 137–142.
19. Eickes, Ch., Piela, P., Davey, J. and Zelenay, P. (2006) Recoverable cathode performance loss in direct methanol fuel cells. *J. Electrochem. Soc.*, **153**, A171–A178.
20. He, C., Qi, Z., Hollett, M. and Kaufman, A. (2002) An electrochemical method to improve the performance of air cathodes and methanol anodes. *Electrochem. Solid-State Lett.*, **5**, A181–A183.

13
Electrocatalysis in the Direct Methanol Alkaline Fuel Cell
Keith Scott and Eileen Yu

13.1
Introduction

The direct methanol fuel cell (DMFC) directly converts the chemical energy of methanol into electrical energy and is one of the most promising alternative power sources for transportation, portable electronics and stationary applications [1]. Unlike hydrogen fuel cells, the ease of using liquid fuels is one of the attractions of the DMFC. In the DMFC, methanol is directly oxidized at the anode in acidic medium using a proton exchange membrane (PEM), such as Nafion (DuPont). Direct methanol fuel cells based on solid polymer electrolytes (SPEs) have been developed to their current status using proton exchange membrane. Obstacles which have restrained their more rapid development and applications include methanol crossover from the anode fuel to the cathode, and the relatively low activity of methanol oxidation electrocatalysts.

The low activity of electrocatalysts is inherent in the overall six-electron oxidation of methanol to carbon dioxide:

$$CH_3OH + 1.5\, O_2 \rightarrow CO_2 + 2H_2O \qquad (13.1)$$

Although there are a number of relatively active electrocatalysts, based on Pt alloys, these catalysts still result in several hundred of millivolts of anode polarization at practical operating current densities. Consequently methanol oxidation performance in PEM-based DMFCs cannot compete with hydrogen oxidation in terms of the delivered fuel cell potential. However it is known that for many reactions, electrocatalysts perform better in alkaline electrolytes, and thus historically there have been some significant investigations of methanol oxidation under alkaline conditions. The electro-oxidation of methanol in an alkaline electrolyte is structure insensitive [1] which, as well as opening up the opportunity for increased activity, can enable the use of non-platinum catalysts, which are significantly cheaper, such as Pd, Ag, Ni [2]. In alkaline electrolyte the DMFC does not suffer from poisoning by chemisorbed methanol fragments, as observed in acid electrolytes.

Electrocatalysis of Direct Methanol Fuel Cells. Edited by Hansan Liu and Jiujun Zhang
Copyright © 2009 WILEY-VCH Verlag GmbH & Co. KGaA, Weinheim
ISBN: 978-3-527-32377-7

In fuel cells the oxygen reduction reaction (ORR) is relatively slow and is responsible for a significant loss in cell voltage when practical current densities are required. The problem of methanol crossover in the DMFC means that methanol oxidation at the cathode inhibits or reduces the effectiveness of the ORR catalysts and results in significantly greater cathode polarization; which further reduces cell voltage in comparison to that expected in the absence of methanol. This crossover is a combined effect of diffusion and drag with protons (+water) through the PEM. Consequently even under open circuit conditions methanol crossover affects the cathode and results in relatively low starting cell potentials. In the field of PEM cells there have been some major developments in membranes (see Chapter 14) which act as barriers to methanol and reduce the impact of crossover on performance. However there is still room for improved membranes that are stable and offer low cost.

With an alkaline electrolyte the reactions at the anode and cathode involve OH^- ions which transfer from the cathode to the anode:

Anode oxidation

$$CH_3OH + 6OH^- \rightarrow 6e^- + CO_2 + 5H_2O \qquad E^o = -0.81 \text{ V} \qquad (13.2)$$

$$\text{Cathode } 3/2 O_2 + 3H_2O + 6e^- \rightarrow 6OH^- \qquad E^0 = 0.402 \text{ V} \qquad (13.3)$$

$$\text{Overall } CH_3OH + 1.5 O_2 \rightarrow CO_2 + 2H_2O \qquad E^o = 1.21 \text{ V} \qquad (13.1)$$

Thus movement of OH^- ions through any barrier/electrolyte which separates the two cell reactions will essentially counteract methanol crossover (Figure 13.1). Thus potentially ionic drag can block the diffusion of methanol and reduce the damaging effect of methanol crossover. Hence there is scope for the use of OH^- ion conducting membranes in the DMFC. Overall therefore there is some significant interest in direct methanol alkaline fuel cells (DMAFC) based on alkaline conducting membranes as depicted in Figure 13.1.

Such DMAFC open up great opportunities for methanol fuel cells in terms of potential reduced costs and improved performance but do present new challenges in membrane and catalyst materials, including performance, durability and lifetime. This chapter examines the issues in the development of alkaline alcohol fuel cells, particularly based on methanol. These issues include electrocatalysis for fuel oxidation and oxygen reduction and the use of anion exchange membranes (AEMs).

13.2
History of Alkaline Methanol Fuel Cells

Alkaline fuel cells (AFCs) were the first fuel cell technology put into practical operation. The use of hydrogen fuel produces high cell efficiencies and cell voltages largely as a result of rapid oxygen reduction in alkaline media. Difficulties with storage and transportation of hydrogen have focused attention onto liquid fuels such as methanol which are cheap and have high volumetric and mass energy densities (6 kWh/kg, 5 kWh/l). As a liquid fuel, the storage and transportation of methanol is

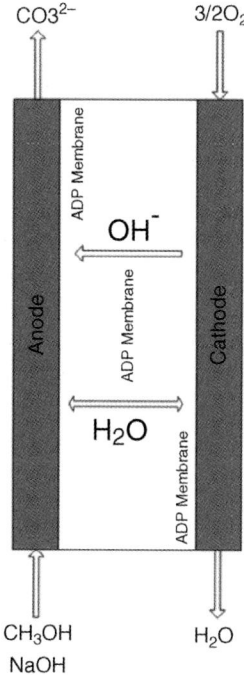

Anode reaction:
$$CH_3OH + 8OH^- \Longrightarrow CO_3^{2-} + 6H_2O + 6e^-$$

Cathode reaction:
$$3/2 O_2 + 3H_2O + 6e^- \Longrightarrow 6OH^-$$

Overall cell reaction:
$$CH_3OH + 3/2\, O_2 + 2OH^- \Longrightarrow CO_3^{2-} + 3H_2O$$

Figure 13.1 Schematic diagram of direct methanol alkaline fuel cell.

less complicated. Methanol could also be supplied by the existing fuel infrastructure. After many years of investigation, methanol has become a popular choice for small fuel cells. As a liquid fuel, methanol's direct use in a fuel cell can eliminate the humidification subsystem, required for hydrogen PEMFC. It also eliminates subsystems using reforming to generate pure hydrogen. Overall, direct oxidation of methanol is ideal and the DMFC is a promising power source because of its high efficiency, very low emissions, low operating temperature (<100 °C) and fast and convenient refueling.

One of the first DMFCs, described in 1955 by Justi and Winsel, operated in alkaline media, used a porous nickel anode and a porous nickel–silver cathode. The first DMFC using a solid polymer membrane was conceived by Hunger [3]. It contained an AEM with porous catalytic electrodes pressed on both sides. However, the performance was very low, only $1\,\text{mA}\,\text{cm}^{-2}$ at 0.25 V at room temperature with methanol and air as the reactants. Investigations of DMFC stacks were carried out by Shell Research and Exxon-Alsthom in the 1960s. Shell used sulfuric acid as the electrolyte, whilst Exxon-Alsthom adopted the alkaline electrolyte.

Some of the earlier work of Vieltich on alkaline methanol fuel cells is reported by Koscher and Kordesch [4] using 10 M KOH and 4.5 M methanol. With Pt (2–5 mg cm^{-2}) cathodes and electrocatalytic carbon anodes, the open-circuit voltage was 0.9 V and at a current density of 2 mA cm^{-2} the cell voltage was 0.6–0.75 V. Improvements in cell design with an unsintered Ag/PbO catalyst for methanol oxidation, and Pt/C/PTFE bonded structure gave single cell performance of 0.6 to 0.7 V at 6 mA cm^{-2}. The cell used a mixture of 6 M methanol and 9 M KOH as anolyte and 9 M KOH as catholyte, which were separated by an anion-exchange membrane. Overall this early work indicated the potential of AFCs using organic fuels; particularly if suitable alkaline conducting electrolytes can be developed.

13.3
Electrocatalysis of Methanol Oxidation in Alkaline Media

The electro-oxidation of methanol has been extensively studied in acid electrolyte. The complexity of the methanol oxidation mechanism has long been considered a difficult problem to solve due to catalytic inefficiency [5]. Compared with hydrogen oxidation, methanol oxidation is several orders of magnitude slower [6]. A general dual-path mechanism for methanol oxidation has been proposed [7, 8]:

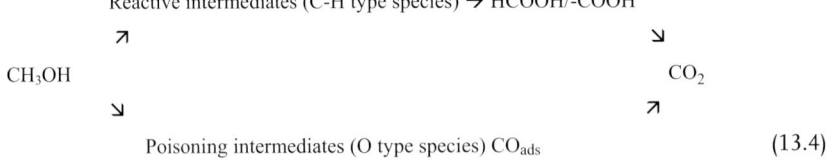

(13.4)

Both pathways need a catalyst; to dissociate the C–H bond and facilitate reaction of the resulting residue with some O-containing species to form CO_2 [6]. On a pure Pt electrode, methanol oxidation involves two processes:

- adsorption and dissociation of methanol molecules;
- dissociation of water, the oxygen donor of the reaction.

Complete electro-oxidation of methanol in acid is difficult to carry out with only Pt catalyst, due to the strongly bound poisoning species formed [9]. Oxidation requires a co-catalyst for more effective dissociation of water. The use of alkaline media was also proposed [10] because catalysts generally perform better in alkaline media [11].

Methanol oxidation in alkaline solutions, on unsupported platinum powder, has shown impressive polarization characteristics compared with those obtained in acid solution [12]. In alkaline solutions, the reaction current densities at certain potentials are at least an order of magnitude greater than in acidic electrolytes. Methanol oxidation in alkaline media on platinum single crystal and polycrystalline electrodes has been investigated by several research groups [13–19]. Unlike in acid, the electrode is not poisoned in alkali because bonding of chemisorbed intermediates on Pt is weak and the amount of possible poisoning species, for example, CO_{ads}, is smaller [8]. However, the use of alkaline solutions has a serious problem of progressive carbonation by CO_2, particularly from methanol oxidation, which influences the oxidation kinetics.

13.3.1
Mechanism of Methanol Oxidation in Alkaline Media

The mechanism of methanol oxidation on platinum in alkaline systems involves formation of absorbed methanolic species and OH on the Pt surface. The mechanism will take place through a series of reaction steps involving successive electron transfer (oxidation) to form adsorbed species which react with adsorbed OH to potentially form carbon dioxide. The reaction mechanism has been written as follows (Equations 13.5) [8]:

$$Pt + OH^- \rightarrow Pt-(OH)_{ads} + e^-$$
$$Pt + (CH_3OH)_{sol} \rightarrow Pt-(CH_3OH)_{ads}$$
$$Pt-(CH_3OH)_{ads} + OH^- \rightarrow Pt-(CH_3O)_{ads} + H_2O + e^-$$
$$Pt-(CH_3O)_{ads} + OH^- \rightarrow Pt-(CH_2O)_{ads} + H_2O + e^-$$
$$Pt-(CH_2O)_{ads} + OH^- \rightarrow Pt-(CHO)_{ads} + H_2O + e^-$$
$$Pt-(CHO)_{ads} + OH^- \rightarrow Pt-(CO)_{ads} + H_2O + e^-$$
$$Pt-(CHO)_{ads} + Pt-(OH)_{ads} + 2OH^- \rightarrow 2Pt + CO_2 + 2H_2O + 2e^-$$
$$Pt-(CHO)_{ads} + Pt-(OH)_{ads} + OH^- \rightarrow Pt + Pt-(COOH)_{ads} + H_2O + e^-$$
$$Pt-(CO)_{ads} + Pt-(OH)_{ads} + OH^- \rightarrow 2Pt + CO_2 + H_2O + e^-$$
$$Pt-(CO)_{ads} + Pt-(OH)_{ads} \rightarrow Pt + Pt-(COOH)_{ads}$$
$$Pt-(COOH)_{ads} + OH^- \rightarrow Pt-(OH)_{ads} + HCOO$$
$$Pt-(COOH)_{ads} + Pt-(OH)_{ads} \rightarrow 2Pt + CO_2 + H_2O \quad (13.5)$$

The rate-determining step is most likely oxidation of the active intermediate $-CHO$ [13–19]. The above mechanism is clearly complex and over recent years significant effort has been made to better understand the methanol oxidation reaction.

13.3.2
Precious Metal Catalysts

For methanol oxidation, the combination of (alloying) elements such as Ru, Mo, Sn, Re, Os, Rh, Pb and Bi with platinum give substantial tolerance to the poisoning species compared to platinum alone (see Chapter 2). Less work appears to have been done on platinum alloys for use in direct alcohol fuel cells in alkaline media. Palladium is a very good electrocatalyst for organic fuel electro-oxidation and is a potential alternative to platinum when alloyed with non-noble metals for use in direct alcohol fuel cells (DAFCs) (see Section 13.7).

The kinetics of methanol electro-oxidation on supported Pt and PtRu catalysts in acid and alkaline solutions have been recently discussed by Lovic [20]. A correlation between the beginning of OH_{ads} adsorption and methanol oxidation on Pt single crystal and nanoparticles was demonstrated. The Ru-containing catalysts shifted the onset potential for methanol oxidation to more negative values; although the effect was less pronounced in alkaline than in acid electrolyte. It was proposed that the rate-determining step in alkaline (and acid) media is the reaction between CO_{ads} and OH_{ads} according to a Langmuir–Hinshelwood mechanism.

The electrochemical oxidation of methanol in NaOH solution has been examined on a thin film $Pt_2 Ru_3/C$ electrode [21]. It was shown that in alkaline solution, the difference in activity between Pt/C and $Pt_2 Ru_3/C$ was significantly smaller than in acid solution. It was proposed that the reaction follows a quasi bifunctional mechanism (Equations 13.6).

$$Pt + CH_3OH_{sol} \rightarrow Pt - (CH_3O)_{ads} + 4H^+ + 4e^-$$
$$Ru + OH^- \rightarrow RuOH_{ads} + e^-$$
$$Pt + OH^- \rightarrow Pt - OH_{ads} e^-$$
$$Pt - CO_{ads} + Ru - OH_{ad} \rightarrow Pt - COOH_{ads} + Ru$$
$$Pt - CO_{ads} + Pt - OH_{ads} \rightarrow Pt - COOH_{ads} + Pt$$
$$Pt - COOH_{ads} + Pt - OH_{ads} \rightarrow 2Pt + CO_2 + H_2O \tag{13.6}$$

The kinetics of methanol oxidation on supported Pt and P/Ru (with nominal Pt:Ru ratios of 2 : 3) catalysts in 0.5 M H_2SO_4 and 0.1 NaOH at 295 and 333 K using a thin-film rotating disc electrode (RDE) have been reported [22]. The activity of Pt and Pt/Ru for methanol oxidation was a strong function of solution pH and temperature: at 333 K a factor of 5 higher than at 295 K. Oxidation kinetics were much better in alkaline than in acid solution; factors of 30 for Pt and 20 for Pt_2Ru_3 at 0.5 V at 333 K. The effect was attributed to pH competitive adsorption of oxygenated species with anions from supporting electrolytes. Negligible differences in kinetics were observed between Pt and Ru rich Pt alloys, presumably due to a slow rate of methanol dehydrogenation on the Ru rich surface and insufficient Pt sites required for dissociative chemisorption of methanol. Kinetic data and parameters indicated that the chemical reaction between adsorbed CO_{ad} and OH_{ad} species could be the rate-limiting step. Based on experimental Tafel slopes of 120 mV^{-1} and reaction orders with respect to methanol and OH^- ions of 0.5, the overall methanol oxidation current density was expressed as Equation 13.7:

$$j = nFk c_{CH_3OH}^{0.5} c_{OH^-}^{0.5} \exp\left(\frac{\alpha F E}{RT}\right) \tag{13.7}$$

assuming that the adsorption of methanol and OH species obey Temkin isotherms.

Cyclic voltammetry (CV) is commonly used to investigate oxidation of organics such as methanol. At a platinum electrode in 1 M methanol in 6 M KOH [12] an anodic peak appears at −0.6 V (vs NHE) with a maximum peak current at −0.04 V. The current then declines until a potential of −0.2 V, where the less active PtO monolayer inhibits methanol oxidation. The following mechanism was proposed (Equations 13.8) [24]:

$$Pt + OH^- \rightarrow Pt - (OH)_{ads} + e^-$$
$$Pt + (CH_3OH)_{sol} \rightarrow Pt - H + (CH_3OH)_{ads}$$
$$Pt - (CH_3OH)_{ads} + Pt - OH_{ads} \rightarrow Pt_2 - (CH_2O)_{ads} + H_2O$$
$$Pt - (CH_2O)_{ads} + Pt - OH_{ads} \rightarrow Pt_3 - (CHO)_{ads} + H_2O$$
$$Pt - (CHO)_{ads} + Pt - OH_{ads} \rightarrow Pt_2 - (CO)_{ads} + 2Pt$$
$$Pt_2 - (CO)_{ads} + Pt - OH_{ads} \rightarrow Pt - (COOH)_{ads} + 2Pt$$
$$Pt - (COOH)_{ads} + Pt - (OH)_{ads} \rightarrow 2Pt + CO_2 + H_2O \quad (13.8)$$

The peak potential range was attributed to the oxidation reactions above. It is suggested that the last four reactions in Equations 13.8 occur in the potential range −0.2 to 0.2 V as indicated by the formation of a potential hump. Residual weakly bonded CHO species are not oxidized at higher potentials as indicated by the absence of a peak in the oxygen evolution potential range. In the reverse sweep a potential peak occurs at around 0.13 V when the PtO layer is reduced. In this region the Pt electrode can oxidize residual weakly bonded CHO species and chemisorbed methanol.

When cyclic voltammograms are obtained in concentrated solutions with equal quantities of KOH and methanol no characteristic peaks are formed. It is suggested that dissociatively chemisorbed OH and methanol cover adjacent sites simultaneously and thus complete oxidation of chemisorbed methanol takes place. Note that very high current densities are experienced in these cyclic voltammograms ($>1\,A\,cm^{-2}$) indicating the high activity of methanol oxidation in alkaline media. This activity is attributed to the high active oxygen concentrations in concentrated KOH. With 1.0 M mixtures of methanol and KOH, oxidation peaks do appear (Figure 13.2) due to incomplete oxidation of CHO species caused by lower available oxygen atoms.

Cyclic voltammetry was used by Mahoharan [23] to investigate the effect of methanol concentration on oxidation. The study showed that at methanol concen-

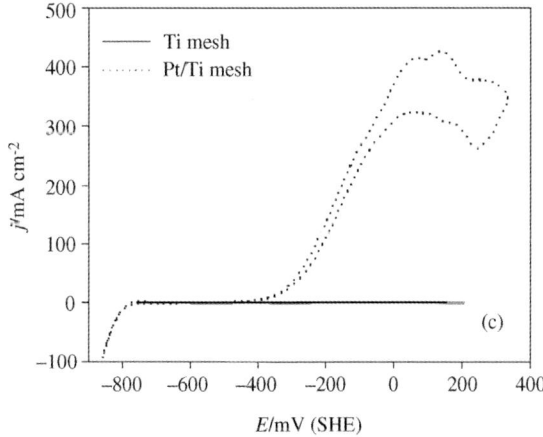

Figure 13.2 Cyclic voltammograms on platinized Ti mesh. 0.5 M NaOH, 2 M MeOH at 60 °C, sweep rate: 20 mV s^{-1}.

trations below 3 M, methanol oxidation was incomplete due to PtO layer formation allowing poisoning CHO_{ads} species to remain on the surface. Similar behavior was reported for Pt-Ru alloys and also for Pd. In the case of palladium, intermediate bridge-bonded CO species were not completely oxidized at all concentrations used.

Verma et al. [24] studied methanol and ethanol oxidation on Pt black using CV. Ethanol produced a single peak at 0.03 V (vs. NHE) and no characteristics of C–C bond breaking were observed. Chemical analysis confirmed production of acetaldehyde and acetic acid. For methanol a broad plateau was produced from −0.4 to 0.6 V and indicated that Pt black was more active for ethanol oxidation than for methanol oxidation.

13.3.3
Non-Precious Metal Catalysts

Nickel is well known as a catalyst due to its surface oxidation properties and is commonly used as an electrocatalyst for both anodic and cathodic reactions in organic synthesis and water electrolysis. Several studies of the electro-oxidation of alcohols on Ni have been reported [25, 26]. Taraszewska and Roslonek [27] found that a glassy carbon/$Ni(OH)_2$ modified electrode was an effective catalyst for the oxidation of methanol. Many organic compounds were found to oxidize at the same potential as that at which the surface of the nickel anode is oxidized [28–30]. Van Effen and Evans [31] found that the oxidation of ethanol in KOH solution involved the formation of a higher valence nickel oxide, which acted as a chemical oxidizing agent. This fact was confirmed [32] by CV, which proved the presence of a mediation process involving the higher oxides and the organic molecules, and by alternating current impedance measurements.

The use of Ni as a catalyst for the electro-oxidation of methanol in alkaline medium was studied using CV by Abdel Rahim et al. [33]. Ni dispersed on graphite showed catalytic activity towards methanol oxidation but massive Ni did not. Ni was dispersed on graphite by electrodeposition from acidic Ni_2SO_4 solution. The catalytic activity of the C/Ni electrodes towards methanol oxidation was found to vary with the amount of electrodeposited Ni. Methanol oxidation starts as Ni oxide is formed on the electrode surface and indicates that the electro-oxidation of methanol is activation controlled, proceeding by direct chemical reaction with NiO(OH) for thin nickel oxides and by charge transfer with the electrode for thick oxides. The accumulation of NiO(OH) had an inhibiting effect on activity which can be counteracted by a period of re-activation in the hydrogen evolution potential region.

A Ni zeolite has been reported with superior methanol oxidation activity to Ni and Pt supported on graphite [34]. This performance was assigned to a lower blocking of active sites than Pt due to repeated use.

A Co-W alloy prepared by electroplating was evaluated by Shobba et al. for methanol oxidation in alkaline media [35]. Heat treatment was shown to improve oxidation performance with open circuit potentials of 0.918 V obtained.

Several studies have examined gold as a possible alternative substrate metal for electro-oxidation of methanol in alkaline media [36–38]. Depending on its concentration, methanol is oxidized, on Au in alkaline media between 0.0 and 0.3 V. The following overall reaction has been proposed [36].

$$CH_3OH + 5OH^- = HCOO^- + 4H_2O + 4e^- \quad (13.11)$$

The role of OH_{ads}^{d-} in methanol oxidation on Au, is represented in the following generalized form:

$$CH_3OH + (Au - OH_{ads}^{\delta-}) = [Au - OH_{ads}^{\delta-} - CH_3OH]$$
$$[Au - OH_{ads}^{\delta-} - CH_3OH] = Au + H_2O + {}^{\bullet}CH_2OH + \delta e^-$$
$${}^{\bullet}CH_2OH + 4OH^{x-} = HCOO^- + 3H_2O + (4x)e^- \quad (13.10)$$

The adsorption characteristics of the intermediate reactants of the above multistep reaction on a gold film electrode in alkaline solutions were studied [39] by combining surface plasmon resonance measurements with Fourier transform electrochemical impedance spectroscopy (FT-EIS). The Au electrode in the alkaline environment is strongly covered with chemisorbed and partially charged hydrated OH^- ions [36]. Under potential activation, methanol reacts with OH^- adsorbates on Au. A subsequent step, most probably involving more solution species of OH leads to formate ions as the final oxidation product. Methanol oxidation showed no significant effects of 'site poisoning' by chemisorbed CO. The OH^- chemisorbed onto Au is reported to act as a stabilizing agent for the surface species of electroactive methanol. Formation and reduction of surface hydroxides and oxides on the Au surface overlap with methanol oxidation, and are rate-limiting factors for methanol oxidation.

Bin Guo et al. studied the electrocatalytic activity of thin gold films electrolessly deposited on fibrous mats of polyacrylonitrile for the oxidation of methanol [40]. The fibrous mats covered with smooth gold films had high conductivities and electrocatalytic activity and can be directly used as electrodes for methanol oxidation due to the high available surface area of gold nanoparticles in the three-dimensional metalized structure.

Synthesis and characterization of electrodeposited Ni–Pd alloy electrodes for methanol oxidation in alkaline electrolyte was studied by Suresh Kumar et al. [41]. Structural characterization of the electrocatalysts showed that the Ni–Pd catalysts were nanocrystalline, single phase, face-centered cubic materials, indicating the formation of complete solid solution in the alloy. The palladium composition of the alloy increased with a decrease in current density as seen in shifts in the X-ray diffraction peaks. The percentage shift in the d-spacing, calculated from X-ray diffraction, was in good agreement with the palladium percentages in the alloy. The electrocatalysts were active for methanol oxidation in alkaline medium.

Studies of oxides of V, Fe, Ni, In, Sn, La and Pb [42] have not generally shown practical activities for methanol oxidation. Graphite-supported perovskite-modified Pt has been reported to have a much higher electrocatalytic activity than smooth Pt or graphite-supported Pt electrodes towards methanol electro-oxidation in 1.0 M NaOH at 25 °C [43]. Lanthanum, strontium oxides have been studied by several researchers: Raghuveer and Vishwanathan [44] investigated the electro-oxidation of methanol on bulk and nanocrystalline $La_{1.8}Sr_{0.2}CuO_4$ in 1 M KOH. Raghuveer et al. [45]

investigated $La_{1.8}Sr_{0.2}CuO_4$, $La_{1.6}Sr_{0.4}CuO_4$ and $La1.9Sr0.1Cu0.9Sb0.1O4$ in 3.0 M KOH plus 2 M CH_3OH. Yu et al. [46] investigated $La0.75Sr0.25Cu\ O3^{-\delta}$ and $La0.75Sr0.25CoO3^{-\delta}$ in 1 M NaOH plus 1 M CH_3OH. All oxides gave reasonable oxidation currents in potential ranges appropriate to fuel cell operation.

Singh et al. studied perovskite-type $La_{2-x}\ Sr_x\ NiO_4$ ($0 \leq x \leq 1$) as anode materials for methanol oxidation in alkaline solutions [47]. The perovskite-type ternary oxides were prepared by a modified citric acid sol–gel route at 600 °C. From CV, chronoamperometry and impedance analyses, electrocatalytic activity of the base oxide ($x=0$) in 1 M KOH plus 1 M CH_3OH at 25 °C was found to increase with x, for oxides with $x=0$, 0.25, 0.5 and 1.0. The perovskite anodes did not exhibit any poisoning by methanol oxidation intermediates/products. The methanol electro-oxidation reaction gave a Tafel slope of $<40\,mV\ decade^{-1}$) on each oxide catalyst, which was independent of the Sr content.

Deshpande et al. [48] studied a variety of ABO_3 and A_2BO_4 (A = Sr, Ce, La and B = Co, Fe, Ni, Pt, Ru) oxides prepared by a solution combustion synthesis and measured their catalytic performance by rapid screening for methanol electro-oxidation. The performance of Sr-based perovskite was comparable to a standard Pt–Ru catalyst. Lan and Mukasyan [49] synthesized perovskites of ABO_3 (A = Ba, Ca, Sr, La; B = Fe, Ru) and electrochemically characterized them for methanol oxidation in DMFCs. The perovskites with ruthenium on the B-site were promising candidates as DMFC oxidation catalysts.

13.3.4
Effect of pH and Electrolyte

Due to the formation of carbonate and bicarbonate during methanol oxidation in aqueous NaOH and KOH solutions the effect of the former species has been investigated. The oxidation of methanol on Pt(111) in carbonate and bicarbonate was found to be slower than in NaOH by Tripokovic [50, 51]. The oxidation was reported to proceed with formation of poisoning species and the main reaction products were formate and CO. However on Pt (100) the oxidation of methanol in NaOH was reported to be less active than in carbonate and bicarbonate solutions. FTIR data showed that CO is weakly adsorbed on Pt(111) compared with Pt(110) and Pt(100) [52].

Figure 13.3 shows cyclic voltammograms recorded for a platinized electrode at 60 °C in 0.5 M solutions of NaOH, Na_2CO_3 and $NaHCO_3$. The peaks in the potential region $-800\,mV < E < -500\,mV$ are associated with the hydrogen adsorption/desorption process, while the anodic peaks that appear at higher potentials ($E > 0\,mV$) are related to a surface oxidation process that is, the formation of platinum oxide. On the negative scan, the peaks in the potential region $-100\,mV < E < 200\,mV$ are associated with the reduction of the platinum oxide. For the different electrolytes used in this study, the shapes of the voltammograms were similar but the peak potentials were shifted more positive, in the order $NaHCO_3 > Na_2CO_3 > NaOH$. The same order was also observed for the measured open circuit potentials which varied depending upon the different OH^- ion concentrations is the associated electrolyte. On the other hand, the potentials observed in Na_2CO_3 and $NaHCO_3$ may have been

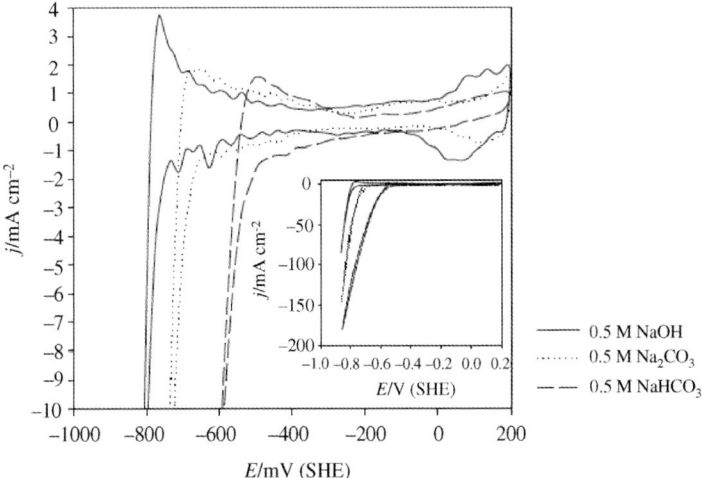

Figure 13.3 Cyclic voltammograms in different alkaline solutions. Sweep rate: 20 mV s^{-1}, $T=60\,°C$.

effected by specific adsorption of carbonate and bicarbonate ions, respectively, since the peak currents for the redox reactions in these electrolytes were lower than in NaOH, that is, the coverage of adsorbed OH could have been reduced due to the adsorption of carbonate and bicarbonate anions. Beden et al. [53] indicated that OH$^-$ was not strongly specifically adsorbed on the electrode surface. Also, it was argued by Tripkovic et al. [54] that strong adsorption of carbonate and bicarbonate on a platinum surface could be anticipated because of the identical geometrical configuration to platinum atom arrangement, especially on Pt (111).

Cyclic voltammograms for methanol oxidation on electrodeposited Pt in 0.5 M solutions of NaOH, Na_2CO_3, and $NaHCO_3$ at 60 °C are shown in Figure 13.4. Methanol oxidation activity in NaOH is significantly higher than in carbonate and bicarbonate media and this most probably reflects a higher coverage of adsorbed OH species (OH_{ads}). A closer examination of the onset potential for methanol oxidation is shown in Figure 13.4b. Compared to the methanol free electrolyte, the magnitude of the hydrogen adsorption/desorption peaks were reduced by the presence of methanol, implying that methanol is adsorbed preferentially on the electrode surface in this potential region. The onset for methanol oxidation occurred at more anodic potentials to those of the hydrogen adsorption/desorption region for all the alkaline solutions, although this was significantly more anodic for Na_2CO_3 and $NaHCO_3$ due to the lower pH in these electrolytes. A contributory factor to the lower activity of the Pt electrode in carbonate and bicarbonate may be due to the lower apparent active surface area (Table 13.1).

Tafel plots for methanol oxidation are shown in Figure 13.5a and b, respectively. The Tafel slopes measured at 20 °C and 60 °C are listed in Table 13.1. A Tafel slope of approximately 110 mV dec^{-1} was measured for methanol oxidation in NaOH, compared to approximately 200 mV dec^{-1} in both Na_2CO_3 and $NaHCO_3$. The variation in the Tafel slopes may again be attributed to carbonate and bicarbonate anions competing with hydroxyl ions and methanol molecules for adsorption sites.

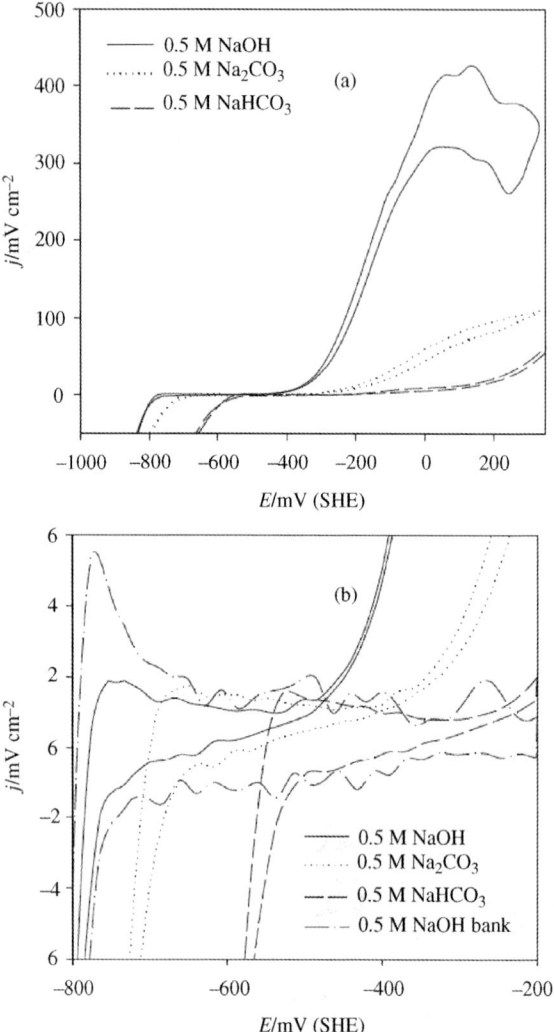

Figure 13.4 Cyclic voltammograms for methanol oxidation in 0.5 M NaOH, Na_2CO_3 and $NaHCO_3$ at 60 °C. Methanol concentration 2 M. Sweep rate: 20 mV s^{-1}. (a) full range (b) potential −860– −200 mV.

13.4
Oxygen Reduction and Methanol Tolerant Electrocatalysts

13.4.1
Oxygen Reduction Mechanism in Alkaline Media

The electrochemical reduction of oxygen in acid and alkaline media has been extensively studied because of its application in fuel cells and batteries [55–58].

Table 13.1 Tafel slopes for the oxidation of 2 M methanol and active areas in various alkaline solutions at 20 °C and 60 °C.

Solutions	60 °C	20 °C	Active area/cm² cm⁻²
0.5 M NaOH	114	109	124.2
0.5 M Na$_2$CO$_3$	199	203	102.5
0.5 M NaHCO$_3$	202	208	62.6

Reduction of oxygen is a kinetically limiting factor in fuel cell performance, notably in hydrogen fuel cells but equally so in organic-based fuel cells. Additionally the influence of fuel on the ORR can be a further performance limiting factor. Effects of various parameters, such as pH [59, 60], pretreatment of the electrode [59], type of electrolytes [59, 61], electrolyte concentrations [62, 63], and temperature [63], on the reaction kinetics have been examined. The most widely accepted mechanism for the oxygen reduction reaction (ORR), first proposed by Damjanovic et al. [64, 65]), is via two parallel reaction paths: the direct four-electron transfer reduction from oxygen to water and the formation of hydrogen peroxide as an intermediate in a two-electron transfer reaction:

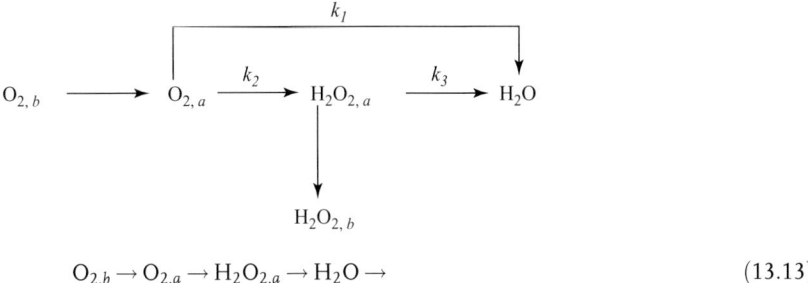

$$O_{2,b} \rightarrow O_{2,a} \rightarrow H_2O_{2,a} \rightarrow H_2O \rightarrow \qquad (13.13)$$

where *a* is an adsorbed state and *b* is the bulk solution.

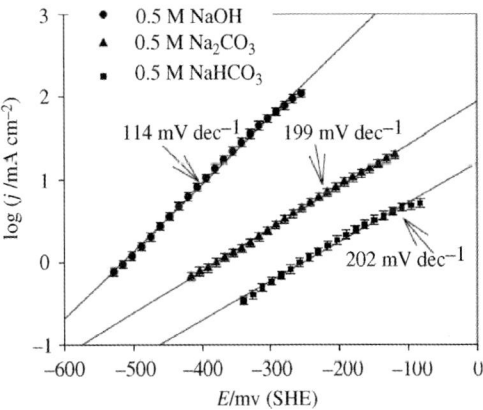

Figure 13.5 Tafel plots for methanol oxidation in 2.0 M methanol with 0.5 M NaOH, Na$_2$CO$_3$ and NaHCO$_3$ at 60 °C.

The reactions in alkaline media can be presented as

$$O_2 + 2H_2O + 4e^- = 4OH^- \tag{13.14}$$

and the path involving hydrogen peroxide as [64]:

$$O_2 + e^- \leftrightarrow O_2^-$$

$$O^{2-} + H_2O = HO_2^- + OH$$

$$OH + e^- \leftrightarrow OH^- \tag{13.15}$$

For oxygen reduction on Pt electrodes, in both acid and alkaline solutions, two Tafel regions were observed [64–66]. At low current density region, the Tafel slope was $-60\,\text{mV dec}^{-1}$ and the reaction order was 0.5 with respect to pH in alkaline solutions. The fractional reaction order was interpreted in terms of the first electrochemical step (reaction 3) as rate determining under a Temkin isotherm, that is, the adsorption of reaction intermediates O_{ads}, OH_{ads} and HO_{2ads} [59].

The rate expression under Temkin conditions of adsorption is:

$$j = nFkc_{O_2}c_{H^+}^{0.5}\exp\left(\frac{-\eta F}{RT}\right) \tag{13.16}$$

In the high current density region, the Tafel slope was $-120\,\text{mV dec}^{-1}$ and the reaction order respect to pH was 0 in alkaline solutions, with H_2O as a reacting species. The adsorption of intermediate species changed to a Langmuir isotherm and Temkin conditions no longer hold. The reaction is first order with respect to O_2 in both acid and alkaline solutions.

It was found that the amount of H_2O_2 formed in the ORR was greater in alkaline than in acid media [64, 65]. In alkaline solutions approximately 80% of the reduction current is via the direct reduction and the remainder of the current goes to the formation of H_2O_2, which suggests a more complicated mechanism.

13.4.1.1 Cyclic Voltammetry of Pt/C and Pt/Ru/C Catalysts in 1 M NaOH

Figure 13.6 illustrates the typical cyclic voltammogram response for carbon-supported Pt and Pt/Ru catalysts in 1 M NaOH solutions at 60 °C. The voltammetric curves are similar to those of bulk electrodes [67, 68] and Pt single crystals [69]. The cyclic voltammograms show three main regions [70, 71]:

(a) hydrogen adsorption/desorption
(b) double layer potential region
(c) surface oxide formation.

The hydrogen region for Pt/C shows several small peaks between $-640\,\text{mV}$ and $-320\,\text{mV}$ (SHE). For Pt/Ru/C it appears as a broader current response at higher cathodic potentials. Schmidt et al. [69] suggested that in alkaline solution, adsorption of OH may have started in the hydrogen region. The small peak immediately after the hydrogen adsorption/desorption peak may be associated with OH adsorption on Pt.

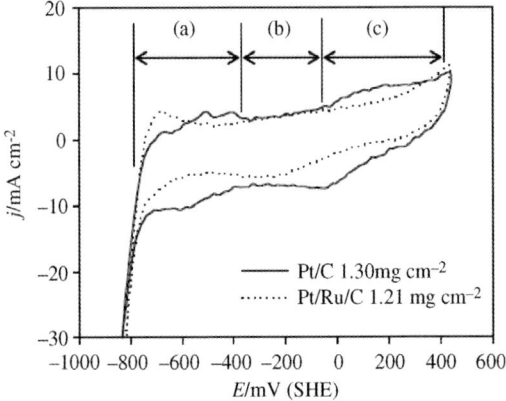

Figure 13.6 Cyclic voltammograms of Pt/C and Pt/Ru/C catalysts in 1 M NaOH blank solutions, sweep rate: 20 mV s^{-1}, T: 60 °C.

The surface oxide region also occurred at more cathodic potentials on Pt-Ru/C (about −238 mV) than on Pt/C (about −50 mV). The potential shift with Pt/Ru catalyst can be explained by adsorption of oxygen-containing species on Ru at significantly more negative potentials than on Pt due to the higher affinity of Ru towards oxygen [67].

Figure 13.7 shows Tafels plots at both 25 °C and 60 °C, which exhibit two linear regions at each temperature. At low overpotentials (ca 70 to 150 mV vs SHE), the Tafel slope was −70 mV dec^{-1}, and at high overpotentials (ca <40 mV vs SHE) it was −202 mV dec^{-1}. In general the Tafel slopes agree with those values reported in the

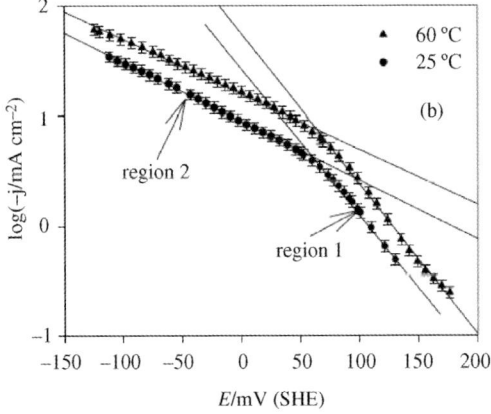

Figure 13.7 Tafel plots of oxygen reduction at 25 and 60 °C Electrode: Pt/C (60 wt.% of Pt) with 5% of PTFE binder with Pt loading of 1.3 mg cm^{-2} on gas diffusion layer: 2 mg cm^{-2} of Ketjen black with 10% PTFE on 20% wetproofed Toray 90 carbon paper. sweep rate: 1 mV s^{-1}.

literature for O$_2$ reduction on Pt: the value for low overpotentials is approximately −60 mV dec^{-1}, and slopes ranging from −120 mV dec^{-1} to over 200 mV dec^{-1} for the higher current density region [63, 72]. Differences in Tafel values may be due to the use of high surface area 'three-dimensional' porous electrodes, which exhibit potential distribution in the structure.

13.4.2
Non-platinum Electrocatalysts

A number of non-platinum electrocatalysts have been studied for ORR in alkaline medium. These include materials based on palladium [73], ruthenium [74], iron-porphyrin [75, 76], nickel, cobalt and nickel-cobalt-spinel catalysts [77–83] and manganese oxides [70–75]. With palladium catalysts, the onset of the reduction wave is typically shifted 50 mV towards more negative potentials in comparison to platinum materials. At the beginning of the reduction wave water is the main product, and at lower potential (higher cathodic overpotentials), hydrogen peroxide becomes the main reaction product [73].

Ruthenium-based catalysts exhibit poor ORR performance because of an oxide film formed at the catalyst surface; as a result of which the electrode becomes inactive towards oxygen reduction [75].

Iron (III) tetramethoxyphenylporphyrins (Fe-TMMP), prepared by adsorption of the Fe-TMMP chloride onto black pearl carbon followed by heat treatment at 1000 °C, have been studied by Gojkovic et al. [75]. The iron-porphyrin catalysts have achieved good activity and stability towards ORR in alkaline media, when calcined at temperatures above about 700 °C. The onset of the reduction wave is close to that obtained with platinum catalysts and the main reaction product is water, although hydrogen peroxide is produced notably at high overpotentials [75, 76]. In oxygen saturated solution the ORR was unaffected by the presence of methanol in both alkaline and acid media. The number of electrons liberated per mole of oxygen was between 3.4 and 4.0.

Nickel- and cobalt-based electrocatalysts have shown poor activity for ORR in alkaline media. Oxygen reduction on their oxides occurs via a two-electron mechanism producing hydrogen peroxide as the main product [77–83]. However, spinels of cobalt and nickel oxides. Ni$_x$Co$_{3-x}$O$_4$ ($0 < x < 1$) showed a reasonable activity towards ORR, although the four-electron reaction mechanism does not occur exclusively and a large amount of hydrogen peroxide was formed [77–79].

Silver-based cathodes have an extensive history for AFCs. Ag nanoparticles deposited on carbon are very easy to prepare and are known to be active towards ORR [84]. Wagner et al. [84] prepared Teflon-bonded Ag catalysts gas diffusion electrodes and studied the performance over extended periods of operation. In 5000 hours of operation electrodes showed up to 15% performance loss. Raney Ag-based cathodes were used by Siemens to develop AFCs [85].

There have been several studies of manganese oxides as oxygen reduction catalysts in alkaline media. Catalyst activity depends upon the type of oxide and often the type of carbon support. The most active were the MnO_2, Mn_2O_3 and MnOOH (manganese oxy-hydroxy) oxides [80, 82, 83]. The ORR mechanism on these catalysts involves the formation of the HO_2^- intermediate via the transfer of two electrons followed by a dismutation reaction of this species into OH^- and O_2. In voltammograms the two observed reduction peaks were assigned to a $2 + 2$ electron reduction mechanism. Mao et al. [86] prepared Mn oxide catalysts by chemical and thermal oxidation and reported that MnOOH had the highest activity of those studied. The kinetics of the dismutation reaction is fast, mainly on the MnOOH catalyst, so that nearly four electrons are exchanged during the ORR. Yang et al. [87] and Verma et al. [88] studied amorphous MnO_2 catalysts and reported high activity; with a four-electron pathway being dominant.

The activity of doped manganese oxides have also been studied for the ORR [75, 82, 86]. The use of nickel as doping agent gave the best activity, with the reduction wave shifted by 100 mV towards higher potentials; which was only 50 mV lower than that at a platinum catalyst. However the mechanism of reaction remains the same as for manganese oxides alone, that is, via the HO_2^- dismutation.

Ponce et al. [89] prepared $Ni_xAl_{1-x}Mn_2O_4$ cathodes by co-precipitation of metal hydroxides and by sol–gel routes from metal propionates. The ORR activity in alkaline media was improved by replacing Al by Ni, that is, with the $NiMn_2O_4$ catalyst.

Meng and Shen [90] evaluated low cost Pt free tungsten carbide nanocrystals promoted Ag composite electrocatalysts for the ORR in alkaline solution using a rotating ring disc electrode (RRDE). The novel Pt-free electrocatalysts gave activities similar to those of Pt-based electrocatalysts; they showed selectivity towards the electroreduction of oxygen and were not influenced by the presence of alcohols such as methanol, ethanol, isopropanol and glycerol. Such electrocatalysts can potentially be applied to AEM alcohol fuel cells.

13.4.3
Mixed Reduction Potential of Methanol with Oxygen

In the DMFC, cathode performance deteriorates because of the depolarizing effect caused by methanol crossover [91]. Most Pt-based cathodes are catalytically active for methanol oxidation under cell operating conditions since methanol oxidation commences at much more negative potential than O_2 reduction. This leads to a mixed cathode potential [92]. As shown in Figures 13.3–13.8, the linear sweep voltammograms shift to more negative potentials as the methanol concentration increased. An oxidation peak can be seen at a potential around 0 mV in the curves for methanol concentrations higher than 0.5 M. Equilibrium potentials were shifted to more negative values as the methanol concentration increased. A decrease in reduction current density can be observed in all potential ranges as the methanol concentration was increased. However, the decrease was relatively small with methanol concentration <0.5 M.

Tafel plots for O_2 reduction with various methanol concentrations are shown in Figure 13.9. Tafel slopes are $-65\,mV\,dec^{-1}$ and $-207\,mV\,dec^{-1}$ for 0.1 M methanol,

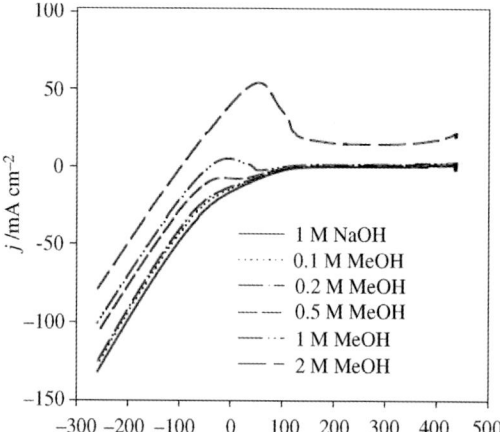

Figure 13.8 Effects of methanol on Oxygen reduction at 60 °C on Pt/C GDE. Pt loading 1.2 mg cm^{-2}, O$_2$ flow rate of 0.5 l min^{-1}, sweep rate 1 mV s^{-1}.

and -48 mV dec^{-1} and -199 mV dec^{-1} for 0.2 M methanol, for region 1 and region 2 respectively. These values agree with the value obtained in 1 M NaOH solution, although the Tafel slope for 0.2 M methanol decreased. A Tafel region was much less evident for O$_2$ reduction in solution mixed with methanol at concentrations higher than 0.5 M. This implies that different electron transfer processes are taking place on the electrode. Three valleys can be observed from the Tafel plot for O$_2$ reduction in the solution with 1 M methanol, which are three equilibrium potentials for O$_2$ reduction and methanol oxidation reactions. Combining the Tafel plot with the linear sweep (Figure 13.8), the first valley at 111 mV is assigned to the onset potential for O$_2$ reduction and between the other two valleys was a methanol oxidation peak. In the 2 M methanol solution, oxidation current dominated the positive potential range; reduction currents did not appear until the potential was -99 mV. Based on the data the 'methanol tolerance' of the Pt/C electrode with Pt loading of 1.2 mg cm^{-2} for O$_2$ reduction is around 0.2 M MeOH in 1 M NaOH.

To solve the problem of methanol crossover in the DMFC methanol-tolerant oxygen reduction catalysts based on alloys of platinum and other metals, for example, Fe, Ni, Co and Ru, have been researched [93–97]. However most of this work is based in acid media and limited work has been reported in alkaline environments. Both Fe-TMMP- and Ag-based electrocatalysts have exhibited good methanol tolerance as ORR electrocatalysts [75, 90]. Demarconnay investigated the electroreduction of dioxygen (ORR) in alkaline medium on Ag/C and Pt/C nanostructured catalysts [98]. The Ag/C catalysts were prepared using a colloidal route to obtain well dispersed catalysts on carbon, with a particle size around 15 nm. A loading of 20 wt.% Ag on carbon was found to give the best mass activity. Using RDE and RRDE methods the ORR kinetics, with the 20 wt.% Ag/C catalyst, was shown to follow almost a four-electron path. Tafel slopes and

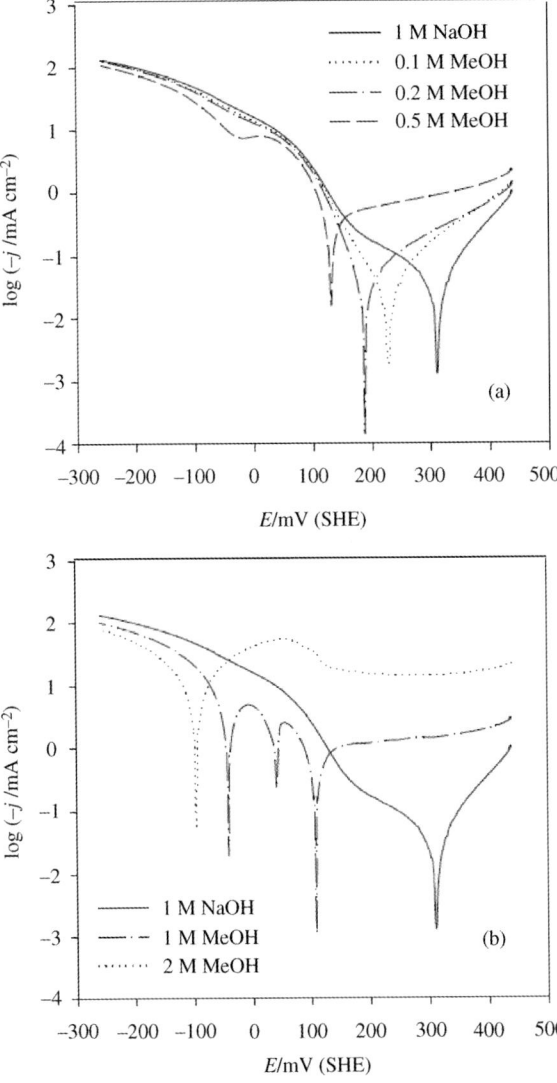

Figure 13.9 Effect of methanol on Tafel plots for oxygen reduction at 60 °C on Pt/C GDE. Pt loading 1.2 mg cm^{-2}, O_2 flow rate of 0.5 l min^{-1}, sweep rate 1 mV s^{-1}. (a) [MeOH] < 1 M, (b) [MeOH] ≥ 1 M.

diffusion limiting current densities for the Ag catalyst were found to be of the same order as for the Pt catalysts, whereas the exchange current density of the Ag/C catalyst, was at least 10 times lower than that of the Pt/C catalyst. The behavior of the Ag/C was affected less by the presence of methanol than was the Pt/C catalyst (Figure 13.10) [98].

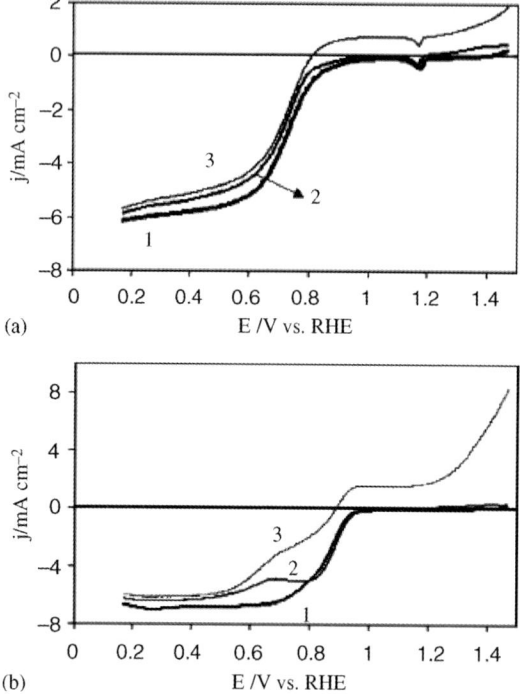

Figure 13.10 Linear sweep polarization curves from RDE. Rotation rate 2500 rpm recorded in O_2-saturated 0.1 M NaOH electrolyte (T = 20 °C, v = 5 mV s^{-1}: (1) methanol-free electrolyte, (2) 0.1 M MeOH, (3) 0.5 M MeOH; (a) on a 20 wt.% Ag/C; (b) on a 20 wt.% Pt/C. [10].

13.5
Direct Methanol Fuel Cells in Alkaline Media

13.5.1
Aqueous Electrolyte Media

Several studies of direct alcohol fuel cells using aqueous alkaline electrolyte have been reported [24]. To date the performance of the fuel cells has not matched that achieved with DMFCs using PEMs. Figure 13.11 shows typical cell polarization curves and electrode potentials obtained for tests performed at 20 °C, 40 °C and 60 °C [99]. Cell performance improved as the temperature increased; with the higher temperature giving a higher open circuit voltage. On the cathode side, the open circuit potential and oxygen reduction polarization increased slightly with temperature. On the anode side, three almost parallel polarization curves were observed which shifted to more negative potential as the temperature increased. The performance of the cell, with an increase in methanol concentration to 4 M, gave power densities at 60 °C of 6 mW cm^{-2}.

13.5 Direct Methanol Fuel Cells in Alkaline Media

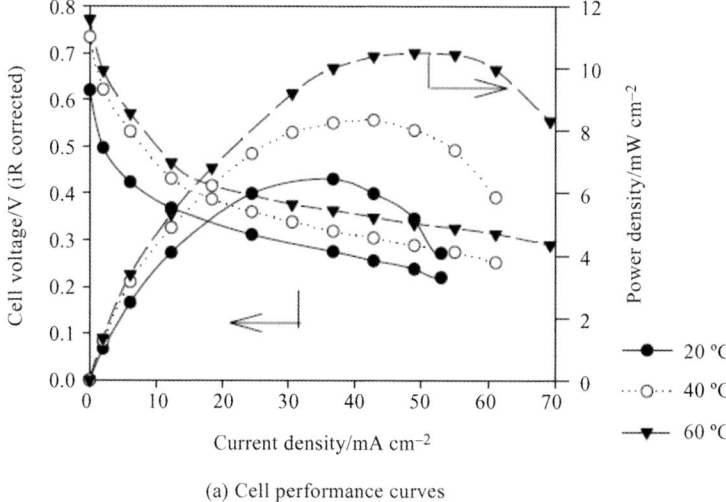

(a) Cell performance curves

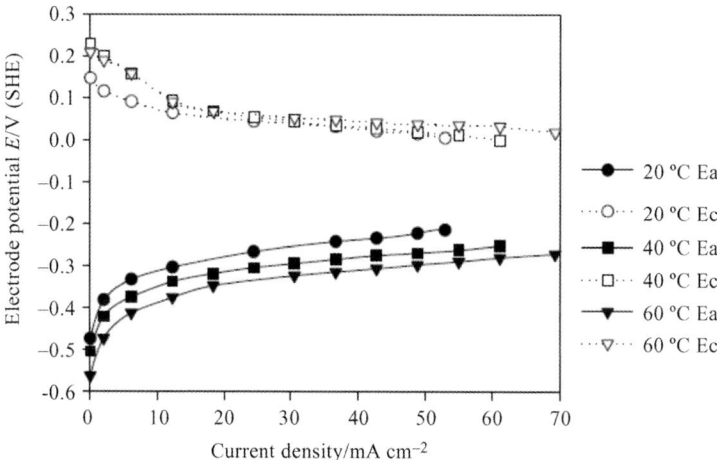

Figure 13.11 Temperature effects on cell performance. MEA: anode: Pt/C (60 wt.% Pt) 2.19 mg cm^{-2} on non-wetproofed Toray 90 carbon paper; cathode: Pt/C (60 wt.% Pt) 2.07 mg cm^{-2} GDL on 20% wetproofed Toray 90 carbon paper. Cell operated at 20 °C, 40 °C and 60 °C with air pressure 1 bar in 2 M methanol and 1 M NaOH.

Yu and Scott [100] also examined alternative supports for the DAMFC; using mini-mesh Ti and an AEM. The Ti electrode was prepared by thermal decomposition and the membrane supported electrode was prepared by impregnation with Pt salt followed by chemical reduction. Typical data is shown in Figure 13.12, where a marginal improvement in performance is seen with the Ti supported electrode.

Methanol oxidation AFCs have been investigated by Verma *et al.* using metal ion impregnated electrodes [101]. With Ag$^+$ impregnated cathodes and Al(III) and Fe

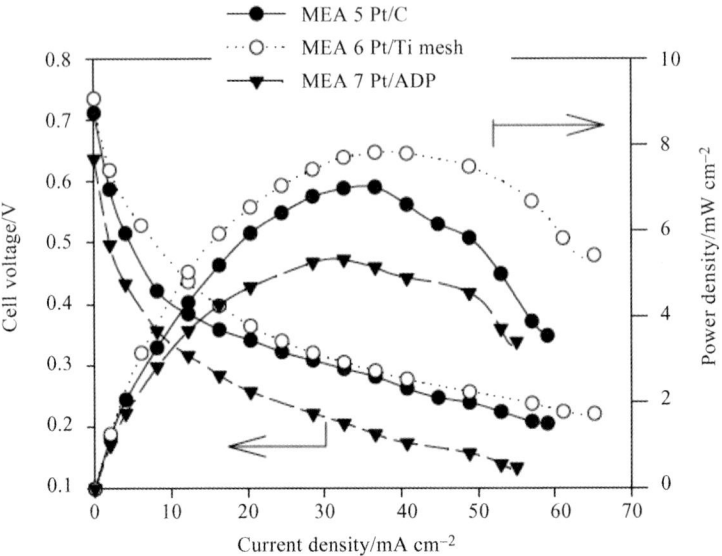

Figure 13.12 Comparison of carbon-supported, Ti mesh supported and membrane supported MEAs at 60 °C (iR corrected). Air pressure 1 bar, 2 M methanol in 1 M NaOH. MEA 5: anode: non-wetproofed Toray + Pt/C (60 wt.%) Pt 1.46 mg cm^{-2}, cathode: standard Pt 2.05 mg cm^{-2}. MEA 6: anode: thermal deposition of Pt (1.24 mg cm^{-2}) on Ti mesh, cathode: standard, Pt 2.15 mg cm^{-2} MEA7: chemical deposition of Pt on both sides, anode: Pt 0.933 mg cm^{-2}, cathode 0.526 mg cm^{-2} with gas diffusion layer.

(III) treated anodes, modest current densities (maximum 54 mA cm^{-2} with Al) were achieved with a 6 M KOH electrolyte. More recently Verma et al. [88] investigated Pt anodes and MnO$_2$ cathodes for direct methanol and ethanol fuel cells. Through investigation of the effect of KOH and alcohol concentrations, maximum power densities of 16.2 and 13.8 mW cm^{-2} were obtained for methanol and ethanol respectively, at 25 °C, using 3 M concentrations of alcohol and electrolyte.

13.5.2
Cationic Exchange Membranes

Polymer electrolyte materials such as Nafion are used commercially for both proton transfer and for sodium ion transfer; in for example the electrolysis of sodium chloride to produce chlorine and sodium hydroxide. Consequently they can be considered as SPE for the DMFC based on the transfer of Na$^+$ from the anode to cathode of the cell. An attraction of these SPEs is that if a Na$^+$ ion form of Nafion was used it offers much greater stability than currently available OH$^-$ ion (anion) conducting membranes. A disadvantage of the Na$^+$ ion form of Nafion is that the ionic conductivity is lower than that of the H$^+$ ion form of Nafion. Thus there will be a compromise between ionic conductivity and electrocatalyst activity at both anode and cathode. These factors affect both the membrane, membrane–catalyst layer interface and the bonded electrocatalyst layers. A schematic diagram of the electrochemical

13.5 Direct Methanol Fuel Cells in Alkaline Media

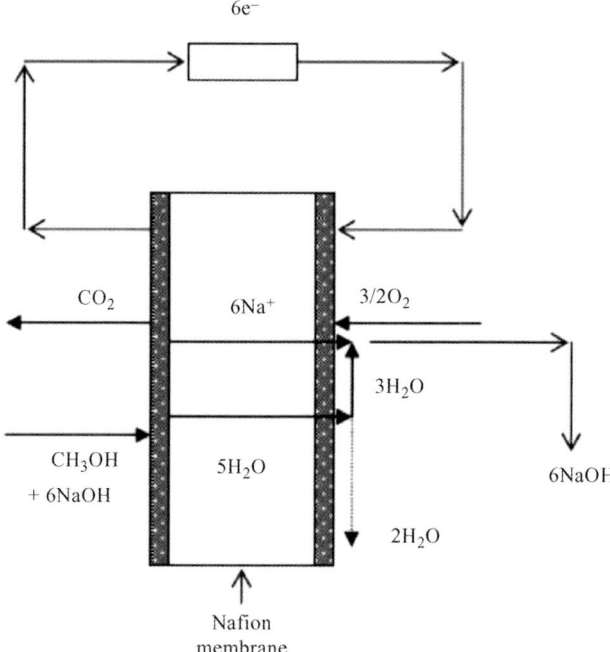

Figure 13.13 Schematic diagram of reactions and transport processes in a DMAFC incorporating a Na$^+$ form Nafion membrane.

reactions and transport processes in such a system is shown in Figure 13.13. The conduction process in this system is by the transport of Na$^+$ ions from anode to cathode. A practical consequence of DMAFCs operating in such a manner is that hydroxide ions produced by oxygen reduction react rapidly with Na$^+$ ions leading to the formation of sodium hydroxide in the cathode side of the cell. In addition, water is transported across the cell to the cathode. This being the case, it is necessary to recycle this component to the anode to maintain an overall sodium ion balance in the system.

The performance of a Na$^+$ conducting Nafion-based DMFC has been studied [102]. Although the open circuit voltage obtained was favorable (about 0.8 V), in comparison to DMFCs operating in acidic media, the polarization of the DMAFC was greater [103]. This greater polarization was due to the greater ionic resistance of the Na$^+$ conduction and to kinetic factors associated with the electrochemical reactions. In the case of the latter, carbonate formation at the anode, by reaction of CO_2 with the supporting electrolyte, is likely to result in concentration polarization, especially at high current densities. It is also possible that the surface coverage of OH species on the electrode is reduced by the specific adsorption of CO_3^{2-} and HCO_3^- ions, hindering the methanol oxidation kinetics [104]. However, it should be noted that in practise, the extent of carbonation of the electrolyte was not large because carbonation of the electrolyte was not a rapid process, with copious CO_2 gas evolution

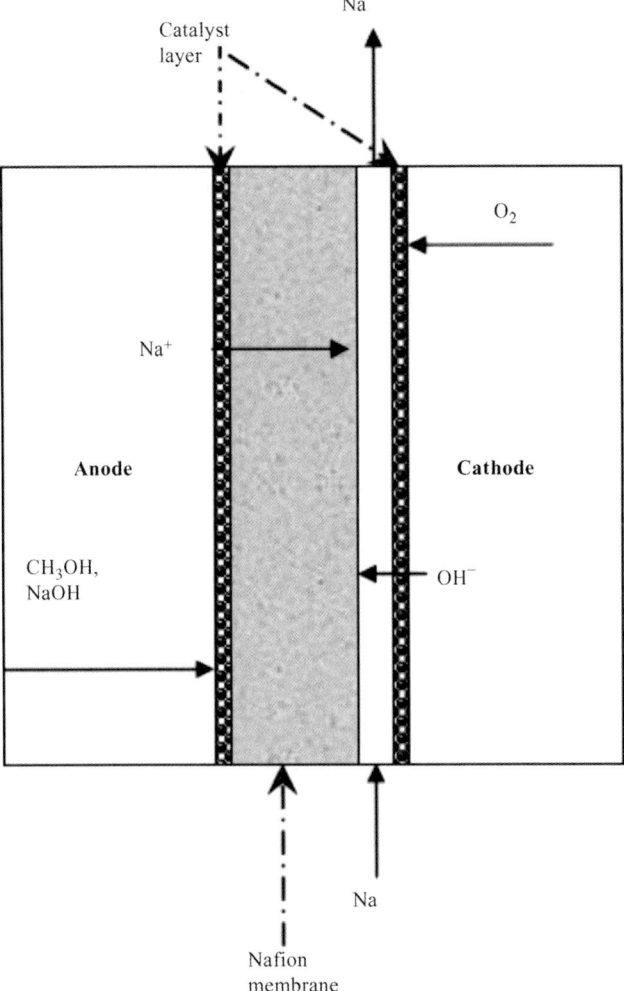

Figure 13.14 Schematic diagram of DMAFC with Nafion membrane incorporating an electrolyte circulating layer in the cathode compartment.

observed during operation. Thus a practical DMAFC could potentially operate with replacement of hydroxide electrolyte after many days of operation.

In addition, in the system used, Na^+ ions that passed from the anode combine with OH^- ions produced at the cathode and the resulting NaOH is removed in the cathode condensate. An alternative method of operation, shown schematically in Figure 13.14, would be to have a third layer of circulating electrolyte between the membrane and the cathode gas diffusion electrode, in which Na^+ ions and OH^- ions combine to form NaOH. Such a concept of electrolyte circulating has been used in the acid form of the DMFC.

Overall a DMFC system based on sodium ion transfer offers limited applications for sustained long term fuel cell use except where electrolyte replacement is acceptable. Alternatively systems based on less basic electrolytes, which also can involve alternative metal cations may be possible.

13.5.3
Invariant Electrolyte Media

The use of non-acid electrolytes in organic fuel cells includes less basic solutions of potassium bicarbonate/carbonate and cesium and rubidium carbonates/bicarbonates. The latter have been used because of their high solubility (>10–15 mol% at 20 °C) which avoided precipitation problems associated with potassium species [105]. Conductivities of the Ce and Rb electrolytes are in the range of 0.05–0.3 S cm^{-1}. Fuel cell tests have been reported on organic fuels such as methanol and ethylene glycol and stable operation of 500 hours duration at 130 °C, with methanol. At temperatures above 100 °C, power densities of some 40 mW cm^{-2} were achieved with methanol vapor using PTFE-bonded platinum black electrodes. In principle, cells can be operated at much higher temperatures. The performance of these cells is quite respectable in view of the advances made in electrode fabrication technology since this work was done.

In cell operation, with methanol, the anode reaction consumes carbonate ions which then react with CO_2 to form bicarbonate which diffuses to the cathode. The hydroxide ions produced by oxygen reduction at the cathode then react with bicarbonate to replenish the carbonate. Overall this maintains the invariance of the electrolyte. A claimed advantage for these carbonate electrolyte cells is that the ORR has a high rate constant; higher than that in OH^- electrolyte [106]. However, the oxygen solubility is lower in the carbonate electrolytes. One limitation is that of OH^- ion accumulation at the cathode in porous electrodes.

The attraction of Ce and Rb carbonate electrolytes is that they avoid the reaction of hydroxide electrolytes (K or Na) with carbon dioxide forming carbonates which can lead to precipitation problems. The major disadvantage is the low conductivity of electrolyte in comparison to KOH electrolytes.

13.6
Direct Alkaline Polymer Electrolyte Membrane Fuel Cells

13.6.1
Anion Exchange Membrane for Methanol Fuel Cells

Due to the obvious limitation of carbon dioxide formation in direct alcohol fuel cells in aqueous KOH or NaOH electrolytes, interest in DMAFC with OH^- ion conducting or anion exchange membranes (AEM) has increased. AEMs are SPE membranes that contain positive ionic groups (e.g., quaternary ammonium (QA) functional groups such as poly-N^+Me_3) and mobile negatively charged anions (e.g., OH^-). However despite the potential advantages of using AEMs in fuel cells, such as reduced materials (membrane electrodes, bipolar plates) costs and improved electrocatalyst activity, there are several important issues to be considered with their development, particularly with the OH^- AEM forms; notably conductivity and stability:

1. The diffusion coefficients and mobilities of OH^- anions are usually less than that of H^+, and QA ionic groups are less dissociated than typical sulfonic acid groups (pKa for sulfonic acid groups are typically -1; pKb for QA groups are around $+4$). Hence, there are concerns that these AEMs will not posses adequate intrinsic conductivities for application in fuel cells.

2. The OH^- anions are effective nucleophiles: if the AEMs contain good leaving groups (e.g., the $-N^+Me_3$ ionic groups themselves) then the chemical stability of the AEMs may be inadequate for use in fuel cells, particularly at elevated temperatures.

3. AEMs are exposed to OH^- ions during fuel cell operation (ORR produces OH^- ion) and during their preparation. Typical AEMs are prepared in their halide form and immersed in aqueous NaOH or KOH.

4. Tolerance of the AEM fuel cells to CO_2 in the cathode supply (carbonation) has not been fully investigated.

Despite these potential issues with AEMs, research to date [107] has shown that AEMs exhibit conductivities that are high enough to allow them to be applied in fuel cells. Also thinner AEMs perform better than thicker AEMs in hydrogen fuel cells due to reduction in ionic resistance and in the supply of stoichiometric reactant water to the cathode ORR catalyst sites. KOH-doped polymer-based AEM fuel cells give the highest power performances to date for H_2/O_2 solid AFCs (about 370 mW cm^{-2} compared to about 90–130 mW cm^{-2} for non-doped analogs) [108]. Additionally AEMs that have been converted to the carbonate (CO_3^{2-}) form, perform as well as the OH^- analogs.

Building upon the positive attributes for AEMs research in DMAFCs has thus focused on preparation of AEMs with lower methanol permeability [109] and lower manufacturing costs than those of Nafion. A number of membrane materials have been reported for application in DMAFC, such as polysiloxane containing QA groups [110]; aminated poly(oxyethylene) methacrylates [111]; quaternized polyethersulfone cardo [112], radiation-grafted poly(vinylidene fluoride) (PVDF) and poly (tetrafluoroethene-co-hexafluoropropylene) (FEP) [113], and quaternized poly(phthalazinon ether sulfone ketone), and so on [114]. However, in many cases these membrane materials are not suitable for further application in DMAFC due to poor stability in hydroxide electrolytes; nonfluorinated polymer membranes have poor fuel-cell-related performance, and fluorinated polymers such as FEP have relatively high cost.

Liang et al. prepared and characterized chloroacetylated poly(2,6-dimethyl-1,4-phenylene oxide) (CPPO) with bromomethylated poly(2,6-dimethyl-1,4-phenylene oxide) (BPPO) blend membranes for potential application in alkaline direct methanol fuel cell [115]. Poly(2,6-dimethyl-1,4-phenylene oxide) (PPO) is a unique material with strong hydrophobicity, a high glass transition temperature ($Tg = 212\,°C$) and hydrolytic stability [116]. Substituted materials have good miscibility since they all have PPO backbones. The PPO blend membranes exhibited high hydroxyl conductivities (0.022–0.032 S cm^{-1} at 25 °C) and low methanol permeability (1.35×10^{-7} to 1.46×10^{-7} cm^2 s^{-1}). A blend membranes with 30–40 wt.% CPPO was recommended for application in DMAFCs because of its low methanol permeability, excellent mechanical properties and comparatively high hydroxyl conductivity [116].

Hong et al. [117] prepared weak-base AEMs, by amination of chlorinated polypropylene with polyethyleneimine (PEI) at low temperatures, which have potential applications in DMAFCs. They used a solvent mixture of acetone and toluene at low temperatures (30–55 °C), yielding colloidal PEI-aminated CPP suspensions and from there AEMs. The chemical structure, chemical composition, microstructure, and thermal stability of CPP and PEI-aminated CPP membranes were characterized by Fourier infrared spectroscopy, elemental analysis, scanning electron microscopy, thermogravimetric analysis, respectively. The PEI-CPP membranes exhibited high ionic conductivities ranging from 0.89×10^{-2} to 1.36×10^{-2} S cm^{-1}, and anion exchange capacities ranging from 7.38 to 9.33 mmol (g dry membrane)$^{-1}$. The water uptake of the PEI-CPP membrane increased whereas the methanol uptake dropped when the degree of amination increased. The PEI-CPP membranes can be further converted into QA-based membranes for DMAFCs.

Quaternized polyethersulfone cardo (QPES-C) AEMs have been prepared for potential applications in DMAFCs [118]. Polyethersulfone cardo was chloromethylated with the complex solution of chloromethylether and zinc chloride. Subsequent reaction with trimethylamine and ion exchange with sodium hydroxide yielded the alkaline AEM. Ionic conductivity and methanol permeability of the membranes were investigated as a function of temperature. Ionic conductivities of QPES-C membrane in 1 M NaOH solution were 4.1×10^{-2} S to 9.2×10^{-2} cm^{-1} and methanol permeabilities were from 5.72×10^{-8} to 1.23×10^{-7} cm^2 s^{-1} over the temperature range 25–70 °C.

An alternative approach to the production of AEMs is the use of suitable doping/reaction of polymer films with KOH. One example is a AEM-based on alkaline-doped polybenzimidazole (PBI) membrane [119]. It has been shown that PBI has a remarkable capacity to concentrate KOH, even concentrations 3 M. The highest conductivity of KOH-doped PBI, of 9 10^{-2} S cm^{-1} at 258 °C, obtained was higher than that obtained previously for H$_2$SO$_4$-doped PBI (5 10^{-2} S cm^{-1}) at 258 °C. PEMFCs based on an alkali-doped PBI membrane were demonstrated, and their characteristics exhibited the same performance as those of PEMFCs based on Nafion 117. Evaluation of such materials is ongoing in the author's laboratory.

13.6.2
Direct Methanol Alkaline Membrane Fuel Cell Performance

The use of AEMs in DMAFCs has been limited and has not yet achieved high performance. Varcoe et al. [107] developed and characterized the QA (as the counter ions bound to the polymer backbone) radiation grafted ETFC, PVDF and FEP alkaline AEM. A maximum power density of 8.5 mW cm^{-2} was obtained [120] in a metal-cation-free methanol/O$_2$ fuel cell with 2–2.5 bar back pressure at 80 °C. In comparison, a peak power density of 130 mW cm^{-2} for the H_2/O_2 fuel cell with the AEM was obtained.

Yang et al. [121, 122] prepared alkaline PVA-based membranes for use in fuel cells. A crosslinked PVA/SSA (10 wt.% sulfosuccinic acid) solid polymer membrane was obtained by a solution casting technique. The membrane was tested in a DMAFC with a MnO_2/CNT cathode and a Ti-base PtRu black anode. The maximum current density of the DMFC was 4.1 mW cm^{-2} at 60 °C and 1 bar pressure in 2 M KOH 2 M CH_3OH solution [122]. Yang et al. also investigated an air-breathing direct methanol fuel cell using a poly(vinyl alcohol)/hydroxyapatite composite polymer membrane [123]. The composite polymer membrane was prepared by solution casting and characterized by thermal gravimetric analysis, X-ray diffraction, scanning electron microscopy, micro-Raman spectroscopy and AC impedance methods. An air-breathing DMFC, comprising an air cathode with MnO_2/BP2000 carbon inks on Ni-foam, an PtRu black on Ti-mesh anode, and the PVA/HAP composite polymer membrane, gave a maximum peak power density in 8 M KOH + 2 M CH_3OH solution of 11.5 mW cm^{-2}. Continuing the PVA theme, Yang investigated a PVA/ TiO = composite polymer membrane for DMAFCs [124]. Glutaraldehyde (GA) was used as a crosslinker for the composite polymer membrane to enhance the chemical, thermal and mechanical stability. The DMAFC using this novel cheap PVA/TiO_2 composite polymer membrane showed good electrochemical performance at ambient temperature and pressure. The maximum peak power density of the alkaline DMFC is about 7.54 mW cm^{-2} at 60 °C and 1 atm.

Bunazawa and Yamazaki [125] studied the influence of the anion ionomer content and a silver cathode catalyst on the performance of alkaline membrane electrode assemblies membrane electrode assembly (MEA) for direct methanol fuel cells (DMFCs). The anion ionomer was an A3-solution, from Tokuyama and was found to enhance the performance of DMAFCs. The ionomer content has been shown (Figure 13.15) to contribute to ohmic polarization of the anode and diffusion polarization of the cathode. The Ag cathode catalyst was suggested as an alternative to Pt catalyst because of its tolerance to methanol crossover, although the power performance of the fuel cell was not high.

Coutanceau et al. [126] used a new kind of anionic membrane, developed by Solvay, to investigate electrocatalysts (both anodic and cathodic) for AFCs with methanol or ethylene glycol at room temperature. For the ORR, catalysts containing silver gave encouraging results. For the anode, platinum-based and platinum-free electrocatalysts were examined. A Pt–Pd catalyst gave power densities of 18–20 mW cm^{-2} with methanol or ethylene glycol at 20 °C, using sodium hydroxide in the electrolyte.

Polybenzimidazole (PBI) membranes are well known to possess excellent endurance both in alkali medium and at high temperature. Although it is an electronic and ionic insulator, there are two imide groups (–N–) in one repeat unit of PBI. Therefore, it can become a proton conductor when doped with the acids, for example, phosphoric acid-doped PBI membrane, which is used for high temperature proton exchange membrane fuel cells (PEMFCs) [128]. Although PBI is a weakly basic polymer, the amine groups (–NH–) in its structure allow it to be an ionic conductor by introducing inorganic hydroxides [119]. However, compared with the acid-doped

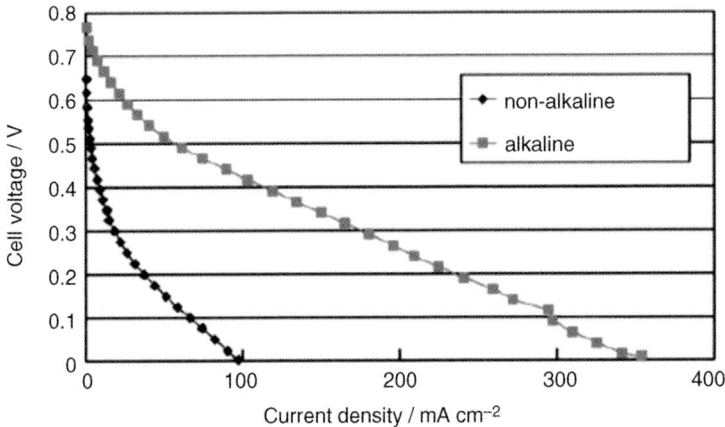

Figure 13.15 I–V curves of MEAs at 80 °C with non-alkaline or alkaline fuel. Anode catalyst: PtRu (Pt loading: 0.75 mg cm^{-2}; Ru loading: 0.58 mg cm^{-2}); cathode catalyst: Pt (Pt loading: 1.4 mg cm^{-2}); ionomer content: 45.4% mass, anode fuel: MeOH (1 M) with and without NaOH (0.5 M); cathode gas: O$_2$ (100% RH). 25 mV min^{-1} from OCV.

PBI, few reports are available about alkali-doped PBI for fuel cells: alkaline PBI membrane was first investigated for a hydrogen fuel cell [119].

Hou et al. [127] have recently used a KOH-doped PBI membrane as an AEM for alkaline direct ethanol fuel cell (ADEFC). The distributions of nitrogen, oxygen and potassium in the membrane were analyzed by XRD and SEM-EDX, respectively. It was reported that free or combined KOH molecules may exist in the PBI matrix, which was helpful for the ionic conductivity of PBI/KOH. The amine groups in PBI may act as proton donors and react with strong alkali although this reaction may take place incompletely. In addition, the water was also adsorbed by the PBI membrane during alkali doping. Some of KOH molecules were probably taken into the polymer by water molecules. Therefore, EDX mapping of potassium and oxygen were higher than that of nitrogen. This result suggested that free or combined KOH molecules by long-distance interaction may exist in the PBI matrix, which aid ionic conductivity of PBI/KOH. According to XRD and SEM-EDX, data two possible combinations between PBI and KOH are possible (1) combination between K$^+$ and $-$NH$-$ in the imidazole ring of PBI as a result of neutralization or interaction; (2) hydrogen bonding between OH$^-$ and N in the imidazole ring of PBI, while K$^+$ was introduced into the polymer connecting with OH$^-$ to form a charge balance, as shown in Figure 13.16.

The ionic conductivity of the membrane was 0.0184 S cm^{-1}, which was much lower than 0.10 S cm^{-1} of H-Nafion. This result may be due to the fact that the transfer number of OH$^-$ is inherently much lower than that of H$^+$, which is only 0.25 times of the transfer number of H$^+$. Ethanol permeability through PBI/KOH was 6.5 × 10^{-7} cm^2 s^{-1}, which was much lower than that of Nafion. Fuel cell tests with 2 M ethanol and 2 M KOH gave open circuit potentials at 75 °C and 90 °C of

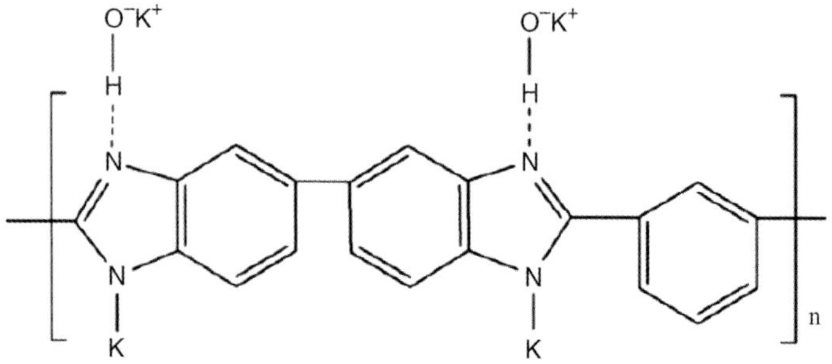

Figure 13.16 Microstructure of the alkali-doped poly[2,2-O-(m-phenylene)-5,5-O-bibenzimidazole], PBI/KOH.

0.923 V and 0.98 V, respectively (Figure 13.17). These results reflect the lower ethanol permeability of PBI/KOH membrane and the anion movement from the cathode to the anode; the electro-osmotic effect. The peak power densities of DEAFC at 75 °C and 90 °C were 49.2 and 61 mW cm^{-2} respectively, some 3 to 6 times previously reported in the literature.

Overall the performance of the DMAFC has not realized its potential, even allowing for the inherent lower ion conductivity in comparison to PEM-based fuel cells. One of the major factors in this is the absence of suitable liquid forms of ionomers that can be used in catalyst layer fabrication. Development of ionomers is required for significant future increases in AFC performance.

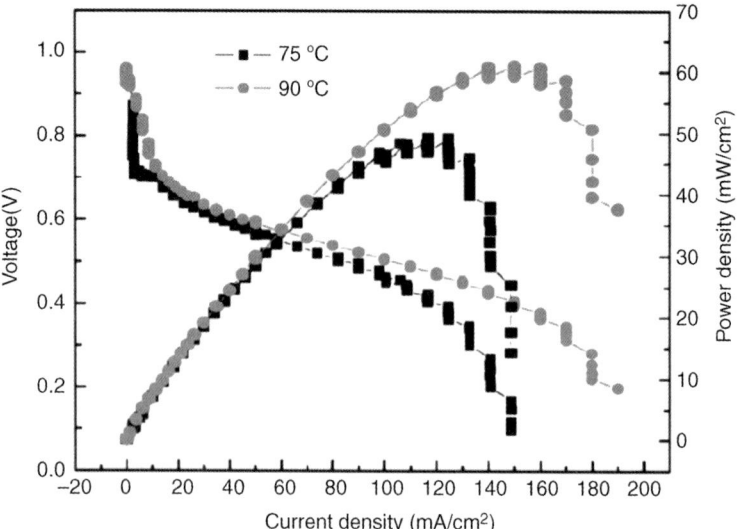

Figure 13.17 The I-V curves of ADEFC with alkali-doped PBI as an electrolyte membrane at 75 °C and 90 °C.

13.6 Direct Alkaline Polymer Electrolyte Membrane Fuel Cells

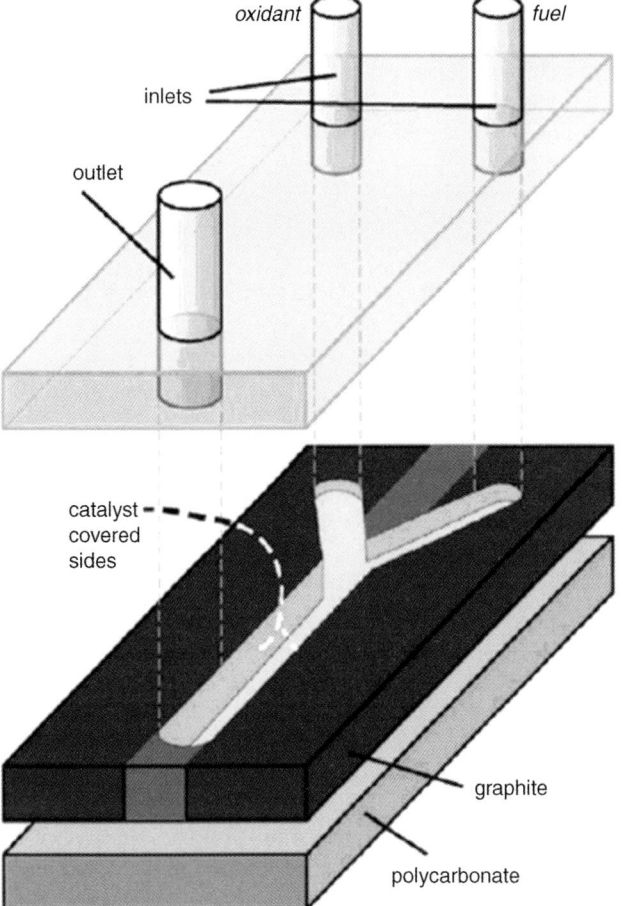

Figure 13.18 Assembly scheme of a membraneless, laminar flow-based fuel cell. Graphite plates as the catalyst support, current collector, and defining structure for the geometry and dimensions of the channel [129].

13.6.3
Membraneless Fuel Cell

For small-scale miniature power applications, fuel cells have been designed to operate without membranes (Figure 13.18). The lack of a membrane (e.g., Nafion) in membraneless, laminar flow-based micro fuel cells (LF-FCs) eliminates several membrane-related issues, for example, fuel crossover, cathode flooding and anode dry-out. Choban *et al.* have explored this concept for alkaline-based methanol fuel cells [129]. They also explored the media flexibility of LF-FCs by working in acidic and alkaline media, and under 'mixed-media' conditions in which the anode is in acidic media whilst the cathode is in alkali, and vice versa. Operating a fuel cell under 'mixed-media' conditions can

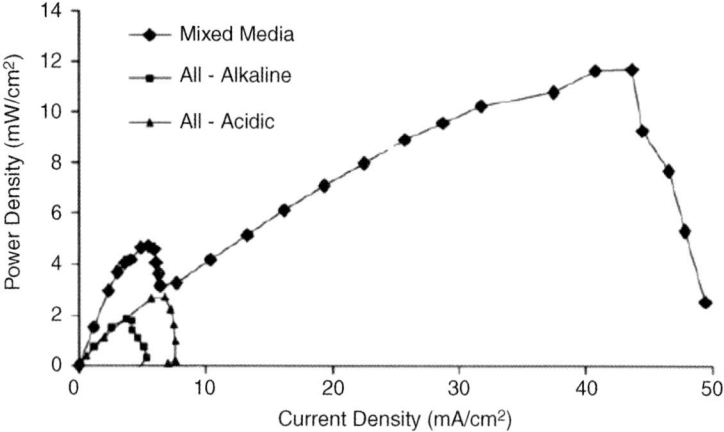

Figure 13.19 Power density curves for overall cell performance of an LF-FC operating in alkaline (1 M KOH), acidic (1 M H_2SO_4), and mixed-media (anode: 1 M KOH and cathode: 1 M H_2SO_4). Fuel stream is 1 M methanol and the oxidant stream an oxygen-saturated solution in the respective media [129].

increase the open cell potential (OCP). For the LF-FC operated in the alkaline anode/acid cathode mixed-media configuration, an OCP of 1.4 V was recorded (cf. theoretical OCP of 2.04 V). The higher OCP (and thus higher potentials under load conditions) resulted in a maximum power density of 12 m/cm^{-2}. The mixed-media LF-FC outperformed the all-acidic and all-alkaline LF-FCs (Figure 13.19). However most of the observed extra power density for this mixed-media LF-FC configuration is supplied by the net consumption of protons at the cathode and hydroxide ions at the anode. Taking into account the consumption of H_2SO4 and KOH a maximum theoretical energy density for this mixed-media LF-FC configuration was calculated at one-fifth of the maximum energy density of all-acidic or all alkaline LF-FCs that consume only methanol.

Sung and Choi have investigated a membraneless microscale fuel cell using non-noble catalysts in alkaline solution [130]. Nickel hydroxide and silver oxide, were employed as anode and cathode catalysts to minimize the effect of crossover of reactants with the membraneless structure. Methanol and hydrogen peroxide were used as fuel and oxidant respectively. A maximum output power density of 28.7 W cm^{-2} was achieved.

Generally a fuel cell without a membrane, with inexpensive catalysts and simple planar structure, enables high design flexibility and easy integration into actual microfluidic systems and portable applications.

13.7
Alkaline Fuel Cells with other Direct Liquid Fuels

Direct ethanol fuel cells (DEFCs) have attracted interest because of ethanol's more attractive features as a fuel than methanol [131]. Ethanol is relatively nontoxic, has a

13.7 Alkaline Fuel Cells with other Direct Liquid Fuels

greater energy density (8.01 kWh kg^{-1}), less volatile than methanol and is thus safer in operation. It can be readily produced in great quantities by the fermentation of sugar-containing raw materials and a range of biomass materials. Development of DEFCs faces several challenges, notably the slow reaction kinetics of ethanol oxidation. The oxidation of ethanol on Pt-based electrocatalysts is more sluggish than that of methanol in acidic media [132]. However PtRuSn/C and PtSn/C are quite efficient electrocatalysts for the oxidation of ethanol [133]. A restriction to the DEFC is a reliance on Pt catalyst especially when high loadings of several mg cm^{-2} are required for a reasonable performance. Pt is very costly and its supply is limited and therefore there is a requirement to develop low Pt loading or Pt-free electrocatalysts. Recent work has shown that the kinetics of ethanol oxidation could be significantly improved in alkaline media even on Pt-free electrocatalysts [134].

Tripovic et al. [135] have studied the mechanisms for oxidation of methanol, ethanol, n-propanol and n-butanol in alkaline media on electrochemically treated Pt single crystal surfaces using CV. Both PtOH and PtO were detected during the surface oxidation on Pt. PtO species is reported to inhibit the oxidation mechanism [135]. A dual parallel path reaction mechanism is proposed, based on RCO$_{ads}$, as a reactive intermediate of the main path, whilst CO$_2$ is the product from oxidation of poisoning species in the other parallel path. Ethanol was reported to give the highest activity of the C—C bonded species, although no rupture of that bond was reported.

Gupta used CV to study ethanol oxidation in alkaline media using CuNi, CuNi/Pt and CuNi/PtRu catalysts [136]. Using Ru in the alloys increased the oxidation kinetics and the formation of acetaldehyde and carbonate, associated with CO$_2$ formation for C—C bond breakage, was reported.

Verma et al. investigated ethanol oxidation on Pt and Pt/Ru in KOH solutions. Oxide layer formation was confirmed with Pt-Ru being more active than Pt black. No C—C bond cleavage was observed [24].

Of alternative electrocatalysts to Pt, palladium has attracted much interest due to its significantly lower cost (approximately $^1/_4$) than Pt. High surface area carbon is usually used as the support to uniformly disperse metal electrocatalysts and consequently try to maximize metal use. However it is known that even with this approach the active surface is still significantly under used. Hence alternative supports such as carbon nanotubes (CNTs) are of interest in fuel cells due to their unique electrical and structural properties [137]. Multiwall carbon nanotubes (MWCNTs) have been used as the support of the cathode electrocatalyst and produced better performance in DEFCs than cathode electrocatalysts supported on carbon black [138]. Activated carbon fiber has also been considered as a catalyst support because of its extended surface area, microporous structure and surface reactivity [139]. The use of gold to promote Pd/C catalysts for 2-propanol electro-oxidation in alkaline media has been studied by Chengwei et al. [140]. Other alcohols have also been examined for fuel cells including propanol, ethylene glycol and glycerol [141–143].

Matsuoka et al. investigated alkaline direct alcohol fuel cells using an OH$^-$ form of AEM [143]. The AEM was a hydrocarbon membrane (AHA Tokuyama) and polyhydric alcohols were ethylene glycol > glycerol > methanol > erythritol > xylitol). A high open circuit voltage of about 800 mV was obtained for a cell using Pt—Ru/C

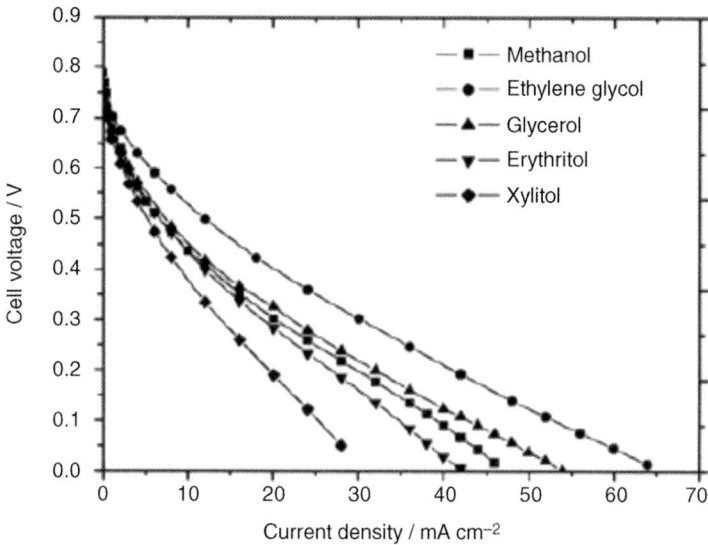

Figure 13.20 Cell voltage curves for alkaline direct alcohol fuel cells using polyhydric alcohols. Anode: Pt–Ru/C; cathode: Pt/C; temperature: 323 K. [143].

(anode) and Pt/C (cathode) at 323 K, which was almost 100–200 mV higher than that for a DMFC using Nafion (Figure 13.20). The maximum power densities in the presence of NaOH electrolyte were in the order of ethylene glycol glycerol > methanol > erythritol > xylitol. Silver catalysts were used as a cathode catalyst to fabricate AFCs, since silver catalysts are almost inactive for oxidation of polyhydric alcohols. Alkaline direct ethylene glycol fuel cells using silver as a cathode catalyst gave excellent performance because higher concentrations of fuel could be supplied to the anode.

13.8
Conclusions

Research to date has demonstrated that direct methanol fuel cells in alkaline media can deliver moderate performance. However much of the research has been conducted in alkaline liquid electrolyte, which brings with it a major disadvantage: long term operation is affected by gradual carbonation of electrolyte due to CO_2 formation from fuel oxidation. Electrocatalysis in alkaline media is potentially more facile that that in acid media. Indeed the potential of using non-Pt catalyst has been demonstrated. Despite significant research on oxygen reduction in alkaline media the performance advantages have not been converted into improved fuel cell performance for methanol fuel cells when compared with PEM-based DMFCs. However a major attraction is that truly methanol-tolerant cathode catalysts can be used and what is needed is to improve the activity of materials such as Ag.

For fuel cell applications where electrolyte replacement is not an option, the use of OH^- ion-conducting membranes is mandatory. Development of such DMAFCs requires more intense research into high conductivity and stable AEMs for significant future increases in AFC performance. In addition, to fully capitalize on the potentially high electrocatalyst activity, fuel cell electrodes require a suitable liquid form of ionomer that can be used in catalyst layer fabrication. Development of ionomers is required for significant future increases in AFC performances and to explore the true catalytic activity of candidate electrodes. Furthermore there is a requirement to study in detail the tolerance of these AEM fuel cells to CO_2.

Alternative alcohol fuels to methanol such as ethylene glycol and ethanol can also be used in DAFCs. The ability of electrocatalysts to be more active, particularly in breaking the C−C double bond is attractive. However as yet the performance of DAFCs has not realised their full potential, even allowing for the inherent lower ion conductivity in comparison to PEM-based fuel cells.

References

1 Adzic, R.R., Aramour, M.C. and Tripkovic, A.V. (1984) Electrochim. Acta, 29, 907.

2 Taraszewska, J. and Roslonek, G. (1994) J. Electroanal. Chem., 364, 209.

3 Lamy, C., Leger, J.M. and Srinivasan, S. (2001) Modern Aspects of Electrochemistry, vol. 34 (eds J.O'.M. Bockris and B.E. Conway), Kluwer Academic/Plenum Press, New York, p. 53.

4 Koscher, G.A. and Kordesch, K. (2003) Handbook of Fuel Cells, vol. 4 (eds W. Vielstich, A., Lamm and H.A. Gasteiger), John Wiley & Sons, Ltd, Chichester, pp. 1123–1129.

5 Wasmus, S. and Kuever, A. (1999) J. Electroanal. Chem., 416, 14.

6 Iwasita, T. (2002) Electrochim. Acta, 47, 3663.

7 Beden, B., Leger, J.M. and Lamy, C. (1992) Modern Aspects of Electrochemistry, vol. 22 (eds J.O'.M. Bockris, B.E. Conway and R.E. White), Plenum Press, New York, p. 97.

8 Breiter, M. (1967) Electrochim. Acta, 12, 1213.

9 Kadirgan, F., Beden, B., Leger, J.M. and Lamy, C. (1981) J. Electroanal. Chem., 125, 89.

10 Parsons, R. and VanderNoot, T. (1988) J. Electroanal. Chem., 257, 9.

11 Adzic, R., Armor, M. and Tripkovic, A. (1984) Electrochim. Acta, 29, 1353.

12 Prabhuram, J. and Manoharan, R. (1998) J. Power Sources, 74, 54.

13 Bagotzky, V.S. and Vassilyev, Yu.B. (1967) Electrochim. Acta, 12, 1323.

14 Lamy, C., Leger, J.M. and Clavilier, J. (1982) J. Electroanal. Chem., 135, 321.

15 Beden, B., Kadirgan, F., Lamy, C. and Leger, J.M. (1982) J. Electroanal. Chem., 142, 171.

16 Morallon, E., Vazquez, J.L. and Aldaz, A. (1990) J. Electroanal. Chem., 288, 217.

17 Tripkovic, A.V., Popovic, K.Dj., Momcilovic, J.D. and Drazic, D.M. (1996) J. Electroanal. Chem., 418, 9.

18 Tripkovic, A.V., Popovic, K.Dj., Momcilovic, J.D. and Drazic, D.M. (1998) J. Electroanal. Chem., 448, 173.

19 Tripkovic, A.V., Popovic, K.Dj., Momcilovic, J.D. and Drazic, D.M. (1998) Electrochim. Acta, 44, 1135.

20 Lović F J.D. (2007) J. Serb. Chem. Soc., 72, 709–712.

21 Tripković, A.V., Popović, K.D., Grgur, B.N., Blizanac, B., Ross, P.N., and

Markovic, N.M., (2002) *Electrochim. Acta*, **47**, 3707–3714.

22 Tripkovic, A.V., Popovic, K.D.J. and Lovic, J.D. (2007) *J. Serb. Chem. Soc.*, **72**, 1095–1101.

23 Manoharan, R. and Prabhuram, J. (2001) *J. Power Sources*, **96**, 220–235.

24 Verma, A. and Basu, S. (2007) Chapter 3, Direct alcohol and borohydride fuel cells, in *Recent Trends in Fuel Cell Science and Technology* (ed. S. Basu), Springer, New York.

25 Vertes, G., Horanyi, G. and Nagi, F. (1971) *Acta Chim. Acad. Sci. Hung.*, **68**, 145.

26 Fleischmann, M., Korinek, K. and Pletcher, D. (1971) *J. Electroanal. Chem.*, **31**, 39.

27 Taraszewska, J. and Roslonek, G. (1994) *J. Electroanal. Chem.*, **364**, 209.

28 Allen, J.R., Florido, A., Young, S.D., Daunert, S. and Bachas, L.G. (1995) *Electroanalysis*, **7**, 710.

29 Bettelheim, A., White, B.A., Raybuck, S.A. and Murray, R.W. (1987) *Inorg. Chem.*, **26**, 1009.

30 Macor, K.A. and Spiro, T.G. (1983) *J. Am. Chem. Soc.*, **105**, 5601.

31 Van Effen, R.M. and Evans, D.H. (1979) *J. Electroanal. Chem. Interfacial Electrochem.*, **103**, 383.

32 Motheo, A.J., Machado, S.A.S., Rabelo, F.J.B. and Santos, J.R. Jr (1994) *J. Braz. Chem. Soc.*, **5**, 161.

33 Abdel Rahim, M.A., Abdel Hameed, R.M. and Khalil, M.W. (2004) *J. Power Sources*, **134**, 160–169.

34 Khalil, M.W., Rahim, M.A.A., Zimmer, A., Hassan, H.B. and Hameed, R.M.A. (2005) *J. Power Sources*, **144**, 35–41.

35 Shobba, T., Mayanna, S.M. and Sequueira, C.A.C. (2002) *J. Power Sources*, **108**, 261–264.

36 Borkowska, Z., Tyhmosiak-Zielinska, A. and Shul, G. (2004) *Electrochim. Acta*, **49**, 1209.

37 Borkowska, Z., Tymosiak-Zielinska, A. and Nowakowski, R. (2004) *Electrochim. Acta*, **49**, 2613.

38 Chug, S.-C., Hamelin, A. and Weaver, M.J. (1991) *J. Phys. Chem.*, **95**, 5560.

39 Assiongbon, K.A. and Roy, D. (2005) *Surf. Sci.*, **594**, 99–119.

40 Guo, Bin, Zhao, Shizhen, Han, Gaoyi and Zhang, Liwei (2008) *Electrochim. Acta*, **53**, 5174–5179.

41 Suresh Kumar, K., Haridoss, Prathap and Seshadri, S.K. (2008) *Surf. Coat. Tech.*, **202**, 1764–1770.

42 Ohmore, T., Nodasaka, K. and Enyo, M. (1990) *J. Electroanal. Chem.*, **281**, 331.

43 Biswas, P.C. and Enyo, M. (1992) *J. Electroanal. Chem.*, **322**, 203.

44 Raghuveer, V. and Viswanathan, B. (2002) *Fuel*, **81**, 2191.

45 Raghuveer, V., Thampi, K.R., Xanthopoulos, N., Mathieu, H.J. and Viswanathan, B. (2001) *Solid State Ionics*, **140**, 263.

46 Yu, H.-C., Fung, K.-Z., Guo, T.-C. and Chang, W.-L. (2004) *Electrochim. Acta*, **50**, 811.

47 Singh, R.N., Sharma, T., Singh, A. Anindita, Mishra, D. and Tiwari, S.K. (2008) *Electrochim. Acta.*, **53**, 2322–2330.

48 Deshpande, K., Mukasyan, A.S. and Varma, A. (2006) *J. Power Sources*, **158**, 60.

49 Lan, A. and Mukasyan, A.S. (2007) *J. Phys. Chem. C*, **111**, 9573.

50 Tripkovic, A.V., Popovic, K.D., Momcilovic, J.D. and Drazic, D.M. (1996) *J. Electroanal. Chem.*, **418**, 9–20.

51 Wang, Y., Li, L., Hu, J. and Xu, B. (2003) *Electrochem. Comm.*, **5**, 662–666.

52 Morallon, E., Rodes, A., Vasquez, J.L. and Perez, J.M. (1995) *J Electroanal. Chem.*, **391**, 149–157.

53 Beden, B., Kadirgan, F., Lamy, C. and Leger, J.M. (1982) *J. Electroanal. Chem.*, **142**, 171.

54 Tripkovic, A.V., Popovic, K.Dj., Momcilovic, J.D. and Drazic, D.M. (1996) *J. Electroanal. Chem.*, **418**, 9.

55 Faubert, G., Guay, D. and Dodelet, J.P. (1998) *J. Electrochem. Soc.*, **145**, 2985.

56 Eikerling, M. and Kornyshev, A.A. (1998) *J. Electroanal. Chem.*, **453**, 89.

57 Fukada, S. (2001) *Energ. Convers. Manage.*, **42**, 1121.

58 Shukla, A.K., Neergat, M., Bera, P., Jayaram, V. and Hegde, M.S. (2001) *J. Electroanal. Chem.*, **504**, 111.
59 Damjanovic, A. and Brusic, V. (1967) *Electrochim. Acta*, **12**, 615.
60 Sepa, D.B., Vojnovic, M.V. and Damjanovic, A. (1981) *Electrochim. Acta*, **26**, 781.
61 Hsueh, K.L., Gonzalez, E.R. and Srinivasan, S. (1983) *Electrochim. Acta*, **28**, 691–697.
62 Hsueh, K.L., Gonzalez, E.R., Srinivasan, S. and Chin, D.T. (1984) *J. Electrochem. Soc.*, **131**, 823.
63 Park, S.M., Ho, S., Aruliah, S., Weber, M.F., Ward, C.A., Venter, R.D. and Srinivasan, S. (1986) *J. Electrochem. Soc.*, **133**, 1641.
64 Damjanovic, A., Genshaw, M.A. and Bockris, J.O'M. (1967) *J. Electrochem. Soc.*, **114**, 1107.
65 Damjanovic, A., Sepa, D.B. and Vojnovic, M.V. (1979) *Electrochim. Acta*, **24**, 887.
66 Damjanovic, A., Genshaw, M.A. and Bockris, J.O'.M. (1967) *J. Electrochem. Soc.*, **114**, 466.
67 Gasteiger, H.A., Markovic, N., Ross, P.N. and Cairns, E.J. (1993) *J. Phys. Chem.*, **97**, 120.
68 Gasteiger, H.A., Markovic, N., Ross, P.N. and Cairns, E.J. (1994) *J. Electrochem. Soc.*, **141**, 1795.
69 Schmidt, T.J., Markovic, N. and Ross, P.N. (2001) *J. Phys. Chem.*, **105**, 12082.
70 Tripkovic, A.V., Popovic, K.D., Grgur, B.N., Blizanac, B., Ross, P.N. and Markovic, N. (2002) *Electrochim. Acta*, **47**, 3707.
71 Lizcano-Valbuena, W.H., Paganin, V.A. and Gonzalez, E.R. (2002) *Electrochim. Acta*, **47**, 3715.
72 Kiros, Y. (1996) *J. Electrochem. Soc.*, **143**, 2152.
73 Yang, Y.-F., Zhou, Y.-H. and Cha, C.-S. (1995) *Electrochim. Acta*, **40**, 2579.
74 Prakash, J. and Joachin, H. (2000) *Electrochim. Acta*, **45**, 2289.
75 Gojkovic, S.L., Gupta, S. and Savinell, R.F. (1999) *J. Electroanal. Chem.*, **462**, 63.
76 Gojkovic, S.L., Gupta, S. and Savinell, R.F. (2000) *Electrochim. Acta*, **45**, 889.
77 Heller-Ling, N., Prestat, M., Gautier, J.-L. and Koenig, J.-F. (1997) *Electrochim. Acta*, **42**, 197.
78 Hu, Y., Tolmachev, Y.V. and Scherson, D.A. (1999) *J. Electroanal. Chem.*, **468**, 64.
79 Rashkova, V., Kitova, S., Konstantinov, I. and Vitanov, T. (2002) *Electrochim. Acta*, **47**, 1555.
80 Matsuki, K. and Kamada, H. (1986) *Electrochim. Acta*, **3**, 13.
81 Ponce, J., Rehspringer, J.-L., Poillerat, G. and Gautier, J.-L. (2001) *Electrochim. Acta*, **46**, 3373.
82 Mao, L., Sotomura, T., Nakatsu, K., Koshiba, N., Zhang, D. and Ohsaka, T. (2002) *J. Electrochem. Soc.*, **149**, 504.
83 Klapste, B., Vondrak, J. and Velicka, J. (2002) *Electrochim. Acta*, **47**, 2365.
84 Wagner, N., Schultze, M. and Gulzow, E. (2004) *J. Power Sources*, **127**, 264–272.
85 Strasser, K. (2003) *Handbook of Fuel Cells—Fundamentals, Technology and Applications*, vol. 4 (eds W. Vielstisch, H.A. Gasteiger and A. Lamm), John Wiley & Sons, Ltd, Chichester.
86 Mao, L., Zhang, D., Sotomura, T., Nakatsu, K., Koshiba, N. and Ohsaka, T. (2003) *Electrochim. Acta*, **48**, 1015.
87 Yang, J. and Xu, J.J. (2003) *Electrochem. Comm.*, **5**, 306–311.
88 Verma, A., Jha, A.K. and Basu, S. (2005) *J. Power Sources*, **2** (141), 30–34.
89 Ponce, J., Rehspringer, J L., Poillerat, G. and Gautier, J L. (1998) *J. Power Sources*, **74**, 3373–3380.
90 Meng, Hui and Shen, Pei Kang, (2006) *Electrochem. Comm.*, **8**, 588–594.
91 Ravikumar, M.K. and Shukla, A.K. (1996) *J. Electrochem. Soc.*, **143**, 2601.
92 Reeve, R.W., Christensen, P.A., Dickinson, A.J., Hamnett, A. and Scott, K. (2000) *Electrochim. Acta*, **45**, 4237.
93 Alonso-Vante, N., Tributsch, H. and Solorza-Feria, O. (1995) *Electrochim. Acta*, **40**, 567.
94 Gupta, S., Tryk, D., Zecevic, S.K., Aldred, W., Guo, D. and Savinell, R.F. (1998) *J. Appl. Electrochem.*, **28**, 673.

95 Reeve, R.W., Christensen, P.A., Hamnett, A., Haydock, S.A. and Roy, S.C. (1998) *J. Electrochem. Soc.*, **45**, 3463.

96 Toda, T., Igarashi, H., Uchida, H. and Watanabe, M. (1999) *J. Electrochem. Soc.*, **146**, 3750.

97 Bron, M., Bogdanoff, P., Fiechter, S., Hilgendorff, M., Radnik, J., Dorbandt, I., Schulenburg, H. and Tributsch, H. (2001) *J. Electroanal. Chem.*, **517**, 85.

98 Demarconnay, L., Coutanceau, C. and Léger, J.-M. (2004) *Electrochim. Acta*, **49**, 4513–4521.

99 Yu, E.H. and Scott, K. (2004) *J. Power Sources*, **137**, 248.

100 Yu, E.H. and Scott, K. (2004) *Electrochem. Commun.*, **6**, 361.

101 Verma, L.K. (2000) *J. Power Sources*, **86**, 464–468.

102 Yu, E., Scott, K. and Reeve, R.W. (2006) *J. Appl. Electrochem.*, **36**, 25–32.

103 Baldauf, M. and Preidel, W. (1999) *J. Power Sources*, **84**, 161.

104 Beden, B., Leger, J.M., Lamy, C. and Bockris, J.O'.M. (1992) *Modern Aspects of Electrochemistry*, vol. 22 (eds B.E. Conway and R.E. White), Plenum Press, New York.

105 Cairns, E.J. and Macdonald, D.I. (1964) *Electrochem. Tech.*, **2**, 65.

106 Sriebel, K.A., McLarnon, F.R. and Cairns, E.J. (1990) *J. Electrochem. Soc.*, **137**, 3351.

107 Varcoe, J.R. and Slade, R.C.T. (2005) *Fuel Cells*, **5**, 187.

108 Xing, B. and Savadogo, O. (2000) *Electrochem. Commun.*, **2**, 697.

109 Guo, Q., Pintauro, P.N., Tang, H. and O'Connor, S. (1999) *J. Membr. Sci.*, **154**, 175.

110 Kang, J.J., Li, W.Y., Lin, Y., Li, X.P., Xiao, X.R. and Fang, S.B. (2004) *Polym. Adv. Technol.*, **15**, 61.

111 Yi, F., Yang, X., Li, Y. and Fang, S. (1999) *Polym. Adv. Technol.*, **10**, 473–475.

112 Li, L. and Wang, Y.X. (2005) *J. Membr. Sci.*, **262**, 1.

113 Slade, Robert C.T., and Varcoe, J.R. (2005) *Solid State Ionics*, **176**, 585–597.

114 Fang, J. and Shen, P.K. (2006) *J. Membr. Sci.*, **285**, 317.

115 Wu, Liang, Xu, Tongwen, Wu, Dan and Zheng, Xin (2008) *J. Membr. Sci.*, **310**, 577–585.

116 Pan, Y., Huang, Y.H., Liao, B. and Cong, G.M. (1996) *J. Appl. Polym. Sci.*, **61**, 1111.

117 Hong, J H., Li, D. and Wang, H. (2008) *J. Membr. Sci.*, **318**, (1-2) 441–444.

118 Lei, Li and Wang, Yuxin, (2005) *J. Membr. Sci.*, **262**, 1–4.

119 Xing, B. and Savadogo, O. (2000) *Electrochem. Comm.*, **2**, 697–702.

120 Danks, T.N., Slade, R.C.T. and Varcoe, J.R. (2002) DMFCs. *J. Mater. Chem.*, **12**, 3371.

121 Wu, G.M., Lin, S.J. and Yang, C.C. (2006) *J. Membr. Sci.*, **275**, 127.

122 Yang, Chun-Chen, Chiu, Shwu-Jer and Chien, Wen-Chen (2006) *J. Power Sources*, **162**, 21–29.

123 Yang, Chun-Chen, Chiu, Shwu-Jer and Lin, Che-Tseng (2008) *J. Power Sources*, **177**, 40–49.

124 Yang, Chun-Chen (2007) *J. Membr. Sci.*, **288**, 51–60.

125 Bunazawa, Hideaki and Yamazaki, Yohtaro (2008) *J. Power Sources*, **182**, 48–51.

126 Coutanceau, C., Demarconnay, L., Lamy, C. and Léger, J.-M. (2006) *J. Power Sources*, **156**, 14–19.

127 Hou, Hongying, Sun, Gongquan, He, Ronghuan, Wu, Zhimou and Sun, Baoying (2008) *J. Power Sources*, **182**, 95–99.

128 Scott, K., Pilditch, S. and Mamlouk, M. (2007) *J. Appl. Electrochem.*, **37**, 1245–1259.

129 Choban, E.R., Spendelow, J.S., Gancs, L., Wieckowski, A. and Kenis, P.J.A. (2005) *Electrochim. Acta*, **50**, 5390–5398.

130 Sunga, Woosuk and Choi, Jin-Woo (2007) *J. Power Sources*, **172**, 198–208.

131 Lamy, C., Lima, A., LeRhun, V., Delime, F., Coutanceau, C. and Léger, J.M. (2002) *J. Power Sources*, **105**, 283.

132 Tremiliosi-Filho, G., Gonzalez, E.R., Motheo, A.J., Belgsir, E.M., Léger, J.M. and Lamy, C. (1998) *J. Electroanal. Chem.*, **444**, 31.

133 Vigier, F., Coutanceau, C., Perrard, A., Belgsir, E.M. and Lamy, C. (2004) *J. Appl. Electrochem.*, **34**, 439.
134 Shen, P.K., Xu, C.W., Zeng, R. and Liu, Y.L. (2006) *Electrochem. Solid-State Lett.*, 9, 39.
135 Tripkovic, A.V., Dj Popovic, K., Momcilovic, J.D. and Drazic, D.M. (2002) *Electrochim. Acta*, **46**, 3163–3173.
136 gupta, S.S., Mahapatra, S.S. and Datta, J. (2004) *J. Power Sources*, **131**, 169–174.
137 Li, W.Z., Liang, C.H., Qiu, J.S., Zhou, W.J., Han, H.M., Wei, Z.B., Sun, G.Q. and Xin, Q. (2002) *Carbon*, **40**, 787.
138 Li, W.Z., Liang, C.H., Zhou, W.J., Qiu, J.S., Zhou, Z.H., Sun, G.Q. and Xin, Q. (2003) *J. Phys. Chem. B*, **1076**, 292.
139 Pereira, M.F.R., Órfão, J.J.M. and Figueiredo, J.L. (2002) *Carbon*, **40**, 2393.
140 Xu, Changwei, Tian, Zhiqun, Chen, Zhaochen and Jiang, San Ping (2008) *Electrochem. Comm.*, **10**, 246–249.
141 Ye, Jianqing, Liu, Jianping Xu, Changwei, Jiang, San Ping and Tong, Yexiang (2007) *Electrochem. Comm.*, **9**, 2760–2763.
142 Demarconnay, L., Brimaud, S., Coutanceau, C. and Léger, J.-M. (2007) *J. Electroanal. Chem.*, **601**, 169–180.
143 Matsuoka, Koji, Iriyama, Yasutoshi, Abe, Takeshi, Matsuoka, Masao and Ogumi, Zempachi. (2005) *J. Power Sources*, **150**, 27–31.

14
Electrocatalysis in Other Direct Liquid Fuel Cells
Sharon L. Blair and Wai Lung (Simon) Law

14.1
Introduction

Direct liquid fuel cells (DLFCs) have been receiving considerable attention over the past decade for portable power applications. Small organic molecules, such as methanol, ethanol and formic acid, are promising alternatives to hydrogen as a fuel for fuel cells [1–4]. Compared with hydrogen, a fuel in its liquid state would make transportation and handling relatively easy, particularly for low power applications. Each of the liquid fuels under consideration has its merits and, in this chapter, specific attributes of fuels other than methanol, especially formic acid, ethanol and hydrazine will be described, with emphasis on electrocatalysis in these systems.

14.1.1
Fuel Characteristics and Theoretical Comparison of Various Fuels

There are a number of liquid fuels that have high theoretical volumetric energy densities. For example, a selection of liquid fuels range from 6280 Wh/L for ethanol to 1930 Wh/L for formic acid at 100% concentration. It is clear that these liquid fuels potentially offer considerable advantage over compressed hydrogen at room temperature with respect to volumetric energy density as illustrated in Table 14.1. Therefore, characteristics of the fuels must be examined to understand how various attributes impact the fuel cell systems using them.

The following sections will describe electrocatalysis using formic acid, ethanol and hydrazine as direct fuels for fuel cells with brief mention of a few other organic fuels. Comparisons will be primarily made against methanol unless otherwise stated.

Electrocatalysis of Direct Methanol Fuel Cells. Edited by Hansan Liu and Jiujun Zhang
Copyright © 2009 WILEY-VCH Verlag GmbH & Co. KGaA, Weinheim
ISBN: 978-3-527-32377-7

Table 14.1 Theoretical properties of various liquid fuels and compressed hydrogen[a].

Fuel	Energy density (Wh/L)[a,b]	E_e° (V)	Reversible energy efficiency (%)
Ethanol	6280	1.14	97
Ethylene glycol	5870	1.22	99
Formic Acid	1930	1.42	106
Hydrazine	5400	1.62	100
Methanol	4820	1.21	97
2-Propanol	7080	1.12	97
Hydrogen	180 (@1000 psi, 25 °C)	1.23	83

Extracted from [2].
[a]Standard conditions, unless stated.
[b]Values for liquid fuels calculated with 100 wt% fuel.

14.2
Electrocatalysis of Direct Formic Acid Fuel Cells

Very high power densities have been achieved in direct formic acid fuel cell (DFAFC) systems. However, similar to the direct methanol fuel cell (DMFC) the DFAFC faces several challenges not normally associated with the traditional polymer electrolyte membrane (PEM) H_2/O_2 system. The most important of these challenges include poor anodic reaction kinetics compared with hydrogen especially for Pt-based catalysts [2], and fuel crossover [5, 6]. The primary challenge of the low anodic reaction kinetics is the tendency for methanol and formic acid oxidation to proceed through a $-CO$ type intermediate, a species which poisons Pt-based catalyst materials [7, 8]. With respect to crossover, methanol or formic acid diffuses from the anode, through the membrane to the cathode. At the cathode, formic acid can react directly with oxygen over the catalyst, creating unwanted heat without producing electricity, thus reducing the overall fuel efficiency of the system. In the case of Pd catalyst, fast anode reaction kinetics has been observed, imparting high fuel cell power densities [9], however, a slow deactivation is observed, resulting in the requirement of occasional regeneration of the catalyst for long term operation [10].

This section considers the challenges associated with electrocatalysis and fuel cell performance for DFAFCs.

Direct formic acid fuel cell operation The anode reaction of the DFAFC is equivalent to the chemical combustion reaction of formic acid as shown in Equation 14.1.

$$HCOOH + \frac{1}{2}O_2 \rightarrow CO_2 + H_2O \tag{14.1}$$

A desire for low temperature operation, along with the limitation that formic acid will not operate in an alkaline fuel cell due to formation of formate salts (which possess significantly lower energy density) via acid–base reactions, restricts the use of formic acid to PEM fuel cells. In the acid-based PEM environment, the anode (Equation 14.2) and cathode (Equation 14.3) reactions for the DFAFC are detailed below. The method of operation, and thus basic cell structure, is similar to that of other PEM-based systems (e.g., hydrogen and methanol fuels).

$$HCOOH \rightarrow CO_2 + 2H^+ + 2e^- \quad \varepsilon^\circ = 0.19 \text{ V} \tag{14.2}$$

$$2H^+ + \frac{1}{2}O_2 + 2e^- \rightarrow H_2O \quad \varepsilon^\circ = 1.23 \text{ V} \tag{14.3}$$

The overall Gibbs free energy change for these combined reactions is 262 kJ/mol, which is equivalent to a reversible standard cell potential of 1.42 V.

Fuel concentration and theoretical energy density Methanol solutions have a theoretical energy density advantage over formic acid solutions on a volumetric basis. The advantage is even greater on a gravimetric basis, owing to the higher gravimetric density of formic acid over methanol. These properties are shown in Figure 14.1.

This lower energy density is partially compensated by the theoretical thermodynamic potential (i.e., 1.42 V for formic acid vs. 1.22 V for methanol) and thermodynamic efficiency of formic acid electro-oxidation which is very high. That is, the reversible energy efficiency for formic acid is 106% compared with 97% for methanol, as shown in Table 14.1.

The lower energy density of formic acid is exemplified by the anode reaction, in that only two electrons are produced per molecule of formic acid, fewer than are produced by other organic molecules such as methane (4 electrons), methanol (6 electrons), ethanol (12 electrons) or hydrazine (4 electrons). Although fewer electrons are produced, there are also fewer steps in the electro-oxidation of formic

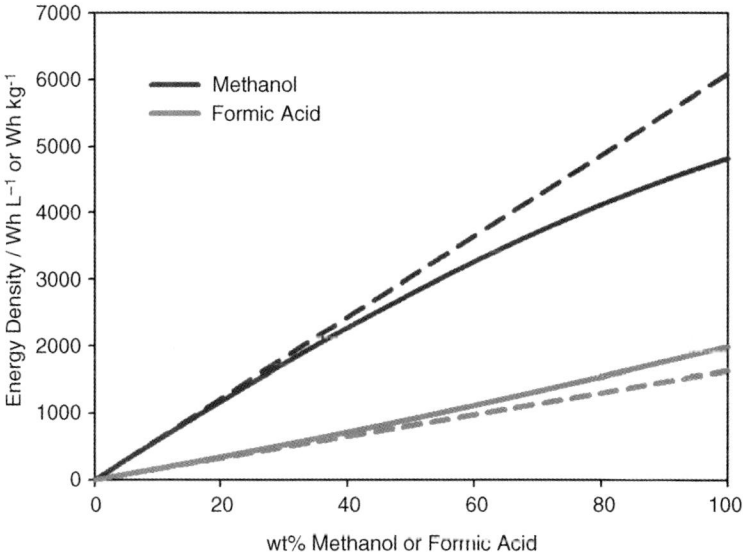

Figure 14.1 Volumetric (—) and gravimetric (---) theoretical free energy density of aqueous formic acid and methanol at 20 °C and 101.3 kPa.

acid, which increases the likelihood for reaction completion, and hence, results in better fuel use. One significant advantage of formic acid is that the presence of two oxygen atoms in its structure means that no water is required at the anode for complete oxidation of formic acid to CO_2. This is in contrast to methanol (Equation 14.4), for instance, that requires water at the anode in a one to one molar ratio for complete oxidation:

$$CH_3OH + H_2O \rightarrow CO_2 + 6H^+ + 6e^- \qquad (14.4)$$

The fundamental need for water in the DMFC restricts operation of the fuel cell to methanol concentrations less than 50 mol% (i.e., 10.6 M or 36 wt%), whereas formic acid can, in theory, be used in the fuel cell without dilution. In practice however, water is usually present in the fuel to meet flammability restrictions (e.g., <85 wt%) and to maintain membrane hydration.

The requirement for water in the anodic reaction of the DMFC is often met by employing novel water management techniques to move water from the cathode of the fuel cell to the anode. This concept is facilitated by the fact that substantial amounts of water are produced in the DMFC (i.e., 3 mol of water per mol of methanol), and that the cathodic reaction in the DMFC proceeds at a much faster rate than the sluggish anodic reaction. These water management techniques however, ultimately lead to more complex balance of plant and, in the case of active water movement, potential increase in system volume as well as parasitic energy losses which are significantly greater than those of DFAFCs. Hence, smaller balance of plant due to simpler anode reaction, coupled with faster reaction kinetics make DFAFCs competitive in very low power applications.

Despite advantages in thermodynamic efficiency, the volumetric energy density of formic acid is only 40% of that of methanol [11]. This requires formic acid to be used at much higher concentrations in the fuel cell. This is possible, and practical, due to the fact that water is not required for the anodic reaction and rates of fuel crossover remain significantly lower [6, 10]. The DFAFC typically exhibits much higher fuel cell efficiency than DMFCs. This results in higher power output as well as lower parasitic energy losses for the DFAFC. Typical fuel cell performance for DFAFCs versus DMFCs with fuels at equivalent energy densities will be discussed in Section 14.2.3.

A compromise between the balance of plant and fuel concentration is made when considering the DFAFC's advantages relative to the DMFC. For example, as the power output requirement increases, the water management balance of plant becomes an increasingly smaller component, while parasitic energy cost relative to the fuel cell system and DMFCs gain a considerable advantage. This limits the DFAFC advantage to low power applications, generally less than 20 W.

14.2.2
Anode Catalysts for DFAFCs

The key distinction between DFAFC and other DLFCs is the anodic oxidation mechanism. The electrocatalysis of formic acid warrants significant attention,

14.2 Electrocatalysis of Direct Formic Acid Fuel Cells

therefore electrocatalysis with both Pt- and Pd-based catalysts will be discussed in detail in this section. In contrast to other small organic molecules such as methanol or ethanol, the oxidation of formic acid on Pt, Pd and their alloys occurs at significantly faster rates than the oxidation of methanol on Pt and Pt alloys, owing to simpler reaction mechanisms. The reaction mechanism on both metal surfaces will be discussed next.

14.2.2.1 Pt-based Anode Catalysts

It is now well established that on Pt metal surfaces, the oxidation of formic acid takes place via a dual pathway mechanism [7, 8]. As exemplified in the case of pure Pt, the direct pathway (Equation 14.5) is a very fast reaction involving reactive intermediates, while the indirect pathway (Equation 14.6) involves site-blocking or poisoning intermediates, similar to the strongly adsorbed –CO intermediate encountered during the electro-oxidation of methanol on Pt.

$$HCOOH \rightarrow CO_2 + 2H^+ + 2e^- \quad \text{(direct pathway)} \quad (14.5)$$

$$HCOOH \rightarrow CO_{ad} + H_2O \rightarrow CO_2 + 2H^+ + 2e^- \quad \text{(indirect pathway)} \quad (14.6)$$

The dual pathway mechanism for formic acid oxidation on Pt-based catalysts is best articulated in the Pt cyclic voltammogram (Figure 14.2) [7]. The first peak in current density, occurring at approximately 0.5–0.6 V versus RHE, represents the direct oxidation pathway and the second peak occurring at 0.9 V versus RHE represents the indirect pathway. Figure 14.2 also demonstrates that the direct pathway is only available on a very active, freshly reduced, Pt surface. When the Pt surface is subjected to HCOOH at open circuit for as little as five minutes,

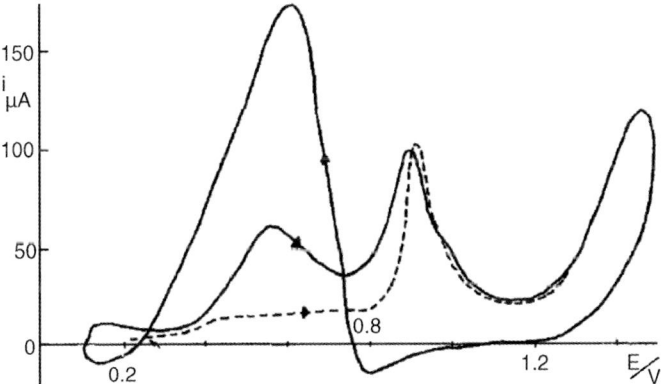

Figure 14.2 Cyclic voltammogram for a smooth Pt bead electrode ($d = 0.15$ mm) at 25 °C in 0.5 M H_2SO_4 containing 0.1 M HCOOH. Scan rate of 140 mV/s. (—) Continuous reading; (---) after 5 minutes at open circuit. Reprinted with permission from Ref. [8] Copyright © 1973 Elsevier.

the surface becomes saturated with strongly adsorbed poisoning intermediates (Equation 14.6) which cannot be removed until the dissociation of water occurs on Pt, at much higher potentials. Thus it is generally agreed that in all practical applications, electro-oxidation of formic acid on pure Pt proceeds predominantly via the indirect oxidation pathway [7, 8, 12–14].

The dual pathway mechanism allows for two routes of platinum-based anode catalyst development for formic acid fuel cells [14] and numerous admetals on Pt have been investigated for formic acid oxidation [15]. In the first route, admetals that allow for the dissociation of water to occur at lower potentials can be employed through a bifunctional mechanism thus facilitating the indirect pathway. In the second, and preferred route, admetals that allow the direct oxidation pathway to proceed uninhibited such as electronic or third body modifications can be employed. Electronic modifiers impart an intrinsic kinetic enhancement of the direct oxidation pathway, whereas third body modifiers create a steric hindrance to formation of the CO_{ad} poisoning species and hence facilitate the direct pathway. In any case, facilitation of the direct pathway is preferred since the direct pathway will demonstrate activity in the potential region of 0–0.4 V versus RHE, which is considered to have the most practical application to the DFAFC [14].

Numerous examples of adatoms to Pt that have been tested and reported in the literature include Pd [9, 16–20], Ru [21–23], Sn [24, 25], Pb [25], Au [14], As [25], Bi [25–27], Sb [25, 28, 29], Ge [25], Se [23], Te [29].

Bifunctional modifiers Examples of bifunctional modifiers to Pt include Ru and Sn [12]. Metals such as Ru or Sn demonstrate a synergistic effect on Pt, resulting in a much higher activity than pure Pt. This higher activity is due to a high rate of dehydrogenation on the Pt surface in addition to Ru or Sn causing effective oxidation of CO_{ad} to CO_2 at lower potentials than Pt, demonstrating the so-called bifunctional mechanism.

The mechanism of formic acid electro-oxidation on PtSn can be represented according to Equations 14.7–14.9, illustrating clearly the bifunctional mechanism. The same mechanism can be used for PtRu.

$$Pt + HCOOH \rightarrow Pt(HCOO)_{ad} + H^+ + e^- \tag{14.7}$$

$$Sn + H_2O \rightarrow Sn(OH)_{ad} + H^+ + e^- \tag{14.8}$$

$$Pt(HCOO)_{ad} + Sn(OH)_{ad} \rightarrow Pt + Sn + CO_2 + H_2O \tag{14.9}$$

PtRu facilitates the oxidation of CO_{ad}, however, it is not preferred as an anode catalyst for DFAFCs. HCOOH interacts strongly with Ru, reducing its effectiveness for the dissociation of water, and the adsorption of HCOOH on Pt is too weak. This combination results in the fact that the CO_{ad} cannot be oxidized at sufficiently low anode potentials (e.g., 0–0.4 V vs. RHE) on PtRu. The difference between PtRu, PtSn and Pt is shown in the cyclic voltammogram scans reported in [12] and illustrated in Figure 14.3. It can be seen from this plot that both Sn and Ru reduce the onset

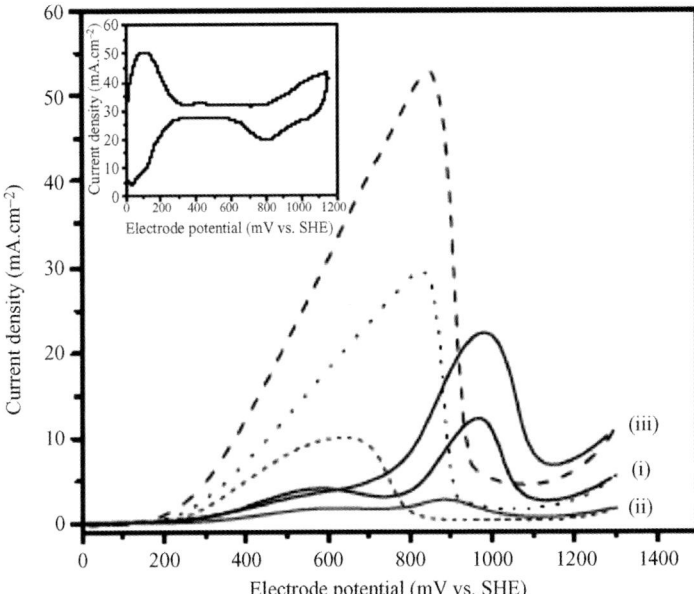

Figure 14.3 Forward (—) and reverse (- - -) scans of cyclic voltammograms of (i) Pt, (ii) PtRu and (iii) PtSn catalysts on Ti mesh in 0.1 M HCOOH in 0.5 M H_2SO_4 at room temperature. Inset shows voltammogram of Pt/Ti in 0.5 M H_2SO_4. Scan rate = 20 mV/s. Reprinted with permission from Ref. [12] Copyright © 2007 J. New Mater. Electrochem. Syst.

potential of H_2O dissociation and as such CO_{ad} oxidation is facilitated (Equations 14.8 and 14.9). PtSn shows a lower onset potential to water dissociation and higher activity in this region, therefore it shows more promise than PtRu, although the overpotential for both catalysts may be too high for practical applications.

Although bifunctional modifiers demonstrate higher activity than pure Pt by facilitating the indirect oxidation pathway, the ideal modifier would be one that facilitates the direct pathway (Equation 14.5) with electronic or third body modifiers. The most promising modifiers to Pt reported include Pd [9, 16–20], Au [14], Sb [25, 26], Pb [30–32], In [32] and Bi [32, 33].

Electronic and third body modifiers Many modifiers to Pt for formic acid oxidation are thought to act as both electronic and third body modifiers in that they act to increase the intrinsic activity of the direct dehydrogenation pathway while at the same time they can also provide a steric hindrance, or geometrical effects that reduce the number of CO_{ad} adsorption sites, hence facilitating the direct pathway. As such, electronic and third body modifiers will be covered together.

Herrero and coworkers [29] reported on poison formation experiments and they found that tellurium adsorbed on the surface inhibits poison formation and accumulation on the surface by a third body mechanism. It was also found that

Table 14.2 Formic acid oxidation on different intermetallic phases.

Electrode	Electrolyte	[HCOOH] (M)	Onset of oxidation (mV)	Peak current potential (mV)[a]	Peak current density ($\mu A/cm^2$)
Pt	H_2SO_4	0.25	150	680	220
Pt	$HClO_4$	0.125	120	650	500
PtBi	H_2SO_4	0.125	−125	550	2400
PtPb	H_2SO_4	0.25	−150	260	8200
PtIn	$HClO_4$	0.125	50	500	930

Extracted from [32].
[a]Potential reported vs. Ag/AgCl.

the catalysis enhancement of formic acid oxidation was due primarily to the inhibition of poison formation.

Casado-Rivera and others [32] reported on the electrocatalytic activities of a wide range of ordered intermetallic phases towards various organic fuels including formic acid. They found the most promising formulations for DFAFCs are PtBi, PtIn and PtPb. Activities are reported in Table 14.2. It can be seen that the onset potential for formic acid oxidation is as much as 300, 275 and 70 mV lower for PtPb, PtBi and PtIn respectively, than that for Pt. Peak current densities were also dramatically higher. Enhancement of the direct dehydrogenation pathway is shown by lower onset potential of formic acid oxidation and higher peak current density at lower potentials, an indication of electronic effect.

Kang and coworkers [34] reported the cyclic voltammograms for Pt and PtBi as shown in Figure 14.4 and determined that Bi adatoms create two modifying effects for HCOOH oxidation. They speculate that the enhanced current, shown in both the low and high potential region (i.e., 300 and 900 mV vs. SCE) is an indication of both intrinsic kinetic enhancement (i.e., electronic effect) of the direct pathway and steric hindrance of the formation of the CO_{ad} poisoning species (i.e., third-body effect).

Formic acid electro-oxidation on PtPd alloys with varying composition, either unsupported or supported on carbon, has been reported by numerous groups [9, 16–20]. Addition of Pd to Pt provides a synergistic electronic effect to Pt. This effect orients the PtPd catalyst towards the direct path of oxidation.

The electro-oxidation of formic acid on PdPt alloys illustrated that while Pd and Pt behave independently, synergistic effects were observed depending on the atomic ratio. Over the entire potential range studied, the greatest synergistic effect was found for the 50:50 Pd:Pt atomic ratio under constant current operation [17], exhibiting 0.55 V anode potential under constant current operation versus DHE, approximately 100 mV lower than pure Pt, which exhibits stable anode performance at approximately 0.65 V versus DHE. Although this demonstrates a dramatic improvement compared with Pt, the overpotential on 50:50 PtPd remains too high for fuel cell applications. A Pt modifier that appears to have significant potential for DFAFCs is Au. Au was originally described as a third body modifier to Pt [35]. However, more

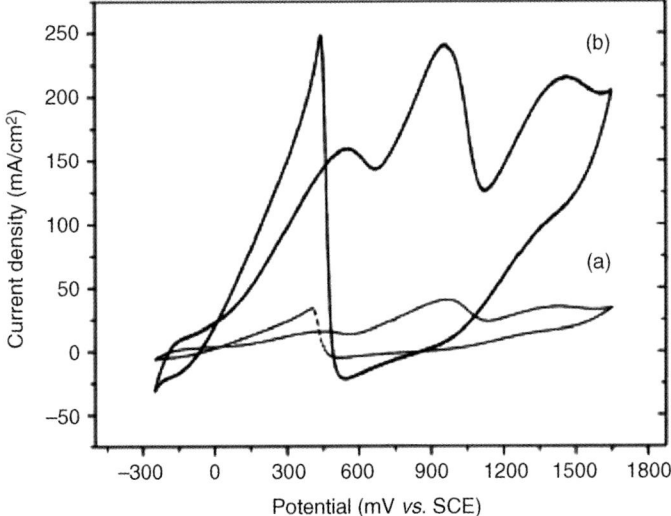

Figure 14.4 Cyclic voltammogram of (a) Pt and (b) $PtBi_m$ in 1.0 M HCOOH/0.5 M H_2SO_4. Scan rate = 50 mV/s. Reprinted with permission from Ref. [34] Copyright © 2006 J. New Mater. Electrochem. Syst.

recently, Choi and coworkers reported on the electro-oxidation of formic acid and methanol on PtAu and PtRu [14]. They indicated that Au modification provides more than just a third body effect.

PtAu was considered to have no positive effect on methanol oxidation because Au itself is inactive towards methanol oxidation and not helpful in removing CO_{ad}, whereas PtRu is preferred for methanol oxidation because of its bifunctional properties towards CO_{ad} oxidation. In contrast to methanol oxidation, PtAu has both a lower onset potential (ca. 0.1 V vs. DHE) and a larger current density for formic acid oxidation than PtRu, as shown in Figure 14.5. It was suggested that the enhancement could be explained by the fact that PtAu produces mainly CO_2 as a final product via a direct dehydrogenation reaction. Thus they conclude that the half replacement of Pt with Au resulted in an enhancement in intrinsic activity for formic acid oxidation. For example, the current densities for the PtRu and PtAu catalysts at 0.4 V are 3.8 and 119.2 mA/cm^2 respectively.

In summary, the most promising Pt-based catalyst formulations for formic acid oxidation observed to date are electronic and/or third-body modifiers, in particular, PtAu, PtBi and PtPb. In other words, admetals that facilitate the direct oxidation pathway to CO_2, avoiding the CO_{ad} intermediate, provide the highest activity in the 0–0.4 V versus RHE range, which is the most practical region for fuel cell operation [21]. Impressive fuel cell data has been reported using PtAu anode catalysts under DFAFC constant current operation for 500 h, indicating stable performance can be obtained at reasonable fuel cell potentials. However, more fuel cell studies are required in order to make an adequate evaluation of the catalysts that have exhibited potential for fuel cell applications.

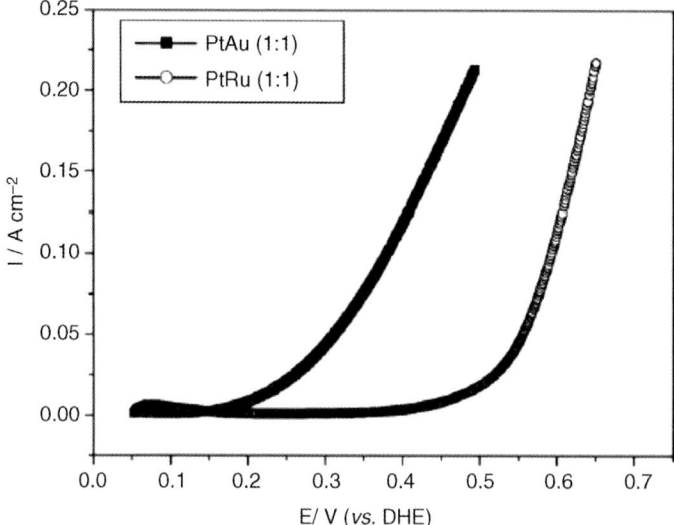

Figure 14.5 Current vs. potential plot for PtAu-based MEA and PtRu-based MEA. The anode was fed with 6 M formic acid and humidified hydrogen fed to the cathode. Reprinted with permission from Ref. [14] Copyright © 2006 Elsevier.

14.2.2.2 Alternative Pt Modifiers

Another novel approach to modifying Pt to facilitate the direct pathway is with the addition of macrocycle compounds such as iron-tetrasulfophthalocyanine (FeTSPc) [15, 36]. Researchers claim there are three factors affecting the promotion of the direct oxidation pathway. First, FeTSPc creates an anion effect to increase the rate of oxidation. Second, the steric hindrance of FeTSPc acts as a third-body modifier preventing formation of the poisoning CO_{ad} species, and finally, FeTSPc on Pt causes a weakening of the adsorption strength of CO_{ad}. Although this approach may take considerable effort to achieve a practical solution for the DFAFC, there are many opportunities for consideration.

14.2.2.3 Pd-Based Anode Catalysts

Although the mechanism for electro-oxidation of formic acid on Pt is known, there is still speculation about the mechanism on Pd. This is due to the fact that no detectable intermediates have been observed on Pd. It is generally agreed that the electro-oxidation of formic acid on Pd proceeds predominantly by the direct pathway through an active intermediate and, in contrast to Pt, no strongly adsorbed intermediates such as CO_{ad} are formed on the catalytic surface [7, 8, 16, 37, 38]. This is observed in the cyclic voltammogram, as shown in Figure 14.6, in that the anodic and cathodic peaks are roughly the same size and shape, thus implying that no strongly adsorbed intermediates are blocking the catalyst surface [7].

On clean palladium surfaces, the electro-oxidation of formic acid proceeds with minimal kinetic losses, rivaling the reaction rate of hydrogen on Pt. This imparts high

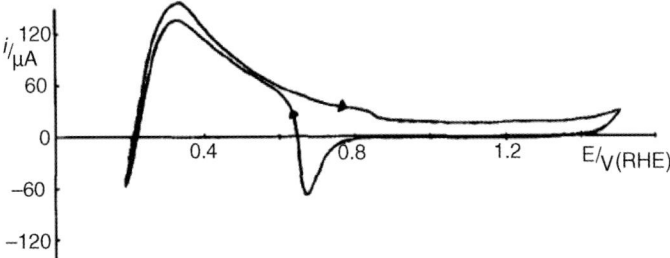

Figure 14.6 Cyclic voltammogram for a smooth Pd wire electrode (10 mm × 0.15 mm) at 25 °C in 0.5 M H_2SO_4 containing 0.1 M HCOOH. Scan rate = 70 mV/s. Reprinted with permission from Ref. [7] Copyright © 1973 Elsevier.

fuel cell efficiency. The oxidation reaction achieves a maximum in activity around 0.4 V versus RHE; however this is followed by a gradual decline in activity. It was shown that there is an easily oxidized intermediate formed on the surface of Pd, although the intermediate has yet to be identified [8].

During constant current operation on a Pd catalyst, a slow deactivation occurs. It has been suggested that the reaction of formic acid on Pd is auto-inhibitive (Equations 14.10–14.12), in that weakly adsorbed reaction intermediates, formed by disproportionation (i.e.,:C(OH)$_2$), cause the apparent decay in activity. Over time, the Pd surface becomes saturated with weakly adsorbed reaction intermediates (each occupying two Pd sites) and their slower rate of reaction (Equation 14.12) becomes the rate-limiting step.

$$Pd + HCOOH \rightarrow Pd(HCOO)_{ad} + H^+ + e^- \qquad (14.10)$$

$$Pd + HCOOH + Pd(HCOO)_{ad} \rightarrow Pd_2(C(OH)_2)_{ad} + CO_2 + H^+ + e^- \qquad (14.11)$$

$$Pd_2(C(OH)_2)_{ad} \rightarrow 2Pd + CO_2 + 2H^+ + 2e^- \qquad (14.12)$$

Alternatively, a more recent proposal that the reaction intermediate, and the intermediate that causes the slow deactivation of Pd, is COOH$_{ad}$, according to the mechanism shown in Equations 14.13–14.15 [39,41]. It was suggested that the rate of transformation to CO_2, according to Equation 14.13, is highest at the surface that is free of COOH$_{ad}$. Particle size effects have been reported by several researchers [40–42]. One postulation is that the smallest nanoparticles have the lowest d-band center, they bind the COOH intermediate less strongly versus larger nanoparticles and thus reduce the surface COOH$_{ad}$ coverage (Equations 14.14 and 14.15). If the rate-limiting step of Equation 14.15 is avoided, due to low COOH$_{ad}$ coverage, then higher rates of direct oxidation of HCOOH to CO_2 (Equation 14.13) are anticipated. This may also help to explain the slow decline in activity, especially with higher concentration formic acid, as there may be more active sites covered with weakly adsorbed COOH.

$$HCOOH_{bulk} \rightarrow CO_2 + 2H^+ + 2e^- \qquad (14.13)$$

$$HCOOH_{ad} \rightarrow (COOH)_{ad} + H^+ + e^- \tag{14.14}$$

$$(COOH)_{ad} \rightarrow CO_2 + H^+ + e^- \text{(rate determining step)} \tag{14.15}$$

Since Pd has demonstrated very high activity, it can be considered as an attractive catalyst for DFAFC applications, especially if the slow decay during constant current operation can be either mitigated or eliminated. However, it is not just catalytic activity that makes Pd appealing. On the basis of cost and gravimetric density alone, Pd offers many advantages as an anode electrocatalyst. Figure 14.7 shows the average monthly trading prices of Pd and Pt, as given by Johnson-Matthey, since 1992. Despite a brief spike in the price of Pd in early 2001, Pt has remained on average more than double the price of Pd over the last 15 years, and nearly four times more expensive in the last 5 years.

In addition, under standard conditions, Pd is 1.75 times less dense than Pt (i.e., 12 vs. 21 g/cm^3), meaning that spherical nanoparticle Pd catalysts of equal mass to Pt catalyst have 75% more available surface area. As a result, a fuel cell that employs Pd in the anode catalyst will benefit in two ways. If two cells contained equal catalyst loading, the Pd-based anode will have higher available surface area for reaction (assuming similar particle sizes) and it will also have a significantly lower cost than a Pt-based anode. Another consideration for Pd catalysts is that they are sensitive to impurities, often present in formic acid sources, such as acetic acid and methyl formate [43]. It is especially important to use a purified fuel source during the characterization of the catalysts.

The emphasis on Pd-based catalyst development, besides particle size studies and the effect of various preparation methods of unsupported and carbon supported Pd [38, 40–42, 44, 45], has been to reduce the decline in activity during operation. This

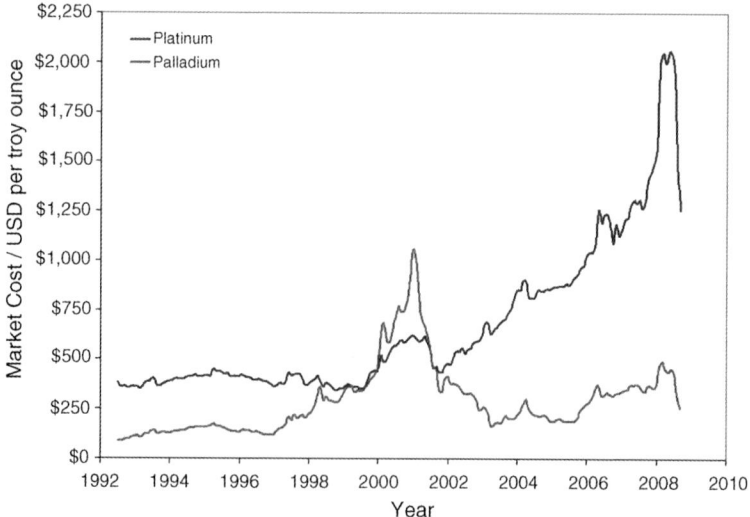

Figure 14.7 Monthly average price of platinum and palladium as listed by Johnson-Matthey. [Data source: http://www.platinum.matthey.com/prices/price_charts.html].

decay in activity results in the requirement that Pd anodes must be regenerated periodically, in order to be considered in a practical system. A reduction in the decay in activity will reduce or preferably eliminate the frequency of required catalyst regeneration [9]. Modifiers to Pd reported in the literature include Ti, Zr, Hf, V, Nb, Ta, Cr, Mo, W and Au [46] as well as PdPt [9, 16–20], PdNi [47], PdAu [16], PdCo/C and PdCoIr/C [48].

The most studied Pd alloy for formic acid oxidation is that of the PdPt system. It has been demonstrated that the addition of even small amounts of Pt to a Pd-based catalyst improves stability, albeit at much higher kinetic overpotentials. As discussed in Section replacing 50% of Pd with Pt stabilizes performance of Pd but at the detriment of its overpotential. Constant current operation of 50:50 Pd:Pt versus DHE was demonstrated to be stable for over 16 hours of operation at 0.55 V versus DHE. This anode potential is far too high to be considered practical for fuel cell applications. However, these results indicate that Pd can be stabilized towards formic acid oxidation.

Casado-Rivera and coworkers [32] reported that ordered intermetallic phases of PdBi and PdSb were inactive towards formic acid oxidation. This is an indication that catalytic activity for Pd can be lost by replacement with certain elements.

Pd modified with Sn has also been reported [12] and the cyclic voltammograms for Pd, PdPt and PdSn catalysts on Ti are illustrated in Figure 14.8. It can be seen

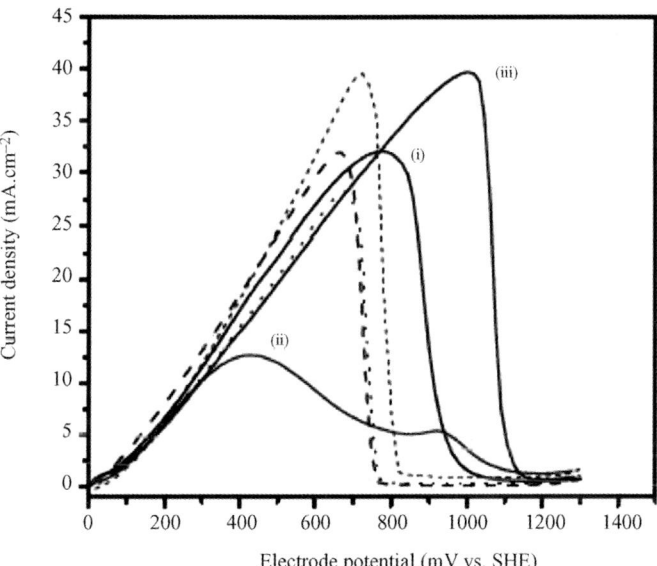

Figure 14.8 Forward (—) and reverse (- - -) scans of cyclic voltammograms of (i) Pd, (ii) PdPt, and (iii) PdSn catalysts thermally formed on titanium mesh in 0.1 M HCOOH/0.5 M H_2SO_4 at room temperature. Scan rate = 20 mV/s. Reprinted with permission from Ref. [12] Copyright © 2007 J. New Mater. Electrochem. Syst.

from this plot that no significant benefit was observed by the addition of Sn to Pd. The group studied this catalyst further by polarizing the electrode at 0.4 V versus DHE and observed a faster decay in current over a 30 min period than Pd alone.

Core shell catalysts using Au core and Pd shell (2 : 1 Au : Pd ratio) demonstrated improved activity and stability towards formic acid oxidation compared with Pd/C over 2 hours of operation [49]. Researchers concluded that the increase in activity was due to the interactions of the Pd shell with the Au core. The thin Pd shell gained electrons from the Au core causing weakening of the adsorptive strengths of the reaction intermediates such as $COOH_{ad}$, facilitating the direct path to CO_2 formation. These results are promising and therefore longer term fuel cell performance of this catalyst would be necessary in order to assess its viability towards the DFAFC.

A series of Pd-M (M = Ti, Zr, Hf, V, Nb, Ta, Cr, Mo, W, Au) catalysts were studied by Masel's group [46] and PdV was identified as the most active catalyst of the series as shown in Figure 14.9. PdV not only demonstrated high activity, but also stability for more than 100 h of operation. These catalysts were prepared by depositing less than a monolayer of Pd on a metal foil support. A further report was made with PdAu formulations in the form of a carbon supported catalyst, which demonstrated promising initial performance, however, long term stability was not observed [50].

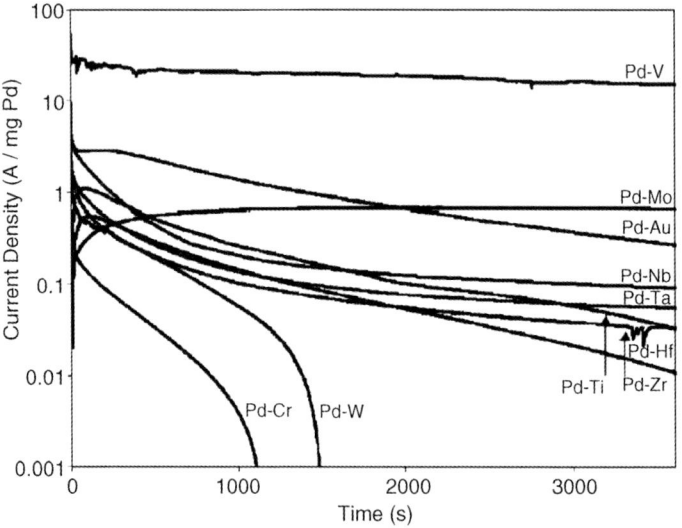

Figure 14.9 Chronoamperometric activity (per mass Pd) of Pd-M catalysts (M = Ti, Zr, Hf, V, Nb, Ta, Cr, Mo, W, Au) with $\theta_{Pd} = 0.6$. The catalysts were immersed in a solution containing 5 M HCOOH and 0.1 M H_2SO_4 at open-cell potential, stepping the potential to 0.3 V vs. RHE and measuring the current as a function of time. Reprinted with permission from Ref. [46] Copyright © 2005 the Electrochemical Society.

Therefore, application of Pd-M catalysts to the formic acid fuel cell has yet to be demonstrated.

Based on the reaction mechanism, stabilization of Pd using bifunctional, electronic or third body modifiers is plausible and the most promising formulations observed appear to be PdAu and PdV. However, to date, a lasting achievement of catalyst stability in a catalyst form that is suited towards use in a DFAFC has not been published.

In the present state of research, the high activity of formic acid oxidation on Pd can be capitalized on in the DFAFC using a modified system operation to maintain or recover palladium activity [9, 46].

14.2.3
Cathode Catalysts for DFAFCs

Reactions at the cathode are essentially the same as that for DMFC and other PEM fuel cell chemistries. As such, the cathode of choice for the DFAFC is Pt, either unsupported (i.e., Pt black) or supported on carbon. Fuel crossover has an impact on cathode performance, but the cathode operates at sufficiently high potentials to ensure that any unreacted formic acid that reaches the cathode is oxidized. As such, the cathode potential remains stable at approximately 0.79 V for over 10 h of constant current operation in a passive DFAFC at 30 °C [1].

It is expected that any improvements to cathode catalysts and/or electrodes for DMFC and other acid-based DLFCs will be applicable to the DFAFC. In the meantime, Pt black and Pt/C are considered the standards for DFAFC cathodes.

14.2.4
Direct Formic Acid Fuel Cell Performance

A combination of low crossover rates and rapid anode reaction kinetics result in very high power densities being reported for direct formic acid fuel cells. DFAFCs are successfully run with up to 60 wt% (15 M) formic acid without showing significant crossover related performance losses under standard active cell operation with both Pd and Pt-based anode catalysts [9, 14]. Furthermore, in the same concentration range, membrane dehydration does not appear to be an issue as determined by high frequency resistance measurements.

Typical DFAFC polarization curves are shown using 10 M (Figure 14.10a) and 15 M (Figure 14.10b) formic acid at 30 °C. The results are given in contrast to DMFC operation, using fuel feeds of equivalent volumetric energy density (i.e., 3.8 and 5.8 M methanol $=$ 10 and 15 M formic acid, respectively). It can be seen by these plots that the DFAFC exhibits much higher fuel cell efficiency than the DMFC using fuels of equivalent energy content at 30 °C, likely due to lower crossover and more efficient anode reactions.

Figures 14.10a and b demonstrate the ability of formic acid fuel cells to operate well at low temperatures and high fuel feed concentrations. When fuel feed con-

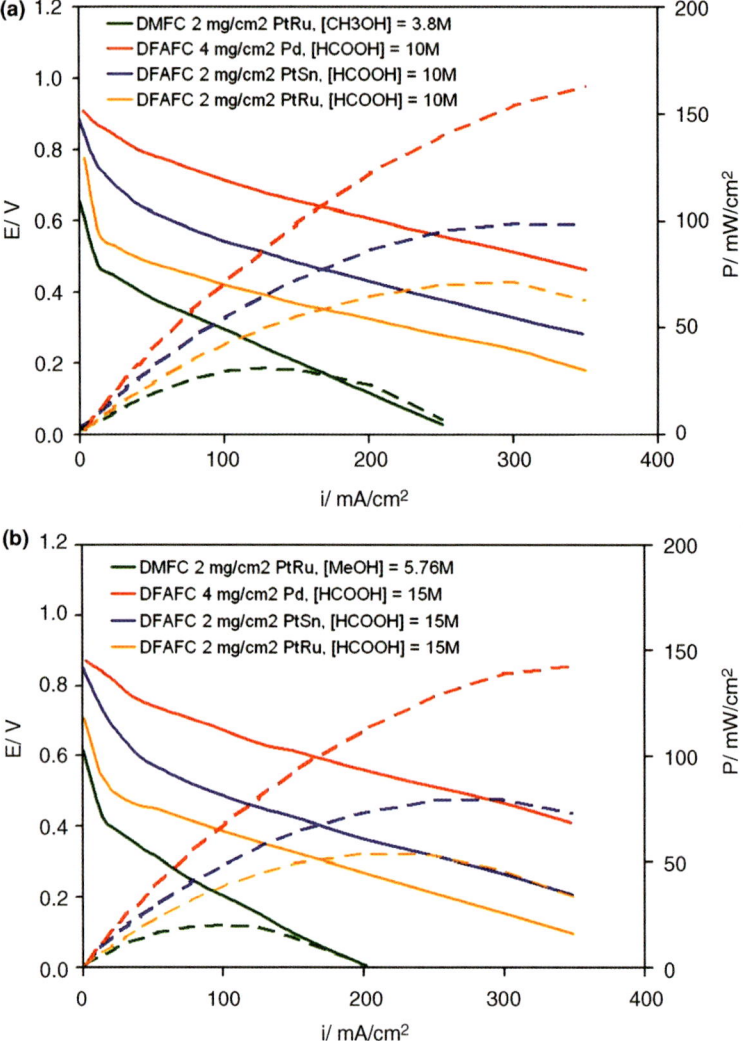

Figure 14.10 Performance of active DFAFC and DMFC at 30 °C with fuels of equivalent energy density (a) 10 M HCOOH; 3.8 M CH_3OH and (b) 15 M HCOOH; 5.8 M CH_3OH. (—) Cell potential (E); (---) Power density (P). Anode feed 0.1 ml/min, cathode (2 mg/cm^2 Pt/C) fed with dry air 200 sccm. [unpublished results courtesy of Tekion Inc.]

centrations are increased 50% from 10 to 15 M, power density performance losses in the low current density range for the DFAFC are minimal, especially when compared with that of methanol (i.e., 8% for formic acid and 32% for methanol using PtRu at 100 mA/cm^2). In contrast, poor anodic reaction rates at 30 °C and methanol crossover for the DMFC results in much lower operating potentials and less tolerance to increases in fuel concentration than the DFAFC. DMFC operation is more favorable

14.2 Electrocatalysis of Direct Formic Acid Fuel Cells

Table 14.3 Reported DFAFC activity for various anode catalysts.

Anode catalyst		Anode conditions		Voltage efficiency at 100 mA/cm^2 (%)
Formulation	Loading (mg/cm^2)	[HCOOH] (M)	T (°C)	
PtAu [14]	3.0	10	30	36
PtBi [34]	3.0	15	60	38
PtRu [51]	3.0	6	60	34
Pd [9]	8.0	15	20	55
Pd/C(40 wt%) [45]	2.4	15	20	45
PdAu/C(20 wt%) [50]	1.2	5	30	51

at higher operating temperatures (e.g., 60 °C) and lower fuel concentration (e.g., 1 M) at the anode.

For both Pt- and Pd-based anode catalysts, the results are particularly impressive at low temperatures due to rapid anodic reaction kinetics. At a low current density of 100 mA/cm^2 and low cell temperatures of 30 to 60 °C, Pd-based catalysts can achieve efficiencies in excess of 50%, whereas Pt-based catalysts report efficiencies between 30 and 40% depending on the catalyst formulation. Representative Pt- and Pd-based catalyst performance, as reported in the literature, is summarized in Table 14.3.

Fuel cell performance for active DFAFCs has been reported using PtAu and PtRu as anode catalysts, and is illustrated in Figure 14.11 [14]. Their effects on Pt as

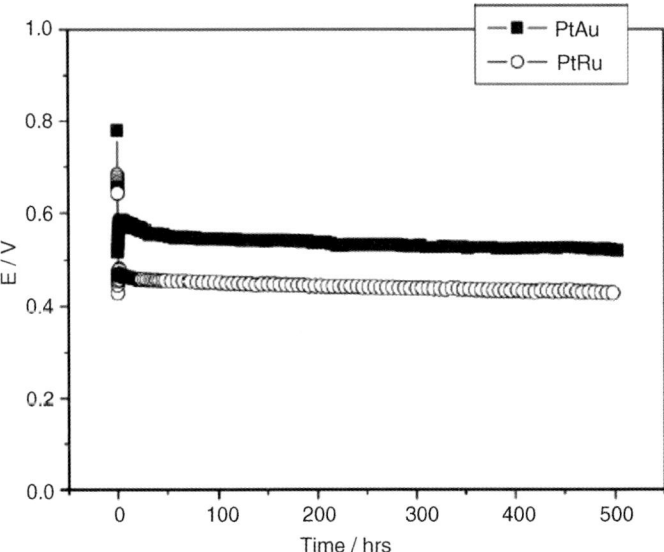

Figure 14.11 Change in cell voltage of PtAu-based MEA and PtRu-based MEA during a 500 h life time test under 100 mA/cm^2. Reprinted with permission from Ref. [14] Copyright © 2006 Elsevier.

discussed in Section , are considered to be bifunctional for PtRu and electronic and third body for PtAu. As can be seen from the fuel cell results at 60 °C, the PtAu outperforms PtRu by about 100 mV. This stable fuel cell performance demonstrates that PtAu can be considered a potentially viable catalyst for DFAFC applications albeit at higher temperatures of operation than those employing Pd.

A distinction between formic acid reaction kinetics on Pd and Pt is also clear in that Pd-based overpotential losses in the kinetic region (i.e., <50 mA/cm^2) are minimal. In operation with Pd catalysts, a very low anodic overpotential for formic acid oxidation has several benefits. Most importantly it allows DFAFCs to be run at high voltage efficiencies (i.e., >50%), while still producing significant amounts of power and waste heat quantities that can easily be managed. The end result is less chance for thermal runaway, provided that system operation can be accommodated to perform occasional catalyst regeneration steps as required to maintain the high activity of Pd [10]. This operation is illustrated in Figure 14.12, which shows a plot of cell potential plotted versus time for an active cell employing Pd black (4 mg/cm^2) at the anode and Pt/C (2 mg Pt/cm^2) with 12 M (ca. 50%) formic acid fuel at a constant cell temperature of 40 °C. The anode catalyst was regenerated periodically as described in the literature [9]. This demonstrates that although Pd shows a decline in activity over time, the performance can be recovered and maintained for long term operation (e.g., 500 h). It also illustrates that the DFAFC can operate with high efficiencies at low temperatures.

A DFAFC portable power system was reported for powering a laptop [51]. This DFAFC was based on a PtRu catalyst. They built a hybrid 15 membrane electrode assembly (MEA) stack capable of producing 30 W at 60 mW/cm^2, with an optimum system efficiency of 0.23. This study demonstrates very stable performance for at

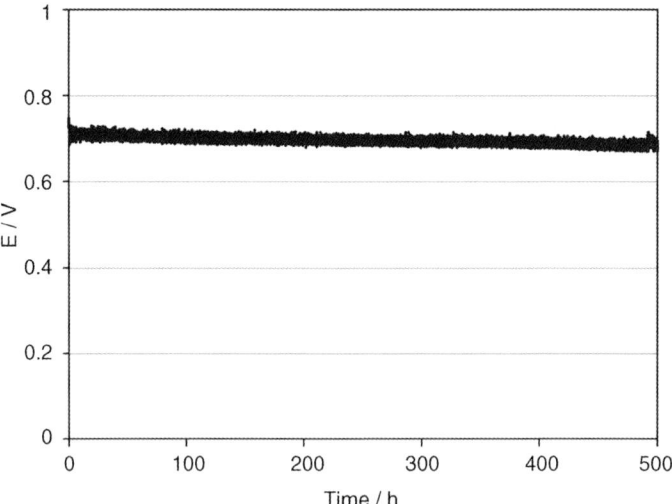

Figure 14.12 Plot of long-term MEA performance (4 mg/cm^2 Pd black for anode, 2 mg/cm^2 Pt/C for cathode). The cell was operated at 40 °C with 12 M formic acid at 0.1 ml/min and 100 sccm air under 100 mA/cm^2. [unpublished results courtesy of Tekion Inc.]

least 500 h of operation. It stands to reason that even more efficient fuel cells are possible with the use of catalysts such as PtAu, PtSn, PtIn or Pb.

Passive cell data has been reported by Ha and coworkers, demonstrating between 147 to 177 mW/cm^2 (with 15 M and 10 M fuel respectively) at 30 °C. The anode catalyst used in this case was Pd black. They demonstrated approximately 10 h of stable operation with 10 M formic acid [1].

The DFAFC results shown in this section illustrate that there is significant potential for development of DFAFCs for low power applications. Commercialization of this technology is being actively pursued by a company named Tekion Inc.

14.2.5
Summary of Electrocatalysis in DFAFCs

The electro-oxidation of formic acid has been studied for decades by numerous researchers. Results to date indicate that there are a few anode catalyst candidates that demonstrate high catalytic efficiencies suitable for DFAFCs at low temperatures. These include either unsupported or carbon supported PtAu, PtBi, PtIn as well as Pd. Active cell DFAFC performance has been demonstrated with PtAu and Pd for 500 h showing impressive performance. More fuel cell performance data would be required however, to obtain adequate comparisons of these catalysts. These candidates (i.e., PtAu, PtBi, PtIn, Pd) show potential for commercialization and there is a possibility that some of these and/or new formulations (provided that they can be commercially produced) will be suitable for low power fuel cell applications.

14.3
Electrocatalysis of Direct Ethanol Fuel Cells

Ethanol is an attractive liquid fuel for direct liquid fuel cells as ethanol has a higher theoretical energy density (6280 Wh/L) than methanol (4820 Wh/L) [2]. Ethanol is non-toxic and it can be produced in large quantities from either biomass or agricultural products. Studies have shown that the electrochemical reactivity of ethanol over Pt-based catalysts was comparable to that of methanol at elevated temperatures [52]. Ethanol also showed a lower permeability through Nafion® membranes than methanol, resulting in a lower fuel cross-over and less effect on cathode performance [53]. These advantages make ethanol a promising alternative fuel source to methanol, the most studied liquid fuel in DLFCs.

The main challenge for direct ethanol fuel cells (DEFCs) is associated with the electro-oxidation of ethanol. Electro-oxidation of ethanol to carbon dioxide involves a 12 electron process and the cleavage of the C—C bond for complete conversion. Due to incomplete oxidation however, the main products of ethanol electro-oxidation are acetic acid and acetaldehyde [54]. This multistep electrochemical process is inefficient for power generation and the intermediate species from the electro-oxidation steps will also lead to poisoning of the fuel cell catalysts, resulting in poor catalytic activities and hence inefficient fuel cell operation.

Direct ethanol fuel cell operation The overall reaction for the electro-oxidation of ethanol is presented in Equation 14.16.

$$C_2H_5OH + 3O_2 \rightarrow 2CO_2 + 3H_2O \tag{14.16}$$

In the acid-based PEM environment, the anode and cathode reactions for the DEFC are presented in Equations 14.17 and 14.18 respectively.

$$C_2H_5OH + 3H_2O \rightarrow 2CO_2 + 12H^+ + 12e^- \quad \varepsilon^\circ = 0.084\,V \tag{14.17}$$

$$12H^+ + 12e^- + 3O_2 \rightarrow 6H_2O \quad \varepsilon^\circ = 1.229\,V \tag{14.18}$$

The overall Gibbs free energy change for the combined reactions is 1367 kJ/mol, equivalent to a reversible standard cell voltage of 1.145 V.

This 12-electron process for the electro-oxidation of ethanol to carbon dioxide leads to the high theoretical energy density for DEFC. Similar to DMFC, water is required at the anode for the complete oxidation of ethanol (3 moles of water per mole of ethanol) hence water management is required in DEFC systems.

The biggest challenge facing the commercialization for DEFC is the low efficiency for the complete electro-oxidation of ethanol to CO_2 at temperatures <100 °C. Incomplete electro-oxidation leads to the formation of other reaction products like acetic acid and acetaldehyde as in Equations 14.19 and 14.20 respectively [55–57]. The formation of these reaction products requires the transfer of 4 and 2 electrons respectively.

$$C_2H_5OH + H_2O \rightarrow CH_3COOH + 4H^+ + 4e^- \tag{14.19}$$

$$C_2H_5OH \rightarrow CH_3CHO + 2H^+ + 2e^- \tag{14.20}$$

These undesired reactions reduce the overall efficiency of the electro-oxidation of ethanol, leading to a low power output for DEFCs. The yields for acetic acid and acetaldehyde are especially high at temperatures <100 °C with Pt anode catalyst [58–60]. The partial oxidation products from the electro-oxidation of ethanol can also be adsorbed onto the Pt surface, leading to rapid poisoning of the fuel cell catalyst [61]. The success of DEFCs thus depends on the identification of anode catalyst candidates which promote the complete electro-oxidation of ethanol to carbon dioxide at temperatures <100 °C. The catalyst will also have to be resistant to poisoning species such as partial oxidation products from ethanol including CO.

14.3.1
Anode Catalysts for DEFCs

Most of the research on the electrocatalysis of ethanol has been on platinum-based anode catalysts, particularly on efforts to enhance the activity of platinum by using binary and ternary electrocatalysts. Similar to the case for DMFC, the function of admetals is two fold. Admetals can promote the bifunctional mechanism, allowing for the dissociation of water to occur at lower potentials. Additionally, admetals can produce electronic modifications to the Pt surface that favor the direct oxidation pathway [21]. The most extensively studied admetals for the electro-oxidation of ethanol are Ru and Sn.

14.3.1.1 Binary Catalysts

PtRu is considered the most active binary catalyst for DMFC. It is also a promising binary catalyst for the electro-oxidation of ethanol in DEFCs. The addition of Ru appears to achieve two functions. First, Ru improves the selectivity for complete ethanol oxidation by favoring the oxidation pathway through the weakly adsorbed species [62]. Second, it promotes the oxidation of strongly-bound adsorbed intermediates, resulting in a higher yield of CO_2 [63]. The catalytic activity of PtRu towards ethanol oxidation was found to be strongly dependent on the Ru content. For current-time curves measured at 0.5 V in 1 M C_2H_5OH at 25 °C, it was observed that a minimum of 20 at.% Ru is required for performance enhancement while the optimum atomic ratio appeared to be at about 40 at.% Ru as shown in Figure 14.13 [64].

Contrary to DMFC, the most active binary catalyst for electro-oxidation of ethanol was found to be PtSn [59, 65, 66]. This is illustrated in a plot of maximum power density for Pt/C, PtRu/C and PtSn/C (Figure 14.14). The increase in the activity for electro-oxidation of ethanol on PtSn can be explained by the bifunctional mecha-

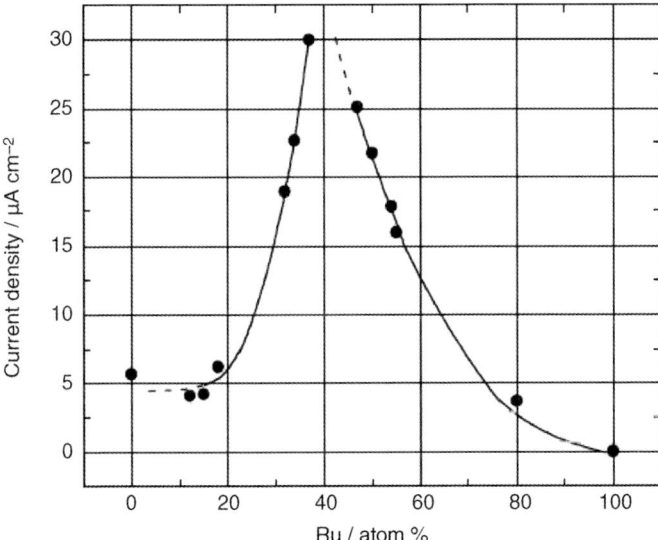

Figure 14.13 Activity towards ethanol electro-oxidation as a function of PtRu composition. The data were obtained from chronoamperometric curves at 25 °C after 30 min of electrode polarization at 0.5 V in 1.0 M C_2H_5OH + 0.5 M H_2SO_4. Reprinted with permission from Ref. [64] Copyright © 2004 Elsevier.

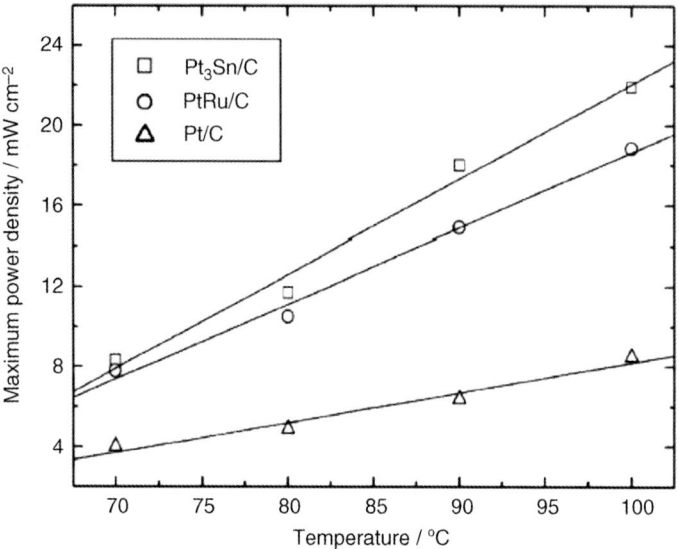

Figure 14.14 Maximum power density of DEFCs with Pt/C, PtRu (1 : 1) and PtSn/C (3 : 1) anodes vs. cell temperature. O_2 pressure: 3 atm; 1 M ethanol. Anode metal loading: 1 mg/cm². Cathode 20 wt.% Pt/C, Pt loading 1 mg/cm². Reprinted with permission from Ref. [66] Copyright © 2006 Elsevier.

nism [67]. When ethanol is adsorbed dissociatively at the platinum sites, acetaldehyde is formed according to Equation 14.21 and 14.22.

$$Pt + CH_3CH_2OH \rightarrow Pt(OCH_2CH_3)_{ad} \text{ or } Pt(CHOHCH_3)_{ad} + H^+ + e^-$$
(14.21)

$$Pt(OCH_2CH_3)_{ad} \text{ or } Pt(CHOHCH_3)_{ad} \rightarrow CHOCH_3 + H^+ + e^-$$
(14.22)

The acetaldehyde is adsorbed on the platinum sites. In the presence of Sn, water can be activated at lower potentials than on Pt (Equation 14.23) and some OH species can be formed at Sn sites. The adsorbed acetaldehyde species can react with the adsorbed OH species to produce acetic acid.

The addition of Sn thus increases the overall rate of the electro-oxidation of ethanol.

$$Pt(COCH_3)_{ad} + Sn(OH)_{ad} \rightarrow Pt + Sn + CH_3COOH$$
(14.23)

Although the addition of Sn improves the activity of the electro-oxidation of ethanol, it also lowers the selectivity for the reaction towards CO_2 (also acetaldehyde) [58]. It is believed that several adjacent platinum sites are required for ethanol to adsorb dissociatively and to break the C–C bond. The addition of Sn to Pt lattice thus discourages C–C bond cleavage on the catalyst surface. This results in the increased production of acetic acid from the electro-oxidation of ethanol on PtSn, leading to a decrease in the faradic efficiency which decreases the overall efficiency of the DEFC.

The optimum Sn content appears to depend on the preparative method for the anode catalyst. When preparing the catalyst using the Bonneman's type coimpregnation reduction method, the optimum composition was found to be in the range of 10–20% Sn by atomic ratio [61]. However in the case when the catalyst was prepared by a modified polyol process, the optimum composition was found in the range 33–40% Sn by atomic ratio [68].

Other binary Pt-M (M = W, Pd, Rh, Re, Mo, Ti, Ce) catalysts were also investigated for the electro-oxidation of ethanol [59, 69–74]. These catalysts have higher activities than pure Pt catalyst but were found to be less active when compared with PtRu and PtSn as anode catalyst for DEFC. Of special note in this group of binary catalysts was PtRh; it was found that the addition of Rh produces a strong decrease in the acetaldehyde yield [75]. CO_2 yield improved in the presence of Rh, leading to the improvement in the faradic efficiency (thus higher overall efficiency of the electro-oxidation for ethanol). The result is promising if continuing research can show that a third metal can be added to PtRh to improve the rate of oxidation of ethanol.

14.3.1.2 Ternary Catalysts

The improvement in activity for the electro-oxidation of ethanol with bimetallic catalysts leads to the study of the addition of a third metal to the catalyst. The goals of the third metal are to further improve the activity for ethanol oxidation, and/or promote the cleavage of the C—C bond (to improve the faradic efficiency) of the reaction. Since the most promising binary anode catalysts for DEFCs are PtSn and PtRu, ternary anode catalyst studies focus on addition of a third metal to these series of binary catalysts.

PtSnRu was found to be among the most active catalyst for the electro-oxidation of ethanol. At the preferred atomic ratio, PtSnRu has higher activity than PtSn for ethanol oxidation. It was found that anode catalyst with PtSnRu ratio between 1:1:0.3 to 1:1:0.4 (with the non-alloyed Ru to non-alloyed Sn ratio approximately at 0.4) appeared to be the best candidate for DEFC (Figure 14.15) [67, 76]. It is believed the RuO_x-SnO_2 interaction promotes the formation of hydroxyl species by dissociating water at a lower potential with respect to PtSn and PtRu catalysts. The modification by the addition of Ru could also weaken the bonding between the hydroxyl species and the catalyst surface, enabling the oxidation of CO and/or acetaldehyde species at lower potentials than either PtSn or PtRu.

However, as the Ru content increases, RuO_x begins to substitute SnO_2 in the PtSn matrix. Pt then has a stronger interaction with Ru than Sn causing a decrease in activity. At a Pt: Sn. Ru ratio of 1:1:1, the activity for electro-oxidation of ethanol was lower than that of Pt: Sn at the ratio of 1:1. It is also important to point out that while the addition of Ru in the Pt-Sn matrix improves the activity of the ethanol electro-oxidation, it did not improve its Faradaic efficiency [67, 77]. The distribution of the oxidation products did not change with the addition of Ru to PtSn, indicating that the modification did not facilitate the C—C bond cleavage. Thus, despite the increase in ethanol electro-oxidation activity, the overall efficiency of the reaction does not improve.

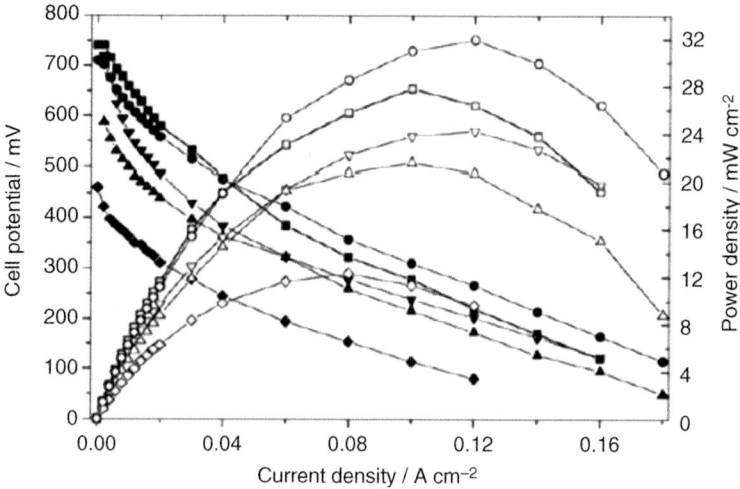

Figure 14.15 Polarization and power density curves in single DEFC with binary PtSn/C (1:1) and ternary PtSnRu/C (1:1:0.3 and 1:1:1) catalysts prepared by the FAM and on commercial Pt/C and PtRu/C by E-TEK as anode electrocatalysts for ethanol oxidation at 110 °C and 3 atm O_2 pressure using a 1 M ethanol solution. Anode metal loading 1 mg/cm^2. Cathode: 20 wt.% Pt/C, Pt loading 1 mg/cm^2 (□) PtSn/C (1:1); (○) PtSnRu/C (1:1:0.3); (△) PtSnRu/C (1:1:1); (□) PtRu (1:1)/C E-TEK; (◊) Pt E-TEK. Full symbols: polarization curves; open symbols: power density curves. Reprinted with permission from Ref. [76] Copyright © 2007 Elsevier.

Besides PtSnRu, ternary catalysts based on either PtRu or PtSn also reported some promising ethanol electro-oxidation properties [59]. Addition of W [68, 78], Mo [68], Ni [79] and Pb [80] to PtRu-based catalysts had all shown improvement when compared with PtRu. For the PtSn-based catalysts, ternary anode catalysts with the addition of Ni [81], and Ru [82] were reported. Spinace et al reported that for Pt-Sn-Ni with molar ratio of 50:40:10 prepared by alcohol-reduction with ethylene glycol as the solvent and reducing agent, the electrocatalyst showed superior performance when compared with PtSn/C for the electro-oxidation of ethanol [81].

14.3.1.3 Anode Catalysts for Alkaline Electrolytes

Alkaline electrolytes are less corrosive; therefore, operating DEFCs in alkaline media enables the use of non-noble metals such as Ni, Fe, Co as well as their oxides as anode catalysts [83]. The mechanism for ethanol electro-oxidation in alkaline medium is not entirely clear but CH_3COO^- was suggested to be the main product of the electro-oxidation [84, 85]. Besides the performance of bimetallic or metal oxide promoted Pt-based anode catalysts, researchers have also investigated non Pt-based anode catalysts [59].

For bimetallic Pt-based anode catalysts, El-Shafei et al studied the ad-atom effect on the electro-oxidation of ethanol in alkaline medium using Pb, Tl and Cd [86]. The results suggested the oxidation rate increased by a factor of about 15 with Pb and Tl as the ad-atoms. A series of metal oxide promoted Pt-based anode catalysts were prepared and their electro-oxidation activities were evaluated. The oxides includes CeO_2 [87, 88], NiO [85, 88], ZrO_2 [89] and MgO [90]. All of these oxide-promoted Pt catalysts revealed higher electrochemical activities for ethanol oxidation in an alkaline medium than Pt catalyst. Xu and coworkers suggested CeO_2 behaves in a similar manner to Ru in a PtRu catalyst in which oxygen containing species could more easily form on the surface of CeO_2. The formation of oxygen-containing species at lower potential can transform CO-like poisoning species on Pt to CO_2, releasing the active sites on Pt for further electrochemical reaction [87].

Palladium was found to be a better catalyst for the electro-oxidation of ethanol in alkaline medium than Pt [88]. As in the case of DFAFC, the potential to substitute Pt with Pd as anode catalyst for DEFC can lead to a reduction of cost for the MEA. Several publications have demonstrated the improvement in electrochemical activities for ethanol oxidation in alkaline media with oxide promoted Pd catalysts. Oxides such as CeO_2 [84, 88], Co_3O_4 [84], Mn_3O_4 [84], NiO [84, 85] and TiO_2 nanotubes [91] have all demonstrated improved electrocatalytic activities and/or poison tolerance towards CO-like species during ethanol oxidation in alkaline media.

Oxide promoted Ru anode catalysts were also reported, and, in particular RuNi [83] and RuNiCo [92] have shown good electrochemical activities for ethanol oxidation in alkaline media.

It is important to point out that these catalysts were in the early stages of research and the actual impact on the performance of these catalysts will have to be verified in a fuel cell.

14.3.2
Cathode Catalysts for DEFCs

Pt/C is traditionally used as the electrocatalyst for oxygen reduction reaction in DEFC due to its high catalytic activity and high chemical stability [93]. When ethanol is used at high concentrations however, ethanol crossover from anode to cathode increases. This results in a decrease of the open circuit potential of the fuel cell due to oxidation of ethanol at the Pt/C cathode. The requirements of DEFC cathodes are similar to those for DMFC. The catalysts require high activity for oxygen reduction and a high ethanol tolerance due to ethanol crossover [59]. Thus, research on cathodes for DEFCs began by adopting suitable cathode candidates for DMFC. Carbon supported PtCo (3 : 1) as a cathode catalyst in DEFCs was studied due to its high oxygen reduction activity and good methanol tolerance, with the presumption that it would also be tolerant of ethanol [94]. Improved DEFC performance was obtained for PtCo/C cathode during operation at 60–100 °C when compared with Pt/C cathode catalyst as shown in Figure 14.16 [94]. PtPd/C (9 : 1) was also found to possess similar oxygen reduction reaction activity but with a higher ethanol tolerance than Pt/C [59].

A recent publication also reported a novel non-noble ternary cathode electrocatalyst for DEFC [93]. NiCoFe/C was demonstrated to selectively reduce oxygen without oxidation of ethanol crossed-over from the anode. The open circuit potentials for the NiCoFe/C cathode were constant at ethanol concentrations of 0.5–5 M while poten-

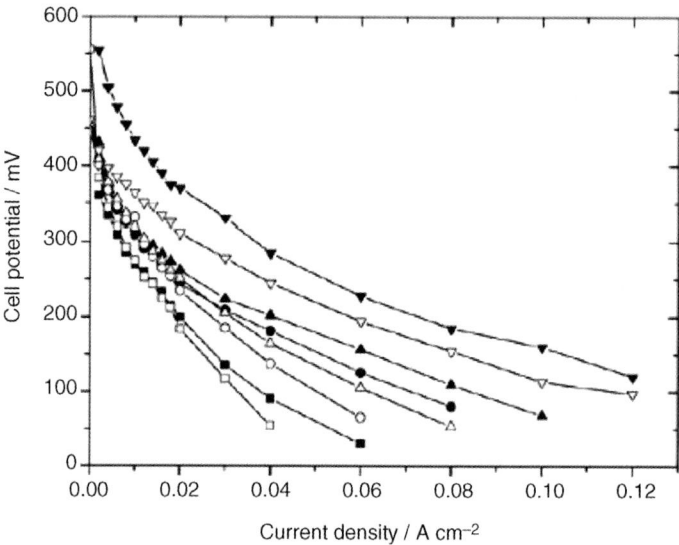

Figure 14.16 Polarization curves in single DEFC with PtCo/C (3:1) and Pt/C electrocatalyst by E-TEK as cathode materials for oxygen reduction at 60, 80, 90 and 100 °C and 3 atm O_2 pressure using a 1 M ethanol solution. Cathode Pt loading 1 mg/cm². Anode: 20 wt.% Pt/C, Pt loading 1 mg/cm². (□) 60 °C; (○) 80 °C; (△) 90 °C; (▽) 100 °C. Open symbols: Pt/C; full symbols: PtCo/C. Currents are expressed in terms of mass activity (numerically equal to the current normalized to the geometric surface area). Reprinted with permission from Ref. [94] Copyright © 2007 Elsevier.

tials for Pt/C decreased as the concentration of ethanol increased over the same range. Due to the similar crossover rate, the high open circuit potentials for NiCoFe/C were attributed to the catalyst's ability to selectively reduce oxygen without undergoing oxidation of ethanol.

14.3.3
Direct Ethanol Fuel Cell Performance

The power density for DEFC is low when compared with DMFC despite ethanol having a higher theoretical energy density than methanol. The lower power density for DEFC is attributed to the slow kinetics of the electro-oxidation of ethanol with Pt-based anode catalyst and the inefficiency of the ethanol electro-oxidation to CO_2. With PtSn and PtSnRu as the anode catalysts, power densities of 50 mW/cm² are readily achieved for DEFC at temperature of about 90 °C [71, 95, 96].

Due to the slow reaction kinetics for the anode electro-oxidation reaction, the power density of the DEFC depends heavily on the cell operating temperature. Figure 14.17 illustrates the effect of operating temperature from 30 °C to 110 °C on power density of the DEFC [71]. Power density from the DEFC remains low until the operating temperature reaches 90 °C. With advancements in anode catalyst research

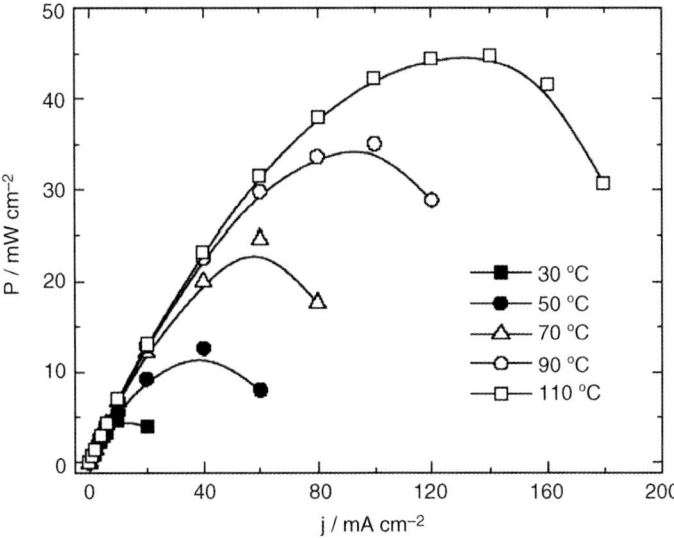

Figure 14.17 Power density P against current density j curves recorded in a single DEFC using a PtSn (100:20)/C anode (30% metal loading) of 5 cm² surface area, at different temperatures (cathode E-TEK, Nafion® 117 membrane, $CH_3CH_2OH = 1\,M$, $P_{CH_3CH_2OH} = 1$ bar, $P_{O_2} = 1$ bar) at various temperatures. Reprinted with permission from Ref. [71] Copyright © 2004 Springer Science.

towards higher activity electro-oxidation catalysts and higher faradic efficiencies, the operating temperature of the DEFC is expected to be reduced.

Typical DEFCs operate with 1–2 M of ethanol. Higher fuel concentration would lead to a reduction in power density due to ethanol crossover [71, 93, 97]. Typical polarization curves for a DEFC operating with 1 and 2 M ethanol using a PtSn (100:20) anode catalyst are illustrated in Figure 14.18 [71]. Operation of the DEFC with 2 M ethanol results in an enhancement in the performance by 20% when compared with operation at 1 M. In the future, operation of the DEFC with higher fuel concentration will be possible through the combination of higher activity anode catalysts, cathode catalysts with higher ethanol tolerance and membranes with lower fuel crossover.

14.3.4
Summary of Electrocatalysis in DEFCs

Direct ethanol fuel cells continue to show promise for DLFCs due to large theoretical energy density, however, to date, power densities do not compete with DMFCs. The most promising catalyst formulations identified so far include either carbon-supported or unsupported PtSn and PtRuSn at operating temperatures of 90 °C or higher demonstrating 50 mW/cm². Further catalyst, membrane and fuel cell development

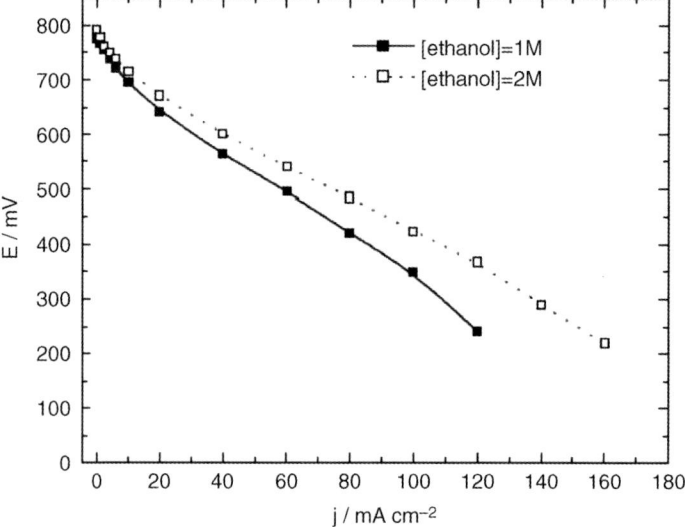

Figure 14.18 Cell voltage (E) against current density (j) curves recorded in a single DEFC using a PtSn(100:20)/C anode (30% metal loading) of 5 cm² surface area at 90 °C for 1 and 2 M ethanol (cathode E-TEK, Nafion® 117 membrane, $P_{CH_3CH_2OH} = 1$ bar, $P_{O_2} = 1$ bar). Reprinted with permission from Ref. [71] Copyright © 2004 Springer Science.

work resulting in higher efficiencies may allow the operating temperatures to decrease below 90 °C. DEFC catalysts are being actively developed for commercialization by Acta S.p.A.

14.4
Electrocatalysis of Direct Hydrazine Fuel Cells

Hydrazine (N_2H_4) has a higher theoretical energy density (5400 Wh/L) than methanol (4820 Wh/L) [2]. Hydrazine is an attractive fuel source for direct liquid fuel cells because it can be readily produced from renewable resources. No greenhouse gases such as CO_2 will be emitted by direct hydrazine fuel cells (DHFCs) because hydrazine is carbon free; only nitrogen and water are produced during the electro-oxidation of hydrazine. Studies also suggest that no catalyst poisoning species are formed during the electro-oxidation of hydrazine, making the direct hydrazine fuel cell system highly desirable with significant advantages over DMFCs [98, 99].

Limitations of the DHFC include the fact that hydrazine is a toxic chemical so special attention is required to design a fuel cell system to ensure that hydrazine will not leak into the environment. DHFCs are also subject to fuel crossover in the form of $N_2H_5^+$, a product from hydrazine hydrolysis in aqueous hydrazine solution [100]. At low current density, DHFC exhibits high voltage, but the cell voltage drops rapidly with an increase in current density due to crossover of $N_2H_5^+$.

Direct hydrazine fuel cell operation The overall reaction for the oxidation of hydrazine in DHFC is presented in Equation 14.24.

$$N_2H_4 + O_2 \rightarrow N_2 + 2H_2O \tag{14.24}$$

The Gibbs free energy change for the overall reaction is 2005 kJ/mol, equivalent to a reversible standard cell voltage of 1.615 V, the highest among common direct liquid fuel cells. Unlike DMFC, water is not required for the anode reaction, thus less balance of plant is required for the fuel cell system resulting in the reduction of overall volume of the system. The overall electro-oxidation for hydrazine is a 4 electron process. DHFCs can operate in acidic or alkaline medium and the anodic reactions for electro-oxidation of hydrazine in acidic and alkaline medium are shown in Equations 14.25 and 14.26 respectively [101].

$$N_2H_5^+ \rightarrow N_2 + 5H^+ + 4e^- \tag{14.25}$$

$$N_2H_4 + 4OH^- \rightarrow N_2 + 4H_2O + 4e^- \tag{14.26}$$

The hydrazine fuel cell was first investigated in the 1970s as an alkaline fuel cell with liquid alkaline electrolyte [102, 103]. Unfortunately, system challenges due to fuel crossover slowed the progress towards commercialization of this technology. Recent advances in fuel cell technology, especially in the area of proton exchange membranes, has led to increased attention towards hydrazine as a fuel for direct liquid fuel cells because of the clean emission of DHFCs as well as high theoretical fuel cell voltage and high energy density.

14.4.1
Anode Catalysts for DHFCs

Research on DHFC electrocatalysts has focused primarily on anode catalysts for the electro-oxidation of hydrazine. In this section, only the development of anode catalysts for hydrazine oxidation will be discussed since the cathode reactions are similar to the DMFC cathode.

Garcia and coworkers proposed that in acidic media, hydrazine electro-oxidation occurs on a Pt surface following a 5-step reaction as shown in Equations 14.27–14.31 [104].

$$\text{Step 1}: N_2H_5^+ \cdot N_2H_{3ad} + 2H^+ + e^- \tag{14.27}$$

$$\text{Step 2}: N_2H_{3ad} \rightarrow N_2H_{2ad} + H^+ + e^- \quad \text{(rate determining)} \tag{14.28}$$

$$\text{Step 3}: N_2H_{2ad} \rightarrow N_2H_{ad} + H^+ + e^- \tag{14.29}$$

$$\text{Step 4}: N_2H_{ad} \rightarrow N_{2ad} + H^+ + e^- \tag{14.30}$$

$$\text{Step 5}: N_{2ad} \rightarrow N_{2(g)} \tag{14.31}$$

The proposed mechanism is similar to that proposed for hydrazine on a gold surface [105]. The difference between Pt and Au is found in the rate determining step in which N_2H_{3ad} oxidizes to N_2H_{2ad}. On a Pt surface, both N atoms are bonded to neighboring Pt sites. On the Au surface, after formation of N_2H_{2ad}, the adsorbate desorbed from the Au surface and continued its oxidation from the solution. The difference is attributed to the incomplete d-band character of the Pt surface atom compared with the d^{10} character of the Au surface site.

Electro-oxidation of hydrazine was also studied in alkaline media (i.e., pH >7) on carbon-based electrodes including modified carbon electrodes [106–108]. These studies suggested that the rate determining step for the electro-oxidation of hydrazine in an alkaline environment is at the first electron transfer step according to Equation 14.32.

$$N_2H_4 + H_2O \rightarrow N_2H_3 + H_3O^+ + e^- \qquad (14.32)$$

The reaction then proceeds following a series of reactions involving a 3-electron transfer shown in Equation 14.33.

$$N_2H_3 + 3H_2O \rightarrow N_2 + 3H_3O^+ + 3e^- \qquad (14.33)$$

14.4.1.1 Noble Metal-based Anode Catalysts

Yamada and coworkers studied a series of noble metals as anode electrocatalysts for DHFCs as listed in Figure 14.19 [109]. Pt, Pd, Rh and Ru were used as the anode catalyst in MEAs tested in a fuel cell environment. It was determined that the trend of cell voltage in the low-current density region was Pt > Pd > Rh = Ru. High fuel cell voltage of over 1 V was obtained for Pt and Pd anode electrocatalysts. Besides electo-

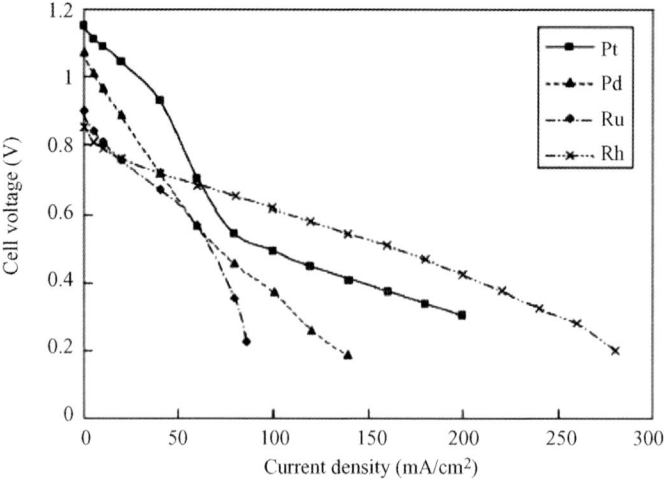

Figure 14.19 Effect of the anode noble catalyst in a direct hydrazine fuel cell. Loading of anode/cathode catalysts = 2/3 mg/cm². Reprinted with permission from Ref. [109] Copyright © 2003 Elsevier.

oxidation, hydrazine also undergoes undesirable catalytic decomposition by the noble metal catalysts according to Equations 14.34 and 14.35.

$$N_2H_4 \rightarrow N_2 + 2H_2 \tag{14.34}$$

$$3N_2H_4 \rightarrow N_2 + 4NH_3 \tag{14.35}$$

The order activity for this catalytic decomposition was determined to be Ru Rh ≫ Pd > Pt. The catalytic decomposition decreases fuel use for the fuel cell as well as lowering the cell voltage due to hydrogen and ammonia generation at anode. The results of the study suggested that Pt was the most effective anode catalyst for operation in DHFC among the noble metal catalysts shown in Figure 14.19.

It was also found that fuel cell voltage decreases rapidly with increasing current density [101, 109]. This voltage drop is likely due to hydrazine crossover rather than the generation of a poisoning species.

14.4.1.2 Non-Noble Metal-based Anode Catalysts

Noble metals such as Pt are expensive and add significantly to the total cost of a fuel cell system. Alternative non-noble metal catalyst development is considered critical to the commercialization of the DHFC system. Nickel-based anode catalysts were used in alkaline hydrazine fuel cell research in the 1970s [110]. Recent research on the electro-oxidation of hydrazine focuses on non-noble metal anode catalysts in an alkaline medium because the acidic medium, either in the fuel and/or electrolyte, limits the use of non-noble metals as anode catalysts for hydrazine electro-oxidation.

Yin and coworkers reported a ZrNi alloy as an anode catalyst, using both Nafion® 117 membrane as well as an anion exchange membrane as the electrolyte [111]. The experimental results confirmed the four-electron process for the electro-oxidation of hydrazine in alkaline medium, as indicated in Equation 14.26. It was observed that the addition of NaOH in hydrazine solution significantly improved the open circuit voltage and cell performance of the DHFC. The performance improvement was associated with the suppression of $N_2H_5^+$ crossover due to the depression of hydrolysis of N_2H_4 in the presence of NaOH. This crossover reduction also contributed to a higher fuel cell power density with the anion exchange membrane compared with a cation exchange membrane such as Nafion®.

With the use of anion exchange polymer electrolytes containing tetraalkylammonium cation groups as pendant groups with a polyolefin main chain, Asazawa and coworkers were able to demonstrate DHFC with both Ni and Co as the anode catalysts [112]. The fuel cell output characteristics exceed those of DMFC. From the performance of Co and Ni anode catalyst, it was found that Co anode catalyst exhibited better performance compared with the Ni anode catalyst. However, in continuous power generation experiments, the performance with a Co anode catalyst decreased sharply after 40 hours due to the conversion of cobalt to cobalt hydroxide during operation, as indicated in Figure 14.20 [112]. Performance with the Ni anode catalyst, on the other hand, exhibited excellent stability for 100 hours of testing and post testing analysis indicated that Ni remained in the metallic state throughout fuel cell operation.

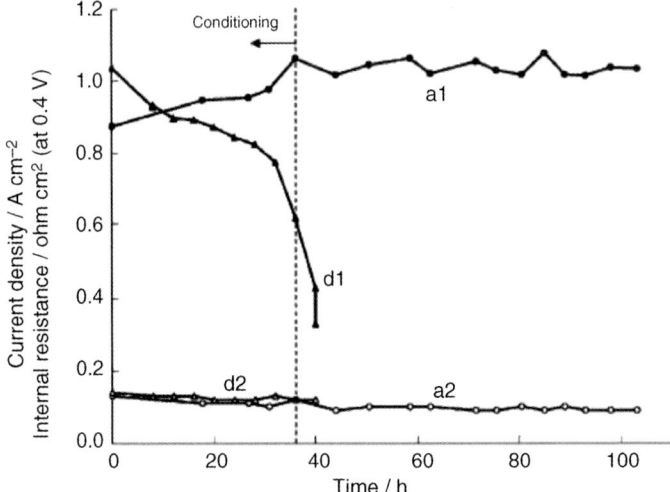

Figure 14.20 Durability test of direct hydrazine anion-exchange PEFCs. Hydrazine was dissolved (0.67 M) in 1 M KOH aqueous solution and supplied at 2 ml/min; O_2 was supplied at 500 ml/min; cell temperature: 80 °C. Current density: a1, d1; internal resistance: a2, d2 (a corresponds to the Ni anode catalyst; d corresponds to the Co anode). Reprinted with permission from Ref. [112] Copyright © 2007 Wiley-Interscience.

14.4.2
Cathode Catalysts for DHFCs

Reactions at the cathode in acidic media are essentially the same as that for DMFC and other PEM fuel cell chemistries. As such, the cathode of choice for the DFAFC is Pt, either unsupported (i.e., Pt black) or supported on carbon [101].

It is expected that any improvements to cathode catalysts and/or electrodes for DMFC and other acid-based DLFCs will be applicable to the DHFC. In the meantime, Pt black and Pt/C are considered the standards for DHFC cathodes in acidic media.

Pt was also used as cathode catalysts for DHFC in alkaline media [111]. In addition to Pt, Asazawa and coworkers demonstrated using Ag/C and colbalt polypyrrole (PPY) composite cathode catalysts in DHFCs [112].

14.4.3
Direct Hydrazine Fuel Cell Performance

Besides the environmental benefit of carbon-free emissions from DHFCs, hydrazine also has a higher energy density than methanol and a higher theoretical cell potential [2]. A fuel cell performance comparison of a DHFC and a DMFC was presented by Yamada and coworkers using Nafion® 117 as the electrolyte [101]. Fuel cell polarization curves illustrating the performance comparison are presented in Figure 14.21.

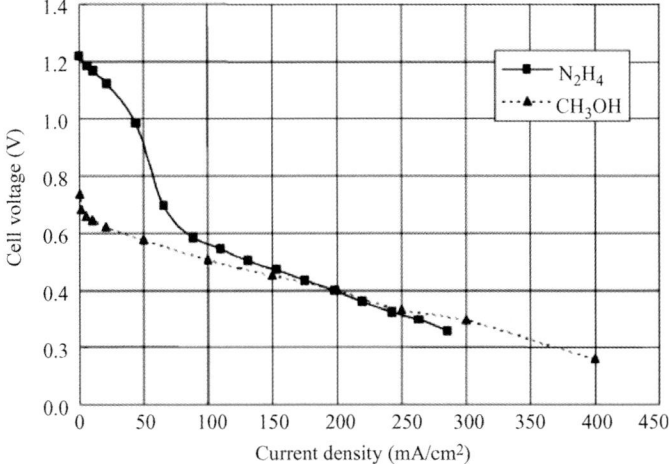

Figure 14.21 Comparison of DHFC vs. DMFC. N_2H_4 (DHFC): anode/cathode catalyst loading = Pt(2)/Pt(3) mg/cm^2, specific surface area of anode catalyst = 7 m^2/g, operating temperature = 80 °C; CH_3OH (DMFC): anode/cathode catalyst loading = Pt(3)Ru(1.5)/Pt (3) mg/cm^2, specific surface area of anode catalyst = 67 m^2/g, operating temperature = 100 °C; common conditions: anode/cathode pressures = 0.2/0.2 Mpa. Reprinted with permission from Ref. [101] Copyright © 2003 Elsevier.

At current densities less than 40 mA/cm^2, DHFCs generate over 1 V, about double that of DMFCs even when the cell is running 20 °C cooler than the DMFC. The higher operating voltage allows the use of fewer cells in the fuel cell stack structure, leading to a potentially more compact fuel cell system design. The rapid decrease of the DHFC cell voltage with increasing current density is likely caused by an increase in hydrazine crossover.

A major benefit of running DHFC in an alkaline medium enables non-noble metal anode catalysts to be used, thus lowering the cost of the catalysts in the fuel cell system. Asazawa and coworkers demonstrated the operation of an alkaline DHFC with a non-platinum catalyst for both anode and cathode [112]. The performance of Ni and Co anode catalysts in DHFC using anion-exchange polymer electrolytes with two types of tetraalkylammonium cation groups (anion A is homogeneous and B is heterogeneous) as pendant groups with a polyolefin main chain, were compared with the performance of DMFC and hydrogen fuel cell in Figure 14.22. The information for each fuel cell configuration is described in Table 14.4.

From the experimental results of Asazawa and coworkers, the power output characteristics of the DHFC were comparable to H_2-O_2 fuel cell and can significantly exceed the performance of the DMFC. These results demonstrate that the DHFC as a direct liquid fuel cell shows great promise when operating in alkaline medium. It should be noted that high power outputs were achieved with alkaline DHFC without Pt-based catalysts. Both Ag/C and cobalt polypyrrole (PPY) composite cathode catalysts were used and high performance was observed in both cases.

Yin and coworkers also demonstrated alkaline DHFCs using a ZrNi alloy and Pt/C as anode and cathode catalysts respectively, combined with an anion exchange

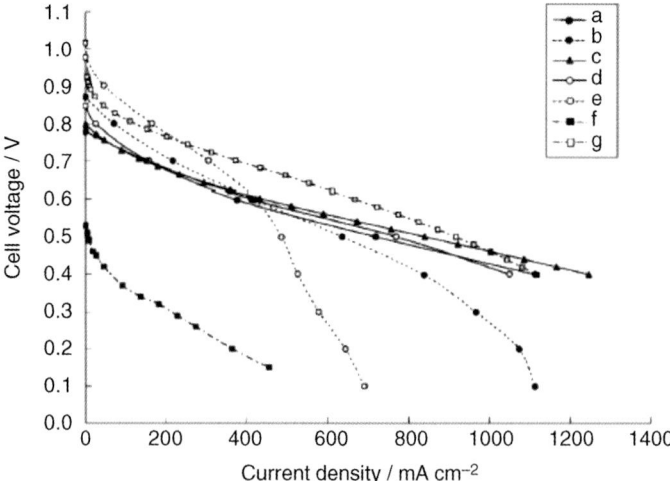

Figure 14.22 Current–voltage performance of DLFCs (cell temperature: 80 °C). Hydrazine was dissolved (0.67 M) in 1 M KOH aqueous solution and supplied at 2 ml/min. (a–e) Hydrazine (+20 kPa); (f) methanol (+100 kPa); (g) hydrogen (+20 kPa). See Table 14.4 for details of samples. Reprinted with permission from Ref. [112] Copyright © 2007 Wiley-Interscience.

membrane [111]. A power density of 84 mW/cm² was achieved, providing another example of high performance DHFCs.

14.4.4
Summary of Electrocatalysis in DHFCs

The early stage performance data from DHFCs is very encouraging. However, a significant research and development effort is required for the successful commercialization of DHFCs. Of significant importance would be the development of anode catalyst formulations that are selective to the electro-oxidation of hydrazine (i.e., avoiding catalytic decomposition). Even more important, however, would be further

Table 14.4 Configuration of samples for Figure 12.22.

Sample	Anode	Cathode	Polymer electrolyte
a	Ni	Ag/C	Anion A
b	Ni	Ag/C	Anion B
c	Ni	Co-PPY-C	Anion A
d	Co	Ag/C	Anion A
e	Co	Ag/C	Anion B
f	PtRu/C	Pt/C	Nafion
g	Pt/C	Pt/C	Nafion

Extracted from Ref. [112] with permission; Copyright © 2007 Wiley-Interscience.

development of anion-exchange membranes to minimize hydrazine crossover. This could potentially lead to a breakthrough for DHFCs, allowing them to maintain high potentials with increasing current density. DHFCs are being actively developed for commercialization by Daihatsu Motor Corporation.

14.5
Other Direct Liquid Fueled Fuel Cells

Besides methanol, ethanol, formic acid and hydrazine, there are other liquid fuels of interest for DLFCs. In this section, some of these fuels will be discussed.

14.5.1
2-Propanol

2-Propanol has an impressive energy density of 7080 Wh/L, almost 1.5 times the energy density of methanol at 4820 Wh/L. This makes it a very attractive fuel for DLFCs [2]. The complete electro-oxidation for 2-propanol is an 18 e^- process (Equation 14.36).

$$CH_3CHOHCH_3 + 5H_2O \rightarrow 3CO_2 + 18H^+ + 18e^- \qquad (14.36)$$

If the direct 2-propanol fuel cell undergoes complete electro-oxidation as shown in Equation 14.36, three times as large current should be observed when compared with the complete oxidation of methanol (i.e., 18 e^- vs. 6 e^- process, respectively) [113]. Performance of direct 2-propanol fuel cells was evaluated by Qi and coworkers [113] and Cao and coworkers [114] employing PtRu as the anode catalyst. The fuel cells exhibit higher open circuit potentials and lower crossover than methanol.

While CO_2 is the primary oxidation product for methanol, acetone is the main product for the electro-oxidation of 2-propanol. The anode catalyst suffers more severe poisoning than catalysts in DMFCs due to acetone as well as other intermediates formed during the oxidation of 2-propanol. In order to consider 2-propanol as a fuel, further poison-resistant catalyst development is required or the anode catalyst surface must be cleaned periodically by an electrical pulse during fuel cell operation as described by Qi [113].

14.5.2
Ethylene Glycol

Ethylene glycol has an energy density of 5870 Wh/L compared with 4820 Wh/L for methanol. The complete electro-oxidation of ethylene glycol to CO_2 is a 10 e^- process illustrated in Equation 14.37 [115].

$$C_2H_6O_2 + 2H_2O \rightarrow 2CO_2 + 10H^+ + 10e^- \qquad (14.37)$$

However, similar to other organic fuels with a C—C bond to cleave during electro-oxidation, the electro-oxidation of ethylene glycol does not go to completion. It is believed that the electro-oxidation of ethylene glycol proceeds via a parallel reaction pathway with numerous intermediate steps that involve the oxidation of the functional OH groups. One suggested pathway does not rupture the C−C bond, while the splitting of the C−C bond with subsequent formation and oxidation of CO_{ad} occurs in the second parallel pathway [116]. This parallel pathway leads to the incomplete oxidation of ethylene glycol, resulting in poisoning intermediates on the catalysts as well as reduction in overall power output. Pt-based catalysts remain the anode catalyst of choice for electro-oxidation and admetals such as Ru and Sn are common choices for performance improvement in the direct ethylene glycol fuel cell (DEGFC). Recent fuel cell results from Livshits and coworkers on a DEGFC operated with triflic acid with PtRu anode catalyst on a nanoporous proton-conducting membrane demonstrated maximum power densities of 80, 285 and 320 mW/cm^2 at 80, 110 and 130 °C respectively [117].

14.5.3
Other Liquid Organic Fuels for Fuel Cells

There are other small organic molecules such as dimethyl ether [118], dimethoxymethane [119], trimethoxymethane [119,120] and tetramethylorthocarbonate [121] that could be of interest as fuels in DLFCs. These molecules are of interest due to their high theoretical energy densities. The complete electro-oxidation of these molecules to CO_2 are at least 12 e$^-$ and up to 24 e$^-$ processes, theoretically leading to potentially high current output in the fuel cells [2]. However, like other small organic fuels that possess C−C bonds, complete electro-oxidation has not been demonstrated with these molecules due to the lack of appropriate anode catalysts. Also, partial oxidation products tend to poison the anode catalysts, leading to the reduction in overall power output of the fuel cell. The future of the commercialization of these organic molecules as fuels for DLFCs will depend upon the development of selective and poison-resistant catalysts for their electro-oxidation.

14.6
Summary

Numerous liquid fuels exhibit potential for use in DLFCs. The main challenges continue to be the development of highly active and poison-tolerant catalysts as well as low crossover membranes. Tremendous achievements have been made in the past decade, and it stands to reason that many of these challenges will be overcome within the next decade. Each of the fuels possess their own attributes and challenges, so depending on the breakthroughs developed, certain DLFCs may be successfully commercialized.

Methanol remains the most common of the liquid organic fuels due to its high energy density and stable performance characteristics. DMFCs, however, suffer from sluggish anode kinetics, fuel crossover and the requirement of water at the anode. Clever water management techniques and system designs enable DMFCs to overcome many of these issues. This fuel cell will be competitive covering a wide power output range, from less than 5 W to kilowatts.

Formic acid, in DFAFCs, exhibits high fuel cell efficiency due to active anode catalysts and lower crossover than methanol. Although DFAFCs possess a lower theoretical energy density, their lower fuel crossover, the existence of efficient catalysts and no requirement for water at the anode add up to a simpler system design. This makes formic acid competitive in low power applications (e.g., <20 W). DFAFCs that use Pd catalysts, however, suffer a slow decay in activity, thus requiring occasional regeneration for practical long term use. DFAFCs that use Pt-based catalysts on the other hand exhibit very stable performance, albeit at higher overpotentials.

Ethanol has a very high theoretical potential and reasonable fuel cell performance has been demonstrated to date. Ethanol also suffers crossover, although to a lesser extent than methanol. One of the main challenges of DEFCs is to achieve reaction completion for this 12 electron process, which requires C–C bond cleavage.

Hydrazine also has a very high theoretical potential, and, due to its simple molecular structure and electro-oxidation reactions, catalyst poisoning is not observed. Another advantage of this fuel cell system is that non-noble metal catalysts can be used in alkaline media, potentially lowering the cost of the fuel cell system dramatically. One major obstacle of this fuel cell appears to be fuel crossover through the membrane. Fuel cell performance results indicate that although high potentials are achieved at low current densities, voltage drops quickly as current density is increased. This rapid performance drop has been attributed to fuel crossover. Besides poor performance at higher current densities, a fuel cell system employing hydrazine must manage the fuel crossover to prevent exposure to the user, due to its toxicity.

Other fuels such as ethylene glycol, 2-propanol, dimethyl ether, dimethoxymethane, trimethoxymethane and tetramethylorthocarbonate can also be considered in DLFCs. All of these fuels possess high theoretical energy densities, but the challenge is to develop poison-resistant catalysts, due to the numerous electro-oxidation steps often leading to poisoning intermediates and oxidation products that can compete for active catalyst sites.

In summary, there are numerous liquid organic fuels that show promise for fuel cell applications. Lower crossover membranes and more active, selective and stable catalysts, specific to each fuel, will facilitate commercialization of DLFCs.

References

1 Ha, S., Dunbar, Z. and Masel, R.I. (2006) *J. Power Sources*, **158**, 129.
2 Qian, W., Wilkinson, D.P., Shen, J., Wang, H. and Zhang, J. (2006) *J. Power Sources*, **154**, 202.
3 Yu, X. and Pickup, P.G. (2008) *J. Power Sources*, **182**, 124.
4 Wasmus, S. and Kuver, A. (1999) *J. Electroanal. Chem.*, **461**, 14.
5 Rhee, Y.-W., Ha, S.Y. and Masel, R.I. (2003) *J. Power Sources*, **117**, 35.
6 Jeong, K.-J., Miesse, C.M., Choi, J.-H., Lee, J., Han, J., Yoon, S.P., Nam, S.W., Lim, T.-H. and Lee, T.G. (2007) *J. Power Sources*, **168**, 119.
7 Capon, A. and Parsons, R. (1973) *J. Electroanal. Chem.*, **44**, 239.

8 Capon, A. and Parsons, R. (1973) J. Electroanal. Chem., **45**, 205.
9 Zhu, Y., Khan, Z. and Masel, R.I. (2005) J. Power Sources, **139**, 15.
10 Ha, S., Larson, R., Zhu, Y. and Masel, R.I. (2004) Fuel Cells, **4**, 337.
11 Wang, X., Hu, J.-M. and Hsing, I.-M. (2004) J. Electroanal. Chem., **562**, 73.
12 Chetty, R. and Scott, K. (2007) J. New Mater. Electrochem. Syst., **10**, 135.
13 Jiang, J. and Kucernak, A. (2002) J. Electroanal. Chem., **520**, 64.
14 Choi, J.-H., Jeong, K.-J., Dong, Y., Han, J., Lim, T.-H., Lee, J.-S. and Sung, Y.-E. (2006) J. Power Sources, **163**, 71.
15 Zhou, X., Liu, C., Liao, J., Lu, T. and Xing, W. (2008) J. Power Sources, **179**, 481.
16 Baldauf, M. and Kolb, D.M. (1996) J. Phys. Chem., **100**, 11375.
17 Blair, S., Lycke, D. and Iordache, C. (2006) ECS Transactions, **3**, 1325.
18 Waszczuk, P., Barnard, T.M., Rice, C., Masel, R.I. and Wieckowski, A. (2002) Electrochem. Comm., **4**, 599.
19 Rice, C., Ha, S., Masel, R.I. and Wieckowski, A. (2003) J. Power Sources, **115**, 229.
20 Li, X. and Hsing, I.-M. (2006) Electrochim Acta, **51**, 3477.
21 Markovi,ć N.M., Gasteiger, H.A. and Ross, P.N. Jr, (1995) Electrochim Acta, **40**, 91.
22 Weber, M., Wang, J.T., Wasmus, S. and Savinell, R.F. (1996) J. Electrochem. Soc., **143**, L158.
23 Zhu, Y., Ha, S.Y. and Masel, R.I. (2004) J. Power Sources, **130**, 8.
24 Watanabe, M., Futuuchi, Y. and Motoo, S. (1985) J. Electroanal. Chem., **191**, 367.
25 Watanabe, M., Horiuchi, M. and Motoo, S. (1988) J. Electroanal. Chem., **250**, 117.
26 Hererro, E., Feliu, J.M. and Aldaz, A. (1994) J. Electroanal. Chem., **368**, 101.
27 Macia, M.D., Herrero, E., Feliu, J.M. and Aldaz, A. (2001) J. Electroanal. Chem., **500**, 498.
28 Yang, Y.Y., Sun, S.G., Gu, Y.J., Zhou, Z.Y. and Zhen, C.H. (2001) Electrochim. Acta, **46**, 4339.
29 Herrero, E., Llorca, M.J., Feliu, J.M. and Aldaz, A. (1995) J. Electroanal. Chem., **383**, 145.
30 Wasmus, S. and Vielstich, W. (1993) J. Electroanal. Chem., **359**, 175.
31 Xia, X.H. and Iwasita, T. (1993) J. Electrochem. Soc., **140**, 2559.
32 Casado-Rivera, E., Volpe, D.J., Alden, L., Lind, C., Downie, C., Vazquez-Alvarez, T., Angelo, A.C.D., Disalvo, F.J. and Abruna, H.D. (2004) J. Am. Chem. Soc., **126**, 4043.
33 Macia, M.D., Hererro, E. and Feliu, J.M. (2003) J. Electroanal. Chem., **554**, 25.
34 Kang, S., Lee, J., Lee, J.K., Chung, S.-Y. and Tak, Y. (2006) J. Phys. Chem. B, **110**, 7270.
35 Rach, E. and Heitbaum, J. (1987) Electrochim. Acta, **32**, 1173.
36 Zhou, X., Xing, W., Liu, C. and Lu, T. (2007) Electrochem. Comm., **9**, 1469.
37 Liu, Z., Hong, L., Tham, M.P., Lim, T.H. and Jiang, H. (2006) J. Power Sources, **161**, 831.
38 Huang, Y., Zhou, X., Liao, J., Liu, C., Lu, T. and Xing, W. (2008) Electrochem. Comm., **10**, 621.
39 Pavese, A. and Solis, V. (1991) J. Electroanal. Chem., **301**, 117.
40 Zhou, W.P., Lewera, A., Larsen, R., Masel, R.I., Bagus, P.S. and Wiekowski, A. (2006) J. Phys. Chem. B, **110**, 13393.
41 Ge, J., Xing, W., Xue, X., Liu, C., Lu, T. and Liao, J. (2007) J. Phys. Chem. C, **111**, 17305.
42 Zhou, W. and Lee, J.Y. (2008) J. Phys. Chem., **112**, 3789.
43 Masel, R.I., Zhu, Y., Khan, Z. and Man, M. (2007) UK Patent No. GB2424650B.
44 Zhang, L., Lu, T., Bao, J., Tang, Y. and Li, C. (2006) Electrochem. Comm., **8**, 1625.
45 Ha, S., Larsen, R. and Masel, R.I. (2005) J. Power Sources, **144**, 28.
46 Larsen, R., Zakzeski, J. and Masel, R.I. (2005) Electrochem. Solid-State Letters, **8**, A291.
47 Shobha, T., Aravinda, C.L., Bera, P., Devi, L.G. and Mayanna, S.M. (2003) Mater. Chem. Phys., **80**, 656.
48 Wang, R., Liao, S. and Ji, S. (2008) J. Power Sources, **180**, 205.

49 Zhou, W. and Lee, J.Y. (2007) *Electrochem. Comm.*, **9**, 1725.
50 Larsen, R., Ha, S., Zakzeski, J. and Masel, R.I. (2006) *J. Power Sources*, **157**, 78.
51 Miesse, C.M., Jung, W.S., Jeong, K.-J., Lee, J.K., Lee, J., Han, J., Yoon, S.P., Nam, S.W., Lim, T.-H. and Hong, S.-A. (2006) *J. Power Sources*, **162**, 532.
52 Wang, J., Wasmus, S. and Savinell, R.F. (1995) *J. Electrochem. Soc.*, **142**, 4218.
53 Song, S., Zhou, W., Liang, Z., Cai, R., Sun, G., Xin, Q., Stergiopoulos, V. and Tsiakaras, P. (2004) *Appl. Catal. B*, **55**, 65.
54 Ghumman, A. and Pickup, P.G. (2008) *J. Power Sources*, **179**, 280.
55 Hitmi, H., Belgsir, E.M., Leger, J.M., Lamy, C. and Lezna, O. (1994) *Electrochim. Acta*, **39**, 407.
56 Iwasita, T. and Pastor, E. (1994) *Electrochim Acta*, **39**, 531.
57 Leger, J.M., Rousseau, S., Coutanceau, C., Hahn, F. and Lamy, C. (2005) *Electrochim Acta*, **50**, 5118.
58 Vigier, F., Rousseau, S., Coutanceau, C., Leger, J.M. and Lamy, C. (2006) *Topics in Catalysis*, **40**, 111.
59 Antolini, E. (2007) *J. Power Sources*, **170**, 1.
60 Song, S.Q., Maragou, V. and Tsiakaras, P. (2007) *J. Fuel Cell Sci. Technol.*, **4**, 203.
61 Lamy, C., Rousseau, S., Belgsir, E.M., Coutanceau, C. and Leger, J.M. (2004) *Electrochim. Acta*, **49**, 3901.
62 Schmidt, V.M., Ianniello, R., Pastor, E. and Gonzalez, S. (1996) *J. Phys. Chem.*, **100**, 17901.
63 Fujiwara, N., Friedrich, K.A. and Stimming, U. (1999) *J. Electroanal. Chem.*, **472**, 120.
64 Camara, G.A., de Lima, R.B. and Iwasita, T. (2004) *Electrochem. Comm.*, **6**, 812.
65 Song, S.Q., Zhou, W.J., Zhou, Z.H., Jiang, L.H., Sun, G.Q., Tsiakaras, P., Xin, Q., Leonditis, V., Kontou, S. and Tsiakaras, P. (2005) *Int. J. Hydrogen Energy*, **30**, 995.
66 Colmati, F., Antolini, E. and Gonzalez, E.R. (2006) *J. Power Sources*, **157**, 98.
67 Rousseau, S., Coutanceau, C., Lamy, C. and Leger, J.M. (2006) *J. Power Sources*, **158**, 18.
68 Zhou, W., Zhou, Z., Song, S., Li, W., Sun, G., Tsiakaras, P. and Xin, Q. (2003) *Appl. Catal. B*, **46**, 273.
69 Zhang, D., Ma, Z., Wang, G., Konstantinov, K., Yuan, X. and Liu, H. (2006) *Electrochem. Solid-State Lett.*, **9**, A423.
70 Zhou, W.J., Li, W.Z., Song, S.Q., Zhou, Z.H., Jiang, L.H., Sun, G.Q., Xin, Q., Poulianitis, K., Kontou, S. and Tsiakaras, P. (2004) *J. Power Sources*, **131**, 217.
71 Vigier, F., Coutanceau, C., Perrard, A., Belgsir, E.M. and Lamy, C. (2004) *J. Appl. Electrochem.*, **34**, 439.
72 dos Anjos, D.M., Kokoh, K.B., Léger, J.M., de Andrade, A.R., Olivi, P. and Tremiliosi-Filho, G. (2006) *J. Appl. Electrochem.*, **36**, 1391.
73 Song, H., Qiu, X., Li, F., Zhu, W. and Chen, L. (2007) *Electrochem. Commun.*, **9**, 1416.
74 Bai, Y., Wu, J., Qiu, X., Xi, J., Wang, J., Li, J., Zhu, W. and Chen, L. (2007) *Appl. Catal. B: Environ.*, **73**, 144.
75 de Souza, J.P.I., Queiroz, S.L., Bergamaski, K., Gonzalez, E.R. and Nart, F.C. (2002) *J. Phys. Chem. B*, **106**, 9825.
76 Antolini, E., Colmati, F. and Gonzalez, E.R. (2007) *Electrochem. Commun.*, **9**, 398.
77 Siné, G., Smida, D., Limat, M., Foti, G. and Comninellis, C.H. (2007) *J. Electrochem. Soc.*, **154**, B170.
78 Oliveira Neto, A., Franco, E.G., Arico, E., Linardi, M. and Gonzalez, E.R. (2004) *J. Eur. Ceram. Soc.*, **23**, 2987.
79 Wang, Z., Yin, G., Zhang, J., Sun, Y. and Shi, P. (2006) *Electrochim. Acta*, **51**, 5691.
80 Li, G. and Pickup, P.G. (2006) *Electrochim. Acta*, **52**, 1033.
81 Spinace, E.V., Linardi, M. and Neto, A. Oliveira (2005) *Electrochem. Commun.*, **7**, 365.
82 Colmati, F., Antolini, E. and Gonzalez, E.R. (2008) *J. Alloys Compds.*, **456**, 264.
83 Tarasevich, M.R., Karichev, Z.R., Bogdanovskaya, V.A., Lubnin, E.N. and Kapustin, A.V. (2005) *Electrochem. Comm.*, **7**, 141.
84 Shen, P.K. and Xu, C. (2006) *Electrochem. Comm.*, **8**, 184.

85 Hu, F., Chen, C., Wang, Z., Wei, G. and Shen, P.K. (2006) *Electrochim. Acta*, **52**, 1087.
86 El-Shafei, A.A., Abd El-Maksoud, S.A. and Moussa, M.N.H. (1992) *J. Electroanal. Chem.*, **336**, 73.
87 Xu, C. and Shen, P.K. (2005) *J. Power Sources*, **142**, 27.
88 Xu, C., Shen, P.K. and Liu, Y. (2007) *J. Power Sources*, **164**, 527.
89 Bai, Y., Wu, J., Xi, J., Wang, J., Zhu, W., Chen, L. and Qin, X. (2005) *Electrochem. Comm.*, **7**, 1087.
90 Xu, C., Shen, P.K., Ji, X., Zeng, R. and Liu, Y. (2005) *Electrochem. Comm.*, **7**, 1305.
91 Hu, F., Ding, F., Song, S. and Shen, P.K. (2006) *J. Power Sources*, **163**, 415.
92 Kim, J. and Park, S. (2003) *J. Electrochem. Soc.*, **150**, E560.
93 Park, N., Shiraishi, T., Kamisugi, K., Hara, Y., Iizuka, K., Kado, T. and Hayase, S. (2008) *J. Appl. Electrochem.*, **38**, 371.
94 Lopes, T., Antolini, E., Colmati, F. and Gonzalez, E.R. (2007) *J. Power Sources*, **164**, 111.
95 Zhou, W.J., Song, S.Q., Li, W.Z., Zhou, Z.H., Sun, G.Q., Xin, Q., Douvartzides, S. and Tsiakaras, P. (2005) *J. Power Sources*, **140**, 50.
96 Ribeiro, J., dos Anjos, D.M., Leger, J.M., Hahn, F., Olivi, P., de Andrade, A.R., Tremiliosi-Filho, G. and Kokoh, K.B. (2008) *J. Appl. Electrochem.*, **38**, 653.
97 Pramanik, H., Wragg, A.A. and Basu, S. (2008) *J. Appl. Electrochem.*, **38**, 1321.
98 Bard, A.J. (1963) *Anal. Chem.*, **35**, 1602.
99 Burke, L.D. and O'Dwyer, K.J. (1989) *Electrochim. Acta*, **34**, 1659.
100 Yamada, K., Yasuda, K., Fujiwara, N., Siroma, Z., Tanaka, H., Miyazaki, Y. and Kobayashi, T. (2003) *Electrochem. Commun.*, **5**, 892.
101 Yamada, K., Asazawa, K., Yasuda, K., Ioroi, T., Tanaka, H., Miyazaki, Y. and Kobayashi, T. (2003) *J. Power Sources*, **115**, 236.
102 Andrew, M.R., Gressler, W.J., Johnson, J.K., Short, R.T. and Williams, K.R. (1972) *J. Appl. Electrochem.*, **2**, 327.
103 Tamura, K. and Kahara, T. (1976) *J. Electrochem. Soc.*, **123**, 776.
104 Garcia, M.D., Marcos, M.L. and Velasco, J.G. (1996) *Electroanalysis*, **8**, 267.
105 Eisner, U. and Gileadi, E. (1970) *J. Electroanal. Chem.*, **28**, 81.
106 Majidi, M.R., Jouyban, A. and Asadpour-Zeynali, K. (2007) *Electrochim. Acta*, **52**, 6248.
107 Adekunle, A.S. and Ozoemena, K. (2008) *J. Solid State Electrochem.*, **12**, 1325.
108 Pournaghi-Azar, M.H. and Sabzi, R. (2003) *J. Electroanal. Chem.*, **543**, 115.
109 Yamada, K., Yasuda, K., Tanaka, H., Miyazaki, Y. and Kobayashi, T. (2003) *J. Power Sources*, **122**, 132.
110 Meibuhr, S.G. (1974) *J. Electrochem. Soc.*, **121**, 1264.
111 Yin, W.X., Li, Z.P., Zhu, J.K. and Qin, H.Y. (2008) *J. Power Sources*, **182**, 520.
112 Asazawa, K., Yamada, K., Tanaka, H., Oka, A., Taniguchi, M. and Kobayashi, T. (2007) *Angew. Chem.*, **46**, 8024.
113 Qi, Z. and Kaufman, A. (2002) *J. Power Sources*, **112**, 121.
114 Cao, D. and Bergens, S.H. (2003) *J. Power Sources*, **124**, 12.
115 Chetty, R. and Scott, K. (2007) *J. Appl. Electrochem.*, **37**, 1077.
116 Wang, H., Zhao, Y., Jusys, Z. and Behm, R.J. (2006) *J. Power Sources*, **155**, 33.
117 Livshits, V. and Peled, E. (2006) *J. Power Sources*, **161**, 1187.
118 Muller, J., Urban, P., Wezel, R., Colbow, K.M. and Zhang, J. (2004) US Patent No. 6,777,116.
119 Narayanan, S.R., Vamos, E., Surampudi, S., Frank, H., Halpert, G., Surya Prakash, G., Smart, M.C., Knieler, R., Olah, G.A., Kosek, J. and Cropley, C. (1997) *J. Electrochem. Soc.*, **144**, 4195.
120 Surampudi, S., Narayanan, S.R., Vamos, E., Frank, H.A., Halpert, G., Olah, G.A. and Prakash, G.S. (1997) US Patent No. 5,599,638.
121 Zhang, J. and Colbow, K.M. (2005) US Patent No. 6,864,001.

Index

a

accelerated durability tests   338
acid-base reaction   39
adapter circuit board   460, 480
– pictures   480
1-adamantanecarboxylic acid (ACA)   101
ad-atoms   13, 14
– electronegativity   13
– influence   14
– theory   13
alkaline anion exchange membranes (AAEMs)   23
alkaline direct alcohol fuel cells   38, 515, 516, 520
– cathode catalysts   38
– cell voltage curves   520
– I-V curves   516
– KOH-doped PBI membrane   515
alkaline electrolytes   21, 520, 550
– anode catalysts   550
– ethanol electro-oxidation mechanism   550
– problem   21
alkaline fuel cells (AFCs)   488, 518–520
– direct liquid fuels   518
alkaline hydrazine fuel cell research   557
– nickel-based anode catalysts   557
alkaline media   20, 21, 490, 491, 498, 506
– advantage   21
– aqueous electrolyte media   506
– cationic exchange membranes   508
– direct methanol fuel cells   506–511
– invariant electrolyte media   511
– methanol oxidation electrocatalysis   490–498
  – electrolyte effect   496
  – non-precious metal catalysts   494–496
  – pH effect   496
– precious metal catalysts   491–494
– oxygen reduction mechanism   498–502
  – catalysts, cyclic voltammetry   500
alkaline methanol fuel cells   488, 490
– history   488–490
alkaline solutions, cyclic voltammograms   497
anion exchange membranes (AEM)   30, 31, 66, 67, 511, 513, 519
– advantage   30, 67, 512
– drawback   67
– issues   512
– quaternized polyethersulfone cardo (QPES-C)   513
– synthesis   31
anion exchange polymer electrolytes   557
anionic membrane direct methanol fuel cells (AMDMFC)   21
anionic polymer electrolytes   30
anode catalysts   121
– atomic force microscopy (AFM)   125
– characterization techniques   121
– energy dispersive X-ray spectroscopy (EDX)   125, 126
– scanning electron microscopy (SEM)   125
– transmission electron microscopy (TEM)   124, 125
anode polarization measurement, design   210
anode recovery mechanism   479
anodic aluminum oxide (AAO)   97
anodic films   257
– surface states, role   257
arc-melted alloys   248
– steady-state voltammetry   248
Auger electron spectroscopy (AES)   122, 134, 135
– low energy ion scattering (LEIS)   134

- primary characterization technique   135
- ultraviolet photoemission spectroscopy (UPS)   135

automated parallel electrochemical synthesis system   172, 174
automated screening method   175–177, 192
- disadvantage   192
automobiles performance, quantum jumps   55
auxiliary power units (APU)   40

b

balance-of-plant (BOP)   221
ball-milling method   433
- catalyst inks preparation   433
Ballard Power Systems Inc. (BPSI)   56
- DMFC system development   56
batteries, portable power sources   165
bifunctional mechanism   97, 115, 166, 231, 547
bifunctional modifiers   533
bifunctional theory   15, 198
bimetallic catalyst   98, 245
- electrocatalytic activity   245
- schematic illustration   98
bimetallic nanoparticles   95, 121
- atomic distribution   127
- EXAFS techniques   131
- NMR spectra   133
- quantitative assessment   127
- structural models   128
- XAS methodology   126, 131
  - data   126
- X-ray diffraction (XRD) technique   121
bimodal mesoporous carbon materials   361
block copolymers, preparation   391
Bönnemann method   273, 275, 298
breathing process   441

c

carbonate electrolyte cells, advantage   511
carbon black   20, 364, 371
- acetylene black   20
- Denka black   364
- Ketjen black   20, 364
- particle structural changes   331
- SEM images   371
- supported catalysts   317, 341
  - catalytic activity   341
- Vulcan XC   20, 364
carbon materials   329, 364
- corrosion, steps   329
- hollow macroporous/mesoporous shell-type, synthesis   362

- structural features   364
carbon nanofibers (CNFs)   92, 94, 317
- herringbone/platelet/tubular   94
- HRTEM images   94
- mesoporous, synthesis   369
carbon nanotubes (CNTs)   247, 316, 317, 320–322, 324–329, 331, 337, 339, 341–348, 432, 436, 519
- carbon cloth-supported catalysts   344
  - electrochemical behavior   344–348
- catalysts   316, 320–322, 324, 326, 331, 337, 339, 341
  - EDX analysis   337
  - electrochemical behavior   331–341
  - preparation methods   316–325
  - properties   316
- growth   341–348
  - carbon cloth   342–344
  - as catalyst supports   341–348
- electrodes   325, 328, 329, 334
  - characteristics   325–331
  - durability   329–331
  - electrochemical behavior   325–328
- functionalization   316
- polarization curves   432
- SEM micrographs   337
carbon powder   262, 263
- Vulcan XC72   262, 263
carbon-supported catalyst   85, 204, 237, 240, 274, 365, 423, 426, 430
- structure   204
- synthetic methods   85–95
- usage   365
- X-ray diffraction patterns   274
catalyst   61, 80, 92, 122, 167, 182, 198, 214, 215, 334, 425, 430, 433, 434, 438
- aggregations   425
- coated membrane   61, 198, 433, 434
- crystallite size   122, 214
  - evaluation   122
- ink formulation   215
- ionomer interaction   425
- layers   216, 339, 341, 430, 434
  - ink-based hot-pressing preparation method   438
  - ink development   216
  - pore distributions   339
- nanostructures   80
- preparation, bichannel equipment   92
- screening methods   182–187
catalyst tolerance   296–306
- in presence of methanol   296
- non Pt-based catalysts   304–306
- PtM/C catalysts behavior   296–300

Index | 569

– transition metal chalcogenides   302–304
– transition metal macrocycles behavior   300–302
catalyzed diffusion medium (CDM)   433, 434
– catalyst layer   434
cathode catalyst   121, 130, 249, 436
– atomic force microscopy (AFM)   125
– characterization techniques   121
– energy dispersive X-ray spectroscopy   125, 126, 249
– layer, optimization   372
– scanning electron microscopy (SEM)   125
– transmission electron microscopy (TEM)   124, 125
– XAS analysis   249
cation-exchange membrane   439
cell potentials   63, 235
cetyltrimethylammonium bromide (CTAB)   96, 99, 369
– assisted microwave synthesis process   369
chalcogenide materials   302
– cluster compound concept   302
charge transfer coefficient   260
charge-transfer resistance, measurement   340
chemical reduction method   439
chemical vapor deposition (CVD) method   369, 436
chloro-alkali cells   33
chronoamperometry   101, 269
– analysis   101
– experiment equipment   269
cobalt tetrasulfonated phthalocyanine (CoTsPc), structure   280
coimpregnation reduction method   549
collection efficiency   149, 153, 269, 272
colloidal method   88–93, 241, 298, 322–323
– drawback   89
– vs. impregnation method   241
– preparing route   323
composite polymer membrane, preparation   514
computational fluid dynamics (CFD) approach   450
constant phase element (CPE)   341
conventional reduction method   239
CO oxidation, IR thermographs   184
copolymers   381, 391, 393, 394, 397, 399–401, 411, 412
– chemical structures   381
– comb-shaped   399, 401, 412
– containing nitrile groups   397
– diblock   381
– 4,4′-dichlorodiphenylsulfone derived   391

– DMFC performance   411
– hydrocarbon   412
– pendant-type copolymer   399
– sulfonated 1,4-bis(4-fluorobenzoyl)benzene derived   394
– sulfonated 4,4′-difluorodiphenyl ketone derived   393
– types   381
co-precipitation method   168
CO_2 sensor method   5, 6
– based crossover measurement   6
crosslinking agent   382
crystal facet effect   82–85
current-voltage polarization   435
cyclic voltammetry (CV)   136, 169, 232, 269, 325, 367, 492, 493
– applications   137
– bulk copper deposition   144
– equipment   269
– potentiodynamic electrochemical measurement   136
– Pt-based catalysts   137–139
  – electrochemical active surface area (ECASA)   140
  – ORR/MOR, qualitative indicator   139
  – particle size   140, 141
  – roughness factor (RF)   140
  – state   137, 138
  – scans for PtRu/carbon cloth   344–347

d
Debye function analysis (DFA)   123, 293
– simulation   293
decal transfer method (DTM) process   45, 433, 434
– MEAs preparation   45
degree of sulfonation (DS)   27, 385
density functional theory (DFT)   82, 120
– calculations   82
diesel engine technology   55
differential electrochemical mass spectrometry (DEMS)   245
diffusion coefficient   32, 147, 154, 271, 454, 512
direct alkaline polymer electrolyte membrane fuel cells   511–518
– direct methanol alkaline membrane fuel cell performance   513–517
– membraneless fuel cell, assembly scheme   517
– methanol fuel cells   511–513
  – anion exchange membrane   511
direct electrode array   175–177
– advantage/disadvantages   177

direct ethanol fuel cells (DEFCs)   518,
   545–554
– anode catalysts   546–551
– cathode catalysts   551, 552
– cell voltage *vs.* current density curves   554
– development, challenges   519
– electrocatalysis   545
   – ethanol   546
– electro-oxidation   545, 546
– performance   552, 553
– polarization curves   550, 552
– power density   548, 550, 552
   – *vs.* current density curves   553
– PtCo/C cathode   551, 552
direct ethylene glycol fuel cell (DEGFC)   562
direct formic acid fuel cell (DFAFC)
   systems   528, 530, 534, 535, 538, 541–545,
   551, 563
– anode catalysts   530, 543
   – bifunctional modifiers   532
   – oxidation mechanism   530
   – Pd-based anode catalysts   531–541
   – Pt modifiers   536
– cathode catalysts   541
   – alkaline media   558
– electrocatalysis   528, 545
– fuel concentration   529
– methanol   542
– performance   541–545
– polarization curves   541, 542
– portable power system   544
– PtRu catalyst   544
direct hydrazine fuel cells (DHFCs)   554–561
– alkaline medium   559
– anion-exchange polymer electrolytes   559
– anode catalysts   555, 556
   – electro-oxidation   555
   – noble metal-based anode catalysts   556
   – non-noble metal-based anode
     catalysts   557
   – effect   556
– cathode catalysts   558
– electrocatalysis   554
– electro-oxidation   555
– fuel cell performance comparison   558, 559
– Gibbs free energy   555
– limitations   554
– performance   558–560
direct liquid fuel cells (DLFCs)   527
– characteristics   527, 528
– current-voltage performance   560
– direct hydrazine fuel cells   554–561
– electrocatalysis   545
– ethylene glycol   561, 562

– organic fuels   562
– 2-propanol   561
direct methanol-hydrogen fuel cells
   (DMHFCs)   466, 469, 470
– experiment   466–470
– functioning mechanism   470–476
direct methanol alkaline fuel cells
   (DMAFC)   487–489, 509, 510, 513
– AEMs, use   513
– CPPO application   513
– with Nafion membrane   510
– schematic diagram   510
– transport processes   489, 509
direct methanol fuel cell (DMFC)   1–12, 18,
   19, 21–25, 27, 29, 34, 37, 39, 40, 47, 49, 53, 55,
   59–65, 67–70, 79, 80, 115, 165, 168, 170, 187,
   189, 190, 197, 198, 200, 203, 212, 215–217,
   220, 222, 227, 231, 248, 257, 302, 303, 315,
   332, 338, 346, 355, 368, 372, 373, 379–381,
   407–409, 417, 419, 424, 428, 431, 433,
   449–454, 456–458, 461, 463–466, 468, 469,
   473, 476, 487–489, 504, 509, 511, 514, 528
– acidic electrolyte, use   22
– advantages   1, 56, 79
– anion exchange membranes   39
– anode   229, 340
   – reaction, water requirement   530
   – catalysts   14, 118, 211, 227, 229–238
   – Nyquist plot   340
   – phase diagram *vs.* activity   229–238
– applications   69, 200
– array   187, 191
   – voltage-current curves   191
– bifunctional activation   476–482
– cell resistance, role   407
– channel equations   454–456
– characterization   1, 8, 170
– commercialization   381
– current density   457, 458, 477
– data analysis   170, 469
– development   10, 39
   – challenges   65
– drawbacks   68
– electrocatalysis   80, 198–203, 487
– electrocatalysts   198–203
   – activity   136
   – adsorptive CO-stripping voltammetry
     141–143
   – combinatorial synthesis methods   165,
     167, 169–175
   – cyclic voltammetry (CV)   136–141
   – development   199–202
   – linear sweep voltammetry (LSV)   149–155
   – role   117

– rotating disk electrode (RDE)
  method   146–148, 149
– underpotential deposition (UPD)
  143–146
– electrochemical device   117
– electrodes layers   2
– electrolyte   217
  – development, technological
    advances   21
  – fuel cells   115
– energy density   9, 68
– equivalent current density   7
– evaluation/data analysis   187
– fuel cell   3, 115
  – efficiency   3, 303
– *in situ* half-cell electrode polarizations   4
– liquid electrolyte   46
– local current distribution (LCD)   449, 461, 463, 468
– mass transport   431
– methanol crossover problem   488, 504
– methanol oxidation reaction (MOR)   14–21, 117–119
  – catalysts   10–14
– micro-channel patterns   63
– plant   64
– mixed-reactant, polarization curves   304
– operation mode   3
  – active/passive   3
  – constant oxygen stoichiometry, polarization curves   462
  – critical air flow rate   464
  – principles   2–4
– oxygen reduction reaction (ORR)   117, 119–121
– performance   4, 61, 231, 249, 332, 380, 409, 542
– polarization curves   4, 19, 29, 37, 302, 338, 464, 466
– polymer electrolyte membranes synthesis   380–412
– proton-conducting electrolyte, drawbacks   29
– proton exchange membranes   379
– quasi-2-D model   456
– single cell electrode polarizations   4
– stacks development   62, 63, 67
  – planar architecture   63
– standard electrolyte membrane   24
– state-of-the-art   10, 14
– technical barriers   165, 197
– techno-economical challenges   65–70
– water recycling   49
– water requirement   530

direct methanol fuel cell (DMFC)
  catalysts   119, 168, 169, 171, 363, 408, 423, 452
– array   170, 171
– colloidal method   168
– co-precipitation   168
– development procedures   168–169, 220–222, 315
– device, polarization curves   5
– electrocatalyst   203
  – characterization   203–215
  – evaluation   203–215
– electrode   433
  – fabrication process   433–436
  – structure   433, 436
– evaluation/data analysis   187
  – combinatorial methods   187–190
– impregnation method   168
– layers   417, 419, 424
  – fabrication   417
  – ionomer in   424
  – membrane electrode assemblies   417
  – optimization   417, 419–433
– MEA   212, 216, 373
  – durability study   212–215
  – potentiodynamic polarization plots   373
  – power density plots   373
  – SEM images   216
– micro-emulsion method   168
– performance   215, 217
– power sources   39, 60
– single cell   18, 189, 368, 372
  – anodic half-cell polarization behavior   18
  – high throughput screening method   189–190
  – polarization curve   19, 368, 372
– system   64, 315, 355, 511
  – design   64
  – Li-ion battery   355
direct methanol fuel cell (DMFC) model   451–459
– assumptions   452
– channel equations   454–456
– goal   451
– methanol stoichiometry   456–459
– through-plane relations   453
direct methanol fuel cell (DMFC) technology   40, 56, 60
– computational fluid dynamics (CFD) models   450
– development   60–65
  – catalyst preparation   61
  – electrode manufacturing   61
  – membrane electrode assemblies   61

– stack hardware/design 62
– portable power sources 40–53
– transportation 53–60
– working characteristics 264
– polarization curves 346
– voltage-current curves 190
direct self-assembly approach 359
disassembled combinatorial electrochemical cell-array 188
dispersion, definition 292
Dreamcar project 59, 60
– DMFC stack development 60
dry production techniques 439
dual parallel path reaction mechanism 519
dual templating approach 362
– schematic diagram 362
dynamic hydrogen electrode (DHE) 4

e

electrocatalyst 11, 19, 79, 85, 116, 122, 126, 215, 202, 293, 306
– anode/cathode 11
– formulations 19
– layer design 215–217
– nanostructures 122, 126
  – characteristics 122, 126–136
– particle size 116
– platinum 120
– properties 85
– role 79
– Scherrer equation 123
– spray conversion reaction platform method 202–203
– TEM images 293
electrochemical active surface area (ECASA) 137, 138, 140
– CO stripping 145
– Cu-UPD method 144, 145
– Pt catalyst 138, 144
electrochemical cell 175, 176, 187, 188
– array 187, 188
– vs. 3-electrode electrochemical cell 175–176
electrochemical energy conversion 121
electrochemical impedance spectroscopy (EIS) 328, 337, 344, 421, 435
– analysis 340–341, 344, 421
electrochemical lithiation process 105
electrochemical methods 8, 58, 177, 182
– AC impedance 177, 182
– chronoamperometry/chronopotentiometry 177, 182
– cyclic voltammetry 177, 182
electrochemical nuclear magnetic resonance (EC-NMR) 133, 134

electrochemical scanning 182
– automated screening/direct electrode array 182
– electrolyte probe screening/indirect electrode array 182
electrochemical synthesis 104–108, 135, 173, 192, 418, 424
– automated system 173
electrode 62, 271, 327
– backing layer, characteristics 62
– electron transfer rate 327
– potential 271
electrode fabrication methods 433, 438–441
– breathing process 441
– chemical reduction method 439
– dry production techniques 439
– electrodeposition 438
– glue method 440
– sedimentation method 440
– sputtering technique 438
electrodeposition method 323–325, 438
– preparing route 325
electrolyte fuel cells 115
electrolyte probe screening methods 179–182, 192
– pen-shaped O_2-electrode 182
electrolytic domains (ED) 462, 467, 472
– voltage losses 472
electron 122, 129, 228, 262, 342, 504
– conducting polymer 262
– donation/back donation mechanism 228
– occupancy 129
– structure modification 82
– transfer process 342, 504
– transitions 129
electro-osmotic drag 6, 25, 30, 32
energy density 529
– methanol solutions 529
– volumetric/gravimetric 529
energy-dispersive X-ray analysis, see energy dispersive X-ray spectroscopy (EDX)
energy-dispersive X-ray spectroscopy (EDX), composition analysis 241
Energy Related Devices Inc. (ERD) 43
– fuel cells development 43
equivalent circuit model 340, 341
ethanol 548
– electro-oxidation 548, 551
  – activity 549, 550
  – alkaline medium 550
  – Pt-based anode catalyst 552
ether-ether-ketone-ketone (EEKK) moieties 394
ethylene glycol (EG) 89, 550

ethylene tetra-fluoroethylene (ETFE)-based membranes 23, 26
European Community project, *see* Morepower project
ex-situ catalyst screening 208
extended X-ray absorption fine structure (EXAFS) 12, 122

f

fabrication process 433–438
face-centered-cubic (fcc) 85, 86, 132, 235, 274
– metals 85, 86
 – surface energy calculations 85
– nanoparticles 132
faradaic efficiency 32, 48, 549
Faraday law 5, 9
feed channels 452
– reactants transport, assumptions 452
Fermi energy level 129, 228
Fick's law 265
formic acid 528, 529, 541
– chemical combustion reaction 528
– dual pathway mechanism 531, 532
– electrocatalytic activities 534
– energy density 529
– fuel cell, Pd-M catalysts 541
– gravimetric density 529
– Pt based catalyst formulations 535
– Pt cyclic voltammogram 531
– Pt modifiers 533
– thermodynamic efficiency 530
formic acid oxidation 531, 534, 536, 545
– intermetallic phases 534
– Pd/Pt alloy 534, 539
Forschungszentrum Julich GmbH (FJG) 45
– DMFC stack development 45
– system 45
fossil fuels, natural gas/coal 1
Fourier infrared spectroscopy 513
Fourier transform electrochemical impedance spectroscopy (FT-EIS) 495
fuel 9, 10
– definition 9
– energy density 9
– volumetric/gravimetric energy density 10
fuel cell 34, 47, 53, 55, 61, 63, 79, 126, 132, 165, 189, 199, 210–212, 296, 315, 379, 380, 437, 518
– air feed-SPE 61
– anodes, SEM micrographs 437
– application 53
– array, characterization 189
– catalyst 126, 132, 199, 212
 – durability testing 212

– preparation methods 199
– electrolyte 380
– evaluation 210–212
– performance 47
– phosphoric acid 34
– powered vehicles 55
– stacks 63
– test set-up 294–296
 – schematic diagram 296
– types 79

g

galvanic domain (GD) 462, 467, 472, 478
– ED interface 469, 471
– voltage losses 472
galvanic replacement reaction 99–104
– advantage 99
– bimetallic clusters 103
– hollow platinum nanochannels 102
– nanoporous platinum 101
– platinum nanospheres/nonotubes 99
gas diffusion electrode (GDE) 203, 434
– preparation, flowchart 434
gas diffusion layer (GDL) 47, 418
gas-phase impregnation 319
Gibbs free energy 3, 529, 546, 555
glassy carbon electrode (GCE) 137
glue method 440
4G mobile phone 356
– load/power demand 356
grafting catalysts, internal structure 426
graft polymers, properties 398
graphite nanofibers, platelet/ribbon type 334

h

hard template 96–97
– disadvantage 97
– usage 96
high angle annular dark field (HAADF) 122
high frequency resistance (HFR) 407, 412
highly ordered pyrolytic graphite (HOPG) 125, 328
– electrodes 328
high-pressure liquid chromatography (HPLC) 264, 296
high resolution electron microscopy (HRTEM) 121
high throughput screening methods 169–171, 175, 250
– electrode arrays 175–182
hydrocarbon membranes 219
– membrane electrode assembly (MEA) 219, 220
 – durability testing 220

– performance comparison   219
– vs. Nafion   219
– PEM   405
– sulfonated copolymers   402
hydrodynamic working electrode, see rotating disk electrode
hydrogen
– adsorption   120, 130, 154, 496
– cell   474–476
  – transition region   474
– charge   144
– consuming devices   55
– coulombic charge   140
– desorption process   138, 140, 496
– fuel cells   487
– oxidation reaction   184, 476
  – IR thermographs   184
– properties   528
hydrophobic teflon polymer, use   44
Hyflon   22, 28, 29, 59
– characteristic   22
– ion membranes   28
  – applications   28
  – based MEA   29

i
impregnation method   87–88, 168, 240, 318–322, 365
– disadvantage   168
– preparing route   318
– problem   88
incipient wetness method   365
indirect electrode array   179–182
induced coupling plasma (ICP)   213
infrared spectroscopy (IR)   122
infrared spectroscopy thermography   183
– principle   183
– screening method   192
inorganic filler, acid-base properties   26
inorganic pore expanding agent, boric acid   362
in-situ infrared reflectance spectroscopy measurements   289
Institute for Fuel Cell Innovation   47
– DMFC stack, designing   47
Institute of Microelectronics   51
– silicon-based micro DMFC development   51
integrated-circuit (IC) technology   55, 63
intermediate temperature solid oxide fuel cells (IT-SOFCs)   24, 33
intrinsic co-catalytic effect   250
intrinsic mechanism, postulates   228
ion exchange capacity (IEC)   381

ion scattering spectroscopy (ISS)   122
iron catalyst, TEM image   343
iron-mediated polyol process   90
– urchin-like Pt agglomerates   90
iron (III) tetramethoxyphenylporphyrins (Fe-TMMP), preparation   502
isoelectric point   319

j
Jet Propulsion Laboratory (JPL)   44
– DMFCs development   44

k
$K_4Fe(CN)_6$, CV curves   327
kinetic current densities, Tafel plots   152
Korea Institute of Energy Research (KIER)   46
– DMFC stack development   46
Korea Institute of Science and Technology (KIST)   50
– MEAs development   50
– micro-DMFCs development   50
Koutecky–Levich equation   150, 152
Koutecky–Levich plots   150, 271

l
laminar flow-based micro fuel cells (LF-FCs)   517, 518
– power density curves   518
– working   517
LAMMI membranes   268
– methanol permeability   268
Langmuir–Hinshelwood mechanism   492
Levich's law   270
ligand effect   81
linear sweep voltammetry (LSV)   136, 149
– diffusion layer thickness   149
– Fick's law   149
– intrinsic kinetics activity   150
liquid alkaline electrolyte   11, 29
– KOH   11
liquid fuels   9, 488, 528
– fed DMFCs   25
– methanol   488
– properties   528
– use   9
lithium ion battery(ies)   69, 165, 197, 222
long-side-chain polymers (LSC)   27
Los Alamos National Laboratory (LANL)   43, 45, 59, 231
– DMFC power source development   43, 45
low-energy electron diffraction (LEED)   134
low-energy ion scattering (LEIS) spectroscopy   12, 236

low temperature co-fired ceramic (LTCC)
  technology 43

m

macrocycle compounds 287, 288, 536
– counter-ion 262
– electropolymerized 287
– iron-tetrasulfophthalocyanine
  (FeTSPc) 536
– monomer, polymerization 261
mass transport 430
– components 430–433
material library fabrication method 174
– evaporation 174
– molecular beam epitaxy 174
– pulsed laser deposition 174
– sputtering 174
material science techniques 66
material spot array 177–179
– advantage 178
MCO, see methanol crossover
membrane's effect 418
membrane electrode assembly (MEA) 2, 26, 45, 48, 50, 51, 58, 59, 169, 187, 198, 211, 213, 248, 315, 334, 347, 364, 370, 373, 402, 406, 409, 417, 427, 429, 433, 436, 440, 459, 514, 515, 544
– annealing 436
– assembly 58
– CDM-based preparation 433
– current-power/current-voltage curves 427
– fabrication 373
– functional layers 459
– I-V curves 515
– microstructured, design 435
– Nafion 115 based, DMFC stack durability test 213
– PEMFC/DMFC, dry production technique 440
– performance 26, 59, 367, 409, 435
– polarization curves 334, 429
– role 417
– schematic diagram 418
– SEM images 347
– types 48
– voltage-current characteristics 406
mesocarbon microbead (MCMB) 431
– supported PtRu catalyst 431
mesoporous carbon (MC) 355–357, 360
– aspects 356–357
– characteristics 360–363
– supported catalyst 363, 364, 366, 367
  – characterization 366, 367
  – fuel cell performance 367–373

– preparation methods 364–366
– synthesis 357–360
mesoporous silica templates 358, 361
– types 358
metal tetra(o-aminophenyl)porphyrin
  (MTAPP), structure 281
metallic nanostructures 93
– microwave-assisted synthesis 93
metallic precursors 319
– $H_2PtCl_6/RuCl_3$ 319
methanol 15, 32, 56, 79, 91, 197, 227, 230, 235, 237, 243, 245, 247, 248, 265, 383, 422, 449, 454, 490, 492, 494, 496–499, 503, 521
– adsorption/dehydrogenation 235
– alternative alcohol fuels 521
– anode polarization measurement 211
– anodic oxidation 243
– crossover 215, 338, 454, 458
  – effect 210
  – flux 458
– diffusion coefficient 32, 265, 454
– electro-oxidation 10, 15, 81, 91, 101, 209, 210, 245, 247, 422, 471, 490, 492, 494, 496
  – activity 13, 14
  – current densities 230
  – cyclic voltammograms 422
  – kinetics 492
  – Tafel slope 496
  – thermodynamic open-circuit potential 471
– fuel cells 24, 38
  – applications 24
– mixed reduction potential 503–506
– permeability vs. proton conductivity 383
– residues, in situ stripping voltammetry 16, 38
– stoichiometry 456
– tolerant cathode catalysts 69, 300, 498–506
  – Chevrel-phase type 69
  – for DMFC 257
  – portable/assisted power units 24
  – transition metal chalcogenides 69
methanol oxidation reaction (MOR) 1, 11, 14–22, 80, 116, 118, 198, 227, 228, 231, 232, 237, 238, 239, 248, 318, 331–337, 366, 417, 419, 422, 449, 492, 497–499
– activation control 22
– activity(ies) 250, 318, 336, 420, 424
– alkaline DMFC systems, anode catalysts for 19, 20
– alternative anode formulations 19
– bifunctional effect 80
– bifunctional model 248
– catalysts 419–423

– activity   12
– computational methods   119
– current density   492
– cyclic voltammograms   497, 498
– deuterium isotope analysis   199
– electrocatalysis   198–199
– electrocatalysts   116
– ligand effect   80
– mass-specific activity, comparison   209
– onset potential dependence ratio   238, 239
– Pt–Ru catalysts   15–19, 419
– RuO_x species promoting effect   15
– Tafel plots   497, 499
microelectromechanical system (MEMS) technology   50
microemulsion method   93–95, 169
– advantage   94, 95
– definition   93
microemulsion reduction technique   240
MicroFuel cell®   43
– array   43, 187–188
– anode design   43
– pore-free electrode   43
microwave (MW)-assisted polyol synthesis method, advantages   91
microwave-irradiated polyol plus annealing (MIPA) synthesis method   240
microwave-plasma-enhanced chemical vapor deposition (MPECVD)   342
modified carbon black (MCB) technology   221
molybdenum species   244
– co-catalytic effect   245
– role   244
More Energy Ltd. (MEL)   46
– direct liquid methanol (DLM) fuel cells development   46
Morepower project   8, 53, 59, 64
– DMFCs development   59
– framework   64
multi-wall carbon nanotubes (MWCNTs)   318, 328, 330, 331, 342, 343, 428, 519
– arc discharge method   328
– vs. CBs, corrosion behavior   330
– chemical vapor deposition technique   328
– components   330
– polyol method   92, 93
– polyoxmetalate-modified   318
– SEM images   343
– structural changes   331

n

Nafion   23, 24, 67, 217, 380, 383, 396, 402, 409
– chemical structure   380
– electrochemical properties   409
– film diffusion current   154
– gel   67
– vs. HC membranes   218
– ionomer   21, 336, 370, 426, 429, 436
  – ionic aggregations   429
– layer resistance   142
– MCO current, comparison   218
Nafion membrane(s)   22, 24, 51, 67, 218, 266, 295, 441
– alternatives to   23
– methanol flow/permeability evaluation   266
– perfluorosulfonic acid   24
– types   218
nanocarbons   364
– graphitic carbon nanocoils   364
– nanofibers/nanotubes   364
nanoparticles   83, 85, 102, 103, 322, 323, 334, 366
– 1-D/2-D anisotropic   85
– catalysts   238
– PtRu/C, thermal crystallization   123
nanoporous platinum   101, 105, 106
nanoporous proton-conducting membrane (NP-PCM)   49, 59
nanoporous structures   106
– template-free electrochemical lithiation/delithiation synthesis   106
nanotubes synthesis, experimental procedure   98
Navier–Stokes equation   147, 149
NEM membranes   266, 267
Nernst correction   473, 474
Nernst equation   143
Nernst's law   472
nickel   494
– catalyst   494
– surface oxidation properties   494
– use   494
non-chlorine-containing precursors, types   88
non-platinum binary array, fluorescence image   179
Nuclepore® filter membrane   43

o

open cell potential (OCP)   518
open circuit voltage (OCV)   48, 52, 264, 298, 336, 368
– measurement   368
optically transparent electrode (OTC)   338
optical screening method   177–179
– disadvantage   178
ordered mesoporous carbon (OMC) material   357, 359–361, 363, 373

– fabrication 363
– frameworks 431
– NCC-1 361
– physical/chemical properties 357
– structural parameters 360
– support, SEM images 371
– synthesis 357–360, 363
  – via direct self-assembly approach 359, 360
  – via nano-casting method 357–359
ordered mesoporous silica (OMS) materials 357
organic fuel cells 511
– non-acid electrolytes, use 511
organic liquid fuels, characteristics 1
organic solvents 289
– 1,2 dichlorobenzene (DCB) 289
– synthesis 289
– xylene 289
oxophilic promoter metal 116
– bifunctional mechanism 115
oxygen molecules 35, 305, 498
– adsorption 141
– chemisorption 35
– containing species 81, 83, 317
– electrochemical reduction 498
– electroreduction current-potential curves 305
– stoichiometry(ies) 463, 464
  – cell polarization curves 463
oxygen reduction reaction (ORR) 33, 35, 66, 80, 103, 119, 249, 257–259, 261, 262, 269, 272, 276, 284–286, 288, 291, 301, 306, 337–340, 366, 417, 423, 454, 488, 498–506
– activity 100
– alternative cathode catalysts 37
– catalysts 423
– current-potential curves 291
– data analysis 269–272
– electrocatalysis process 33–39, 306
– electrocatalysts 261
  – non-noble metal 35–37, 261, 502
– kinetics 258
– parameters 301
– mass activity 153
– methanol effect 504, 505
– nanostructured catalysts 272–296
  – synthesis/characterizations 272
– paths 499
– performance 66
– at platinum electrode in DMFC 258–261
– polarization curve(s) 103, 276, 284, 286, 288, 301, 339
– polymer, role 285
– Pt-based catalysts 35–37, 152

– reaction mechanism 258, 259, 337
– Tafel plots 263, 501
– thermodynamics 258

p
palladium 538
– catalysts 502, 538
– chronoamperometric activity 540
– cyclic voltammograms 539
  – forward/reverse scans 539
– monthly average price 538
particle size distribution (PSD) 207
particle size effect 81–82, 537
pattern recognition methodology 232
polymer electrolyte fuel cells (PEFCs) 558
– direct hydrazine anion-exchange, durability test 558
perfluorinated polymers 25
– advantage 25
– types 25
perfluorosulfonic acid (PFSA) 22, 24, 66, 165
– membrane 22, 24
– polymer electrolyte 66
perfluorosulfonic membranes 26, 27, 31
– conductivity/performance 27
periodical activation effect 478, 479
phosphoric acid fuel cells (PAFCs) 33
platinum 34, 36, 80, 82, 89, 227, 229, 272, 277, 367, 538
– anisotropic growth 89
– based electrocatalyst, designing 87
– based electrodes 272–278
– carbon activity 151
– crystal structure 229
– cubic particles, molecular dynamic simulations 83
– drawbacks 80
– electrocatalytic activity 34
– electrochemical oxygen reduction reaction 135
– electronic state 277
  – *in situ* XANES/EXAFS experiments 277
– gold nanoparticles, synergistic catalytic effect 245
– lattice parameter, usage 206
– multipods 90
  – morphologies 90
  – synthesis 90
– MWCNT 322, 339
  – HRTEM/TEM images 322
– nanostructure 125
  – four-armed 91
  – TEM images 91
  – morphological control 125

– SEM observations   125
– OMC-300 preparation method   371
  – modified incipient wetness method   371
– ORR activity   82, 102
– properties   100, 367
– surface, oxide film growth   478
– XRD patterns   36
platinum alloys   238, 242
– activity evaluation   242–246
    – Pt-based binary catalysts   243–246
    – ternary Pt-Ru-based catalysts   246–248
  – Au alloy   535
    – anode catalysts   543
    – cell voltage   543
    – current vs. potential plot   536
    – methanol oxidation   535
  – Pd alloys   534
    – formic acid electro-oxidation   534
  – preparation methods   238–242
    – supported catalysts   240–242
    – unsupported catalysts   238–240
platinum-based catalysts   36, 44, 154, 166, 183, 240, 243–246, 272–278, 291, 355, 369, 371, 423
– agglomeration effect   154
– colloid methods   240
– EDX/XRD analyses   276
– electrochemical characterization   275–278
– functions   166
– physicochemical characterizations   274
– Raney type   277
– structural parameters   275
– synthesis by colloidal/carbonyl complex route   273
– voltammogram(s)   276, 277, 278
platinum nanoparticles   83, 102, 103, 322, 323, 334, 366
– catalytic activity   83
– cyclic voltammetry curves   100
– d orbital vacancies   102
– formation   90
– modification   103
– SEM images   366
– size histogram   322
platinum nanotubes   99–101
– electron diffraction pattern   101
– SEM images   100, 101
– TEM image   101
platinum-ruthenium (PtRu) alloy   128, 419
– CNT clusters   316
– IR measurements   128, 439
– nanoparticles   97, 134
  – synthesis   93

platinum–ruthenium (PtRu) catalysts   15–19, 89, 199, 214, 229–233, 248, 250, 345, 368, 370, 419, 420, 423, 425
– bimetallic catalysts   133
– black catalyst   217
– catalytic activity   206, 233, 246, 547
– ethanol electro-oxidation activity   547
– ex-situ Ru leaching amount   214
– high-throughput screening methods   250
– HRTEM/XRD analyses data   241
– KB crystallite size   205, 207, 208
  – particle size distribution   208
  – pore size distribution   207
  – TEM   205
– oxygen-containing species generation   419
– preparation methods   201
– SEM/TEM image   345
– stability   249
  – alloying degree, role   249
  – in DMFC environment   248–250
– structural/electrochemical properties   231
– structure/activity matrix   232
– structure-activity relationship   420
– thin-film libraries   186
– X-ray photoelectron spectra   17
platinum–ruthenium (PtRu) nanoparticles   88, 124, 133, 370, 372
– binary alloy   124
– electron diffraction pattern   124
  – face centered cubic (fcc)   124
– NMR spectrum   134
– thermal-treatment strategy   131
– XAS characterization   130
platinum–ruthenium (PtRu) nanowire network   422
poison formation experiments   533
polarization
– behavior   36
– curves   5, 249, 408
  – anodic/cathodic   5
poly(arylene ether ketone)s (PAEKs) copolymers   379, 386
– sulfonated/unsulfonated structure   386
– sulfonation reaction   386, 388
poly(arylene ether)s   385, 397, 400
– pendant sulfonated groups   400, 401
– poly(arylene ether ether ketone) (PEEK)   385
– poly(arylene ether sulfone) (PES)   379, 385, 386, 401
– structure   389
polybenzimidazole (PBI) membranes   514
poly(diallyldimethylammonium chloride) (PDDA)   318

polyethyleneimine (PEI) 513
– CPP membranes, water uptake 513
poly(ethylene oxide) (PEO) 22
polymer 30, 31, 285, 379
– electrolyte DMFC, open circuit voltage 4
– electrolyte materials, Nafion 508
– matrix, polyaniline 263
– membrane 30
– carbon nanotubes 317
– types 379
polymer electrolyte fuel cell technology 227
polymer electrolyte membranes (PEM)s 379, 380–385, 407, 410, 528
– aliphatic polymers, synthesis 380
– fuel cell 116, 249
– performance 403–412
– properties 380, 406, 409
– Pt/Ru-composite 439
– state-of-the-art 380
– sulfonated poly(aryl ether) copolymers synthesis 385–403
  – copolymerization 388–397
  – post-sulfonation of polymers 385–388
  – strategies 397–402
polynuclear cluster compound 289
– ^{13}C NMR analysis 289
polyol synthesis method 89, 239, 240
poly(phthalazinone ether sulfone ketone) (PPESK) 386
– structure/sulfonation reaction 387
polyvinyl alcohols (PVA)s 379
– membranes 381
poly(vinylpyrrolidone) (PVP) 317
– stabilizing agent 89
polytetrafluoroethylene (PTFE) 61
– based membranes, chemical structures 382
pore-forming agent 432
porous carbon plate (PCP) 48
porous carbon spheres, BET analysis 431
porous hollow platinum nanospheres 102
– synthetic routes 102
porous materials 356, 363
– classification 356
– limiting factor, primary particle size 363
porphyrin compounds 282
– based electrodes 287
– ESR signal characteristic 282
– π-π* transition 282
post-deposition annealing treatment 174, 192
potassium ferrocyanide technique 153
potential conductors, see carbon nanotubes (CNTs)
potential of zero charge (PZC) 319

printed circuit board (PCB) 51, 460, 480
proton-conducting ionomer impregnation 58
proton conductivity 58, 383, 384, 403–405, 421
proton exchange membrane (PEM) 30, 117, 249, 260, 264, 383, 433, 441, 487
– catalysts loading, breathing process 441
– degradation mechanisms 30
– methanol crossover determination procedures 264
– Nafion 487
– water uptake 383
proton exchange membrane fuel cells (PEMFCs) 1, 24, 35, 56, 228, 260, 261, 315, 417
– drawback 56, 260
– polarizations curves 261
– types 417
protonic electrolytes 21, 66
– based direct methanol fuel cell 3, 5, 7, 30
Pt-Co bimetallic system 155
– catalysts, carbon activity 151
Pt-Ni bimetallic system 155
– alloy catalysts, voltammograms 297, 298
PtRu/C catalyst 91, 92, 95, 131, 205, 209, 210, 223, 232
– anodic treatment 232
– carbon supported 87
– redox couples, standard electrical potentials 92
– SAR for 232
– TEM image 205
Pt-Sn system 11, 532
– catalysts 233–235
  – activity 234
  – formic acid method 234, 532
  – cyclic voltammograms 533
  – forward/reverse scans 533
pulse electrodeposition method 104
3-pyridin-2yl-(4,5,6) triazolo-(1,5-a) pyridine (PTP) complex 178

q
quadrupole mass spectrometry (QMS) 184, 185
quasi bifunctional mechanism 492
quasi-2-D models 451. see also direct methanol fuel cell (DMFC) model

r
radiochemical grafting technique 25, 53
– application 25
rechargeable battery 40
reducing agents, types 87

resorcinol-formaldehyde (RF) resin   359
reticulated vitreous carbon (RVC)   437
reversible hydrogen electrode (RHE)   5, 269
rotating disk electrode (RDE)   136, 146, 169, 263, 492, 506
– 3-electrode cell   146
– electrolytic solutions, diffusion   147
– equation   146
– experiment   151, 269
    – electrochemical set-up   269
    – Levich line   151
    – LSV   147, 151
– linear sweep polarization curves   506
– mass transport correction   152
– Navier–Stokes equation   146
– Pine instrument setup, schematic diagram   147
rotating ring-disk electrode (RRDE)   136, 148, 503
– current-potential curve   148
– electrode/solution interface   148
– LSV experiments   153, 269
    – electrochemical set-up   269
– roughness factor (RF)   140
– thin-film   154
ruthenium
– based catalysts, ORR performance   502
– chalcogenide-based low temperature synthesis   290
    – flow chart   290
– leaching experiment   213

S
Samsung Advanced Institute of Technology (SAIT)   46
– DMFC cell development   46
scanning electrochemical microscope (SECM)   185–187
– electrochemical imaging technique   185
– electrochemical setup   186
– experiments   186
– usage   185
scanning electron microscopy (SEM)   122
scanning mass spectrometer   183–185
– principle   185
scanning QMS system, photograph   185
Scherrer's equation   123, 275, 334, 366
secondary ion mass spectrometry (SIMS)   122
sedimentation method   440
selected area diffraction (SAD) pattern   344
semiconductor technology   69
semi-interpenetrating network (SIPN) membranes   382
short-side-chain (SSC) polymer   27

silver catalysts, use   520
single channel cell   421, 476–480
– current-voltage curves   421
single/multiple-channel potentiostat   191
single-walled carbon nanotubes (SWCNTs)   323, 331, 337, 338
– oxygen/carbon atomic ratios   331
– porosity   338
six-electron transfer reaction   331. see also methanol oxidation reaction (MOR)
small organic fuels   84
– ethanol/formic acid   84
Sn negative effect, see methanol adsorption/dehydrogenation
sodium dodecylbenzene sulfonate (SDBS)   323
soft-templates   96, 357
– amphiphilic block copolymers   357
– surfactants   357
soft-templating approach, see direct self-assembly approach
sol-gel type procedure   22
– tetraethyl orthosilicate (TEOS) precursor, use   22
solid polymer electrolytes (SPEs)   289, 487
– fuel cells   62
solid polymer membrane   489
Solvay membranes   59, 267
– methanol permeability   267
– vs. Nafion membranes   267
spherical agglomerates, cross-section image   204
spherical platinum clusters   104–105
spillover effect   141
spray conversion reaction (SCR)   200
– anode catalyst, performance   212
– cathode catalysts, comparison   220, 221
– electrocatalysts   203, 207
    – development platform   203
    – structure   207
    – types   207
– platform   202
– supported catalysts   206, 209, 213, 217
    – high resolution TEM results   206
    – PtRu catalyst   206, 209, 213, 217
sputtering technique   44, 438
sputter system, chamber lay out   175
square-shaped cell   480–482
– activation   480–482
– power-current/voltage-current characteristics   482
stabilizing agents   88
standard hydrogen electrode (SHE)   258
state-of-the-art cathode catalysts   33–39

state-of-the-art DMFC electrolytes 24–32
– alkaline membranes 29–32
– development aspects 24
– high temperature application membrane 26–29
– performance/efficiency, crossover effects 32
– proton conducting membranes 24–26
styrene, radiation-grafting 31
sulfonated aromatic polymers, types 397
sulfonated hydrocarbon polymer membranes 410
sulfonated monomers 388–397, 400
– aromatic nucleophilic substitution polycondensation 389
– direct copolymerization 388
– poly(arylene ether ketone)s 391
– poly(arylene ether nitrile)s 396
– poly(arylene ether sulfone)s 390
– structure 390, 392
sulfonated poly(arylene ether nitrile) copolymers 405
– structures 405
sulfonated poly(ether ether ketone) (S-PEEK) 23, 53, 267–269
sulfonation reaction 385–388
sulfonic acid-based membranes 25
– water uptake properties 25
sulfosuccinic acid (SSA) 382, 514
surface oxidation process 496
surfactant molecule 94
– protected monodisperse nanoparticles 94
– role 94

t
Tafel slope 151, 152, 155, 259, 260, 271, 501
– extraction 152
– kinetic parameter 151
Tekion Inc. 50
– air breathing DMFC development 50
Temkin adsorption isotherm 7, 259, 500
temperature programmed reduction (TPR) 135
templates 95
– based synthesis 95–99
– types 95
tetraethylammonium perchlorate (TEAP) electrolyte 285
tetrahexahedral platinum nanostructures 106–108
– electrochemical preparation 108
thin film rotating disc electrode (TFRDE) catalyst characterization methodology 208–210

– advantage 208
thin gold films, electrocatalytic activity 495
through-plane model 451, 453
tip generation-substrate collection (TG-SC) 186
transition metal 37, 261, 289, 304
– chalcogenide 38
  – catalysts/electrodes 289–294
  – electrocatalytic activity 304
  – metal chalcogenides synthesis 289
  – physicochemical characterizations 292–294
– electrode activity/selectivity/stability 282–289
  – preparation/characterization 282
– macrocycles 166
  – characterization/synthesis 278–289
  – phthalocyanines synthesis 279
  – porphyrins synthesis 280
  – structure 279
– sulfides 38
– tetraazaannulenes synthesis 281
transition region (TR) 469
– hydrogen transport mechanism 469
transmission electron microscopy (TEM) 121, 292
– images 292
– use 292
trifluoromethanesulfonic acid (TFMFSA) 22
two-electrode cells 181
– batteries 181
two-electron pathway 287, 338

u
ultrasonicating, see ball-milling method
ultrasonic spray pyrolysis (USP) 431
uncompressed graphite felt (UGF) 437
underpotential deposition (UPD) 143
– of copper 137
unsupported platinum nanostructures 95
– synthetic methods 95–108

v
Vegard's law 123
vinylbenzylchloride (VBC) 32
– radiation grafting 32
voltage efficiency, definition 9
Voltalab PGZ-402 potentiostats 269
Vulcan XC-72 carbon 87, 273, 282, 290, 295

w
Waseda University 50
– µDMFC development 50
water drag number 218, 219

water-in-oil reverse microemulsion system 240
working electrode array 175, 176, 179

x

X-ray absorption near-edge structure (XANES) 102, 122
– particle size effect 129
X-ray absorption spectroscopy (XAS) 122, 126–132, 249
X-ray diffraction (XRD) 122, 240, 274, 292–294, 515
– lattice parameter comparison 206
– patterns 233
– spectroscopy data 206
– use 292
X-ray photoelectron spectroscopy (XPS) 12, 122, 132, 133, 244, 291, 292, 421
– data 236
– photoelectric effect 132
X-ray radiation 132

z

ZrNi alloy, anode catalyst 557, 559